K&E Series on Knots and Everything — Vol. 29

QUANTUM INVARIANTS

A Study of Knots, 3-Manifolds, and Their Sets

Tomotada Ohtsuki
Tokyo Institute of Technology
Japan

World Scientific
Singapore • New Jersey • London • Hong Kong

Published by

World Scientific Publishing Co. Pte. Ltd.
5 Toh Tuck Link, Singapore 596224
USA office: 27 Warren Street, Suite 401-402, Hackensack, NJ 07601
UK office: 57 Shelton Street, Covent Garden, London WC2H 9HE

British Library Cataloguing-in-Publication Data
A catalogue record for this book is available from the British Library.

Series on Knots and Everything — Vol. 29
QUANTUM INVARIANTS
A Study of Knots, 3-Manifolds, and Their Sets

ISBN-13 978-981-02-4675-4
ISBN-10 981-02-4675-7

Preface

In the 1980s low-dimensional topology encountered mathematical physics, and their interactions yielded infinitely many new invariants, called quantum invariants, of knots and 3-manifolds. This monograph provides an extensive and self-contained presentation of quantum and related invariants. This wealth of invariants gives us a new viewpoint in 3-dimensional topology: a study of the sets of knots and 3-manifolds, as well as a study of individual knots and 3-manifolds.

This monograph is addressed to readers with a basic knowledge of topology, such as homology groups and fundamental groups, as well as some acquaintance with basic abstract algebra.

When we tie a strand, a knot appears. Since ancient times many knots have been invented and contrived for practical and decorative purposes (see [TuGr96]). Tying strands in different ways, different knot types appear, albeit with significant redundancy. Thus we are led to ask: how various are the knot types? From the mathematical viewpoint knots are quite complicated and fascinating objects, even when we study them with all the mathematical approaches developed so far. The study of knots and knot types as mathematical objects is called *knot theory.*

There were attempts, say [Tai1898] (see also [TuGr96, Prz92, Epp99] for expositions of the history of knot theory), to classify knot types in the 19th century, though the mathematical fundamentals required to proceed with the classification rigorously were far from sufficiently developed in that age. This situation was changed by the creation of the discipline of topology at the beginning of the 20th century. Mathematical fundamentals of topology, such as homology groups, were established in the early 20th century, and subsequently "classical" knot theory, based on algebraic topology, developed rapidly (see [CrFo77, Rol90] for expositions of "classical" knot theory). The Alexander polynomial, discovered in the 1920s, was one of main achievements in "classical" knot theory. This polynomial is an isotopy invariant of knots, where by an *isotopy invariant* we mean a map from the set of knot types to a well-known set, such as a polynomial ring.

A dramatic transformation of the study of knot invariants arose from the discovery of the Jones polynomial in 1984, which was soon followed by a flood of discoveries of infinitely many new invariants of knots. They were derived from interactions between knot theory and various other fields; the Jones polynomial was defined by using the theory of operator algebras, some other of the new invariants were defined by using solutions of the Yang-Baxter equation in statistical mechanics, and others still were introduced by using representations of quantum groups and by using solutions of the Knizhnik-Zamolodchikov equation, related to quantum field theory in theoretical physics, which organized them all into *quantum invariants* of knots.

— ⋆ —

Another main topic of this monograph is the study of invariants of 3-manifolds.

Until the 19th century 3-dimensional spaces appearing in mathematics and physics were mainly straight spaces (copies of the Euclidean space), but, since the beginning of the 20th century, when the theory of relativity disclosed that the real space is a globally curved space (possibly having a different topological type than the Euclidean space in the astronomical scale), curved spaces (manifolds) have become fundamental in the study of mathematics and physics of the 20th century. Furthermore, from the mid-20th-century solution of the classification problem of higher-dimensional (dim ≥ 5) manifolds (an historic achievement of differential topology), up to the present day, the study of low-dimensional manifolds has been the main stream of manifold theory.

Quantum theory, another great physical theory initiated at the beginning of the 20th century, asserts that an atomic particle moving in a space can be described in mathematics, not by a moving point, but by a wave function on the whole of the space. In the 1980s, various marvellous results were yielded to low-dimensional topology by the new methodology, which studies topological properties of a given space by analyzing the structure of "the space of wave functions on the underlying space" in a suitably defined quantum field theory (*i.e.*, the space of solutions of a certain differential equation, which is canonical in some physical sense). In particular in the late 1980s Witten considered a quantum field theory which depends only on the topological type of an underlying 3-manifold, and proposed that the partition function of the theory provides a "topological invariant" of the 3-manifold, where by a *topological invariant* we mean a map of the set of topological types of 3-manifolds to a well-known set. Motivated by Witten's proposal, infinitely many invariants of 3-manifolds, called *quantum invariants*, have been constructed rigorously in various mathematical ways.

— ⋆ —

So, in the 1980s we obtained infinitely many new invariants (quantum invariants) of knots and 3-manifolds. For these quantum invariants to be useful they needed to be organized, which was achieved with the introduction of the Kontsevich invariant (resp. the LMO invariant) and the theory of Vassiliev invariants (resp. finite type invariants) in the 1990s. Further, the Kontsevich invariant (resp. the LMO invariant) gives a map of the set of knots (resp. integral homology 3-spheres) into a lattice of infinite rank. It is now conjectured that the image under the map classifies knots (resp. integral homology 3-spheres). The set of knots (resp. integral homology 3-spheres) would be identified with a subset of the lattice if this conjecture was true.

— ⋆ —

In "classical" topology, where the study focuses, in the main, on individual knots (resp. 3-manifolds), an invariant was often regarded just as a tool to distinguish knots (resp. 3-manifolds). This situation has changed since the discoveries of these many new invariants in the 1980-90s. In the sense that an invariant of knots (resp. 3-manifolds) gives a partition (*i.e.*, a rough classification) of the set of knots (resp. 3-manifolds), the study of such a wealth of invariants is, in effect, a study of these sets. In this way, we are now ready to study the *sets* of knots and 3-manifolds.

Number theory has developed, over the course of its long history, from a study of numbers to be, also, a study of the *set* of numbers, and in this way has obtained many profound results on the numbers themselves. Can knots and 3-manifolds be to topology as numbers are to number theory? That is, we expect future developments in knot theory (resp. 3-manifold theory) to arise as much from the study of the *set* of knots (resp. 3-manifolds) as from the study of the individuals of those sets themselves. If we could find appropriate structures for those sets, then they could be studied as the "spaces of knots and 3-manifolds", and invariants (such as quantum invariants) of knots and 3-manifolds could be studied by analyzing appropriate functions on those spaces. The author hopes that this monograph will serve as a first step towards such a future study of the sets of knots and 3-manifolds.

This monograph consists of two parts. The former part (Chapter 1–7) is concerned with invariants of knots and links; see Figure 0.1 for the main knot invariants discussed in this monograph and the chapters related to them respectively. The latter part (Chapter 8–11) is concerned with invariants of 3-manifolds; see Figure 0.2 for the main 3-manifold invariants discussed in this monograph and the chapters related to them respectively.

In Chapters 1–3 the Jones and Alexander polynomials, a modern and a classical invariant of knots, are constructed from various directions by introducing knot diagrams in Chapter 1, braids in Chapter 2, and tangles in Chapter 3. The constructions of them discussed in Chapters 2 and 3 reduce the definition of the invariants of knots into invariants of "elements" (elementary tangle diagrams). In particular,

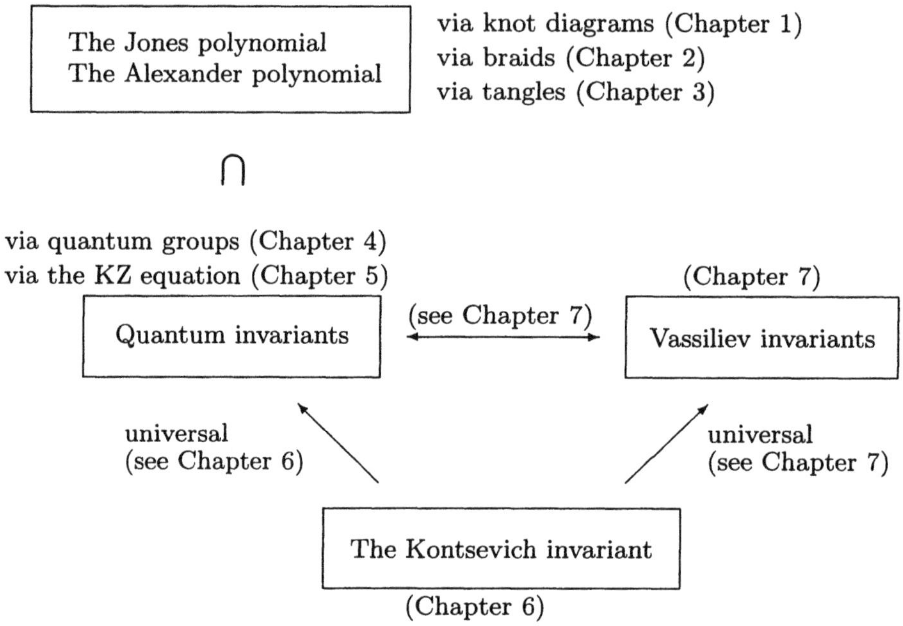

Figure 0.1 Invariants of knots and related chapters of this monograph. See also Figure 7.4 for a concrete description of this figure.

an invariant of an elementary tangle diagram with a single crossing is given by an R matrix (a solution of the Yang-Baxter equation). Such a construction is general enough to yield, not only the Jones and Alexander polynomials, but also many other invariants of knots. In fact, this construction associates an invariant of knots to every R-matrix.

As shown in Chapter 4, many R matrices systematically arise from representations of ribbon Hopf algebras, whose typical examples are quantum groups. We call invariants defined by using such R matrices derived from quantum groups *quantum invariants.* They are reformulated, in Chapter 5, by using monodromy along solutions of the KZ equation.

In Chapters 6 and 7 we discuss two approaches to control these infinitely many quantum invariants. One approach is to unify all quantum invariants into the Kontsevich invariant, which is introduced in Chapter 6 by using monodromy along solutions of the "universal" KZ equation. By definition the Kontsevich invariant is universal among quantum invariants. The other approach is to characterize all quantum invariants with a property. Vassiliev invariants are defined by such a characteristic property in Chapter 7. With the introduction of the Kontsevich invariant and the theory of Vassiliev invariants, the quantum invariants become well-organized. Further, via a descending series of equivalence relations among

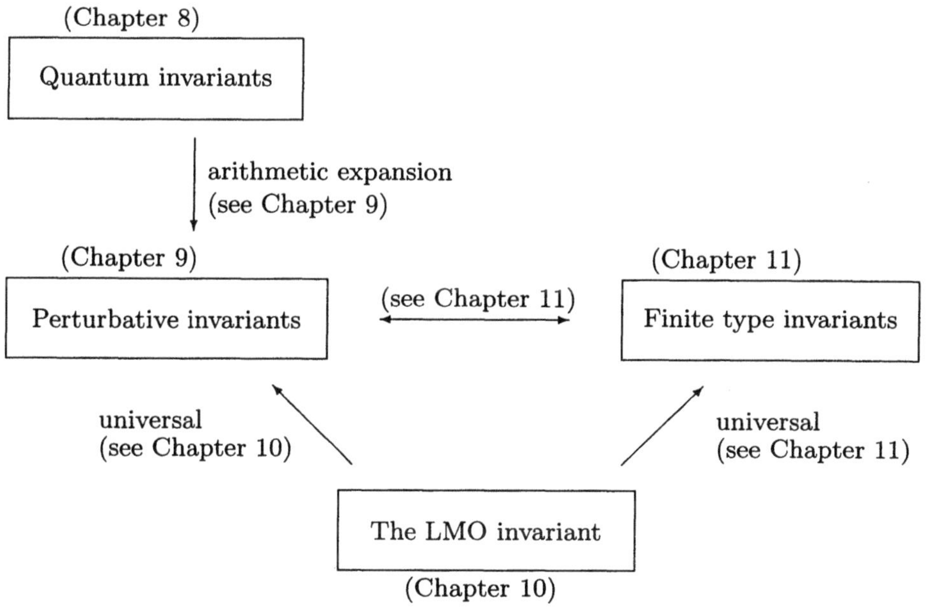

Figure 0.2 Invariants of 3-manifolds and related chapters of this monograph. See also Figure 11.3 for a concrete description of this figure.

knots given by Vassiliev invariants, the Kontsevich invariant gives a map of the set of knots into a lattice of infinite rank, which is now conjectured to classify knots.

Quantum invariants of 3-manifolds were originally proposed by Witten by the Chern-Simons path integral in mathematical physics (see Appendix F). In Chapter 8 we mathematically define quantum invariants of a 3-manifold to be appropriate linear sums of quantum invariants of a framed link, which gives a surgery presentation of the 3-manifold; such a definition was predicted by the operator formalism of the Chern-Simons path integral. Further, motivated by the perturbative expansion of the Chern-Simons path integral, we define a perturbative invariant to be a certain number-theoretical limit of quantum invariants in Chapter 9. This is a power series invariant, while quantum invariants of 3-manifolds are $\mathbb{C}$-valued.

In Chapters 10 and 11 we introduce the LMO invariant and finite type invariants which play similar roles among the perturbative invariants as the Kontsevich invariant and Vassiliev invariants play among knot invariants. In Chapter 10 we define the LMO invariant by picking up information in the Kontsevich invariant which is invariant under the Kirby moves. We also show the universality of the LMO invariant among perturbative invariants. Further, these arguments are reviewed by using a formal Gaussian integral, called the Aarhus integral. In Chapter 11 we introduce finite type invariants of integral homology 3-spheres, and show that they satisfy similar properties as those of Vassiliev invariants. Further, via a descending series of equivalence relations among integral homology 3-spheres given by finite type invariants, the LMO invariant gives a map of the set of integral homology 3-spheres into a lattice of infinite rank, which is now conjectured to classify integral homology 3-spheres.

The author tremendously and deeply appreciates Andrew Kricker, who gave him many valuable suggestions and comments in each page of the whole of the draft, which really improved readability of many English and mathematical expressions in the draft.

The author would like to thank Tatsuya Yagi, Ayumu Inoue, Tatsuhiro Yamakawa, Kentarou Kikuchi, Naosuke Okuda, Eri Hatakenaka, and Tomohide Yamada for reading early drafts carefully and pointing out many corrections.

The author is sincerely grateful to many helpful comments given by Hitoshi Murakami for Chapters 1–6, Kazuo Habiro for Appendix E, and Toru Gocho and Jorgen Ellegaard Andersen for Appendix F. He also wishes to thank Jun Murakami and Toshitake Kohno for comments and suggestions, Dylan Paul Thurston for comments on configuration space integrals, and Kouji Kodama for calculations of the colored Jones polynomials.

Tomotada Ohtsuki
Tokyo, October 2001

Contents

Chapter 1

Knots and polynomial invariants

In knot theory we study knots and knot types (*i.e.*, isotopy classes of knots) as mathematical objects. We often deal with knots by depicting them in a plane; the depicted pictures of knots are called knot diagrams. Further, we describe isotopy of knots by some moves among knot diagrams, which are called the Reidemeister moves. Then, the set of isotopy classes of knots can be identified with the quotient set of knot diagrams modulo the Reidemeister moves. A remarkable advantage of such identification is that although knots are topological objects in a 3-dimensional space they can be treated as combinatorial objects, like graphs embedded in a plane. In this way we can reduce studies of knots to studies of the combinatorics of knot diagrams.

A main topic in knot theory is the study of invariants of knots. An invariant is a map from the set of knots to a (well-known) set, such as a polynomial ring, such that two isotopic knots have the same image by the map. By using invariants of knots we can distinguish isotopy classes of knots concretely. A typical way of constructing an invariant is to construct a function of the set of knot diagrams, in a combinatorial way, such that it is unchanged under the Reidemeister moves. We introduce the Jones polynomial in such a way in Section 1.2.

The historic discovery of the Jones polynomial took place in the middle of 1980s, relatively recent in the history of knot theory. In the 1980s many invariants of knots, what are called quantum invariants, were discovered in active interaction between low dimensional topology and mathematical physics. The Jones polynomial can be regarded as the simplest quantum invariant. On the other hand a most classical invariant of knots is the Alexander polynomial, which was discovered in the 1920s, relatively early in the history of knot theory. It is defined in a classical way, using the homology of the infinite cyclic covering space of a knot complement. We introduce the Alexander polynomial in Section 1.3.

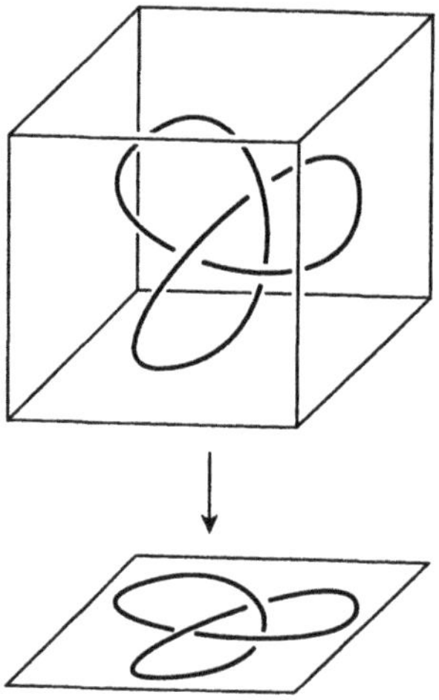

Figure 1.2 A knot and a diagram of it

approximation of the immersed annulus we can express the annulus as the union of linear small triangles in $\mathbb{R}^3$ after small perturbation of the annulus. Hence, the deformation between K and K' can be expressed as the composition of successive steps such that each of the steps is the move such that one edge (resp. the union of two edges) of a small triangle is replaced by the union of the other two edges (resp. the other edge). By classifying the phenomena that can happen at such a move we obtain the RI, RII and RIII moves and the mirror images of the RI and RIII move. Further, the mirror images of the RI and RIII moves can be obtained as sequences of the RII and RIII moves, as below.

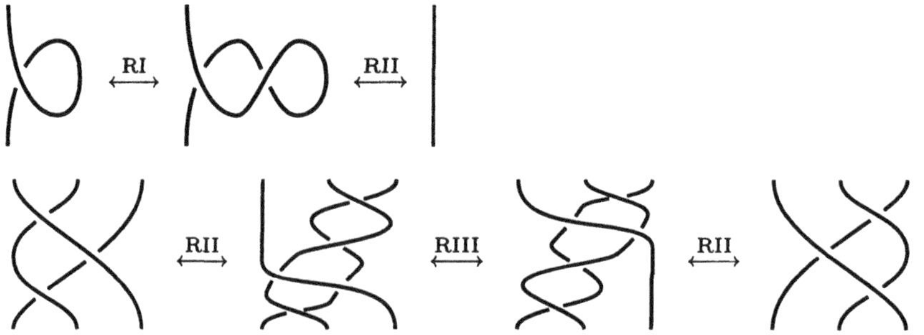

Hence, the isotopy can be expressed by a sequence of the Reidemeister moves; for a detailed proof, see, *e.g.*, [BuZi85]. □

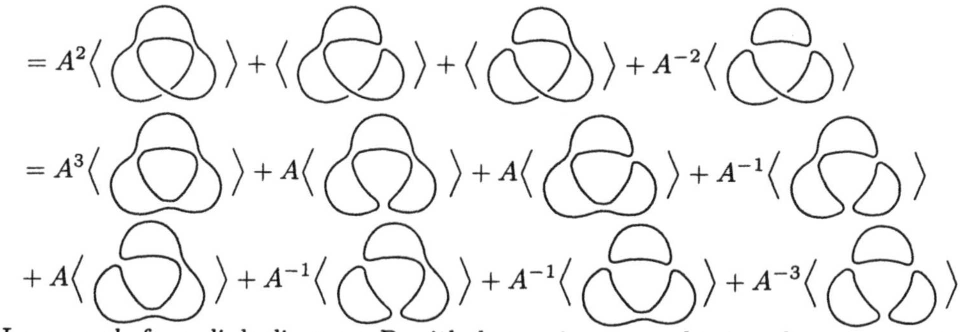

In general, for a link diagram D with k crossings, we obtain a linear sum of 2^k diagrams without crossings. Note that the linear sum is obtained independently of the order of expansion of the k crossings. Further, since a diagram without crossings is the disjoint union of loops, we obtain the value of its bracket recursively by (1.6) and (1.7), *i.e.*, the bracket of a diagram consisting of l disjoint loops has the value $(-A^2 - A^{-2})^l$. Hence, we obtain $\langle D \rangle$ from the above linear sum of 2^k diagrams by replacing each diagram, say, l loops, by $(-A^2 - A^{-2})^l$. For example, for a trefoil knot diagram, we have that

$$\begin{aligned} &= A^3(-A^2 - A^{-2})^2 + A(-A^2 - A^{-2}) + A(-A^2 - A^{-2}) \\ &\quad + A^{-1}(-A^2 - A^{-2})^2 + A(-A^2 - A^{-2}) + A^{-1}(-A^2 - A^{-2})^2 \\ &\quad + A^{-1}(-A^2 - A^{-2})^2 + A^{-3}(-A^2 - A^{-2})^3 \\ &= (-A^2 - A^{-2})(-A^5 - A^{-3} + A^{-7}). \end{aligned} \tag{1.8}$$

Lemma 1.4. The Kauffman bracket satisfies the formulae

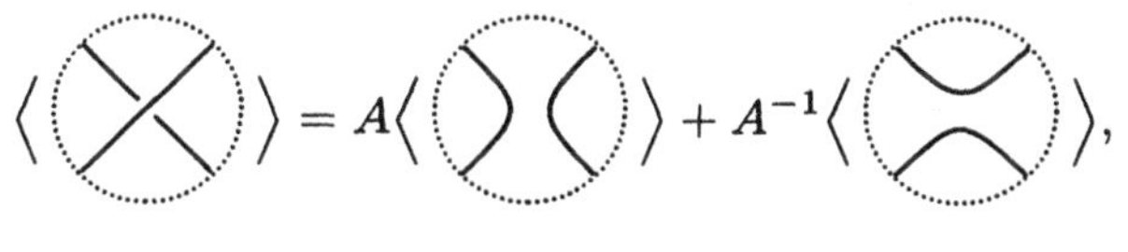

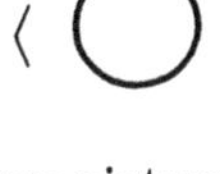 $D\rangle = (-A^2 - A^{-2})\langle D \rangle$ for any diagram D,

where the three pictures in the first formula imply three diagrams which are identical outside the dotted circles and differ only inside the dotted circles, as shown in the pictures.

The first formula of the lemma is called the *skein relation* of the Kauffman bracket. It is the same formula as a defining relation (1.5) and they have the following different sense; the defining relation (1.5) is a recursive formula to expand all crossings repeatedly to define the Kauffman bracket, while the skein relation is a relation among the Kauffman bracket of three diagrams which have been already defined.

Proof of Lemma 1.4. From the definition of the Kauffman bracket, we expand all crossings of the diagram, say k crossings, on the left hand side of the first formula, to obtain a linear sum of 2^k diagrams without crossings. We split the linear sum into two parts, according to which of the two terms in the expansion of crossing in the dotted circle each final term arises from. Then, each part is a linear sum of 2^{k-1} diagrams without crossings, and each part corresponds to a term on the right hand side of the first formula. Hence, we obtain the first formula.

We obtain the second formula by a similar argument using (1.6). □

To obtain an isotopy invariant of links from the Kauffman bracket of their diagrams, it is sufficient to show the invariance of the Kauffman bracket under the RI, RII and RIII moves by Theorem 1.1. Let us verify the invariance under each of the moves.

Firstly, is  equal to 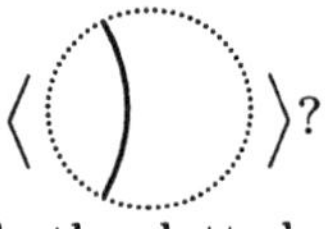? Here, the two pictures indicate two diagrams which only differ inside the dotted circles, as shown. By Lemma 1.4 we have that

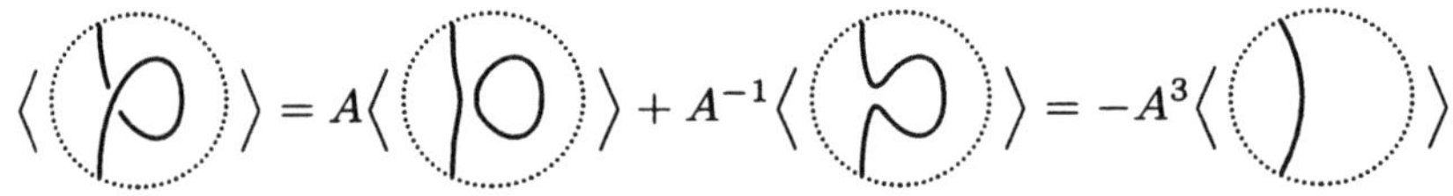

(1.9)

In the same way, we find that

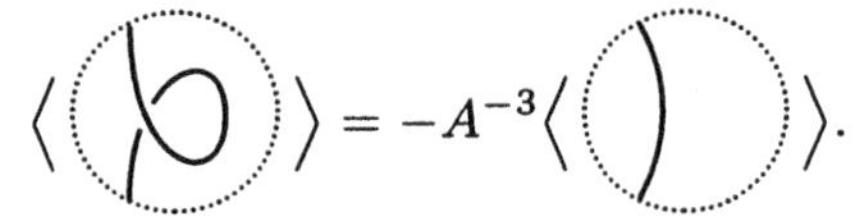

(1.10)

Hence, unfortunately, $\langle D\rangle$ does change under the RI move on a diagram D.

For simplicity we omit the dotted circles in the following.

Secondly, is  equal to 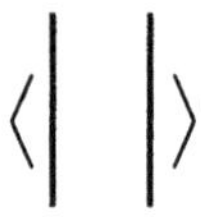? By Lemma 1.4 we have that

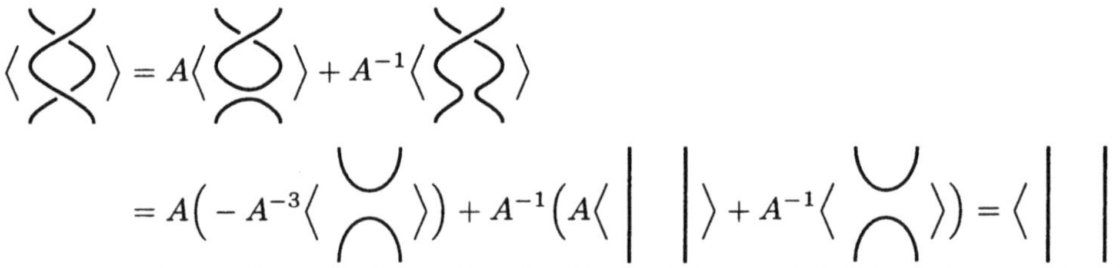

where we obtain the second equality by (1.10). Hence, $\langle D\rangle$ is invariant under the RII move on a diagram D.

Thirdly, is  equal to ? By Lemma 1.4 we have the

following two formulae

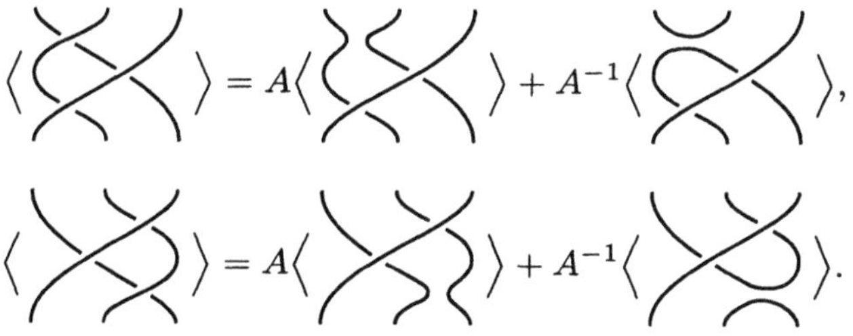

The first terms on the right hand sides of the two formulae are equal, because their diagrams are isotopic. Further, the second terms are equal, because each of them is equal to 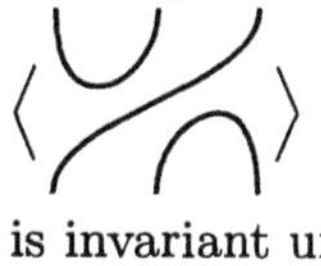 by the invariance under the RII move, as shown above. Hence, $\langle D \rangle$ is invariant under the RIII move on a diagram D.

Summarizing the above arguments, we showed that $\langle D \rangle$ is invariant under the RII and RIII moves, though $\langle D \rangle$ is not invariant under the RI move. To deal with this, we have two options; the first option is to modify $\langle D \rangle$ by "writhe" of D, and the second option is to replace links with "framed links", and regard the Kauffman bracket as an isotopy invariant of framed links.

Modifying the Kauffman bracket by writhe

We proceed by the first option as follows. For an oriented diagram D we define the *writhe* of D by

$$\begin{aligned} w(D) = \; & (\text{the number of positive crossings of } D) \\ & - (\text{the number of negative crossings of } D) \; . \end{aligned}$$

Modifying the Kauffman bracket with the writhe we obtain an isotopy invariant of oriented links as follows.

Theorem 1.5. Let L be an oriented link, and D an oriented diagram of L. Then,

$$(-A^3)^{-w(D)} \langle D \rangle \tag{1.11}$$

is invariant under the RI, RII and RIII moves, where $\langle D \rangle$ is the Kauffman bracket of D with its orientation forgotten. In particular, it is an isotopy invariant of L.

It is an elementary exercise to show that the values of (1.11) belong to the polynomial ring $\mathbb{Z}[A^2, A^{-2}]$. Putting $A^2 = t^{-1/2}$ we denote the quotient of the invariant (1.11) by $(-A^2 - A^{-2})$ by $V_L(t)$ and call it the *Jones polynomial* of an oriented link L, *i.e.*, we set

$$V_L(t) = (-A^2 - A^{-2})^{-1} (-A^3)^{-w(D)} \langle D \rangle \Big|_{A^2 = t^{-1/2}} \in \mathbb{Z}[t^{1/2}, t^{-1/2}].$$

For example, for the trivial knot K_0,

$$V_{K_0}(t) = 1.$$

Further, for the trefoil knot $K_{\overline{3}_1}$,

$$V_{K_{\overline{3}_1}}(t) = (-A^3)^{-3}(-A^5 - A^{-3} + A^{-7})\Big|_{A^2=t^{-1/2}} = t + t^3 - t^4$$

from the computation (1.8). By similar computations for the other knots in Figure 1.1 we have that

$$\begin{aligned} V_{K_{4_1}}(t) &= t^2 - t + 1 - t^{-1} + t^{-2} \\ V_{K_{\overline{5}_1}}(t) &= -t^7 + t^6 - t^5 + t^4 + t^2 \\ V_{K_{5_2}}(t) &= t^{-1} - t^{-2} + 2t^{-3} - t^{-4} + t^{-5} - t^{-6}. \end{aligned}$$

It is also possible to obtain them by using computer software such as in [Kod]; see also [Kaw+90] for a list of values of various invariants.

Proof of Theorem 1.5. Let us verify the invariance of (1.11) under the RI, RII and RIII moves.

We show the invariance under the RI move as follows. Let D be a link diagram and D' a diagram obtained from D by adding a positive crossing by the RI move. Then, from the definition of the writhe, $w(D') = w(D) + 1$. Further, by (1.9), we have that $\langle D' \rangle = -A^3 \langle D \rangle$. Hence,

$$(-A^3)^{-w(D)} \langle D \rangle = (-A^3)^{-w(D')} \langle D' \rangle$$

which implies the invariance of (1.11) under the RI move with a positive crossing. We obtain the invariance under the RI move with a negative crossing similarly, by using (1.10).

The invariance of (1.11) under the RII move is easily obtained from the invariance of each of $w(D)$ and $\langle D \rangle$ under the RII move.

We show the invariance of (1.11) under the RIII move as follows. Since $\langle D \rangle$ is invariant under the RIII move as shown before, it is sufficient to show that $w(D)$ is invariant. We consider a correspondence between the sets of three crossings of both sides of the RIII move, such as

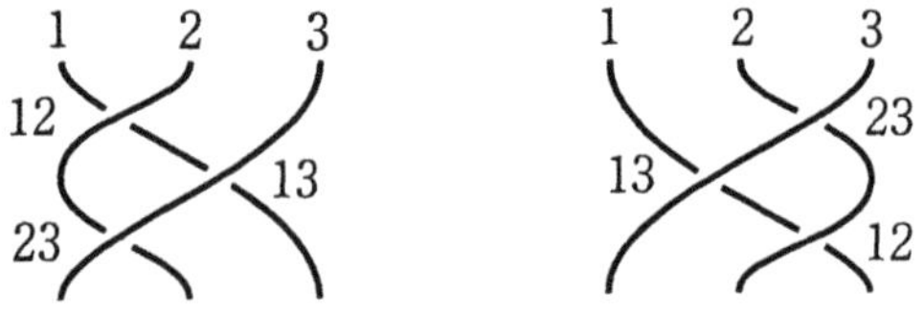

where the crossings with the same label correspond. Since the crossings labeled by "ij" are crossings between the strands labeled by i and j, the corresponding two crossings have the same sign for any orientation of the three strands. Hence, the writhes of both sides are equal. Therefore, we obtain the invariance of (1.11) under the RIII move. □

Proposition 1.6. The Jones polynomial satisfies the following relation, called the *skein relation* of the Jones polynomial,

$$t^{-1}V_{L_+}(t) - tV_{L_-}(t) = (t^{1/2} - t^{-1/2})V_{L_0}(t).$$

Here, L_+, L_-, and L_0 are three oriented links, which are identical except for a ball, where they differ as shown in the pictures.

L_+ L_- L_0

Proof. Let $D_\pm$ and D_0 be diagrams of $L_\pm$ and L_0. By using Lemma 1.4 twice, we obtain

$$A\langle D_+\rangle - A^{-1}\langle D_-\rangle = (A^2 - A^{-2})\langle D_0\rangle. \tag{1.12}$$

Hence, noting the equalities $w(D_0) = w(D_+) - 1 = w(D_-) + 1$, we have that

$$A^4\cdot(-A^3)^{-w(D_+)}\langle D_+\rangle - A^{-4}\cdot(-A^3)^{-w(D_-)}\langle D_-\rangle = -(A^2 - A^{-2})\cdot(-A^3)^{-w(D_0)}\langle D_0\rangle.$$

By the above definition of the Jones polynomial we obtain the required formula. □

Remark 1.7. A relation amongst the values of an invariant on a number of links which differ by local pictures, as for L_+, L_- and L_0, is called a *skein relation*; Proposition 1.6 is an example of a skein relation; for skein relations of other invariants see Propositions 2.6 and 4.18 and the relation (6.36). Suppose that an invariant satisfies a skein relation. Then, often the value of an invariant of each oriented link is uniquely determined by the skein relation and the value of the trivial knot. For example, the Jones polynomial is uniquely characterized by

$$\begin{cases} t^{-1}V_{L_+}(t) - tV_{L_-}(t) = (t^{1/2} - t^{-1/2})V_{L_0}(t), \\ V_O(t) = 1, \end{cases}$$

where L_+, L_- and L_0 are as in Proposition 1.6 and O denotes the trivial knot. In fact, the value of each oriented link can be computed by using the above two relations. Instead of giving a complete proof of this, we shall demonstrate it for the case of the trefoil knot. We consider a resolution of the trefoil knot by the skein relation, as below.

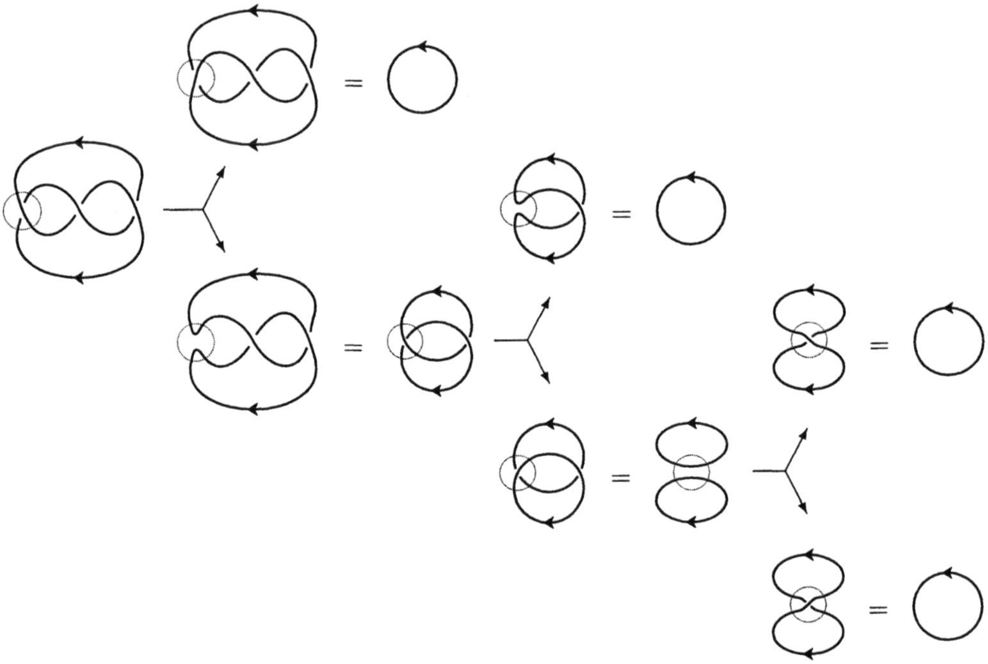

In the picture a triple of dotted circles implies a triple of L_+, L_- and L_0 in Proposition 1.6. We compute the Jones polynomial of the links in the picture backward from the trivial knots along the resolution. We have that

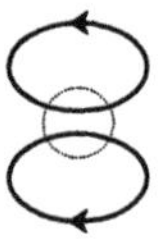

$$= \frac{1}{t^{1/2} - t^{-1/2}} \Big(t^{-1}$$

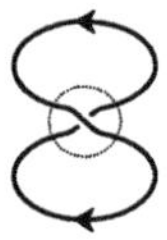

$$- t$$

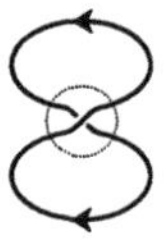

$$\Big)$$

$$= \frac{1}{t^{1/2} - t^{-1/2}} (t^{-1} - t) = -t^{1/2} - t^{-1/2},$$

where, for simplicity, by a picture of a link we mean its Jones polynomial. Further, by using the above relation, we have that

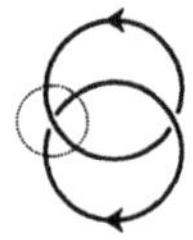

$$= t^2$$

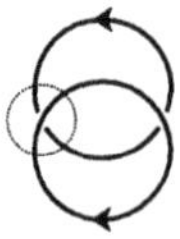

$$+ t(t^{1/2} - t^{-1/2})$$

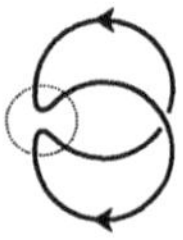

$$= t^2(-t^{1/2} - t^{-1/2}) + t(t^{1/2} - t^{-1/2}) = -t^{5/2} - t^{1/2}.$$

Hence, we obtain the Jones polynomial of the trefoil knot as

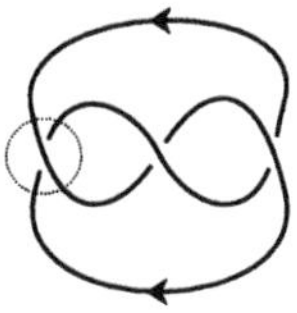 $= t^2$ 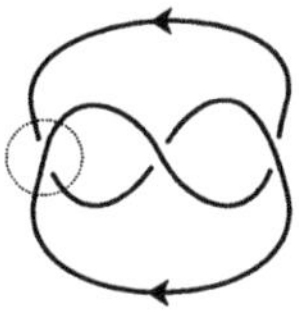 $+ t(t^{1/2} - t^{-1/2})$ 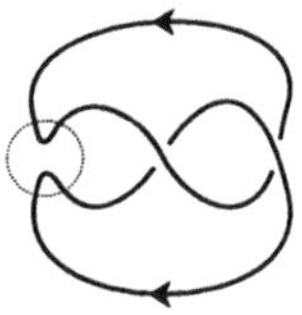

$$= t^2 + t(t^{1/2} - t^{-1/2})(-t^{5/2} - t^{1/2}) = -t^4 + t^3 + t.$$

It is left to the reader to show that the Jones polynomial of any oriented link can be computed by the skein relation in the above way (for a complete proof of that, see, *e.g.*, [PrSo97]).

The Kauffman bracket as an invariant of framed links

Next we consider the second option to obtain an invariant from the Kauffman bracket, introducing the notion of a framed link. A *framed link* is the image of an embedding of a disjoint union of annuli into $\mathbb{R}^3$. The *underlying link* of a framed link is the link obtained by restricting an annulus $S^1 \times [0,1]$ to its center line $S^1 \times \{1/2\}$. The *framing* of a component of a framed link is the isotopy class of framed knots whose underlying knots are equal to the component. We often express the framing of a component of a framed link by the linking number of the boundary components of the embedded annulus. Here, we let the two boundary components be oriented in the same direction of the component of a framed link. The *blackboard framing* of a diagram on $\mathbb{R}^2$ is the framing parallel to $\mathbb{R}^2$. Any framed link is expressed by a diagram by blackboard framing; see Figure 1.7.

Figure 1.7 A framed link and a diagram of it by blackboard framing

The following theorem is obtained as a corollary of Theorem 1.1.

Theorem 1.8. Let L and L' be two framed links, and D and D' diagrams of them by blackboard framings. Then, L is isotopic to L' if and only if D is related to D' by a sequence of isotopies of $\mathbb{R}^2$ and the $\mathcal{RI}$, RII and RIII moves in Figure 1.8.

Sketch of the proof. It is trivial to show that, if D is related to D' by a sequence of the moves, then L is isotopic to L'.

Conversely, suppose that L is isotopic to L'. Then, by forgetting the framings D and D' are related by a sequence of isotopies of $\mathbb{R}^2$ and RI, RII and RIII moves by Theorem 1.1. Modifying the sequence we obtain a sequence of isotopies of $\mathbb{R}^2$ and the $\mathcal{RI}$, RII and RIII moves using the fact that D and D' have the same framing.

We omit the details; it is left to the reader, as an exercise, to complete the proof. □

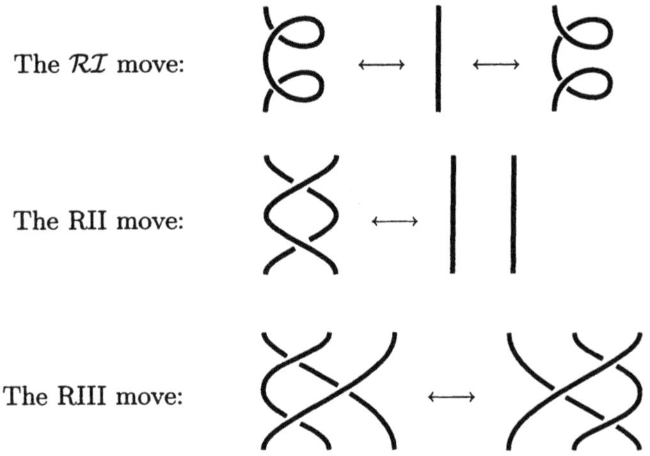

Figure 1.8 The Reidemeister moves for diagrams of framed links

Corollary 1.9. Let L, L', D and D' be as in Theorem 1.8. Further, we regard D and D' as diagrams on $S^2 = \mathbb{R}^2 \cup \{\infty\}$. Then, L is isotopic to L' if and only if D is related to D' by a sequence of isotopies of S^2 and RII and RIII moves.

Proof. We obtain the $\mathcal{RI}$ move as a sequence of isotopies of S^2 and RII, RIII moves as shown in Figure 1.9. Hence, we obtain the corollary by Theorem 1.8. □

Theorem 1.10. Let L be a framed link, and let D be a diagram of L by blackboard framing. Then, the Kauffman bracket $\langle D \rangle$ is an isotopy invariant of L.

We denote the invariant also by $\langle L \rangle$ and call it the *Kauffman bracket* of a framed link L.

Proof of Theorem 1.10. As shown before, $\langle D \rangle$ is invariant under the RII and RIII moves on a diagram D. Hence, by Corollary 1.9 it is an isotopy invariant of a framed link L. □

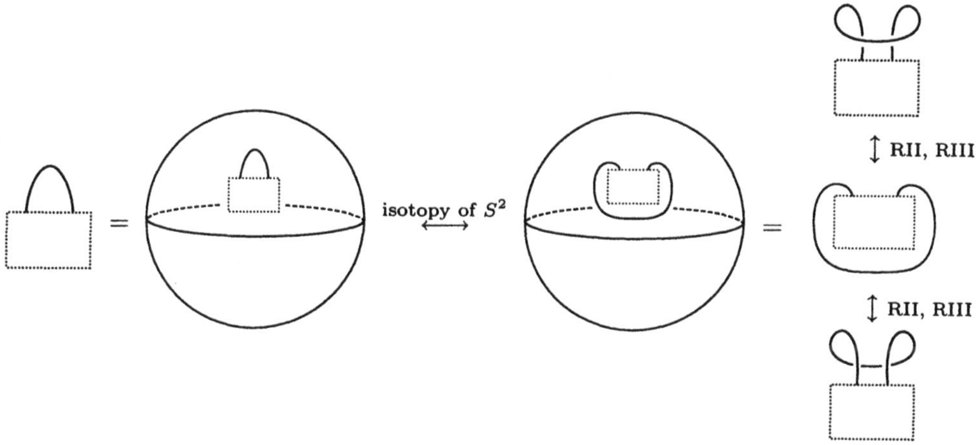

Figure 1.9 The $\mathcal{RI}$ move in Figure 1.8 is derived from isotopy of S^2 and RII and RIII moves.

It is an elementary exercise to show that the Jones polynomial of a knot K can be alternatively defined to be $\langle \mathcal{K} \rangle$ for the framed knot $\mathcal{K}$ which is the knot K with 0 framing, putting $A^4 = t^{-1}$.

1.3 The Alexander polynomial

The Alexander polynomial is the most classical polynomial invariant of knots. It was discovered by Alexander [Ale23] in the 1920s, early in the history of topology, using the homology of the infinite cyclic cover of a knot complement. Further, the homology is presented by the Seifert matrix of the knot. In this section we introduce the Alexander polynomial by computing the homology with the Seifert matrix. There are also other approaches (obtained in the 1960s) to reconstruct the Alexander polynomial; see [CrFo77] for the free differential calculus of Fox and see [Con70] for the skein relation of the Alexander polynomial. See also [Gor78] for a classical history of the Alexander polynomial.

A *Seifert surface* of a knot is a compact oriented surface embedded in $\mathbb{R}^3$ such that the boundary of the surface is equal to the knot. Every knot has a Seifert surface; in fact we can find a Seifert surface of a given knot as follows. We take a diagram of the knot, and replace each crossing of the diagram with two disjoint strands as  . This results in a disjoint union of loops. Further, we consider disjoint discs in $\mathbb{R}^3$ which are bounded by the loops respectively, and attach bands with positive and negative half twist corresponding to positive and negative crossings of the original diagram.

The union of the discs and the bands gives a surface bounded by the original knot. It is a Seifert surface; see Figure 1.10 for graphical expression of this construction.

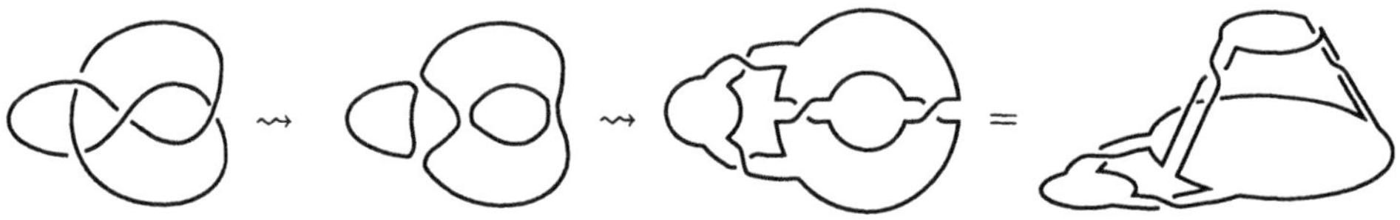

Figure 1.10 Finding a Seifert surface of a knot. As for the figure-eight knot (the left picture), we find its Seifert surface as the union of three discs and four bands between the discs (the right picture).

Let M be the 3-manifold obtained from $S^3 - K$ by cutting it along F. As sections of the cutting we have F^+ and F^- in the boundary of M. Let M_i $(i \in \mathbb{Z})$ be copies of M, and $F_i^{\pm}$ the copies of $F^{\pm}$ in M_i. The infinite cyclic cover $\widetilde{S^3 - K}$ is obtained as the union of M_i gluing F_i^+ and F_{i+1}^-; see Figure 1.11. The covering transformation t acts on $\widetilde{S^3 - K}$ taking M_i to M_{i+1}. By this action $H_1(\widetilde{S^3 - K})$ is a $\mathbb{R}[t, t^{-1}]$ module, where we consider the homology groups with $\mathbb{R}$ coefficients in this section.

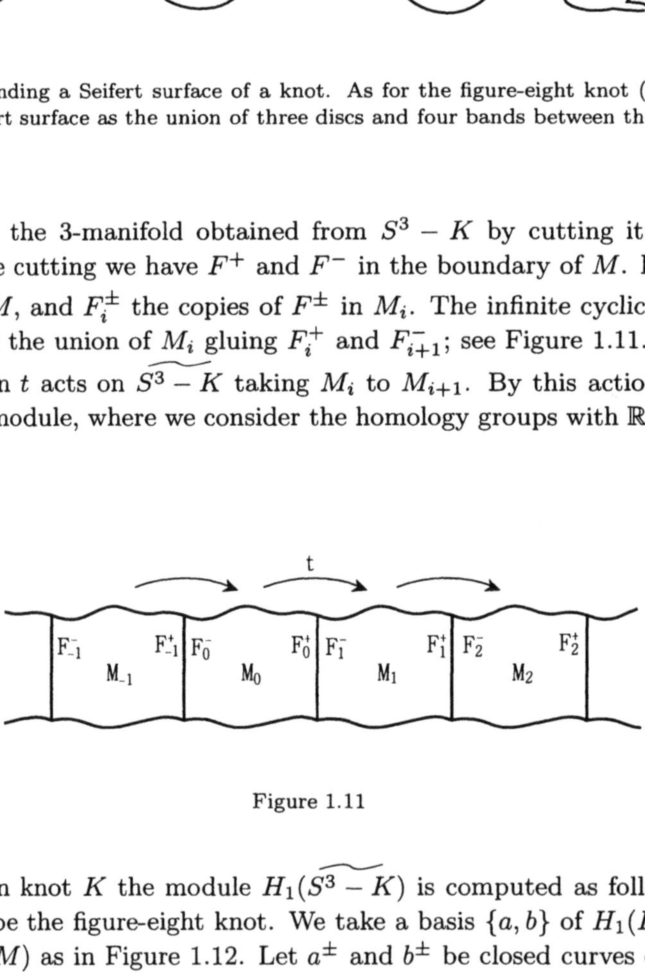

Figure 1.11

For a given knot K the module $H_1(\widetilde{S^3 - K})$ is computed as follows. For simplicity let K be the figure-eight knot. We take a basis $\{a, b\}$ of $H_1(F)$ and a basis $\{\alpha, \beta\}$ of $H_1(M)$ as in Figure 1.12. Let $a^{\pm}$ and $b^{\pm}$ be closed curves obtained from the curves a and b by pushing in the normal directions of F; we take the direction "+" as shown in Figure 1.12. Further, we regard $a^{\pm}$ and $b^{\pm}$ as curves in $F^{\pm} \subset M$. As elements in $H_1(M)$ we have that

$$a^+ = \alpha + \beta, \qquad a^- = \alpha, \qquad b^+ = -\beta, \qquad b^- = \alpha - \beta.$$

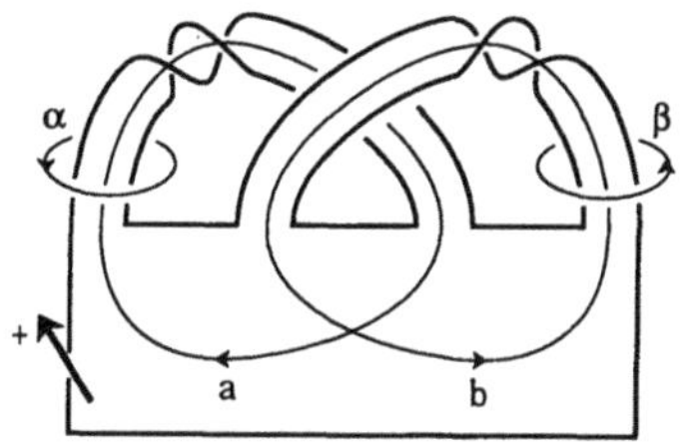

Figure 1.12 A Seifert surface F of the figure-eight knot. $\{a,b\}$ is a basis of $H_1(F)$ and $\{\alpha,\beta\}$ is a basis of $H_1(S^3-F)$.

We present these formulae by

$$\begin{pmatrix} a^+ \\ b^+ \end{pmatrix} = V \begin{pmatrix} \alpha \\ \beta \end{pmatrix}, \qquad \begin{pmatrix} a^- \\ b^- \end{pmatrix} = {}^tV \begin{pmatrix} \alpha \\ \beta \end{pmatrix},$$

by introducing the matrix V by

$$V = \begin{pmatrix} \mathrm{lk}(a^+,a) & \mathrm{lk}(a^+,b) \\ \mathrm{lk}(b^+,a) & \mathrm{lk}(b^+,b) \end{pmatrix} = \begin{pmatrix} 1 & 1 \\ 0 & -1 \end{pmatrix}.$$

Further, since $\widetilde{S^3-K}$ is obtained by gluing F_i^+ and F_{i+1}^-, the $\mathbb{R}[t,t^{-1}]$ module $H_1(\widetilde{S^3-K})$ is presented by

$$\mathrm{span}_{\mathbb{R}}\left\{ \begin{matrix} \alpha_i, \beta_i, a_i^\pm, b_i^\pm, \\ \text{for } i \in \mathbb{Z} \end{matrix} \right\} \Big/ \left(\begin{pmatrix} a_i^+ \\ b_i^+ \end{pmatrix} = V \begin{pmatrix} \alpha_i \\ \beta_i \end{pmatrix}, \begin{pmatrix} a_i^- \\ b_i^- \end{pmatrix} = {}^tV \begin{pmatrix} \alpha_i \\ \beta_i \end{pmatrix}, \begin{pmatrix} a_i^+ \\ b_i^+ \end{pmatrix} = \begin{pmatrix} a_{i+1}^- \\ b_{i+1}^- \end{pmatrix} \right),$$

where the action of t is given by $tx_i = x_{i+1}$ for $x_i = \alpha_i, \beta_i, a_i^\pm, b_i^\pm$. By decreasing the number of generators using the relations and the action of t we have that

$$H_1(\widetilde{S^3-K}) \cong \mathrm{span}_{\mathbb{R}}\left\{ t^i \begin{pmatrix} \alpha_0 \\ \beta_0 \end{pmatrix} \text{ for } i \in \mathbb{Z} \right\} \Big/ \mathrm{span}_{\mathbb{R}}\left\{ t^i (t\,{}^tV - V) \begin{pmatrix} \alpha_0 \\ \beta_0 \end{pmatrix} \text{ for } i \in \mathbb{Z} \right\}.$$

The numerator on the right hand side is regarded as

$$\mathrm{span}_{\mathbb{R}[t,t^{-1}]}\{\alpha_0, \beta_0\} \cong \mathbb{R}[t,t^{-1}]^2,$$

where we mean by $\mathbb{R}[t,t^{-1}]^2$ a free $\mathbb{R}[t,t^{-1}]$ module with the basis $\{\alpha_0,\beta_0\}$. Further, the denominator is regarded as

$$\mathrm{span}_{\mathbb{R}[t,t^{-1}]}\left\{ (t\,{}^tV - V) \begin{pmatrix} \alpha_0 \\ \beta_0 \end{pmatrix} \right\} \cong (t\,{}^tV - V)\mathbb{R}[t,t^{-1}]^2,$$

which is a submodule of $\mathbb{R}[t,t^{-1}]^2$ spanned by multiples of $(t\,{}^tV - V)$. Hence, we have the following expression

$$H_1(\widetilde{S^3 - K}) \cong \mathbb{R}[t,t^{-1}]^2 \Big/ \Big((t\,{}^tV - V)\mathbb{R}[t,t^{-1}]^2\Big)$$

for the figure-eight knot K. This formula implies that the matrix $(t\,{}^tV - V)$ is a presentation matrix for the $\mathbb{R}[t,t^{-1}]$ module $H_1(\widetilde{S^3 - K})$. For a general knot K we obtain almost the same presentation of $H_1(\widetilde{S^3 - K})$ in the same way as follows. Any Seifert surface F of genus g of a knot K is expressed (up to isotopy) as in Figure 1.13. We take a basis $\{a_1, b_1, \cdots, a_g, b_g\}$ of $H_1(F)$ and a basis $\{\alpha_1, \beta_1, \cdots, \alpha_g, \beta_g\}$ of $H_1(S^3 - F)$ as in Figure 1.13.

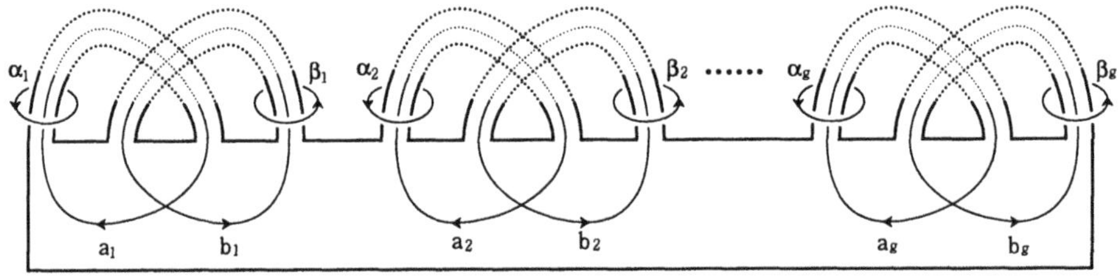

Figure 1.13 Any Seifert surface of genus g can be expressed (up to isotopy) as shown in the picture.

Further, we give a matrix V by

$$V = \begin{pmatrix} \mathrm{lk}(a_1^+, a_1) & \mathrm{lk}(a_1^+, b_1) & \cdots & \cdots & \mathrm{lk}(a_1^+, a_g) & \mathrm{lk}(a_1^+, b_g) \\ \mathrm{lk}(b_1^+, a_1) & \mathrm{lk}(b_1^+, b_1) & \cdots & \cdots & \mathrm{lk}(b_1^+, a_g) & \mathrm{lk}(b_1^+, b_g) \\ \cdots & \cdots & \cdots & \cdots & \cdots & \cdots \\ \cdots & \cdots & \cdots & \cdots & \cdots & \cdots \\ \mathrm{lk}(a_g^+, a_1) & \mathrm{lk}(a_g^+, b_1) & \cdots & \cdots & \mathrm{lk}(a_g^+, a_g) & \mathrm{lk}(a_g^+, b_g) \\ \mathrm{lk}(b_g^+, a_1) & \mathrm{lk}(b_g^+, b_1) & \cdots & \cdots & \mathrm{lk}(b_g^+, a_g) & \mathrm{lk}(b_g^+, b_g) \end{pmatrix}.$$

We call V a *Seifert matrix* of the Seifert surface F. The matrix V is a presentation matrix of the bilinear form $H_1(F) \times H_1(F) \to \mathbb{R}$ which takes (x, y) to $\mathrm{lk}(x^+, y)$. In the same way as the above case of the figure-eight knot, we obtain a presentation of the $\mathbb{R}[t,t^{-1}]$ module $H_1(\widetilde{S^3 - K})$ by

$$H_1(\widetilde{S^3 - K}) \cong \mathbb{R}[t,t^{-1}]^{2g} \Big/ \Big((t\,{}^tV - V)\mathbb{R}[t,t^{-1}]^{2g}\Big).$$

Moreover, since $\mathbb{R}[t,t^{-1}]$ is a principal ideal domain, it is known (see, *e.g.*, [HaHa70]) that the matrix $(t\,{}^tV - V)$ can be adjusted to a diagonal matrix as

$$P(t\,{}^tV - V)Q = \begin{pmatrix} f_1(t) & & & 0 \\ & f_2(t) & & \\ & & \ddots & \\ 0 & & & f_{2g}(t) \end{pmatrix} \qquad (1.13)$$

for some matrices P and Q with unit determinants in $\mathbb{R}[t,t^{-1}]$, where the series $\{f_i(t)\}$ satisfies that $f_i(t)$ is divisible by $f_{i-1}(t)$ for any i. Further, the series $\{f_i(t)\}$ is uniquely determined (unique up to multiple of units in $\mathbb{R}[t,t^{-1}]$) by the module $H_1(\widetilde{S^3 - K})$. In particular,

$$H_1(\widetilde{S^3 - K}) \cong \bigoplus_i \mathbb{R}[t,t^{-1}]/(f_i(t)).$$

Theorem 1.11. Let K be a knot and V a Seifert matrix of a Seifert surface of K. Set

$$\Delta_K(t) = \det(t^{1/2}\,{}^tV - t^{-1/2}V) \in \mathbb{Z}[t,t^{-1}]. \qquad (1.14)$$

Then, it is an isotopy invariant of K.

We call the invariant of the theorem the *Alexander polynomial* of a knot K. For example, for the knots in Figure 1.1 we have values of the Alexander polynomials of them as shown in the beginning of Section 1.1.

Proof. By the expression (1.13) we have that

$$\prod_i f_i(t) \quad \sim \quad \det(t\,{}^tV - V) \quad \sim \quad \Delta_K(t)$$

where the notation "$\sim$" implies equality up to a multiple of a unit ct^i in $\mathbb{R}[t,t^{-1}]$. Since $\prod_i f_i(t)$ is uniquely determined up to a multiple of a unit by the module $H_1(\widetilde{S^3 - K})$ as mentioned above, so is $\Delta_K(t)$. Hence, the set $\{ct^i\Delta_K(t) \mid i \in \mathbb{Z}, c \in \mathbb{R} - \{0\}\}$ is uniquely determined by K.

From the definition of $\Delta_K(t)$, we have that $\Delta_K(t) = \Delta_K(t^{-1})$, noting that the size of a Seifert matrix is always even. Further, by Lemma 1.12 below, $\Delta_K(1) = 1$. Since the element $f(t)$ in the above set satisfying the properties $f(t) = f(t^{-1})$ and $f(1) = 1$ is unique, the Alexander polynomial $\Delta_K(t)$ is uniquely determined by K as the unique polynomial with the required properties in the above set. □

Lemma 1.12. For a Seifert matrix V of a Seifert surface of a knot, $\det({}^tV - V) = 1$.

Proof. Recall that a Seifert matrix V is a presentation matrix of the bilinear form $s : H_1(F;\mathbb{Z}) \times H_1(F;\mathbb{Z}) \to \mathbb{Z}$ given by $s(x,y) = \mathrm{lk}(x^+, y)$. Further, we consider the bilinear form ι given by

$$\iota(x,y) = s(y,x) - s(x,y),$$

noting that ${}^tV - V$ is a presentation matrix of ι. From the definitions of s and ι, we have that

$$\iota(x,y) = \mathrm{lk}(y^+,x) - \mathrm{lk}(x^+,y) = \mathrm{lk}(x,y^+) - \mathrm{lk}(x,y^-) = \mathrm{lk}(x,y^+ - y^-).$$

This is equal to the intersection number of x and $y \times I$ in $\mathbb{R}^3$, where I is a small interval such that the boundary of $y \times I$ is the disjoint union of y^+ and y^-. Further, it is equal to the intersection number of x and y on F. Therefore, the bilinear form ι is equal to the intersection form on $H_1(F,\mathbb{Z})$. In particular, for the basis $\{a_1, b_1, \cdots, a_g, b_g\}$ in Figure 1.13 its presentation matrix ${}^tV - V$ is equal to the block sum of g copies of $\begin{pmatrix} 0 & -1 \\ 1 & 0 \end{pmatrix}$. Hence, its determinant is equal to 1. □

For an oriented link L we define the Alexander polynomial $\Delta_L(t)$ in the same way as above. In this case a Seifert surface of L is a connected oriented surface whose boundary is L. With a Seifert matrix V of a Seifert surface of L we define $\Delta_L(t)$ by (1.14). In the case of links, there is a technical difficulty to show the invariance of $\Delta_L(t)$ unlike the case of knots, since $\Delta_L(1) = 0$ for a link L with at least two components. See [Kau81] for a proof of the invariance based on this definition of the Alexander polynomial of links. We give, later in Theorem 3.12, another alternative definition of the Alexander polynomial of links with a proof of its invariance.

Chapter 2

Braids and representations of the braid groups

Braids were introduced by Artin in the 1920s. Links are expressed by braids. Further, braids make braid groups. The set of (isotopy classes of) links can be identified with a quotient set of the union of the braid groups (see Section 2.1). Thus, by using braids, we can reduce studies of the set of knots and links to studies of algebraic objects. A remarkable advantage of such an identification is that we can use the group structure to study them.

We give two kinds of representations of braid groups in Sections 2.2 and 2.3. Of one kind are those representations derived from R matrices (which are solutions of the Yang-Baxter equation). They are representations on the tensor product $V^{\otimes n}$ of n copies of a vector space V. From a certain (modified) trace of such representations we obtain isotopy invariants of links. The Jones and Alexander polynomials of links can be recovered from such representations. Another kind of representation is the Burau representation. It is a representation on $\mathbb{C}^n$. The determinant of matrices obtained from the Burau representation yields another construction of the Alexander polynomial of links. Thus, the Alexander polynomial is obtained from two different representations of braid groups; we show the relation between them in Appendix C.1.

2.1 Braids and braid groups

A *braid* in n strands is a union of n strands embedded in $\mathbb{R}^2 \times [0,1]$ such that its boundary is the set $\{1, 2, \cdots, n\} \times \{0\} \times \{0,1\}$ in $\mathbb{R}^2 \times [0,1]$ and such that no strand has a critical point with respect to the vertical coordinate; see, for example, Figure 2.1. Two braids are *isotopic* if they are related by an isotopy of $\mathbb{R}^2 \times [0,1]$, preserving its boundary and the vertical coordinate.

Any braid in n strands is obtained from σ_i's and σ_i^{-1}'s ($i = 1, 2, \cdot, n-1$) by attaching together vertically. For example, the braid in Figure 2.1 is presented (up to isotopy) by

$$\sigma_1 \sigma_3 \sigma_2^{-1} \sigma_4^{-1} \sigma_1 \sigma_4^{-1} \sigma_2 \sigma_3^{-1},$$

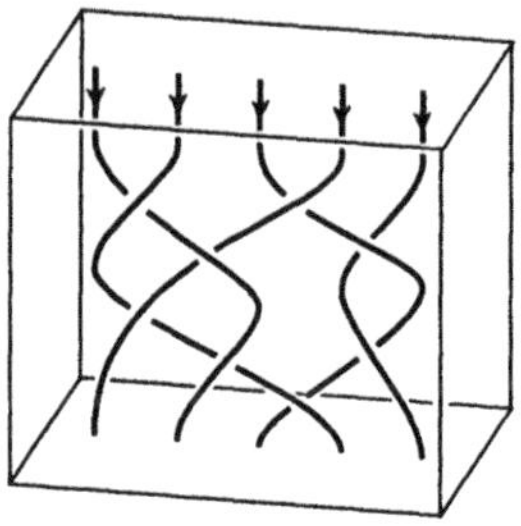

Figure 2.1 A braid in 5 strands

where this product of generators implies the braid obtained by attaching the generators from the top to the bottom.

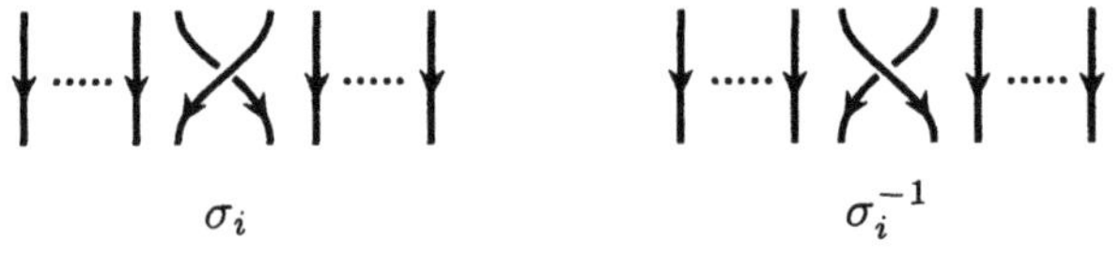

Figure 2.2 σ_i's generate braids. σ_i (resp. σ_i^{-1}) has a positive (resp. negative) crossing between the i-th and $(i+1)$-th strands.

The set of the isotopy classes of braids in n strands becomes a group with respect to the product of attaching braids vertically. We call the group the *braid group* in n strands and denote it by B_n. It is known that the braid group B_n has the following presentation,

$$\begin{aligned} \text{generators} \quad &: \quad \sigma_1, \sigma_2, \cdots, \sigma_{n-1}, \\ \text{relations} \quad &: \quad \sigma_i \sigma_j = \sigma_j \sigma_i \quad (|i-j| \geq 2), \\ & \qquad \sigma_i \sigma_{i+1} \sigma_i = \sigma_{i+1} \sigma_i \sigma_{i+1} \quad (i = 1, 2, \cdots, n-2). \end{aligned}$$

It is easy to show that the relations of the braid group are derived from isotopy and Reidemeister moves. Conversely, it is not so easy to show that the above relations are sufficient; it can be shown by applying the same argument as in the proof of Reidemeister's theorem (Theorem 1.1) to isotopy of braids. We omit a detailed argument; see [Bir74, PrSo97] for complete arguments.

The *closure* of a braid is the link obtained from the braid by connecting upper ends and lower ends respectively as shown in Figure 2.3. The following theorem assures us that such an expression always exists.

Theorem 2.1 (Alexander). Any (oriented) link is isotopic to the closure of some braid (with downward orientation).

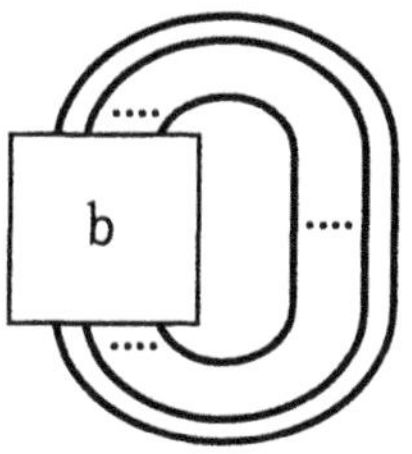

Figure 2.3 The closure of a braid b

Sketch of the proof. For simplicity, we will only sketch a proof for the case of the figure-eight knot shown in Figure 2.4. As in the figure, we mark in the diagrams of the knot with a cross. We choose a part of the diagram with clockwise orientation with respect to the mark; the chosen part is shown by dotted line in the second picture in the figure. We then move the part by isotopy of $\mathbb{R}^3$ so that the part becomes counter-clockwise. Then, repeating this procedure if necessary, we obtain a diagram whose strand is counter-clockwise everywhere. Further, by isotopy of $\mathbb{R}^2$, moving the area surrounded by the thin line in the third picture to a box depicted by thin line in the last picture, the diagram becomes isotopic to the closure of a braid. For detailed proofs see [Bir74, PrSo97].

See also [Yam87, Vog90, Kaw+90, PrSo97] for another algorithm, using "Seifert circles"; this algorithm is economical in the number of extra crossings which have to be introduced. □

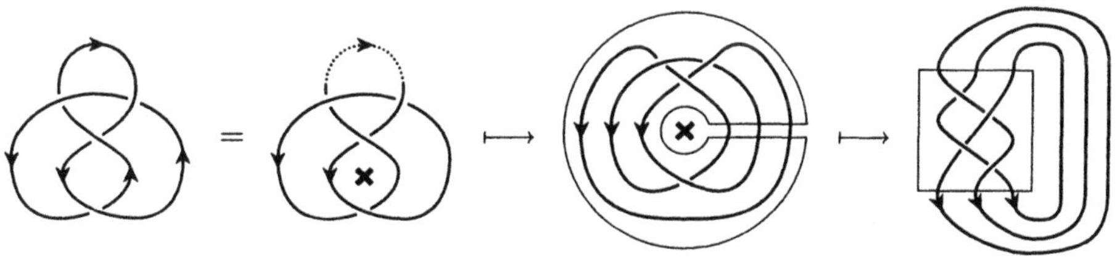

Figure 2.4 The figure-eight knot (the left picture) is isotopic to the closure of some braid (the right picture).

We consider the following map

$$\bigcup_{n=1}^{\infty} B_n \longrightarrow \{\text{oriented links}\}/\text{isotopic},$$

which takes a braid to its closure. Theorem 2.1 implies that this map is surjective. Further, the kernel of the map can be described, by the following theorem.

Theorem 2.2 (Markov). Let b and b' be two braids, and L and L' their closures. Then, L is isotopic to L' as oriented links if and only if b is related to b' by a sequence of the following MI and MII moves,

$$\text{The MI move:} \qquad ab \longleftrightarrow ba \qquad \text{for any } a, b \in B_n,$$
$$\text{The MII move:} \qquad b\sigma_n \longleftrightarrow b \longleftrightarrow b\sigma_n^{-1} \qquad \text{for any } b \in B_n,$$

where we regard the b of $b\sigma_n^{\pm 1}$ in the MII move as the braid in B_{n+1} obtained from the original b by adding a trivial $(n+1)$-st strand.

We call the MI and MII moves the *Markov moves*.

Theorem 2.2 implies the following equality

$$\{\text{oriented links}\}/\text{isotopic} = \Big(\bigcup_{n=1}^{\infty} B_n\Big)/\text{the MI, MII moves}.$$

Note that the left hand side is topological while the right hand side is algebraic. Through the equation we can reduce studies of the set of isotopy classes of links to studies of the right hand side of the equation.

Sketch of the proof of Theorem 2.2. If two braids are related by each of the MI and MII moves, their closures are isotopic as

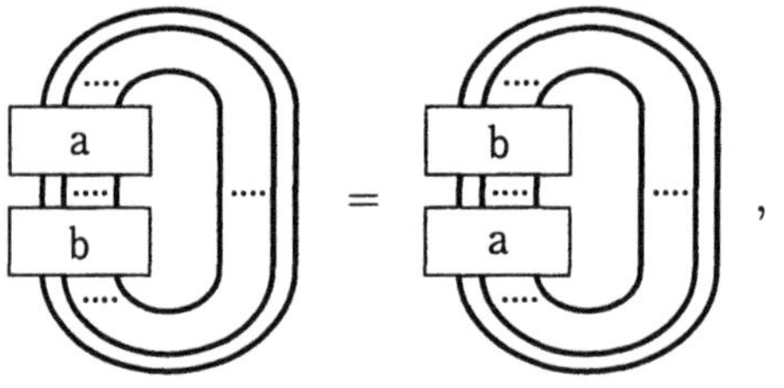

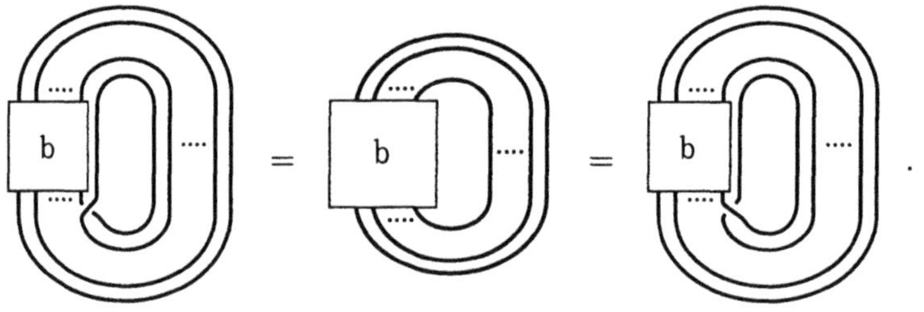

Conversely, if the closures of two braids are isotopic, then the closures are related by a sequence of Reidemeister moves by Theorem 1.2. By modifying the sequence to another sequence consisting of closed braids, it is shown that the original two braids are related by a sequence of Markov moves. We omit a detailed argument; see [Bir74] for a complete proof. □

2.2 Representations of the braid groups via R matrices

In this section we introduce representations of the braid groups obtained from R matrices and reconstruct the Jones polynomial via such representations.

We review trace of linear maps. Let V be a vector space over $\mathbb{C}$, $V^\star$ its dual vector space, and let $\mathrm{End}(V)$ denote the set of endomorphisms of V, *i.e.*, the set of linear maps $V \to V$. There is a natural identification between $V^\star \otimes V$ and $\mathrm{End}(V)$ such that $f \otimes x \in V^\star \otimes V$ is identified with the map $(y \mapsto f(y)x) \in \mathrm{End}(V)$. The trace on $\mathrm{End}(V)$ is the following composed linear map

$$\mathrm{trace} : \mathrm{End}(V) \longrightarrow V^\star \otimes V \longrightarrow \mathbb{C},$$

where the first map is the above identification and the second map is the contraction defined by $f \otimes x \mapsto f(x)$ for $f \in V^\star$ and $x \in V$. With respect to a basis $\{e_i\}$ of V, a linear map $A \in \mathrm{End}(V)$ is presented by $A(e_i) = \sum_j A_i^j e_j$ with a matrix (A_j^i). The trace of A is equal to the sum of the diagonal entries of the matrix, *i.e.*, $\mathrm{trace}(A) = \sum_i A_i^i$. Further, let V_1 and V_2 be vector spaces over $\mathbb{C}$. For an endomorphism in $\mathrm{End}(V_1 \otimes V_2)$ we define trace_i of the endomorphism to be the trace with respect to the i-th entry as

$$\begin{aligned}
\mathrm{trace}_1 : \mathrm{End}(V_1 \otimes V_2) &= (V_1 \otimes V_2)^\star \otimes (V_1 \otimes V_2) \\
&= V_2^\star \otimes V_1^\star \otimes V_1 \otimes V_2 \overset{\text{contraction}}{\longrightarrow} V_2^\star \otimes V_2 = \mathrm{End}(V_2), \\
\mathrm{trace}_2 : \mathrm{End}(V_1 \otimes V_2) &= (V_1 \otimes V_2)^\star \otimes (V_1 \otimes V_2) \\
&= V_2^\star \otimes V_1^\star \otimes V_1 \otimes V_2 \overset{\text{contraction}}{\longrightarrow} V_1^\star \otimes V_1 = \mathrm{End}(V_1),
\end{aligned}$$

where the contractions are the contractions $V_i^\star \otimes V_i \to \mathbb{C}$ for $i = 1, 2$ respectively. With respect to the bases $\{e_i\}$ and $\{e_i'\}$ of V_1 and V_2, a linear map $A \in \mathrm{End}(V_1 \otimes V_2)$ is presented by $A(e_i \otimes e_j) = \sum_{k,l} A_{ij}^{kl} e_k \otimes e_l$. The trace_1 and trace_2 are presented by

$$\begin{aligned}
\Big(\mathrm{trace}_1(A)\Big)(e_i') &= \sum_{j,k} A_{ki}^{kj} e_j', \\
\Big(\mathrm{trace}_2(A)\Big)(e_i) &= \sum_{j,k} A_{ik}^{jk} e_j.
\end{aligned}$$

For example, when V is a 2-dimensional vector space with a basis $\{e_0, e_1\}$, a linear map $A \in \mathrm{End}(V \otimes V)$ is presented by a matrix

$$A = \begin{pmatrix} A_{00}^{00} & A_{00}^{01} & A_{00}^{10} & A_{00}^{11} \\ A_{01}^{00} & A_{01}^{01} & A_{01}^{10} & A_{01}^{11} \\ A_{10}^{00} & A_{10}^{01} & A_{10}^{10} & A_{10}^{11} \\ A_{11}^{00} & A_{11}^{01} & A_{11}^{10} & A_{11}^{11} \end{pmatrix}$$

with respect to the basis $\{e_0 \otimes e_0,\ e_0 \otimes e_1,\ e_1 \otimes e_0,\ e_1 \otimes e_1\}$ of $V \otimes V$. By using the above matrix, trace_1 and trace_2 are presented by

$$\text{trace}_1(A) = \begin{pmatrix} A^{00}_{00} + A^{10}_{10} & A^{01}_{00} + A^{11}_{10} \\ A^{00}_{01} + A^{10}_{11} & A^{01}_{01} + A^{11}_{11} \end{pmatrix},$$

$$\text{trace}_2(A) = \begin{pmatrix} A^{00}_{00} + A^{01}_{01} & A^{10}_{00} + A^{11}_{01} \\ A^{00}_{10} + A^{01}_{11} & A^{10}_{10} + A^{11}_{11} \end{pmatrix}.$$

The n-th *symmetric group* $\mathfrak{S}_n$ is the group of bijections of the set $\{1, 2, \cdots, n\}$ to itself. There is a natural surjective homomorphism $B_n \to \mathfrak{S}_n$ defined by $\sigma_i \mapsto s_i$, where s_i denotes the permutation of i and $i+1$. Let V be a 2-dimensional vector space over $\mathbb{C}$ with a basis $\{e_0, e_1\}$. We consider the representation $\psi_n : \mathfrak{S}_n \to \text{End}(V^{\otimes n})$, where $V^{\otimes n}$ denotes the tensor product of n copies of V, such that $\psi_n(s_i)$ exchanges the i-th and $(i+1)$-th entries of $V^{\otimes n}$, *i.e.*, we put

$$\psi_n(s_i) = (\text{id}_V)^{\otimes(i-1)} \otimes P \otimes (\text{id}_V)^{\otimes(n-i-1)}$$

where the linear map $P : V \otimes V \to V \otimes V$ is defined by $P(x \otimes y) = y \otimes x$ for $x, y \in V$. With respect to the basis $\{e_0 \otimes e_0,\ e_0 \otimes e_1,\ e_1 \otimes e_0,\ e_1 \otimes e_1\}$ of $V \otimes V$ the linear map P is presented by

$$P = \begin{pmatrix} 1 & 0 & 0 & 0 \\ 0 & 0 & 1 & 0 \\ 0 & 1 & 0 & 0 \\ 0 & 0 & 0 & 1 \end{pmatrix} \in \text{End}(V \otimes V).$$

By composing this with the natural homomorphism $B_n \to \mathfrak{S}_n$ we obtain the representation $\psi_n : B_n \to \text{End}(V^{\otimes n})$ defined by

$$\psi_n(\sigma_i) = (\text{id}_V)^{\otimes(i-1)} \otimes P \otimes (\text{id}_V)^{\otimes(n-i-1)}.$$

Modifying the above representation we try to obtain a representation $\psi_n : B_n \to \text{End}(V^{\otimes n})$ defined by

$$\psi_n(\sigma_i) = (\text{id}_V)^{\otimes(i-1)} \otimes R \otimes (\text{id}_V)^{\otimes(n-i-1)} \tag{2.1}$$

for some linear map $R : V \otimes V \to V \otimes V$ different from P, say, a linear map obtained from P by perturbation. To obtain such a representation ψ_n of the braid group B_n from the map R, the map ψ_n is required to satisfy the following relations,

$$\psi_n(\sigma_i \sigma_j) = \psi_n(\sigma_j \sigma_i), \qquad |i - j| \geq 2, \tag{2.2}$$

$$\psi_n(\sigma_i \sigma_{i+1} \sigma_i) = \psi_n(\sigma_{i+1} \sigma_i \sigma_{i+1}). \tag{2.3}$$

Let us consider what is required of a map R so that ψ_n will satisfy the above relations. Such a map ψ_n given in (2.1) always satisfies (2.2). To obtain (2.3) the matrix R is required to satisfy the relation

$$(R \otimes \text{id}_V)(\text{id}_V \otimes R)(R \otimes \text{id}_V) = (\text{id}_V \otimes R)(R \otimes \text{id}_V)(\text{id}_V \otimes R).$$

We call this equation the *Yang-Baxter equation*, and call a solution of it an *R matrix*. For an R matrix we obtain a representation ψ_n of the braid group B_n by (2.1).

Using the notation $R(e_i \otimes e_j) = \sum R_{ij}^{kl} e_k \otimes e_l$, we consider R matrices which satisfy the property that $R_{ij}^{kl} = 0$ unless $i + j = k + l$, *i.e.*, the matrix R preserves the sum of subscripts of the basis. We call this property *charge conservation.*

Let us compute an R matrix with the property of charge conservation in the simple case that V is a 2-dimensional vector space with a basis $\{e_0, e_1\}$. With respect to the basis $\{e_0 \otimes e_0,\ e_0 \otimes e_1,\ e_1 \otimes e_0,\ e_1 \otimes e_1\}$ of $V \otimes V$ we put

$$R = \begin{pmatrix} a & 0 & 0 & 0 \\ 0 & b & c & 0 \\ 0 & d & e & 0 \\ 0 & 0 & 0 & f \end{pmatrix}.$$

By charge conservation of the R matrix the linear maps $R \otimes \mathrm{id}_V$ and $\mathrm{id}_V \otimes R$ preserve the four subspaces of $V \otimes V \otimes V$ spanned by the basis $\{e_0 \otimes e_0 \otimes e_0\}$, $\{e_0 \otimes e_0 \otimes e_1,\ e_0 \otimes e_1 \otimes e_0,\ e_1 \otimes e_0 \otimes e_0\}$, $\{e_0 \otimes e_1 \otimes e_1,\ e_1 \otimes e_0 \otimes e_1,\ e_1 \otimes e_1 \otimes e_0\}$ and $\{e_1 \otimes e_1 \otimes e_1\}$ respectively. By regarding $V \otimes V \otimes V$ as the direct sum of the four subspaces, the linear maps are presented by

$$R \otimes \mathrm{id}_V = (a) \oplus \begin{pmatrix} a & 0 & 0 \\ 0 & b & c \\ 0 & d & e \end{pmatrix} \oplus \begin{pmatrix} b & c & 0 \\ d & e & 0 \\ 0 & 0 & f \end{pmatrix} \oplus (f),$$

$$\mathrm{id}_V \otimes R = (a) \oplus \begin{pmatrix} b & c & 0 \\ d & e & 0 \\ 0 & 0 & a \end{pmatrix} \oplus \begin{pmatrix} f & 0 & 0 \\ 0 & b & c \\ 0 & d & e \end{pmatrix} \oplus (f).$$

Hence,

$$(R \otimes \mathrm{id}_V)(\mathrm{id}_V \otimes R) = (a^2) \oplus \begin{pmatrix} ab & ac & 0 \\ bd & be & ac \\ d^2 & de & ae \end{pmatrix} \oplus \begin{pmatrix} bf & bc & c^2 \\ df & be & ce \\ 0 & df & ef \end{pmatrix} \oplus (f^2).$$

Further,

$$\begin{aligned} &(R \otimes \mathrm{id}_V)(\mathrm{id}_V \otimes R)(R \otimes \mathrm{id}_V) \\ &= (a^3) \oplus \begin{pmatrix} a^2b & abc & ac^2 \\ abd & b^2e + acd & bce + ace \\ ad^2 & bde + ade & cde + ae^2 \end{pmatrix} \oplus \begin{pmatrix} b^2f + bcd & bcf + bce & c^2f \\ bdf + bde & cdf + be^2 & cef \\ d^2f & def & ef^2 \end{pmatrix} \oplus (f^3), \end{aligned}$$

$$\begin{aligned} &(\mathrm{id}_V \otimes R)(R \otimes \mathrm{id}_V)(\mathrm{id}_V \otimes R) \\ &= (a^3) \oplus \begin{pmatrix} ab^2 + bcd & abc + bce & ac^2 \\ abd + bde & acd + be^2 & ace \\ ad^2 & ade & a^2e \end{pmatrix} \oplus \begin{pmatrix} bf^2 & bcf & c^2f \\ bdf & b^2e + cdf & bce + cef \\ d^2f & bde + def & cde + e^2f \end{pmatrix} \oplus (f^3). \end{aligned}$$

Since they are required to be equal, we have that

$$\begin{aligned} &b(cd+ab-a^2)=0, \quad b(cd+bf-f^2)=0,\\ &e(cd+ae-a^2)=0, \quad e(cd+ef-f^2)=0,\\ &bce=0, \quad bde=0, \quad be(b-e)=0. \end{aligned}$$

It is interesting to consider the case that $b=0$ and $e\neq 0$. In this case,

$$a^2-ae=cd=f^2-ef.$$

Hence,

$$e=a-\frac{cd}{a}, \qquad (a-f)(a+f-e)=0.$$

Corresponding to the two cases $a-f=0$ and $a+f-e=0$, we obtain the following two R matrices,

$$R=\begin{pmatrix} a & 0 & 0 & 0\\ 0 & 0 & c & 0\\ 0 & d & a-cd/a & 0\\ 0 & 0 & 0 & a \end{pmatrix}, \quad \begin{pmatrix} a & 0 & 0 & 0\\ 0 & 0 & c & 0\\ 0 & d & a-cd/a & 0\\ 0 & 0 & 0 & -cd/a \end{pmatrix}. \tag{2.4}$$

We obtain the Jones and Alexander polynomials from the first and second R matrices respectively. We discuss the first R matrix below, to give a reconstruction of the Jones polynomial, while we discuss the second R matrix in Section 2.3.

After some normalization the first R matrix in (2.4) is presented by

$$R=\begin{pmatrix} t^{1/2} & 0 & 0 & 0\\ 0 & 0 & t & 0\\ 0 & t & t^{1/2}-t^{3/2} & 0\\ 0 & 0 & 0 & t^{1/2} \end{pmatrix} \in \mathrm{End}(V\otimes V).$$

The above R matrix induces a representation ψ_n of the braid group B_n by (2.1), as mentioned before. We try to make an invariant of a link L by taking the trace of $\psi_n(b)$ where b is a braid whose closure is isotopic to L. To obtain an isotopy invariant of L, we require the invariance of $\mathrm{trace}\big(\psi_n(b)\big)$ under the MI and MII moves. The invariance under the MI move is obtained from the property of trace that $\mathrm{trace}(AB)=\mathrm{trace}(BA)$ holds for any A and B. To obtain the invariance under the MII move we require the equality $\mathrm{trace}_2 R^{\pm 1}=1$. However, a concrete computation shows that the equality fails unfortunately. Hence, we consider the following modification. Using the linear map $h=\begin{pmatrix} t^{-1/2} & 0\\ 0 & t^{1/2}\end{pmatrix}\in \mathrm{End}(V)$, we have that

$$\mathrm{trace}_2\Big((\mathrm{id}_V \otimes h)\cdot R\Big) = \mathrm{trace}_2 \begin{pmatrix} 1 & 0 & 0 & 0 \\ 0 & 0 & t^{3/2} & 0 \\ 0 & t^{1/2} & 1-t & 0 \\ 0 & 0 & 0 & t \end{pmatrix} = \begin{pmatrix} 1 & 0 \\ 0 & 1 \end{pmatrix} = \mathrm{id}_V,$$

$$\mathrm{trace}_2\Big((\mathrm{id}_V \otimes h)\cdot R^{-1}\Big) = \mathrm{trace}_2 \begin{pmatrix} t^{-1} & 0 & 0 & 0 \\ 0 & 1-t^{-1} & t^{-1/2} & 0 \\ 0 & t^{-1} & 0 & 0 \\ 0 & 0 & 0 & 1 \end{pmatrix} = \begin{pmatrix} 1 & 0 \\ 0 & 1 \end{pmatrix} = \mathrm{id}_V.$$

That is,

$$\mathrm{trace}_2\Big((\mathrm{id}_V \otimes h)\cdot R^{\pm 1}\Big) = \mathrm{id}_V.$$

Hence,

$$\begin{aligned} \mathrm{trace}\big(h^{\otimes(n+1)} \cdot \psi_{n+1}(\sigma_n^{\pm 1} b)\big) &= \mathrm{trace}\big(h^{\otimes(n+1)} \cdot (\mathrm{id}_V^{\otimes(n-1)} \otimes R^{\pm 1}) \cdot \psi_{n+1}(b)\big) \\ &= \mathrm{trace}\big(h^{\otimes n} \cdot \psi_n(b)\big). \end{aligned}$$

This formula implies that $\mathrm{trace}\big(h^{\otimes n} \cdot \psi_n(b)\big)$ is invariant under the MII move (up to the MI move). Further, by charge conservation of R,

$$R \cdot (h \otimes h) = (h \otimes h) \cdot R.$$

This formula implies that $\psi_n(b)$ commutes with $h^{\otimes n}$. Hence,

$$\begin{aligned} \mathrm{trace}\Big(h^{\otimes n} \cdot \psi_n(b_1 b_2)\Big) &= \mathrm{trace}\Big(h^{\otimes n} \cdot \psi_n(b_1)\psi_n(b_2)\Big) = \mathrm{trace}\Big(\psi_n(b_2) \cdot h^{\otimes n} \cdot \psi_n(b_1)\Big) \\ &= \mathrm{trace}\Big(h^{\otimes n} \cdot \psi_n(b_2)\psi_n(b_1)\Big) = \mathrm{trace}\Big(h^{\otimes n} \cdot \psi_n(b_2 b_1)\Big). \end{aligned}$$

Hence, $\mathrm{trace}\big(h^{\otimes n} \cdot \psi_n(b)\big)$ is invariant under the MI move. Therefore, we have

Theorem 2.3. Let L be an oriented link and b a braid whose closure is isotopic to L. Then, for the above representation ψ_n of B_n and the linear map h,

$$\mathrm{trace}\big(h^{\otimes n} \cdot \psi_n(b)\big) \tag{2.5}$$

is invariant under the MI and MII moves. In particular, by Theorem 2.2, it is an isotopy invariant of L. Further, it is equal to $(t^{1/2} + t^{-1/2})$ times the Jones polynomial $V_L(t)$ of L.

The theorem gives an alternative definition of the *Jones polynomial.*

Proof. The invariance of (2.5) under the Markov moves is shown above. We show its equality to the Jones polynomial as follows. By Remark 1.7 it is sufficient to verify that (2.5) satisfies the same skein relation as that of the Jones polynomial in Proposition 1.6. In fact,

$$\begin{aligned}
&t^{-1}R - tR^{-1} \\
&= t^{-1}\begin{pmatrix} t^{1/2} & 0 & 0 & 0 \\ 0 & 0 & t & 0 \\ 0 & t & t^{1/2}-t^{3/2} & 0 \\ 0 & 0 & 0 & t^{1/2} \end{pmatrix} - t\begin{pmatrix} t^{-1/2} & 0 & 0 & 0 \\ 0 & t^{-1/2}-t^{-3/2} & t^{-1} & 0 \\ 0 & t^{-1} & 0 & 0 \\ 0 & 0 & 0 & t^{-1/2} \end{pmatrix} \\
&= (t^{-1/2}-t^{1/2})\begin{pmatrix} 1 & 0 & 0 & 0 \\ 0 & 1 & 0 & 0 \\ 0 & 0 & 1 & 0 \\ 0 & 0 & 0 & 1 \end{pmatrix} = (t^{-1/2}-t^{1/2})\mathrm{id}_{V\otimes V}.
\end{aligned}$$

Since R, R^{-1} and $\mathrm{id}_{V\otimes V}$ correspond to a positive crossing, a negative crossing and two parallel strands respectively, (2.5) satisfies the same skein relation as that of the Jones polynomial. □

2.3 Burau representation of the braid groups

In this section we reconstruct the Alexander polynomial via two representations of the braid group B_n; one is the Burau representation and the other is the representation defined by (2.1) with the second R matrix in (2.4).

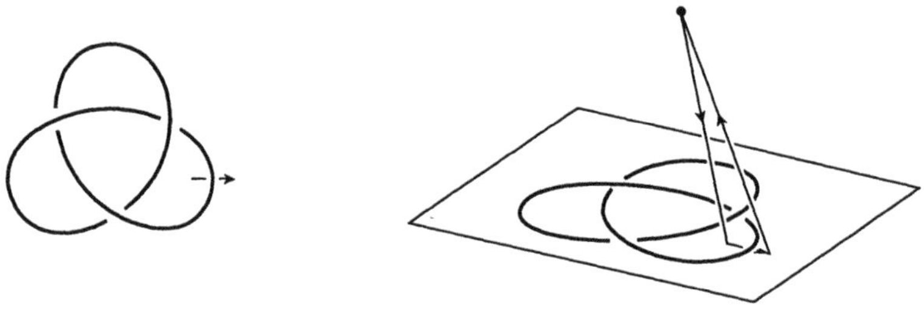

Figure 2.5 An arrow under a strand of a diagram, as shown in the left picture, denotes the cycle in the right picture. The upper point in the right picture is a fixed base point.

Burau representation

Let b be a braid, and L the link obtained as the closure of b. We compute the $\mathbb{R}[t, t^{-1}]$ module $H_1(\widetilde{S^3 - L})$ along the braid b as follows. Consider the diagram of L naturally determined by the closure of b, as in the picture in Figure 2.3.

Around a crossing of the diagram we consider 4 cycles $x, y, x', y' \in H_1(\widetilde{S^3 - L})$,

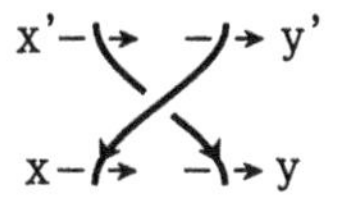

,

where each arrow implies a cycle as in Figure 2.5, fixing a base point p in $S^3 - L$. In $H_1(S^3 - L)$ the 4 cycles are related by

$$\begin{aligned} y' &= x, \\ x' + y' &= x + y. \end{aligned} \tag{2.6}$$

Let M be the 3-manifold obtained from $S^3 - L$ by cutting along a Seifert surface. As mentioned in Section 1.3 the universal cyclic cover $\widetilde{S^3 - L}$ is constructed as the union of copies M_i ($i \in \mathbb{Z}$) of M. Let p_i be the copy of p in M_i. We take the lift x_i of a cycle x in $M_i \cup M_{i+1}$, which is a path from p_i to p_{i+1}, see Figure 2.6. Further, we take the lifts of other cycles in the same way. Then, as a lift of (2.6), we have that

$$\begin{aligned} y_i' &= x_i, \\ x_i' + y_{i+1}' &= x_i + y_{i+1}, \end{aligned}$$

among cycles in $\widetilde{S^3 - L}$. Noting that the action of t is given by $tx_i = x_{i+1}, \cdots$, the above formulae are presented by

$$\begin{pmatrix} x_i' \\ y_i' \end{pmatrix} = \begin{pmatrix} 1-t & t \\ 1 & 0 \end{pmatrix} \begin{pmatrix} x_i \\ y_i \end{pmatrix}. \tag{2.7}$$

Similarly, for 4 cycles x, y, x', y' around a negative crossing,

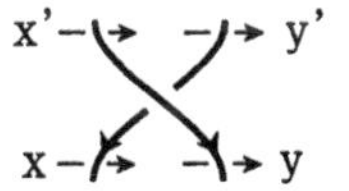

,

we have the following formula for the lifts x_i, y_i, x_i', y_i' of the 4 cycles,

$$\begin{pmatrix} x_i' \\ y_i' \end{pmatrix} = \begin{pmatrix} 0 & 1 \\ t^{-1} & 1-t^{-1} \end{pmatrix} \begin{pmatrix} x_i \\ y_i \end{pmatrix} = \begin{pmatrix} 1-t & t \\ 1 & 0 \end{pmatrix}^{-1} \begin{pmatrix} x_i \\ y_i \end{pmatrix}.$$

For simplicity, we assume that b is a braid in 3 strands. We consider 6 cycles around each of the generators σ_1 and σ_2 of the braid group B_3 as below.

σ_1 σ_2

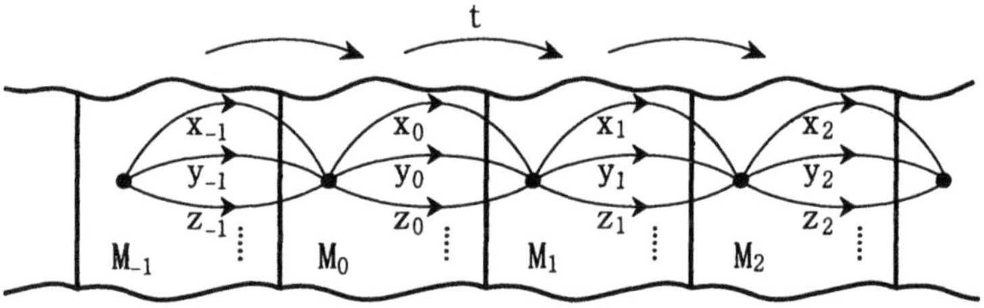

Figure 2.6 Lifts of cycles in $\widetilde{S^3 - L}$

By applying (2.7) we obtain the following relations between the lifts $x_i, y_i, z_i, x'_i, y'_i, z'_i$ of the 6 cycles,

$$\begin{pmatrix} x'_i \\ y'_i \\ z'_i \end{pmatrix} = \begin{pmatrix} 1-t & t & 0 \\ 1 & 0 & 0 \\ 0 & 0 & 1 \end{pmatrix} \begin{pmatrix} x_i \\ y_i \\ z_i \end{pmatrix} \quad \text{for the lifts of the 6 cycles around } \sigma_1, \tag{2.8}$$

$$\begin{pmatrix} x'_i \\ y'_i \\ z'_i \end{pmatrix} = \begin{pmatrix} 1 & 0 & 0 \\ 0 & 1-t & t \\ 0 & 1 & 0 \end{pmatrix} \begin{pmatrix} x_i \\ y_i \\ z_i \end{pmatrix} \quad \text{for the lifts of the 6 cycles around } \sigma_2. \tag{2.9}$$

Further, for a braid b in 3 strands we consider the 6 cycles as

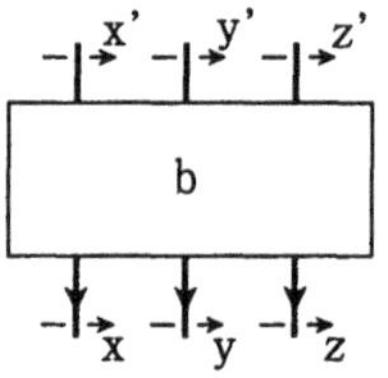

.

By applying the above two formulae to generators in b we obtain the following relation for the lifts of the 6 cycles,

$$\begin{pmatrix} x'_i \\ y'_i \\ z'_i \end{pmatrix} = \phi_3(b) \begin{pmatrix} x_i \\ y_i \\ z_i \end{pmatrix},$$

where ϕ_3 is the representation of B_3 defined by

$$\phi_3(\sigma_1) = \begin{pmatrix} 1-t & t & 0 \\ 1 & 0 & 0 \\ 0 & 0 & 1 \end{pmatrix}, \quad \phi_3(\sigma_2) = \begin{pmatrix} 1 & 0 & 0 \\ 0 & 1-t & t \\ 0 & 1 & 0 \end{pmatrix}.$$

Further, when taking the closure of b we have the following additional relations,

$$x_i = x'_i, \qquad y_i = y'_i, \qquad z_i = z'_i.$$

Hence,

$$\Big(I - \phi_3(b)\Big)\begin{pmatrix} x_i \\ y_i \\ z_i \end{pmatrix} = 0.$$

We regard the braid b as in $D^2 \times [0,1]$ and regard its closure L as in $D^2 \times S^1 \subset S^3$. The relative homology group $H_1(\widetilde{D^2 \times S^1} - L, \cup_{i\in\mathbb{Z}}\{p_i\})$ is spanned by a lift of $\{\text{point}\} \times S^1$ in $D^2 \times S^1$ and the above lifts x_i, y_i and z_i, with the above relation. Therefore, it can be presented by

$$\text{span}_{\mathbb{R}}\left\{\begin{array}{l}\text{a lift of}\\ \{\text{point}\} \times S^1\end{array}\right\} \oplus \text{span}_{\mathbb{R}}\{x_i, y_i, z_i \text{ for } i \in \mathbb{Z}\} \Big/ \Big(I - \phi_3(b)\Big)\begin{pmatrix} x_i \\ y_i \\ z_i \end{pmatrix} = 0 \text{ for } i \in \mathbb{Z}.$$

We put $\Lambda = \mathbb{R}[t, t^{-1}]$. Noting that the action of Λ is trivial on a lift of $\{\text{point}\} \times S^1$ and is given by $tx_i = x_{i+1}$, $ty_i = y_{i+1}$ and $tz_i = z_{i+1}$, the above homology group is presented by the following Λ module,

$$\Lambda/(1-t) \;\oplus\; \text{span}_\Lambda\{x_0, y_0, z_0\} \Big/ \text{span}_\Lambda\left\{\Big(I - \phi_3(b)\Big)\begin{pmatrix} x_0 \\ y_0 \\ z_0 \end{pmatrix}\right\}.$$

We write the above module as

$$H_1(\widetilde{D^2 \times S^1} - L, \bigcup_{i\in\mathbb{Z}}\{p_i\}) \;\cong\; \Big(\Lambda/(1-t)\Big) \oplus \Big(\Lambda^3 \Big/ \Big(I - \hat{\phi}_3(b)\Big)\Lambda^3\Big),$$

where Λ^3 denotes the direct sum of 3 copies of Λ. Further,

$$\begin{aligned} H_1(\widetilde{D^2 \times S^1} - L) &\cong \text{kernel}\Big(H_1(\widetilde{D^2 \times S^1} - L, \bigcup_{i\in\mathbb{Z}}\{p_i\}) \xrightarrow{\partial} H_0(\bigcup_{i\in\mathbb{Z}}\{p_i\})\Big) \\ &\cong \Big(\Lambda/(1-t)\Big) \oplus \Big(\widehat{\Lambda^3} \Big/ \Big(I - \hat{\phi}_3(b)\Big)\widehat{\Lambda^3}\Big), \end{aligned}$$

where $\widehat{\Lambda^3}$ denotes the subspace of Λ^3 consisting of $(f_1(t), f_2(t), f_3(t))$ with $f_1(t) + f_2(t) + f_3(t) = 0$ and $\hat{\phi}_3$ denotes the restriction of ϕ_3 to $\widehat{\Lambda^3}$, noting that ϕ_3 preserves the subspace $\widehat{\Lambda^3}$. Since we have that

$$\begin{pmatrix} x_i' - y_i' \\ y_i' - z_i' \end{pmatrix} = \begin{pmatrix} -t & 0 \\ 1 & 1 \end{pmatrix}\begin{pmatrix} x_i - y_i \\ y_i - z_i \end{pmatrix} \quad \text{for } \sigma_1$$

$$\begin{pmatrix} x_i' - y_i' \\ y_i' - z_i' \end{pmatrix} = \begin{pmatrix} 1 & t \\ 0 & -t \end{pmatrix}\begin{pmatrix} x_i - y_i \\ y_i - z_i \end{pmatrix} \quad \text{for } \sigma_2$$

by (2.8) and (2.9), the representation $\hat{\phi}_3$ is presented by

$$\hat{\phi}_3(\sigma_1) = \begin{pmatrix} -t & 0 \\ 1 & 1 \end{pmatrix}, \qquad \hat{\phi}_3(\sigma_2) = \begin{pmatrix} 1 & t \\ 0 & -t \end{pmatrix}.$$

Since $\widetilde{S^3 - L}$ is obtained as the union of $\widetilde{S^3 - D^2 \times S^1}$ and $\widetilde{D^2 \times S^1 - L}$ along $\widetilde{\partial D^2 \times S^1}$, we have the following exact sequence,

$$\xrightarrow{0} H_1(\widetilde{\partial D^2 \times S^1}) \longrightarrow H_1(\widetilde{S^3 - D^2 \times S^1}) \oplus H_1(\widetilde{D^2 \times S^1 - L}) \longrightarrow H_1(\widetilde{S^3 - L}) \xrightarrow{0}.$$

Since $\widetilde{\partial D^2 \times S^1}$ in $\widetilde{S^3 - L}$ is homeomorphic to the disjoint union of 3 copies of $S^1 \times \mathbb{R}$, we have that $H_1(\widetilde{\partial D^2 \times S^1}) \cong \Lambda/(1-t^3)$ as Λ module. Further, since $\widetilde{S^3 - D^2 \times S^1}$ is homeomorphic to the disjoint union of 3 copies of $D^2 \times \mathbb{R}$, we have that $H_1(\widetilde{S^3 - D^2 \times S^1}) = 0$. Hence, we have the exact sequence

$$0 \longrightarrow \Lambda/(1-t^3) \longrightarrow \big(\Lambda/(1-t)\big) \oplus \left(\widehat{\Lambda^3} \Big/ \big(I - \hat{\phi}_3(b)\big)\widehat{\Lambda^3}\right) \longrightarrow H_1(\widetilde{S^3 - L}) \longrightarrow 0.$$

Since the Alexander polynomial $\Delta_L(t)$ is the determinant of a presentation matrix of the Λ module $H_1(\widetilde{S^3 - L})$, we have that

$$\Delta_L(t) \ \sim\ \frac{1-t}{1-t^3} \det\big(I - \hat{\phi}_3(b)\big)$$

where the notation "$\sim$" implies equality up to multiplication by a unit of Λ.

In general, we define the representation ϕ_n of the braid group B_n by

$$\phi_n(\sigma_i) = \underbrace{(1) \oplus \cdots \oplus (1)}_{i-1} \oplus \begin{pmatrix} 1-t & t \\ 1 & 0 \end{pmatrix} \oplus \underbrace{(1) \oplus \cdots \oplus (1)}_{n-i-1} \tag{2.10}$$

for each generator $\sigma_i \in B_n$, where the block sum is defined by $A \oplus B = \begin{pmatrix} A & 0 \\ 0 & B \end{pmatrix}$. We call ϕ_n the *(non-reduced) Burau representation* of the braid group B_n. Then, B_n acts on Λ^n through the representation ϕ. Further, the action preserves the submodule $\widehat{\Lambda^n}$ of Λ^n consisting of $(f_1(t), f_2(t), \cdots, f_n(t))$ with $f_1(t) + f_2(t) + \cdots + f_n(t) = 0$. We denote by $\hat{\phi}_n(b)$ the restriction of $\phi_n(b)$ on $\widehat{\Lambda^n}$, and call $\hat{\phi}_n$ the *(reduced) Burau representation* of the braid group B_n. In the same way as the above case of B_3 we obtain

Theorem 2.4. Let L be an oriented link and b a braid whose closure is isotopic to L. Then, the Alexander polynomial $\Delta_L(t)$ of L is determined by the reduced Burau representation $\hat{\phi}_n$ as

$$\Delta_L(t) \ \sim\ \frac{1-t}{1-t^n} \det\big(I - \hat{\phi}_n(b)\big),$$

where the notation "$\sim$" implies equality up to multiplication by a unit of $\mathbb{R}[t, t^{-1}]$.

The theorem gives an alternative definition of the *Alexander polynomial* of knots, *i.e.*, the Alexander polynomial of a knot K can be uniquely determined by selecting a representative of $\big((1-t)/(1-t^n)\big)\det\big(I-\hat{\phi}_n(b)\big)$ (by multiplying by some unit of $\mathbb{R}[t,t^{-1}]$) with the properties $\Delta_K(1)=1$ and $\Delta_K(t)=\Delta_K(t^{-1})$.

Representation with R matrix

In the remainder of this section we give an alternative construction of the Alexander polynomial via another representation ψ_n of the braid group B_n, defined by (2.1) with the second R matrix in (2.4).

Let V be a 2-dimensional vector space with a basis $\{e_0, e_1\}$. Recall the representation $\psi_n : B_n \to \mathrm{End}(V^{\otimes n})$ given by

$$\psi_n(\sigma_i) = (\mathrm{id}_V)^{\otimes(i-1)} \otimes R \otimes (\mathrm{id}_V)^{\otimes(n-i-1)}$$

for each generator $\sigma_i \in B_n$, where R is an R matrix. Here we consider the second R matrix in (2.4). After some normalization the R matrix is presented by

$$R = \begin{pmatrix} t^{-1/2} & 0 & 0 & 0 \\ 0 & 0 & 1 & 0 \\ 0 & 1 & t^{-1/2}-t^{1/2} & 0 \\ 0 & 0 & 0 & -t^{1/2} \end{pmatrix} \in \mathrm{End}(V \otimes V).$$

Further, we put

$$h = \begin{pmatrix} t^{1/2} & 0 \\ 0 & -t^{1/2} \end{pmatrix} \in \mathrm{End}(V).$$

Then,

$$\mathrm{trace}_2\Big((\mathrm{id}_V \otimes h)\cdot R\Big) = t^{1/2}\mathrm{trace}_2\Big(\begin{pmatrix} 1 & 0 \\ 0 & 1 \end{pmatrix} \otimes \begin{pmatrix} 1 & 0 \\ 0 & -1 \end{pmatrix} \cdot R\Big)$$
$$= t^{1/2}\mathrm{trace}_2 \begin{pmatrix} t^{-1/2} & 0 & 0 & 0 \\ 0 & 0 & -1 & 0 \\ 0 & 1 & t^{-1/2}-t^{1/2} & 0 \\ 0 & 0 & 0 & t^{1/2} \end{pmatrix} = \begin{pmatrix} 1 & 0 \\ 0 & 1 \end{pmatrix}.$$

Moreover,

$$\mathrm{trace}_2\Big((\mathrm{id}_V \otimes h)\cdot R^{-1}\Big) = t^{1/2}\mathrm{trace}_2\Big(\begin{pmatrix} 1 & 0 \\ 0 & 1 \end{pmatrix} \otimes \begin{pmatrix} 1 & 0 \\ 0 & -1 \end{pmatrix} \cdot R^{-1}\Big)$$
$$= t^{1/2}\mathrm{trace}_2 \begin{pmatrix} t^{1/2} & 0 & 0 & 0 \\ 0 & t^{-1/2}-t^{1/2} & -1 & 0 \\ 0 & 1 & 0 & 0 \\ 0 & 0 & 0 & t^{-1/2} \end{pmatrix} = \begin{pmatrix} 1 & 0 \\ 0 & 1 \end{pmatrix}.$$

That is,

$$\operatorname{trace}_2\Big((\mathrm{id}_V \otimes h)\cdot R^{\pm 1}\Big) = \mathrm{id}_V. \tag{2.11}$$

Further, by charge conservation of R, we have that

$$R\cdot (h\otimes h) = (h\otimes h)\cdot R. \tag{2.12}$$

Like the case of the construction of the Jones polynomial in Theorem 2.3, the above two formulae imply that $\operatorname{trace}\big(h^{\otimes n}\cdot\psi_n(b)\big)$ is invariant under the MI and MII moves. However, it is an immediate corollary of Proposition 3.10 that $\operatorname{trace}\big(h^{\otimes n}\cdot\psi_n(b)\big)$ is always equal to 0. To obtain a non-trivial invariant from this representation we consider the modification that we do not take the trace with respect to the first entry of $V^{\otimes n}$, as in the following theorem.

Theorem 2.5. Let L be an oriented link and $b\in B_n$ a braid whose closure is isotopic to L. Then, for the above representation ψ_n and the linear map h the following equation holds for some scalar c,

$$\operatorname{trace}_{2,3,\cdots,n}\Big((1\otimes h^{\otimes(n-1)})\psi_n(b)\Big) = c\cdot \mathrm{id}_V,$$

where $\operatorname{trace}_{2,3,\cdots,n}$ denotes the trace on the $2,3,\cdots,n$-th entries of $V^{\otimes n}$. Further, c is an isotopy invariant of L. Moreover, c is equal to the Alexander polynomial $\Delta_L(t)$ of L.

Theorem 2.5 gives a reconstruction of the *Alexander polynomial* via a representation of B_n with an R matrix.

The isotopy invariance of the theorem is a special case of the isotopy invariance of Theorem 3.12. See the proof of Theorem 3.12 for the invariance. The key of the proof is to verify the relations (2.11) and (2.12), which has been done above. The proof of Theorem 3.12 provides an additional relation to complete the proof of the theorem.

We show the proof of the equality of the invariant and the Alexander polynomial in Appendix C.1. The idea of the proof is as follows. By regarding t as a complex parameter the Burau representation ϕ_n given in (2.10) is regarded as a representation into $\mathrm{End}(W_n)$, where W_n is an n-dimensional complex vector space. For a braid $b\in B_n$, let the eigenvalues of $\phi_n(b)$ be $\alpha_1,\cdots,\alpha_n$. Then,

$$\det\Big(I-\phi_n(b)\Big) = \prod_{1\le i\le n}(1-\alpha_i).$$

Further, it is shown that the exterior algebra $\bigwedge W_n$ is isomorphic to $V^{\otimes n}$ by an isomorphism which is equivariant with respect to the actions of B_n on both spaces. Hence, the trace of the action of B_n on $V^{\otimes n}$ is presented by

$$\operatorname{trace}\bigg(\begin{pmatrix}1 & 0\\ 0 & -1\end{pmatrix}^{\otimes n}\cdot\psi_n(b)\bigg) = 1-\sum_{1\le i\le n}\alpha_i+\sum_{1\le i<j\le n}\alpha_i\alpha_j-\cdots+(-1)^n\alpha_1\alpha_2\cdots\alpha_n,$$

where the signs on the right hand side are derived from $\begin{pmatrix} 1 & 0 \\ 0 & -1 \end{pmatrix}$ on the left hand side. The key of the proof is that the above two formulae are equal. Unfortunately, both of them vanish because an eigenvalue, say α_1, is equal to 1. The equality in Theorem 2.5 is proved by "removing" the eigenvalue $\alpha_1 = 1$ from the above two formulae; see Appendix C.1 for a detailed argument.

Proposition 2.6. The Alexander polynomial satisfies the following relation

$$\Delta_{L_+}(t) - \Delta_{L_-}(t) = -(t^{1/2} - t^{-1/2})\Delta_{L_0}(t)$$

where L_+, L_- and L_0 are three oriented links, which are identical except for a ball, where they differ as shown. We call this relation the *skein relation* of the Alexander polynomial.

L_+ L_- L_0

Proof. For the R matrix we used above, we have that

$$R - R^{-1} = \begin{pmatrix} t^{-1/2} & 0 & 0 & 0 \\ 0 & 0 & 1 & 0 \\ 0 & 1 & t^{-1/2} - t^{1/2} & 0 \\ 0 & 0 & 0 & -t^{1/2} \end{pmatrix} - \begin{pmatrix} t^{1/2} & 0 & 0 & 0 \\ 0 & t^{1/2} - t^{-1/2} & 1 & 0 \\ 0 & 1 & 0 & 0 \\ 0 & 0 & 0 & -t^{-1/2} \end{pmatrix}$$

$$= -(t^{1/2} - t^{-1/2}) \begin{pmatrix} 1 & 0 & 0 & 0 \\ 0 & 1 & 0 & 0 \\ 0 & 0 & 1 & 0 \\ 0 & 0 & 0 & 1 \end{pmatrix} = -(t^{1/2} - t^{-1/2}) \cdot \mathrm{id}_V \otimes \mathrm{id}_V.$$

Since L_+, L_- and L_0 correspond to R, R^{-1} and $\mathrm{id}_V \otimes \mathrm{id}_V$ in the construction of the representation ψ_n, we obtain the required relation. □

Chapter 3

Operator invariants of tangles via sliced diagrams

Let us start with a physical interpretation of operator invariants. Consider a universe consisting of particles moving in a plane. A link in $\mathbb{R}^3$ describes a "history" of this universe, in the following sense. Consider Figure 3.7. The vertical coordinate is regarded as the "time" of this universe (increasing as one moves upward), and the intersections of the link with some plane at some fixed time describe the "state" of this universe at that time. These "time-slices" are represented by horizontal lines in the diagram. As such a time-slice sweeps upward, the particles are born, evolve, and are annihilated.

The quantum field theory of such a universe gives a physical interpretation of operator invariants. In the theory we associate a vector space to each physical state of the universe, where a physical state is a configuration of particles in a plane in this case. Further, we associate operators between vector spaces to time transitions of states. The quantum field theory is described by such operators. In particular, when a quantum field theory does not depend on the metric of the space, the theory is called a topological quantum field theory. In this case we obtain an isotopy invariant of tangles as an operator between vector spaces; this is called an "operator invariant".

In this chapter, based on such a physical picture, we introduce sliced diagrams of tangles. Further, we introduce the Turaev moves among them, which play a role similar to the Reidemeister moves among link diagrams and the Markov moves among braids. By using sliced diagrams of tangles we give rigorous constructions of operator invariants of tangles. In particular, we obtain reconstructions of the Jones and Alexander polynomials as operator invariants.

3.1 Tangles and their sliced diagrams

In this section we introduce sliced diagrams by expressing tangles by elementary tangle diagrams. To describe isotopy classes of tangles (in particular, links) we introduce the Turaev moves among sliced diagrams.

A *tangle* is a compact 1-manifold properly embedded in $\mathbb{R}\times\mathbb{R}\times[0,1]$ such that the boundary of the embedded 1-manifold is a set of distinct points in $\{0\}\times\mathbb{R}\times\{0,1\}$. Two tangles are called *isotopic* if they are related by an isotopy of $\mathbb{R}^2\times[0,1]$ fixing its boundary points. A *framed tangle* is a tangle with a framing of each component; see, for example, Figure 3.1.

Figure 3.1 A tangle and a framed tangle

A *tangle diagram* is a diagram of a tangle in $\mathbb{R}\times[0,1]$. A tangle diagram (in particular a link diagram) consists of crossings, critical points and vertical paths connecting these points, where a "critical point" is a critical point of the tangle diagram in $\mathbb{R}\times[0,1]$ with respect to the height function $\mathbb{R}\times[0,1]\to[0,1]$ and a "vertical path" implies a curve embedded in $\mathbb{R}^2$ such that the tangent line at each point of the curve is not horizontal. In other words any tangle diagram can be expressed, up to isotopy, as a union of copies of elementary tangle diagrams shown in Figure 3.3 and vertical paths between them; see Figure 3.2 for such expressions of a link diagram.

For such expressions of a tangle diagram in $\mathbb{R}\times[0,1]$ we consider an isotopy of $\mathbb{R}\times[0,1]$ preserving such expression *i.e.*, isotopy which moves copies of elementary tangle diagrams in the expression by translation without rotation and moves vertical paths whilst keeping them vertical; see, for example, Figure 3.2. We call such an isotopy an isotopy preserving crossings and critical points. The difference between a usual isotopy of $\mathbb{R}^2$ and an isotopy preserving crossings and critical points is that such an isotopy can not rotate neighborhoods of crossings and critical points whereas a usual isotopy can rotate them. Hence, the difference is described by the moves (3.4) and (3.5) shown in Figure 3.4, where the moves (3.4) describe rotations of a neighborhood of a critical point and the moves (3.5) describes that of a crossing. Thus, we have the following symbolic representation,

$$\begin{aligned}&\{\text{tangle diagrams}\}/\text{isotopy of } \mathbb{R}\times[0,1]\\ &= \{\text{tangle diagrams}\}\Big/ \begin{array}{l}\text{the moves (3.4) and (3.5) and}\\ \text{isotopy preserving crossings and critical points.}\end{array}\end{aligned} \tag{3.1}$$

Note that, while the moves (3.5) are moves such that a vertical path moves over a maximal critical point of a diagram, we obtain the moves such that a vertical path

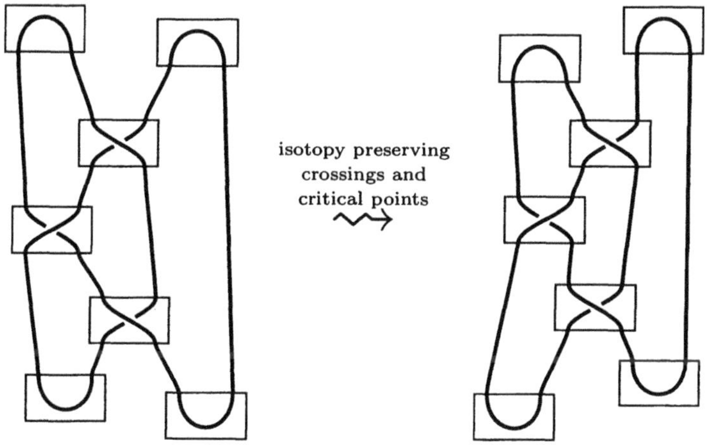

Figure 3.2 Two expressions of a link diagram by elementary diagrams and vertical paths. The two expressions in the figure are related by isotopy preserving crossings and critical points.

moves over a minimal critical point by the moves (3.4) and (3.5) as below.

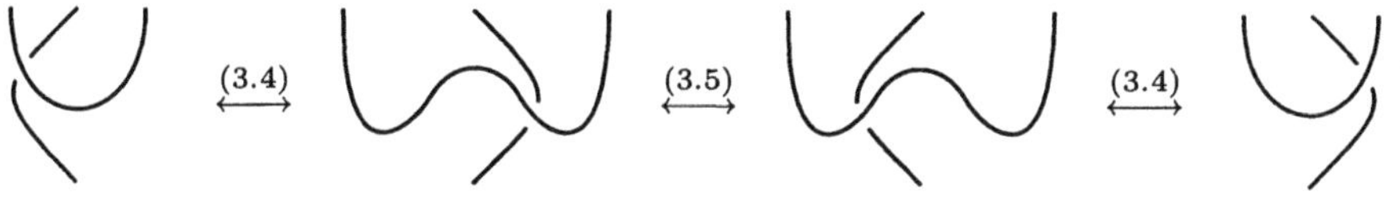

Any tangle diagram (in particular any link diagram) can be expressed, up to isotopy, as a composition of tensor products of copies of elementary tangle diagrams as shown in Figure 3.7, where we define the *tensor product* and the *composition* of two tangle diagrams as

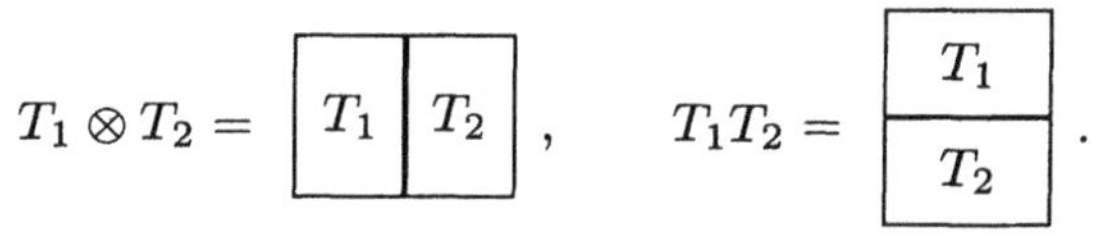

Note that the composition of T_1 and T_2 is defined only if the numbers of the lower ends of T_1 and the upper ends of T_2 are equal to each other.

Figure 3.3 The *elementary tangle diagrams*

As a special case of such a decomposition, we consider a tangle diagram sliced by horizontal lines such that each domain between adjacent horizontal lines has either a single crossing or a single critical point as shown in Figure 3.7. We call such a diagram a *sliced diagram*. Each domain between slices in a sliced diagram can be presented as a tensor product of a number of vertical lines and either a crossing or a critical point.

Since an isotopy preserving crossings and critical points can be expressed by the moves (3.2) and (3.3) among sliced digrams, as a modification of the equality (3.1) we have that

$$\begin{aligned}&\{\text{tangle diagrams}\}/\text{isotopy of } \mathbb{R}\times[0,1]\\ &=\{\text{sliced tangle diagrams}\}/\text{the moves (3.2), (3.3), (3.4) and (3.5)},\end{aligned}$$

where the moves (3.2)–(3.5) are in Figure 3.4. Further, by considering the quotients of both sides by the Reidemeister moves, we obtain the following equality by Theorem 1.1,

$$\{\text{tangles}\}\ /\text{isotopy of } \mathbb{R}^2\times[0,1] = \ \{\text{sliced tangle diagrams}\}\ /\text{the Turaev moves},$$

where the *Turaev moves* are the moves shown in Figure 3.4. In other words, we obtain the following theorem.

Theorem 3.1 ([Tur88, Tur89, FrYe89]). Two sliced tangle diagrams express the same (isotopic) tangle if and only if the two sliced diagrams are related by a sequence of the Turaev moves (see Figure 3.4).

In particular, by regarding links as tangles with no ends, we have that

$$\{\text{links}\}/\text{isotopy of } \mathbb{R}^3 = \ \{\text{sliced link diagrams}\}\ /\text{the Turaev moves}.$$

Remark 3.2. Regarding "the Reidemeister moves" in the statement of (3.6), the Reidemeister moves for sliced diagrams are defined from the original Reidemeister moves by fixing sliced structure in both sides of the original Reidemeister moves in an arbitrary way. For example, we have the following two slicings of the Reidemeister move RI as

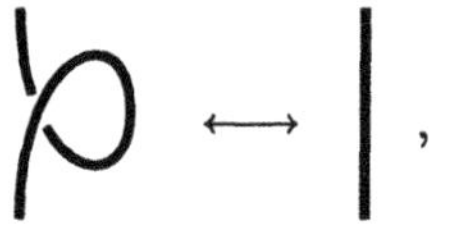
(3.7)

and alternatively 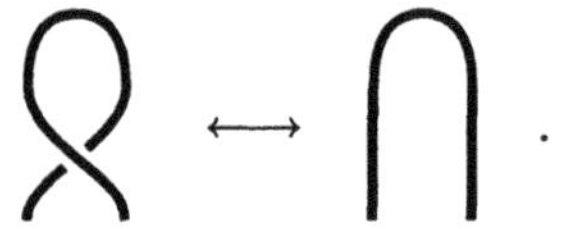 (3.8)

These moves are equivalent up to the moves (3.4) and (3.5); for example, the move (3.8) can be derived from the move (3.7) as

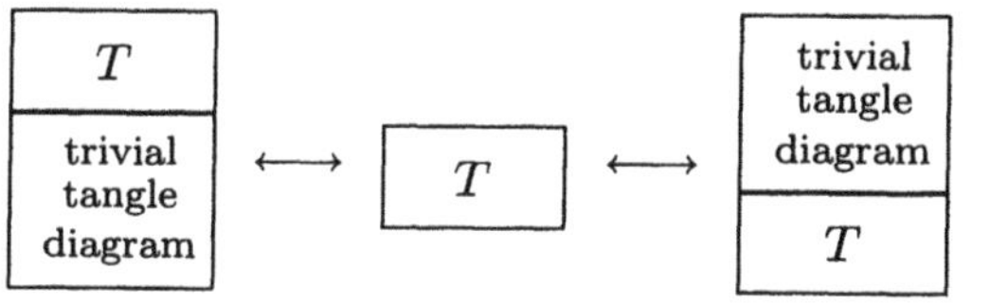

(3.2)

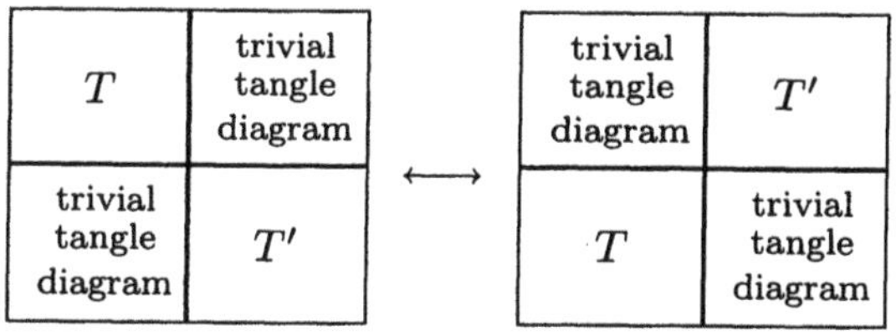

(3.3)

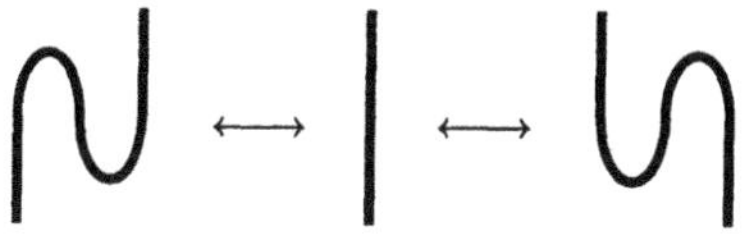

(3.4)

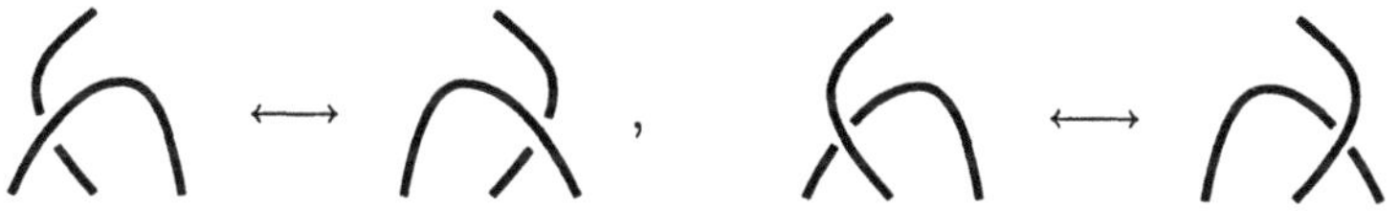

(3.5)

The Reidemeister moves in $\begin{cases} \text{Figure 1.3 for diagrams of tangles} \\ \text{Figure 1.8 for diagrams of framed tangles} \end{cases}$ (3.6)

Figure 3.4 The Turaev moves for sliced diagrams. A *trivial tangle diagram* is a tangle diagram consisting of disjoint vertical lines.

An *oriented tangle* and an *oriented tangle diagram* are a tangle and a tangle diagram with each component oriented. Like the above case of tangle diagrams any oriented tangle diagram can be expressed up to isotopy as a composition of tensor products of copies of the elementary oriented tangle diagrams shown in Figure 3.5 . Note that each crossing of an oriented tangle diagram is rotated by an isotopy of $\mathbb{R} \times [0,1]$ in such a way that the two strands becomes downward in a neighborhood of the crossing, before we decompose the diagram into elementary diagrams. For example, we rotate crossings as

Though such a rotation is not unique for a given crossing, all possible rotations are related by the move (3.11).

Analogously to the case of an unoriented sliced tangle diagram, an *oriented sliced tangle diagram* is defined to be an oriented tangle diagram with horizontal lines such

Figure 3.5 The *oriented elementary tangle diagrams*

that each part of the diagram between two adjacent horizontal lines consists of a disjoint union of vertical lines and one of the elementary oriented tangle diagrams shown in Figure 3.10. Similarly to the unoriented case we have that

$$\begin{aligned}&\{\text{oriented tangle diagrams}\}/\text{isotopy of } \mathbb{R}\times[0,1]\\&=\{\text{oriented sliced tangle diagrams}\}/\text{the moves (3.9), (3.10) and (3.11)},\end{aligned}$$

where these moves are the first three moves among the Turaev moves shown in Figure 3.6. We obtain the other Turaev moves from the Reidemeister moves for oriented diagrams in Figure 1.6 by fixing slice structures; note that as mentioned in Remark 3.2 we can choose such sliced structures in an arbitrary way, up to the moves (3.9), (3.10) and (3.11). Taking the quotients of both sides of the above equation by the Reidemeister moves we obtain the following equation by Theorem 1.2,

$$\begin{aligned}&\{\text{oriented tangles}\}/\text{isotopy of } \mathbb{R}^2\times[0,1]\\&=\{\text{oriented sliced tangle diagrams}\}/\text{the Turaev moves}.\end{aligned}$$

In other words, we obtain the following theorem.

Theorem 3.3 ([Tur88, Tur89, FrYe89]). Two oriented sliced tangle diagrams express the same (isotopic) oriented tangle if and only if the two sliced diagrams are related by the Turaev moves shown in Figure 3.6.

By replacing the move (3.12) in the Turaev moves in Figure 3.6 with

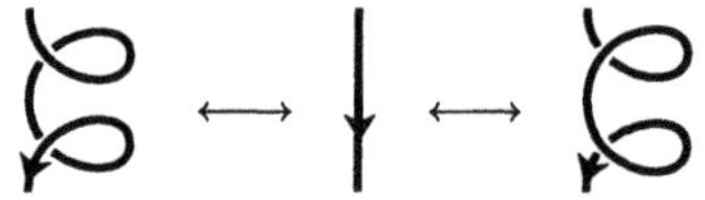

(3.16)

we obtain the moves for sliced diagrams of oriented framed tangles which describe isotopy classes of the tangles in the same way as in Theorem 3.3.

3.2 Operator invariants of unoriented tangles

In our quantum field theory of 2-dimensional space, we regard the horizontal plane $\mathbb{R}^2 \times \{\text{point}\}$ in $\mathbb{R}^2 \times [0,1]$ as the universe and regard a section of a tangle by such a horizontal plane as a set of particles in the universe. By moving a horizontal plane upward, we observe moving particles in the universe, regarding the vertical

The moves (3.2) and (3.3) (3.9)

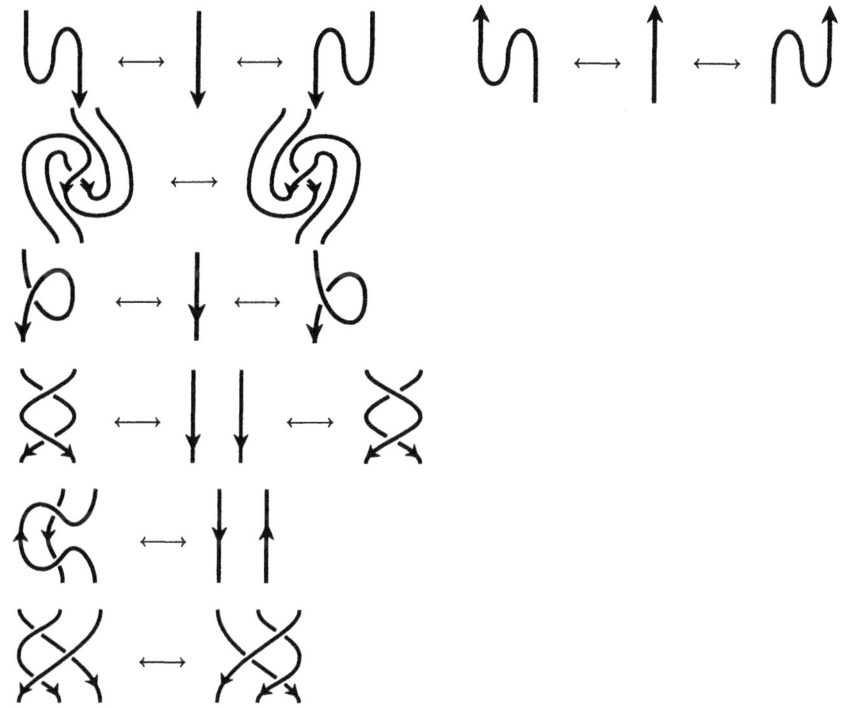

(3.10)

(3.11)

(3.12)

(3.13)

(3.14)

(3.15)

Figure 3.6 The Turaev moves for oriented sliced diagrams

coordinate $[0, 1]$ as the time of the universe. Quantum field theory suggests that our theory would be described by using linear maps associated to tangles between vector spaces associated to sections of tangles by horizontal planes. Further, to obtain isotopy invariants of tangles, we require topological invariance of the theory; this is a key idea of "topological quantum field theory". Following this principle, we introduce operator invariants of tangles in this section. The Kauffman bracket (which will be shown to be equal to the quantum (sl_2, V) invariant) will be reconstructed as an operator invariant.

Let V be a vector space over $\mathbb{C}$. For a sliced tangle diagram D with i upper ends and j lower ends we introduce the bracket $[D] \in \mathrm{Hom}(V^{\otimes i}, V^{\otimes j})$ as follows. By definition a sliced diagram has horizontal lines such that there is exactly one of elementary tangle diagrams between each pair of adjacent horizontal lines, as in Figure 3.7. We associate the vector space V to each point of the section of the diagram by a horizontal line, and take the tensor product of the vector spaces along each such horizontal line. We associate the tensor product to each such horizontal line, with the proviso that $\mathbb{C}$ is associated to a horizontal line which does not

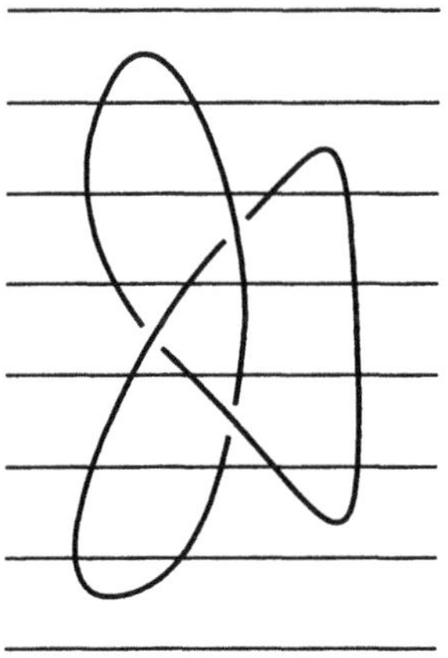

$$
\begin{array}{c}
\mathbb{C} \\
\uparrow n \\
V \otimes V \\
\uparrow \mathrm{id}_V \otimes \mathrm{id}_V \otimes n \\
V \otimes V \otimes V \otimes V \\
\uparrow \mathrm{id}_V \otimes R^{-1} \otimes \mathrm{id}_V \\
V \otimes V \otimes V \otimes V \\
\uparrow R \otimes \mathrm{id}_V \otimes \mathrm{id}_V \\
V \otimes V \otimes V \otimes V \\
\uparrow \mathrm{id}_V \otimes R^{-1} \otimes \mathrm{id}_V \\
V \otimes V \otimes V \otimes V \\
\uparrow \mathrm{id}_V \otimes \mathrm{id}_V \otimes u \\
V \otimes V \\
\uparrow u \\
\mathbb{C}
\end{array}
$$

Figure 3.7 A sliced link diagram and its operator invariant

intersect the link. (see 3.7.) Further, we consider linear maps between the tensor products associated to the horizontal lines as follows. From the definition of a sliced tangle diagram, in each domain between two adjacent horizontal lines there is one of the elementary tangle diagrams in Figure 3.8. We associate id_V to each vertical line. We associate an invertible endomorphism of $V \otimes V$, denoted by R, to the elementary diagram of a positive crossing, and the inverse map R^{-1} to the diagram of a negative crossing. We also associate linear maps, denoted by n and u, to the elementary diagrams of maximal and minimal critical points respectively, as shown in Figure 3.8. Further, we consider the composition of tensor products of the linear maps as shown in Figure 3.7. We define the bracket $[D] \in \mathrm{Hom}(V^{\otimes i}, V^{\otimes j})$ to be the composition. In particular, the bracket $[D]$ of a sliced link diagram D belongs to $\mathrm{End}(\mathbb{C}) = \mathbb{C}$ (the isomorphism on an element $f \in \mathrm{End}(\mathbb{C})$ given by $f(1)$).

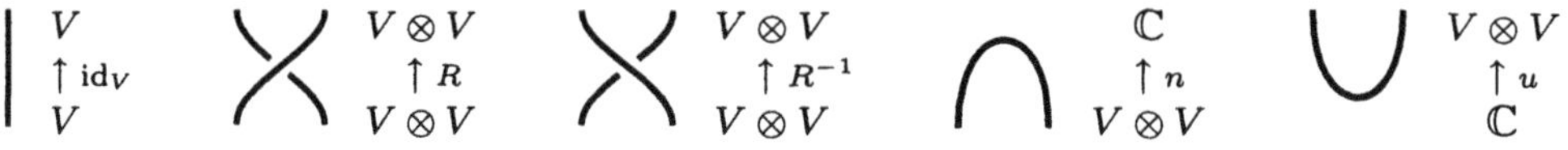

Figure 3.8 The linear maps associated to the elementary tangle diagrams

The bracket $[D]$ is invariant under the moves (3.2) and (3.3) by construction. For the bracket to be an isotopy invariant of links, there are further conditions, corresponding to the other moves in Figure 3.4, that must be satisfied.

Let us consider the conditions on the maps n and u that will arise from the move

(3.4). By computing the bracket of both sides of (3.4) the condition is presented by

$$(n \otimes \mathrm{id}_V)(\mathrm{id}_V \otimes u) = \mathrm{id}_V = (\mathrm{id}_V \otimes n)(u \otimes \mathrm{id}_V). \tag{3.17}$$

This relation is computed as follows. With respect to a basis $\{e_i\}$ of V, the linear maps $n \in \mathrm{Hom}(V \otimes V, \mathbb{C})$ and $u \in \mathrm{Hom}(\mathbb{C}, V \otimes V)$ are presented by

$$n(e_i \otimes e_j) = n_{ij},$$
$$u(1) = \sum_{i,j} u^{ij} e_i \otimes e_j,$$

for some scalars n_{ij} and u^{ij}. Using this notation, the first term of (3.17) is computed as

$$e_k \overset{\mathrm{id}_V \otimes u}{\longmapsto} \sum_{i,j} u^{ij} e_k \otimes e_i \otimes e_j \overset{n \otimes \mathrm{id}_V}{\longmapsto} \sum_{i,j} n_{ki} u^{ij} e_j.$$

This is equal to e_k if the matrices (u^{ij}) and (n_{ki}) are the inverses of each other. Further, for the last term of (3.17) we have that

$$e_k \overset{u \otimes \mathrm{id}_V}{\longmapsto} \sum_{i,j} u^{ij} e_i \otimes e_j \otimes e_k \overset{\mathrm{id}_V \otimes n}{\longmapsto} \sum_{i,j} u^{ij} n_{jk} e_i.$$

This is equal to e_k if the above condition holds. Therefore, the maps n and u satisfy (3.17) if the matrices (n_{ij}) and (u^{ij}) are the inverses of each other. In particular, for a given non-degenerate bilinear form $n \in \mathrm{Hom}(V \otimes V, \mathbb{C})$, the map $u \in \mathrm{Hom}(\mathbb{C}, V \otimes V)$ is uniquely determined by (3.17) such that the matrix (u^{ij}) is the inverse of the matrix (n_{ij}). For example, when V is a 2-dimensional vector space with a basis $\{e_0, e_1\}$, the linear maps n and u are presented by

$$n = \begin{pmatrix} n_{00} & n_{01} & n_{10} & n_{11} \end{pmatrix}, \qquad u = \begin{pmatrix} u^{00} \\ u^{01} \\ u^{10} \\ u^{11} \end{pmatrix},$$

with respect to the basis $\{e_0 \otimes e_0,\ e_0 \otimes e_1,\ e_1 \otimes e_0,\ e_1 \otimes e_1\}$ of $V \otimes V$. For a given n, the map u is determined by

$$\begin{pmatrix} u^{00} & u^{10} \\ u^{01} & u^{11} \end{pmatrix} = \begin{pmatrix} n_{00} & n_{10} \\ n_{01} & n_{11} \end{pmatrix}^{-1}, \tag{3.18}$$

and such n and u satisfy (3.17).

Let R be an invertible endomorphism of $V \otimes V$ and n a non-degenerate bilinear form on $V \otimes V$. Setting the map u as above, to satisfy the relation (3.17), we obtain the bracket $[D]$ of a sliced diagram D as mentioned before. We call $[D]$ the bracket associated with the maps R and n.

Theorem 3.4 ([Tur88, Tur89, FrYe89]). Let T be a tangle and D a sliced diagram of T. If an invertible endomorphism R of $V \otimes V$ and a non-degenerate bilinear form n on $V \otimes V$ satisfy

$$(\mathrm{id}_V \otimes n)(R^{\pm 1} \otimes \mathrm{id}_V) = (n \otimes \mathrm{id}_V)(\mathrm{id}_V \otimes R^{\mp 1}), \tag{3.19}$$

$$n \cdot R = n, \tag{3.20}$$

$$(R \otimes \mathrm{id}_V)(\mathrm{id}_V \otimes R)(R \otimes \mathrm{id}_V) = (\mathrm{id}_V \otimes R)(R \otimes \mathrm{id}_V)(\mathrm{id}_V \otimes R), \tag{3.21}$$

then the bracket $[D]$ associated with the maps R and n is an isotopy invariant of T.

We denote the invariant by $[T]$ and call it the *operator invariant* of a tangle T associated with the linear maps R and n. In particular, the operator invariant $[L]$ of a link L belongs $\mathrm{End}(\mathbb{C}) = \mathbb{C}$.

Proof. By Theorem 3.1 it is sufficient to show that the bracket of sliced diagrams associated with the maps R and n is invariant under the Turaev moves in Figure 3.4.

The invariance under the moves (3.2), (3.3) and (3.4) are obtained by the construction of the bracket $[D]$.

We obtain the invariance under the first move of (3.5) as

[] $= (\mathrm{id}_V \otimes n)(R \otimes \mathrm{id}_V) = (n \otimes \mathrm{id}_V)(\mathrm{id}_V \otimes R^{-1}) =$ []

by using the relation (3.19). The invariance under the second move of (3.5) is obtained in the same way.

The invariance under the Reidemeister moves is obtained by (3.20), (3.21), and the equality $R \cdot R^{-1} = \mathrm{id}_V \otimes \mathrm{id}_V$. □

By construction, an operator invariant preserves the structure of tensor product and composition as

$$[T_1 \otimes T_2] = [T_1] \otimes [T_2],$$
$$[T_1 \cdot T_2] = [T_1] \cdot [T_2].$$

Modifying Theorem 3.4 we obtain an operator invariant of framed tangles by the following theorem.

Theorem 3.5. Let T be a framed tangle and D a sliced diagram presenting T by blackboard framing. If an invertible endomorphism R of $V \otimes V$ and a non-degenerate bilinear form n on $V \otimes V$ satisfy the relations (3.19), (3.21) and

$$n \cdot R = c \cdot n, \tag{3.22}$$

for some non-zero scalar c, then the bracket $[D]$ associated with the maps R and n is an isotopy invariant of T.

The Kauffman bracket as an operator invariant

Let V be a 2-dimensional vector space with a basis $\{e_0, e_1\}$. With respect to the basis $\{e_0 \otimes e_0,\ e_0 \otimes e_1,\ e_1 \otimes e_0,\ e_1 \otimes e_1\}$ of $V \otimes V$ we put

$$R = \begin{pmatrix} A & 0 & 0 & 0 \\ 0 & 0 & A^{-1} & 0 \\ 0 & A^{-1} & A - A^{-3} & 0 \\ 0 & 0 & 0 & A \end{pmatrix} \in \mathrm{End}(V \otimes V), \qquad (3.23)$$
$$n = \begin{pmatrix} 0 & A & -A^{-1} & 0 \end{pmatrix} \in \mathrm{Hom}(V \otimes V, \mathbb{C}).$$

The linear map u is determined by (3.18) as

$$u = \begin{pmatrix} 0 \\ -A \\ A^{-1} \\ 0 \end{pmatrix} \in \mathrm{Hom}(\mathbb{C}, V \otimes V).$$

Let us verify that the above R and n satisfy the relations of Theorem 3.5. Firstly, to compute (3.19) we regard $V \otimes V \otimes V$ as the direct sum of four subspaces W_1, W_2, W_3 and W_4 spanned by the basis $\{e_0 \otimes e_0 \otimes e_0\}$, $\{e_0 \otimes e_0 \otimes e_1,\ e_0 \otimes e_1 \otimes e_0,\ e_1 \otimes e_0 \otimes e_0\}$, $\{e_0 \otimes e_1 \otimes e_1,\ e_1 \otimes e_0 \otimes e_1,\ e_1 \otimes e_1 \otimes e_0\}$ and $\{e_1 \otimes e_1 \otimes e_1\}$ respectively. We have that

$$\mathrm{id}_V \otimes n = \begin{pmatrix} 0 \\ 0 \end{pmatrix} \oplus \begin{pmatrix} A & -A^{-1} & 0 \\ 0 & 0 & 0 \end{pmatrix} \oplus \begin{pmatrix} 0 & 0 & 0 \\ 0 & A & -A^{-1} \end{pmatrix} \oplus \begin{pmatrix} 0 \\ 0 \end{pmatrix},$$
$$R \otimes \mathrm{id}_V = (A) \oplus \begin{pmatrix} A & 0 & 0 \\ 0 & 0 & A^{-1} \\ 0 & A^{-1} & A - A^{-3} \end{pmatrix} \oplus \begin{pmatrix} 0 & A^{-1} & 0 \\ A^{-1} & A - A^{-3} & 0 \\ 0 & 0 & A \end{pmatrix} \oplus (A),$$

where these lie in $\bigoplus_{i=1}^4 \mathrm{Hom}(W_i, V)$ and $\bigoplus_{i=1}^4 \mathrm{End}(W_i)$ respectively. Hence,

$$(\mathrm{id}_V \otimes n)(R \otimes \mathrm{id}_V) = \begin{pmatrix} 0 \\ 0 \end{pmatrix} \oplus \begin{pmatrix} A^2 & 0 & -A^{-2} \\ 0 & 0 & 0 \end{pmatrix} \oplus \begin{pmatrix} 0 & 0 & 0 \\ 1 & A^2 - A^{-2} & -1 \end{pmatrix} \oplus \begin{pmatrix} 0 \\ 0 \end{pmatrix}.$$

By a similar computation it is verified that $(n \otimes \mathrm{id}_V)(\mathrm{id}_V \otimes R^{-1})$ is equal to the above formula. The version of the relation (3.19) with the sign set to "$-$" is verified similarly. Secondly, the Yang-Baxter equation (3.21) is derived from the fact that the above R is a special case of the first R matrix in (2.4). Thirdly, (3.22) is verified as follows,

$$n \cdot R = \begin{pmatrix} 0 & -A^{-2} & A^{-4} & 0 \end{pmatrix} = -A^{-3} \cdot n,$$

where we put the scalar c in (3.22) to be $-A^{-3}$.

Therefore, the linear maps R and n given in (3.23) satisfy the assumptions of Theorem 3.5. Hence, we obtain

Theorem 3.6. There exists the operator invariant $[T]$ of a framed tangle T determined by the linear maps R and n given in (3.23). For a framed link L the operator invariant $[L] \in \mathrm{End}(\mathbb{C}) = \mathbb{C}$ is equal to the Kauffman bracket of L.

The theorem gives a reconstruction of the *Kauffman bracket* via sliced diagrams.

Proof. We verify that the operator invariant satisfies the defining relations of the Kauffman bracket as follows,

$A\big[$ 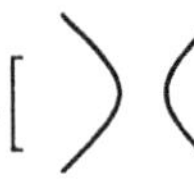 $\big] + A^{-1}\big[$  $\big] = A(\mathrm{id}_{V\otimes V}) + A^{-1}(u \cdot n)$

$$= A\begin{pmatrix} 1 & 0 & 0 & 0 \\ 0 & 1 & 0 & 0 \\ 0 & 0 & 1 & 0 \\ 0 & 0 & 0 & 1 \end{pmatrix} + A^{-1}\begin{pmatrix} 0 & 0 & 0 & 0 \\ 0 & -A^2 & 1 & 0 \\ 0 & 1 & -A^2 & 0 \\ 0 & 0 & 0 & 0 \end{pmatrix} = R = \big[$$  $\big],$

$\big[$  $\big] = n \cdot u = -A^2 - A^{-2}.$

Hence, this operator invariant of a link is equal to the Kauffman bracket. □

3.3 Operator invariants of oriented tangles

In this section, we construct operator invariants of *oriented* tangles in a similar way as we constructed operator invariants of unoriented tangles in the previous section. The Jones polynomial and the Alexander polynomial are reconstructed as operator invariants of oriented tangles.

Before defining operator invariants of oriented tangles we introduce some notation. Let V_1 and V_2 be two vector spaces. Regard an endomorphism A of $V_1 \otimes V_2$ as lying in the following tensor product,

$$A \in \mathrm{End}(V_1 \otimes V_2) = (V_1 \otimes V_2)^\star \otimes (V_1 \otimes V_2) = V_2^\star \otimes V_1^\star \otimes V_1 \otimes V_2.$$

We denote by $A^{\circlearrowright}$ and $A^{\circlearrowleft}$ the linear maps obtained from A by cyclic permutations of the entries of the tensor product $V_2^\star \otimes V_1^\star \otimes V_1 \otimes V_2$ as

$$A^{\circlearrowright} \in \mathrm{Hom}(V_2 \otimes V_2^\star, V_1^\star \otimes V_1) = (V_2 \otimes V_2^\star)^\star \otimes (V_1^\star \otimes V_1) = V_2 \otimes V_2^\star \otimes V_1^\star \otimes V_1,$$
$$A^{\circlearrowleft} \in \mathrm{Hom}(V_1^\star \otimes V_1, V_2 \otimes V_2^\star) = (V_1^\star \otimes V_1)^\star \otimes (V_2 \otimes V_2^\star) = V_1^\star \otimes V_1 \otimes V_2 \otimes V_2^\star.$$

For simplicity we put $V = V_1 = V_2$. When A is presented by a matrix (A_{kl}^{ij}) with respect to a basis $\{e_i\}$ of V, such that $A(e_k \otimes e_l) = \sum_{i,j} A_{kl}^{ij} e_i \otimes e_j$, it is presented by an element of the tensor product by $A = \sum_{i,j,k,l} A^{ij}_{kl} e^\star_l \otimes e^\star_k \otimes e_i \otimes e_j$. Hence, $A^{\circlearrowright}$ is given by $A^{\circlearrowright} = \sum_{i,j,k,l} A^{ij}_{kl} e_j \otimes e^\star_l \otimes e^\star_k \otimes e_i$. Therefore, as a linear map $A^{\circlearrowright}$ is given by $A^{\circlearrowright}(e_l \otimes e^\star_j) = \sum_{k,i} A^{ij}_{kl} e^\star_k \otimes e_i$. When we present $A^{\circlearrowright}$ by a matrix $(A^{\circlearrowright}{}^{ij}_{kl})$, we have that $A^{ij}_{kl} = A^{\circlearrowright}{}^{ki}_{lj}$. In the same way, for the linear map $A^{\circlearrowleft}$, we have that $A^{ij}_{kl} = A^{\circlearrowleft}{}^{jl}_{ik}$. For example, when V is a 2-dimensional vector space with a basis $\{e_0, e_1\}$, an endomorphism A of $V \otimes V$ is presented by a matrix

$$A = \begin{pmatrix} A^{00}_{00} & A^{01}_{00} & A^{10}_{00} & A^{11}_{00} \\ A^{00}_{01} & A^{01}_{01} & A^{10}_{01} & A^{11}_{01} \\ A^{00}_{10} & A^{01}_{10} & A^{10}_{10} & A^{11}_{10} \\ A^{00}_{11} & A^{01}_{11} & A^{10}_{11} & A^{11}_{11} \end{pmatrix}.$$

Then, $A^{\circlearrowright}$ and $A^{\circlearrowleft}$ are presented by

$$A^{\circlearrowright} = \begin{pmatrix} A^{00}_{00} & A^{10}_{00} & A^{00}_{10} & A^{10}_{10} \\ A^{01}_{00} & A^{11}_{00} & A^{01}_{10} & A^{11}_{10} \\ A^{00}_{01} & A^{10}_{01} & A^{00}_{11} & A^{10}_{11} \\ A^{01}_{01} & A^{11}_{01} & A^{01}_{11} & A^{11}_{11} \end{pmatrix}, \qquad A^{\circlearrowleft} = \begin{pmatrix} A^{00}_{00} & A^{00}_{01} & A^{01}_{00} & A^{01}_{01} \\ A^{00}_{10} & A^{00}_{11} & A^{01}_{10} & A^{01}_{11} \\ A^{10}_{00} & A^{10}_{01} & A^{11}_{00} & A^{11}_{01} \\ A^{10}_{10} & A^{10}_{11} & A^{11}_{10} & A^{11}_{11} \end{pmatrix}.$$

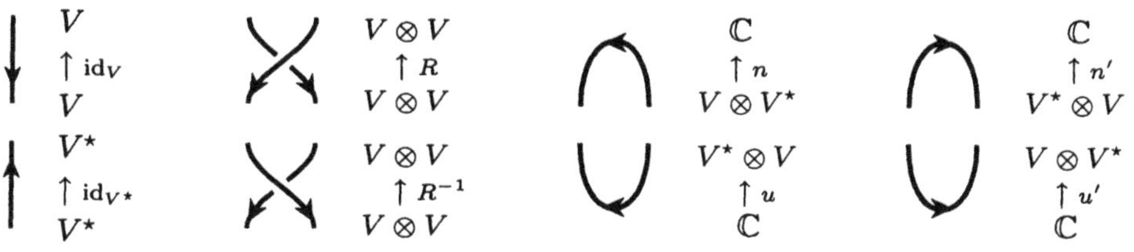

Figure 3.9 The linear maps associated to the oriented elementary tangle diagrams.

To define a bracket $[D]$ of an oriented tangle diagram D we begin with certain definitions of linear maps associated to elementary oriented tangle diagrams as follows. Let V be a vector space. We consider two invertible endomorphisms $R \in \mathrm{End}(V \otimes V)$ and $h \in \mathrm{End}(V)$. In Figure 3.9, we associate certain linear maps constructed from the endomorphisms R and h to certain elementary oriented tangle diagrams. Here, the maps n and n' are defined by $n(x \otimes f) = f(hx)$ and $n'(f \otimes x) = f(x)$ for $x \in V$ and $f \in V^\star$. Further, the maps u and u' are defined by $u(1) = \sum_i e^\star_i \otimes (h^{-1} e_i)$ and $u'(1) = \sum_i e_i \otimes e^\star_i$ where $\{e_i\}$ is a basis of the vector space V and $\{e^\star_i\}$ the dual basis to $\{e_i\}$.

Further, in the same way as the bracket was defined in the unoriented case, we define the bracket $[D]$ of an oriented sliced tangle diagram D determined by the endomorphisms R and h to be the composition of tensor products of copies of the linear maps associated to elementary diagrams, as is illustrated in Figure 3.10.

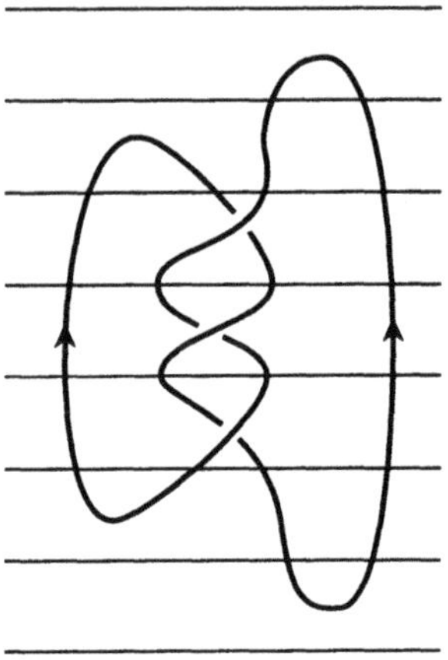

$$\begin{array}{c} \mathbb{C} \\ \uparrow n \\ V \otimes V^\star \\ \uparrow n' \otimes \mathrm{id}_V \otimes \mathrm{id}_{V^\star} \\ V^\star \otimes V \otimes V \otimes V^\star \\ \uparrow \mathrm{id}_{V^\star} \otimes R \otimes \mathrm{id}_{V^\star} \\ V^\star \otimes V \otimes V \otimes V^\star \\ \uparrow \mathrm{id}_{V^\star} \otimes R \otimes \mathrm{id}_{V^\star} \\ V^\star \otimes V \otimes V \otimes V^\star \\ \uparrow \mathrm{id}_{V^\star} \otimes R \otimes \mathrm{id}_{V^\star} \\ V^\star \otimes V \otimes V \otimes V^\star \\ \uparrow u \otimes \mathrm{id}_V \otimes \mathrm{id}_{V^\star} \\ V \otimes V^\star \\ \uparrow u' \\ \mathbb{C} \end{array}$$

Figure 3.10 An oriented sliced link diagram and its operator invariant

For example, we obtain the bracket of the following tangle diagram as

$$= (n' \otimes \mathrm{id}_V \otimes \mathrm{id}_{V^\star})(\mathrm{id}_{V^\star} \otimes R^{-1} \otimes \mathrm{id}_{V^\star})(\mathrm{id}_{V^\star} \otimes \mathrm{id}_V \otimes u') = (R^{-1})^{\circlearrowleft},$$

where we obtain the second equality by the following computation,

$$\begin{array}{rcl} e_i^\star \otimes e_k & \overset{\mathrm{id}_{V^\star} \otimes \mathrm{id}_V \otimes u'}{\longmapsto} & \sum_l e_i^\star \otimes e_k \otimes e_l \otimes e_l^\star \\ & \overset{\mathrm{id}_{V^\star} \otimes R^{-1} \otimes \mathrm{id}_{V^\star}}{\longmapsto} & \sum_{i',j,l} {R^{-1}}^{i'j}_{kl} e_i^\star \otimes e_{i'} \otimes e_j \otimes e_l^\star \\ & \overset{n' \otimes \mathrm{id}_V \otimes \mathrm{id}_{V^\star}}{\longmapsto} & \sum_{j,l} {R^{-1}}^{ij}_{kl} e_j \otimes e_l^\star = (R^{-1})^{\circlearrowleft}(e_i^\star \otimes e_k). \end{array}$$

Similarly, we have that

$$\begin{aligned} &= (\mathrm{id}_{V^\star \otimes V} \otimes n)(\mathrm{id}_{V^\star} \otimes R \otimes \mathrm{id}_{V^\star})(u \otimes \mathrm{id}_{V \otimes V^\star}) \\ &= \Big((\mathrm{id}_V \otimes h) \cdot R \cdot (h^{-1} \otimes \mathrm{id}_V)\Big)^{\circlearrowleft}. \end{aligned}$$

Composing the above two formulae of the bracket we have that

$$= (R^{-1})^{\circlearrowleft} \cdot \Big((\mathrm{id}_V \otimes h) \cdot R \cdot (h^{-1} \otimes \mathrm{id}_V)\Big)^{\circlearrowleft}.$$

Theorem 3.7 ([Tur88, Tur89, FrYe89]). Let T be an oriented tangle and D a sliced diagram of T. If invertible endomorphisms $R \in \mathrm{End}(V \otimes V)$ and $h \in \mathrm{End}(V)$

satisfy the following relations,

$$R \cdot (h \otimes h) = (h \otimes h) \cdot R, \tag{3.24}$$

$$\operatorname{trace}_2\Big((\mathrm{id}_V \otimes h) \cdot R^{\pm 1}\Big) = \mathrm{id}_V, \tag{3.25}$$

$$(R^{-1})^{\circlearrowleft} \cdot \Big((\mathrm{id}_V \otimes h) \cdot R \cdot (h^{-1} \otimes \mathrm{id}_V)\Big)^{\circlearrowleft} = \mathrm{id}_V \otimes \mathrm{id}_{V^\star}, \tag{3.26}$$

$$(R \otimes \mathrm{id}_V)(\mathrm{id}_V \otimes R)(R \otimes \mathrm{id}_V) = (\mathrm{id}_V \otimes R)(R \otimes \mathrm{id}_V)(\mathrm{id}_V \otimes R), \tag{3.27}$$

then the bracket $[D]$ is an isotopy invariant of T.

We denote the invariant by $[T]$ and call it the *operator invariant* of T associated with the linear maps R and h. In particular, the operator invariant $[L]$ of an oriented link L belongs to $\mathrm{End}(\mathbb{C}) = \mathbb{C}$.

We showed in Chapter 2 that a braid representation is derived from the relations (3.24), (3.25) and (3.27). Thus, the above theorem implies that we need an additional relation (3.26) to obtain an invariant of, not only links, but also tangles.

Proof of Theorem 3.7. By Theorem 3.3 it is sufficient to verify that $[D]$ is invariant under the Turaev moves for oriented diagrams given in Figure 3.6.

Firstly, the invariance under (3.9) is derived from the construction of the operator invariant. Secondly, the invariance under the first move of (3.10) is obtained as

[] $= (\mathrm{id}_V \otimes n')(u' \otimes \mathrm{id}_V) = \mathrm{id}_V =$ [],

where the second equality is obtained by

$$e_i \overset{u' \otimes \mathrm{id}_V}{\longmapsto} \sum_j e_j \otimes e_j^\star \otimes e_i \overset{\mathrm{id}_V \otimes n'}{\longmapsto} \sum_j \Big(e_j^\star(e_i)\Big) e_j = e_i,$$

noting that $e_j^\star(e_i) = 1$ if $j = i$, 0 otherwise. The invariance under the other moves of (3.10) is obtained similarly. Thirdly, we obtain the invariance under the moves (3.11) and (3.12) by the relations (3.24) and (3.25) respectively. Fourthly, the invariance under the moves (3.13) are derived from the relations $R \cdot R^{-1} = \mathrm{id}_V \otimes \mathrm{id}_V = R^{-1} \cdot R$. Fifthly, the invariance under the move (3.14) is derived from the relation (3.26), noting the computation before the theorem. Lastly, we obtain the invariance under the move (3.15) by the Yang-Baxter equation (3.27). □

Modifying the proof of Theorem 3.7 we have

Theorem 3.8. Let T be an oriented framed tangle and D a sliced diagram presenting T by blackboard framing. If the invertible endomorphisms $R \in \mathrm{End}(V \otimes V)$ and $h \in \mathrm{End}(V)$ satisfy the relations (3.24), (3.26), (3.27) and

$$\operatorname{trace}_2\Big((\mathrm{id}_V \otimes h) \cdot R^{\pm 1}\Big) = c^{\pm 1} \cdot \mathrm{id}_V, \tag{3.28}$$

for some scalar c, then the bracket $[D]$ is an isotopy invariant of T.

The Jones polynomial as an operator invariant

Let V be a 2-dimensional vector space over $\mathbb{C}$ with a basis $\{e_0, e_1\}$. As in Section 2.2 we put $R \in \mathrm{End}(V \otimes V)$ and $h \in \mathrm{End}(V)$ by

$$R = \begin{pmatrix} t^{1/2} & 0 & 0 & 0 \\ 0 & 0 & t & 0 \\ 0 & t & t^{1/2} - t^{3/2} & 0 \\ 0 & 0 & 0 & t^{1/2} \end{pmatrix}, \qquad h = \begin{pmatrix} t^{-1/2} & 0 \\ 0 & t^{1/2} \end{pmatrix}, \qquad (3.29)$$

with a complex parameter t. It has already been verified, in Section 2.2, that these linear maps R and h satisfy the relations (3.24), (3.25) and (3.27). We verify that they also satisfy (3.26) as follows. We have that

$$R^{-1} = \begin{pmatrix} t^{-1/2} & 0 & 0 & 0 \\ 0 & t^{-1/2} - t^{-3/2} & t^{-1} & 0 \\ 0 & t^{-1} & 0 & 0 \\ 0 & 0 & 0 & t^{-1/2} \end{pmatrix} \in \mathrm{End}(V \otimes V).$$

Hence,

$$(R^{-1})^{\circlearrowleft} = \begin{pmatrix} t^{-1/2} & 0 & 0 & t^{-1/2} - t^{-3/2} \\ 0 & 0 & t^{-1} & 0 \\ 0 & t^{-1} & 0 & 0 \\ 0 & 0 & 0 & t^{-1/2} \end{pmatrix} \in \mathrm{Hom}(V^\star \otimes V, V \otimes V^\star). \quad (3.30)$$

Further,

$$\Big((\mathrm{id}_V \otimes h) \cdot R \cdot (h^{-1} \otimes \mathrm{id}_V)\Big) = \begin{pmatrix} t^{1/2} & 0 & 0 & 0 \\ 0 & 0 & t & 0 \\ 0 & t & t^{-1/2} - t^{1/2} & 0 \\ 0 & 0 & 0 & t^{1/2} \end{pmatrix} \in \mathrm{End}(V \otimes V).$$

Hence,

$$\Big((\mathrm{id}_V \otimes h) \cdot R \cdot (h^{-1} \otimes \mathrm{id}_V)\Big)^{\circlearrowleft} = \begin{pmatrix} t^{1/2} & 0 & 0 & t^{-1/2} - t^{1/2} \\ 0 & 0 & t & 0 \\ 0 & t & 0 & 0 \\ 0 & 0 & 0 & t^{1/2} \end{pmatrix} \in \mathrm{Hom}(V \otimes V^\star, V^\star \otimes V).$$

Since the above matrix is the inverse of the matrix (3.30), *i.e.*, of $(R^{-1})^{\circlearrowleft}$, the maps R and h satisfy the relation (3.26).

Theorem 3.9. Let L be an oriented link. Since the R and h given in (3.29) satisfy the conditions of Theorem 3.7, the operator invariant $[L] \in \mathbb{C}$ is an isotopy invariant of L. Further, the invariant is equal to the Jones polynomial $V_L(t)$.

This theorem gives an alternative definition of the *Jones polynomial* as an operator invariant.

Proof. We show the equality of this invariant and the Jones polynomial as follows. Let b be a braid in n strands whose closure is isotopic to L. Comparing the definitions of the operator invariant and the representation ψ_n in Section 2.2 we have that

$$[b] = \psi_n(b).$$

Further, let D be the sliced diagram of the closure of b obtained from the picture of the closure, given in Figure 2.3, by adding horizontal lines. Then, from the construction of the operator invariant, we have that

$$[L] = \text{trace}\Big(h^{\otimes n} \cdot [b]\Big).$$

Further, by Theorem 2.3, the right hand side, putting $[b] = \psi_n(b)$, is equal to the Jones polynomial. Hence, the invariant $[L]$ is equal to the Jones polynomial. □

The Alexander polynomial as an operator invariant

Let V be a 2-dimensional vector space over $\mathbb{C}$ with a basis $\{e_0, e_1\}$. As in Section 2.3 we put $R \in \text{End}(V \otimes V)$ and $h \in \text{End}(V)$ by

$$R = \begin{pmatrix} t^{-1/2} & 0 & 0 & 0 \\ 0 & 0 & 1 & 0 \\ 0 & 1 & t^{-1/2} - t^{1/2} & 0 \\ 0 & 0 & 0 & -t^{1/2} \end{pmatrix}, \qquad h = \begin{pmatrix} t^{1/2} & 0 \\ 0 & -t^{1/2} \end{pmatrix}, \tag{3.31}$$

with a complex parameter t. It has already been verified, in Section 2.3, that these linear maps R and h satisfy the relations (3.24), (3.25) and (3.27). We verify that they also satisfy (3.26) as follows. We have that

$$R^{-1} = \begin{pmatrix} t^{1/2} & 0 & 0 & 0 \\ 0 & t^{1/2} - t^{-1/2} & 1 & 0 \\ 0 & 1 & 0 & 0 \\ 0 & 0 & 0 & -t^{-1/2} \end{pmatrix} \in \text{End}(V \otimes V).$$

Hence,

$$(R^{-1})^{\circlearrowleft} = \begin{pmatrix} t^{1/2} & 0 & 0 & t^{1/2} - t^{-1/2} \\ 0 & 0 & 1 & 0 \\ 0 & 1 & 0 & 0 \\ 0 & 0 & 0 & -t^{-1/2} \end{pmatrix} \in \text{Hom}(V^* \otimes V, V \otimes V^*). \tag{3.32}$$

Further,

$$\Big((\mathrm{id}_V \otimes h)\cdot R\cdot (h^{-1}\otimes \mathrm{id}_V)\Big) = \begin{pmatrix} t^{-1/2} & 0 & 0 & 0 \\ 0 & 0 & 1 & 0 \\ 0 & 1 & t^{1/2}-t^{-1/2} & 0 \\ 0 & 0 & 0 & -t^{1/2} \end{pmatrix} \in \mathrm{End}(V\otimes V).$$

Hence,

$$\Big((\mathrm{id}_V\otimes h)\cdot R\cdot (h^{-1}\otimes \mathrm{id}_V)\Big)^{\circlearrowleft} = \begin{pmatrix} t^{-1/2} & 0 & 0 & t^{1/2}-t^{-1/2} \\ 0 & 0 & 1 & 0 \\ 0 & 1 & 0 & 0 \\ 0 & 0 & 0 & -t^{1/2} \end{pmatrix} \in \mathrm{Hom}(V\otimes V^\star, V^\star\otimes V).$$

Since the above matrix is the inverse of the matrix (3.32), *i.e.*, of $(R^{-1})^{\circlearrowleft}$, the maps R and h satisfy the relation (3.26).

Since the linear maps R and h satisfy the relations of Theorem 3.7 we have the operator invariant of oriented tangles associated with the maps R and h. Unlike the above case of the Jones polynomial, the operator invariant vanishes for links, as stated by the following proposition.

Proposition 3.10. For any oriented link L, the operator invariant $[L]$ associated with the linear maps R and h given in (3.31) is equal to 0.

A *1-tangle* is a tangle whose ends are an upper end and a lower end. A *2-tangle* is a tangle whose ends are two upper ends and two lower ends. Further, an *oriented 1-tangle* (resp. an oriented *2-tangle*) is a 1-tangle (resp. a 2-tangle), which has been oriented, with its ends oriented downward.

Proof of Proposition 3.10. Let T be a 1-tangle obtained from L by cutting a component of L. Then, L is isotopic to the closure of T, where the *closure* of a 1-tangle is defined by connecting the two ends of the 1-tangle as in Figure 3.11. By the construction of the operator invariant we have that

$$[L] = \mathrm{trace}\Big(h\cdot [T]\Big).$$

Since $[T]$ is equal to a scalar multiple of id_V by Lemma 3.11, the above formula is equal to 0 for the map h given in (3.31). □

Lemma 3.11. For a 1-tangle T the operator invariant $[T]$ associated with the linear maps R and h given in (3.31) is presented by

$$[T] = c\cdot \mathrm{id}_V,$$

for some scalar c.

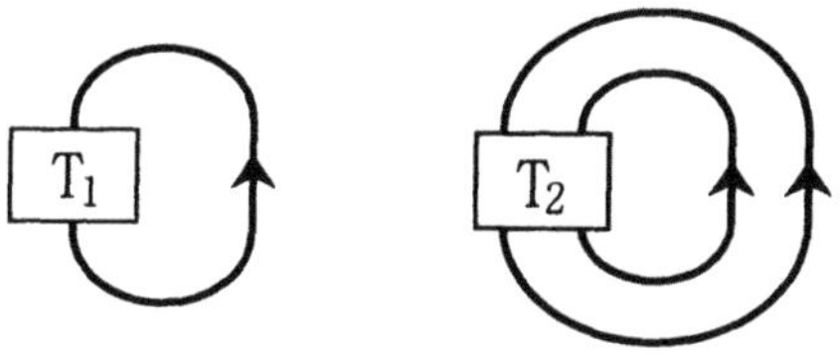

Figure 3.11 The closures of a 1-tangle T_1 and a 2-tangle T_2

The lemma can be obtained as an immediate conclusion of the fact that $[T]$ is an intertwiner with respect to the action of the quantum group $U_\varsigma(sl_2)$ on an irreducible representation V as explained later in Section 4.5 (see also Proposition 4.10). Here we give another, direct, proof based on the construction of operator invariants given in this section.

Proof of Lemma 3.11. By charge conservation, the operator $[T]$ takes the form,

$$[T] = \begin{pmatrix} a & 0 \\ 0 & b \end{pmatrix} \in \mathrm{End}(V).$$

Further, by Theorem 3.7, we have that

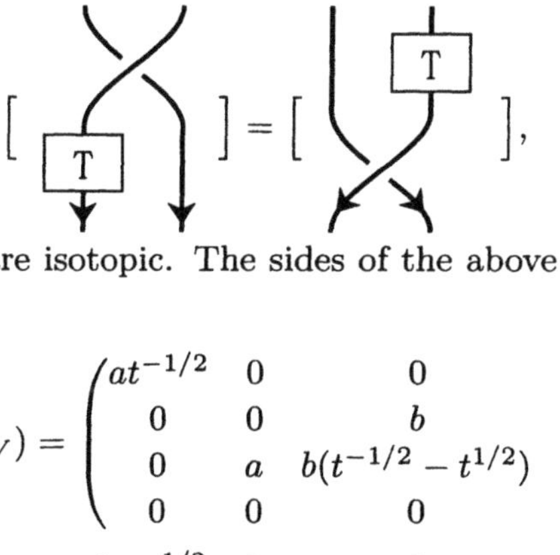

since these two tangles are isotopic. The sides of the above formula are computed respectively as

$$R \cdot ([T] \otimes \mathrm{id}_V) = \begin{pmatrix} at^{-1/2} & 0 & 0 & 0 \\ 0 & 0 & b & 0 \\ 0 & a & b(t^{-1/2} - t^{1/2}) & 0 \\ 0 & 0 & 0 & -bt^{1/2} \end{pmatrix},$$

$$(\mathrm{id}_V \otimes [T]) \cdot R = \begin{pmatrix} at^{-1/2} & 0 & 0 & 0 \\ 0 & 0 & b & 0 \\ 0 & a & a(t^{-1/2} - t^{1/2}) & 0 \\ 0 & 0 & 0 & -bt^{1/2} \end{pmatrix}.$$

By comparing these two matrices we obtain $a = b$. This implies the lemma. □

The following theorem gives an alternative definition of the *Alexander polynomial* as an operator invariant.

Theorem 3.12. Let L be an oriented link and T a 1-tangle whose closure is isotopic to L. Then, the operator invariant $[T]$ associated with the linear maps R and h given in (3.31) is an isotopy invariant of L. Further, $[T]$ is presented by the Alexander polynomial $\Delta_L(t)$ of L as

$$[T] = \Delta_L(t) \cdot \mathrm{id}_V. \tag{3.33}$$

Proof. Let L_1 and L_2 be two isotopic oriented links and T_1 and T_2 two 1-tangles whose closures are isotopic to L_1 and L_2 respectively. We show $[T_1] = [T_2]$ as follows.

The 1-tangle T_1 (resp. T_2) is obtained from L_1 (resp. L_2) by cutting a component of L_1 (resp. L_2). If the cut components of L_1 and L_2 are related by the isotopy between L_1 and L_2, then T_1 and T_2 are isotopic and we obtain $[T_1] = [T_2]$ by Theorem 3.7, completing the proof in this case. Otherwise, T_1 and T_2 are obtained from some 2-tangle T by closing the second and the first strands respectively, as below.

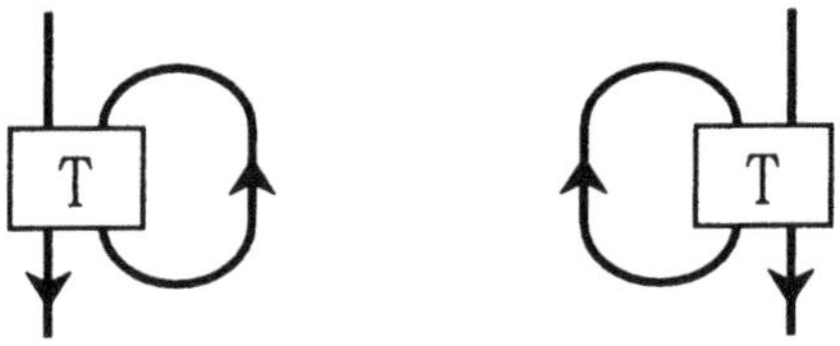

T_1 T_2

We can obtain $[T_1] = [T_2]$ as an immediate conclusion of Lemma A.17 using the fact that $[T]$ is an intertwiner with respect to the action of the quantum group $U_\varsigma(sl_2)$ on V. Here we give another direct proof below, based on the construction of the operator invariant in this section.

By charge conservation the operator invariant $[T]$ of the 2-tangle T is presented by

$$[T] = \begin{pmatrix} a & 0 & 0 & 0 \\ 0 & b & c & 0 \\ 0 & d & e & 0 \\ 0 & 0 & 0 & f \end{pmatrix} \in \mathrm{End}(V \otimes V).$$

We have the following equality by isotopy invariance of the operator invariant,

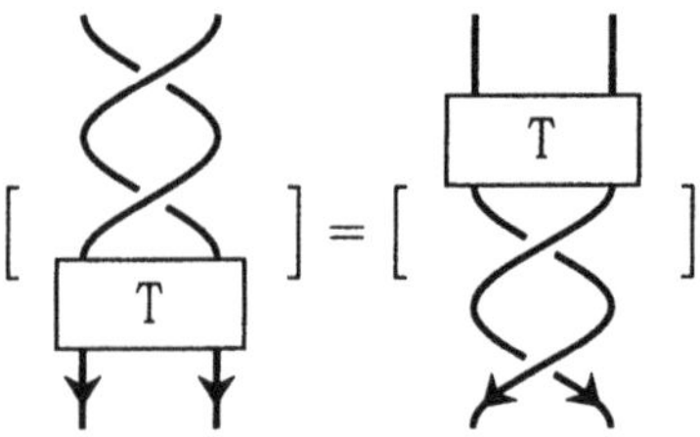

The difference of the two sides is computed as

$$R^2 \cdot [T] - [T] \cdot R^2$$
$$= (t^{1/2} - t^{-1/2}) \begin{pmatrix} 0 & 0 & 0 & 0 \\ 0 & c-d & b-e-c(t^{1/2}-t^{-1/2}) & 0 \\ 0 & -b+e+d(t^{1/2}-t^{-1/2}) & d-c & 0 \\ 0 & 0 & 0 & 0 \end{pmatrix}.$$

Hence,

$$c = d, \qquad b - e = c(t^{1/2} - t^{-1/2}).$$

Further, we consider the following equality,

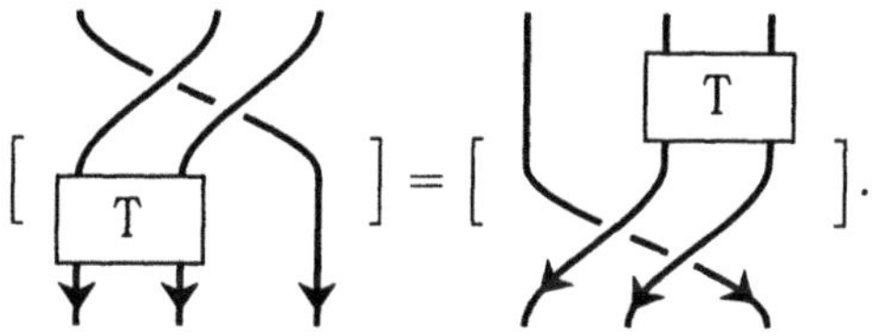

By regarding $V \otimes V \otimes V$ as the direct sum of subspaces with the basis $\{e_0 \otimes e_0 \otimes e_0\}$, $\{e_0 \otimes e_0 \otimes e_1,\ e_0 \otimes e_1 \otimes e_0,\ e_1 \otimes e_0 \otimes e_0\}$, $\{e_0 \otimes e_1 \otimes e_1,\ e_1 \otimes e_0 \otimes e_1,\ e_1 \otimes e_1 \otimes e_0\}$ and $\{e_1 \otimes e_1 \otimes e_1\}$ respectively, we have the following presentations, setting $d = c$,

$$[T] \otimes \mathrm{id}_V = (a) \oplus \begin{pmatrix} a & 0 & 0 \\ 0 & b & c \\ 0 & c & e \end{pmatrix} \oplus \begin{pmatrix} b & c & 0 \\ c & e & 0 \\ 0 & 0 & f \end{pmatrix} \oplus (f),$$
$$\mathrm{id}_V \otimes [T] = (a) \oplus \begin{pmatrix} b & c & 0 \\ c & e & 0 \\ 0 & 0 & a \end{pmatrix} \oplus \begin{pmatrix} f & 0 & 0 \\ 0 & b & c \\ 0 & c & e \end{pmatrix} \oplus (f),$$

and

$$(R \otimes \mathrm{id}_V)(\mathrm{id}_V \otimes R)$$
$$= (t^{-1}) \oplus \begin{pmatrix} 0 & t^{-1/2} & 0 \\ 0 & 0 & t^{-1/2} \\ 1 & t^{-1/2} - t^{1/2} & t^{-1} - 1 \end{pmatrix} \oplus \begin{pmatrix} 0 & 0 & 1 \\ -t^{1/2} & 0 & t^{-1/2} - t^{1/2} \\ 0 & -t^{1/2} & t-1 \end{pmatrix} \oplus (t).$$

Hence,

$$\frac{1}{t^{1/2} - t^{-1/2}} \Big((R \otimes \mathrm{id}_V)(\mathrm{id}_V \otimes R)(\mathrm{id}_V \otimes [T]) - (\mathrm{id}_V \otimes [T])(R \otimes \mathrm{id}_V)(\mathrm{id}_V \otimes R) \Big)$$
$$= (0) \oplus \begin{pmatrix} 0 & 0 & 0 \\ 0 & 0 & 0 \\ 0 & a-b-ct^{-1/2} & at^{-1/2} - c - et^{-1/2} \end{pmatrix} \oplus \begin{pmatrix} 0 & 0 & 0 \\ 0 & 0 & b - ct^{1/2} - f \\ 0 & 0 & c - et^{1/2} + ft^{1/2} \end{pmatrix} \oplus (0).$$

Therefore,

$$a - b = ct^{-1/2} = e - f,$$
$$a - e = ct^{1/2} = b - f.$$

Further, by the construction of the operator invariant, we have that

$$[T_1] = \mathrm{trace}_2\Big((\mathrm{id}_V \otimes h) \cdot [T]\Big) = t^{1/2}\mathrm{trace}_2 \begin{pmatrix} a & 0 & 0 & 0 \\ 0 & -b & -c & 0 \\ 0 & d & e & 0 \\ 0 & 0 & 0 & -f \end{pmatrix}$$
$$= t^{1/2} \begin{pmatrix} a-b & 0 \\ 0 & e-f \end{pmatrix} = \begin{pmatrix} c & 0 \\ 0 & c \end{pmatrix}.$$

Similarly, we have that

$$[T_2] = \mathrm{trace}_1\Big([T] \cdot (h^{-1} \otimes \mathrm{id}_V)\Big) = t^{-1/2} \begin{pmatrix} a-e & 0 \\ 0 & b-f \end{pmatrix} = \begin{pmatrix} c & 0 \\ 0 & c \end{pmatrix}.$$

Hence, $[T_1] = [T_2]$. Therefore, the operator invariant $[T]$ of the theorem is an isotopy invariant of L.

We obtain the equality (3.33) as follows. Let b be a braid in n strands whose closure is isotopic to L. Let T be a 1-tangle obtained from b by closing the $2, 3, \cdots, n$-th strands. Comparing the definitions of the operator invariant and the representation ψ_n in Section 2.3 we have that

$$[T] = \mathrm{trace}_{2,3,\cdots,n}\Big((1 \otimes h^{\otimes(n-1)})\psi_n(b)\Big). \tag{3.34}$$

Further, by Lemma 3.11 the operator invariant $[T]$ is presented by

$$[T] = c \cdot \mathrm{id}_V,$$

for some scalar c. As mentioned in Theorem 2.5, the right hand side of (3.34) is equal to $\Delta_L(t) \cdot \mathrm{id}_V$, which is proved in Appendix C.1. Hence, we obtain the equality (3.33). □

Chapter 4

Ribbon Hopf algebras and invariants of links

Quantum groups were discovered independently by Drinfel'd [Dri85] and by Jimbo [Jim85] in the middle 1980s, as quantizations of Lie groups. Roughly speaking, a quantum group can be regarded as a non-commutative deformation of the function algebra of a Lie group.

There is a natural Hopf algebra structure on the function algebra A of a group G; for simplicity, we consider the case of a finite group as follows. The Hopf algebra structure of A is given by three maps named the comultiplication, the antipode and the counit. The *comultiplication* $\Delta : A \to A \otimes A$ is the pull back of the multiplication $G \times G \to G$. The *antipode* $S : A \to A$ is the pull back of the map $G \to G$ taking inverses of elements. The *counit* $\varepsilon : A \to \mathbb{C}$ is the pull back of the inclusion $\{e\} \to G$ where e denotes the unit of G.

The Hopf algebra structure of a quantum group is given as a deformation of such a Hopf algebra structure of the (suitably defined) function algebra of a Lie group. Thus a quantum group is *not* a group (despite its naming) but rather a Hopf algebra. Further, a quantum group is a ribbon Hopf algebra, which is a Hopf algebra with additional properties which are sufficient to construct invariants of framed links from it. The aim of this chapter is to introduce invariants of framed links and tangles, derived from ribbon Hopf algebras, and investigate such invariants for the case of the quantum group $U_q(sl_2)$. The invariants of framed links and tangles derived from quantum groups are called *quantum invariants*.

Such invariants derived from ribbon Hopf algebras are related to the operator invariants introduced in the previous chapter, as follows. We obtain R matrices as intertwiners with respect to the action of a ribbon Hopf algebra. By using such R matrices we obtain operator invariants of tangles, as mentioned in Section 3.3. Further, we show in Sections 4.4 and 4.5 that the operator invariants of the Jones and Alexander polynomials are derived from 2-dimensional representations of the quantum group $U_q(sl_2)$, at generic q and at a root of unity q respectively.

There is a structure similar to a Hopf algebra structure on the set of links (or suitable tangles). The *comultiplication* Δ, the *antipode* S and the *counit* ε acting on a component of a link are graphically given as

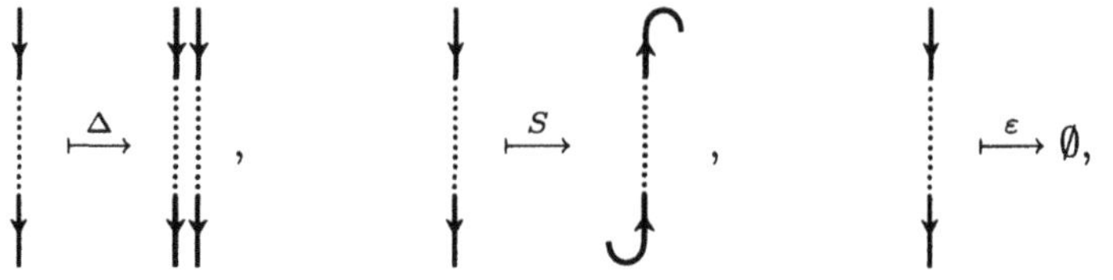

where Δ takes a strand to the union of two parallel copies of the strand, S takes a strand to the same strand with opposite orientation with additional curves at its ends as shown in the above picture and ε removes a strand. Motivated by such structures, we give graphical computations of Hopf algebras in Section 4.1.

4.1 Ribbon Hopf algebras

In this section we introduce, in order, Hopf algebras, quasi-triangular Hopf algebras, and ribbon Hopf algebras, and show properties of these algebras that will be useful in construction of link invariants that will take place in the following sections. Most results in this section are originally due to [Dri89a]. Here, we obtain these results by graphical computations.

Hopf algebra

An algebra A over $\mathbb{C}$, with a unit element 1, equipped a homomorphism $\Delta : A \to A \otimes A$, an anti-homomorphism $S : A \to A$, and a homomorphism $\varepsilon : A \to \mathbb{C}$, is called a *Hopf algebra* if the maps satisfy the following relations,

$$(\Delta \otimes \mathrm{id}) \circ \Delta = (\mathrm{id} \otimes \Delta) \circ \Delta, \tag{4.1}$$
$$(\varepsilon \otimes \mathrm{id}) \circ \Delta = \mathrm{id}, \tag{4.2}$$
$$(\mathrm{id} \otimes \varepsilon) \circ \Delta = \mathrm{id}, \tag{4.3}$$
$$m \circ (S \otimes \mathrm{id}) \circ \Delta = i \circ \varepsilon, \tag{4.4}$$
$$m \circ (\mathrm{id} \otimes S) \circ \Delta = i \circ \varepsilon, \tag{4.5}$$

where the map $m : A \otimes A \to A$ is the product of the algebra A and the map $i : \mathbb{C} \to A$ is given by $i(1) = 1$. We call the map Δ the *comultiplication*, the map S the *antipode* and the map ε the *counit*.

When we define a map of A to $A \otimes A \otimes A$ by a composition of two comultiplications, we have two ways $(\Delta \otimes \mathrm{id}) \circ \Delta$ and $\Delta \circ (\Delta \otimes \mathrm{id})$ to make the map. The relation (4.1) guarantees that the two ways give the same result. We denote the map $(\Delta \otimes \mathrm{id}) \circ \Delta$ (or equivalently the map $(\mathrm{id} \otimes \Delta) \circ \Delta$) by $\Delta^{(2)}$. Further, we define the map $\Delta^{(n)} : A \to A^{\otimes(n+1)}$ recursively by $\Delta^{(n)} = (\Delta^{(n-1)} \otimes \mathrm{id}) \circ \Delta$. The relation (4.1) guarantees that the map $\Delta^{(n)}$ can be defined independently of the choice of way to compose n comultiplications.

For the following computations of elements of a Hopf algebra, it is convenient to present the relations (4.4) and (4.5) in a graphical way by

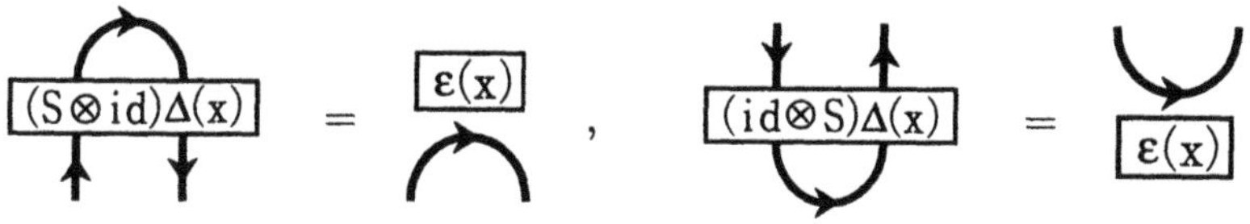

(4.6)

where x is any element of A. In the pictures of (4.6) each strand implies a copy of the Hopf algebra A. Further, for each strand we consider the product of the elements on the strand along the orientation of the strand. For example, we regard the third picture as follows:

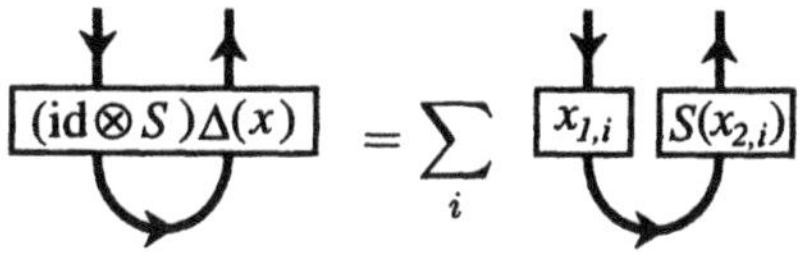

$$= \sum_i x_{1,i} S(x_{2,i}) = \big(m \circ (\mathrm{id} \otimes S) \circ \Delta\big)(x),$$

writing $\Delta(x) = \sum_i x_{1,i} \otimes x_{2,i}$. The equalities of (4.6) imply that such products along the corresponding strands on the two sides of such a formula are equal.

Quasi-triangular Hopf algebra

For a Hopf algebra A we consider an invertible element $\mathcal{R} \in A \otimes A$ satisfying

$$(P \circ \Delta)(x) = \mathcal{R}\Delta(x)\mathcal{R}^{-1}, \quad \text{for any } x \in A, \tag{4.7}$$

$$(\Delta \otimes \mathrm{id})(\mathcal{R}) = \mathcal{R}_{13}\mathcal{R}_{23}, \tag{4.8}$$

$$(\mathrm{id} \otimes \Delta)(\mathcal{R}) = \mathcal{R}_{13}\mathcal{R}_{12}, \tag{4.9}$$

where the map P is the permutation given by $P(x \otimes y) = y \otimes x$ and we put $\mathcal{R}_{12} = \mathcal{R} \otimes 1$, $\mathcal{R}_{23} = 1 \otimes \mathcal{R}$ and $\mathcal{R}_{13} = \sum \alpha_i \otimes 1 \otimes \beta_i$ putting $\mathcal{R} = \sum \alpha_i \otimes \beta_i$. We call such a pair $(A, \mathcal{R})$ a *quasi-triangular Hopf algebra* and $\mathcal{R}$ a *universal R matrix*.

The relation (4.7) implies, in a graphical way, that  and  commute for any $x \in A$, *i.e.*, the following equality holds for any $x \in A$,

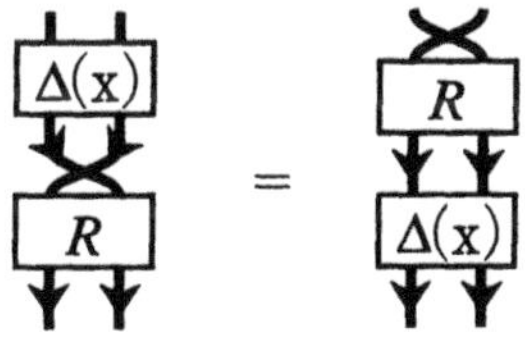

(4.10)

The relations (4.8) and (4.9) imply that $(\Delta \otimes \mathrm{id})(\mathcal{R})$ and $(\mathrm{id} \otimes \Delta)(\mathcal{R})$ are equal to compositions of two copies of 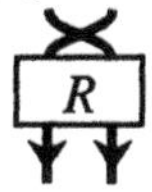 respectively *i.e.*, the relations (4.8) and

(4.9) are graphically shown as

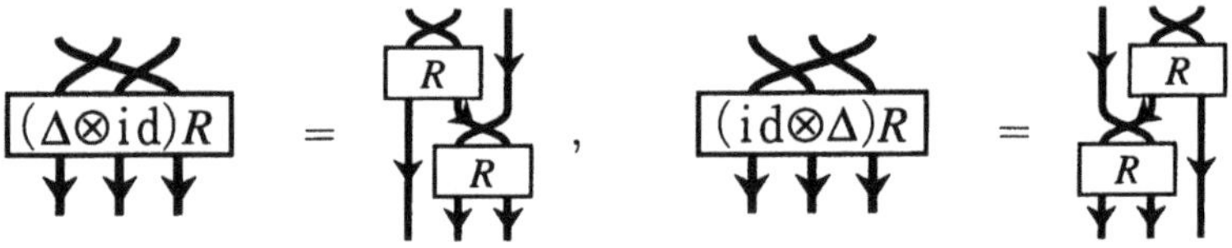

(4.11)

For a quasi-triangular Hopf algebra $(A, \mathcal{R})$, putting $\mathcal{R} = \sum \alpha_i \otimes \beta_i$, we define the element $u \in A$ by $u = \sum S(\beta_i)\alpha_i \in A$. We present the definition of u in a graphical way by

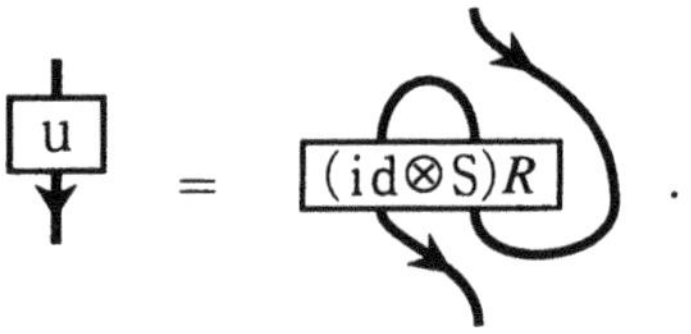

(4.12)

See also the end of Section 4.2 for a similar graphical presentation of u.

Proposition 4.1 ([Dri89a]). A Hopf algebra A with an invertible element $\mathcal{R} \in A \otimes A$ satisfying the relation (4.7) has the following properties,

$$S^2(x) = uxu^{-1} \quad \text{for any } x \in A, \tag{4.13}$$

$$u^{-1} = \sum S^{-1}(\beta'_i)\alpha'_i, \tag{4.14}$$

where we put $\mathcal{R}^{-1} = \sum \alpha'_i \otimes \beta'_i$.

Proof. We give a graphical proof of (4.13) as follows. By (4.10) we have that

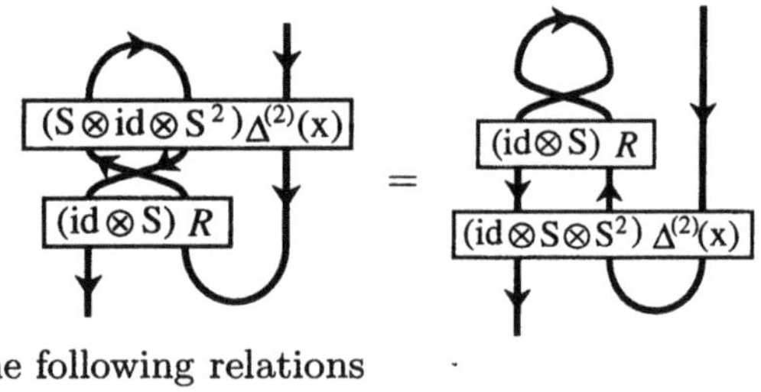

(4.15)

Further, we have the following relations

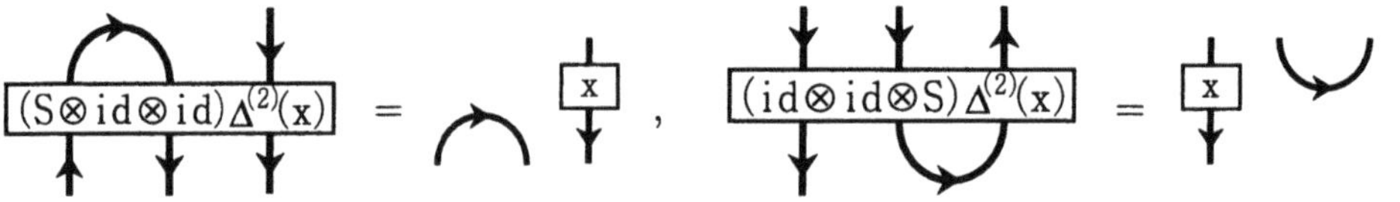

where we obtain the equalities by applying the relations (4.6) to the left hand sides of the above formulae and by using (4.2) and (4.3). By applying the first and second relations of the above formula to the left and right hand sides of (4.15) respectively we have that

LHS of (4.15) = 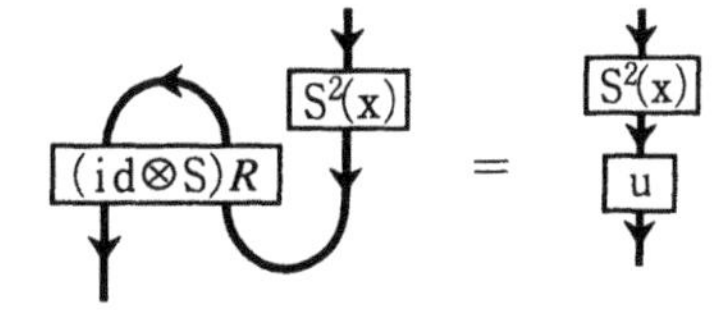

,

RHS of (4.15) = 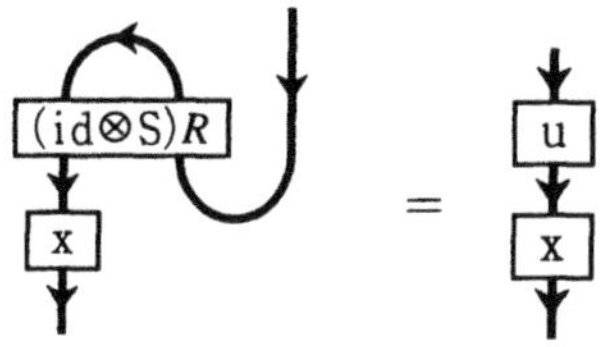

,

where the second equality of each of the above formulae is derived from the definition (4.12) of u. Hence, we obtain $S^2(x)u = ux$ and the proof of (4.13) below guarantees the existence of u^{-1}, completing the proof of (4.13).

Further, we can give a graphical proof of (4.14) as follows. From the relation $S^2(x)u = ux$ shown above, we have that

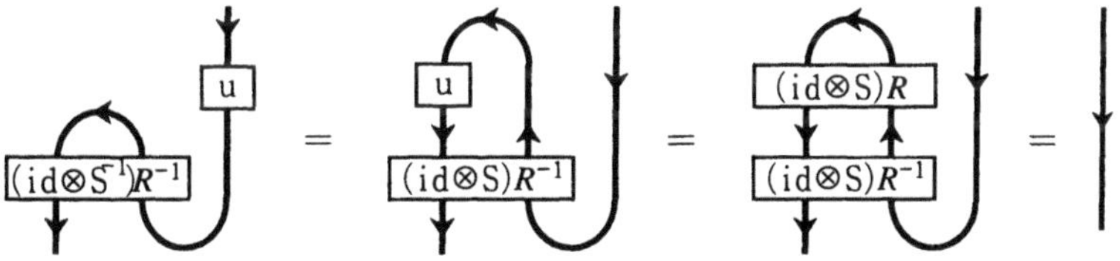

,

where the second equality is derived from the definition (4.12) of u. The above formula implies the relation (4.14). □

The following proposition will be used when we define the universal A invariant in the next section.

Proposition 4.2 ([Dri89a]). A quasi-triangular Hopf algebra $(A, \mathcal{R})$ has the following properties,

$$\mathcal{R}_{12}\mathcal{R}_{13}\mathcal{R}_{23} = \mathcal{R}_{23}\mathcal{R}_{13}\mathcal{R}_{12}, \tag{4.16}$$
$$(\varepsilon \otimes \mathrm{id})\mathcal{R} = 1 = (\mathrm{id} \otimes \varepsilon)\mathcal{R}, \tag{4.17}$$
$$(S \otimes \mathrm{id})\mathcal{R} = \mathcal{R}^{-1} = (\mathrm{id} \otimes S^{-1})\mathcal{R}, \tag{4.18}$$
$$(S \otimes S)\mathcal{R} = \mathcal{R}, \tag{4.19}$$
$$u \otimes u \cdot \mathcal{R} = \mathcal{R} \cdot u \otimes u. \tag{4.20}$$

In particular, the invertibility of $\mathcal{R}$ is derived from (4.18). Recall that we put $\mathcal{R}_{12} = \mathcal{R} \otimes 1$, $\mathcal{R}_{23} = 1 \otimes \mathcal{R}$ and $\mathcal{R}_{13} = \sum \alpha_i \otimes 1 \otimes \beta_i$ putting $\mathcal{R} = \sum \alpha_i \otimes \beta_i$.

The relation (4.16) is called the *quantized Yang-Baxter equation*; it is graphically shown as

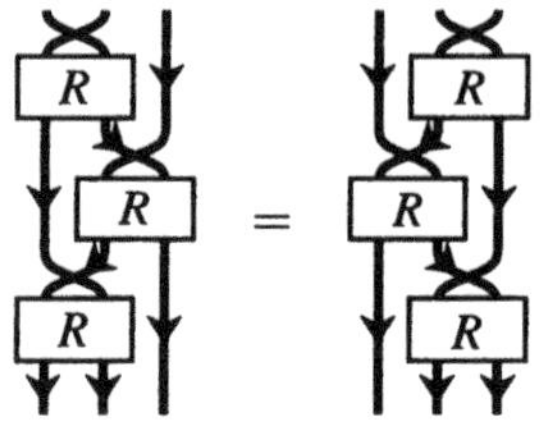

(4.21)

Proof of Proposition 4.2. Firstly, we show a proof of the relation (4.16) (alternatively its graphical presentation (4.21)) as

LHS of (4.21) =

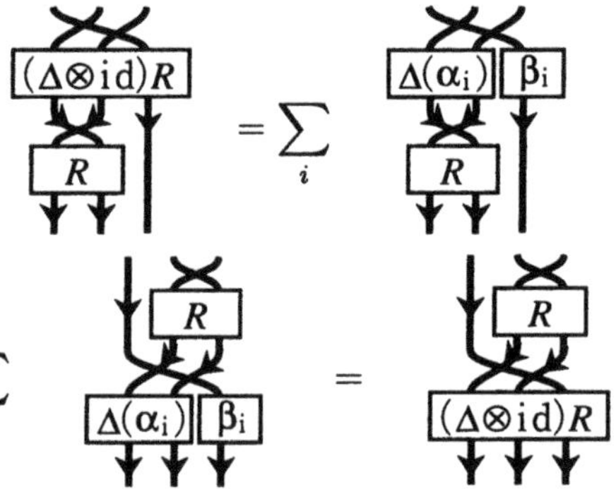

= RHS of (4.21)

where the first and the last equalities are derived from the defining relations (4.11) of a quasi-triangular Hopf algebra, and the second and the fourth equalities employ the notation $\mathcal{R} = \sum_i \alpha_i \otimes \beta_i$, and the third equality is obtained by (4.10).

Secondly, we prove (4.17) as follows. By applying the counit ε to the left strand of the first formula of (4.11), we have that

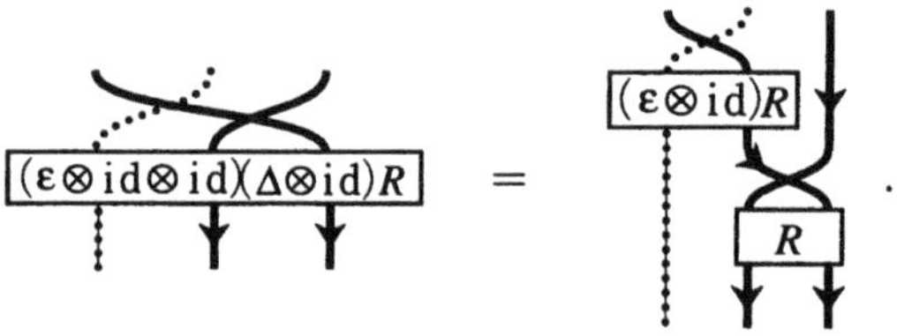

By (4.2) the left hand side of the above formula is equal to 

. Hence, we obtain the first relation of (4.17). Similarly, by applying the counit ε to the right strand of the second relation of (4.11) we obtain the second relation of (4.17).

Thirdly, we obtain (4.18) as follows. From the first relation of (4.11) we have that

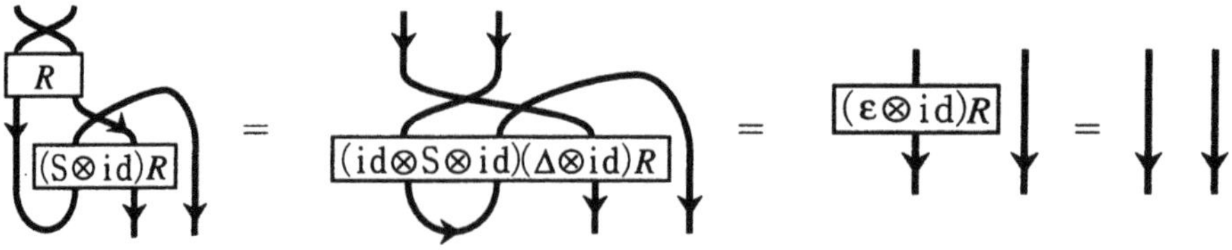

where the second and the third equalities are derived from the second relation of (4.6) and the first equality of (4.17) respectively. The above formula implies the first equality of (4.18). Similarly, the second equality of (4.18) can be obtained from the second relation of (4.11) by using the second equality of (4.17).

Fourthly, we obtain (4.19) immediately from the formula (4.18) by comparing the first and the last term of (4.18).

Finally, the relation (4.20) is obtained as follows,

$$u \otimes u \cdot R = (S^2 \otimes S^2)\mathcal{R} \cdot u \otimes u = R \cdot u \otimes u,$$

where we obtain the first equality by (4.13), and obtain the second equality by applying (4.19) twice. □

The following proposition will be used when we show properties of the universal A invariant in the next section.

Proposition 4.3 (see **[Dri89a]**). The element u of a quasi-triangular Hopf algebra satisfies

$$\begin{aligned}
&\Delta(u) = (u \otimes u) \cdot (\mathcal{R}_{21}\mathcal{R})^{-1}, \\
&\Delta\big(S(u)\big) = \big(S(u) \otimes S(u)\big) \cdot (\mathcal{R}_{21}\mathcal{R})^{-1}, \\
&\varepsilon(u) = 1.
\end{aligned}$$

Proof. The third formula is obtained by (4.17).

The first formula is obtained as follows. By (4.9) and (4.19) we have that

$$\begin{aligned}
&\big(\mathrm{id} \otimes (\Delta \circ S)\big)\mathcal{R} = (S^{-1} \otimes \Delta)\mathcal{R} = (S^{-1} \otimes \mathrm{id} \otimes \mathrm{id})(\mathcal{R}_{13}\mathcal{R}_{12}) \\
&= (S^{-1} \otimes \mathrm{id} \otimes \mathrm{id})\mathcal{R}_{12} \cdot (S^{-1} \otimes \mathrm{id} \otimes \mathrm{id})\mathcal{R}_{13} \\
&= (\mathrm{id} \otimes S \otimes \mathrm{id})\mathcal{R}_{12} \cdot (\mathrm{id} \otimes \mathrm{id} \otimes S)\mathcal{R}_{13} = (\mathrm{id} \otimes S \otimes S)(\mathcal{R}_{12}\mathcal{R}_{13}) \\
&= \Big(\mathrm{id} \otimes \big(P \circ (S \otimes S)\big)\Big)(\mathcal{R}_{13}\mathcal{R}_{12}) = \Big(\mathrm{id} \otimes \big(P \circ (S \otimes S) \circ \Delta\big)\Big)\mathcal{R},
\end{aligned}$$

where $p : A \otimes A \to A \otimes A$ denotes the permutation given by $x \otimes y \mapsto y \otimes x$. Hence, $\Delta \circ S$ is equal to $P \circ (S \otimes S) \circ \Delta$ when they are acting on the second entry of $\mathcal{R}$. Thus, the two sides of the first formula are equal to the corresponding sides of the following formula.

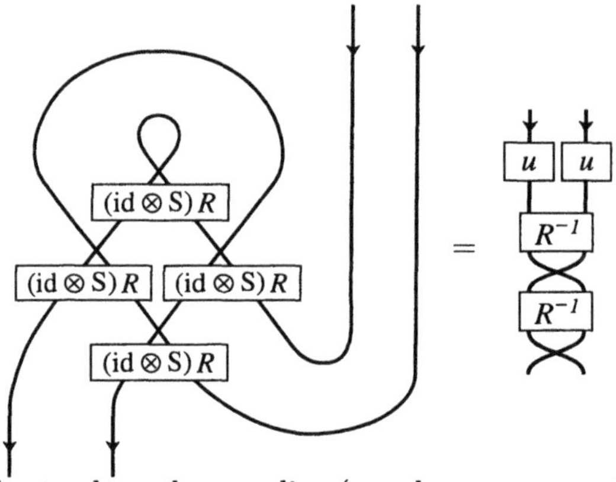

(4.22)

It is left to the reader to show the equality (see the comment after the proof).

The second formula of the proposition is obtained similarly (see also the comment below). □

Using the universal A invariant $Q^{A;\star}$, defined in the next section, the formula (4.22) can be presented by

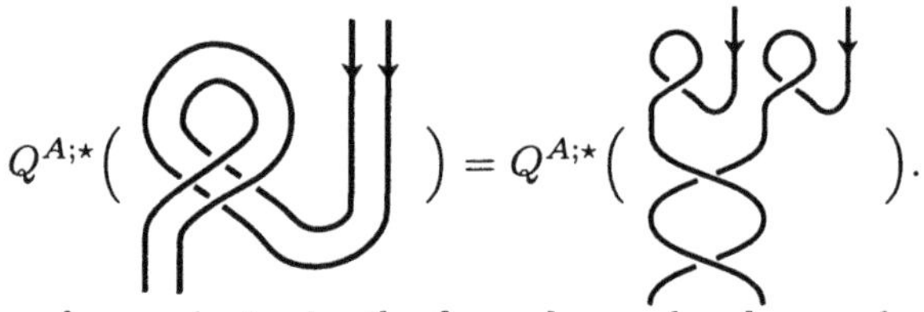

Since these two tangles are isotopic, the formula can be shown along the lines of the proof of isotopy invariance of the universal A invariant (Theorem 4.5), noting that Proposition 4.3 is not used in the proof of isotopy invariance. In a similar way, the second formula of Proposition 4.3 can be shown by using the following equality,

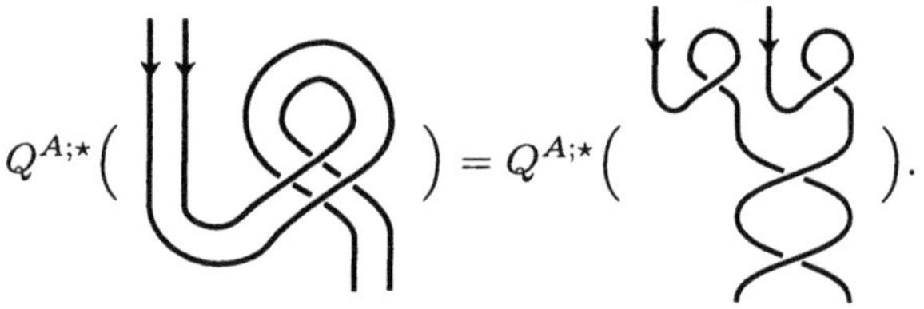

The following proposition motivates the definition of a ribbon Hopf algebra (see the ends of this and the next sections).

Proposition 4.4 (see **[Dri89a]**). For the element u (given in (4.12)) of a quasi-triangular Hopf algebra $(A, \mathcal{R})$, we put $v_2 = S(u)u$. Then, v_2 is central in A and satisfies the following relations,

$$\Delta(v_2) = (v_2 \otimes v_2) \cdot (\mathcal{R}_{21}\mathcal{R})^{-2},$$

$$S(v_2) = v_2,$$

$$\varepsilon(v_2) = 1,$$

where we put $\mathcal{R}_{21} = \sum_i \beta_i \otimes \alpha_i$ (using the notation $\mathcal{R} = \sum_i \alpha_i \otimes \beta_i$).

Proof. That the element v_2 is central follows from

$$v_2 x = S(u)ux = S(u)S^2(x)u = S\big(S(x)u\big)u = S\big(uS^{-1}(x)\big)u = xS(u)u = xv_2$$

for any $x \in A$, where we obtain the equalities by using (4.13).

The first formula of the proposition is obtained

$$\begin{aligned}\Delta(v_2) &= \Delta\big(S(u)\big)\Delta(u) = \big(S(u) \otimes S(u)\big) \cdot (\mathcal{R}_{21}\mathcal{R})^{-1} \cdot (u \otimes u) \cdot (\mathcal{R}_{21}\mathcal{R})^{-1} \\ &= (v_2 \otimes v_2) \cdot (\mathcal{R}_{21}\mathcal{R})^{-2},\end{aligned}$$

where we obtain the second equality by Proposition 4.3, and obtain the third equality by (4.20).

The second formula of the proposition is obtained

$$S(v_2) = S\big(S(u)u\big) = S(u)S^2(u) = S(u)u = v_2,$$

where we obtain the third equality by (4.13).

The third formula of the proposition is obtained

$$\varepsilon(v_2) = \varepsilon\big(S(u)\big)\varepsilon(u) = 1,$$

by Proposition 4.3, noting that $\varepsilon \circ S = \varepsilon$ from the definition of a Hopf algebra. □

Ribbon Hopf algebra

We consider a quasi-triangular Hopf algebra $(A, \mathcal{R})$ with a special element $v \in A$ such that

$$v \text{ is central in } A, \tag{4.23}$$

$$v^2 = S(u)u, \tag{4.24}$$

$$\Delta(v) = (v \otimes v) \cdot (\mathcal{R}_{21}\mathcal{R})^{-1}, \tag{4.25}$$

$$S(v) = v, \tag{4.26}$$

$$\varepsilon(v) = 1. \tag{4.27}$$

As in [ReTu90] we call a triple $(A, \mathcal{R}, v)$ satisfying the above five relations a *ribbon Hopf algebra*. Thus we obtain a ribbon Hopf algebra from a quasi-triangular Hopf algebra by fixing a square root v of $S(u)u$. Note that the relations of v^2 induced by the above relations of v are the same as the relations in Proposition 4.4. See the end of the next section for a reason why we need, not only the notion of a quasi-triangular Hopf algebra, but also the notion of a ribbon Hopf algebra, to construct invariants of framed links.

4.2 Invariants of links in ribbon Hopf algebras

In this section, for a ribbon Hopf algebra $(A, \mathcal{R}, v)$, we introduce the universal A invariant of a link L, denoted by $Q^{A;\star}(L)$, which is defined to be in the algebra; the universal invariant was introduced in [Law89, Law90], see also [Oht93].

We give the values of $Q^{A;\star}$ for oriented elementary tangle diagrams as

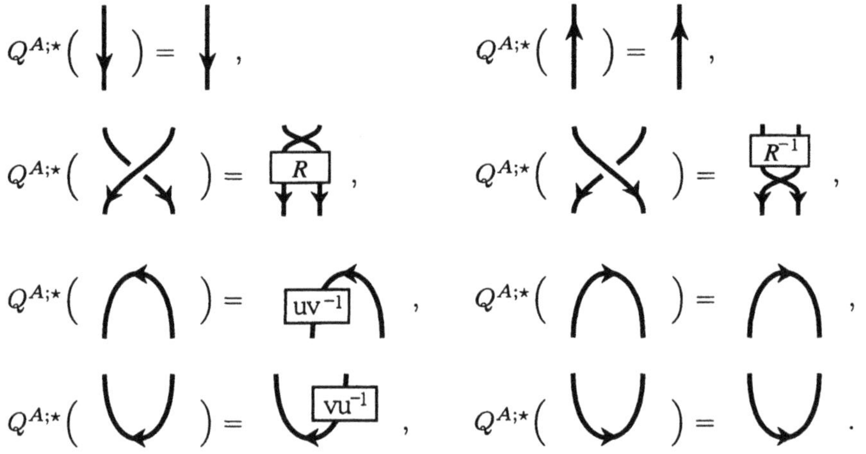

Further, for a diagram D of a framed link, we take a decomposition of D into elementary tangle diagrams (as in Section 3.1); we then glue the values of the invariant on these elementary diagrams according to D to obtain a graphical construction of $Q^{A;\star}(D)$; see, *e.g.*, Figure 4.1. When D is a diagram of an l-component framed link, we select a location on each component to mark, and then $Q^{A;\star}(D)$ is defined to be the element of $(A/I)^{\otimes l}$ that arises by taking the product of the elements of A on each strand, along the orientation of the strand starting from the marking on the strand, where I is the vector subspace of A spanned by $xy - yx$ for any $x, y \in A$. Note that we must quotient by the subspace I to account for the ambiguity in the choice of the marking of each component. The following theorem states that $Q^{A;\star}$ gives an isotopy invariant of framed links.

Theorem 4.5. For a diagram D of a framed link L with l components the value $Q^{A;\star}(D) \in (A/I)^{\otimes l}$ is an isotopy invariant of the framed link L. We call this invariant the *universal A invariant* of L, and denote it by $Q^{A;\star}(L)$.

Proof. By Theorem 3.3 it is sufficient to show that $Q^{A;\star}$ is invariant under the Turaev moves for oriented tangles presented in Figure 3.6, using the move (3.16) for oriented framed tangles instead of (3.12).

The invariance of $Q^{A;\star}$ under the moves (3.9) and (3.10) immediately follows from the definition of $Q^{A;\star}$.

The invariance of $Q^{A;\star}$ under the move (3.11) is derived from the commutativity of R and $u \otimes u$ shown in (4.20) and the fact that v is central in A.

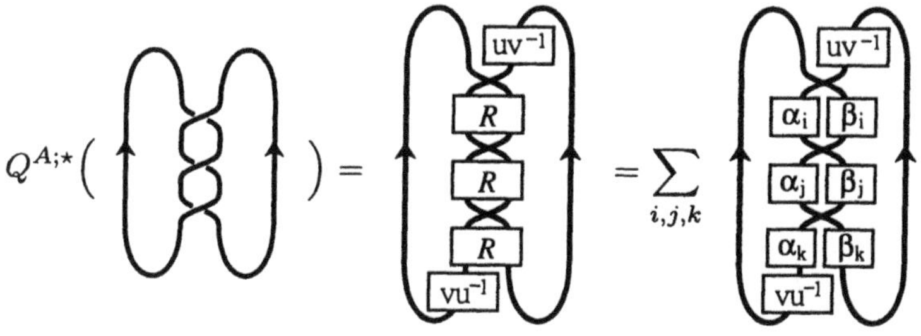

$$= \sum_{i,j,k} \beta_i \alpha_j \beta_k u v^{-1} \alpha_i \beta_j \alpha_k v u^{-1} \in A/I$$

Figure 4.1 An example of the calculation of $Q^{A;\star}$. Making the marking at the left maximal critical point gives the expression for $Q^{A;\star}$ appearing in the second line.

We show the invariance of $Q^{A;\star}$ under the move (3.16) as follows. We can define the value of $Q^{A;\star}$ on n full twists of a strand to be v^{-n} times the value $Q^{A;\star}$ of the original strand, since

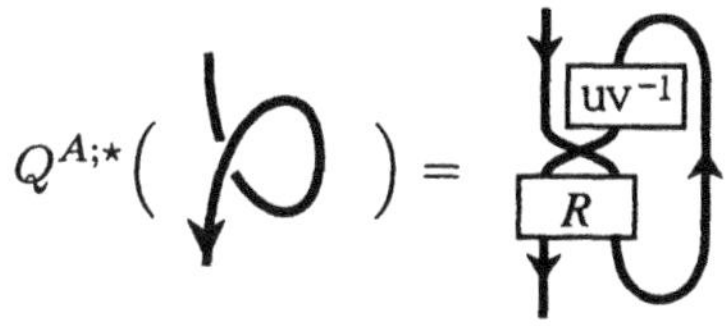

$$= \sum_i \beta_i u v^{-1} \alpha_i$$

$$= v^{-1} u \sum S^{-2}(\beta_i)\alpha_i = v^{-1} = Q^{A;\star}\Big($$  $$\Big)$$

where the picture in the last term denotes the trivial diagram with a full $+1$ twist. Here we obtain the first two equalities from the definition of the universal A invariant, and the third equality from the relation (4.13) and the fact that v is central in A. Further, by (4.14) and (4.18) we have the equality $u^{-1} = \sum_i S^{-2}(\beta_i)\alpha_i$; this equality induces the fourth equality of the above formula. Similarly,

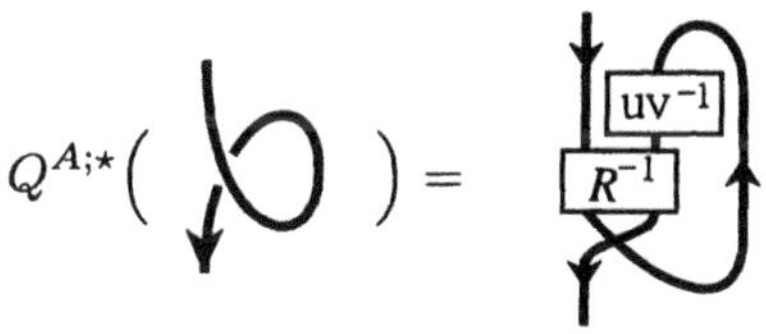

$$= \sum_i \alpha_i' u v^{-1} \beta_i' = v^{-1} u \sum_i S^{-2}(\alpha_i')\beta_i'$$

$$= v^{-1} u \sum S^{-1}(\alpha_i)\beta_i = v^{-1} u S(u) = v = Q^{A;\star}\Big($$  $$\Big),$$

where the picture in the last term denotes the trivial strand with a full -1 twist. Here we obtain the third equality from (4.13) and the fact that v is central, and the fourth equality from the relation (4.18). Further, from the definition of u and two uses of the relation (4.19), we have that $S(u) = \sum_i S(\alpha_i)S^2(\beta_i) = \sum_i S^{-1}(\alpha_i)\beta_i$; this induces the fifth equality of the above formula. By the above formulae we obtain the invariance of $Q^{A;\star}$ under the move (3.16).

We obtain the invariance of $Q^{A;\star}$ under the move (3.13) as follows:

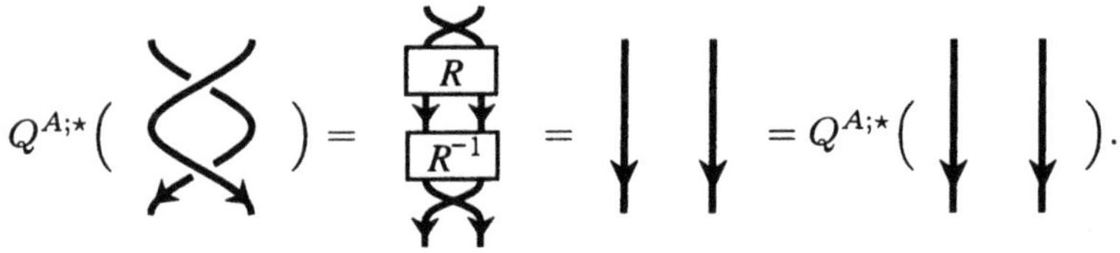

Further, we obtain the invariance of $Q^{A;\star}$ under the move (3.14) as follows:

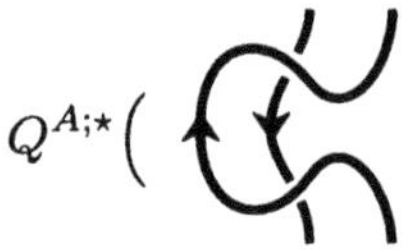

$$= \sum_{i,j} \beta'_j\beta_i \otimes uv^{-1}\alpha_i vu^{-1}\alpha'_j = \sum_{i,j} \beta'_j\beta_i \otimes S^2(\alpha_i)\alpha'_j$$

$$= \sum_{i,j} \beta'_j\beta_i \otimes S\Big(S^{-1}(\alpha_i)S(\alpha'_j)\Big)$$

$$= (\mathrm{id}\otimes S)\Big((\mathrm{id}\otimes S^{-1})\mathcal{R}^{-1}\cdot(\mathrm{id}\otimes S)\mathcal{R}\Big) = (\mathrm{id}\otimes S)(\mathcal{R}\cdot\mathcal{R}^{-1})$$

$$= (\mathrm{id}\otimes S)(1\otimes 1) = 1\otimes 1 = Q^{A;\star}\Big(\ \Big), \tag{4.28}$$

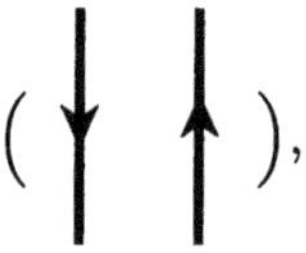

using (4.13) and (4.18) in the computation.

The invariance of $Q^{A;\star}$ under the move (3.15) is derived from the quantized Yang-Baxter equation (4.16) (or equivalently its graphical presentation (4.21)) by

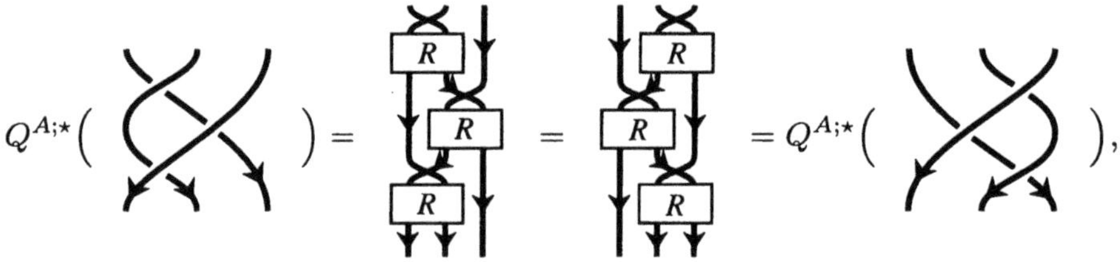

completing the proof of the theorem. □

Proposition 4.6. Let $C \cup L$ be a framed link of $(l+1)$ component, where C denotes a component of the link. We consider the universal A invariant $Q^{A;\star}(C \cup L) \in (A/I)^{\otimes(l+1)}$. We denote the maps induced by Δ, S and ε on A/I also by the same notations.

(1) Let $C^{(2)} \cup L$ be the link obtained from $C \cup L$ by replacing the component C with two parallel copies $C^{(2)}$ of C. Then,

$$Q^{A;\star}(C^{(2)} \cup L) = (\Delta \otimes \mathrm{id}^{\otimes l})(Q^{A;\star}(C \cup L)).$$

(2) Let $\overline{C} \cup L$ be the link obtained from $C \cup L$ by reversing the orientation of C. Then,

$$Q^{A;\star}(\overline{C} \cup L) = (S \otimes \mathrm{id}^{\otimes l})(Q^{A;\star}(C \cup L)).$$

(3) We have that

$$Q^{A;\star}(L) = (\varepsilon \otimes \mathrm{id}^{\otimes l})(Q^{A;\star}(C \cup L)).$$

Proof. The proposition is proved by verifying the formulae for each of the elementary oriented tangle diagrams. We show only (1) for simplicity; it is left to the reader to show (2) and (3).

We show (1) for the elementary diagrams of oriented tangles, *i.e.*, diagrams of positive and negative crossings and maximal and minimal critical points with left and right orientations (see Figure 3.5). We obtain (1) for the elementary diagram of a positive crossing from the defining relations (4.8) and (4.9) of a quasi-triangular Hopf algebra. Further, we obtain that of a negative crossing by the inverses of the relations. Furthermore, we obtain (1) for the elementary diagrams of critical points by

$$\Delta(uv^{-1}) = uv^{-1} \otimes uv^{-1},$$

which is obtained from Proposition 4.3 and a defining relation (4.25) of a ribbon Hopf algebra. By gluing the above cases we obtain (1) for all tangles, in particular for all framed links. □

Why do we need ribbon Hopf algebras?

If we did not have the notion of a ribbon Hopf algebra, then what trouble would arise when we tried to define invariants of framed links? Let us try to construct a universal A invariant of framed links, not for a ribbon Hopf algebra A, but for a quasi-triangular Hopf algebra A.

As a working hypothesis, assume that the following assignments define a universal A invariant $Q^{A;\star}$:

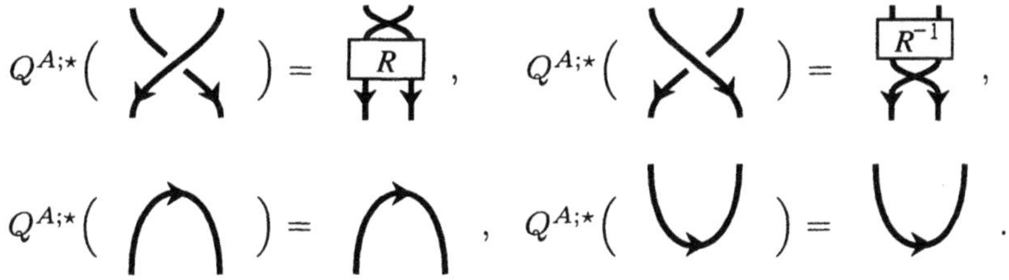

(4.29)

Also assume the following principle, like Proposition 4.6 (2):

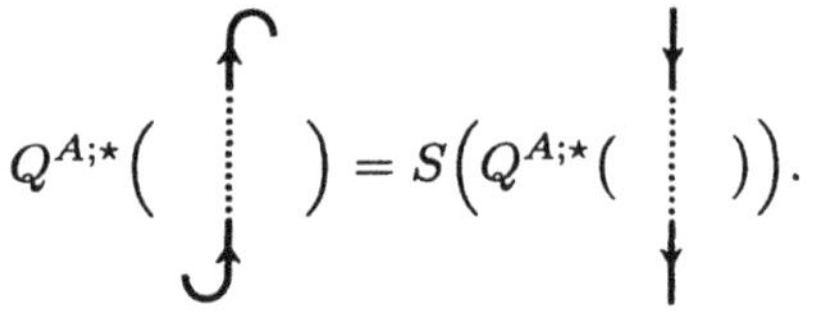

Then,

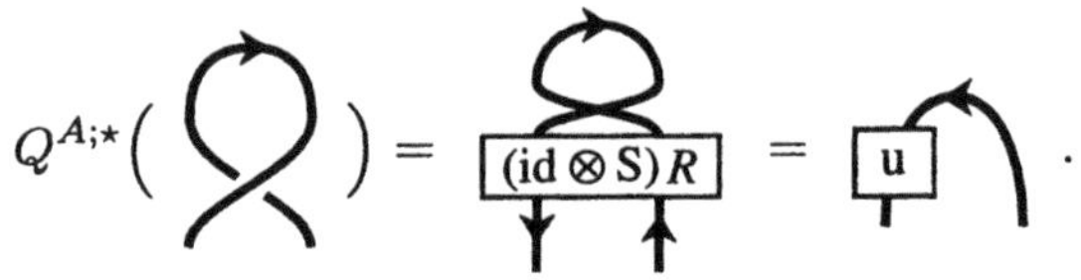

Similarly,

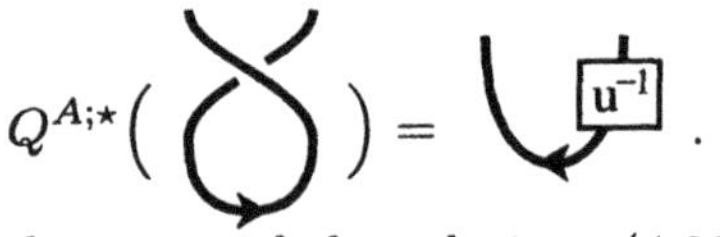

Regarding the above two relations and the relations (4.29) as the defining relations of the universal A invariant $Q^{A;\star}$ we obtain a universal A invariant of framed links whose components have odd framings; it is left to the reader to show a complete proof of that. For the universal A invariant we have that

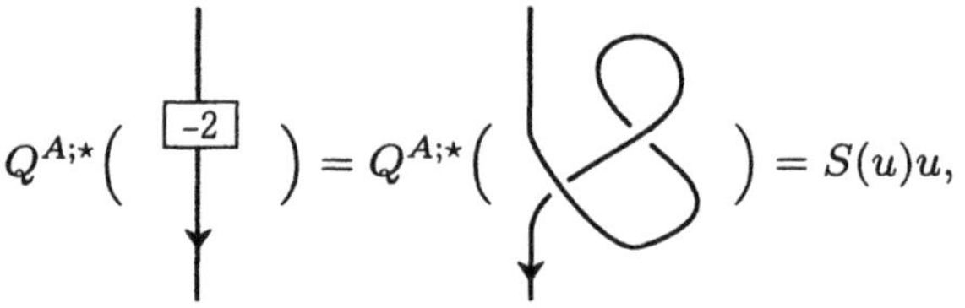

where the boxed "-2" denotes -2 full twists of the strand and we obtain the second equality from the above definition of $Q^{A;\star}$ and from properties of a quasi-triangular Hopf algebra.

In order to extend the universal A invariant to an invariant of all framed links we need an element v to be the value of $Q^{A;\star}(-1$ full twist) to be v. By the above formula, v must be a square root of $S(u)u$. By fixing such v in a quasi-triangular Hopf algebra we obtain the definition of a ribbon Hopf algebra mentioned in the previous section. This is a reason why we need, not only a quasi-triangular Hopf algebra, but also a ribbon Hopf algebra to construct the invariant of framed links.

4.3 Operator invariants of tangles derived from ribbon Hopf algebras

When equipped with a representation of a ribbon Hopf algebra, the universal A invariant induces an operator invariant of tangles. We introduce such operator invariants and show that such operator invariants recover from the universal A invariant. A remarkable property of such an operator invariant is that the value is an intertwiner with respect to the action of the ribbon Hopf algebra.

Let A be a ribbon Hopf algebra. For a vector space V we call a homomorphism $\rho : A \to \mathrm{End}(V)$ a *representation* of A on V. When a representation ρ of A on V is given, we also say that the algebra A *acts* on V through ρ, there is an *action* of A on V through ρ and V is an A *module* through ρ. A representation of A on V (and an A module V) is called *irreducible* if no proper subspace of V is preserved by ρ, *i.e.*, if a subspace W of V is preserved by $\rho(a)$ for any $a \in A$, then $W = \{0\}$ or V. It is known as *Schur's lemma* that for an irreducible representation ρ and a central element $a \in A$, the image $\rho(a)$ is equal to $c \cdot \mathrm{id}_V$ for some scalar c.

For an irreducible representation ρ of A on V we consider the two endomorphisms $R \in \mathrm{End}(V \otimes V)$ and $h \in \mathrm{End}(V)$ given by

$$R = P \circ \big((\rho \otimes \rho)(\mathcal{R})\big), \qquad h = \rho(uv^{-1}), \tag{4.30}$$

where P denotes the map defined by $P(x \otimes y) = y \otimes x$. By presenting $\mathcal{R} = \sum_i \alpha_i \otimes \beta_i$, the above map R is alternatively given by

$$R(x \otimes y) = \sum_i \rho(\beta_i) y \otimes \rho(\alpha_i) x.$$

For example, we obtain the operator invariant of the following tangle associated with the above maps R and h as

$[$  $] = (\mathrm{id}_V \otimes n)(R \otimes \mathrm{id}_{V^*})(\mathrm{id}_V \otimes u') = \rho\left(Q^{A;*}(\right.$  $\left.)\right) = \rho(v^{-1})$ (4.31)

where we obtain the second equality from the following computation

$$x \overset{\mathrm{id}_V \otimes u'}{\longmapsto} \sum_i x \otimes e_i \otimes e_i^*$$

$$\overset{R \otimes \mathrm{id}_{V^*}}{\longmapsto} \sum_{i,j} \beta_j e_i \otimes \alpha_j x \otimes e_i^*$$

$\overset{\mathrm{id}_V \otimes n}{\longmapsto} \sum_{i,j} \beta_j \cdot e_i^*(uv^{-1}\alpha_j x) e_i = \sum_j \beta_j uv^{-1} \alpha_j x = Q^{A;*}($  $)x,$

where, for simplicity, we have omitted the appearances of the notation ρ. Similarly,

$[$  $] = \rho\left(Q^{A;*}(\right.$  $\left.)\right) = \rho(v),$ (4.32)

$[$  $] = (\rho \otimes \rho^*)\left(Q^{A;*}(\right.$  $\left.)\right).$ (4.33)

Further,

$[$  $] = P_{13} \circ \rho \otimes \rho \otimes \rho(\mathcal{R}_{23}\mathcal{R}_{13}\mathcal{R}_{12}),$

$= P_{13} \circ \rho \otimes \rho \otimes \rho(\mathcal{R}_{12}\mathcal{R}_{13}\mathcal{R}_{23}) = [$  $],$ (4.34)

where we obtain the second equality from the quantized Yang-Baxter equation (4.16).

Theorem 4.7. The two endomorphisms R and h given in (4.30) satisfy the relations of Theorem 3.8. Hence, by Theorem 3.8, we obtain the operator invariant of oriented framed tangles associated with the endomorphisms R and h.

We denote the operator invariant of a tangle T by $Q^{A;V}(T)$ and call it the *operator invariant* derived from a representation V of a ribbon Hopf algebra A.

Proof. Since $\mathcal{R}$ and uv^{-1} are invertible the maps R and h given in (4.30) are also invertible. We verify that the maps R and h satisfy the relations (3.24), (3.28), (3.26) and (3.27) as follows.

Firstly, the relation (3.24) is derived from the relation (4.20) and the fact that v is central.

Secondly, we obtain (3.28) as follows. By the above computations (4.31) and (4.32) the left hand side of (3.28) is equal to $\rho(v^{\pm 1})$. Since $v^{\pm 1}$ is central in A, $\rho(v^{\pm 1})$ is equal to $c^{\pm 1} \cdot \mathrm{id}_V$ for some scalar $c^{\pm 1}$ by Schur's lemma. This implies the relation (3.28).

Thirdly, we obtain (3.26) by applying the relation (4.28) to the above computation (4.33).

Finally, the Yang-Baxter equation (3.27) follows from (4.34). □

It is left to the reader to prove the following example.

Example 4.8. Let T be a framed 1-tangle and K the closure of T. Then,

$$Q^{A;V}(K) = \mathrm{trace}\big(\rho(uv^{-1}) \cdot Q^{A;V}(T)\big).$$

For a linear map $f : V \to V$ we call $\mathrm{trace}\big(\rho(uv^{-1}) \cdot f\big)$ the *quantum trace* of f. The above example says that $Q^{A;V}(K)$ is equal to the quantum trace of $Q^{A;V}(T)$. In particular, the quantum trace of the identity map of V is called the *quantum dimension* of V. As a special case of the above example, the invariant $Q^{A;V}$ of the trivial knot with 0 framing is equal to the quantum dimension of V.

The operator invariant $Q^{A;V}$ recovers from the universal A invariant $Q^{A;\star}$

When an algebra A acts on a vector space V over $\mathbb{C}$ through a representation $\rho : A \to \mathrm{End}(V)$, we define a linear map $\mathrm{trace}_V : A \to \mathbb{C}$ by $\mathrm{trace}_V(a) = \mathrm{trace}\rho(a)$

and call it the *trace* on V. Let I be the vector subspace of A spanned by $ab - ba$ for any $a, b \in A$. Since trace has the property $\mathrm{trace}(AB) = \mathrm{trace}(BA)$ in general, the trace on V induces a map $A/I \to \mathbb{C}$; this map is also denoted by trace_V.

Theorem 4.9. Let L be an oriented framed link with l components. The operator invariant $Q^{A;V}(L)$ of L derived from a representation of a ribbon Hopf algebra A on V is recovered from the universal A invariant $Q^{A;*}(L) \in (A/I)^{\otimes l}$ as follows:

$$Q^{A;V}(L) = (\mathrm{trace}_V)^{\otimes l}\Big(Q^{A;*}(L)\Big).$$

We introduce some notations for the following proof. Let $\{e_i\}_{i=1,2,\cdots,n}$ be a basis of the vector space V. A linear map $A \in \mathrm{Hom}(V^{\otimes n}, V^{\otimes m})$ will be presented by a tensor $A^{j_1,\cdots,j_m}_{i_1,\cdots,i_n}$ so that $A(e_{i_1} \otimes \cdots \otimes e_{i_n}) = \sum_{j_1,\cdots,j_m} A^{j_1,\cdots,j_m}_{i_1,\cdots,i_n} e_{j_1} \otimes \cdots \otimes e_{j_m}$.

Proof. We introduce a certain "state sum" construction, and prove the theorem by showing that each side of the required formula is equal to it.

To start, present a diagram of L as the composition of some tensor products of copies of elementary oriented tangle diagrams. We call part of a strand going between crossings or critical points an *edge*. We associate an element of $\{1, 2, \cdots, n\}$ to each edge, and call the element a *label* of the edge. A *state* is a choice of a label for every edge. We associate a *weight* to every elementary diagram in the decomposition as follows:

$w\Big($ 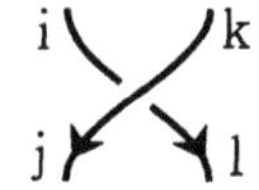 $\Big) = R^{ik}_{jl}$, $w\Big($ 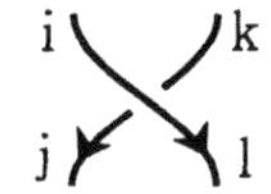 $\Big) = (R^{-1})^{ik}_{jl}$,

$w\Big($  $\Big) = n_{ij}$, $w\Big($  $\Big) = n'_{ij}$,

$w\Big($  $\Big) = u^{ij}$, $w\Big($  $\Big) = u'^{ij}$,

where the right hand sides are entries in the matrices presenting the linear maps appearing in the definition of the operator invariant (see Figure 3.9). By definition, the latter four maps are presented by

$$n_{ij} = h^j_i, \qquad n'_{ij} = \delta_{ij}, \qquad u^{ij} = (h^{-1})^j_i, \qquad u'^{ij} = \delta_{ij}, \tag{4.35}$$

with entries of the linear map h according to the definition of the operator invariant, where we put $\delta_{ij} = 1$ if $i = j$, 0 otherwise. The *state sum* is defined by

$$\sum_S \prod_E w(E; S), \tag{4.36}$$

where the sum is taken over all states S and the product is taken over all elementary diagrams E in the link diagram.

For example, we consider a state of the following diagram of a Hopf link L.

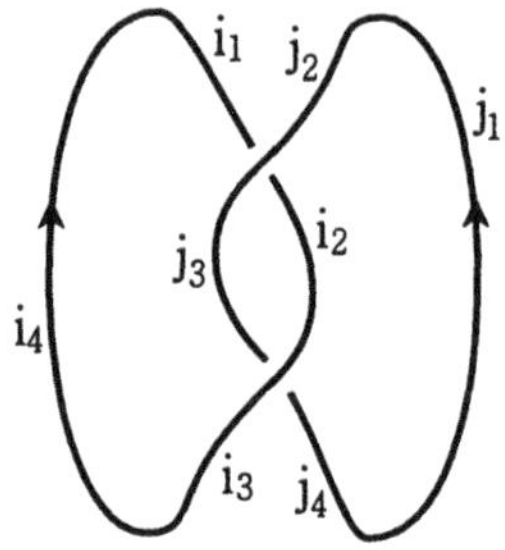

For this link diagram the state sum is given by

$$(4.36) = \sum n'_{i_4 i_1} n_{j_2 j_1} R^{i_1 j_2}_{j_3 i_2} R^{j_3 i_2}_{i_3 j_4} u^{i_4 i_3} u'^{j_4 j_1}. \tag{4.37}$$

Let us show, for this example, that this state sum is equal to each side of the required formula.

By (4.35) we have that

$$(4.36) = \sum h^{j_1}_{j_2} R^{i_1 j_2}_{j_3 i_2} R^{j_3 i_2}_{i_3 j_1} (h^{-1})^{i_3}_{i_1}.$$

Further, from the definition of the operator invariant derived from a representation of a ribbon Hopf algebra, we have that

$$R^{ij}_{kl} = \sum_m \rho(\alpha_m)^j_k \rho(\beta_m)^i_l, \qquad h^i_j = \rho(uv^{-1})^i_j,$$

where we put $\mathcal{R} = \sum_m \alpha_m \otimes \beta_m$ for the universal R matrix $\mathcal{R}$. Hence,

$$\begin{aligned}(4.36) &= \sum \rho(\beta_k)^{i_1}_{i_2} \rho(\alpha_l)^{i_2}_{i_3} \rho(vu^{-1})^{i_3}_{i_1} \rho(uv^{-1})^{j_1}_{j_2} \rho(\alpha_k)^{j_2}_{j_3} \rho(\beta_l)^{j_3}_{j_1} \\ &= \sum_{k,l} \mathrm{trace}\Big(\rho(\beta_k)\rho(\alpha_l)\rho(vu^{-1})\Big) \mathrm{trace}\Big(\rho(uv^{-1})\rho(\alpha_k)\rho(\beta_l)\Big) \\ &= \sum_{k,l} \mathrm{trace}_V(\beta_k \alpha_l v u^{-1}) \mathrm{trace}_V(uv^{-1}\alpha_k \beta_l).\end{aligned}$$

Further, from the definition of the universal A invariant, we have that

$Q^{A;\star}(L) = \sum_{k,l}$ 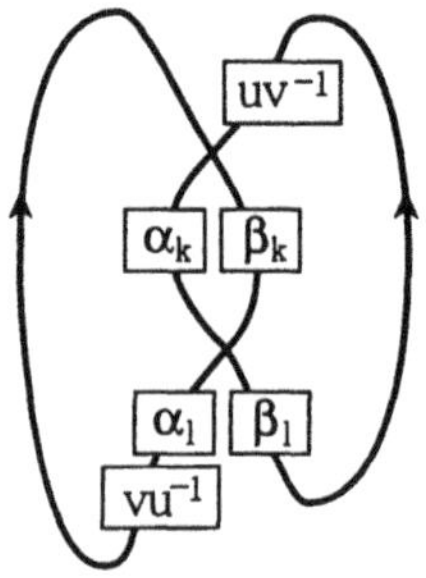 $= \sum \beta_k \alpha_l v u^{-1} \otimes uv^{-1} \alpha_k \beta_l.$

Hence, the state sum is equal to the right hand side of the required formula.

We construct the operator invariant of the link diagram as follows:

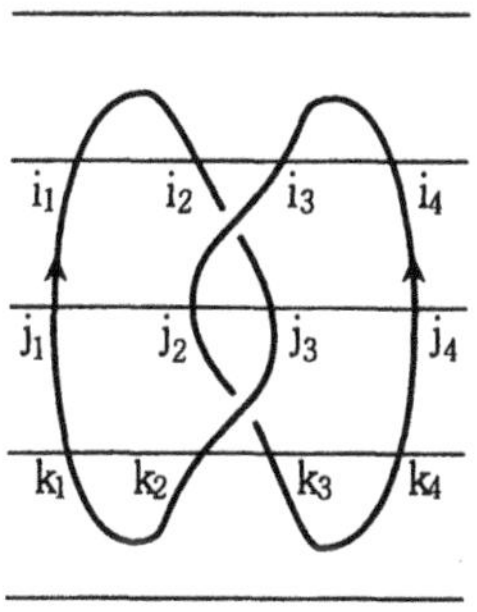

$$\begin{array}{c} \mathbb{C} \\ \uparrow n' \otimes n \\ V^* \otimes V \otimes V \otimes V^* \\ \uparrow \mathrm{id}_{V^*} \otimes R \otimes \mathrm{id}_{V^*} \\ V^* \otimes V \otimes V \otimes V^* \\ \uparrow \mathrm{id}_{V^*} \otimes R \otimes \mathrm{id}_{V^*} \\ V^* \otimes V \otimes V \otimes V^* \\ \uparrow u \otimes u' \\ \mathbb{C} \end{array}$$

Expressing the above product in indicial notation, we have that

$$\begin{aligned} Q^{A;V}(L) &= (n' \otimes n)(\mathrm{id}_{V^*} \otimes R \otimes (\mathrm{id}_{V^*})(\mathrm{id}_{V^*} \otimes R \otimes (\mathrm{id}_{V^*})(u \otimes u') \\ &= \sum (n' \otimes n)_{i_1 i_2 i_3 i_4} (\mathrm{id}_{V^*} \otimes R \otimes (\mathrm{id}_{V^*})^{i_1 i_2 i_3 i_4}_{j_1 j_2 j_3 j_4} \\ &\qquad \times (\mathrm{id}_{V^*} \otimes R \otimes (\mathrm{id}_{V^*})^{j_1 j_2 j_3 j_4}_{k_1 k_2 k_3 k_4} (u \otimes u')^{k_1 k_2 k_3 k_4} \\ &= \sum n'_{i_1 i_2} n_{i_3 i_4} R^{i_2 i_3}_{j_2 j_3} R^{j_2 j_3}_{k_2 k_3} u^{i_1 k_2} u'^{k_3 i_4}. \end{aligned}$$

This is equal to the state sum in (4.37). Hence, the state sum is also equal to the left hand side of the required formula. □

The operator invariant $Q^{A;V}$ is an intertwiner

For two representations $\rho_V : A \to \mathrm{End}(V)$ and $\rho_{V'} : A \to \mathrm{End}(V')$ of a ribbon Hopf algebra A, the *tensor representation* $\rho_{V \otimes V'} : A \to \mathrm{End}(V \otimes V')$ of ρ_V and $\rho_{V'}$ is defined by

$$\rho_{V \otimes V'} : A \xrightarrow{\Delta} A \otimes A \xrightarrow{\rho_V \otimes \rho_{V'}} \mathrm{End}(V) \otimes \mathrm{End}(V') \cong \mathrm{End}(V \otimes V').$$

The A module $V \otimes V'$ obtained from the tensor representation $\rho_{V \otimes V'}$ is called the *tensor module* of the A modules V and V'. Using the notation $\Delta(a) = \sum_i a_{1,i} \otimes a_{2,i}$ for $a \in A$, the tensor representation is presented by $\rho_{V \otimes V'}(a) = \sum_i \rho_V(a_{1,i}) \otimes \rho_{V'}(a_{2,i})$. Further, for the dual vector space V^* of V, we define the *dual representation* $\rho_{V^*} : A \to \mathrm{End}(V^*)$ by $\rho_{V^*}(a) = \rho_V(S(a))^*$, where $\rho_V(S(a))^*$ denotes the pull back by $\rho_V(S(a))$, as follows:

$$\Big(\rho_{V^*}(a)\Big)(f) : V \xrightarrow{\rho_V(S(a))} V \xrightarrow{f} \mathbb{C}.$$

Since the antipode is an anti-homomorphism, the dual representation ρ_{V^*} is a homomorphism, as shown by the following computation,

$$\Big(\rho_{V^*}(a_1 a_2)\Big)(f) = f \circ \rho_V\Big(S(a_1 a_2)\Big) = f \circ \rho_V\Big(S(a_2) S(a_1)\Big)$$

$$= f \circ \rho_V\big(S(a_2)\big)\rho_V\big(S(a_1)\big) = \big(\rho_{V^\star}(a_2)\big)(f) \circ \rho_V\big(S(a_1)\big) = \big(\rho_{V^\star}(a_1)\rho_{V^\star}(a_2)\big)(f).$$

The A module $V^\star$ is called the *dual module* of V. Furthermore, we define the *unit representation* $\rho_{\mathbb{C}}$ on $\mathbb{C}$ by $\rho_{\mathbb{C}} : A \xrightarrow{\varepsilon} \mathbb{C} = \mathrm{End}(\mathbb{C})$. The A module $\mathbb{C}$ obtained from $\rho_{\mathbb{C}}$ is called the *unit module.* It follows from the definition of Δ and ε, that the unit representation is a unit among representations of A with respect to tensor product.

When we have two representations $\rho_{W_i} : A \to \mathrm{End}(W_i)$ on vector spaces W_i $(i = 1, 2)$, a linear map $f : W_1 \to W_2$ is called an *intertwiner* if the map f is equivariant with respect to the actions of A on W_1 and W_2, *i.e.*, the following diagram is commutative for any $a \in A$.

$$\begin{array}{ccc} W_2 & \xrightarrow{\rho_{W_2}(a)} & W_2 \\ {\scriptstyle f}\uparrow & & \uparrow{\scriptstyle f} \\ W_1 & \xrightarrow{\rho_{W_1}(a)} & W_1 \end{array}$$

A remarkable property of the operator invariant derived from a representation ρ_V of a ribbon Hopf algebra A is that the operator invariant of any oriented tangle is an intertwiner with respect to the actions of A. Let us verify the property for some simple tangles as follows.

For example,

$Q^{A;V}($  $) = R \in \mathrm{End}(V \otimes V),$

where the R matrix is given by $R = P \circ \big((\rho_V \otimes \rho_V)(\mathcal{R})\big)$ as in the definition of $Q^{A;V}$. By the relation (4.7), $(P \circ \Delta)(a)\mathcal{R} = \mathcal{R}\Delta(a)$ for any $a \in A$. By sending both sides by $\rho_V \otimes \rho_V$, we have that

$$\big((\rho_V \otimes \rho_V) \circ P \circ \Delta\big)(a) \cdot (\rho_V \otimes \rho_V)(\mathcal{R}) = (\rho_V \otimes \rho_V)(\mathcal{R}) \cdot \big((\rho_V \otimes \rho_V) \circ \Delta\big)(a).$$

Further, by composing the permutation P with the left of the formula, we have that

$$\big((\rho_V \otimes \rho_V) \circ \Delta\big)(a) \cdot R = R \cdot \big((\rho_V \otimes \rho_V) \circ \Delta\big)(a)$$

from the definition of R. Moreover, from the definition of the tensor representation, we obtain

$$\rho_{V\otimes V}(a) \cdot R = R \cdot \rho_{V\otimes V}(a).$$

This formula implies that the following diagram is commutative.

$$\begin{array}{ccc} V \otimes V & \xrightarrow{\rho_{V\otimes V}(a)} & V \otimes V \\ {\scriptstyle R}\uparrow & & \uparrow{\scriptstyle R} \\ V \otimes V & \xrightarrow{\rho_{V\otimes V}(a)} & V \otimes V \end{array}$$

Hence, the linear map R is an intertwiner with respect to the action of A on $V \otimes V$.

Next, we verify that the operator invariant of the following tangle is an intertwiner,

$$Q^{A;V}\Big(\quad$$  $$\quad\Big) = n' \in \mathrm{Hom}(V^\star \otimes V, \mathbb{C}),$$

recall that the map n' is defined by $n'(f \otimes x) = f(x)$ for $f \in V^\star$ and $x \in V$. We show the following diagram is commutative for any $a \in A$.

$$\begin{array}{ccc} \mathbb{C} & \xrightarrow{\rho_{\mathbb{C}}(a)} & \mathbb{C} \\ {\scriptstyle n'}\uparrow & & \uparrow{\scriptstyle n'} \\ V^\star \otimes V & \xrightarrow{\rho_{V^\star \otimes V}(a)} & V^\star \otimes V \end{array}$$

The clockwise route of the diagram is computed as

$$f \otimes x \overset{n'}{\longmapsto} f(x) \overset{\rho_{\mathbb{C}}(a)}{\longmapsto} \varepsilon(a) f(x),$$

from the definition of the unit representation $\rho_{\mathbb{C}}$. On the other hand, writing $\Delta(a) = \sum_i a_{1,i} \otimes a_{2,i}$, the counter-clockwise route is computed as follows:

$$\begin{aligned} f \otimes x \quad & \overset{\rho_{V^\star \otimes V}(a)}{\longmapsto} \quad \textstyle\sum_i \rho_{V^\star}(a_{1,i}) f \otimes \rho_V(a_{2,i}) x = \sum_i \rho_V\Big(S(a_{1,i})\Big)^\star f \otimes \rho_V(a_{2,i}) x \\ & \overset{n'}{\longmapsto} \quad \textstyle\sum_i f\Big(\rho_V(S(a_{1,i}))\rho_V(a_{2,i})x\Big) = \sum_i f\Big(\rho_V(S(a_{1,i})a_{2,i})x\Big) \\ & \qquad = f\Big(\varepsilon(a)x\Big) = \varepsilon(a) f(x), \end{aligned}$$

by using defining relations of Δ and S. Hence, the above diagram is commutative, which implies that the map n' is an intertwiner.

Further, we verify that the operator invariant of the following diagram is an intertwiner,

$$Q^{A;V}\Big(\quad$$  $$\quad\Big) = n \in \mathrm{Hom}(V \otimes V^\star, \mathbb{C}),$$

recalling that the map n is defined by $n(x \otimes f) = f\big(\rho_V(uv^{-1})x\big)$ for $x \in V$ and $f \in V^\star$. We show that the following diagram is commutative for any $a \in A$.

$$\begin{array}{ccc} \mathbb{C} & \xrightarrow{\rho_{\mathbb{C}}(a)} & \mathbb{C} \\ {\scriptstyle n}\uparrow & & \uparrow{\scriptstyle n} \\ V \otimes V^\star & \xrightarrow{\rho_{V \otimes V^\star}(a)} & V \otimes V^\star \end{array}$$

The clockwise route of the diagram is computed as follows:

$$x \otimes f \overset{n}{\longmapsto} f\Big(\rho_V(uv^{-1})x\Big) \overset{\rho_{\mathbb{C}}(a)}{\longmapsto} \varepsilon(a) f\Big(\rho_V(uv^{-1})x\Big).$$

On the other hand, writing $\Delta(a) = \sum_i a_{1,i} \otimes a_{2,i}$, the counter-clockwise route is computed as follows:

$$\begin{aligned} x \otimes f \overset{\rho_{V \otimes V^*}(a)}{\longmapsto} & \textstyle\sum_i \rho_V(a_{1,i})x \otimes \rho_{V^*}(a_{2,i})f = \sum_i \rho_V(a_{1,i})x \otimes \rho_V\Big(S(a_{2,i})\Big)^* f \\ \overset{n}{\longmapsto} & \textstyle\sum_i f\Big(\rho_V\big(S(a_{2,i})uv^{-1}a_{1,i}\big)x\Big) = \varepsilon(a) f\Big(\rho_V(uv^{-1})x\Big), \end{aligned}$$

where we obtain the last equality by the following computation,

$$\begin{aligned} \sum_i S(a_{2,i})uv^{-1}a_{1,i} &= \sum_i uv^{-1}S^{-1}(a_{2,i})a_{1,i} = \sum_i uv^{-1}S^{-1}\Big(S(a_{1,i})a_{2,i}\Big) \\ &= uv^{-1}S^{-1}\Big(\varepsilon(a)\cdot 1\Big) = \varepsilon(a)uv^{-1}. \end{aligned}$$

Here we obtain the first equality from (4.13) and the fact that v is central. Further, we obtain the third equality by a defining relation of Δ and S. Hence, we have shown that the above diagram is commutative. Therefore, the map n is an intertwiner.

In general we have the following theorem.

Proposition 4.10. We consider the operator invariant $Q^{A;V}(T)$ of an oriented framed tangle T derived from a representation of a ribbon Hopf algebra A on V. Then, $Q^{A;V}(T)$ is an intertwiner with respect to the action of A.

Proof. Since an operator invariant of a tangle can be expressed as the composition of the tensor product of some operator invariants of elementary tangle diagrams as shown in Figure 3.10, it is sufficient to show the proposition for each elementary oriented tangle diagram shown in Figure 3.9. We have already verified it for three of the elementary diagrams in the above computations. We now verify it for the remaining elementary diagrams as follows.

Consider the diagram  ; its operator invariant is equal to R^{-1}. Since R is an intertwiner, as verified above, so is R^{-1}.

Consider the diagrams  and  . Set

$Q^{A;V}\Big($  $\Big) = u'$, $\qquad Q^{A;V}\Big($  $\Big) = u$,

as in Figure 3.9. By isotopy invariance of the operator invariant we have that

$(\mathrm{id}_V \otimes n')(u' \otimes \mathrm{id}_V) = Q^{A;V}\Big($  $\Big) = Q^{A;V}\Big($  $\Big) = \mathrm{id}_V$,

$(n \otimes \mathrm{id}_V)(\mathrm{id}_V \otimes u) = Q^{A;V}\Big($  $\Big) = Q^{A;V}\Big($  $\Big) = \mathrm{id}_V$.

Since n and n' are intertwiners, as verified above, so are u and u'. $\square$

Let $T = T_1 \cup \cdots \cup T_l$ be an oriented framed tangle with l components. We associate representations $V_1, \cdots, V_l$ of a ribbon Hopf algebra A to the components $T_1, \cdots, T_l$ of the tangle respectively. We modify the construction of the operator invariant in Figure 3.10 by letting V_i instead of V correspond to the component T_i. Instead of the R matrix given by $R = \big(P \circ (\rho \otimes \rho)\big)(\mathcal{R}) \in \mathrm{End}(V \otimes V)$ we consider the following map

$V_i \quad V_j$

$$\begin{array}{c} V_j \otimes V_i \\ \uparrow R_{ij} \\ V_i \otimes V_j \end{array}$$

where we put $R_{ij} = \big(P \circ (\rho_{V_i} \otimes \rho_{V_j})\big)(\mathcal{R})$. Further, instead of the maps n, n', u and u' in Figure 3.9 we consider the maps defined in the same way but using V_i instead of V. Using these maps we define the operator invariant of $T = T_1 \cup \cdots \cup T_l$ associated with the representations $V_1, \cdots, V_l$. We denote the invariant by $Q^{A;V_1,\cdots,V_l}(T)$.

Modifying the proofs of Theorem 4.9 and Proposition 4.10 we obtain the following two theorems.

Theorem 4.11. For an oriented framed link L with l components we have that

$$Q^{A;V_1,\cdots,V_l}(L) = (\mathrm{trace}_{V_1} \otimes \cdots \otimes \mathrm{trace}_{V_l})\Big(Q^{A;*}(L)\Big).$$

Proposition 4.12. For an oriented framed tangle T with l components the invariant $Q^{A;V_1,\cdots,V_l}(T)$ is an intertwiner with respect to the action of A.

By applying Theorem 4.11 to Proposition 4.6 we obtain the following proposition.

Proposition 4.13. With the same notation as in Proposition 4.6 we have that

$$\begin{aligned} &Q^{A;V_0,V_0',V_1,\cdots,V_l}(C^{(2)} \cup L) = Q^{A;V_0 \otimes V_0',V_1,\cdots,V_l}(C \cup L), \\ &Q^{A;V_0,V_1,\cdots,V_l}(\overline{C} \cup L) = Q^{A;V_0^*,V_1,\cdots,V_l}(C \cup L), \\ &Q^{A;V_1,\cdots,V_l}(L) = Q^{A;\mathbb{C},V_1,\cdots,V_l}(C \cup L). \end{aligned}$$

4.4 The quantum group $U_q(sl_2)$ at a generic q

In this section, we introduce the quantum group $U_q(sl_2)$ as a typical example of a ribbon Hopf algebra. The operator invariant derived from a representation V_n of $U_q(sl_2)$ defines the quantum (sl_2, V_n) invariant of oriented framed links. The Jones polynomial is reconstructed by modifying such an operator invariant.

The *Lie algebra* sl_2 is defined to be the vector space over $\mathbb{C}$ spanned by the basis E, F and H equipped with the bracket $[\ ,\ ] : sl_2 \otimes sl_2 \to sl_2$ given by

$$[H, E] = 2E, \qquad [H, F] = -2F, \qquad [E, F] = H.$$

We can enlarge the Lie algebra sl_2 to obtain the *universal enveloping algebra* $U(sl_2)$ of sl_2, defined to be the algebra over $\mathbb{C}$, with the unit element 1, generated by the generators E, F and H subject to the following relations

$$HE - EH = 2E, \qquad HF - FH = -2F, \qquad EF - FE = H.$$

Note that the vector subspace of $U(sl_2)$ spanned by E, F and H can be identified with sl_2, where the bracket is given by the commutator. The algebra $U(sl_2)$ is the universal one among algebras including sl_2 in this way.

Further, perturbing $U(sl_2)$ by a complex parameter q, called a *quantum parameter*, we obtain the quantum group $U_q(sl_2)$, as follows. The *quantum group* $U_q(sl_2)$ is defined to be the algebra over $\mathbb{C}$, with the unit element 1, generated by the generators K, K^{-1}, E, and F, subject to the following relations

$$K \cdot K^{-1} = K^{-1} \cdot K = 1,$$

$$KE = qEK, \qquad KF = q^{-1}FK, \qquad EF - FE = \frac{K - K^{-1}}{q^{1/2} - q^{-1/2}},$$

where in this section we suppose that q is not equal to a root of unity *i.e.*, $q^n \neq 1$ for any $n \in \mathbb{Z}$. The topology of $U_q(sl_2)$ is the topology of the power series ring in an indeterminate $\hbar$, putting $q = e^{\hbar}$.

Note that the universal enveloping algebra $U(sl_2)$ is recovered from $U_q(sl_2)$ by taking the limit $q \to 1$, where we regard K as $q^{H/2}$. There is a defining relation of $U_q(sl_2)$ corresponding to the relation $HE = E(H+2)$, as follows:

$$KE = e^{\hbar H/2}E = Ee^{\hbar(H+2)/2} = qEK,$$

in the completion of $U_q(sl_2)$ with respect to the topology of $U_q(sl_2)$. Similarly, it follows that a defining relation $KFK^{-1} = q^{-1}F$ of $U_q(sl_2)$ is equivalent to the relation $HF = F(H-2)$ in the completion of $U_q(sl_2)$. The third defining relations, of the form $EF - FE = \cdots$, are different between $U(sl_2)$ and $U_q(sl_2)$, even in the completion of $U_q(sl_2)$. The third relation for $U_q(sl_2)$ is the relation $EF - FE = [H]$, where $[H]$ denotes the "quantum H" which is defined to be $(q^{H/2} - q^{-H/2})/(q^{1/2} - q^{-1/2})$ in the completion of $U_q(sl_2)$. The "quantum H" becomes the original H in the limit $q \to 1$. In this sense, $U(sl_2)$ is recovered from $U_q(sl_2)$ in the limit $q \to 1$.

$U_q(sl_2)$ is called a quantum "group", though it is actually an algebra. The reason, for its being so named, is as follows. In general, a quantization of a space X is obtained by the following procedure. We take the function algebra on X and regard that as the primary structure in the characterization of a space; the original X thus becomes a secondary concept associated to its function algebra. When we obtain a (suitable) non-commutative algebra as a perturbation of this function algebra, we call this non-commutative algebra a *quantization* of X. In our case, roughly speaking, the universal enveloping algebra $U(sl_2)$ can be regarded as the algebra of Taylor expansions of elements of the function algebra on the Lie group

$SL_2(\mathbb{C})$. Hence, we call its non-commutative perturbation $U_q(sl_2)$ the quantum "group" in the sense that we regard it as the quantization of this Lie group.

There is a natural Hopf algebra structure on the function algebra of a group G, with the comultiplication, the antipode, and the counit, defined by the pull-back of the multiplication $G \times G \to G$, the map $G \to G$ taking inverses of elements, and the inclusion $\{e\} \to G$ respectively. Motivated by the above Hopf algebra structure we introduce the Hopf algebra structure for $U(sl_2)$ by the comultiplication $\Delta : U(sl_2) \to U(sl_2) \otimes U(sl_2)$, the antipode $S : U(sl_2) \to U(sl_2)$, and the counit $\varepsilon : U(sl_2) \to \mathbb{C}$, defined by

$$\Delta(X) = X \otimes 1 + 1 \otimes X, \qquad S(X) = -X, \qquad \varepsilon(X) = 0$$

for any $X \in sl_2$. Further, as perturbations of the above maps, we have the Hopf algebra structure for the quantum group $U_q(sl_2)$, with the comultiplication $\Delta : U_q(sl_2) \to U_q(sl_2) \otimes U_q(sl_2)$, the antipode $S : U_q(sl_2) \to U_q(sl_2)$, and the counit $\varepsilon : U_q(sl_2) \to \mathbb{C}$, defined by

$$\begin{aligned} &\Delta(K^{\pm 1}) = K^{\pm 1} \otimes K^{\pm 1}, && S(K^{\pm 1}) = K^{\mp 1}, && \varepsilon(K^{\pm 1}) = 1, \\ &\Delta(E) = E \otimes K + 1 \otimes E, && S(E) = -EK^{-1}, && \varepsilon(E) = 0, \\ &\Delta(F) = F \otimes 1 + K^{-1} \otimes F, && S(F) = -KF, && \varepsilon(F) = 0. \end{aligned}$$

It can be verified by elementary computation that $U_q(sl_2)$ becomes a Hopf algebra when equipped with the above maps Δ, S, and ε.

Further, the quantum group $U_q(sl_2)$ becomes a quasi-triangular Hopf algebra when given the universal R matrix

$$\mathcal{R} = q^{H \otimes H/4} \exp_q \Big((q^{1/2} - q^{-1/2}) E \otimes F \Big), \tag{4.38}$$

where, to be precise, the element $q^{H \otimes H/4}$ does not belong to $U_q(sl_2) \otimes U_q(sl_2)$; the element belongs to the completion of $U_q(sl_2) \otimes U_q(sl_2)$ with respect to the degree in $\hbar$, putting $q = e^{\hbar}$. Further, we define the *q-exponential map* $\exp_q$ by

$$\exp_q(x) = \sum_{n=0}^{\infty} \frac{q^{n(n-1)/4}}{[n]!} x^n,$$

where we define the *quantum integer* by $[n] = (q^{n/2} - q^{-n/2})/(q^{1/2} - q^{-1/2})$; for example, $[1] = 1$, $[2] = q^{1/2} + q^{-1/2}$, $[3] = q + 1 + q^{-1}$ and so on. Further, we define its factorial by $[n]! = [n][n-1] \cdots [1]$. It is proved in Appendix A.1 that $U_q(sl_2)$ forms a quasi-triangular Hopf algebra with the above universal R matrix $\mathcal{R}$. The infinite sum of the q-exponential map in (4.38) converges in the topology of the power series ring of $\hbar$. Further, when acting on finite dimensional modules of $U_q(sl_2)$ given below, this infinite sum is a finite sum. The inverse of the above universal R matrix is given by

$$\mathcal{R}^{-1} = \exp_{q^{-1}} \Big((q^{-1/2} - q^{1/2}) E \otimes F \Big) q^{-H \otimes H/4}. \tag{4.39}$$

We verify that this really is the inverse of the above $\mathcal{R}$ in Appendix A.1.

Furthermore, as usual with a quasi-triangular Hopf algebra, we define the element u to be $\sum_i S(\beta_i)\alpha_i$ putting $\mathcal{R} = \sum_i \alpha_i \otimes \beta_i$. In this case u is calculated as follows (see Appendix A.1):

$$u = q^{-H^2/4} \sum_{n=0}^{\infty} q^{3n(n-1)/4} \frac{(q^{-1/2} - q^{1/2})^n}{[n]!} F^n K^{-n} E^n.$$

Further, we find that

$$v = K^{-1}u = q^{-H^2/4} \sum_{n=0}^{\infty} q^{n(3n+1)/4} \frac{(q^{-1/2} - q^{1/2})^n}{[n]!} F^n K^{-n-1} E^n. \tag{4.40}$$

Theorem 4.14. For the elements $\mathcal{R}$ and v given above, the triple $(U_q(sl_2), \mathcal{R}, v)$ forms a ribbon Hopf algebra.

We give a proof of this theorem in Appendix A.1.

Proceeding, by Theorem 4.5, we obtain the universal $U_q(sl_2)$ invariant of a framed oriented link L for the ribbon Hopf algebra $(U_q(sl_2), \mathcal{R}, v)$. We call this invariant the *quantum* $(sl_2;\star)$ *invariant* of L and denote it by $Q^{sl_2;\star}(L) \in (U_q(sl_2)/I)^{\otimes l}$, where I denotes the vector subspace of $U_q(sl_2)$ spanned by $ab - ba$ for $a, b \in U_q(sl_2)$.

Example 4.15. Let L be a framed link and $\overline{L}$ its mirror image. The quantum $(sl_2;\star)$ invariants of L and $\overline{L}$ are related by

$$Q^{sl_2;\star}(\overline{L}) = \iota\big(Q^{sl_2;\star}(L)\big),$$

where ι is the involutive endomorphism of $U_q(sl_2)$ defined by

$$\iota(q) = q^{-1}, \quad \iota(K) = K, \quad \iota(H) = -H, \quad \iota(E) = KF, \quad \iota(F) = EK^{-1}.$$

Proof. It is shown by concrete computation that the involution ι takes the universal R matrix to its inverse, as follows:

$$(\iota \otimes \iota)(\mathcal{R}) = P(\mathcal{R}),$$

where P denotes the permutation defined by $P(x \otimes y) = y \otimes x$. Further,

$$\iota(uv^{-1}) = \iota(K) = K = uv^{-1}.$$

The above two formulae relate $Q^{A;\star}(T)$ and $Q^{A;\star}(\overline{T})$ for each elementary tangle T and its mirror image $\overline{T}$ with respect to a mirror parallel to the plane in which the diagram is. The example follows. □

The quantum (sl_2, V_n) invariant

As a vector space V_n is equal to $\mathbb{C}^n$. It is known (see, *e.g.*, [Hum72]) that the irreducible n-dimensional representation ρ_{V_n} of sl_2 and $U(sl_2)$ on V_n is given by

$$\rho_{V_n}(E) = \begin{pmatrix} 0 & n-1 & & & \text{\Large 0} \\ & 0 & n-2 & & \\ & & \ddots & \ddots & \\ & & & 0 & 1 \\ \text{\Large 0} & & & & 0 \end{pmatrix},$$

$$\rho_{V_n}(F) = \begin{pmatrix} 0 & & & & \text{\Large 0} \\ 1 & 0 & & & \\ & 2 & 0 & & \\ & & \ddots & \ddots & \\ \text{\Large 0} & & & n-1 & 0 \end{pmatrix},$$

$$\rho_{V_n}(H) = \begin{pmatrix} n-1 & & & & \text{\Large 0} \\ & n-3 & & & \\ & & n-5 & & \\ & & & \ddots & \\ \text{\Large 0} & & & & -(n-1) \end{pmatrix}.$$

Further, perturbing the above representation by the quantum parameter q we obtain the irreducible n-dimensional representation $\rho_{V_n} : U_q(sl_2) \longrightarrow \mathrm{End}(V_n)$, given by

$$\rho_{V_n}(E) = \begin{pmatrix} 0 & [n-1] & & & \text{\Large 0} \\ & 0 & [n-2] & & \\ & & \ddots & \ddots & \\ & & & 0 & [1] \\ \text{\Large 0} & & & & 0 \end{pmatrix},$$

$$\rho_{V_n}(F) = \begin{pmatrix} 0 & & & & \text{\Large 0} \\ [1] & 0 & & & \\ & [2] & 0 & & \\ & & \ddots & \ddots & \\ \text{\Large 0} & & & [n-1] & 0 \end{pmatrix},$$

$$\rho_{V_n}(K) = \begin{pmatrix} q^{(n-1)/2} & & & & \text{\Large 0} \\ & q^{(n-3)/2} & & & \\ & & q^{(n-5)/2} & & \\ & & & \ddots & \\ \text{\Large 0} & & & & q^{-(n-1)2} \end{pmatrix},$$

see, *e.g.*, [Kas94, ChPr95, KiMe91]. Note that the *quantum dimension* of V_n (which is the trace of $\rho_{V_n}(K)$) is equal to the quantum integer $[n]$, while the dimension of V_n is equal to n. Let us verify that the above matrices satisfy the last defining relation of $U_q(sl_2)$ as follows. For the standard basis $\{e_i\}_{i=1,2,\cdots,n}$ of $\mathbb{C}^n$ the representation ρ_{V_n} is alternatively given by

$$\rho_{V_n}(E)e_i = \begin{cases} [n-i+1]e_{i-1} & \text{if } i > 1, \\ 0 & \text{if } i = 1, \end{cases}$$

$$\rho_{V_n}(F)e_i = \begin{cases} [i]e_{i+1} & \text{if } i < n, \\ 0 & \text{if } i = 1, \end{cases}$$

$$\rho_{V_n}(K)e_i = q^{(n-2i+1)/2}.$$

Hence,

$$\begin{aligned} \rho_{V_n}(EF)e_i &= [n-i][i]e_i, \\ \rho_{V_n}(FE)e_i &= [i-1][n-i+1]e_i, \\ \rho_{V_n}(\frac{K-K^{-1}}{q^{1/2}-q^{-1/2}}) &= [n-2i+1]e_i. \end{aligned}$$

Since the equality $[n-i][i]-[i-1][n-i+1] = [n-2i-1]$ holds for any n and i, the representation ρ_{V_n} satisfies the last defining relation of $U_q(sl_2)$. A key step in the above computation was the process by which formulae for $\rho_{V_n}(EF)e_i$ and $\rho_{V_n}(FE)e_i$ were obtained; note that we used the following fact:

$$\begin{aligned} &[n-i][i] = 0 && \text{when } i = n, \\ &[i-1][n-i+1] = 0 && \text{when } i = 1. \end{aligned} \tag{4.41}$$

By Theorem 4.7 we obtain an operator invariant of a framed oriented link L, derived from a representation V_n of a ribbon Hopf algebra $U_q(sl_2)$. We call this invariant the *quantum* (sl_2, V_n) *invariant* of L and denote it by $Q^{sl_2;V_n}(L) \in \mathbb{Z}[q^{\pm 1/2}]$.

Further, as a generalization as mentioned in the previous section, we obtain the operator invariant of a framed oriented link $L = L_1 \cup \cdots \cup L_l$ associated with representations $V_{n_1}, \cdots, V_{n_l}$ of $U_q(sl_2)$. We call this invariant the *quantum* $(sl_2; V_{n_1}, \cdots, V_{n_l})$ *invariant* of L and denote it by $Q^{sl_2;V_{n_1},\cdots,V_{n_l}}(L) \in \mathbb{Z}[q^{\pm 1/2}]$.

Example 4.16. The quantum $(sl_2; V_{n_1}, \cdots, V_{n_l})$ invariant $Q^{sl_2;V_{n_1},\cdots,V_{n_l}}(L)$ of a framed link L does not depend on the orientation of each component of L. Furthermore, by fixing an isomorphism between V_n and $V_n^\star$, the quantum invariant induces the operator invariant of unoriented tangles.

Proof. By Proposition 4.13, the quantum invariant of L with a different orientation is equal to the corresponding quantum invariant of L, where the involved representations are replaced by their duals. Further, it is known, see [Kas94, ChPr95, KiMe91], that the irreducible n-dimensional representation V_n of $U_q(sl_2)$ is isomorphic to its dual representation $V_n^\star$. Therefore, the value of $Q^{sl_2;V_{n_1},\cdots,V_{n_l}}(L)$ is independent of orientations of the components of L.

Furthermore, fix an isomorphism between V_n and $V_n^\star$. In the definition of an operator invariant of oriented tangles we use V_n and $V_n^\star$. Replace such $V_n^\star$ with V_n by using the isomorphism. Then, we obtain an operator invariant of unoriented tangles. □

Example 4.17. Let L be a framed link with l components. Then, the quantum $(sl_2; V_{n_1}, \cdots, V_{n_l})$ invariant of the mirror image $\overline{L}$ of L is the "conjugate" of the invariant of L as

$$Q^{sl_2;V_{n_1},\cdots,V_{n_l}}(\overline{L}) = \overline{Q^{sl_2;V_{n_1},\cdots,V_{n_l}}(L)} \in \mathbb{Z}[q^{\pm 1/2}],$$

where the over line on the right hand side implies the conjugate which is the involution of $\mathbb{Z}[q^{\pm 1/2}]$ exchanging q and q^{-1}; *i.e.*, we put $\overline{f(q)} = f(q^{-1})$ for a polynomial $f(q)$ in $q^{\pm 1/2}$.

Proof. We define the matrix S by

$$S = \begin{pmatrix} 0 & & & 1 \\ & & q^{\frac{n-1}{2}} & \\ & q^{\frac{n-1}{2}+\frac{n-3}{2}} & & \\ \cdots & & & \\ q^{\frac{n-1}{2}} & & & 0 \end{pmatrix},$$

i.e., the $(i, n-i)$ entry of S is $q^{(n-i+1)(i-1)/2}$ for $1 \le i \le n$ and the other entries are 0. Then, for the involutive endomorphism ι in Example 4.15, we have that

$$\rho_{V_n}(\iota(x)) = S^{-1} \cdot \rho_{V_n}(x) \cdot S,$$

for any $x \in U_q(sl_2)$. Hence, computing $Q^{sl_2;V_{n_1},\cdots,V_{n_l}}(\overline{L})$ by Example 4.15 and Theorem 4.11, we obtain the required formula (observing that ι takes q to q^{-1}). □

The quantum (sl_2, V) invariant and the Kauffman bracket

Let V be the vector representation of sl_2, *i.e.*, we let V denote V_2. Setting $n = 2$ in the representation ρ_{V_n} given above, we find that

$$\rho_V(E) = \begin{pmatrix} 0 & 1 \\ 0 & 0 \end{pmatrix}, \qquad \rho_V(F) = \begin{pmatrix} 0 & 0 \\ 1 & 0 \end{pmatrix}, \qquad \rho_V(K) = \begin{pmatrix} q^{1/2} & 0 \\ 0 & q^{-1/2} \end{pmatrix}.$$

Through this representation ρ_V we obtain the R matrix from the universal R matrix $\mathcal{R}$ as follows. Since $\rho_V(E^2) = \rho_V(F^2) = 0$, we have that

$$\begin{aligned}
&\rho_V \otimes \rho_V\Big(\exp_q\big((q^{1/2} - q^{-1/2})E \otimes F\big)\Big) \\
&\quad = \rho_V(1) \otimes \rho_V(1) + (q^{1/2} - q^{-1/2})\rho_V(E) \otimes \rho_V(F) \\
&\quad = \begin{pmatrix} 1 & 0 \\ 0 & 1 \end{pmatrix} \otimes \begin{pmatrix} 1 & 0 \\ 0 & 1 \end{pmatrix} + (q^{1/2} - q^{-1/2}) \begin{pmatrix} 0 & 1 \\ 0 & 0 \end{pmatrix} \otimes \begin{pmatrix} 0 & 0 \\ 1 & 0 \end{pmatrix} \\
&\quad = \begin{pmatrix} 1 & 0 & 0 & 0 \\ 0 & 1 & q^{1/2} - q^{-1/2} & 0 \\ 0 & 0 & 1 & 0 \\ 0 & 0 & 0 & 1 \end{pmatrix}.
\end{aligned}$$

Further,

$$\begin{aligned}
\rho_V \otimes \rho_V(q^{H \otimes H/4}) &= \rho_V \otimes \rho_V\Big(\exp(\frac{\hbar}{4} H \otimes H)\Big) = \exp\Big(\frac{\hbar}{4}\rho_V(H) \otimes \rho_V(H)\Big) \\
&= \exp\left(\frac{\hbar}{4}\begin{pmatrix} 1 & 0 \\ 0 & -1 \end{pmatrix} \otimes \begin{pmatrix} 1 & 0 \\ 0 & -1 \end{pmatrix}\right) = \begin{pmatrix} q^{1/4} & 0 & 0 & 0 \\ 0 & q^{-1/4} & 0 & 0 \\ 0 & 0 & q^{-1/4} & 0 \\ 0 & 0 & 0 & q^{1/4} \end{pmatrix}.
\end{aligned}$$

Hence, the R matrix and the matrix h are given by

$$\begin{aligned}
R = P \circ (\rho_V \otimes \rho_V)(\mathcal{R}) &= \begin{pmatrix} q^{1/4} & 0 & 0 & 0 \\ 0 & 0 & q^{-1/4} & 0 \\ 0 & q^{-1/4} & q^{1/4} - q^{-3/4} & 0 \\ 0 & 0 & 0 & q^{1/4} \end{pmatrix}, \\
h = \rho_V(uv^{-1}) = \rho_V(K) &= \begin{pmatrix} q^{1/2} & 0 \\ 0 & q^{-1/2} \end{pmatrix}.
\end{aligned} \tag{4.42}$$

We refer to the operator invariant that is obtained from these matrices according to the construction given in Section 3.3, as the *quantum* (sl_2, V) *invariant* $Q^{sl_2;V}(L)$ of a framed link L.

Proposition 4.18. The quantum (sl_2, V) invariant satisfies the following *skein relation*

$$q^{1/4} Q^{sl_2;V}(L_+) - q^{-1/4} Q^{sl_2;V}(L_-) = (q^{1/2} - q^{-1/2}) Q^{sl_2;V}(L_0),$$

where L_+, L_- and L_0 are three framed links which only differ in a ball, as follows:

L_+ $\qquad$ L_- $\qquad$ L_0

Proof. The proposition is an immediate consequence of the following equation,

$$q^{1/4}R - q^{-1/4}R^{-1} = (q^{1/2} - q^{-1/2})\mathrm{id}_{V\otimes V},$$

which can be verified for the R matrix given in (4.42). □

Recall that $\langle L\rangle$ denotes the Kauffman bracket of a framed link L, as described in Section 1.2. The following theorem gives a reconstruction of the *quantum* (sl_2, V) *invariant* by the Kauffman bracket. For a similar relation see Appendix H.1.

Theorem 4.19. Let L be a framed link. Then,

$$Q^{sl_2;V}(L) = (-1)^{\#L+f(L)}\langle L\rangle\Big|_{A=q^{1/4}}$$

where $\#L$ denotes the number of the components of L and $f(L)$ denotes the sum of the framings of the components of L.

See Appendix H.1 for the proof of this theorem.

The Jones polynomial

We compute $\rho_V(v)$:

$$\begin{aligned}\rho_V(v) &= \rho_V(q^{-H^2/4})\rho_V(K^{-1} + q(q^{-1/2} - q^{1/2})FK^{-2}E)\\ &= \begin{pmatrix} q^{-1/4} & 0 \\ 0 & q^{-1/4}\end{pmatrix}\begin{pmatrix} q^{-1/2} & 0 \\ 0 & q^{-1/2}\end{pmatrix} = q^{-3/4}\cdot \mathrm{id}_V.\end{aligned}$$

Normalizing the R matrix in (4.42) with the scalar $q^{-3/4}$, we find that

$$q^{-3/4}R\Big|_{q=t^{-1}} = \begin{pmatrix} t^{1/2} & 0 & 0 & 0 \\ 0 & 0 & t & 0 \\ 0 & t & t^{1/2} - t^{3/2} & 0 \\ 0 & 0 & 0 & t^{1/2}\end{pmatrix}.$$

Such a normalization is useful to obtain an invariant, not of framed links, but of (unframed) links, just as we normalized by writhe in Theorem 1.5. Further, we find that

$$h = \rho_V(uv^{-1}) = \rho_V(K)\Big|_{q=t^{-1}} = \begin{pmatrix} t^{-1/2} & 0 \\ 0 & t^{1/2}\end{pmatrix}.$$

These are just the matrices (3.29), which gave the construction of the *Jones polynomial* as an operator invariant in Section 3.3.

4.5 The quantum group $U_\zeta(sl_2)$ at a root of unity ζ

In this section we formulate the quantum group $U_\zeta(sl_2)$ at a root of unity ζ, as a ribbon Hopf algebra. Though the formulation is similar to that of $U_q(sl_2)$, the element v of this ribbon Hopf algebra is different from the v of $U_q(sl_2)$ at a generic q. The Alexander polynomial can be reconstructed as an operator invariant derived from a representation of $U_\zeta(sl_2)$. Further, we introduce the multi-variable Alexander polynomial as an operator invariant associated with certain representations of $U_\zeta(sl_2)$.

Let r be an integer ≥ 2 and set $\zeta = \exp(2\pi\sqrt{-1}/r)$. The *quantum group* $U_\zeta(sl_2)$ at a root of unity ζ is defined to be the algebra over $\mathbb{C}$, with the unit element 1, generated by the generators $K^{\pm 1}, E$ and F subject to the following relations

$$K \cdot K^{-1} = K^{-1} \cdot K = 1,$$

$$KE = \zeta EK, \qquad KF = \zeta^{-1} FK, \qquad EF - FE = \frac{K - K^{-1}}{\zeta^{1/2} - \zeta^{-1/2}},$$

$$E^r = F^r = 0.$$

The universal R matrix $\mathcal{R}$ is defined in $U_\zeta(sl_2) \otimes U_\zeta(sl_2)$ by

$$\mathcal{R} = \zeta^{H \otimes H/4} \exp_\zeta^{(<r)} \Big((\zeta^{1/2} - \zeta^{-1/2}) E \otimes F \Big), \tag{4.43}$$

where we define the *truncated q-exponential map* $\exp_q^{(<r)}$ by

$$\exp_q^{(<r)}(x) = \sum_{0 \leq n < r} \frac{q^{n(n-1)/4}}{[n]!} x^n.$$

Further, we regard $\zeta^{H \otimes H/4}$ as a formal symbol satisfying the same properties as those of $q^{H \otimes H/4}$ at $q = \zeta$. The inverse $\mathcal{R}^{-1}$ of the above $\mathcal{R}$ is given by

$$\mathcal{R}^{-1} = \exp_{\zeta^{-1}}^{(<r)} \Big((\zeta^{-1/2} - \zeta^{1/2}) E \otimes F \Big) \zeta^{-H \otimes H/4}.$$

It is shown in Appendix A.2 that the pair $(U_\zeta(sl_2), \mathcal{R})$ forms a quasi-triangular Hopf algebra. Furthermore, as usual with a quasi-triangular Hopf algebra, we put $u = \sum_i S(\beta_i)\alpha_i$ where we have written $\mathcal{R} = \sum_i \alpha_i \otimes \beta_i$. We calculate

$$u = \zeta^{-H^2/4} \sum_{n=0}^{r-1} \zeta^{3n(n-1)/4} \frac{(\zeta^{-1/2} - \zeta^{1/2})^n}{[n]!} F^n K^{-n} E^n.$$

Moreover, we set

$$v = K^{r-1} u.$$

Note that this definition of v is different from that of v of the ribbon Hopf algebra of $U_q(sl_2)$ at a generic q.

Theorem 4.20. For the above $\mathcal{R}$ and v, the triple $(U_\zeta(sl_2), \mathcal{R}, v)$ forms a ribbon Hopf algebra.

We give a proof of this theorem in Appendix A.2.

Finite dimensional representations of $U_\zeta(sl_2)$

We now introduce the representations ρ_λ of $U_\zeta(sl_2)$. These representations, which depend on a complex parameter λ, act on the vector space $\mathbb{C}^r$ (we denote the corresponding $U_\zeta(sl_2)$ module by V_λ), as follows:

$$\rho_\lambda(E) = \begin{pmatrix} 0 & [r-1+\lambda] & & & \\ & 0 & [r-2+\lambda] & & \\ & & \ddots & \ddots & \\ & & & 0 & [1+\lambda] \\ & & & & 0 \end{pmatrix},$$

$$\rho_\lambda(F) = \begin{pmatrix} 0 & & & & \\ [1] & 0 & & & \\ & [2] & 0 & & \\ & & \ddots & \ddots & \\ & & & [r-1] & 0 \end{pmatrix},$$

$$\rho_\lambda(K) = \begin{pmatrix} \zeta^{(r-1+\lambda)/2} & & & & \\ & \zeta^{(r-3+\lambda)/2} & & & \\ & & \zeta^{(r-5+\lambda)/2} & & \\ & & & \ddots & \\ & & & & \zeta^{(-r+1+\lambda)/2} \end{pmatrix}.$$

We can verify that this definition gives a representation of $U_\zeta(sl_2)$ by elementary computation. At the heart of the computation are the following facts,

$$[r-i+\lambda][i] = 0 \quad \text{when } i = r,$$
$$[i-1][r-i+1+\lambda] = 0 \quad \text{when } i = 1.$$

Compare these facts with (4.41), which were important in the analogous computation for $U_q(sl_2)$ for generic q.

The Alexander polynomial

In particular, the Alexander polynomial arises as the invariant derived from the above representation in the case $r = 2$, as follows. In this case we find that

$$\rho_\lambda(E) = \begin{pmatrix} 0 & [1+\lambda] \\ 0 & 0 \end{pmatrix}, \qquad \rho_\lambda(F) = \begin{pmatrix} 0 & 0 \\ 1 & 0 \end{pmatrix}, \qquad \rho_\lambda(K) = \begin{pmatrix} \zeta^{(1+\lambda)/2} & 0 \\ 0 & \zeta^{(-1+\lambda)/2} \end{pmatrix}.$$

Through this representation ρ_λ we obtain the R matrix from the universal R matrix $\mathcal{R}$ as follows. Since $\rho_\lambda(E^2) = \rho_\lambda(F^2) = 0$, we have that

$$\begin{aligned}
&(\rho_\lambda \otimes \rho_\lambda)\Big(\exp_\zeta^{(<2)} \big((\zeta^{1/2} - \zeta^{-1/2}) E \otimes F\big)\Big) \\
&= \rho_\lambda(1) \otimes \rho_\lambda(1) + (\zeta^{1/2} - \zeta^{-1/2}) \rho_\lambda(E) \otimes \rho_\lambda(F) \\
&= \begin{pmatrix} 1 & 0 \\ 0 & 1 \end{pmatrix} \otimes \begin{pmatrix} 1 & 0 \\ 0 & 1 \end{pmatrix} + (\zeta^{1/2} - \zeta^{-1/2}) \begin{pmatrix} 0 & [1+\lambda] \\ 0 & 0 \end{pmatrix} \otimes \begin{pmatrix} 0 & 0 \\ 1 & 0 \end{pmatrix} \\
&= \begin{pmatrix} 1 & 0 & 0 & 0 \\ 0 & 1 & \zeta^{(1+\lambda)/2} - \zeta^{-(1+\lambda)/2} & 0 \\ 0 & 0 & 1 & 0 \\ 0 & 0 & 0 & 1 \end{pmatrix}.
\end{aligned}$$

Since $\rho_\lambda(H) = \begin{pmatrix} 1+\lambda & 0 \\ 0 & -1+\lambda \end{pmatrix}$, we find that

$$(\rho_\lambda \otimes \rho_\lambda)(\zeta^{H \otimes H/4}) = \begin{pmatrix} \zeta^{(\lambda+1)^2/4} & 0 & 0 & 0 \\ 0 & \zeta^{(\lambda^2-1)/4} & 0 & 0 \\ 0 & 0 & \zeta^{(\lambda^2-1)/4} & 0 \\ 0 & 0 & 0 & \zeta^{(\lambda-1)^2/4} \end{pmatrix}.$$

Hence,

$$\begin{aligned}
R &= P \circ \big((\rho_\lambda \otimes \rho_\lambda)(\mathcal{R})\big) \\
&= \zeta^{(\lambda^2-1)/4} \begin{pmatrix} \zeta^{(\lambda+1)/2} & 0 & 0 & 0 \\ 0 & 0 & 1 & 0 \\ 0 & 1 & \zeta^{(1+\lambda)/2} - \zeta^{-(1+\lambda)/2} & 0 \\ 0 & 0 & 0 & \zeta^{(1-\lambda)/2} \end{pmatrix}.
\end{aligned}$$

Further, we calculate

$$\begin{aligned}
&\rho_\lambda(v) = \rho_\lambda(Ku) = \rho_\lambda(K) \rho_\lambda(\zeta^{-H^2/4}) \rho_\lambda\Big(1 + (\zeta^{-1/2} - \zeta^{1/2}) F K^{-1} E\Big) \\
&= \begin{pmatrix} \zeta^{(\lambda+1)/2} & 0 \\ 0 & \zeta^{(\lambda-1)/2} \end{pmatrix} \begin{pmatrix} \zeta^{-(\lambda+1)^2/4} & 0 \\ 0 & \zeta^{-(\lambda-1)^2/4} \end{pmatrix} \begin{pmatrix} 1 & 0 \\ 0 & \zeta^{-(\lambda+1)} \end{pmatrix} = \zeta^{(1-\lambda^2)/4} \cdot \mathrm{id}_{V_\lambda},
\end{aligned}$$

noting $\zeta^2 = 1$. We put $t^{1/2} = \zeta^{-(1+\lambda)/2}$. By normalizing the R matrix by the scalar on the right hand side of the above formula, we find that

$$\zeta^{(1-\lambda^2)/4} \cdot R = \begin{pmatrix} t^{-1/2} & 0 & 0 & 0 \\ 0 & 0 & 1 & 0 \\ 0 & 1 & t^{-1/2} - t^{1/2} & 0 \\ 0 & 0 & 0 & -t^{1/2} \end{pmatrix}.$$

Further, we find that

$$h = \rho_\lambda(uv^{-1}) = \rho_\lambda(K^{-1}) = \begin{pmatrix} t^{1/2} & 0 \\ 0 & -t^{1/2} \end{pmatrix}.$$

The above two matrices are just the matrices (3.31), which gave the reconstruction of the *Alexander polynomial* as an operator invariant in Section 3.3.

The multi-variable Alexander polynomial

By computing $(\rho_{\lambda_1} \otimes \rho_{\lambda_2})(\mathcal{R})$ for two representations ρ_{λ_1} and ρ_{λ_2} in the same way as above, we obtain the following R matrix with two parameters t_1 and t_2,

$$R_{t_1t_2} = \begin{pmatrix} t_1^{-1/4}t_2^{-1/4} & 0 & 0 & 0 \\ 0 & 0 & t_1^{-1/4}t_2^{1/4} & 0 \\ 0 & t_1^{1/4}t_2^{-1/4} & (t_1^{-1/4} - t_1^{3/4})t_2^{-1/4} & 0 \\ 0 & 0 & 0 & -t_1^{1/4}t_2^{1/4} \end{pmatrix}.$$

We consider a link $L = L_1 \cup \cdots \cup L_l$ whose l components are labeled by l variables $t_1, \cdots, t_l$. We obtain an invariant of such a labeled link by using the above R matrix, as follows. Let T be an 1-tangle obtained from L by cutting the ith component L_i. We consider the operator invariant of the labeled tangle T whose components are associated with representations $\rho_{\lambda_1}, \cdots, \rho_{\lambda_l}$ of $U_\zeta(sl_2)$ at $\zeta = -1$, as in Section 4.3. In the definition of the operator invariant, the R matrix $R_{t_it_j}$ is associated to a crossing of two strands labeled respectively by t_i and t_j. Since the operator invariant of T is an intertwiner, it takes the form $c_i \cdot \mathrm{id}_{V_{\lambda_i}}$ for some scalar c_i. This scalar c_i is an isotopy invariant of T. Further, when we cut another component L_j to obtain T from L, we have another scalar c_j in the same way as above. By Lemma A.17 the relation between them is given by

$$c_i(t_j^{1/2} - t_j^{-1/2}) = c_j(t_i^{1/2} - t_i^{-1/2})$$

in a similar way as in the proof of Theorem 3.12. Hence, $c_i/(t_i^{1/2} - t_i^{-1/2})$ does not depend on i, and it is an isotopy invariant of a link L. We call it the *multi-variable Alexander polynomial* of labeled oriented links. See [MuJ93, ADO92] for detailed arguments.

Chapter 5

Monodromy representations of the braid groups derived from the Knizhnik-Zamolodchikov equation

The Knizhnik-Zamolodchikov equation [KnZa84] (the KZ equation) was found as a differential equation satisfied by the partition functions in the Wess-Zumino-Witten model of a conformal field theory (see Appendix F.4). Here a conformal field theory is a quantum field theory on a 2-dimensional space which is conformally invariant; this implies that such a theory does not depend on the metric of the space in some sense. As mentioned in the introduction of Chapter 3 a tangle (in particular, a braid) can be regarded as a time evolution of a set of moving particles in a plane from the physical viewpoint. Further, in quantum field theory, such a time evolution can be described by an operator between vector spaces. Hence, a braid determines a linear map in the theory.

Motivated by this physical interpretation, we construct representations of braid groups derived from the KZ equation in this chapter. We consider the space of sets of n distinct points in a plane which is called the *configuration space* of n points in a plane. The KZ equation is formulated as a differential equation on the configuration space. We regard a braid in n strands as a path in the configuration space. By lifting such paths along (local) solutions of the KZ equation we obtain representations of the braid group in n strands. We call such representations *monodromy representations* of the braid group along solutions of the KZ equation.

Kohno [Koh87, Koh88] first pointed out the fact that such monodromy representations are equivalent to representations derived from quantum groups mentioned in the previous chapter, and hence, quantum invariants are obtained from monodromy representations along solutions of the KZ equation. Further, motivated by Kohno's work, Drinfel'd [Dri89b, Dri90] gave a complete proof of this fact for any pair of a Lie algebra and its representation using the uniqueness of the category of quasi-Hopf algebras.

In Section 5.1 we show that solutions of the KZ equation induces the monodromy representations of braid groups. We give explicit presentations of the monodromy representations for generators of the braid groups in Sections 5.2 and 5.3. Based on these presentations, we introduce quasi-bialgebras (a weak notion of quasi-Hopf algebras) in Section 5.4. Further, in Section 5.5 we explain the equivalence between the monodromy representations and the representations derived from the quantum groups. For detailed expositions for these topics see also [Kas94, EFK98, ChPr95].

5.1 Representations of braid groups derived from the KZ equation

The Knizhnik-Zamolodchikov equation (KZ equation) is a total differential equation on a configuration space. In this section, by regarding a braid as a path in the configuration space, we give a representation of a braid group by monodromy along solutions of the KZ equation.

We begin with a review of some basic formulae of the theory of Lie algebras.

Let $\mathfrak{g}$ be a semi-simple Lie algebra over $\mathbb{C}$. The *adjoint representation* $\mathrm{ad} : \mathfrak{g} \to \mathrm{End}(\mathfrak{g})$ is defined by $(\mathrm{ad}(X))(Y) = [X, Y]$. The map ad is a representation because of the identity $\mathrm{ad}([X,Y]) = \mathrm{ad}(X)\mathrm{ad}(Y) - \mathrm{ad}(Y)\mathrm{ad}(X)$, which is derived from the Jacobi identity of the Lie algebra $\mathfrak{g}$. Further, the *Killing form* $B : \mathfrak{g} \otimes \mathfrak{g} \to \mathbb{C}$ is defined by $B(X,Y) = \mathrm{trace}(\mathrm{ad}(X)\mathrm{ad}(Y))$. It is a symmetric bilinear form on $\mathfrak{g}$. It is known that for any semi-simple Lie algebra its Killing form is non-degenerate; see, *e.g.*, [Hum72]. Let $\{I_\mu\}$ be an orthonormal basis of $\mathfrak{g}$ with respect to the Killing form, *i.e.*, $B(I_\mu, I_\nu)$ is equal to 0 if $\mu \neq \nu$ and 1 if $\mu = \nu$. For the dual basis $\{I_\mu^\star\}$ of the dual vector space $\mathfrak{g}^\star$ of $\mathfrak{g}$, the Killing form B can be regarded as the element $B = \sum_\mu I_\mu^\star \otimes I_\mu^\star \in \mathfrak{g}^\star \otimes \mathfrak{g}^\star$. By using the dual basis we have that

$$\sum_\mu B(X, I_\mu) I_\mu = \sum_\mu I_\mu^\star(X) \cdot I_\mu = X. \tag{5.1}$$

The Killing form B satisfies the associativity in the sense $B([X,Y],Z) = B(X,[Y,Z])$. This follows from the equality of the following two formulae,

$$\begin{aligned}
B\Big([X,Y],Z\Big) &= \mathrm{trace}\Big(\mathrm{ad}([X,Y])\mathrm{ad}(Z)\Big) \\
&= \mathrm{trace}\Big(\mathrm{ad}(X)\mathrm{ad}(Y)\mathrm{ad}(Z) - \mathrm{ad}(Y)\mathrm{ad}(X)\mathrm{ad}(Z)\Big), \\
B\Big(X,[Y,Z]\Big) &= \mathrm{trace}\Big(\mathrm{ad}(X)\mathrm{ad}([Y,Z])\Big) \\
&= \mathrm{trace}\Big(\mathrm{ad}(X)\mathrm{ad}(Y)\mathrm{ad}(Z) - \mathrm{ad}(X)\mathrm{ad}(Z)\mathrm{ad}(Y)\Big),
\end{aligned}$$

where the right hand sides of the above two formulae are equal by the property $\mathrm{trace}(AB) = \mathrm{trace}(BA)$ of trace.

We define the element $\tau \in \mathfrak{g} \otimes \mathfrak{g}$ to be the dual element of the element $B \in \mathfrak{g}^\star \otimes \mathfrak{g}^\star$ with respect to the orthonormal structure derived from the Killing form $B : \mathfrak{g} \otimes \mathfrak{g} \to \mathbb{C}$. We call τ the *invariant 2-tensor* of $\mathfrak{g} \otimes \mathfrak{g}$; for the case $\mathfrak{g} = sl_N$ we re-define the invariant 2-tensor τ to be a scalar multiple of the dual element of B later. By using the orthonormal basis $\{I_\mu\}$ the invariant 2-tensor τ is presented by $\tau = \sum_\mu I_\mu \otimes I_\mu$. Further, the invariant 2-tensor τ commutes with $X \otimes 1 + 1 \otimes X$

for any $X \in \mathfrak{g}$, since we obtain the commutativity as follows:

$$\begin{aligned}\sum_{\mu}[X, I_\mu] \otimes I_\mu &= \sum_{\mu,\nu} B\Big(I_\nu, [X, I_\mu]\Big) I_\nu \otimes I_\mu \\ &= \sum_{\mu,\nu} B\Big([I_\nu, X], I_\mu\Big) I_\nu \otimes I_\mu = \sum_{\nu} I_\nu \otimes [I_\nu, X],\end{aligned} \tag{5.2}$$

where the first and last equalities are derived from (5.1) and the second equality is derived from the associativity of the Killing form B. Hence, the invariant 2-tensor τ is invariant under two actions on $\mathfrak{g} \otimes \mathfrak{g}$; one is the action of the tensor product of two copies of the adjoint representation of $\mathfrak{g}$, and the other is the action of the symmetric group $\mathfrak{S}_2$ interchanging the two entries of $\mathfrak{g} \otimes \mathfrak{g}$.

We denote by $U(\mathfrak{g})$ the *universal enveloping algebra* of $\mathfrak{g}$, which is defined to be the algebra over $\mathbb{C}$, with the unit element 1, generated by a basis $\{X_i\}$ of $\mathfrak{g}$ subject to the relations $X_iX_j - X_jX_i = [X_i, X_j]$; see Section 4.4 for the definition in the case $\mathfrak{g} = sl_2$. Note that the vector subspace of $U(\mathfrak{g})$ spanned by $\{X_i\}$ can be identified with $\mathfrak{g}$, where the bracket is given by the commutator. The algebra $U(\mathfrak{g})$ is the universal one among algebras including $\mathfrak{g}$ in this way, and hence it is defined independently of the choice of a basis of $\mathfrak{g}$.

We define the *Casimir element* $C \in U(\mathfrak{g})$ to be the image of τ under the multiplication $U(\mathfrak{g}) \otimes U(\mathfrak{g}) \to U(\mathfrak{g})$. Using the orthonormal basis of $\mathfrak{g}$ the Casimir element C is presented by $C = \sum_\mu I_\mu I_\mu$. Taking the images of both sides of (5.2) by the multiplication of $U(\mathfrak{g})$ we find that $\sum_\mu [X, I_\mu] I_\mu = \sum_\mu I_\mu [I_\mu, X]$. It follows from this formula that C is central in $U(\mathfrak{g})$.

The invariant 2-tensor τ, as an element of $U(\mathfrak{g}) \otimes U(\mathfrak{g})$, can be presented, in terms of the Casimir element, by $\tau = \frac{1}{2}(\Delta(C) - C \otimes 1 - 1 \otimes C)$, since $\Delta(I_\mu) = I_\mu \otimes 1 + 1 \otimes I_\mu$, where the comultiplication $\Delta : U(\mathfrak{g}) \to U(\mathfrak{g}) \otimes U(\mathfrak{g})$ is defined by $\Delta(X) = X \otimes 1 + 1 \otimes X$ for any $X \in \mathfrak{g}$.

For distinct i and j with $1 \le i, j \le n$, we define τ_{ij} by taking τ to the i, jth factors of $U(\mathfrak{g})^{\otimes n}$, *i.e.*, we put

$$\tau_{ij} = \sum_\mu 1 \otimes \cdots \otimes 1 \otimes I_\mu \otimes 1 \otimes \cdots \otimes 1 \otimes I_\mu \otimes 1 \otimes \cdots \otimes 1 \in U(\mathfrak{g})^{\otimes n}, \tag{5.3}$$

where the two I_μ's are located at the ith and jth factors of the tensor product $U(\mathfrak{g})^{\otimes n}$.

Lemma 5.1. We have that

$$[\tau_{ij}, \tau_{kl}] = 0, \qquad \text{for distinct } i, j, k, l, \tag{5.4}$$

$$[\tau_{ij}, \tau_{ik} + \tau_{jk}] = 0, \qquad \text{for distinct } i, j, k. \tag{5.5}$$

Proof. The first formula of the lemma is immediately obtained from the definition of τ_{ij}.

We prove the second formula as follows. Since the Casimir element C is central, $(\Delta\otimes \mathrm{id})\tau$ commutes with each of $\Delta(C)\otimes 1$, $C\otimes 1\otimes 1$ and $1\otimes C\otimes 1$. Further,

$$(\Delta\otimes \mathrm{id})\tau = \sum_\mu \Delta(I_\mu)\otimes I_\mu = \sum_\mu (I_\mu\otimes 1 + 1\otimes I_\mu)\otimes I_\mu = \tau_{13}+\tau_{23},$$

$$\frac{1}{2}\Big(\Delta(C)\otimes 1 - C\otimes 1\otimes 1 - 1\otimes C\otimes 1\Big) = \tau_{12}.$$

The commutativity of the above two formulae implies the second formula of the lemma for $(i,j,k) = (1,2,3)$. By replacing the subscripts we obtain the second formula for all (i,j,k). □

In the case of the Lie algebra sl_2 the bracket is given by

$$[H,E]=2E, \qquad [H,F]=-2F, \qquad [E,F]=H,$$

with respect to a basis $\{E,H,F\}$ of sl_2. Hence, with respect to this basis, the adjoint representation of sl_2 is given by

$$\mathrm{ad}(E) = \begin{pmatrix} 0 & -2 & 0\\ 0 & 0 & 1\\ 0 & 0 & 0\end{pmatrix}, \qquad \mathrm{ad}(F) = \begin{pmatrix} 0 & 0 & 0\\ -1 & 0 & 0\\ 0 & 2 & 0\end{pmatrix}, \qquad \mathrm{ad}(H) = \begin{pmatrix} 2 & 0 & 0\\ 0 & 0 & 0\\ 0 & 0 & -2\end{pmatrix}.$$

Therefore, the Killing form B is equal to the bilinear form given by the symmetric matrix

$$\begin{pmatrix} 0 & 0 & 4\\ 0 & 8 & 0\\ 4 & 0 & 0\end{pmatrix}$$

with respect to the basis of sl_2. Hence, the dual element of $B\in (sl_2)^*\otimes(sl_2)^*$ is given by

$$(E\ \ H\ \ F)\otimes \begin{pmatrix} 0 & 0 & 4\\ 0 & 8 & 0\\ 4 & 0 & 0\end{pmatrix}^{-1}\begin{pmatrix} E\\ H\\ F\end{pmatrix} = \frac{1}{4}E\otimes F + \frac{1}{8}H\otimes H + \frac{1}{4}F\otimes E.$$

We define the *invariant 2-tensor* τ of $sl_2\otimes sl_2$ to be, not the above element, but the following scalar multiple of it:

$$\tau = E\otimes F + F\otimes E + \frac{1}{2}H\otimes H.$$

Further, the Casimir element C is given by

$$C = EF + FE + \frac{1}{2}H^2.$$

In particular, $\tau = \frac{1}{2}(\Delta(c) - c\otimes 1 - 1\otimes c)$. Conventionally, the canonical invariant 2-tensor of $sl_N\otimes sl_N$ is defined to be a scalar multiple of the dual of the Killing form, as follows. Putting $V=\mathbb{C}^N$, we define the *vector representation* $\rho_V : sl_N \to$

End(V) to be that representation which takes a matrix in sl_N to the same matrix in End(V). Further, we consider the bilinear form $B' : sl_N \otimes sl_N \to \mathbb{C}$ defined by $B'(X,Y) = \mathrm{trace}(\rho_V(X)\rho_V(Y))$. It is known that B' is a scalar multiple of the Killing form. Regarding b' in $(sl_N)^* \otimes (sl_N)^*$ we define the *invariant 2-tensor* $\tau \in sl_N \otimes sl_N$ to be the dual element of B'. In particular, when $N = 2$, we obtain the presentation of the invariant 2-tensor τ of $sl_2 \otimes sl_2$ given above.

Let X_n be the *configuration space* of n distinct points in $\mathbb{C}$, *i.e.*, we put

$$X_n = \{(z_1, z_2, \cdots, z_n) \in \mathbb{C}^n \mid z_i \neq z_j \text{ for any } i \neq j\},$$

and fix a base point $x_0 \in X_n$. A path $\gamma : [0,1] \to X_n$ such that $\gamma(0) = \gamma(1) = x_0$ can be regarded as a pure braid in n strands, as follows. For each $t \in [0,1]$, writing $\gamma(t) = (z_1(t), \cdots, z_n(t)) \in X_n$, the points $(z_1(t), t), \cdots, (z_n(t), t)$ are plotted in $\mathbb{C} \times [0,1]$; see Figure 5.1. Here a *pure braid* is a braid such that the i-th upper end point is connected to the i-th lower end point by a strand of the braid, for each i. Since a homotopy of a path in X_n implies an isotopy of the corresponding pure braid, the correspondence gives an isomorphism between the fundamental group $\pi_1(X_n)$ and the pure braid group P_n in n strands, *i.e.*, we have that

$$\pi_1(X_n) \cong P_n. \tag{5.6}$$

Here the *pure braid group* P_n is the subgroup of the braid group B_n in n strands consisting of pure braids. The nth symmetric group $\mathfrak{S}_n$ acts on X_n by interchanging the n coordinates $z_1, \cdots, z_n$ of X_n. We have an isomorphism between the fundamental group $\pi_1(X_n/\mathfrak{S}_n)$ of the quotient space $X_n/\mathfrak{S}_n$ and the braid group B_n in n strands, *i.e.*, we have that

$$\pi_1(X_n/\mathfrak{S}_n) \cong B_n. \tag{5.7}$$

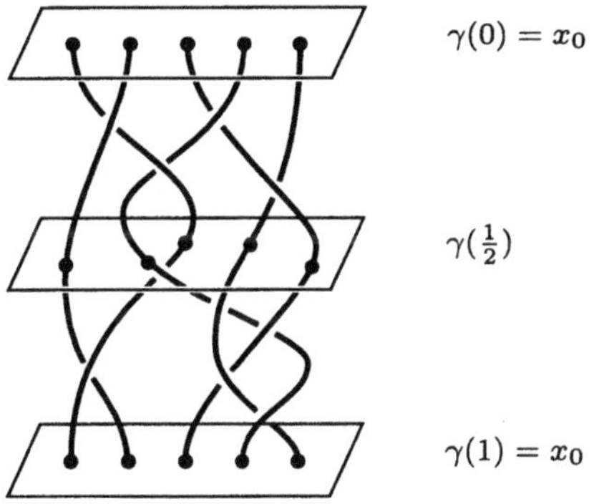

Figure 5.1 A path $\gamma : [0,1] \to X_n$ can be regarded as a pure braid in n strands.

Let V be a vector space on which $U(\mathfrak{g})$ acts. Then, the element $\tau_{ij} \in U(\mathfrak{g})^{\otimes n}$ given in (5.3) acts on $V^{\otimes n}$. For a smooth function $W : X_n \to V^{\otimes n}$ taking $(z_1, \cdots, z_n)$ to $W(z_1, \cdots, z_n)$, we consider the following system of the partial differential equations

$$\frac{\partial W}{\partial z_i} = \frac{\hbar}{2\pi\sqrt{-1}} \sum_{\substack{1\leq j\leq n \\ j\neq i}} \frac{\tau_{ij}}{z_i - z_j} W, \qquad (i = 1, 2, \cdots, n), \tag{5.8}$$

with a complex parameter $\hbar$. This can alternatively be written as a total differential equation

$$dW = \frac{\hbar}{2\pi\sqrt{-1}} \sum_{1\leq i<j\leq n} \tau_{ij} \frac{dz_i - dz_j}{z_i - z_j} W. \tag{5.9}$$

We call this equation *the Knizhnik-Zamolodchikov equation* (the *KZ equation*). Define a $U(sl_2)^{\otimes n}$-valued 1-form τ_n on X_n

$$\tau_n = \frac{\hbar}{2\pi\sqrt{-1}} \sum_{1\leq i<j\leq n} \tau_{ij} \frac{dz_i - dz_j}{z_i - z_j}. \tag{5.10}$$

Using this 1-form the KZ equation can be written $dW = \tau_n W$.

Proposition 5.2. Let A be a non-commutative algebra acting on a vector space V', and let α be an A-valued 1-form on a space X. We consider the total differential equation $(d - \alpha)W = 0$ for functions $W : X \to V'$. If the equality

$$d\alpha - \alpha \wedge \alpha = 0 \tag{5.11}$$

holds, then the total differential equation has a unique local solution W in a neighborhood of any point $x_0 \in X$ with an arbitrarily given initial value $v = W(x_0) \in V'$.

In the terminology of differential geometry $d - \alpha$ is called a *connection* on the trivial vector bundle over X with the fiber V'. For a 1-form α on X the relation (5.11) is called the *integrability condition* for the connection. A connection satisfying the integrability condition is called *flat*. The proposition implies that a flat connection gives a local section of the vector bundle by integration along the connection; see also Figure 5.2 for a graphical explanation of the integrability condition.

Proof of Proposition 5.2. For any point x in a neighborhood of x_0 we consider a path $\gamma : [0,1] \to X$ such that $\gamma(0) = x_0$ and $\gamma(1) = x$. The pull back $\gamma^*\alpha$ of the 1-form α becomes a 1-form on $[0,1]$ and we put $\gamma^*\alpha = w(t)dt$ with a function $w(t) \in A$ for a coordinate $t \in [0,1]$. For a function $f : [0,1] \to V'$ we consider the following ordinary differential equation

$$\frac{df(t)}{dt} = w(t)f(t), \qquad f(0) = v.$$

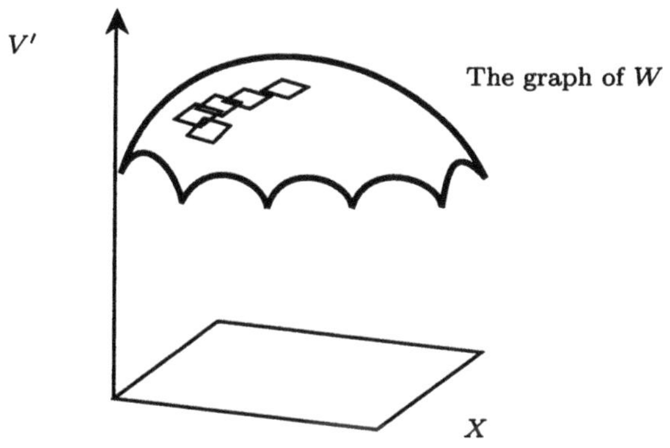

Figure 5.2 The integrability condition (5.11) of the differential equation $(d-\alpha)W = 0$ implies the existence of a local solution W whose graph has the family of tangent spaces given by the differential equation in advance.

Denote the solution of this by $f_\gamma(t)$, and denote the vector $f_\gamma(1)$ by $W(x)$. If the value $W(x)$, at each point x in a neighborhood of x_0, does not depend on the choice of γ, then $W(x)$ gives a solution of the original differential equation $(d-\alpha)W = 0$, because $f_\gamma(t)$ is a solution of $(d-\gamma^\star\alpha)f = 0$. In the remainder of the proof we show that $f_\gamma(1)$ does not depend on the choice of a path γ from x_0 to x.

Let γ_0 and γ_1 be two paths from x_0 to x lying in a neighborhood of x_0. Since the paths are in a neighborhood of x_0, they are homotopic. Hence, we have a map $\psi : [0,1] \times [0,1] \to X$ such that $\gamma_0([0,1]) = \psi\Big((\{0\} \times [0,1]) \cup ([0,1] \times \{1\})\Big)$ and $\gamma_1([0,1]) = \psi\Big(([0,1] \times \{0\}) \cup (\{1\} \times [0,1])\Big)$. We put $\psi^\star\alpha = p(x,y)dx + q(x,y)dy$ for a coordinate $(x,y) \in [0,1] \times [0,1]$. The integrability condition (5.11) can be reexpressed in these terms

$$\psi^\star(d\alpha - \alpha\wedge\alpha) = \Big(-\frac{\partial p(x,y)}{\partial y} + \frac{\partial q(x,y)}{\partial x} - p(x,y)q(x,y) + q(x,y)p(x,y)\Big)dxdy = 0. \tag{5.12}$$

Further, let $P(x,y)$ and $Q(x,y)$ denote the (respective) solution of the following (essentially ordinary) differential equations

$$\frac{\partial P(x,y)}{\partial x} = p(x,y)P(x,y), \qquad P(0,y) = 1, \tag{5.13}$$

$$\frac{\partial Q(x,y)}{\partial y} = q(x,y)Q(x,y), \qquad Q(x,0) = 1. \tag{5.14}$$

Then, we obtain the equality

$$\frac{\partial Q(x,y)}{\partial x} = p(x,y)Q(x,y) - Q(x,y)p(x,0) \tag{5.15}$$

as follows. Putting

$$f(x,y) = \frac{\partial Q(x,y)}{\partial x} - p(x,y)Q(x,y) + Q(x,y)p(x,0),$$

it follows from (5.14) and (5.12) that $f(x,y)$ is characterized by the following (essentially ordinary) differential equation

$$\frac{\partial f(x,y)}{\partial y} = q(x,y)f(x,y), \qquad f(x,0) = 0.$$

Hence, $f(x,y) = 0$; this implies (5.15). We consider a path γ_T, for $0 \le T \le 1$, whose image is equal to $\psi\Big(([0,T] \times \{0\}) \cup (\{T\} \times [0,1]) \cup ([T,1] \times \{1\})\Big)$. Note that the above paths γ_0 and γ_1 are obtained from γ_T by putting $T = 0, 1$ respectively. By factoring $f_{\gamma_T}(1)$ into the contributions of the integrals along $[0,T] \times \{0\}$, $\{T\} \times [0,1]$ and $([T,1] \times \{1\})$, we find that

$$f_{\gamma_T}(1) = P(1,1)P(T,1)^{-1}Q(T,1)P(T,0).$$

Noting

$$\frac{dP(T,1)^{-1}}{dT} = -P(T,1)^{-1}p(T,1),$$

we have that

$$\begin{aligned}
\frac{df_{\gamma_T}(1)}{dT} &= P(1,1)\frac{d(P(T,1)^{-1})}{dT}Q(T,1)p(T,0) \\
&\quad + P(1,1)P(T,1)^{-1}\frac{dQ(T,1)}{dT}P(T,0) \\
&\quad + P(1,1)P(T,1)^{-1}Q(T,1)\frac{dP(T,0)}{dT} = 0.
\end{aligned}$$

Hence, $f_{\gamma_T}(1)$ is constant with respect to T, and we deduce that $f_{\gamma_0}(1) = f_{\gamma_1}(1)$. Therefore, $f_\gamma(1)$ does not depend on the choice of a path γ in a neighborhood of x_0. □

Let us verify that the KZ equation has a local solution by checking the integrability condition (5.11) for the 1-form τ_n given in (5.10). It is trivial that the first term $d\tau_n$ vanishes from the definition of τ_n. We show the equality $\tau_n \wedge \tau_n = 0$ as follows. Setting

$$u_{ij} = \frac{dz_i - dz_j}{z_i - z_j}, \tag{5.16}$$

we present $\tau_n \wedge \tau_n$ by

$$\tau_n \wedge \tau_n = \left(\frac{\hbar}{2\pi\sqrt{-1}}\right)^2 \sum_{i<j,\ k<l} \tau_{ij}\tau_{kl}u_{ij} \wedge u_{kl}.$$

We split the above sum into three sums depending on the cardinality of the set of indices $\{i, j, k, l\}$; the cardinality can be 2, 3 or 4. In the case of cardinality 2, the sum vanishes by $u_{ij} \wedge u_{ij} = 0$. In the case of cardinality 4, the sum vanishes by the relation $u_{ij} \wedge u_{kl} = -u_{kl} \wedge u_{ij}$ and the commutativity (5.4) of τ_{ij} and τ_{kl} for distinct i, j, k, l. In the remaining case of cardinality 3, we have that

$$\begin{aligned}
&\sum_{\substack{i<j,\ k<l \\ |\{i,j,k,l\}|=3}} \tau_{ij}\tau_{kl}u_{ij} \wedge u_{kl} \\
&= \sum_{\substack{i<j,\ i<k \\ j\neq k}} \tau_{ij}\tau_{ik}u_{ij} \wedge u_{ik} + \sum_{\substack{i<k,\ j<k \\ i\neq j}} \tau_{ik}\tau_{jk}u_{ik} \wedge u_{jk} + \sum_{i<j<k} (\tau_{ij}\tau_{jk}u_{ij} \wedge u_{jk} + \tau_{jk}\tau_{ij}u_{jk} \wedge u_{ij}) \\
&= \sum_{i<j<k} ([\tau_{ij}, \tau_{ik}]u_{ij} \wedge u_{ik} + [\tau_{ik}, \tau_{jk}]u_{ik} \wedge u_{jk} + [\tau_{ij}, \tau_{jk}]u_{ij} \wedge u_{jk}).
\end{aligned}$$

Since, according to the relation (5.5),

$$[\tau_{ij}, \tau_{ik}] = [\tau_{ik}, \tau_{jk}] = -[\tau_{ij}, \tau_{jk}],$$

the summand of the above summation is equal to

$$[\tau_{ij}, \tau_{ik}](u_{ij} \wedge u_{ik} + u_{ik} \wedge u_{jk} - u_{ij} \wedge u_{jk}).$$

The following lemma asserts that this vanishes, completing our verification that the 1-form τ_n given in (5.10) satisfies the integrability condition (5.11).

Lemma 5.3 (Arnold's lemma **[Arn69]**). For the 1-forms u_{ij} defined in (5.16),

$$u_{ij} \wedge u_{jk} + u_{jk} \wedge u_{ki} + u_{ki} \wedge u_{ij} = 0. \tag{5.17}$$

Arnold [Arn69] showed that the cohomology ring of the configuration space X_n is generated by the 1-forms u_{ij} subject to the relation (5.17).

Proof of Lemma 5.3. The left hand side of (5.17) is equal to

$$\begin{aligned}
&u_{ij} \wedge u_{jk} + (\text{cyclic permutations of the subscripts } i, j, k) \\
&= \frac{dz_i \wedge dz_j + dz_j \wedge dz_k + dz_k \wedge dz_i}{(z_i - z_j)(z_j - z_k)} + (\text{cyclic permutations}) \\
&= \Big((z_k - z_i) + (\text{cyclic permutations})\Big) \frac{dz_i \wedge dz_j + dz_j \wedge dz_k + dz_k \wedge dz_i}{(z_i - z_j)(z_j - z_k)(z_k - z_i)} = 0.
\end{aligned}$$

Hence, we obtain (5.17). □

We now consider a closed path $\gamma : [0, 1] \to X_n$ such that $\gamma(0) = \gamma(1) = x_0$ for a base point x_0. For a vector $v \in V^{\otimes n}$, just as in the proof of Proposition 5.2, we obtain a vector $f_\gamma(1) \in V^{\otimes n}$ which is invariant under homotopies of γ. This case differs from the previous, in that we now consider a global path, whereas in the previous case we only considered a path in a neighborhood of x_0. We now consider the construction of this vector $f_\gamma(1)$. Consider a local solution W of the KZ equation with the initial value $W(\gamma(0)) = v$. By considering local solutions W of the KZ equation continuously along the path γ, we obtain a lift of γ on the graph of W as shown in Figure 5.3. As the image of the lift of γ in $V^{\otimes n}$ we obtain the path $f_\gamma : [0,1] \to V^{\otimes n}$ with the initial value $f_\gamma(0) = v$. The path f_γ is presented by $f_\gamma(t) = W(\gamma(t))$ in a small neighborhood of each point in $[0,1]$ though this formula is available only in a local sense, since there does not exist a global solution W on the whole of X_n. Note, however, that the vector $f_\gamma(1)$ does not change under homotopies of γ, since the vector does not change under a local perturbation of the path γ; the reader can verify the situation graphically in Figure 5.3. For a closed path γ in X_n with the base point x_0 we consider the map $V^{\otimes n} \to V^{\otimes n}$ taking $v = f_\gamma(0)$ to $f_\gamma(1)$. We denote this map by $\rho_{\mathrm{KZ}}(\gamma)$; this is linear, because the KZ equation is linear. By the reason mentioned above, the linear map $\rho_{\mathrm{KZ}}(\gamma)$ does not change under homotopies of γ. Since a homotopy class of γ expresses a pure braid as in (5.6), ρ_{KZ} induces a representation of the pure braid group,

$$\rho_{\mathrm{KZ}} : P_n \to \mathrm{End}(V^{\otimes n}), \tag{5.18}$$

where we denote this representation also by ρ_{KZ}, and call it the *monodromy representation* of the pure braid group P_n along solutions of the KZ equation.

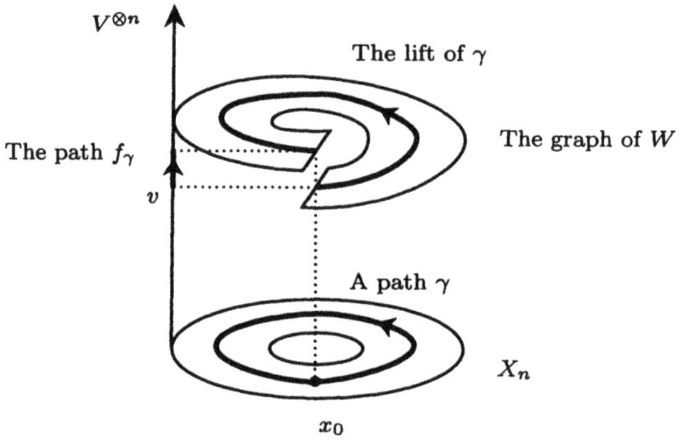

Figure 5.3 By taking the lift of a path γ in X_n along a local solution W of the KZ equation we obtain the path f in $V^{\otimes n}$.

We consider the action of the nth symmetric group $\mathfrak{S}_n$ on $V^{\otimes n}$ by permutations of the n entries of $V^{\otimes n}$. We also consider the action of $\mathfrak{S}_n$ on the configuration space X_n by permutations of the n coordinates of X_n, as mentioned before. For a closed path γ in the quotient space $X_n/\mathfrak{S}_n$, we take the lift $\tilde{\gamma}$ of γ to X_n so that $\tilde{\gamma}$ is a path from x_0 to σx_0, for some $\sigma \in \mathfrak{S}_n$. For a vector $v \in V^{\otimes n}$, similarly as above, we consider the vector $f_{\tilde{\gamma}}(1) \in V^{\otimes n}$. Further, we consider the linear map taking $v = f_{\tilde{\gamma}}(0)$ to $f_{\tilde{\gamma}}(1)$; we denote the map by $\rho_{\mathrm{KZ}}(\gamma)$. Noting that a homotopy class of γ expresses a braid as in (5.7), we obtain the representation

$$\rho_{KZ} : B_n \longrightarrow \mathrm{End}(V^{\otimes n}). \tag{5.19}$$

We call it the *monodromy representation* of the braid group B_n along solutions of the KZ equation.

5.2 Computing monodromies of the KZ equation

In this section we compute monodromies of the KZ equation (5.8) on the configuration space X_n of n points in $\mathbb{C}$ when $n = 2, 3, 4$, and obtain concrete presentations of the monodromy representations (5.19) of the braid groups B_n in these cases.

The case of $n = 2$

When $n = 2$, the KZ equation (5.9) reduces to the differential equation

$$dW = \frac{\hbar\tau}{2\pi\sqrt{-1}} \frac{dz_1 - dz_2}{z_1 - z_2} W$$

for a function $W : X_2 \to V \otimes V$ with variables $(z_1, z_2) \in X_2$. It is easy to see that any local solution of the above differential equation can be presented by

$$W(z_1, z_2) = (z_2 - z_1)^{\hbar\tau/2\pi\sqrt{-1}} v \tag{5.20}$$

for some vector $v \in V \otimes V$ in a local region, say, $\{(z_1, z_2) \in \mathbb{C}^2 \mid z_2 - z_1 \in \mathbb{C} - (-\infty, 0]\}$. Here, on the right hand side of (5.20), we regard z^A as $\exp(A \log z)$; note that the function $\log z$ can be defined as a single-valued function in the region $\mathbb{C} - (-\infty, 0]$.

Let γ be a path in X_2 given by $\gamma(t) = (0, e^{2\pi\sqrt{-1}t})$ for $t \in [0, 1]$. The monodromy of the local solution (5.20) of the KZ equation along the path γ is locally presented by $W(0, e^{2\pi\sqrt{-1}t}) = e^{\hbar\tau t} v$. By taking the variable t from 0 to 1, the vector v is taken to $e^{\hbar\tau} v$. Hence, the representation $\rho_{KZ} : P_2 \to \mathrm{End}(V \otimes V)$ of (5.18) is given by

$$\rho_{KZ}(\sigma^2) = e^{\hbar\tau}$$

where σ^2 denotes the generator of P_2. Further, by considering the path in $X_2/\mathfrak{S}_2$ given by $(0, e^{\pi\sqrt{-1}t})$ instead of the above path γ, it can be seen that the representation $\rho_{KZ} : B_2 \to \mathrm{End}(V \otimes V)$ of (5.19) is given by

$$\rho_{KZ}(\sigma) = P \circ e^{\hbar\tau/2}$$

where $P : V \otimes V \to V \otimes V$ denotes the permutation of the two entries of $V \otimes V$, which is derived from the action of $\mathfrak{S}_2$ which interchanges the two entries of $V \otimes V$ in the definition of ρ_{KZ}.

The case of $n = 3$

When $n = 3$, the KZ equation (5.9) is written

$$dW = \frac{\hbar}{2\pi\sqrt{-1}}\Big(\tau_{12} d\log(z_1 - z_2) + \tau_{13} d\log(z_1 - z_3) + \tau_{23} d\log(z_2 - z_3)\Big)W. \quad (5.21)$$

Further, putting

$$W(z_1, z_2, z_3) = (z_3 - z_1)^{\hbar(\tau_{12}+\tau_{13}+\tau_{23})/2\pi\sqrt{-1}}\tilde{G}(z_1, z_2, z_3),$$

the above differential equation becomes

$$d\tilde{G} = \frac{\hbar}{2\pi\sqrt{-1}}\Big(\tau_{12} d\log\frac{z_2 - z_1}{z_3 - z_1} + \tau_{23} d\log(\frac{z_2 - z_1}{z_3 - z_1} - 1)\Big)\tilde{G}. \quad (5.22)$$

More generally we consider the differential equation

$$dG = \frac{1}{2\pi\sqrt{-1}}\Big(A d\log z + B d\log(z - 1)\Big)G \quad (5.23)$$

for a function $G(z) \in \mathbb{C}\langle\langle A, B\rangle\rangle$, where $\mathbb{C}\langle\langle A, B\rangle\rangle$ denotes the ring of formal power series over $\mathbb{C}$ in non-commutative two variables, A and B. Any solution $\tilde{G}$ of (5.22) would be obtained as $\tilde{G}(z_1, z_2, z_3) = G\big((z_2 - z_1)/(z_3 - z_1)\big)$ putting $A = \hbar\tau_{12}$ and $B = \hbar\tau_{23}$, if we could find a solution G of the above differential equation

Lemma 5.4. There exists unique solutions $G_{(\bullet\bullet)\bullet}$ and $G_{\bullet(\bullet\bullet)}$ of the differential equation (5.23) of the forms

$$G_{(\bullet\bullet)\bullet}(z) = f(z) z^{A/2\pi\sqrt{-1}}$$
$$G_{\bullet(\bullet\bullet)}(z) = g(1 - z)(1 - z)^{B/2\pi\sqrt{-1}}$$

where $f(z)$ and $g(z)$ are analytic functions with $f(0) = g(0) = 1 \in \mathbb{C}\langle\langle A, B\rangle\rangle$ defined in a neighborhood of $0 \in \mathbb{C}$.

Proof. Consider the first formula of the lemma. The function $G_{(\bullet\bullet)\bullet}$ is a solution if the function f satisfies

$$\frac{df}{dz} - \frac{1}{z}[\overline{A}, f] = -\frac{\overline{B}}{1 - z} f,$$

where we put $\overline{A} = A/2\pi\sqrt{-1}$ and $\overline{B} = B/2\pi\sqrt{-1}$. This differential equation can be solved by the method of successive approximation, as follows. Setting $f(z) = 1 + \sum_{k=1}^{\infty} f_k z^k$ with $f_k \in \mathbb{C}\langle\langle A, B\rangle\rangle$,

$$k f_k - [\overline{A}, f_k] = -\overline{B}(1 + f_1 + \cdots + f_{k-1}).$$

Since the operator $k - \mathrm{ad}(\overline{A})$ (where we set $\big(\mathrm{ad}(\overline{A})\big)(f_k) = [\overline{A}, f_k]$) has the inverse $\sum_{i=0}^{\infty} \mathrm{ad}(\overline{A})^i / k^{i+1}$, we obtain the f_k successively by using

$$f_k = \sum_{i=0}^{\infty} \frac{1}{k^{i+1}} \mathrm{ad}(\overline{A})^i\Big(- \overline{B}(1 + f_1 + \cdots + f_{k-1})\Big).$$

For example, $f_1 = -\overline{B} - [\overline{A}, \overline{B}] - [\overline{A}, [\overline{A}, \overline{B}]] - \cdots$. Further, since a power series solution of a differential equation implies an analytic solution, we obtain the first formula of the lemma.

The second formula of the lemma can be obtained similarly. □

Since both of $G_{(\bullet\bullet)\bullet}$ and $G_{\bullet(\bullet\bullet)}$ are non-zero solutions of the linear differential equation (5.23), they are related by a constant multiple in $\mathbb{C}\langle\langle A, B\rangle\rangle$. Hence, we can define the power series $\varphi_{\text{KZ}}(A, B) \in \mathbb{C}\langle\langle A, B\rangle\rangle$ by

$$G_{(\bullet\bullet)\bullet} = G_{\bullet(\bullet\bullet)}\varphi_{\text{KZ}}(A, B). \tag{5.24}$$

See Appendix D.2 for the concrete computation of $\varphi_{\text{KZ}}(A, B)$.

Recall that Lemma 5.4 gives two particular solutions of (5.23) and that they induce solutions W of the KZ equation (5.21); concretely we obtain the following two local solutions of the KZ equation expressed with the functions f and g of Lemma 5.4:

$$W_{(\bullet\bullet)\bullet}(z_1, z_2, z_3) = G_{(\bullet\bullet)\bullet}\Big(\frac{z_2 - z_1}{z_3 - z_1}\Big)(z_3 - z_1)^{\hbar(\tau_{12}+\tau_{13}+\tau_{23})/2\pi\sqrt{-1}} \tag{5.25}$$

$$= f\Big(\frac{z_2 - z_1}{z_3 - z_1}\Big)(z_2 - z_1)^{\hbar\tau_{12}/2\pi\sqrt{-1}}(z_3 - z_1)^{\hbar(\tau_{13}+\tau_{23})/2\pi\sqrt{-1}}, \tag{5.26}$$

$$W_{\bullet(\bullet\bullet)}(z_1, z_2, z_3) = G_{\bullet(\bullet\bullet)}\Big(\frac{z_3 - z_2}{z_3 - z_1}\Big)(z_3 - z_1)^{\hbar(\tau_{12}+\tau_{13}+\tau_{23})/2\pi\sqrt{-1}} \tag{5.27}$$

$$= g\Big(\frac{z_3 - z_2}{z_3 - z_1}\Big)(z_3 - z_2)^{\hbar\tau_{23}/2\pi\sqrt{-1}}(z_3 - z_1)^{\hbar(\tau_{12}+\tau_{13})/2\pi\sqrt{-1}}, \tag{5.28}$$

putting $A = \hbar\tau_{12}$ and $B = \hbar\tau_{23}$. Since $\tau_{12} + \tau_{13} + \tau_{23}$ commutes with each of τ_{12} and τ_{23}, by (5.24) the above two solutions are related by

$$W_{(\bullet\bullet)\bullet} = W_{\bullet(\bullet\bullet)}\varphi_{\text{KZ}}(\hbar\tau_{12}, \hbar\tau_{23}). \tag{5.29}$$

We call $\varphi_{\text{KZ}}(\hbar\tau_{12}, \hbar\tau_{23})$ the *Drinfel'd associator* of $U(\mathfrak{g})^{\otimes 3}[[\hbar]]$; see also Appendix D for associators.

We consider the representation $\rho_{\text{KZ}} : P_3 \to \text{End}(V^{\otimes 3})$ of (5.18) for the base point $(0, \varepsilon, 1) \in X_3$ as follows, where ε is a constant with $0 < \varepsilon < 1$. Let γ be the path in X_3 given by $\gamma(t) = (0, \varepsilon e^{2\pi\sqrt{-1}t}, 1) \in X_3$. By (5.25),

$$W_{(\bullet\bullet)\bullet}(0, \varepsilon e^{2\pi\sqrt{-1}t}, 1) = f(\varepsilon e^{2\pi\sqrt{-1}t})\varepsilon^{\hbar\tau_{12}/2\pi\sqrt{-1}}e^{\hbar\tau_{12}t}.$$

Hence, by taking the variable t from 0 to 1, the monodromy along the path γ takes the vector $f(\varepsilon)\varepsilon^{\hbar\tau_{12}/2\pi\sqrt{-1}}v$ to $f(\varepsilon)\varepsilon^{\hbar\tau_{12}/2\pi\sqrt{-1}}e^{\hbar\tau_{12}}v$. Therefore,

$$\rho_{\text{KZ}}(\sigma_1^2) = f(\varepsilon)\varepsilon^{\hbar\tau_{12}/2\pi\sqrt{-1}}e^{\hbar\tau_{12}}\varepsilon^{-\hbar\tau_{12}/2\pi\sqrt{-1}}f(\varepsilon)^{-1}. \tag{5.30}$$

Further, let γ_1 be a path from $(0, \varepsilon, 1)$ to $(0, 1 - \varepsilon, 1)$ along the real line, γ_2 the path given by $\gamma_2(t) = (0, 1 - \varepsilon e^{2\pi\sqrt{-1}}, 1)$ and γ_3 a path from $(0, 1 - \varepsilon, 1)$ back to $(0, \varepsilon, 1)$.

By (5.25) the monodromy along the path γ_1 takes the vector $f(\varepsilon)\varepsilon^{\hbar\tau_{12}/2\pi\sqrt{-1}}$ to $f(1-\varepsilon)(1-\varepsilon)^{\hbar\tau_{12}/2\pi\sqrt{-1}}$, which is equal to $g(\varepsilon)\varepsilon^{\hbar\tau_{23}/2\pi\sqrt{-1}}\Phi_{\mathrm{KZ}}$ by (5.29), putting $\Phi_{\mathrm{KZ}} = \varphi_{\mathrm{KZ}}(\hbar\tau_{12}, \hbar\tau_{23})$. Hence, the monodromy along γ_1 is given by

$$g(\varepsilon)\varepsilon^{\hbar\tau_{23}/2\pi\sqrt{-1}}\Phi_{\mathrm{KZ}}\varepsilon^{-\hbar\tau_{12}/2\pi\sqrt{-1}}f(\varepsilon)^{-1}. \tag{5.31}$$

Further, in a similar way as the above case of γ, the monodromy along γ_2 is given by

$$g(\varepsilon)e^{\hbar\tau_{23}/2\pi\sqrt{-1}}e^{\hbar\tau_{23}}e^{-\hbar\tau_{23}/2\pi\sqrt{-1}}g(\varepsilon)^{-1}.$$

Furthermore, the monodromy along γ_3 is the inverse of (5.31). We write $\rho_{\mathrm{KZ}}(\sigma_2^2)$ as the product of three monodromies along γ_1, γ_2 and γ_3 as

$$\rho_{\mathrm{KZ}}(\sigma_2^2) = f(\varepsilon)\varepsilon^{\hbar\tau_{12}/2\pi\sqrt{-1}}\Phi_{\mathrm{KZ}}^{-1}e^{\hbar\tau_{23}}\Phi_{\mathrm{KZ}}\varepsilon^{-\hbar\tau_{12}/2\pi\sqrt{-1}}f(\varepsilon)^{-1}. \tag{5.32}$$

By taking conjugates of (5.30) and (5.32) we redefine the map $\rho_{\mathrm{KZ}} : P_3 \to \mathrm{End}(V^{\otimes 3})$ to be

$$\begin{aligned}\rho_{\mathrm{KZ}}(\sigma_1^2) &= e^{\hbar\tau_{12}},\\ \rho_{\mathrm{KZ}}(\sigma_2^2) &= \Phi_{\mathrm{KZ}}^{-1}e^{\hbar\tau_{23}}\Phi_{\mathrm{KZ}}.\end{aligned}$$

Further, by using the half of γ and γ_2 instead of γ and γ_2 respectively it is shown that the representation (5.19) of the braid group B_3 is given by

$$\begin{aligned}\rho_{\mathrm{KZ}}(\sigma_1) &= P_{12}\circ e^{\hbar\tau_{12}/2},\\ \rho_{\mathrm{KZ}}(\sigma_2) &= \Phi_{\mathrm{KZ}}^{-1}(P_{23}\circ e^{\hbar\tau_{12}/2})\Phi_{\mathrm{KZ}},\end{aligned} \tag{5.33}$$

where $P_{ij} : V^{\otimes 3} \to V^{\otimes 3}$ denotes the permutation between the ith and the jth entries of $V^{\otimes 3}$.

The case of $n = 4$

When $n = 4$, the KZ equation (5.9) is written

$$dW = \overline{\hbar}\sum_{1\le i<j\le 4}\tau_{ij}d\log(z_i - z_j)W, \tag{5.34}$$

where we put $\overline{\hbar} = \hbar/2\pi\sqrt{-1}$.

Lemma 5.5 ([Dri90], see also [Kas94, Lemma XIX.8.1]). There exist unique solutions of (5.34) of the following five forms,

$$W_{((\bullet\bullet)\bullet)\bullet} = f_1\left(\frac{z_2-z_1}{z_3-z_1}, \frac{z_3-z_1}{z_4-z_1}\right)(z_2-z_1)^{\overline{\hbar}\tau_{12}}(z_3-z_1)^{\overline{\hbar}(\tau_{13}+\tau_{23})}(z_4-z_1)^{\overline{\hbar}(\tau_{14}+\tau_{24}+\tau_{34})},$$
$$W_{(\bullet(\bullet\bullet))\bullet} = f_2\left(\frac{z_3-z_2}{z_3-z_1}, \frac{z_3-z_1}{z_4-z_1}\right)(z_3-z_2)^{\overline{\hbar}\tau_{23}}(z_3-z_1)^{\overline{\hbar}(\tau_{12}+\tau_{13})}(z_4-z_1)^{\overline{\hbar}(\tau_{14}+\tau_{24}+\tau_{34})},$$
$$W_{\bullet((\bullet\bullet)\bullet)} = f_3\left(\frac{z_3-z_2}{z_4-z_2}, \frac{z_4-z_2}{z_4-z_1}\right)(z_3-z_2)^{\overline{\hbar}\tau_{23}}(z_4-z_2)^{\overline{\hbar}(\tau_{24}+\tau_{34})}(z_4-z_1)^{\overline{\hbar}(\tau_{12}+\tau_{13}+\tau_{14})},$$
$$W_{\bullet(\bullet(\bullet\bullet))} = f_4\left(\frac{z_4-z_3}{z_4-z_2}, \frac{z_4-z_2}{z_4-z_1}\right)(z_4-z_3)^{\overline{\hbar}\tau_{34}}(z_4-z_2)^{\overline{\hbar}(\tau_{23}+\tau_{24})}(z_4-z_1)^{\overline{\hbar}(\tau_{12}+\tau_{13}+\tau_{14})},$$
$$W_{(\bullet\bullet)(\bullet\bullet)} = f_5\left(\frac{z_2-z_1}{z_4-z_1}, \frac{z_4-z_3}{z_4-z_1}\right)(z_2-z_1)^{\overline{\hbar}\tau_{12}}(z_4-z_3)^{\overline{\hbar}\tau_{34}}(z_4-z_1)^{\overline{\hbar}(\tau_{13}+\tau_{14}+\tau_{23}+\tau_{24})},$$

where $f_i(u,v)$'s are analytic functions defined in a neighborhood of $(0,0) \in \mathbb{C}^2$ with $f_i(0,0) = 1$.

Proof. We obtain the first formula of the lemma as follows. By putting

$$W(z_1,z_2,z_3,z_4) = (z_4-z_1)^{\overline{\hbar}\sum_{1\le i<j\le 4}\tau_{ij}}\tilde{G}(z_1,z_2,z_3,z_4) \tag{5.35}$$

the KZ equation (5.34) is rewritten as

$$d\tilde{G} = \overline{\hbar}\sum_{1\le i<j\le 4}\tau_{ij}\,d\log\frac{z_i-z_j}{z_4-z_1}\tilde{G}. \tag{5.36}$$

Further, by putting $u = (z_2-z_1)/(z_3-z_1)$ and $v = (z_3-z_1)/(z_4-z_1)$, the above equation becomes

$$dG = \overline{\hbar}\Big(\tau_{12}d\log uv + \tau_{13}d\log v + \tau_{23}d\log(uv-v) + \tau_{24}d\log(uv-1) + \tau_{34}d\log(v-1)\Big)G \tag{5.37}$$

for a function $G = G(u,v)$. To be precise we obtain any solution $\tilde{G}$ of (5.36) from a solution G of (5.37) as $\tilde{G}(z_1,z_2,z_3,z_4) = G((z_2-z_1)/(z_3-z_1),(z_3-z_1)/(z_4-z_1))$. Further, by putting

$$G(u,v) = f(u,v)u^{\overline{\hbar}\tau_{12}}v^{\overline{\hbar}(\tau_{12}+\tau_{13}+\tau_{23})}$$

the above differential equation reduces to the following differential equation

$$df - \overline{\hbar}\frac{du}{u}[\tau_{12},f] - \overline{\hbar}\frac{dv}{v}[\tau_{12}+\tau_{13}+\tau_{23},f] = -\Big(\tau_{23}\frac{du}{1-u} + \tau_{24}\frac{vdu+udv}{1-uv} + \tau_{34}\frac{dv}{1-v}\Big)\hbar f.$$

This total differential equation is rewritten as

$$\frac{\partial f}{\partial u} - \frac{\overline{\hbar}}{u}[\tau_{12},f] = -\Big(\frac{\tau_{23}}{1-u} + \frac{\tau_{24}v}{1-uv}\Big)\hbar f,$$
$$\frac{\partial f}{\partial v} - \frac{\overline{\hbar}}{v}[\tau_{12}+\tau_{13}+\tau_{23},f] = -\Big(\frac{\tau_{34}}{1-v} + \frac{\tau_{24}u}{1-uv}\Big)\hbar f.$$

Similarly as in the proof of Lemma 5.4, by the method of successive approximation putting $f(u,v) = \sum_{k,l\geq 0} f_{k,l}u^k v^l$ with $f_{0,0} = 1$, we can obtain the solution of the above system of differential equations. $\square$

Lemma 5.6 ([Dri90], see also [Kas94, Lemma XIX.8.2]). We have the following relations among the solutions in Lemma 5.5,

$$\begin{aligned}
W_{((\bullet\bullet)\bullet)\bullet} &= W_{(\bullet(\bullet\bullet))\bullet}\varphi_{\mathrm{KZ}}(\hbar\tau_{12}, \hbar\tau_{23}),\\
W_{(\bullet(\bullet\bullet))\bullet} &= W_{\bullet((\bullet\bullet)\bullet)}\varphi_{\mathrm{KZ}}\big(\hbar(\tau_{12}+\tau_{13}), \hbar(\tau_{24}+\tau_{34})\big),\\
W_{\bullet((\bullet\bullet)\bullet)} &= W_{\bullet(\bullet(\bullet\bullet))}\varphi_{\mathrm{KZ}}(\hbar\tau_{23}, \hbar\tau_{34}),\\
W_{((\bullet\bullet)\bullet)\bullet} &= W_{(\bullet\bullet)(\bullet\bullet)}\varphi_{\mathrm{KZ}}\big(\hbar(\tau_{13}+\tau_{23}), \hbar\tau_{34}\big),\\
W_{(\bullet\bullet)(\bullet\bullet)} &= W_{\bullet(\bullet(\bullet\bullet))}\varphi_{\mathrm{KZ}}\big(\hbar\tau_{12}, \hbar(\tau_{23}+\tau_{24})\big).
\end{aligned}$$

Proof. Since both $W_{((\bullet\bullet)\bullet)\bullet}$ and $W_{(\bullet(\bullet\bullet))\bullet}$ are non-zero solutions of the same linear differential equation, they are related by a constant multiple, say, $W_{((\bullet\bullet)\bullet)\bullet} = W_{(\bullet(\bullet\bullet))\bullet}\Phi'$. By Lemma 5.5,

$$\begin{aligned}
&f_1\Big(\frac{z_2-z_1}{z_3-z_1}, \frac{z_3-z_1}{z_4-z_1}\Big)(z_2-z_1)^{\hbar\tau_{12}}(z_3-z_1)^{\hbar(\tau_{13}+\tau_{23})}\\
&= f_2\Big(\frac{z_3-z_2}{z_3-z_1}, \frac{z_3-z_1}{z_4-z_1}\Big)(z_3-z_2)^{\hbar\tau_{23}}(z_3-z_1)^{\hbar(\tau_{12}+\tau_{13})}\Phi'.
\end{aligned}$$

By taking the limit $z_4 \to \infty$, we have that

$$\begin{aligned}
&f_1\Big(\frac{z_2-z_1}{z_3-z_1}, 0\Big)(z_2-z_1)^{\hbar\tau_{12}}(z_3-z_1)^{\hbar(\tau_{13}+\tau_{23})}\\
&= f_2\Big(\frac{z_3-z_2}{z_3-z_1}, 0\Big)(z_3-z_2)^{\hbar\tau_{23}}(z_3-z_1)^{\hbar(\tau_{12}+\tau_{13})}\Phi'.
\end{aligned}$$

Since the functions on both sides are solutions of the KZ equations on X_3 of the forms (5.25) and (5.27), they must coincide with $W_{(\bullet\bullet)\bullet}$ and $W_{\bullet(\bullet\bullet)}$ respectively. Hence, by (5.29), Φ' must be equal to $\varphi_{\mathrm{KZ}}(\hbar\tau_{12}, \hbar\tau_{23})$. Therefore, we obtain the first formula of the lemma.

We now sketch a proof of the second formula of the lemma. Just as above, we write $W_{(\bullet(\bullet\bullet))\bullet} = W_{\bullet((\bullet\bullet)\bullet)}\Phi''$. By Lemma 5.5,

$$\begin{aligned}
&f_2\Big(\frac{z_3-z_2}{z_3-z_1}, \frac{z_3-z_1}{z_4-z_1}\Big)(z_3-z_1)^{\hbar(\tau_{12}+\tau_{13})}(z_4-z_1)^{\hbar(\tau_{14}+\tau_{24}+\tau_{34})}\\
&= f_3\Big(\frac{z_3-z_2}{z_4-z_2}, \frac{z_4-z_2}{z_4-z_1}\Big)(z_4-z_2)^{\hbar(\tau_{24}+\tau_{34})}(z_4-z_1)^{\hbar(\tau_{12}+\tau_{13}+\tau_{14})}\Phi''.
\end{aligned}$$

Putting $z_2 = z_3$ and replacing z_4 with z_3 we have that

$$\begin{aligned}
&f_2\Big(0, \frac{z_2-z_1}{z_3-z_1}\Big)(z_2-z_1)^{\hbar(\tau_{12}+\tau_{13})}(z_3-z_1)^{\hbar(\tau_{14}+\tau_{24}+\tau_{34})}\\
&= f_3\Big(0, \frac{z_3-z_2}{z_3-z_1}\Big)(z_3-z_2)^{\hbar(\tau_{24}+\tau_{34})}(z_3-z_1)^{\hbar(\tau_{12}+\tau_{13}+\tau_{14})}\Phi''.
\end{aligned}$$

Applying id $\otimes \Delta \otimes$ id to the formula obtained from (5.26), (5.28), (5.29), we have that

$$f\Big(\frac{z_2 - z_1}{z_3 - z_1}\Big)(z_2 - z_1)^{\hbar(\tau_{12}+\tau_{13})}(z_3 - z_1)^{\hbar(\tau_{14}+\tau_{24}+\tau_{34})}$$
$$= g\Big(\frac{z_3 - z_2}{z_3 - z_1}\Big)(z_3 - z_2)^{\hbar(\tau_{24}+\tau_{34})}(z_3 - z_1)^{\hbar(\tau_{12}+\tau_{13}+\tau_{14})}\varphi_{\mathrm{KZ}}\big(\hbar(\tau_{12} + \tau_{13}), \hbar(\tau_{24} + \tau_{34})\big).$$

Comparing the above two formulae we find that

$$\Phi'' = \varphi_{\mathrm{KZ}}\big(\hbar(\tau_{12} + \tau_{13}), \hbar(\tau_{24} + \tau_{34})\big),$$

which implies the second formula of the lemma.

We obtain the other formulae of the lemma similarly. See [Kas94, Lemma XIX.8.2] for a detailed proof. □

By similar arguments as in the cases of $n = 2, 3$ we obtain the monodromy representation $\rho_{\mathrm{KZ}} : B_4 \to \mathrm{End}(V^{\otimes 4})$, by using Lemma 5.6, as

$$\rho_{\mathrm{KZ}}(\sigma_1) = P_{12} \circ e^{\hbar\tau_{12}/2},$$
$$\rho_{\mathrm{KZ}}(\sigma_2) = (\Phi_{\mathrm{KZ}}^{-1} \otimes 1)(P_{23} \circ e^{\hbar\tau_{12}/2})(\Phi_{\mathrm{KZ}} \otimes 1),$$
$$\rho_{\mathrm{KZ}}(\sigma_3) = (\Delta \otimes 1 \otimes 1)\Phi_{\mathrm{KZ}}^{-1} \cdot (P_{23} \circ e^{\hbar\tau_{12}/2}) \cdot (\Delta \otimes 1 \otimes 1)\Phi_{\mathrm{KZ}}.$$

Thus we have obtained concrete presentations of the monodromy representation ρ_{KZ} in the cases $n = 2, 3, 4$. To describe the monodromy representation $\rho_{\mathrm{KZ}} : B_n \to \mathrm{End}(V^{\otimes n})$ for general n we introduce the notion of a parenthesized braid in the next section.

5.3 Combinatorial reconstruction of the monodromy representations

Recall that in Lemma 5.5 we have solutions, say, $W_{((\bullet\bullet)\bullet)\bullet}$ and $W_{(\bullet(\bullet\bullet))\bullet}$ of the KZ equation. They are solutions at $(0, \varepsilon^2, \varepsilon, 1)$ and $(0, \varepsilon - \varepsilon^2, \varepsilon, 1)$, where ε is in a neighborhood of 0. It is convenient to consider their limits at $\varepsilon \to 0$ and regard the solutions as solutions around the limits. Though the limits are the same in $\mathbb{R}^4$, to distinguish such limits we consider a suitable compactification of the configuration space of n distinct points; for simplicity, n distinct points in $\mathbb{R}$. Further, we consider a path between the limits in the boundary of the compactification of the configuration space. In this section we introduce sliced parenthesized braids corresponding to paths in the boundary of the configuration space. By using parenthesized braids we give a combinatorial reconstruction of the monodromy representation ρ_{KZ} of the braid group B_n.

Compactification of configuration spaces

Let Y_n be the configuration space of n distinct points in $\mathbb{R}$, *i.e.*, we put

$$Y_n = \{(t_1, t_2, \cdots, t_n) \in \mathbb{R}^n \mid t_i \neq t_j \text{ for any } i \neq j\}.$$

The nth symmetric group $\mathfrak{S}_n$ acts on Y_n by interchanging the orders of n points. The quotient space by this action is presented by

$$Y_n/\mathfrak{S}_n = \{(t_1, t_2, \cdots, t_n) \in \mathbb{R}^n \mid t_1 < t_2 < \cdots < t_n\}.$$

Further, we consider the action of an affine transformation $t \mapsto at + b$ on the space $Y_n/\mathfrak{S}_n$. Let Y'_n be the quotient space by this action, which can be presented by

$$Y'_n = \{(t_2, \cdots, t_{n-1}) \in \mathbb{R}^{n-2} \mid 0 < t_2 < \cdots < t_{n-1} < 1\}.$$

When $n = 3$, we find that

$$Y'_3 = \{t_2 \in \mathbb{R} \mid 0 < t_2 < 1\}.$$

We define a compactification of Y'_3 as

$$\overline{Y'_3} = \{t_2 \in \mathbb{R} \mid 0 \leq t_2 \leq 1\}.$$

We denote the two 0-cells of $\overline{Y'_3}$ by $(\bullet\bullet)\bullet$ and $\bullet(\bullet\bullet)$, corresponding to $t_2 = 0$ and 1 respectively, and call $(\bullet\bullet)\bullet$ and $\bullet(\bullet\bullet)$ *parenthesized sets* of 3 dots. Further, we denote the (oriented) 1-cell of $\overline{Y'_3}$ by 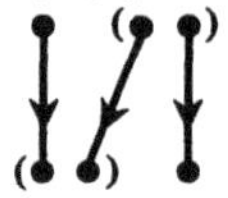 which is oriented from $\bullet(\bullet\bullet)$ to $(\bullet\bullet)\bullet$. The 1-cell with the opposite orientation is denoted by 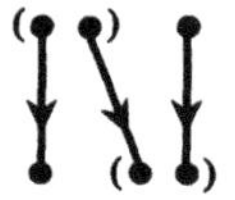. We express a path in $\overline{Y'_3}$ as a composition of copies of 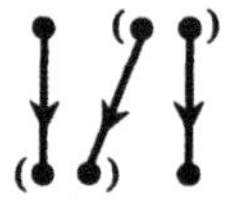 and 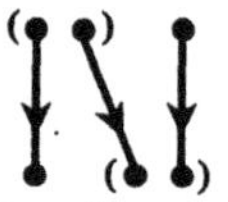. Two homotopic paths are related by the relations (5.39) in Figure 5.6, where we regard $\circ$ in the relations as $\bullet$ at present, though we regard $\circ$ in a more general sense later.

In the case $n = 4$,

$$Y'_4 = \{(t_2, t_3) \in \mathbb{R}^2 \mid 0 < t_2 < t_3 < 1\}.$$

We let $\overline{Y'_4}$ be the closure of Y'_4 embedded in $\mathbb{R}^2 \times \mathbb{R}^2$ as

$$\{(t_2, t_3; s_1, s_2) \in \mathbb{R}^2 \times \mathbb{R}^2 \mid 0 < t_2 < t_3 < 1, \quad s_1 = \frac{t_2}{t_3}, \quad s_2 = \frac{t_3 - t_2}{1 - t_2}\}.$$

This $\overline{Y'_4}$ is a closed pentagon obtained as a real blow up of the closed triangle

$$\{(t_2, t_3) \in \mathbb{R}^2 \mid 0 \leq t_2 \leq t_3 \leq 1\}$$

at $(t_2, t_3) = (0,0)$ and $(1,1)$. As in Figure 5.4, we denote the five 0-cells of $\overline{Y'_4}$ by •(•(••)), (••)(••), ((••)•)•, •((••)•), and (•(••))• corresponding to the limits of $(t_2, t_3) = (1-\varepsilon, 1-\varepsilon^2), (\varepsilon, 1-\varepsilon), (\varepsilon^2, \varepsilon), (1-\varepsilon, 1-\varepsilon+\varepsilon^2)$ and $(\varepsilon, \varepsilon+\varepsilon^2)$ in $Y'_4 \subset \overline{Y'_4}$ at $\varepsilon \to 0$, and call them *parenthesized sets* of 4 dots. We denote the five 1-cells of $\overline{Y'_4}$ by

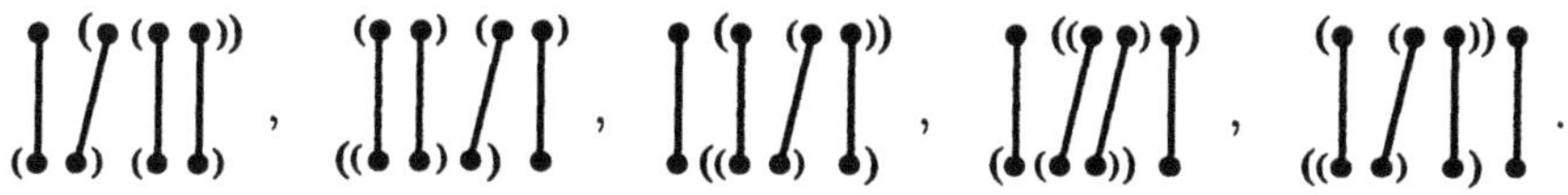
(5.38)

The 2-cell of $\overline{Y'_4}$ is expressed by the relation (5.40) in Figure 5.6, where we regard the mark ∘ there as • in this case. We express a path in the 1-skeleton of $\overline{Y'_4}$ as a composition of copies of 1-cells (5.38) and their inverses. Here the *1-skeleton* denotes the union of 0-cells and 1-cells. If two paths in the 1-skeleton of $\overline{Y'_4}$ are homotopic in $\overline{Y'_4}$, the corresponding compositions of 1-cells are related by the relations (5.39) and (5.40).

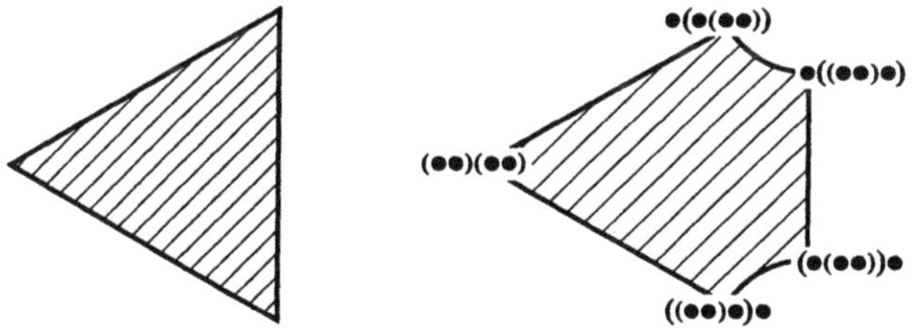

Figure 5.4 An open triangle Y'_4 and its compactification $\overline{Y'_4}$. The 2-cell of $\overline{Y'_4}$ corresponds to the pentagon relation (5.40).

In the case $n = 5$,

$$Y'_5 = \{(t_2, t_3, t_4) \in \mathbb{R}^3 \mid 0 < t_2 < t_3 < t_4 < 1\}.$$

This is an open tetrahedra as shown in the left picture of Figure 5.5. Let $\overline{Y'_5}$ be a compactification of Y'_5 shown in the right picture of Figure 5.5, which is obtained by iterating real blow ups of the closed tetrahedra

$$\{(t_2, t_3, t_4) \in \mathbb{R}^3 \mid 0 \le t_2 \le t_3 \le t_4 \le 1\}.$$

The pentagonal 2-cells of $\overline{Y'_5}$ correspond to the pentagon relation (5.40), where the copies of  in the relation denote the bundle of two parallel vertical lines. The square 2-cells of $\overline{Y'_5}$ correspond to the relation (5.41) in Figure 5.6, where either of

T_1, T_2 or T_3 in the relation denotes 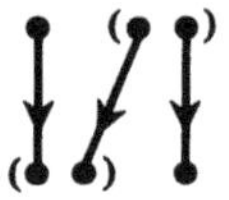 in this case.

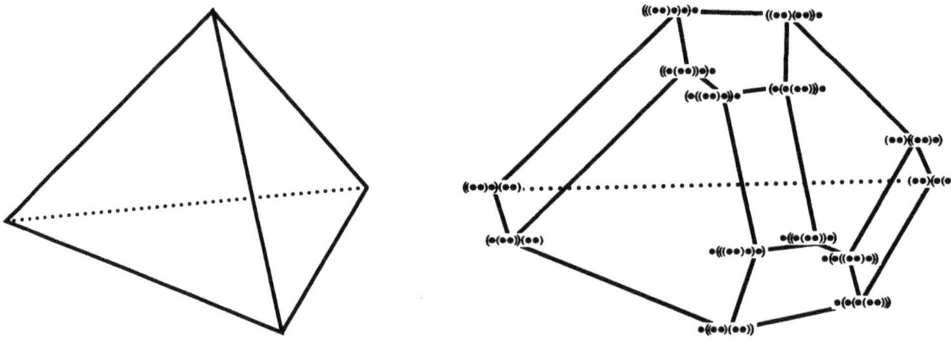

Figure 5.5 An open tetrahedra Y_5' and its compactification $\overline{Y_5'}$. The pentagonal and square 2-cells on the boundary of $\overline{Y_5'}$ correspond to the pentagon relation (5.40) and the relation (5.41) respectively.

In the case of general n the compactification $\overline{Y_n'}$ of Y_n' is described as follows. A 0-cell of $\overline{Y_n'}$ is given by a parenthesized set of n dots, where a *parenthesized set of n dots* is recursively obtained from a parenthesized set of $n-1$ dots by replacing one • by (••); recall that the parenthesized sets of 3 dots are (••)• and •(••). A *trivial parenthesized bundle* of vertical lines is the direct product of a vertical interval and a parenthesized set of dots such as

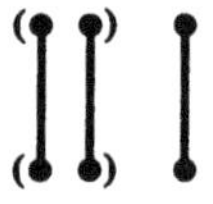 , 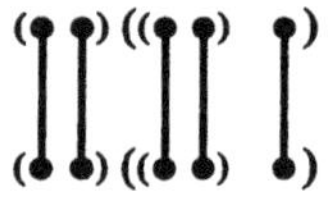 .

Further, an *elementary parenthesized bundle* of vertical lines is a parenthesized bundle of vertical lines obtained by putting trivial parenthesized bundles of vertical lines at both sides of either of 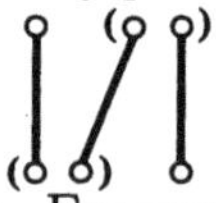 or 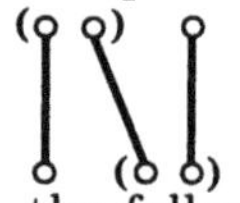 where each  denotes a trivial bundle of vertical lines. For example, the following diagram is such a diagram.

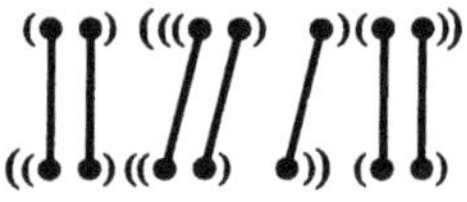 .

A 1-cell of $\overline{Y_n'}$ (with an orientation) is expressed by an elementary parenthesized bundle of n vertical lines. Further, 2-cells of $\overline{Y_n'}$ are expressed by the relations

(5.40), (5.41) and (5.42) in Figure 5.6, where each  in the three relations (5.40), (5.41) and (5.42) denotes a trivial parenthesized bundle of vertical lines. Further, either of the T_i's in each of the relations (5.41) and (5.42) denotes an elementary parenthesized bundle of vertical lines.

We express a path in the 1-skeleton of $\overline{Y'_n}$ as a composition of copies of elementary parenthesized bundles of vertical lines.

Proposition 5.7. Any two compositions of elementary parenthesized bundle of vertical lines with the same end sets of parenthesized dots are related by a finite sequence of the relations (5.39), (5.40), (5.41) and (5.42) in Figure 5.6.

This proposition is equivalent to the coherence theorem of the category of monoids [Mac71]. It is later extended to Theorem 5.8, which implies the coherence theorem for a braided tensor category [MoSe89], see also [Kas94, BaKi01].

Proof of Proposition 5.7. Two such compositions give two paths in the 1-skeleton of $\overline{Y'_n}$ with the same end sets of parenthesized dots. Hence, we obtain a loop as the union of the two paths in the 1-skeleton of $\overline{Y'_n}$. Since $\overline{Y'_n}$ is homeomorphic to an $(n-2)$-ball, it is simply connected, as is the 2-skeleton of $\overline{Y'_n}$. Hence, the loop must be null-homotopic in the 2-skeleton of $\overline{Y'_n}$. This implies that the two composition of the proposition are related by a finite sequence of relations corresponding to 2-cells of $\overline{Y'_n}$, which are the relations of the proposition. □

Parenthesized braids

A *parenthesized braid* in n strands is a braid in n strands such that each end of the braid is a parenthesized set of n dots. Some simple examples of parenthesized braids are

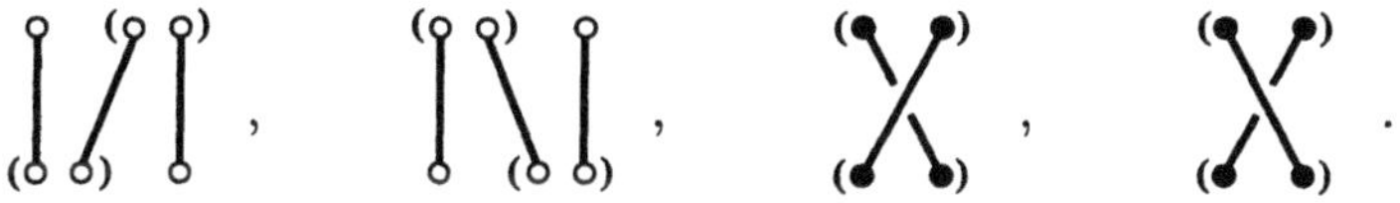

A *sliced parenthesized braid* is a composition of tensor products of elementary parenthesized braids and trivial parenthesized bundles of vertical lines. Any parenthesized braid is expressed by a sliced parenthesized braid.

Theorem 5.8. Fix a parenthesized set of n dots. Then, two sliced parenthesized braids whose ends are equal to the fixed parenthesized set of n dots express isotopic braids with their parentheses forgotten, if and only if they are related by a sequence of the moves in Figure 5.6.

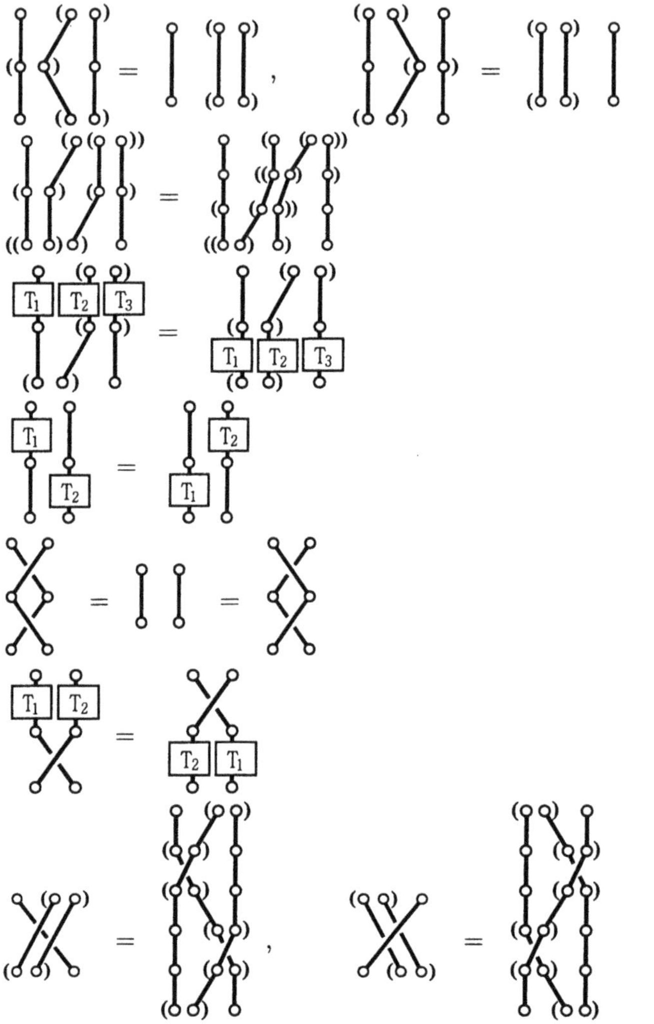

(5.39)

(5.40)

(5.41)

(5.42)

(5.43)

(5.44)

(5.45)

Figure 5.6 The moves for sliced parenthesized braids. Each T_i in the relations denotes any elementary parenthesized braid. We call the relations (5.40) and (5.45) the *pentagon* and *hexagon relations* respectively.

This theorem has the following symbolic representation,

$$B_n \cong \left\{ \begin{array}{l} \text{sliced parenthesized braids} \\ \text{whose ends are equal to} \\ \text{the fixed parenthesized set of } n \text{ dots} \end{array} \right\} \Big/ \text{the relations in Figure 5.6.}$$

Proof of Theorem 5.8. It is clear that, if two sliced parenthesized braids are related by the moves in Figure 5.6, then they express isotopic braids.

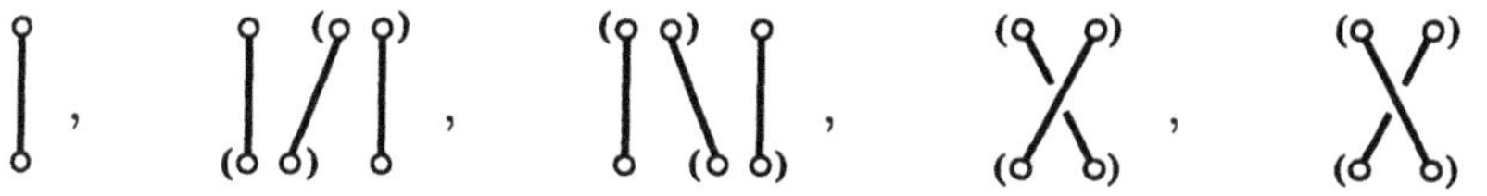

Here each  denotes a trivial parenthesized bundle of vertical lines.

Figure 5.7 The *elementary parenthesized braids.*

Conversely, suppose that two sliced parenthesized braids express isotopic braids. Then, they are related by the moves of Proposition 5.7 and the defining relations

$$\sigma_i \sigma_j = \sigma_j \sigma_i, \qquad (|i-j| \geq 2), \tag{5.46}$$

$$\sigma_i \sigma_{i+1} \sigma_i = \sigma_{i+1} \sigma_i \sigma_{i+1} \tag{5.47}$$

of the braid groups. The moves of Proposition 5.7 are given in Figure 5.6. Then, it is sufficient to show that the relations (5.46) and (5.47) are derived from these moves. Firstly, the relation (5.46) is obtained by (5.41). Secondly, the relation (5.47) is obtained

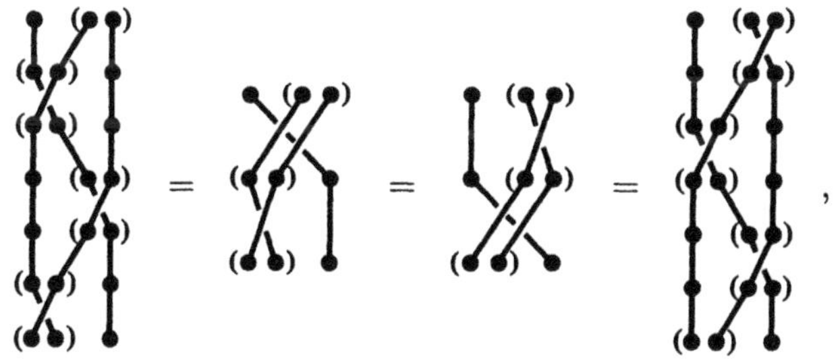

where the second equality is derived from (5.44) and the first and third equalities are derived from (5.45). □

Combinatorial reconstruction of the monodromy representations

As in the previous section we put $\Phi_{\mathrm{KZ}} = \varphi_{\mathrm{KZ}}(\hbar\tau_{12}, \hbar\tau_{23})$. Further, we put $\mathcal{R}_{\mathrm{KZ}} = \exp(\hbar\tau/2)$.

In the case of $n = 3$, by using parenthesized braids, the monodromy representation $\rho_{\mathrm{KZ}} : B_3 \to \mathrm{End}(V^{\otimes 3})$ given in (5.33) is presented by

$\rho_{\mathrm{KZ}}($ 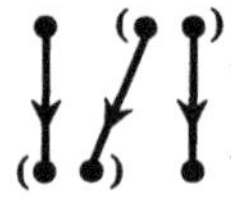 $) = \Phi_{\mathrm{KZ}},$

$\rho_{\mathrm{KZ}}($ 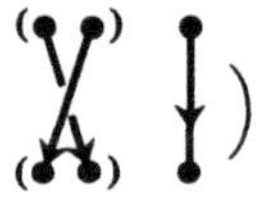 $) = (P \circ \mathcal{R}_{\mathrm{KZ}}) \otimes 1,$

$\rho_{\mathrm{KZ}}($ 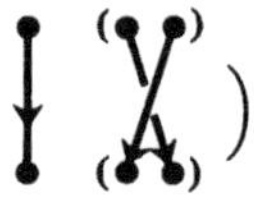 $) = 1 \otimes (P \circ \mathcal{R}_{\mathrm{KZ}}),$

where P denotes the permutation of two entries of $V \otimes V$.

In the case $n = 4$ we can express the monodromy representation $\rho_{\mathrm{KZ}} : B_4 \to \mathrm{End}(V^{\otimes 4})$ as follows. The first and second formulae of Lemma 5.6 imply that

$\rho_{\mathrm{KZ}}($ 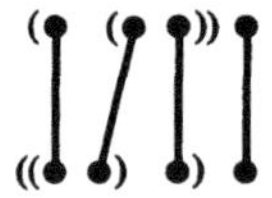 $) = \Phi_{\mathrm{KZ}} \otimes 1,$

$\rho_{\mathrm{KZ}}($ 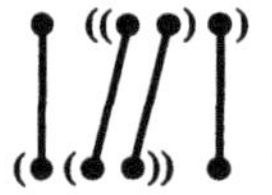 $) = (\mathrm{id} \otimes \Delta \otimes \mathrm{id})\Phi_{\mathrm{KZ}}.$

The values of ρ_{KZ} on the other parenthesized braids in (5.38) are defined similarly, to be Φ_{KZ} followed by appropriate applications of the comultiplication Δ and tensor products with 1. Furthermore, the values of ρ_{KZ} on the generators of B_4 are given by a tensor product of $P \circ \mathcal{R}$ and two 1's such as

$\rho_{\mathrm{KZ}}($ 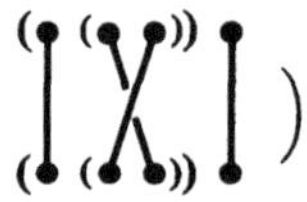 $) = 1 \otimes (P \circ \mathcal{R}_{\mathrm{KZ}}) \otimes 1.$

Extending the above cases of small n, we obtain the following explicit presentation of the monodromy representation $\rho_{\mathrm{KZ}} : B_n \to \mathrm{End}(V^{\otimes n})$, for general n. We present ρ_{KZ} for two simple parenthesized braids by

$\rho_{\mathrm{KZ}}($ 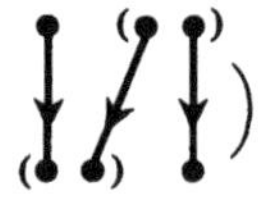 $) = \Phi_{\mathrm{KZ}},$

$\rho_{\mathrm{KZ}}($  $) = P \circ \mathcal{R}_{\mathrm{KZ}}.$

We extend these formulae by taking tensor products with 1 and by actions of the comultiplication Δ as in the above cases of $n = 3, 4$. When we add a trivial parenthesized braid beside a given parenthesized braid, we take an appropriate tensor product with 1 to obtain the value of ρ_{KZ}. Further, when we replace a strand of a given braid with two parallel copies of the strand, we apply the comultiplication Δ

to the corresponding factor to obtain the value of ρ_{KZ}. For example,

$\rho_{\mathrm{KZ}}\Big($ 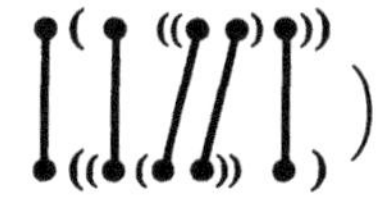 $\Big) = 1 \otimes \Big((\mathrm{id} \otimes \Delta \otimes \mathrm{id})\Phi_{\mathrm{KZ}}\Big),$

$\rho_{\mathrm{KZ}}\Big($ 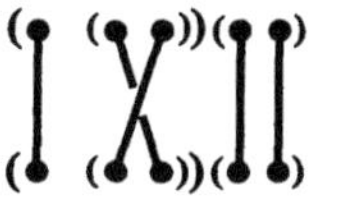 $\Big) = 1 \otimes (P \circ \mathcal{R}) \otimes 1 \otimes 1.$

By composing such values of parenthesized braids we obtain a combinatorial reconstruction of the monodromy representation $\rho_{\mathrm{KZ}} : B_n \to \mathrm{End}(V^{\otimes n})$. Key data of the reconstruction are the elements Φ_{KZ} and $\mathcal{R}_{\mathrm{KZ}}$. They are reformulated in a quasi-triangular quasi-bialgebra in the next section.

5.4 Quasi-triangular quasi-bialgebra

In this section we introduce the notion of a quasi-triangular quasi-bialgebra. The notion has been extended to the notions of a quasi-triangular quasi-Hopf algebra and a ribbon quasi-Hopf algebra, though we omit them in this section; for quasi-triangular quasi-Hopf algebra see [Dri89b] and for ribbon quasi-Hopf algebra see [AlCo92]. Further, we show that $U(\mathfrak{g})[[\hbar]]$, equipped with the elements Φ_{KZ} and $\mathcal{R}_{\mathrm{KZ}}$ given in the previous section, forms a quasi-triangular quasi-bialgebra.

The notion of a bialgebra is obtained from the notion of a Hopf algebra by forgetting the antipode as follows. Let A be an associative algebra over $\mathbb{C}$, with a unit element 1. If there are algebra homomorphisms $\Delta : A \to A \otimes A$ and $\varepsilon : A \to \mathbb{C}$ satisfying the following relations,

$$(\mathrm{id} \otimes \Delta) \circ \Delta = (\Delta \otimes \mathrm{id}) \circ \Delta, \tag{5.48}$$

$$(\varepsilon \otimes \mathrm{id}) \circ \Delta = \mathrm{id} = (\mathrm{id} \otimes \varepsilon) \circ \Delta, \tag{5.49}$$

then we call (A, Δ, ε) a *bialgebra*. We call Δ the *comultiplication* and ε the *counit* of the bialgebra like the terminology used for Hopf algebras.

Further, the notion of a quasi-bialgebra is obtained from the notion of a bialgebra by replacing coassociativity (5.48) of the algebra with quasi-coassociativity, as follows. Let A be an associative algebra over $\mathbb{C}$, with a unit element 1. If there are algebra homomorphisms $\Delta : A \to A \otimes A$, $\varepsilon : A \to \mathbb{C}$ and an invertible element $\Phi \in A \otimes A \otimes A$ satisfying the following relations,

$$(\mathrm{id} \otimes \Delta)(\Delta(x)) = \Phi \cdot (\Delta \otimes \mathrm{id})(\Delta(x)) \cdot \Phi^{-1}, \tag{5.50}$$

$$(\mathrm{id} \otimes \mathrm{id} \otimes \Delta)\Phi \cdot (\Delta \otimes \mathrm{id} \otimes \mathrm{id})\Phi = 1 \otimes \Phi \cdot (\mathrm{id} \otimes \Delta \otimes \mathrm{id})\Phi \cdot \Phi \otimes 1, \tag{5.51}$$

$$(\varepsilon \otimes \mathrm{id}) \circ \Delta = \mathrm{id} = (\mathrm{id} \otimes \varepsilon) \circ \Delta, \tag{5.52}$$

$$(\mathrm{id} \otimes \varepsilon \otimes \mathrm{id})\Phi = 1 \otimes 1, \tag{5.53}$$

then we call $(A, \Delta, \varepsilon, \Phi)$ a *quasi-bialgebra*. We call Δ the *comultiplication*, ε the *counit* and Φ an *associator* of the quasi-bialgebra. Further, we call the relation (5.51) the *pentagon relation* for the associator Φ. Note that the relations (5.51)–(5.53) imply that

$$(\varepsilon \otimes \mathrm{id} \otimes \mathrm{id})\Phi = (\mathrm{id} \otimes \mathrm{id} \otimes \varepsilon)\Phi = 1 \otimes 1.$$

We consider the algebra $U(\mathfrak{g})[[\hbar]]$ which is the power series ring of $\hbar$ with coefficients in the universal enveloping algebra $U(\mathfrak{g})$ of a Lie algebra $\mathfrak{g}$. As usual we define the comultiplication Δ and the counit ε by $\Delta(X) = X \otimes 1 + 1 \otimes X$ and $\varepsilon(X) = 0$ for any $X \in \mathfrak{g}$. Further, we consider the Drinfel'd associator $\Phi_{\mathrm{KZ}} = \varphi_{\mathrm{KZ}}(\hbar\tau_{12}, \hbar\tau_{23})$ given in (5.29).

Lemma 5.9. $(U(\mathfrak{g})[[\hbar]], \Delta, \varepsilon, \Phi_{\mathrm{KZ}})$ forms a quasi-bialgebra.

Proof. We verify that $(U(\mathfrak{g})[[\hbar]], \Delta, \varepsilon, \Phi_{\mathrm{KZ}})$ satisfies the defining relations (5.50)–(5.53) of a quasi-bialgebra.

We show (5.50) as follows. In Section 5.1 we showed that τ commutes with $\Delta(X)$ for any $X \in \mathfrak{g}$. Hence, τ_{12} commutes with $(\Delta \otimes \mathrm{id})\Delta(X)$ and τ_{23} commutes with $(\mathrm{id} \otimes \Delta)\Delta(X)$. Therefore, Φ_{KZ} commutes with $(\Delta \otimes \mathrm{id})\Delta(X)$ for any $X \in \mathfrak{g}$. Further, from the definition of Δ, $(\Delta \otimes \mathrm{id}) \circ \Delta = (\mathrm{id} \otimes \Delta) \circ \Delta$. Hence, we obtain (5.50).

We show (5.51) as follows. By Lemma 5.6 two solutions $W_{((\bullet\bullet)\bullet)\bullet}$ and $W_{\bullet(\bullet(\bullet\bullet))}$ of the KZ equation are related by each side of (5.51) for the associator Φ_{KZ}. Hence, (5.51) is solved by Φ_{KZ}.

We obtain (5.52) from the definitions of Δ and ε.

We obtain (5.53) as follows. Since $\tau = \sum I_\mu \otimes I_\mu$, as in Section 5.1, we find that $(\mathrm{id} \otimes \varepsilon \otimes \mathrm{id})\tau_{12} = (\mathrm{id} \otimes \varepsilon \otimes \mathrm{id})\tau_{23} = 0$ from the definition of ε. Since $\varphi_{\mathrm{KZ}}(0,0) = 1$, we obtain (5.53). $\square$

Let $(A, \Delta, \varepsilon, \Phi)$ be a quasi-bialgebra. If there is an invertible element $\mathcal{R} \in A \otimes A$ satisfying the following relations

$$(P \circ \Delta)(x) = \mathcal{R}\Delta(x)\mathcal{R}^{-1} \tag{5.54}$$
$$(\Delta \otimes \mathrm{id})\mathcal{R} = \Phi_{312}\mathcal{R}_{13}\Phi_{132}^{-1}\mathcal{R}_{23}\Phi \tag{5.55}$$
$$(\mathrm{id} \otimes \Delta)\mathcal{R} = \Phi_{231}^{-1}\mathcal{R}_{13}\Phi_{213}\mathcal{R}_{12}\Phi^{-1}, \tag{5.56}$$

then $(A, \Delta, \varepsilon, \Phi, \mathcal{R})$ is called *quasi-triangular*. The relations (5.55) and (5.56) are called the *hexagon relations*. Here the map P denotes the permutation given by $P(x \otimes y) = y \otimes x$ and we put $\mathcal{R}_{12} = \mathcal{R} \otimes 1$, $\mathcal{R}_{23} = 1 \otimes \mathcal{R}$ and $\mathcal{R}_{13} = \sum \alpha_i \otimes 1 \otimes \beta_i$, where $\mathcal{R} = \sum \alpha_i \otimes \beta_i$. Further, Φ_{ijk} denotes Φ after the permutation of the entries of $A \otimes A \otimes A$ putting p_l, q_l and r_l at the ith, jth and kth entries respectively, where $\Phi = \sum_l p_l \otimes q_l \otimes r_l$. For example, $\Phi_{312} = \sum_l q_l \otimes r_l \otimes p_l$.

For the quasi-bialgebra $(U(\mathfrak{g})[[\hbar]], \Delta, \varepsilon, \Phi_{\mathrm{KZ}})$ in Lemma 5.9 we put $\mathcal{R}_{\mathrm{KZ}} = \exp(\hbar\tau/2)$ for the invariant 2-tensor $\tau \in U(\mathfrak{g})^{\otimes 2}$ dual to the Killing form. Then we have

Lemma 5.10. $(U(\mathfrak{g})[[\hbar]], \Delta, \varepsilon, \Phi_{\mathrm{KZ}}, \mathcal{R}_{\mathrm{KZ}})$ forms a quasi-triangular quasi-bialgebra.

Sketch of the proof. We verify that $(U(\mathfrak{g})[[\hbar]], \Delta, \varepsilon, \Phi_{\mathrm{KZ}}, \mathcal{R}_{\mathrm{KZ}})$ satisfies the defining relations (5.54), (5.55) and (5.56) of a quasi-triangular quasi-bialgebra.

We show (5.54) as follows. The comultiplication Δ of $U(\mathfrak{g})[[\hbar]]$ is co-commutative, *i.e.*, $P \circ \Delta = \Delta$ from the definition of the Δ of $U(\mathfrak{g})[[\hbar]]$. Further, since the invariant 2-tensor τ is presented by $\tau = \frac{1}{2}(\Delta(C) - C \otimes 1 - 1 \otimes C)$ for a central element C of $U(\mathfrak{g})$, the invariant 2-tensor τ commutes with $\Delta(x)$ for any $x \in U(\mathfrak{g})$. Hence, so does $\mathcal{R}$. Thus, (5.54) is obtained.

A sketch of the proof of the hexagon relation (5.55) is as follows. We consider the local solutions $W_{(12)3}$ and $W_{3(12)}$ of the KZ equation on the configuration space X_3 of 3 points z_1, z_2, z_3 such that their asymptotic behaviors at $|z_2 - z_1| \ll |z_3 - z_1| < 1$ are given by

$$W_{(12)3}(z_1, z_2, z_3) \sim (z_2 - z_1)^{\hbar \tau_{12}} (z_3 - z_1)^{\hbar(\tau_{13} + \tau_{23})},$$
$$W_{3(12)}(z_1, z_2, z_3) \sim (z_2 - z_1)^{\hbar \tau_{12}} (z_2 - z_3)^{\hbar(\tau_{13} + \tau_{23})}.$$

They have the following relation

$$W_{(12)3}(z_1, z_2, z_3) = W_{3(12)}(z_1, z_2, z_3) \Phi_{312} \mathcal{R}_{13} \Phi_{132}^{-1} \mathcal{R}_{23} \Phi,$$

which is obtained by computing the monodromy along the path in X_3 corresponding to the parenthesized braid on the right side of the following equation.

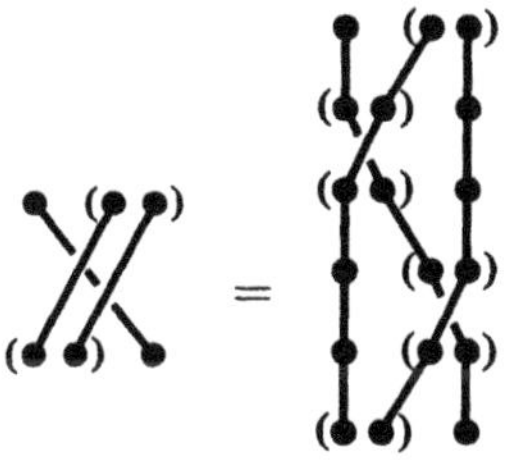

Further, by computing the monodromy along the path given by the parenthesized braid on the left side of the equation, we have that

$$W_{(12)3}(z_1, z_2, z_3) = W_{3(12)}(z_1, z_2, z_3)(\Delta \otimes \mathrm{id})\mathcal{R}.$$

Hence, we obtain (5.55).

The relation (5.56) is obtained similarly.

See [Dri90] and [Kas94, Section XIX.8] for detailed proofs. □

Proposition 5.11. Let $(A, \Delta, \varepsilon, \Phi, \mathcal{R})$ be a quasi-triangular quasi-bialgebra and V an A module. Then, they induce a representation of the braid group B_n into $\mathrm{End}(V^{\otimes n})$.

Proof. We denote the action of A on V by $\rho_V : A \to \mathrm{End}(V^{\otimes n})$. The representation

ρ of B_n is obtained as follows. We put

$\rho\Big($  $\Big) = \rho_V^{\otimes 3}(\Phi),$

$\rho\Big($  $\Big) = P \circ ((\rho \otimes \rho)(\mathcal{R})).$

Extending these formula by the comultiplication Δ and tensor products with 1, just as in the previous setion, we obtain the required representation ρ. □

5.5 Relation to braid representations derived from the quantum group

The aim of this section is to show Theorem 5.16 which asserts the equivalence between two representations of the braid group B_n; one is the monodromy representation ρ_{KZ} along solutions of the KZ equation and the other is the representation provided by the universal R matrix of the quantum group. We follow [Kas94] in this section. For details on topics in this section see [Kas94, EFK98, ChPr95].

Let $(A, \Delta, \varepsilon, \Phi, \mathcal{R})$ be a quasi-triangular quasi-bialgebra. For an invertible element $f \in A \otimes A$ we define $\Delta^f(x)$, Φ^f, and $\mathcal{R}^f$ obtained by *twisting* Δ, Φ, and $\mathcal{R}$ by f by setting

$$\begin{aligned}
\Delta^f(x) &= f \cdot \Delta(x) \cdot f^{-1}, \\
\Phi^f &= f_{23} \cdot (\mathrm{id} \otimes \Delta) f \cdot \Phi \cdot (\Delta \otimes \mathrm{id}) f^{-1} \cdot f_{12}^{-1}, \\
\mathcal{R}^f &= f_{21} \cdot \mathcal{R} \cdot f^{-1},
\end{aligned}$$

where f_{ij} denotes f acting in the ith and jth entries of $A \otimes A \otimes A$. In a similar way as in Chapter 4 they are shown graphically as

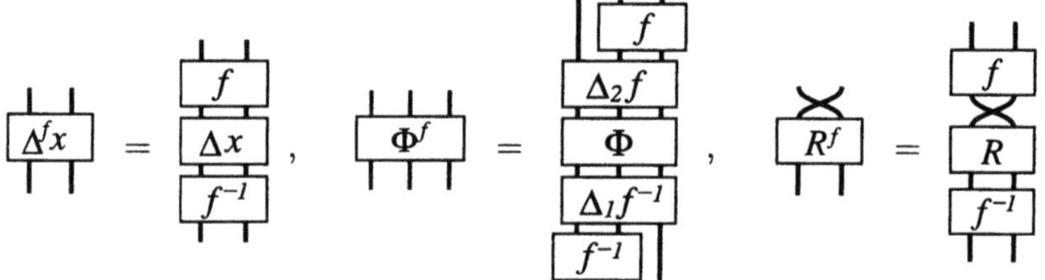

(5.57)

where Δ_i denotes Δ acting on the ith strand.

Proposition 5.12. For a quasi-triangular quasi-bialgebra $(A, \Delta, \varepsilon, \Phi, \mathcal{R})$, the algebra $(A, \Delta^f, \varepsilon, \Phi^f, \mathcal{R}^f)$ obtained by twisting by f also forms a quasi-triangular quasi-bialgebra.

Proof. It is sufficient to show that $(A, \Delta^f, \varepsilon, \Phi^f, \mathcal{R}^f)$ satisfies the defining relations of a quasi-triangular quasi-bialgebra. We show (5.50) and (5.54); it is left to the reader to show the other relations.

The relation (5.50) is shown graphically as

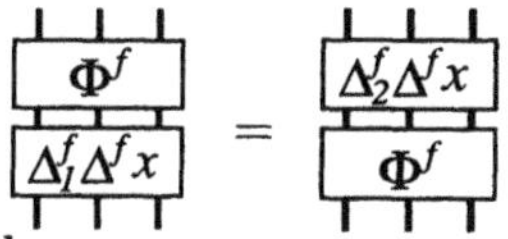

(5.58)

The left hand side is computed as

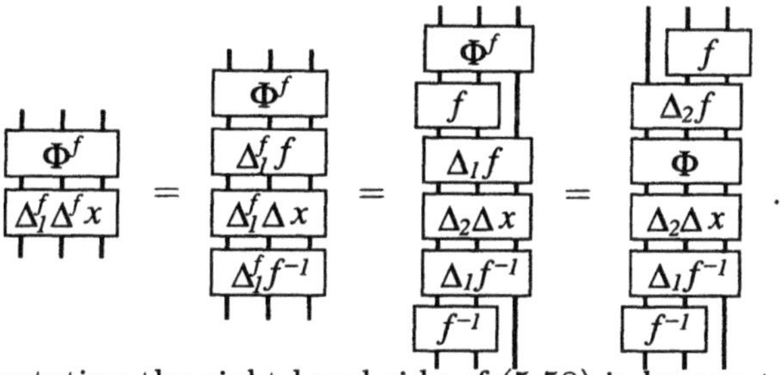

By a similar computation the right hand side of (5.58) is known to be equal to the right hand side of the above formula. Hence, we obtain (5.50).

As for (5.54), we have that

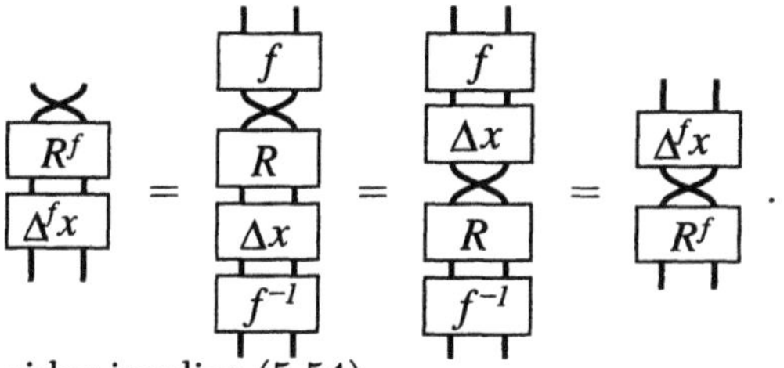

The equality of both sides implies (5.54). □

When the associator Φ is equal to 1, the notion of a quasi-triangular quasi-bialgebra becomes the notion of a quasi-triangular bialgebra. As mentioned in the previous chapter the quantum group $U_q(\mathfrak{g})$ is a quasi-triangular Hopf algebra (hence, is a quasi-triangular bialgebra) with the comultiplication, the counit and the universal R matrix of $U_q(\mathfrak{g})$, which we denote by Δ_q, ε_q and $\mathcal{R}_q$ here. Therefore, $(U_q(\mathfrak{g}), \Delta_q, \varepsilon_q, 1, \mathcal{R}_q)$ forms a quasi-triangular quasi-bialgebra. Further, $(U(\mathfrak{g})[[\hbar]]$, Δ, ε, Φ_{KZ}, $\mathcal{R}_{\mathrm{KZ}})$ is also a quasi-triangular quasi-bialgebra, as shown in Lemma 5.10.

Proposition 5.13. $(U(\mathfrak{g})[[\hbar]], \Delta, \varepsilon, \Phi_{\mathrm{KZ}}, \mathcal{R}_{\mathrm{KZ}})$ is obtained from $(U_q(\mathfrak{g}), \Delta_q, \varepsilon_q, 1, \mathcal{R}_q)$ by twisting by some $f \in U(\mathfrak{g})^{\otimes 2}[[\hbar]]$, regarding the algebra $U_q(\mathfrak{g})$ as $U(\mathfrak{g})[[\hbar]]$ putting $q = e^{\hbar}$.

Drinfel'd showed, not only the equivalence between the two quasi-triangular quasi-bialgebras, but also the equivalence among any (suitable) quasi-triangular quasi-bialgebra structures. We sketch a proof below, following [Kas94]; for detailed proofs see also [Dri89b, Dri90] and their exposition [ChPr95].

Sketch of the proof of Proposition 5.13. It is known (see, *e.g.*, [Kas94, ChPr95]) that there exists a unique algebra isomorphism $\alpha : U_q(\mathfrak{g}) \to U(\mathfrak{g})[[\hbar]]$ such that α is congruent to the identity map of $\mathfrak{g}$ modulo $\hbar$, putting $q = e^{\hbar}$. Further, we have that

$$\alpha(v) = e^{-\hbar C/2}, \tag{5.59}$$

where v is the canonical central element of $U_q(\mathfrak{g})$ such that $(U_q(\mathfrak{g}), \mathcal{R}_q, v)$ forms a ribbon Hopf algebra, and C is the Casimir element of $U(\mathfrak{g})$, which is also central in $U(\mathfrak{g})$. (The two sides of (5.59) are equal to the invariants of -1 full twist in $U_q(\mathfrak{g})$ and $U(\mathfrak{g})[[\hbar]]$ respectively. Hence, they should be equal.) The equation (5.59) is proved by showing that the two sides have the same action on all highest weight modules (for details, see [Dri89a, Proposition 5.1]). For example, when $\mathfrak{g} = sl_2$, the center of $U_q(\mathfrak{g})$ (resp. $U(\mathfrak{g})[[\hbar]]$) is generated by v (resp. C), and it is shown by elementary computations that v and $e^{-\hbar C/2}$ act on the n-dimensional irreducible module V_n by the multiple of the same scalar $q^{-(n^2-1)/4}$; this implies (5.59) in this case (see [Dri89a]).

By the algebra isomorphism α, the quasi-triangular quasi-bialgebra $(U_q(\mathfrak{g}), \Delta_q, \varepsilon_q, 1, \mathcal{R}_q)$ is isomorphic to $(U(\mathfrak{g})[[\hbar]], \Delta', \varepsilon', 1, (\alpha\otimes\alpha)\mathcal{R}_q)$ for some Δ' and ε'. The proposition is proved by twisting it in 3 steps; firstly, to $(U(\mathfrak{g})[[\hbar]], \Delta, \varepsilon, \Phi', \mathcal{R}')$ for some Φ' and $\mathcal{R}'$, and secondly, to $(U(\mathfrak{g})[[\hbar]], \Delta, \varepsilon, \Phi'', \mathcal{R}_{\mathrm{KZ}})$ for some Φ'', and thirdly, to $(U(\mathfrak{g})[[\hbar]], \Delta, \varepsilon, \Phi_{\mathrm{KZ}}, \mathcal{R}_{\mathrm{KZ}})$. To show these steps, we will use (5.59) and the fact that v satisfies a defining relation (4.25) of a ribbon Hopf algebra for the universal R matrix $\mathcal{R}_q$ for $U_q(\mathfrak{g})$; these are the only properties of $U_q(\mathfrak{g})$ which we use in the remainder of this proof.

The first step is shown in Lemma 5.15 below, which says that there exists some f' such that $\Delta = \Delta'^{f'}$. Putting $\mathcal{R}' = \big((\alpha\otimes\alpha)\mathcal{R}_q\big)^{f'}$ and $\Phi' = 1^{f'}$, we complete this step.

The second step is shown as follows. It is sufficient to show that $\mathcal{R}_{\mathrm{KZ}}$ (which was defined to be $e^{\hbar\tau/2}$) can be obtained from $\mathcal{R}'$ by twisting by some f'', keeping Δ unchanged. Putting $f'' = \big(\mathcal{R}'(\mathcal{R}'_{21}\mathcal{R}')^{-1/2}\big)^{1/2}$, it follows from some computation (see [Kas94, Proposition XVI.6.2]) that $\mathcal{R}'^{f''}$ is symmetric, where we call $\rho \in U(\mathfrak{g})^{\otimes 2}[[\hbar]]$ *symmetric* if ρ is invariant under the exchange of the two entries of $U(\mathfrak{g})^{\otimes 2}$. Then,

$$\begin{aligned}
(\mathcal{R}'^{f''})^2 &= (\mathcal{R}'^{f''})_{21}\mathcal{R}'^{f''} = \big((\alpha\otimes\alpha)(\mathcal{R}_q)_{21}\big)^{f''_{21}f'_{21}}\big((\alpha\otimes\alpha)\mathcal{R}_q\big)^{f''f'} \\
&= f''f'(\alpha\otimes\alpha)\big((\mathcal{R}_q)_{21}\mathcal{R}_q\big)f'^{-1}f''^{-1} \\
&= f''f'(\alpha\otimes\alpha)\big(\Delta_q(v^{-1})\cdot(v\otimes v)\big)f'^{-1}f''^{-1} \\
&= f''f'\Delta^{f'^{-1}}(e^{\hbar C/2})\cdot(e^{-\hbar C/2}\otimes e^{-\hbar C/2})f'^{-1}f''^{-1} \\
&= f''\Delta(e^{\hbar C/2})f'(e^{-\hbar C/2}\otimes e^{-\hbar C/2})f'^{-1}f''^{-1} \\
&= \Delta(e^{\hbar C/2})\cdot(e^{-\hbar C/2}\otimes e^{-\hbar C/2}) = e^{\hbar(\Delta(C)-C\otimes 1-1\otimes C)/2} = e^{\hbar\tau},
\end{aligned}$$

where we used (4.25), (5.59), the centrality of C, and the definitions of the notations. Hence, $\mathcal{R}^{f''} = e^{\hbar\tau/2}$. Putting $\Phi'' = \Phi'^{f''}$, we complete this step.

The third step is shown as follows (just as in the proof of Proposition D.3). It is sufficient to show that Φ_{KZ} is obtained from Φ'' by twisting by some f''', keeping Δ and $\mathcal{R}_{\mathrm{KZ}}$ unchanged. Since $\Delta = \Delta^f$ and $\mathcal{R}_{\mathrm{KZ}} = \mathcal{R}_{\mathrm{KZ}}^f$ for any symmetric f in general, we seek a symmetric f''' such that $\Phi_{\mathrm{KZ}} = \Phi''^{f'''}$. We give an inductive construction of such a f''': assuming that $\Phi_{\mathrm{KZ}} \equiv \Phi''^{f_n}$ modulo $\hbar^n$ for some symmetric f_n, we find f_{n+1} (putting $f_{n+1} = (1+\hbar^n \dot{f})f_n$ for some $\dot{f}$) which satisfies that $\Phi_{\mathrm{KZ}} \equiv \Phi''^{f_{n+1}}$ modulo $\hbar^{n+1}$. (Then, the limit of f_n gives a required f'''.) Putting $\Phi_{\mathrm{KZ}} = \Phi''^{f_n} + \hbar^n \dot{\Phi}$, it is sufficient to find $\dot{f}$ satisfying that

$$\dot{\Phi} = 1 \otimes \dot{f} - (\Delta \otimes \mathrm{id})\dot{f} + (\mathrm{id} \otimes \Delta)\dot{f} - \dot{f} \otimes 1.$$

This is rewritten as $d_2 \dot{f} = \dot{\Phi}$, by using the map $d_2 : U(\mathfrak{g})^{\otimes 2} \to (\mathfrak{g})^{\otimes 3}$ defined by

$$d_2\eta = 1 \otimes \eta - (\Delta \otimes \mathrm{id})\eta + (\mathrm{id} \otimes \Delta)\eta - \eta \otimes 1.$$

Further, the pentagon relations of Φ_{KZ} and Φ''^{f_n} deduce the equation,

$$1 \otimes \dot{\Phi} - (\Delta \otimes \mathrm{id} \otimes \mathrm{id})\dot{\Phi} + (\mathrm{id} \otimes \Delta \otimes \mathrm{id})\dot{\Phi} - (\mathrm{id} \otimes \mathrm{id} \otimes \Delta)\dot{\Phi} + \dot{\Phi} \otimes 1 = 0.$$

This is rewritten as $d_3\dot{\Phi} = 0$, by using the map $d_3 : U(\mathfrak{g})^{\otimes 3} \to U(\mathfrak{g})^{\otimes 4}$ defined by

$$d_3\xi = 1 \otimes \xi - (\Delta \otimes \mathrm{id} \otimes \mathrm{id})\xi + (\mathrm{id} \otimes \Delta \otimes \mathrm{id})\xi - (\mathrm{id} \otimes \mathrm{id} \otimes \Delta)\xi + \xi \otimes 1.$$

It follows, from the vanishing of (a certain modification of) the cohomology group kernel(d_3)/image(d_2), that there exists $\dot{f}$ such that $d_2\dot{f} = \dot{\Phi}$. Further, we can replace $\dot{f}$ with a symmetric one, by using the relations,

$$\dot{\Phi}_{321} = -\dot{\Phi}, \qquad \dot{\Phi} + \dot{\Phi}_{231} + \dot{\Phi}_{312} = 0,$$

where the first relation is derived from Lemma 5.14 below, and the second relation is derived from the hexagon relations of Φ_{KZ} and Φ''^{f_n}. Hence, we obtain a required f'''. For a detailed proof of this step, see [Kas94, Theorem XVIII.8.1]. □

Lemma 5.14. Let Φ be an associator of a quasi-triangular quasi-bialgebra $(U(\mathfrak{g})[[\hbar]], \Delta, \varepsilon, \mathcal{R}_{\mathrm{KZ}}, \Phi)$. Then, $\Phi_{321} = \Phi^{-1}$.

Proof. The lemma is obtained, just as in the proof of Lemma D.5; see also [Kas94, Lemma XVIII.8.2]. It is left to the reader to give a detailed proof. □

Let us review definition of the first cohomology of Lie algebra $\mathfrak{g}$ with coefficients in a $\mathfrak{g}$ module V. We put

$$C^0(\mathfrak{g}, V) = V,$$
$$C^1(\mathfrak{g}, V) = \mathrm{Hom}(\mathfrak{g}, V),$$
$$C^2(\mathfrak{g}, V) = \mathrm{Hom}(\wedge^2\mathfrak{g}, V),$$

where $\mathrm{Hom}(W,V)$ denotes the vector space of linear maps of W to V (which are not necessarily $\mathfrak{g}$-equivariant). By using the action of $\mathfrak{g}$ on V, we define the coboundary operator d by

$$(dv)(X) = Xv,$$
$$(d\psi)(X,Y) = -\psi([X,Y]) + X\psi(Y) - Y\psi(X),$$

for $v \in C^0(\mathfrak{g},V)$ and $\psi \in C^1(\mathfrak{g},V)$. Then, $d^2 = 0$. We define the first cohomology of $\mathfrak{g}$ by

$$H^1(\mathfrak{g},V) = \frac{\mathrm{kernel}\big(d : C^1(\mathfrak{g},V) \to C^2(\mathfrak{g},V)\big)}{\mathrm{image}\big(d : C^0(\mathfrak{g},V) \to C^1(\mathfrak{g},V)\big)}$$

It is known (see [Kas94, Proposition XVIII.3.1]) that $H^1(\mathfrak{g},V) = 0$ for any non-trivial simple $\mathfrak{g}$ module V; in particular, $H^1(\mathfrak{g},U(\mathfrak{g})^{\otimes 2}) = 0$ when we regard $U(\mathfrak{g})^{\otimes 2}$ as a $\mathfrak{g}$ module by $Xu = [\Delta(X),u]$ for $X \in \mathfrak{g}$ and $u \in U(\mathfrak{g})^{\otimes 2}$.

By using the vanishing of this cohomology, we obtain the following lemma, which implies the rigidity of quasi-bialgebra structure of $U(\mathfrak{g})[[\hbar]]$.

Lemma 5.15. Let $(U(\mathfrak{g})[[\hbar]],\Delta,\varepsilon)$ and $(U(\mathfrak{g})[[\hbar]],\Delta',\varepsilon')$ be quasi-bialgebras such that their comultiplications and counits are congruent modulo $\hbar$ to the comultiplication and the unit of $U(\mathfrak{g})$. Then, $(U(\mathfrak{g})[[\hbar]],\Delta,\varepsilon)$ can be obtained from $(U(\mathfrak{g})[[\hbar]],\Delta',\varepsilon')$ by twisting by some $f \in U(\mathfrak{g})^{\otimes 2}[[\hbar]]$ such that $f \equiv 1$ modulo $\hbar$.

Sketch of the proof. Assuming $\Delta' \equiv \Delta$ modulo $\hbar^n$, we show that $\Delta' \equiv \Delta^{f_n}$ modulo $\hbar^{n+1}$ for some $f_n = 1 + \hbar^n \dot{f}$. (Then, we obtain the element f with $\Delta' = \Delta^f$ from the inductive limit of the product $\cdots f_3 f_2 f_1$.) We require that

$$\Delta'(X) \equiv (1+\hbar^n \dot{f})\Delta(X)(1+\hbar^n \dot{f})^{-1} \quad \text{modulo } \hbar^{n+1}$$

for some $\dot{f} \in U(\mathfrak{g})^{\otimes 2}$. The above relation implies that

$$\big(\Delta(X) - \Delta'(X)\big)^{(n)} = X\dot{f},$$

where $\alpha^{(n)}$ denotes the coefficient of $\hbar^n$ in α. It can be shown by an elementary computation that the map $X \mapsto \big(\Delta(X) - \Delta'(X)\big)^{(n)}$ is a 1-cocycle. Since $H^1(\mathfrak{g},U(\mathfrak{g})^{\otimes 2}) = 0$, the 1-cocycle is equal to the coboundary of some 0-cochain. Putting the 0-cochain to be $\dot{f}$, we obtain the above formulae. Hence, we obtain the element f with $\Delta' = \Delta^f$.

When the comultiplications are equal, the counits are also equal from the definitions of comultiplication and counit. Hence, $(U(\mathfrak{g})[[\hbar]],\Delta',\varepsilon')$ is obtained from $(U(\mathfrak{g})[[\hbar]],\Delta,\varepsilon)$ by twisting by the above f.

For detailed proofs see [Kas94, ChPr95]. □

Let V be a $U(\mathfrak{g})$ module. Then, $V[[\hbar]]$ is a $U(\mathfrak{g})[[\hbar]]$ module. Further, we have a $U_q(\mathfrak{g})$ module obtained from V by perturbing it by a quantum parameter

q. We denote the module that results from putting $q = e^{\hbar}$ also by $V[[\hbar]]$. By Proposition 5.11, the quasi-triangular quasi-bialgebras $(U(\mathfrak{g})[[\hbar]], \Delta, \varepsilon, \Phi_{\mathrm{KZ}}, \mathcal{R}_{\mathrm{KZ}})$ and $(U_q(\mathfrak{g}), \Delta, \varepsilon, 1, \mathcal{R})$ induce representations of the braid group B_n on $V^{\otimes n}[[\hbar]]$. We denote them by ρ_{KZ} and $\rho_{\mathcal{R}}$ respectively.

We now establish the following remarkable theorem, which was originally pointed out by Kohno, and has completely been proved by Drinfel'd.

Theorem 5.16 (Drinfel'd-Kohno theorem). For any semi-simple Lie algebra $\mathfrak{g}$ and a $U(\mathfrak{g})$ module V, the two representations ρ_{KZ} and $\rho_{\mathcal{R}}$ of the braid group B_n are equivalent, *i.e.*, there exists an automorphism u of $V^{\otimes n}[[\hbar]]$ such that $\rho_{\mathrm{KZ}}(b) = u\rho_{\mathcal{R}}(b)u^{-1}$ for any $b \in B_n$.

See also [Kas94, EFK98, ChPr95] for details of this theorem.

Proof. We denote the action of $U(\mathfrak{g})$ on V by $\rho_V : U(\mathfrak{g}) \to \mathrm{End}(V)$. We also denote the action of $U(\mathfrak{g})[[\hbar]]$ on $V[[\hbar]]$ by ρ_V (in an abuse of notation). By Proposition 5.13 $(U(\mathfrak{g})[[\hbar]], \Delta, \varepsilon, \Phi_{\mathrm{KZ}}, \mathcal{R}_{\mathrm{KZ}})$ is obtained from $(U_q(\mathfrak{g}), \Delta, \varepsilon, 1, \mathcal{R})$ by twisting by some $f \in U(\mathfrak{g})^{\otimes 2}[[\hbar]]$. Hence,

$\rho_{\mathrm{KZ}}\Big($ 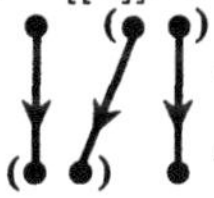 $\Big) = \rho_V^{\otimes 3}(1 \otimes f) \cdot \rho_V^{\otimes 3}((\mathrm{id} \otimes \Delta)f) \cdot \rho_{\mathcal{R}}\Big($  $\Big)$,

$\times\, \rho_V^{\otimes 3}((\Delta \otimes \mathrm{id})f^{-1}) \cdot \rho_V^{\otimes 3}(f^{-1} \otimes 1)$

$\rho_{\mathrm{KZ}}\Big($  $\Big) = \rho_V^{\otimes 2}(f) \cdot \rho_{\mathrm{KZ}}\Big($  $\Big) \cdot \rho_V^{\otimes 2}(f^{-1})$.

Therefore, ρ_{KZ} and $\rho_{\mathcal{R}}$ are intertwined by the action of f on $V[[\hbar]]$. For example, when we regard B_4 as the group of parenthesized braids whose ends are equal to $((\bullet\bullet)\bullet)\bullet$, the automorphism u is given explicitly by

$$u = \rho_V^{\otimes 4}\Big((f \otimes 1 \otimes 1) \cdot ((\Delta \otimes \mathrm{id})f \otimes 1) \cdot ((\Delta \otimes \mathrm{id} \otimes \mathrm{id})(\Delta \otimes \mathrm{id})f)\Big).$$

Thus we obtain the theorem. $\square$

Chapter 6

The Kontsevich invariant

For each simple Lie algebra $\mathfrak{g}$ and each representation R of $\mathfrak{g}$ we have the quantum $(\mathfrak{g}, R)$ invariant of links, as described in Chapter 4. Historically speaking, this seemed to imply that, since there are infinitely many Lie algebras and representations of them, we had obtained infinitely many isotopy invariants of links, following the discovery of the Jones polynomial. For quantum invariants to be useful (say, in the study of the classification problem of knots and links) we needed a better understanding of the set of quantum invariants. How could we control so many invariants? An answer was provided with the introduction of the *Kontsevich invariant* [Kon93], which is the invariant unifying all quantum invariants of links.

The Kontsevich invariant is obtained as follows. As mentioned in the previous chapter, the quantum $(\mathfrak{g}, R)$ invariant of links can be constructed by using the monodromy of solutions of the Knizhnik-Zamolodchikov equation (the KZ equation) associated with the Lie algebra $\mathfrak{g}$ and its representation R. To obtain the universal invariant among quantum invariants, we consider the "universal" KZ equation (called the *formal KZ equation* in this chapter), obtained from the KZ equation by replacing $\mathfrak{g}$ with a dashed line and R a solid line, introducing Jacobi diagrams which consist of dashed and solid lines. Jacobi diagrams have universal properties among the pairs $(\mathfrak{g}, R)$, in some sense, and we can consider the substitution of $\mathfrak{g}$ and R into the dashed and solid lines of Jacobi diagrams, which construction is called a *weight system*. The formal KZ equation is formulated in a space of Jacobi diagrams. By using the monodromy along solutions of the formal KZ equation we construct the Kontsevich invariant of links whose values belong to a space of Jacobi diagrams. The quantum $(\mathfrak{g}, R)$ invariant is recovered from the Kontsevich invariant through the weight system which substitutes $\mathfrak{g}$ and R into Jacobi diagrams.

In Section 6.1 we introduce Jacobi diagrams and describe their basic properties. In Section 6.2 we formulate the formal KZ equation in a space of Jacobi diagrams, and construct the Kontsevich invariant using the monodromy along solutions of the formal KZ equation. In Section 6.3 we introduce quasi-tangles to give a combinatorial reformulation of the Kontsevich invariant, similar to be constructions of Sections 5.3 and 5.4. This combinatorial construction of the Kontsevich invariant appears in Section 6.4, modifying the invariant to an invariant of framed links, which is called the *framed Kontsevich invariant* (by J. Murakami and T. Le [LeMu95b]) in this chapter. (For combinatorial constructions of the Kontsevich invariant see also [AlFr95, Piu95a].) Based on the combinatorial reconstruction we show properties of the (framed) Kontsevich invariant in Section 6.5. Further, we show the universality of the framed Kontsevich invariant among quantum invariants of framed links in Section 6.6.

6.1 Jacobi diagrams

In this section we introduce Jacobi diagrams. As shown in Section 6.6 we can "substitute" any pair of a Lie algebra $\mathfrak{g}$ and its representation R into Jacobi diagrams. Hence, Jacobi diagrams have universal properties among the pairs $(\mathfrak{g}, R)$. We also show basic properties of Jacobi diagrams in this section. See also [Bar95b] for this topic.

A *uni-trivalent graph* is a graph, every vertex of which is either univalent or trivalent, where a vertex of a graph is *univalent* (resp. *trivalent*) if there is one edge (resp. there are three edges) of the graph adjacent to the vertex. A *vertex-oriented* trivalent vertex is a trivalent vertex such that a cyclic order of the three edges around the trivalent vertex is fixed. Let X be a compact oriented (possibly empty) 1-manifold with (possibly empty) boundary. A *Jacobi diagram** on X is the manifold X together with a uni-trivalent graph such that univalent vertices of the graph are distinct points on X and each trivalent vertex is vertex-oriented; see Figure 6.1 for examples of Jacobi diagrams. We regard homeomorphic Jacobi diagrams (by homeomorphisms whose restrictions to X are isotopic to the identity map of X) as the same one. In figures we draw X by solid lines and the uni-trivalent graphs by dashed lines, in such a way that each trivalent vertex is vertex-oriented in the counterclockwise order. We define the *degree* of a Jacobi diagram to be half the number of univalent and trivalent vertices of the uni-trivalent graph of the Jacobi diagram. A component of a uni-trivalent graph with no trivalent vertex, *i.e.*, a dashed interval, is called a *chord*. A Jacobi diagram whose uni-trivalent graph is a disjoint union of chords is called a *chord diagram*. The degree of a chord diagram is equal to the number of chords.

We denote by $\mathcal{A}(X)$ the quotient vector space over $\mathbb{C}$ spanned by Jacobi diagrams on X subject to the AS, IHX, and STU relations; see Figure 6.2 for the definition of these relations. We call $\mathcal{A}(X)$ the *space of Jacobi diagrams* on X. We denote by $\mathcal{A}(X)^{(d)}$ the vector subspace of $\mathcal{A}(X)$ spanned by the Jacobi diagrams of degree d. We denote the completion of $\mathcal{A}(X)$ with respect to the degree also by $\mathcal{A}(X)$ (abusing notation).

*A Jacobi diagram is also called a *web diagram* or a *trivalent diagram* in some literatures. In physics this is often called a *Feynman diagram*.

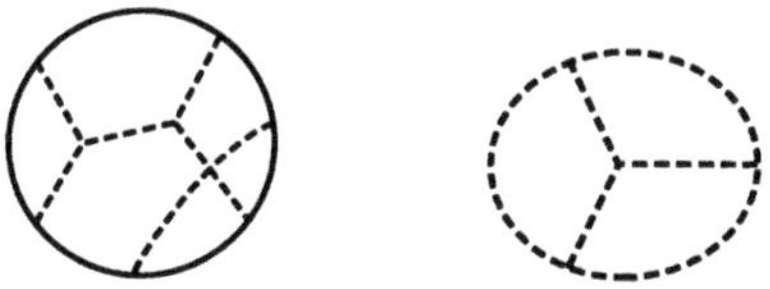

Figure 6.1 A Jacobi diagram on S^1 (the left picture) and a Jacobi diagram on ϕ (the right picture). For a Jacobi diagram on X, the uni-trivalent graph of the Jacobi diagram is depicted by dashed lines and X by solid lines. A trivalent vertex is vertex-oriented in the counterclockwise order. The uni-trivalent graph of the left Jacobi diagram has 2 trivalent vertices and 6 univalent vertices; note that there is no vertex of the uni-trivalent graph at the crossing of dashed lines in the picture.

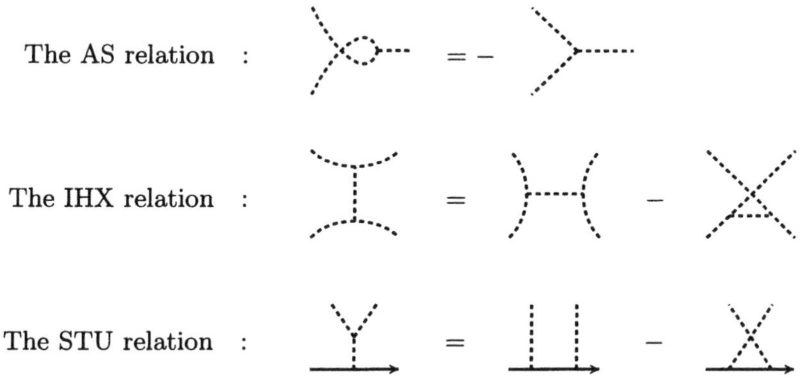

Figure 6.2 Definition of the AS, IHX, and STU relations

Let us note the usage of the STU relation. When we apply the STU relation, we implicitly consider the cyclic order of three lines around a univalent vertex of a Jacobi diagram, given by the order of the outward-oriented solid line, the dashed line, and the inward-oriented solid line. At each univalent vertex in Figure 6.2 this cyclic order is counterclockwise. Note that, when this cyclic order is clockwise in a picture, the STU relation should be applied as follows,

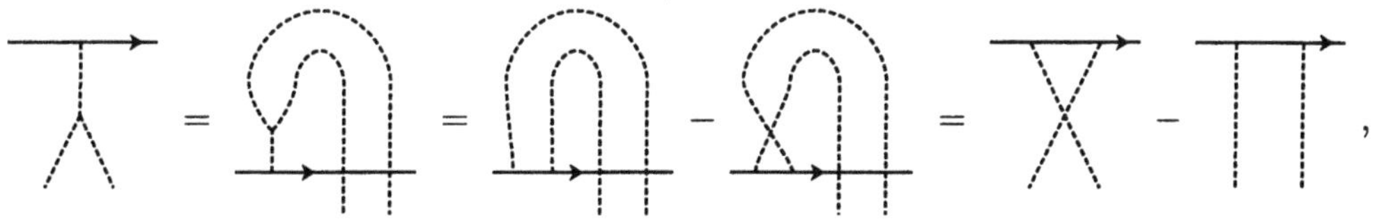

where we obtain the second equality by the STU relation in Figure 6.2. This relation can alternatively be obtained from the STU relation in Figure 6.2 by taking mirror image with respect to a horizontal line, noting that the mirror image of a trivalent vertex gives rise to the change of sign by the AS relation.

Let $C \sqcup X$ be a compact oriented 1-manifold such that C is a component of the 1-manifold and X is the rest of it. We introduce the comultiplication $\Delta_{(C)}$, the antipode $S_{(C)}$, and the counit $\varepsilon_{(C)}$ of Jacobi diagrams on $C \sqcup X$ acting on the component C as follows. Firstly, the *comultiplication* $\Delta_{(C)} : \mathcal{A}(C \sqcup X) \to \mathcal{A}(C^{(2)} \sqcup X)$, where $C^{(2)}$ denotes the union of two parallel copies of C, is defined by

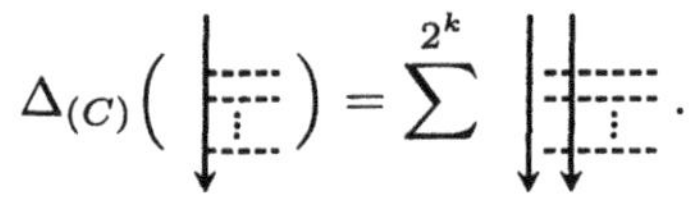

Here, C is depicted as a vertical line on the left hand side, and k is the number of univalent vertices on the vertical line. The sum on the right hand side runs over the 2^k possible configurations that arise by choosing for each dashed edge one of the two solid vertical lines to connect to. Secondly, the *antipode* $S_{(C)} : \mathcal{A}(C \sqcup X) \to \mathcal{A}(\overline{C} \sqcup X)$, where $\overline{C}$ denotes the component C with the opposite orientation, is defined by

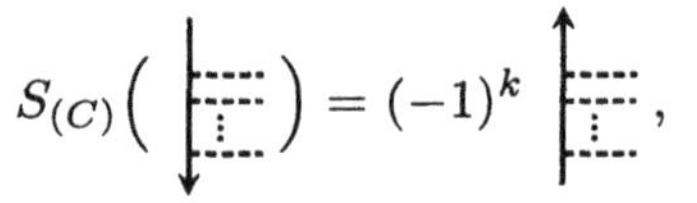

where k is the number of univalent vertices on the vertical line on the left. Thirdly, the *counit* $\varepsilon_{(C)} : \mathcal{A}(C \sqcup X) \to \mathcal{A}(X)$ is defined as follows. For a Jacobi diagram D on $C \sqcup X$, we put $\varepsilon_{(C)}(D) = 0$ if there exist univalent vertices on C, and $\varepsilon_{(C)}(D) = D - C$ otherwise. Moreover, the map ${\Delta_{(C)}}^{(n)} : \mathcal{A}(C \sqcup X) \to \mathcal{A}(C^{(n+1)} \sqcup X)$, where $C^{(n+1)}$ denotes the union of $n+1$ parallel copies of the component C, is defined recursively by ${\Delta_{(C)}}^{(n)} = {\Delta_{(C_1)}}^{(n-1)} \circ \Delta_{(C)}$ where C_1 denotes one of copies of C.

Proposition 6.1 ([Bar97]). Let $C \sqcup X$ be an oriented 1-manifold whose component C is homeomorphic to an interval. Let D_1 be a Jacobi diagram on $C \sqcup X$.
(1) Let D_2 be a Jacobi diagram on an interval. We consider two Jacobi diagrams obtained from D_1 by gluing the interval of D_2 to two ends of C respectively. Then, the two Jacobi diagrams are equal in the space of Jacobi diagrams. This is graphically shown as

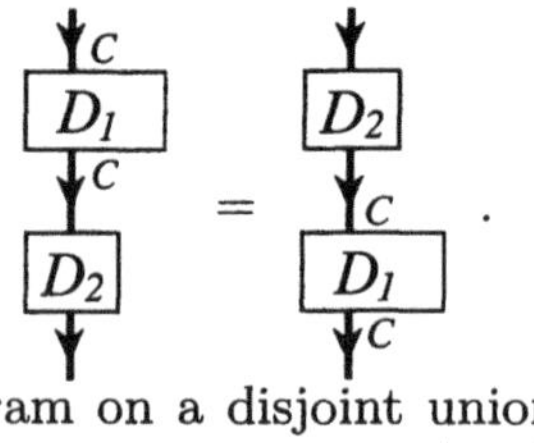

(2) Let D_2 be a Jacobi diagram on a disjoint union of n intervals. Recall that ${\Delta_{(C)}}^{(n)}(D_1)$ is a sum of Jacobi diagrams on $C^{(n+1)}$, where $C^{(n+1)}$ denotes the union of $n+1$ parallel copies of the interval C. We consider two sums of Jacobi

diagrams obtained from ${\Delta_{(C)}}^{(n)}(D_1)$ by gluing the disjoint union of n intervals of D_2 to the upper and the lower ends of $C^{(n+1)}$ respectively. Then, the two Jacobi diagrams are equal in the space of Jacobi diagrams. This is graphically shown as

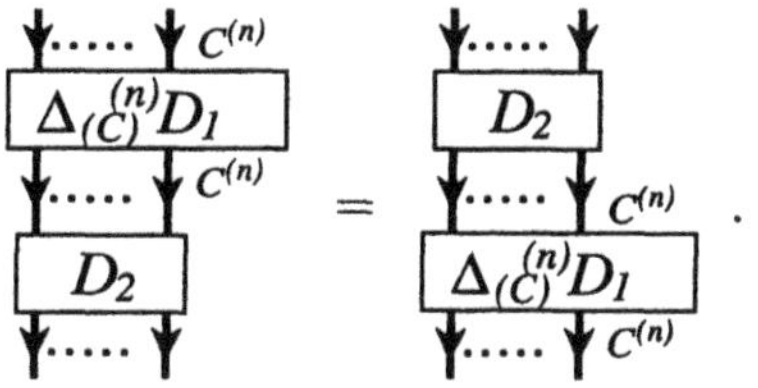

Proof. We prove (1) as follows. A dashed horizontal line can pass over a Jacobi diagram on a solid vertical interval, in the following way,

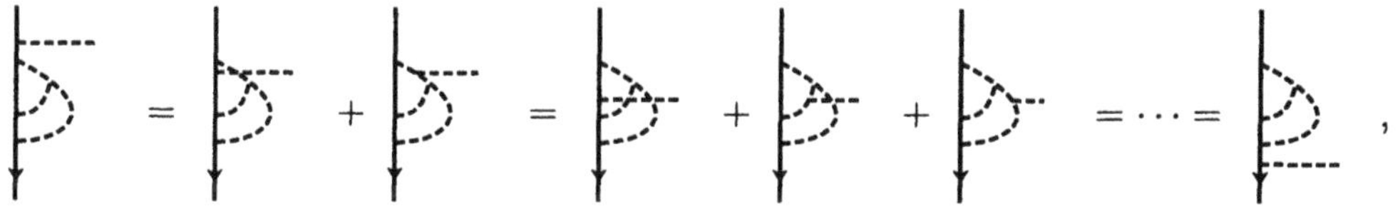

where the equalities are derived from the STU and IHX relations. This implies, in general, that a dashed horizontal line commutes with any Jacobi diagram D_2 on a solid vertical line. This observation can be applied repeatedly to sweep, one by one, each leg of D_1 over D_2. This obtains part (1) of the proposition.

For simplicity, we focus on the $n = 1$ case of (2). Similarly as in the above case, we have that

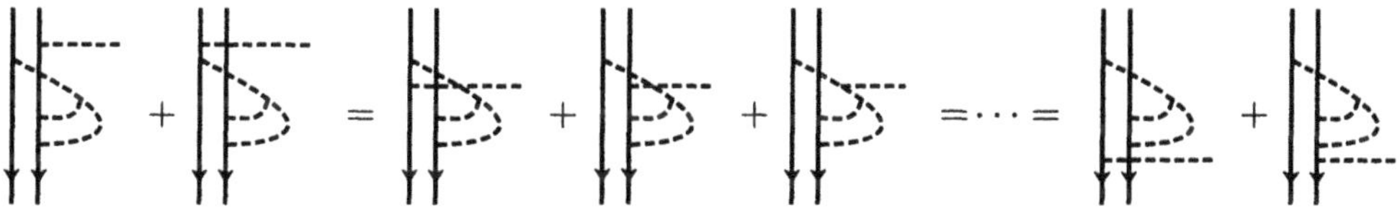

by the STU and IHX relations. This implies, in general for any Jacobi diagram D_2 on $\downarrow\downarrow$, that the sum over the lifts of a horizontal dashed edge to parallel copies of a vertical solid edge commutes with any diagram on those parallel copies, *i.e.*, $D_2 \in \mathcal{A}(\downarrow\downarrow)$. Since the diagram $\Delta_{(C)}(D_1)$ consists of such sums in a neighborhood of the components $C^{(2)}$, we obtain (2) of the proposition. □

The formulae in the following proposition are similar to the defining relations (4.4) and (4.5) of a Hopf algebra.

Proposition 6.2. Let $C \sqcup X$ be as in Proposition 6.1. For any Jacobi diagram D on $C \sqcup X$, graphically shown as 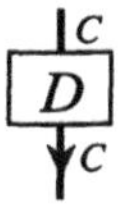

, we have the following relations,

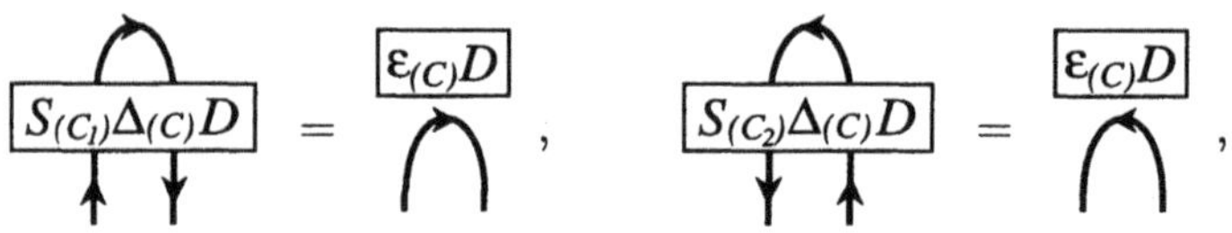

where C_1 and C_2 denote the first and second copies of C respectively.

Proof. We show the first relation as follows. If D has no univalent vertices on C, then the relation is easily obtained. Otherwise, we present D by 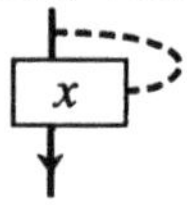

for some x. Then,

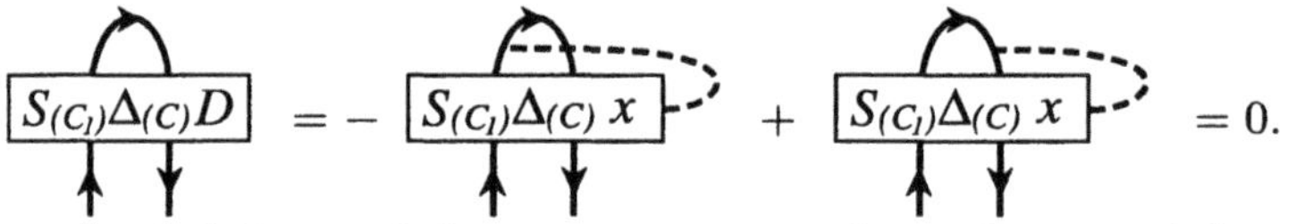

Further, from the definition of the counit $\varepsilon_{(C)}$, we have that $\varepsilon_{(C)}(x) = 0$. Hence, we obtain the first relation of the proposition.

The second relation is obtained similarly. □

Proposition 6.3.
(1) There exists an isomorphism between the spaces $\mathcal{A}(\downarrow)$ and $\mathcal{A}(S^1)$, which is induced by attaching the two end points of $\downarrow$ to obtain S^1.
(2) The space $\mathcal{A}(\downarrow)$ has a commutative algebra structure whose product is given by attaching two intervals vertically at their end points. Further, $\mathcal{A}(S^1)$ has a commutative algebra structure whose product is given by the connected sum of the copies of S^1.
(3) Let X be a compact oriented 1-manifold. We fix a component C of X. Then, we have an action of $\mathcal{A}(S^1)$ on $\mathcal{A}(X)$, given by taking the connected sum of S^1 to the component C. That is, $\mathcal{A}(X)$ is an $\mathcal{A}(S^1)$-module.

Proof. By attaching the two end points of $\downarrow$, we have a linear map $\mathcal{A}(\downarrow) \to \mathcal{A}(S^1)$. This map is an isomorphism, because we have the inverse map $\mathcal{A}(S^1) \to \mathcal{A}(\downarrow)$ obtained by cutting S^1 at a point to obtain $\downarrow$. This inverse map is well defined, that is, independent of the choice of the cutting point, because a dashed univalent vertex can pass over the cutting point as follows:

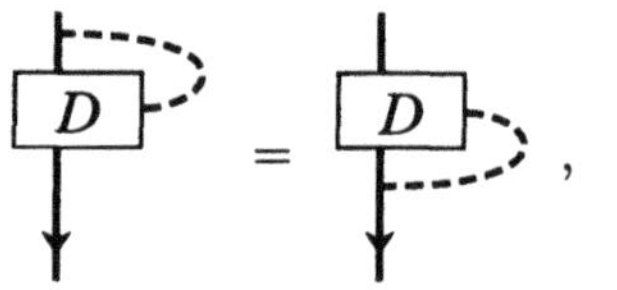

where we obtain this equality by the same argument as in the proof of Proposition 6.1 (1).

We show (2) as follows. We have an algebra structure of $\mathcal{A}(\downarrow)$ by attaching two intervals vertically at their end points to obtain an interval. This algebra is commutative, because two Jacobi diagrams on a solid interval commute with each other by Proposition 6.1 (1); the commutativity is graphically shown, for example, as

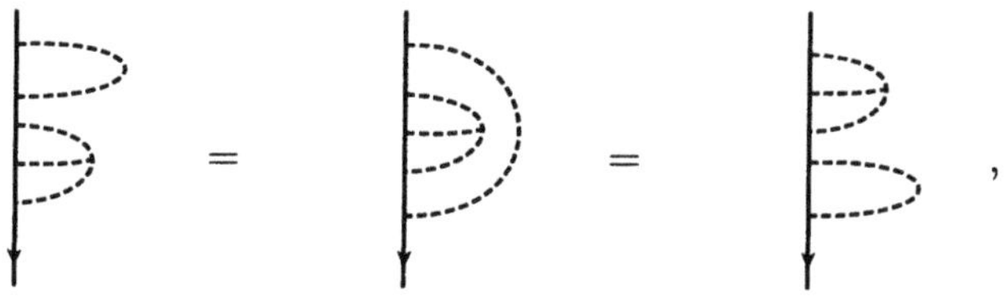

where each equality is derived from the same argument as in the proof of Proposition 6.1 (1). Further, by the isomorphism between $\mathcal{A}(\downarrow)$ and $\mathcal{A}(S^1)$, we have a commutative algebra structure on $\mathcal{A}(S^1)$. The product of $\mathcal{A}(S^1)$ is graphically shown, for example, as

This connected sum is well defined, because of the following equality,

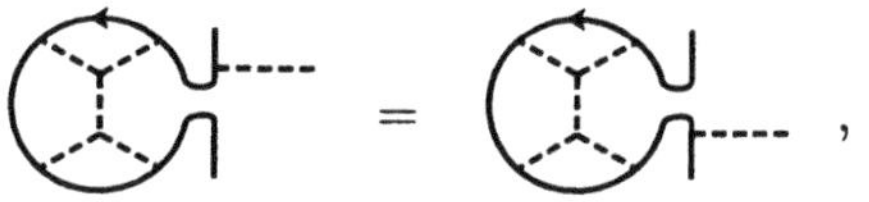

which can be obtained by the same argument as appears in the proof of Proposition 6.1 (1).

The above equality also guarantees that there exists a well-defined action of $\mathcal{A}(S^1)$ on $\mathcal{A}(X)$ by taking the connected sum of the S^1 to a fixed component C of X. □

Similarly as in Proposition 6.3 (2), the space $\mathcal{A}\big(\underbrace{\downarrow\downarrow\cdots\downarrow}_{n}\big)$ has an algebra structure whose product is given by composing two copies of $\underbrace{\downarrow\downarrow\cdots\downarrow}_{n}$ vertically along their end points, though this algebra is not commutative for $n > 1$ unlike the case $n = 1$.

6.2 The Kontsevich invariant derived from the formal KZ equation

In Section 5.1 we introduced the monodromy representation of the braid groups along solutions of the KZ equation. We gave a presentation of the monodromy by an iterated integral. Motivated by this presentation, we now introduce the Kontsevich invariant of links taking values in a space of Jacobi diagrams in this section.

Following the notations of Section 5.1, we denote by V a vector space on which a Lie algebra $\mathfrak{g}$ acts, and by X_n the configuration space of n distinct points in $\mathbb{C}$, *i.e.*, we put

$$X_n = \{(z_1, z_2, \cdots, z_n) \in \mathbb{C}^n \mid z_i \neq z_j \text{ for any } i \neq j\}.$$

We consider a closed path $\gamma : [0,1] \to X_n$. Further, as in Figure 5.3, we consider the path $f : [0,1] \to V^{\otimes n}$ which arises from the lift of γ along solutions of the KZ equation. Let us review the construction of the path f. As in the proof of Proposition 5.2 the path f is determined by the differential equation,

$$\frac{df(t)}{dt} = w(t)f(t), \tag{6.1}$$

where $w(t)$ is the function such that $w(t)dt$ is the 1-form on $[0,1]$ obtained from the KZ 1-form τ_n on X_n given in (5.10) by pulling back by the path $\gamma : [0,1] \to X_n$. To obtain a concrete presentation of the function w, we put $\gamma(t) = \big(z_1(t), \cdots, z_n(t)\big) \in X_n$, for $t \in [0,1]$. Since $f(t)$ and $\gamma(t)$ are related by $f(t) = W\big(\gamma(t)\big)$ where W is a local solution of the KZ equation, we have that

$$\begin{aligned} \frac{df}{dt} &= \sum_{1 \le i \le n} \frac{\partial W}{\partial z_i} \frac{dz_i}{dt} \\ &= \sum_{1 \le i \le n} \frac{\hbar}{2\pi\sqrt{-1}} \sum_{\substack{1 \le j \le n \\ j \neq i}} \frac{\tau_{ij}}{z_i - z_j} f \frac{dz_i}{dt} \\ &= \frac{\hbar}{2\pi\sqrt{-1}} \sum_{1 \le i < j \le n} \frac{d}{dt} \log(z_i - z_j) \tau_{ij} f, \end{aligned}$$

where we obtain the second equality by the KZ equation (5.8). Hence, we have the following presentation of the function $w : [0,1] \to U(\mathfrak{g})^{\otimes n}$:

$$w(t) = \frac{\hbar}{2\pi\sqrt{-1}} \sum_{1 \le i < j \le n} \frac{d}{dt} \log\big(z_i(t) - z_j(t)\big) \tau_{ij}. \tag{6.2}$$

Given the above function w, we can obtain a presentation of the solution of the differential equation (6.1) by an iterated integral, as shown in the following proposition; for expositions of iterated integrals see [Che75, Kas94] and references in [Kas94, Section XIX.10].

Proposition 6.4. Let A be a non-commutative algebra acting on a vector space V'. For a given function $w : [0,1] \to A$ we consider the differential equation (6.1) for a function $f : [0,1] \to V'$ with an initial value $f(0) = \alpha \in V'$. Then, the solution of the differential equation is presented by the following iterated integral:

$$f(t) = \Big(1 + \sum_{m=1}^{\infty} \int_{0 \le t_1 \le \cdots \le t_m \le t} w(t_m) \cdots w(t_1) dt_1 \cdots dt_m\Big) \alpha. \tag{6.3}$$

Proof. By putting

$$I_m(t) = \int_{0 \leq t_1 \leq \cdots \leq t_m \leq t} w(t_m) \cdots w(t_1) dt_1 \cdots dt_m,$$

we have that

$$I_m(t) = \int_0^t w(t_m) I_{m-1}(t_m) dt_m.$$

This formula implies that $I_m(t)$ is an integral function of the function $w(t)I_{m-1}(t)$. Hence,

$$\frac{d}{dt} I_m(t) = w(t) I_{m-1}(t).$$

This formula implies that

$$f(t) = \Big(1 + \sum_{m=1}^{\infty} I_m\Big)\alpha$$

is the solution of (6.1). The above formula gives the required solution (6.3). □

Hence, by Proposition 6.4, the path f in Figure 5.3 is presented as in (6.3) with the function w in (6.2). Therefore, as discussed in Section 5.1, the following sum

$$1 + \sum_{m=1}^{\infty} \int_{0 \leq t_1 \leq \cdots \leq t_m \leq 1} w(t_m) \cdots w(t_1) dt_1 \cdots dt_m$$

gives the monodromy of the path γ along solutions of the KZ equation. Therefore, for a closed path γ in X_n, the monodromy representation (5.18) of the pure braid group P_n is alternatively defined to be the representation which takes γ to the above series.

To obtain the Kontsevich invariant we reformulate the above argument in the space of Jacobi diagrams on $\underbrace{\downarrow\downarrow \cdots \downarrow}_{n}$. This reformulation factors out the dependence on Lie algebras and their representations.

Let τ_{ij} be the chord diagram on $\underbrace{\downarrow\downarrow \cdots \downarrow}_{n}$ consisting of just one chord connecting the ith and jth vertical lines of $\underbrace{\downarrow\downarrow \cdots \downarrow}_{n}$. Recall that the space $\mathcal{A}\Big(\underbrace{\downarrow\downarrow \cdots \downarrow}_{n}\Big)$ forms an algebra with the product derived from composing copies of $\underbrace{\downarrow\downarrow \cdots \downarrow}_{n}$ vertically as mentioned before. The unit 1 of the algebra is given by the Jacobi diagram $\underbrace{\downarrow\downarrow \cdots \downarrow}_{n}$ with empty uni-trivalent graph. Motivated by the KZ equation (5.9), we consider the following total differential equation of a function $W : X_n \to \mathcal{A}\big(\underbrace{\downarrow\downarrow\cdots\downarrow}_{n}\big)$,

$$dW = \frac{1}{2\pi\sqrt{-1}} \sum_{1\leq i<j\leq n} \tau_{ij} \frac{dz_i - dz_j}{z_i - z_j} W.$$

We call this equation the *formal KZ equation*.

Let us verify that the formal KZ equation satisfies the integrability condition. The first thing is to check the relations (5.4) and (5.5) of Lemma 5.1 for the present τ_{ij}; the other part of this check is the same as the check of integrability of the KZ equation in Section 5.1. It is trivial that the present τ_{ij} satisfies (5.4), which is graphically shown as

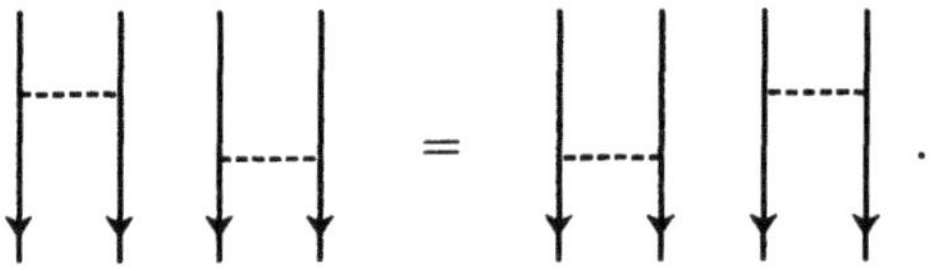

The relation (5.5) for the present τ_{ij} is graphically shown as

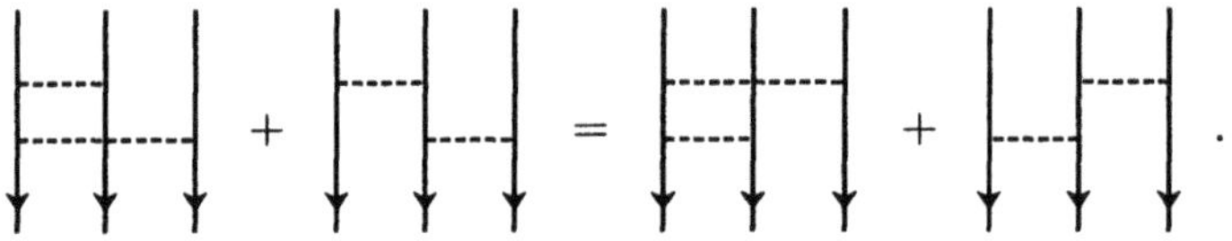

We obtain this relation by Proposition 6.1 (2). Hence, the formal KZ equation satisfies the integrability condition.

Therefore, similarly as discussed in Section 5.1, the monodromy along solutions of the formal KZ equation induces a monodromy representation of the braid group. Let γ be a closed path in X_n, and put $\gamma(t) = \big(z_1(t), \cdots, z_n(t)\big)$ as before. Similarly as in (6.2), we put the function $w : [0,1] \to \mathcal{A}\big(\underbrace{\downarrow\downarrow\cdots\downarrow}_{n}\big)$ to be

$$w(t) = \frac{1}{2\pi\sqrt{-1}} \sum_{1\leq i<j\leq n} \frac{d}{dt} \log\big(z_i(t) - z_j(t)\big)\tau_{ij} \in \mathcal{A}\big(\underbrace{\downarrow\downarrow\cdots\downarrow}_{n}\big),$$

where we do not use the indeterminate $\hbar$ in the above definition of w (unlike the definition in (6.2)). By the same argument as above, the following series of iterated integrals

$$1 + \sum_{m=1}^{\infty} \int_{0\leq t_1\leq\cdots\leq t_m\leq 1} w(t_m)\cdots w(t_1)dt_1\cdots dt_m \in \mathcal{A}\big(\underbrace{\downarrow\downarrow\cdots\downarrow}_{n}\big) \qquad (6.4)$$

gives the monodromy of γ along solutions of the formal KZ equation. Similarly as in Section 5.1, we obtain a monodromy representation

$$\rho_{\mathrm{KZ}} : P_n \longrightarrow \mathcal{A}\big(\underbrace{\downarrow\downarrow\cdots\downarrow}_{n}\big) \tag{6.5}$$

of the pure braid group P_n, which takes a path γ to the series (6.4).

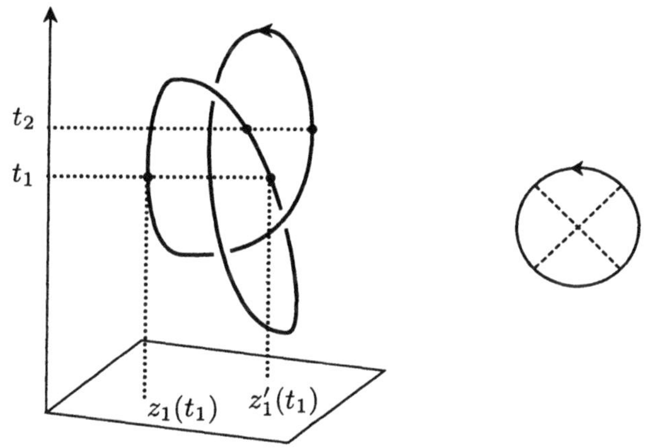

Figure 6.3 A configuration P of a knot in $\mathbb{C} \times \mathbb{R}$ (the left picture) and the chord diagram D_P of P (the right picture). For the configuration P in the left picture the number $\#P_{\uparrow}$ is equal to 3.

Motivated by the above monodromy representation along solutions of the formal KZ equation, we introduce the Kontsevich invariant of a knot as follows.

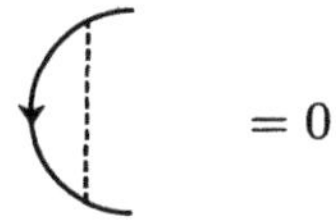

Figure 6.4 The FI relation. This relation states that a Jacobi diagram with an isolated chord is equal to 0, where an *isolated chord* is a chord such that there is no univalent vertex of the uni-trivalent graph on the solid line between two ends of the chord.

We consider an oriented knot in $\mathbb{C} \times \mathbb{R}$ such that the knot has finitely many critical points with respect to the height $\mathbb{R}$. Let $t_1, t_2, \cdots, t_m$ be parameters in $\mathbb{R}$ with $t_1 \leq t_2 \leq \cdots \leq t_m$. For each t_i we choose two points of K at the height t_i and connect the two points with a horizontal chord. We call a set of such a choice for each t_i a *configuration* P and denote the corresponding chord diagram by D_P; see

Figure 6.3 for an example of a configuration. We define $Z'(K) \in \mathcal{A}(S^1)/\mathrm{FI}$ by

$$Z'(K) = \sum_{m=0}^{\infty} \frac{1}{(2\pi\sqrt{-1})^m}$$
$$\times \int_{t_1 \le t_2 \le \cdots \le t_m} \sum_P (-1)^{\#P_\uparrow} \Big(\prod_{i=1}^{m} \frac{d}{dt_i} \log\big(z_i(t) - z_i'(t)\big) \Big) D_P dt_1 dt_2 \cdots dt_m,$$

where the second sum runs over all possible configurations P, and $\#P_\uparrow$ denotes the number of end points of chords such that the knot is oriented upwards at the end points. Here, we define the FI relation as shown in Figure 6.4. Note that, if K was a braid, then the above iterated integral would give the iterated integral (6.4). Further, we put

$$Z(K) = \frac{Z'(K)}{Z'(U)^M} \in \mathcal{A}(S^1)/\mathrm{FI},$$

and call it the *Kontsevich invariant* of an oriented knot K, where M is the number of maximal points of K, and U is the following knot,

$U =$  .

In the same way as in the knot case, the *Kontsevich invariant* $Z(L)$ of an oriented link L is defined to be in $\mathcal{A}(\sqcup^l S^1)/\mathrm{FI}$.

We omit the proof of the isotopy invariance of the Kontsevich invariant $Z(L)$; see [Bar95b] for the proof. In fact, $Z(L) \in \mathcal{A}(\sqcup^l S^1)/\mathrm{FI}$ can be obtained from the framed Kontsevich invariant $\hat{Z}(L) \in \mathcal{A}(\sqcup^l S^1)$ (introduced in Section 6.4) by introducing the FI relation. (It is left to the reader to show that.) Hence, the proof of the isotopy invariance of $Z(L)$ can be obtained from the isotopy invariance of $\hat{Z}(L)$, which was shown in Section 6.4.

6.3 Quasi-tangles and their sliced diagrams

In Section 5.3 we introduced the notion of a parenthesized braid as a refinement of the notion of a braid. This was done to describe the combinatorial construction of the monodromy representation of the braid groups along solutions of the KZ equation. In this section we introduce the notion of a quasi-tangle as a refinement of the notion of a tangle, to describe the combinatorial construction of the Kontsevich invariant.

A *quasi-tangle* is a tangle in $\mathbb{R} \times \mathbb{R} \times [0,1]$ whose ends are parenthesized sets of dots in $\{0\} \times \mathbb{R} \times \{0,1\}$; see, *e.g.*, Figure 6.5. Recall that a parenthesized set of n dots is recursively obtained from a parenthesized set of $n-1$ dots by replacing one of • with (••); for example, the parenthesized sets of 3 dots are (••)• and •(••), and the parenthesized sets of 4 dots are •(•(••)), (••)(••), ((••)•)•, •((••)•), and (•(••))•.

Two quasi-tangles are called *isotopic* if they are isotopic as tangles with their ends fixed as parenthesized sets of dots. A *framed quasi-tangle* is a quasi-tangle with each component framed (like a framed tangle). We mean by a *quasi-tangle diagram* the projection of a tangle in the plane like a tangle diagram. The framing of a framed quasi-tangle is usually expressed in a diagram of it by the blackboard framing, just like a diagram of a framed tangle.

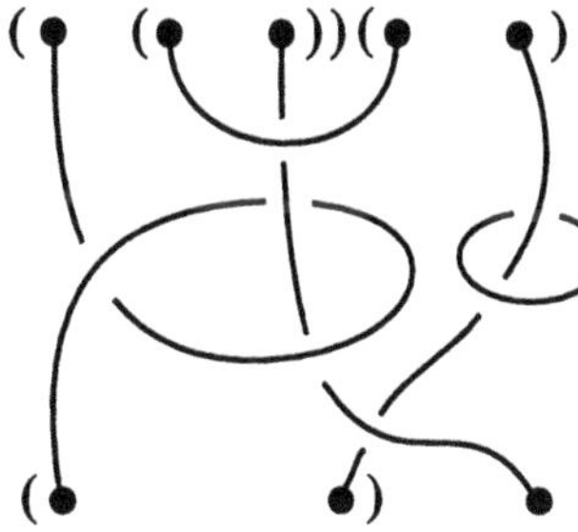

Figure 6.5 An example of a quasi-tangle

Figure 6.6 The elementary quasi-tangle diagrams.

The elementary quasi-tangle diagrams shown in Figure 6.6 are simple examples of quasi-tangles. As in Section 5.3 each  in Figure 6.6 denotes a trivial parenthesized bundle of vertical lines such as

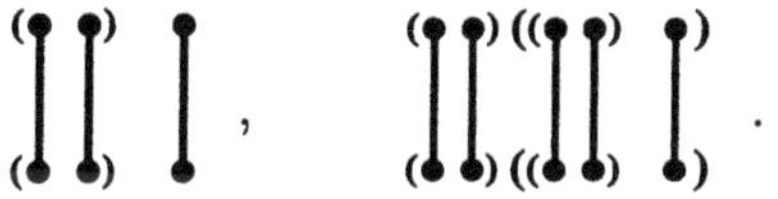

Further, we define the *tensor product* and the *composition* of two quasi-tangle diagrams by

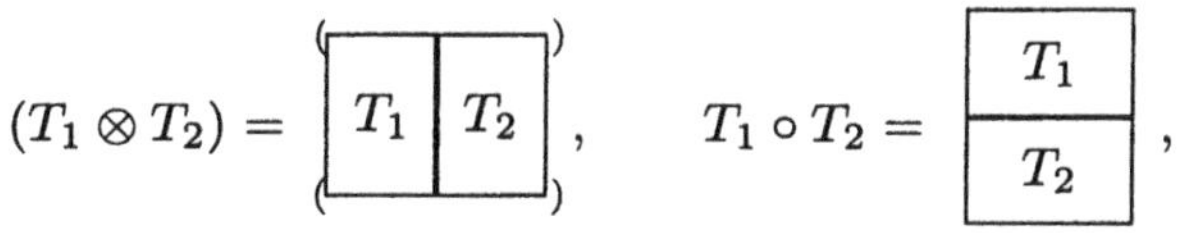

where the composition $T_1 \circ T_2$ is defined if the lower end of T_1 is equal to the upper end of T_2 as a parenthesized set of dots. A *sliced quasi-tangle diagram* is a composition of tensor products of copies of elementary quasi-tangle diagrams. Any quasi-tangle diagram is expressed as a sliced quasi-tangle diagram; see, *e.g.*, Figure 6.7. For a given quasi-tangle there are always, and many, sliced quasi-tangle diagrams expressing it. To relate them to each other, we consider the moves among sliced quasi-tangle diagrams shown in Figure 6.8.

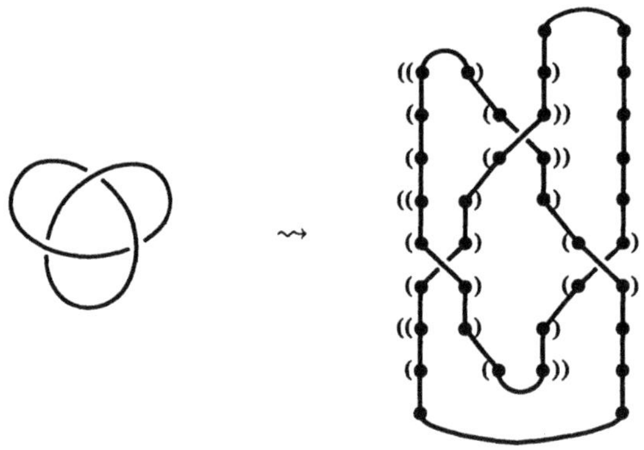

Figure 6.7 A knot diagram (the left picture) is expressed as a sliced quasi-tangle diagram (the right picture).

Theorem 6.5. Two sliced quasi-tangle diagrams express isotopic quasi-tangles if and only if the two sliced diagrams are related by a finite series of the moves shown in Figure 6.8.

This theorem has the following symbolic representation,

$$\begin{aligned}&\{\text{sliced quasi-tangle diagrams}\}/\text{the moves in Figure 6.8}\\ &= \{\text{quasi-tangles}\}/\text{isotopy of } \mathbb{R}^2 \times [0,1].\end{aligned} \tag{6.10}$$

In particular, restricting the above equality to links, the statement becomes

$$\begin{aligned}&\{\text{sliced quasi-tangle diagrams of links}\}/\text{the moves in Figure 6.8}\\ &= \{\text{links}\}/\text{isotopy of } \mathbb{R}^3.\end{aligned}$$

Remark 6.6. The statement of Theorem 6.5 can exactly be fixed by arbitrarily choosing a sliced quasi-tangle diagram expressing each side of the moves (6.7)–(6.9) in Figure 6.8. For example, we regard the second move of (6.8) as the following move,

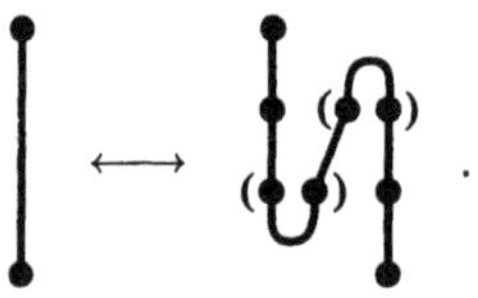

The moves in Figure 5.6

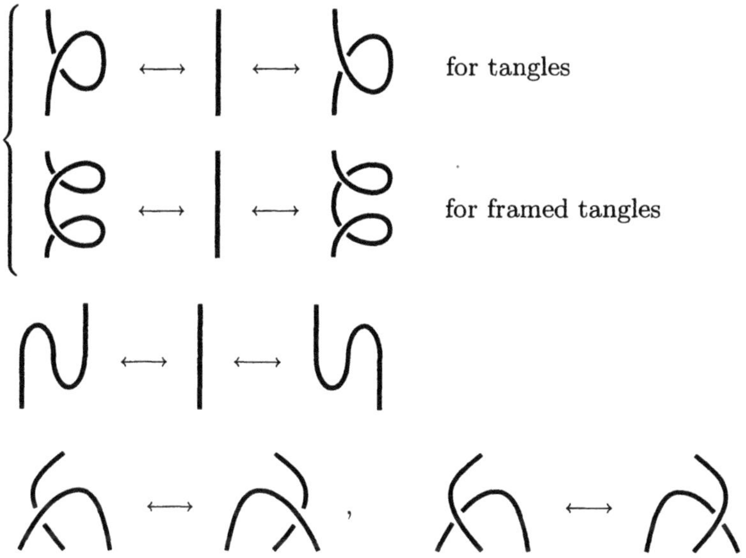

for tangles (6.6)

for framed tangles (6.7)

(6.8)

(6.9)

Figure 6.8 The moves among sliced quasi-tangle diagrams. We regard each of the moves (6.7)–(6.9) as a move between two sliced quasi-tangle diagrams which express the two sides of the move; see Remark 6.6.

If we chose another sliced quasi-tangle diagram expressing each side of this move, all such choices are equivalent by the moves in Figure 5.6, by the same reason as mentioned in Remark 3.2. In this sense, the statement of Theorem 6.5 does not depend on such choices, and it is not necessary to fix such choices of sliced quasi-tangle diagrams which express both sides of the moves (6.7)–(6.9), though they are moves among sliced quasi-tangle diagrams.

Proof of Theorem 6.5. Similarly as in the proofs of Proposition 5.7 and Theorem 5.8, we have that

$$\begin{aligned}&\{\text{sliced quasi-tangle diagrams}\}/\text{some of the moves in Figure 5.6}\\ &= \{\text{quasi-tangle diagrams}\}/\text{isotopy of } \mathbb{R} \times [0,1].\end{aligned}$$

Further, the Turaev moves in Figure 3.4 follow from the moves in Figure 6.8. By applying Theorem 3.1 to the above equality, we obtain equality (6.10); this implies the theorem. □

6.4 Combinatorial definition of the framed Kontsevich invariant

In this section we give a combinatorial definition of the framed Kontsevich invariant of framed quasi-tangles using sliced quasi-tangle diagrams as introduced in the previous section. In particular, it is an invariant of framed links. Furthermore, the original Kontsevich invariant of the underlying (unframed) link is recovered from the framed Kontsevich invariant by introducing the FI relation.

Before defining the framed Kontsevich invariant, we must introduce an associator. An *associator*, denoted by Φ, is defined to be an invertible group-like element in the space $\mathcal{A}(\downarrow\downarrow\downarrow)$ satisfying that $\varepsilon_2\Phi = 1 \in \mathcal{A}(\downarrow\downarrow)$ and the following relations,

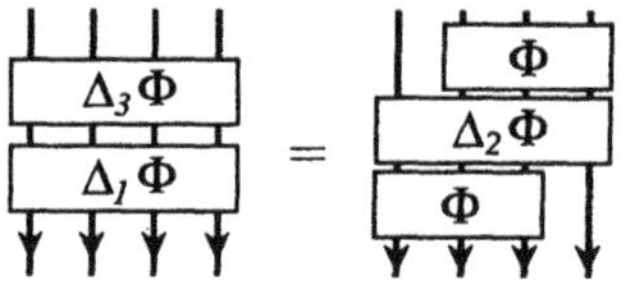

(6.11)

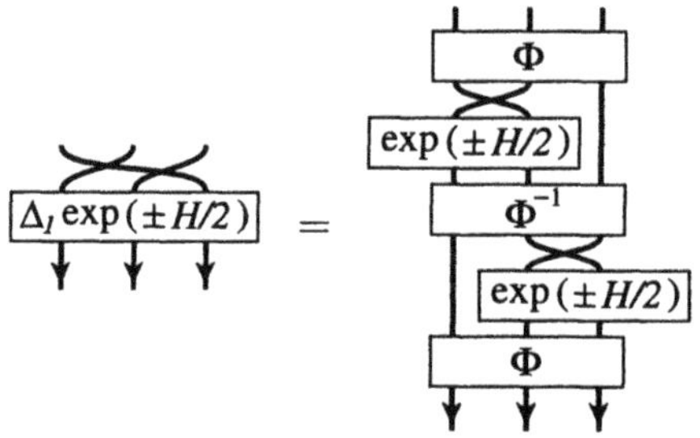

where $H =$  . (6.12)

Here, the map ε_i and Δ_i denote ε and Δ acting on the i-th strand, and we define the notion of group-like later in the next section. The relations (6.11) and (6.12) are called the *pentagon* and *hexagon relations* of an associator Φ. Note that the relations $\varepsilon_1\Phi = \varepsilon_3\Phi = 1$ are derived from the relation $\varepsilon_2\Phi = 1$ and the pentagon relation (6.11). Further, the following relation is derived from the hexagon relation (6.12),

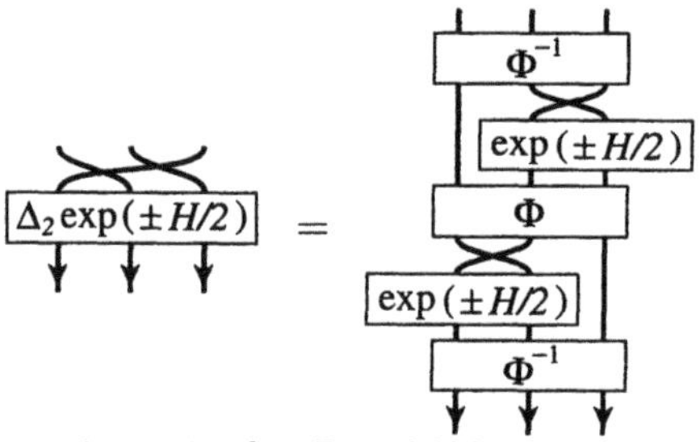

(6.13)

A first example of an associator is the *Drinfel'd associator* given by

$\Phi_{KZ} = \varphi_{KZ}($  $) \in \mathcal{A}(\downarrow\downarrow\downarrow)$

for the power series φ_{KZ} defined in (5.24). See Appendix D.2 for the proof that this is an associator. In Appendix D we exhibit some concrete calculations of some associators in low degrees. To define the framed Kontsevich invariant $\hat{Z}$ an associator Φ must be chosen. (Although $\hat{Z}(T)$ will, in general, depend on the choice of Φ, Theorem 6.7 will show that $\hat{Z}(L)$ is independent of the choice in the case that L is a link.) Further, we define the element ν by

$$\nu = \Big($$ 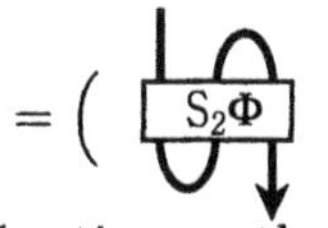

$$\Big)^{-1},$$

where S_2 denotes the antipode S acting on the second strand. It is left to the reader to show that ν is uniquely determined, independently of the choice of Φ; (this can be shown by Theorem 6.7).

Let X be a compact oriented 1-manifold, and let T be an oriented framed quasi-tangle obtained by equipping an embedding (as a quasi-tangle) of X into $\mathbb{R}^2 \times [0,1]$ with a framing. We fix a sliced quasi-tangle diagram expressing the quasi-tangle T. We define the framed Kontsevich invariant $\hat{Z}(T) \in A(X)$ by gluing $\hat{Z}(E)$ for elementary quasi-tangle diagrams E in the sliced quasi-tangle diagram, where we give $\hat{Z}(E)$ for each elementary quasi-tangle E as follows.

Firstly, we give $\hat{Z}(E)$ for the following six elementary quasi-tangle diagrams with particular orientations,

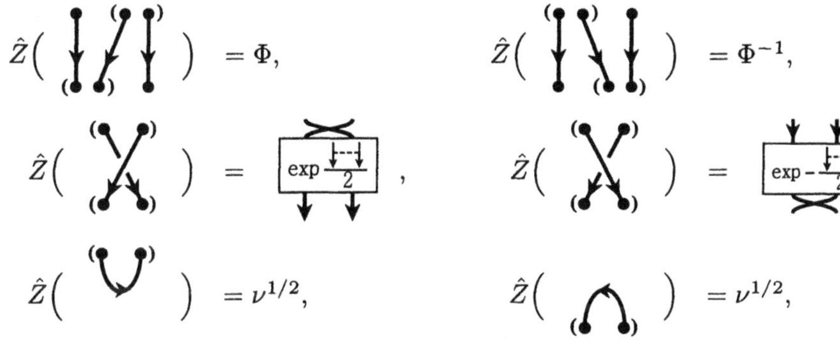

where Φ and ν are defined above. Further, the right hand side of the third formula implies the following expansion:

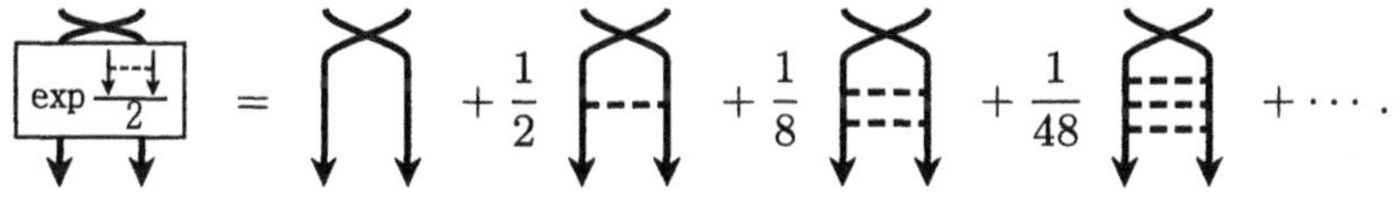

Secondly, we give $\hat{Z}(E)$ for the other elementary quasi-tangle diagrams with particular orientations. We define $\hat{Z}\Big($ 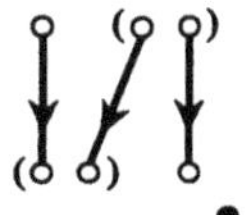 $\Big)$ and $\hat{Z}\Big($ 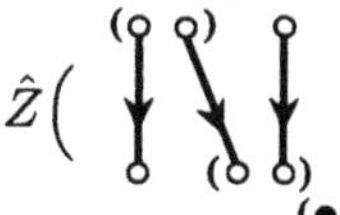 $\Big)$ by successively applying the comultiplication Δ to $\hat{Z}\Big($ 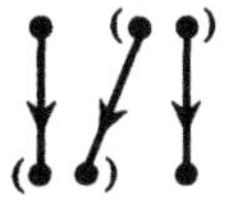 $\Big)$ and $\hat{Z}\Big($ 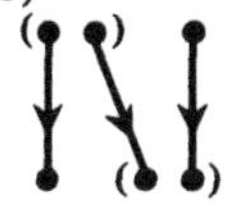 $\Big)$ as follows; recall that each  denotes a parenthesized bundle of vertical lines. When a diagram E' is obtained from E by replacing a component C of E with the union of two parallel copies of C, we define $\hat{Z}(E')$ to be $\Delta_{(C)}\hat{Z}(E)$. For example,

$\hat{Z}\Big($ 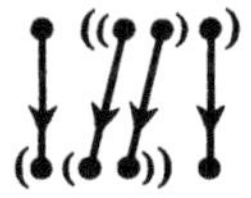 $\Big) = \Delta_2\hat{Z}\Big($ 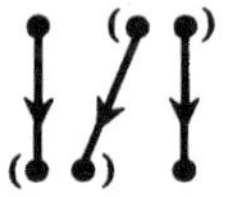 $\Big) = \Delta_2\Phi.$

We define $\hat{Z}\Big($  $\Big)$ and $\hat{Z}\Big($  $\Big)$ similarly, by successively applying the comultiplication Δ to $\hat{Z}\Big($  $\Big)$ and to $\hat{Z}\Big($  $\Big)$.

Thirdly, we give $\hat{Z}(E)$ for elementary quasi-tangle diagrams E with the other orientations. We define $\hat{Z}(E)$ by successively applying the antipode S to the values of $\hat{Z}$ already defined above. When a diagram E'' is obtained from a diagram E by reversing the orientation on some component C of E, we define $\hat{Z}(E'')$ to be $S_{(C)}\hat{Z}(E)$. For example,

$\hat{Z}\Big($ 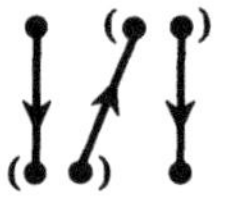 $\Big) = S_2\hat{Z}\Big($ 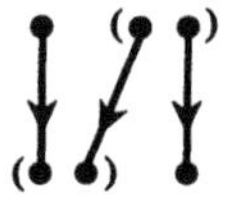 $\Big) = S_2\Phi.$

Lastly, we define $\hat{Z}(T)$ for an oriented framed quasi-tangle T with a fixed sliced quasi-tangle diagram expressing T by composing the $\hat{Z}(E)$ of the elementary quasi-tangle diagrams E in the sliced quasi-tangle diagram.

Theorem 6.7. For a fixed associator $\Phi \in \mathcal{A}(\downarrow\downarrow\downarrow)$, $\hat{Z}(T)$ is an isotopy invariant of a framed oriented quasi-tangle T. In particular, for a framed oriented link L, $\hat{Z}(L)$ is an isotopy invariant of L, which does not depend on the choice of an associator Φ.

We call $\hat{Z}(T)$ the *framed Kontsevich invariant* of a quasi-tangle T. The coefficients of the framed Kontsevich invariant $\hat{Z}(L)$ of a framed link L are rational, because of the existence of an associator with rational coefficients (see Appendix D.1).

The framed Kontsevich invariant $\hat{Z}(L) \in \mathcal{A}(\sqcup^l S^1)$ of a framed oriented link L modulo the FI relation is equal to the Kontsevich invariant $Z(L) \in \mathcal{A}(\sqcup^l S^1)/\mathrm{FI}$ of L defined in Section 6.2.

Proof. By Theorem 6.5 it is sufficient to show that $\hat{Z}(T)$ is invariant under the moves in Figure 6.8, *i.e.*, under the moves (6.7), (6.8), (6.9) and the moves (5.39)–(5.45) in Figure 5.6.

The invariance of $\hat{Z}$ under the move (5.39) is derived from the relations $\Phi \cdot \Phi^{-1} = \Phi^{-1} \cdot \Phi =$ 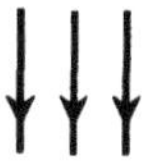 . The invariance under the move (5.43) is obtained, since 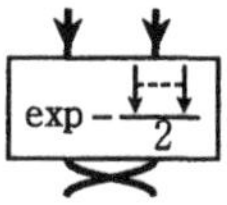 is the inverse of 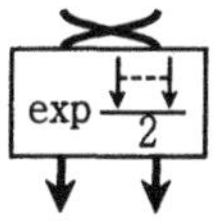 .

The invariance under the moves of the pentagon relation (5.40) and the hexagon relations (5.45) of parenthesized braids are obtained by the pentagon relation (6.11) and the hexagon relations (6.12) of the associator Φ, respectively.

The invariance of $\hat{Z}$ under the moves (5.41) and (5.44) is obtained from the definition of $\hat{Z}$ and Proposition 6.1 (2).

The relation (5.42) is obtained from the definition of $\hat{Z}$.

The invariance of $\hat{Z}$ under the moves (6.7) is obtained from the following changes of $\hat{Z}$:

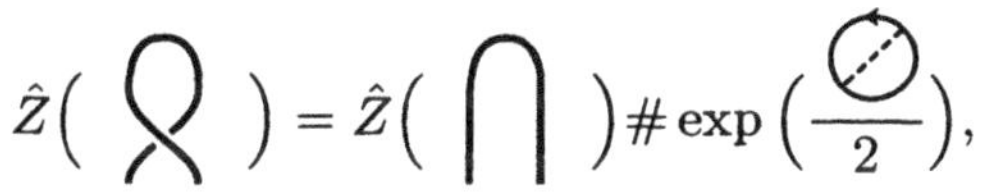

(6.14)

(6.15)

where the notation "#" denotes the connected sum of Jacobi diagrams along solid lines; we showed that the connected sum of Jacobi diagrams is well defined in Proposition 6.3 (3). (To be precise, we need the invariance under the moves (6.8), shown below, to complete this part.)

The invariance under the moves (6.8) is shown as follows. We obtain the invariance of $\hat{Z}$ under the second move of (6.8) by

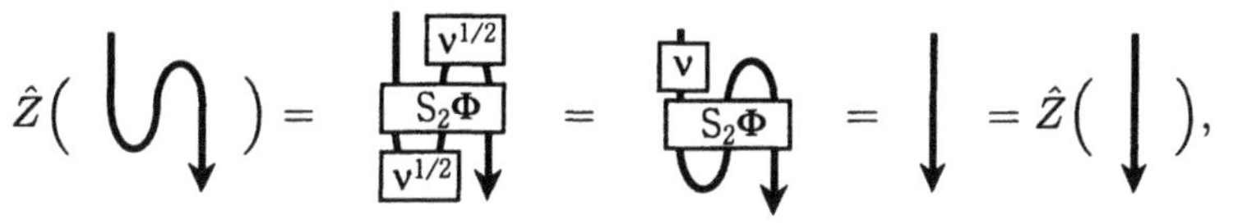

(6.16)

where we obtain the second equality by Proposition 6.1 (1). Further, we obtain the invariance under the first move of (6.8) as follows. By the invariance of $\hat{Z}$ under the pentagon relation we have that

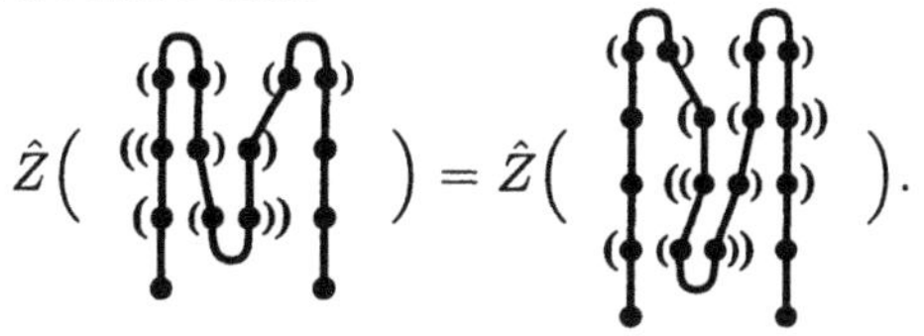

(6.17)

Both sides of the above formula are computed as

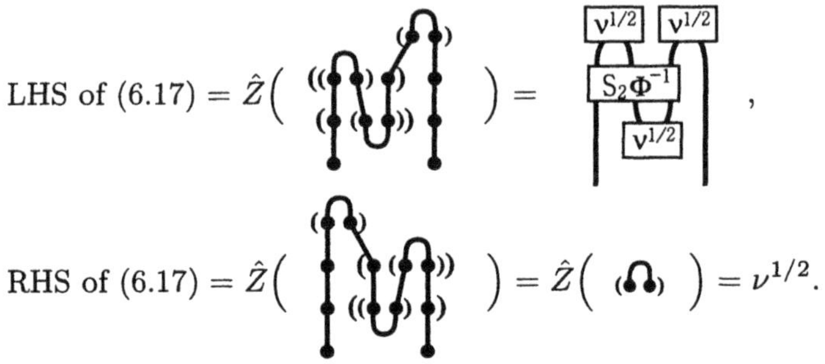

Comparing the above two formulae we obtain the invariance of $\hat{Z}$ under the first move of (6.8).

The invariance of $\hat{Z}$ under the moves (6.9) is obtained from the hexagon relations (6.12) of the associator Φ as follows. We have that

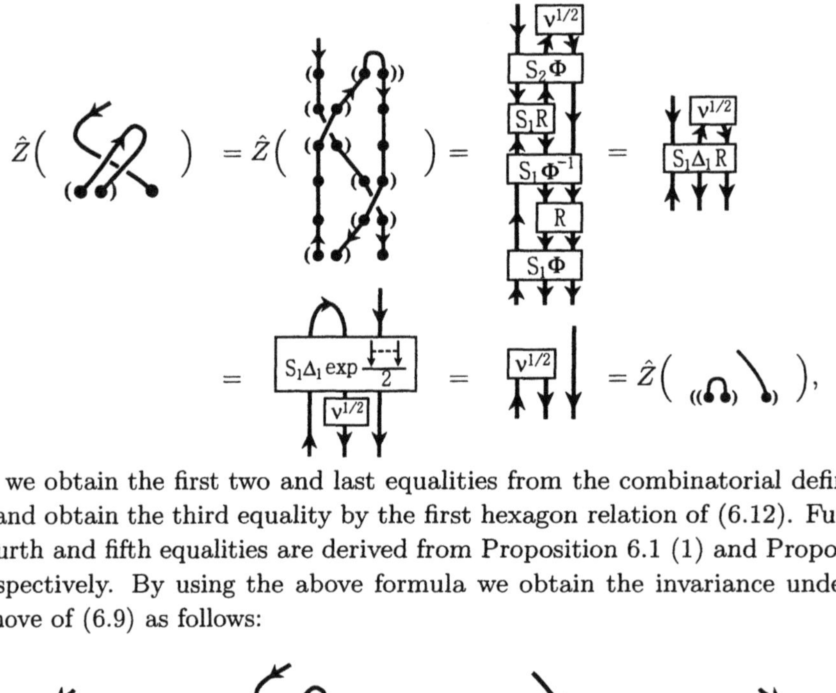

where we obtain the first two and last equalities from the combinatorial definition of $\hat{Z}$, and obtain the third equality by the first hexagon relation of (6.12). Further, the fourth and fifth equalities are derived from Proposition 6.1 (1) and Proposition 6.2 respectively. By using the above formula we obtain the invariance under the first move of (6.9) as follows:

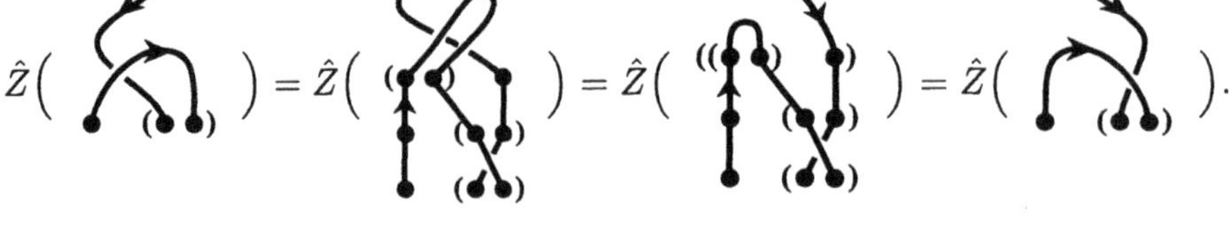

We obtain invariance under the second moves of (6.9) similarly.

For a framed link L, the value of $\hat{Z}(L)$ does not depend on the choice of an associator Φ, for the following reason. Proposition D.6 says that any other associator Φ' can be obtained from Φ by twisting it by some $f \in \mathcal{A}(\downarrow\downarrow)$; for the definition of twisting see Appendix D.1. Hence, when we change associators, $\hat{Z}(T)$ of a quasi-tangle T is changed by an action of f which is localized at (and depends only on) the parenthesized sets of dots at the ends of T. (See the proof of Theorem 5.16 for a similar argument.) Therefore, $\hat{Z}(L)$ is unchanged for a link L, since a link has no ends. □

6.5 Properties of the framed Kontsevich invariant

In this section we show some properties of the framed Kontsevich invariant, based on the combinatorial definition of it given in the previous section.

The framed Kontsevich invariant $\hat{Z}$ commutes with $\Delta_{(C)}$, $S_{(C)}$, and $\varepsilon_{(C)}$.

Proposition 6.8 ([LeMu97, Theorem 5.1]). Let C be a circle, and X a disjoint union of circles. Let $K \cup L$ be a framed link obtained from an embedding of $C \sqcup X$ by adding framing information, where K denotes the component corresponding to the embedding of C. We consider the framed Kontsevich invariant $\hat{Z}(K \cup L) \in \mathcal{A}(C \sqcup X)$.
(1) Let $K^{(2)} \cup L$ be the link obtained from $K \cup L$ by replacing the component K with two parallel copies $K^{(2)}$ of K along the given framing of K. Then,

$$\hat{Z}(K^{(2)} \cup L) = \Delta_{(C)}\big(\hat{Z}(K \cup L)\big). \tag{6.18}$$

(2) Let $\overline{K} \cup L$ be the link obtained from $K \cup L$ by reversing the orientation of K. Then,

$$\hat{Z}(\overline{K} \cup L) = S_{(C)}\big(\hat{Z}(K \cup L)\big). \tag{6.19}$$

(3) We have that

$$\hat{Z}(L) = \varepsilon_{(C)}\big(\hat{Z}(K \cup L)\big). \tag{6.20}$$

As in Chapter 4 we define the *comultiplication* $\Delta_{(C)}$, the *antipode* $S_{(C)}$, and the *counit* $\varepsilon_{(C)}$ of a component of a link by

$$\begin{aligned}
\Delta_{(C)}(K \cup L) &= K^{(2)} \cup L,\\
S_{(C)}(K \cup L) &= \overline{K} \cup L,\\
\varepsilon_{(C)}(K \cup L) &= L,
\end{aligned}$$

with notations as in the proposition. Then, the proposition implies that the framed Kontsevich invariant $\hat{Z}$ commutes with $\Delta_{(C)}$, $S_{(C)}$, and $\varepsilon_{(C)}$.

Proof of Proposition 6.8. The first step is to factor $K \cup L$ into elementary tangle diagrams. By the combinatorial construction of $\hat{Z}$, we can then prove the proposition by proving an appropriate version of it for each elementary quasi-tangle diagram. Parts (6.19) and (6.20) follow easily in this way. Now consider the relation (6.18).

For the diagrams 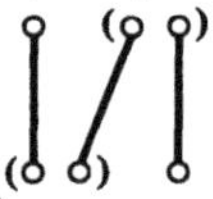 and 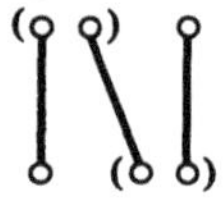 we obtain (6.18) immediately from the definition of $\hat{Z}$.

For the diagrams  and  we obtain (6.18) by the hexagon relations (6.12).

The remainder of this proof concerns the elementary tangle diagrams  and . The sides of (6.18) for the diagram  are computed as follows:

$\Delta\Big(\hat{Z}($  $)\Big) = \Delta(\nu^{1/2}),$

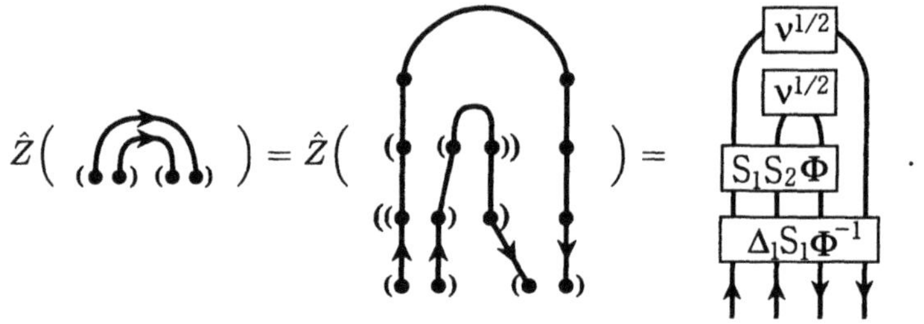

If the above formulae were equal, then the proof would be complete, but unfortunately they are not equal in general, though they might be equal for a particular choice of an associator Φ in the definition of $\hat{Z}$. We let n denote the error which relates the above two expressions, that is,

$n =$ 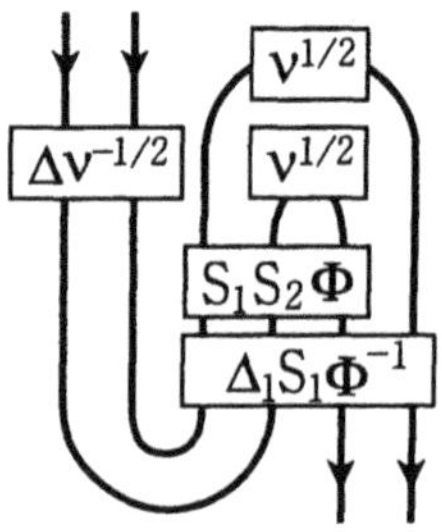 $\in \mathcal{A}(\downarrow\downarrow).$

Then,

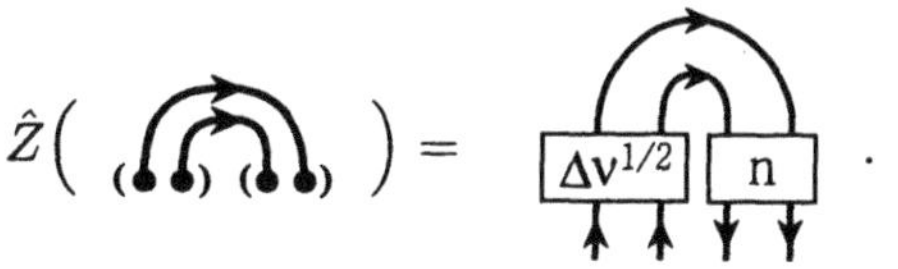

(6.21)

Further, we show that the error relating the two sides of (6.18), in the case of the diagram  , is also equal to n, as follows. We have that

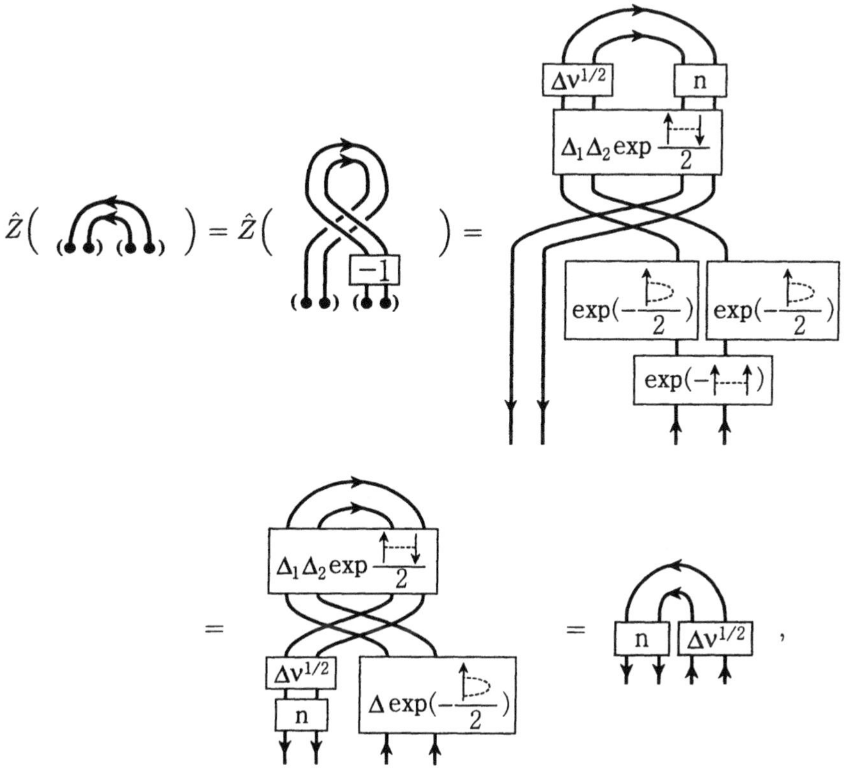

(6.22)

where "−1" in the second picture denotes −1 full twist. Here, we obtain the third equality by Proposition 6.1 (2), and obtain the last equality by applying the comultiplication Δ to the following equation,

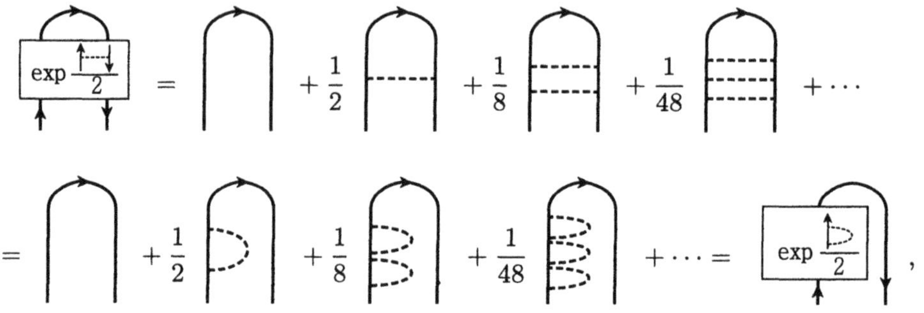

where we obtain the second equality by Proposition 6.1 (1). Moreover, we put

$u =$ 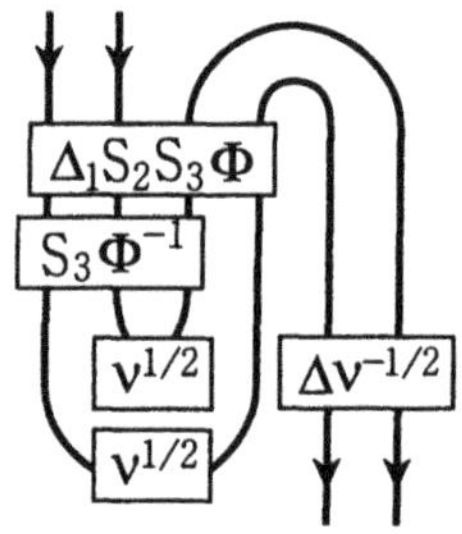 $\in \mathcal{A}(\downarrow\downarrow)$.

Then, by a similar argument as above, we have that

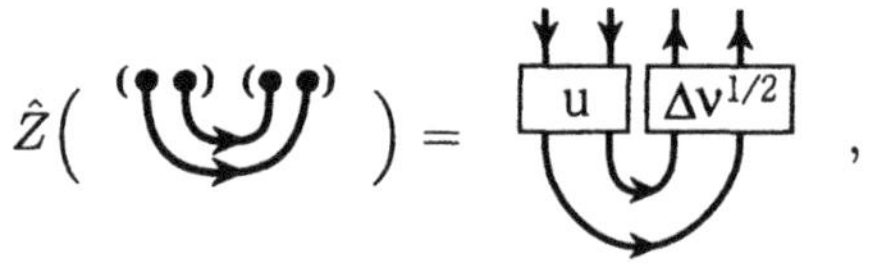 (6.23)

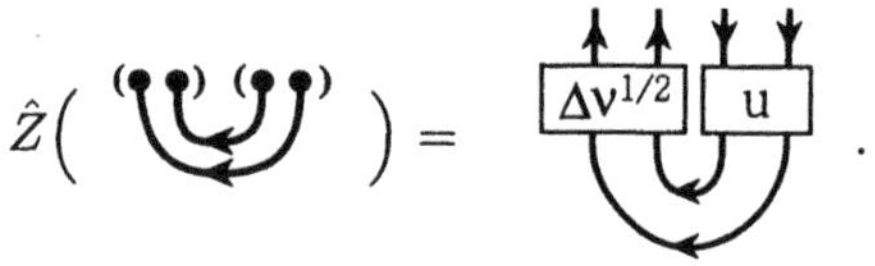 (6.24)

Further, u and n are related as

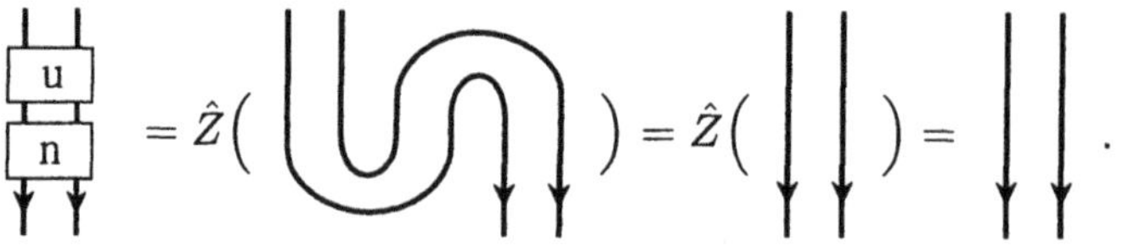 (6.25)

Similarly, we obtain

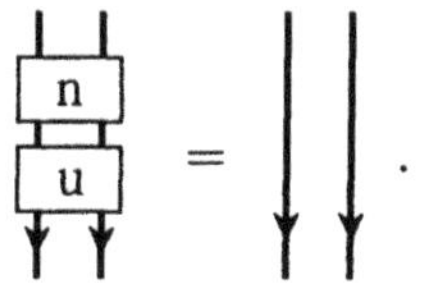

(6.26)

As mentioned in Section 3.1 any oriented link $K \cup L$ can be presented by some sliced tangle diagram which is a composition of tensor products of copies of elementary oriented tangle diagrams in Figure 3.5. Further, since we consider a link in this case, the diagram is equal to such a composition of copies of the following diagrams.

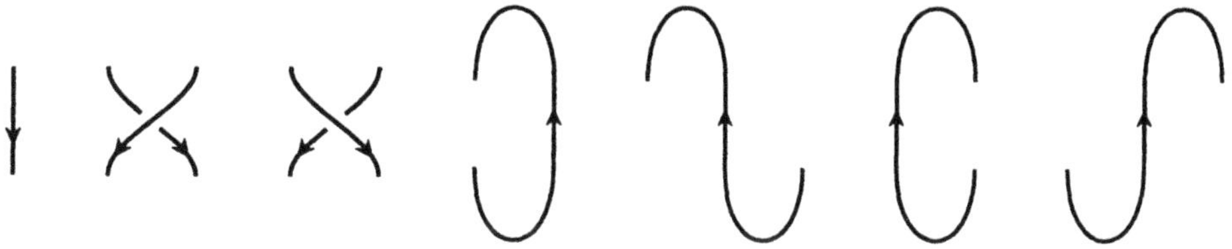

We have already verified that the first three diagrams satisfy (6.18). Further, the fourth diagram satisfies (6.18) by (6.22) and (6.24), since the errors n and u cancel by (6.25). It is verified similarly that the last three diagram satisfy (6.18). □

The framed Kontsevich invariant $\hat{Z}(L)$ of a link L is group-like.

We define the map

$$\hat{\Delta} : \mathcal{A}(\sqcup^l S^1) \to \mathcal{A}(\sqcup^l S^1) \otimes \mathcal{A}(\sqcup^l S^1)$$

as follows; see also [Bar95b] for its definition. For a Jacobi diagram D, we define $\hat{\Delta}(D)$ to be the sum of $D' \otimes D''$ where D' runs over all Jacobi diagrams obtained as a subset of D by removing some of the dashed connected components and D'' is the Jacobi diagram consisting of solid circles and the other components of dashed components. For example,

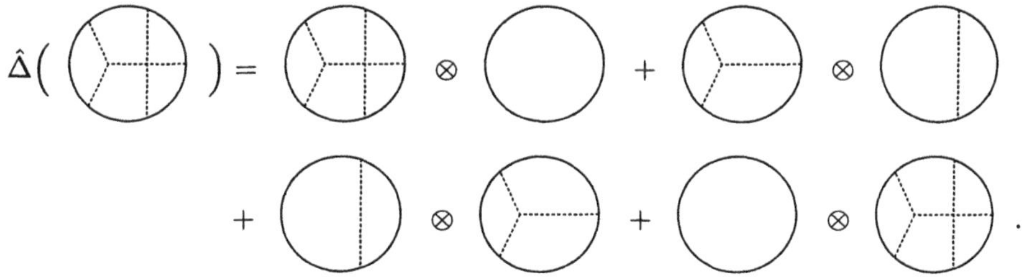

The map $\hat{\Delta}$ behaves similarly as the comultiplication in a Hopf algebra. We call an element $\alpha \in \mathcal{A}(\sqcup^l S^1)$ *group-like* if $\hat{\Delta}(\alpha) = \alpha \otimes \alpha$, like the terminology in the theory of Hopf algebras.

Proposition 6.9 ([LMO98, Theorem 3.7]). For an oriented framed link L with l components, the value of $\hat{Z}(L)$ is group-like in $\mathcal{A}(\sqcup^l S^1)$, that is,

$$\hat{\Delta}\Big(\hat{Z}(L)\Big) = \hat{Z}(L) \otimes \hat{Z}(L).$$

Proof. Let X_1 and X_2 be two copies of $\sqcup^l S^1$. For a Jacobi diagram D on $X_1 \sqcup X_2$ we define $p(D)$ by

$$p(D) = \begin{cases} 0, & \text{if } X_1 \text{ and } X_2 \text{ are connected by the uni-trivalent graph of } D, \\ D_1 \otimes D_2, & \text{otherwise,} \end{cases}$$

where in the latter case D is equal to the disjoint union of some Jacobi diagrams D_1 on X_1 and D_2 on X_2. By its linear extension we obtain a map $p : \mathcal{A}(X_1 \sqcup X_2) \to \mathcal{A}(X_1) \otimes \mathcal{A}(X_2)$. From the definition of $\hat{\Delta}$, we have that $\hat{\Delta} = p \circ \Delta_{(C_1)} \circ \cdots \circ \Delta_{(C_l)}$, where C_i denotes the ith component of $\sqcup^l S^1$. Therefore, by Proposition 6.8 (1),

$$\hat{\Delta}(\hat{Z}(L)) = \big(p \circ \Delta_{(C_1)} \circ \cdots \circ \Delta_{(C_l)}\big)\Big(\hat{Z}(L)\Big) = p\big(\hat{Z}(L^{(2)})\big), \tag{6.27}$$

where $L^{(2)}$ denotes the union of 2 parallel copies of all components of L.

Let $L \sqcup L$ be the split union of two copies of L. We show that

$$p\big(\hat{Z}(L^{(2)}) - \hat{Z}(L \sqcup L)\big) = 0 \tag{6.28}$$

as follows. We obtain $L^{(2)}$ from $L \sqcup L$ by crossing changes between the first copy of L and the second copy of L. At each of the crossing changes, the difference between the two values of $\hat{Z}$ is obtained by substituting the following expression into the composition which calculates $\hat{Z}$:

$\exp\Big(\frac{1}{2}$ 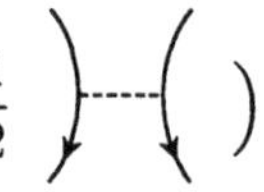 $\Big) - \exp\Big(-\frac{1}{2}$ 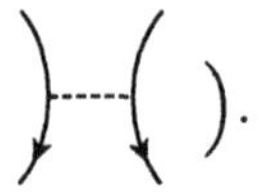 $\Big).$

Hence, the difference is presented by a linear sum of Jacobi diagrams with chords connecting X_1 and X_2. Since they vanish by the map p, we obtain (6.28).

By (6.27) and (6.28) we have that $\hat{\Delta}\big(\hat{Z}(L)\big) = p\big(\hat{Z}(L \sqcup L)\big)$. Further, from the definition of p, we have that $p\big(\hat{Z}(L \sqcup L)\big) = \hat{Z}(L) \otimes \hat{Z}(L)$. Hence, we obtain the proposition. □

The framed Kontsevich invariant $\hat{Z}(K)$ of a knot K

In this subsection we restrict our attention to the framed Kontsevich invariant of knots.

In this case the framed Kontsevich invariant $\hat{Z}(K)$ of a knot K is valued in the space $\mathcal{A}(S^1)$. Here, $\mathcal{A}(S^1)$ forms a Hopf algebra with the multiplication given by the connected sum, and with the comultiplication given by $\hat{\Delta}$. The unit of the algebra $\mathcal{A}(S^1)$ is given by $1 =$  . We call an element α in $\mathcal{A}(S^1)$ *primitive* if α satisfies that $\hat{\Delta}(\alpha) = \alpha \otimes 1 + 1 \otimes \alpha$, like the terminology of a Hopf algebra. It is known (see, *e.g.*, [Abe80]) that any non-zero group-like element of a graded Hopf algebra can be presented by the exponential image of some primitive element. In the present case we have

Lemma 6.10. For any primitive element $\alpha \in \mathcal{A}(S^1)$, $\exp(\alpha)$ is group-like. Conversely, for any non-zero group-like element $\beta \in \mathcal{A}(S^1)$ there exists a primitive element α satisfying that $\beta = \exp(\alpha)$.

Proof. For a primitive element α,

$$\hat{\Delta}\big(\exp(\alpha)\big) = \exp\big(\hat{\Delta}(\alpha)\big) = \exp(\alpha \otimes 1 + 1 \otimes \alpha) = \exp(\alpha) \otimes \exp(\alpha).$$

Hence, $\exp(\alpha)$ is group-like.

Let β be a non-zero group-like element in $\mathcal{A}(S^1)$. We construct α satisfying that $\beta = \exp(\alpha)$ by the following inductive construction with respect to the degrees of Jacobi diagrams. Suppose that we have a primitive element $\alpha^{(<d)} \in \mathcal{A}(S^1)^{(<d)}$ satisfying that $\exp(\alpha^{(<d)})$ is congruent to β modulo $\mathcal{A}(S^1)^{(\geq d)}$. Let $\alpha^{(d)}$ be the degree d part of $\beta \exp(-\alpha^{(<d)})$. Then, by the above assumption, $\beta \exp(-\alpha^{(<d)})$ is congruent to $1 + \alpha^{(d)}$ modulo $\mathcal{A}(S^1)^{(>d)}$. Further, since $\beta \exp(-\alpha^{(<d)})$ is group-like, $\hat{\Delta}(1 + \alpha^{(d)})$ is congruent to $(1 + \alpha^{(d)}) \otimes (1 + \alpha^{(d)})$ modulo $\mathcal{A}(S^1)^{(>d)}$. By picking up the degree d part of this congruence, we have that $\hat{\Delta}(\alpha^{(d)}) = \alpha^{(d)} \otimes 1 + 1 \otimes \alpha^{(d)}$. Hence, $\alpha^{(d)}$ is primitive. Putting $\alpha^{(\leq d)} = \alpha^{(<d)} + \alpha^{(d)}$, this is a primitive element in $\mathcal{A}(S^1)^{(\leq d)}$ satisfying that $\exp(\alpha^{(\leq d)})$ is congruent to β modulo $\mathcal{A}(S^1)^{(>d)}$. Repeating this inductive procedure we obtain a primitive element α satisfying that $\beta = \exp(\alpha)$. □

We denote by $\mathcal{A}(S^1)_{\text{prim}}$ the vector subspace of $\mathcal{A}(S^1)$ spanned by the primitive elements in $\mathcal{A}(S^1)$, and call it the *space of primitive Jacobi diagrams* on S^1. By Proposition 6.9, $\hat{Z}(K)$ is group-like. In particular, when K is the trivial knot, the element ν (which is equal to $\hat{Z}$ of the trivial knot) is group-like. Hence, so is $\hat{Z}(K)\#\nu^{-1}$. Therefore, there exists a primitive element $z(K)$ by Lemma 6.10, which satisfies that $\hat{Z}(K)\#\nu^{-1} = \exp\big(z(K)\big)$. Further, since $z(K)$ is uniquely determined by $\hat{Z}(K)$, $z(K)$ is an isotopy invariant of K in $\mathcal{A}(S^1)_{\text{prim}}$. We call it the *primitive Kontsevich invariant* of a framed oriented knot K.

Let us investigate the space $\mathcal{A}(S^1)_{\text{prim}}$.

Lemma 6.11. The space $\mathcal{A}(S^1)_{\text{prim}}$ is equal to the vector subspace of $\mathcal{A}(S^1)$ spanned by Jacobi diagrams on S^1 with connected uni-trivalent graphs.

Thus, we call a Jacobi diagram on S^1 *primitive* if its uni-trivalent graph is connected.

Proof of Lemma 6.11. A Jacobi diagram with a connected uni-trivalent graph is primitive, from the definition of primitive elements. It is an exercise for the reader to show that the algebra $\mathcal{A}(S^1)$ is generated by Jacobi diagrams with connected uni-trivalent graphs; it is illustrated, for example, as

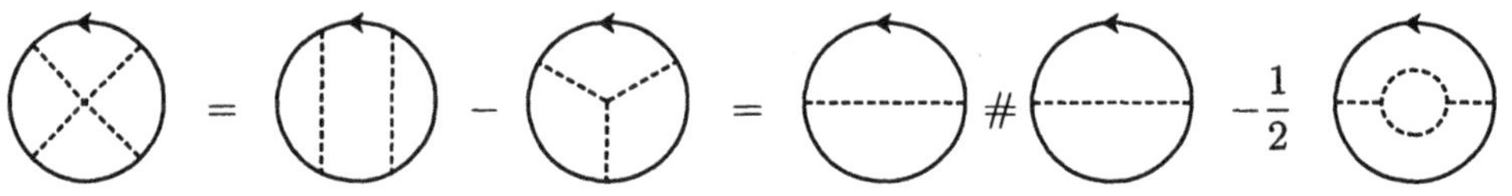

Further, it follows from the definition of primitive elements that a polynomial in Jacobi diagrams with connected uni-trivalent graphs is primitive if and only if the polynomial is equal to a linear sum of Jacobi diagrams with connected uni-trivalent graphs. Hence, the subspace of primitive elements is spanned by Jacobi diagrams with connected uni-trivalent graphs. □

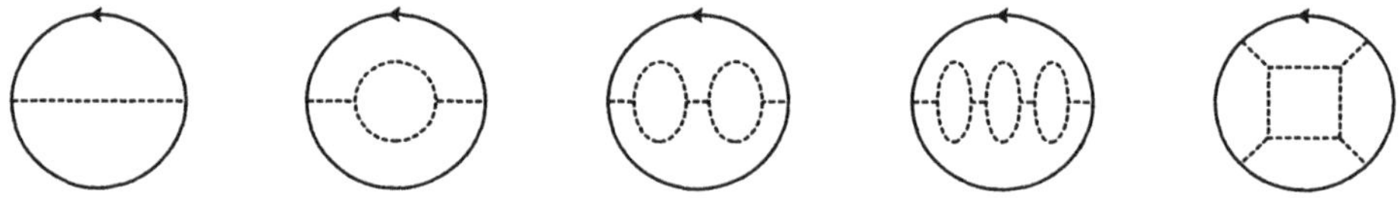

Figure 6.9 A basis of the degree ≤ 4 subspace of the space $\mathcal{A}(S^1)_{\text{prim}}$

Let us investigate Jacobi diagrams on S^1 with connected uni-trivalent graphs in low degrees. By the IHX and STU relations we have the following relations,

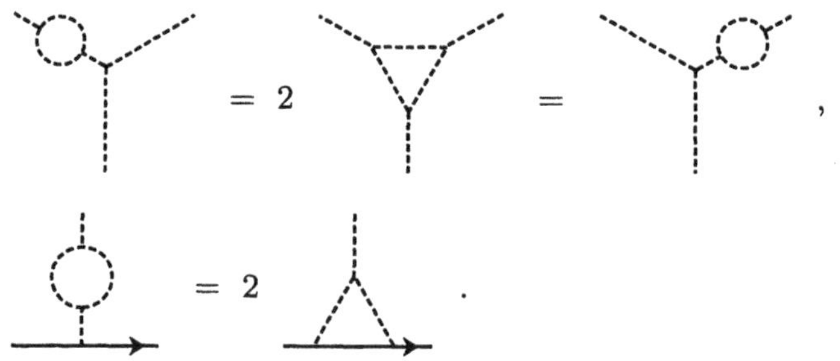

By the relations we can replace a triangle of a Jacobi diagram with a dashed bubble. Further, by the first relation a dashed bubble can move beyond a trivalent vertex. It is left to the reader to show that the set of the Jacobi diagrams in Figure 6.9 is a basis of the degree ≤ 4 subspace of $\mathcal{A}(S^1)_{\text{prim}}$, by using the above relations.

By using the basis in Figure 6.9, the primitive Kontsevich invariant of a framed knot K is presented by

$z(K) = a_1(K)$ 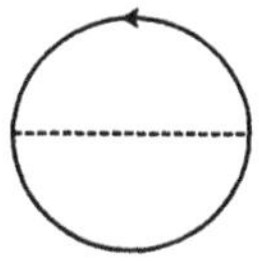 $+ a_2(K)$ 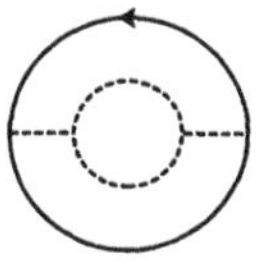 $+ a_3(K)$ 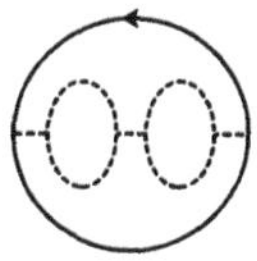

$+ a_4(K)$ 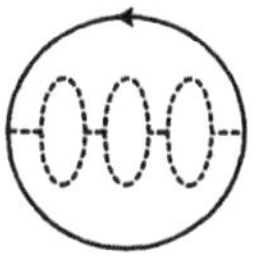 $+ a_4'(K)$ 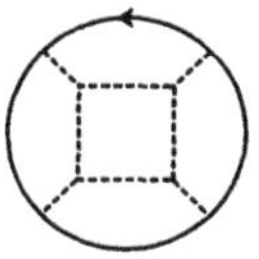 $+$ (terms of degree ≥ 5),

with scalar knot invariants $a_1(K), \cdots, a_4(K)$ and $a_4'(K)$. It is left to the reader to show that $a_1(K)$ is equal to half the framing of K, and $a_2(K)$ is equal to the second coefficient of the Conway polynomial, where the *Conway polynomial* $\nabla_L(z)$ of an oriented link L is defined by $\nabla_L(t^{1/2} - t^{-1/2}) = \Delta_L(t)$ with the Alexander polynomial $\Delta_L(t)$. These scalar invariants are primitive Vassiliev invariants defined later in Section 7.3; see Theorem 7.6.

6.6 Universality of the Kontsevich invariant among quantum invariants

In this section we show that for any pair $(\mathfrak{g}, R)$ the quantum $(\mathfrak{g}, R)$ invariant is recovered from the framed Kontsevich invariant through a map called a weight system. This implies the universality of the framed Kontsevich invariant among the quantum invariants of framed links.

Weight system

Let $\mathfrak{g}$ be a simple Lie algebra, R an irreducible representation of $\mathfrak{g}$, and X a disjoint union of circles. In this subsection we define the linear map $\mathcal{A}(X) \to \mathbb{C}$ obtained by "substituting" $\mathfrak{g}$ into the dashed lines of Jacobi diagrams and R into the solid lines of Jacobi diagrams. The map is denoted by $W_{\mathfrak{g},R}$ and is called the weight system; see also [Bar95b] for its definition. For a Jacobi diagram D we define the value $W_{\mathfrak{g},R}(D)$ in the following two ways.

The first way of defining $W_{\mathfrak{g},R}(D)$ is as follows. Like the definition of operator invariants in Chapter 3 we put the Jacobi diagram D in $\mathbb{R}^2$ and slice it by horizontal lines such that each part between two adjacent horizontal lines includes one of the pictures in (6.29). To each intersection of each horizontal line with a dashed edge, a downward solid edge, and an upward solid edge, we associate, respectively, a copy of $\mathfrak{g}$, a copy of R, and a copy of $R^\star$. Further, we associate the tensor product of all these copies along each horizontal line, with that horizontal line. In particular, we associate the trivial representation $\mathbb{C}$ to a horizontal line which has

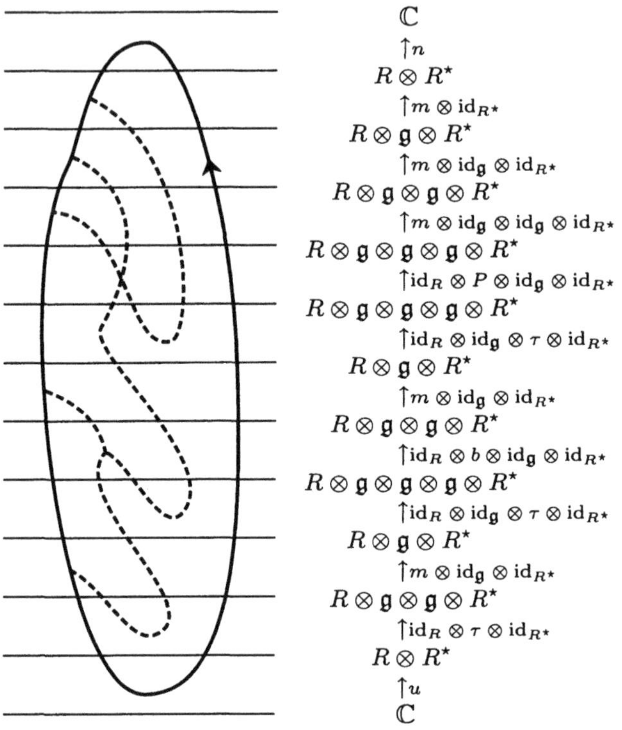

Figure 6.10 The weight system $W_{\mathfrak{g},R}$ of a Jacobi diagram D is defined to be the image of 1 by the composition of the linear maps given in (6.29); for example, as shown above. The map P in the above linear maps denotes the permutation of two entries.

no intersection with the Jacobi diagram. We give a linear map between two tensor products corresponding to two adjacent horizontal lines as follows,

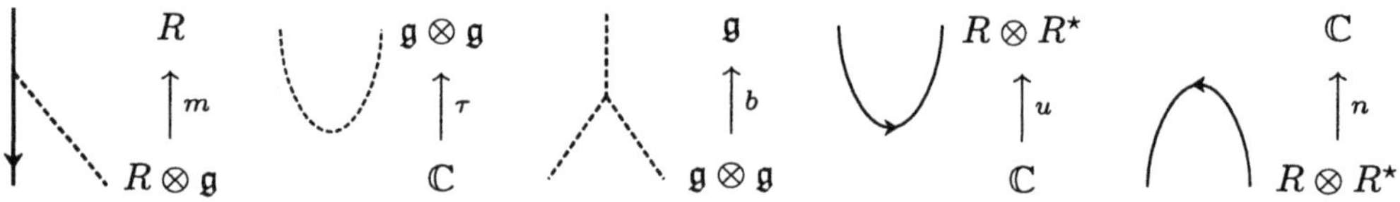

(6.29)

where the first map m is the action of $\mathfrak{g}$ on R regarding R as a right $\mathfrak{g}$ module, the second map τ is the map taking 1 to the invariant 2-tensor τ of $\mathfrak{g} \otimes \mathfrak{g}$ (which is defined to be the dual of the Killing form; see Section 5.1 for its definition), the third map b is the Lie bracket, the fourth map u is the map taking 1 to $\sum_i e_i \otimes e_i^\star$ for a basis $\{e_i\}$ of R and its dual basis $\{e_i^\star\}$ of $R^\star$, and the last map n is given by $n(x \otimes f) = f(x)$ for $x \in R$ and $f \in R^\star$. By composing the linear maps, we obtain a linear map $\mathbb{C} \to \mathbb{C}$ as shown in Figure 6.10. We define $W_{\mathfrak{g},R}(D)$ to be the image of 1 under the composed map $\mathbb{C} \to \mathbb{C}$.

The second way of defining $W_{\mathfrak{g},R}(D)$ is as follows. Let $\{\mathfrak{g}_a\}_{a\in I}$ be an orthonormal basis with respect to the Killing form on $\mathfrak{g}$, and let f_{bcd} be the structure constants of $\mathfrak{g}$ with respect to the basis $\{\mathfrak{g}_a\}$, *i.e.*, $[\mathfrak{g}_b, \mathfrak{g}_c] = \sum_d f_{bcd}\mathfrak{g}_d$. By the associativity of the Killing form shown in Section 5.1, we have a symmetry $f_{bcd} = f_{cdb} = f_{dbc}$. For a Jacobi diagram D, we consider labeling each dashed edge by an element of I. We associate the structure constant f_{bcd} to a dashed trivalent vertex such that the three edges around the vertex are labeled by b, c, and d in the opposite order of the vertex-orientation of the trivalent vertex, and associate the basis element $\mathfrak{g}_a$ to a univalent dashed vertex which adjoins an edge labeled by a. We define the weight of such a labeling to be the product of all these associated structure constants times the trace in R of all the labels on the univalent vertices, as read along the opposite orientation of the solid line. We define $W_{\mathfrak{g},R}(D)$ to be the sum over all labelings of their weights. For example,

$$W_{\mathfrak{g},R}\Big(\ \text{[diagram]}\ \Big) = \sum_{a,b,c,d\in I} f_{dcb}\mathrm{trace}\Big(\rho_R(\mathfrak{g}_d)\rho_R(\mathfrak{g}_c)\rho_R(\mathfrak{g}_a)\rho_R(\mathfrak{g}_b)\rho_R(\mathfrak{g}_a)\Big), \quad (6.30)$$

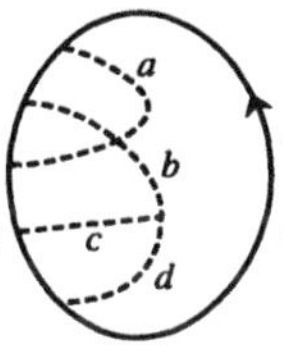

where ρ_R denotes the representation $\mathfrak{g} \to \mathrm{End}(R)$.

Lemma 6.12. $W_{\mathfrak{g},R}(D)$ is well defined for each Jacobi diagram D.

Proof. We show that the above two definitions give the same value $W_{\mathfrak{g},R}(D)$, for simplicity, for the Jacobi diagram D in Figure 6.10.

In addition to the maps in (6.29) we consider the following map t

$$\begin{array}{c} \mathbb{C} \\ \uparrow t \\ \mathfrak{g}\otimes\mathfrak{g}\otimes\mathfrak{g} \end{array} \qquad (6.31)$$

given by $t(X \otimes Y \otimes Z) = B([X,Y],Z)$ for the Killing form B. The map t is alternatively given by $t(\mathfrak{g}_b\otimes\mathfrak{g}_c\otimes\mathfrak{g}_d) = f_{bcd}$ with the structure constants in the second definition of $W_{\mathfrak{g},R}(D)$. Further, it can be shown by an elementary computation that the composition of the following maps is equal to the map b in (6.29).

$$\begin{array}{c} \mathfrak{g} \\ \uparrow \mathrm{id}_{\mathfrak{g}}\otimes t \\ \mathfrak{g}\otimes\mathfrak{g}\otimes\mathfrak{g}\otimes\mathfrak{g} \\ \uparrow \tau\otimes\mathrm{id}_{\mathfrak{g}}\otimes\mathrm{id}_{\mathfrak{g}} \\ \mathfrak{g}\otimes\mathfrak{g} \end{array}$$

In the first definition of $W_{\mathfrak{g},R}(D)$ we temporarily regard the map b as presented by the composition of t and τ as above. Further, instead of the picture in Figure 6.10

we consider the picture below and compute the composed map $\mathbb{C} \to \mathbb{C}$ by using the maps τ, t, and m as follows.

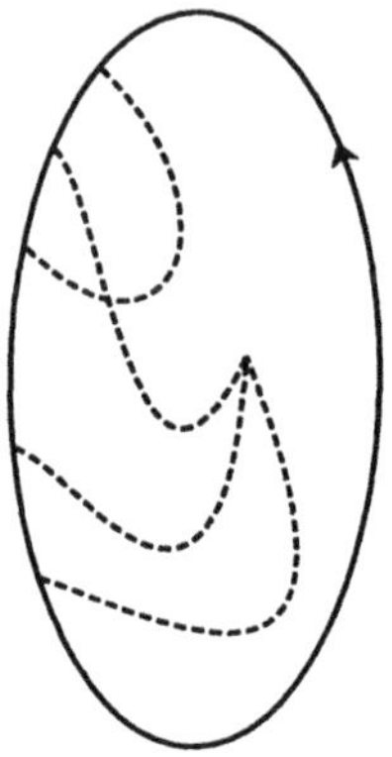

By using the orthonormal basis $\{\mathfrak{g}_a\}$ the map τ is given by $\tau(1) = \sum_a \mathfrak{g}_a \otimes \mathfrak{g}_a$. We obtain the right hand side of (6.30) as the image of the composition of the following linear maps,

$$
\begin{aligned}
1 &\longmapsto \sum_i e_i \otimes e_i^\star \\
&\longmapsto \sum_{i,c,d} e_i \otimes \mathfrak{g}_d \otimes \mathfrak{g}_c \otimes \mathfrak{g}_c \otimes \mathfrak{g}_d \otimes e_i^\star \\
&\longmapsto \sum_{i,c,d} e_i \rho_R(\mathfrak{g}_d)\rho_R(\mathfrak{g}_c) \otimes \mathfrak{g}_c \otimes \mathfrak{g}_d \otimes e_i^\star \\
&\longmapsto \sum_{i,b,c,d} e_i \rho_R(\mathfrak{g}_d)\rho_R(\mathfrak{g}_c) \otimes \mathfrak{g}_b \otimes \mathfrak{g}_b \otimes \mathfrak{g}_c \otimes \mathfrak{g}_d \otimes e_i^\star \\
&\longmapsto \sum_{i,b,c,d} f_{dcb} \cdot e_i \rho_R(\mathfrak{g}_d)\rho_R(\mathfrak{g}_c) \otimes \mathfrak{g}_b \otimes e_i^\star \\
&\longmapsto \sum_{i,a,b,c,d} f_{dcb} \cdot e_i \rho_R(\mathfrak{g}_d)\rho_R(\mathfrak{g}_c) \otimes \mathfrak{g}_b \otimes \mathfrak{g}_a \otimes \mathfrak{g}_a \otimes e_i^\star \\
&\longmapsto \sum_{i,a,b,c,d} f_{dcb} \cdot e_i \rho_R(\mathfrak{g}_d)\rho_R(\mathfrak{g}_c)\rho_R(\mathfrak{g}_a)\rho_R(\mathfrak{g}_b)\rho_R(\mathfrak{g}_a) \otimes e_i^\star \\
&\longmapsto \sum_{a,b,c,d} f_{dcb} \operatorname{trace}\Big(\rho_R(\mathfrak{g}_d)\rho_R(\mathfrak{g}_c)\rho_R(\mathfrak{g}_a)\rho_R(\mathfrak{g}_b)\rho_R(\mathfrak{g}_a)\Big).
\end{aligned}
$$

Hence, the two definition of $W_{\mathfrak{g},R}(D)$ give the same value.

From the second way of defining $W_{\mathfrak{g},R}(D)$ it follows that $W_{\mathfrak{g},R}(D)$ is well defined, once an orthonormal basis $\{\mathfrak{g}_a\}$ of $\mathfrak{g}$ has been chosen. Further, as shown above, its resulting value can also be expressed by the first way of defining $W_{\mathfrak{g},R}(D)$. As we can see in Figure 6.10, the first way is independent of the choice of a basis of $\mathfrak{g}$. It follows from the second way and this independence that $W_{\mathfrak{g},R}(D)$ is well defined. □

Recall that the space $\mathcal{A}(X)$ is defined to be the vector space spanned by Jacobi diagrams on X subject to the AS, IHX, and STU relations shown in Figure 6.2.

Proposition 6.13. The map $W_{\mathfrak{g},R} : \mathcal{A}(X) \to \mathbb{C}$ is well defined.

We call the map the *weight system* derived from the substitution of a Lie algebra $\mathfrak{g}$ and its representation R into Jacobi diagrams.

Proof. We show that $W_{\mathfrak{g},R}(D)$ is invariant under the AS, IHX, and STU relations.

Firstly, we obtain invariance under the AS relation by the anti-symmetry of the Lie bracket, $[X,Y] = -[Y,X]$.

Secondly, we obtain invariance under the IHX relation as follows. We take each side of the IHX relation by the weight system.

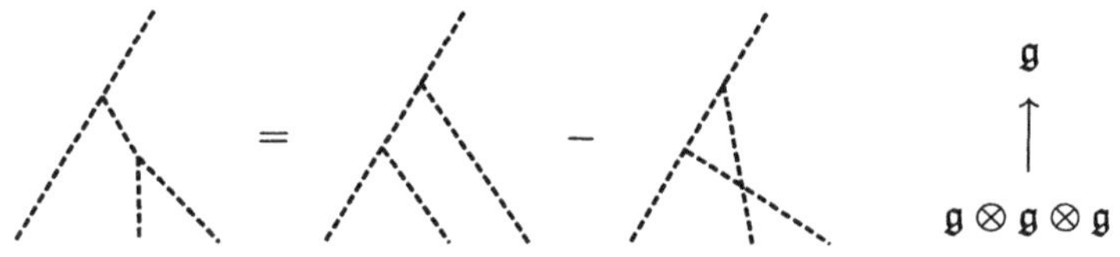

Then, corresponding to both sides of the IHX relation, we have the following two linear maps,

$$X \otimes Y \otimes Z \longmapsto [X,[Y,Z]],$$
$$X \otimes Y \otimes Z \longmapsto [[X,Y],Z] - [[X,Z],Y],$$

for $X, Y, Z \in \mathfrak{g}$. The above two images are equal by the Jacobi identity,

$$[[X,Y],Z] + [[Y,Z],X] + [[Z,X],Y] = 0,$$

of the Lie algebra $\mathfrak{g}$. Hence, we obtain invariance under the IHX relation.

Lastly, we obtain invariance under the STU relation by computing both sides of the following relation,

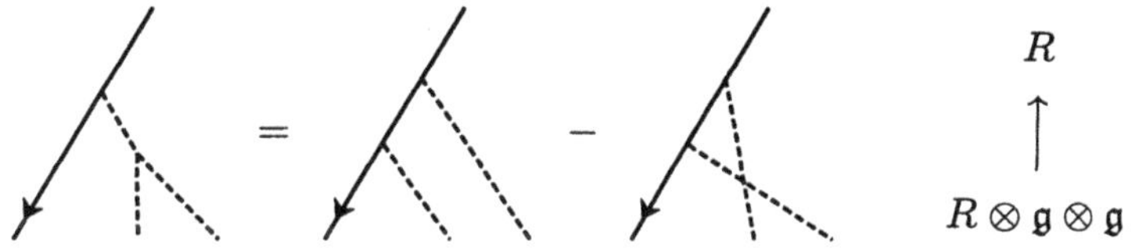

by using the defining relation of representations

$$\rho_R([X,Y]) = \rho_R(X)\rho_R(Y) - \rho_R(Y)\rho_R(X),$$

noting that we regard R as a right $\mathfrak{g}$ module in the definition of weight systems. □

If X is not closed (for example, as in Figure 6.11) then the value $W_{\mathfrak{g},R}(D)$, for a Jacobi diagram D on X, is defined to be a linear map between the two representations of $\mathfrak{g}$ corresponding to the top and the bottom horizontal lines, as in

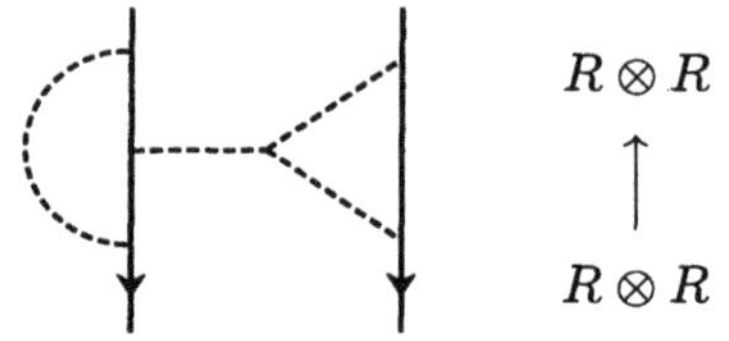

Figure 6.11 The weight system of a Jacobi diagram on a compact 1-manifold with boundary is an intertwiner.

the first definition of $W_{\mathfrak{g},R}$. The linear map $W_{\mathfrak{g},R}(D)$ is an intertwiner, where we call a linear map between two representations compatible with the action of $\mathfrak{g}$ an *intertwiner*, because in the first definition of the weight system each map in (6.29) is an intertwiner. We show another proof of the fact that $W_{\mathfrak{g},R}(D)$ is an intertwiner as follows. For simplicity, we consider a Jacobi diagram D on an interval. In the same way as in the proof of Proposition 6.3 we obtain the following equality.

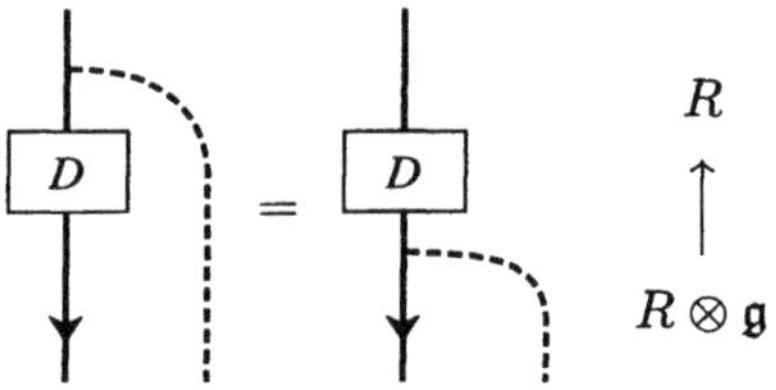

The image of this equality by the weight system $W_{\mathfrak{g},R}$ implies that $W_{\mathfrak{g},R}(D)$ is an intertwiner.

As an exercise let us compute some examples of the weight system for the Lie algebra sl_2. We have that

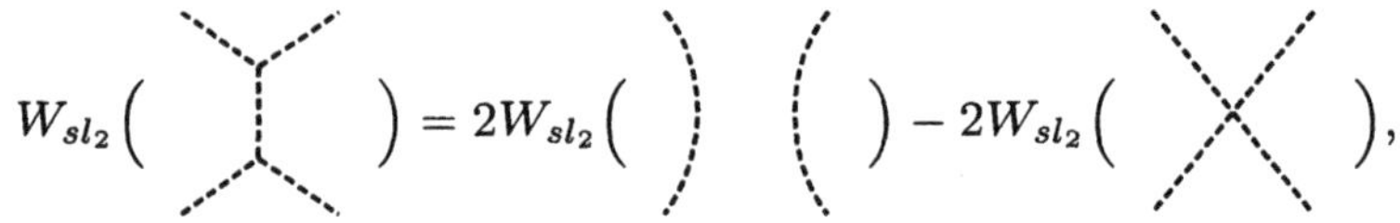

(6.32)

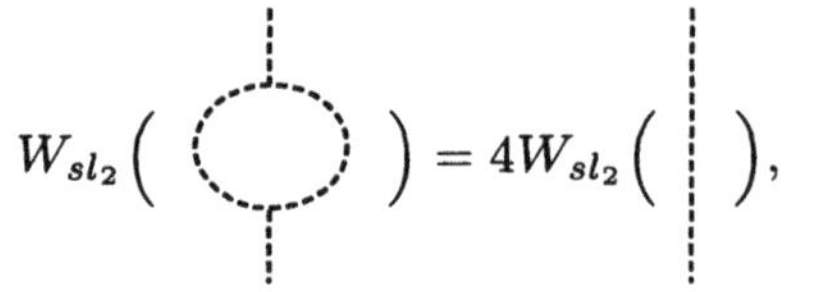

(6.33)

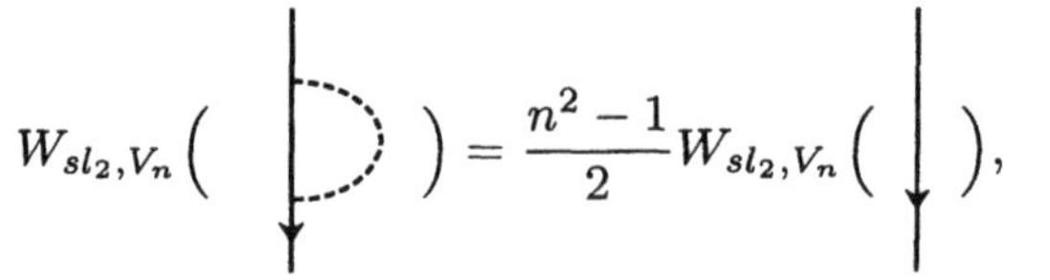

(6.34)

where we show a proof of the first formula later, in Lemma 9.4, and we obtain the second formula from the third formula by putting $n = 3$, noting that the adjoint representation of sl_2 is isomorphic to the 3-dimensional irreducible representation V_3 of sl_2. Further, the third formula is obtained by

$$\rho_{V_n}(C) = \rho_{V_n}\Big(EF + FE + \frac{1}{2}H^2\Big) = \frac{n^2-1}{2} \cdot \mathrm{id}_{V_n}. \tag{6.35}$$

By using the above relations we obtain the following values for the Jacobi diagrams in Figure 6.9,

$W_{sl_2,V_n}\Big($ 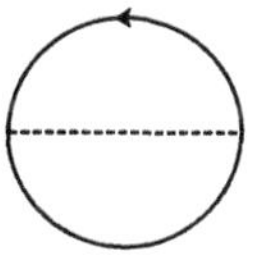 $\Big) = \dfrac{n^3-n}{2}$, $W_{sl_2,V_n}\Big($ 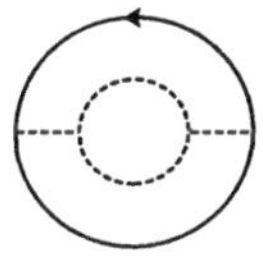 $\Big) = 2(n^3-n)$,

$W_{sl_2,V_n}\Big($ 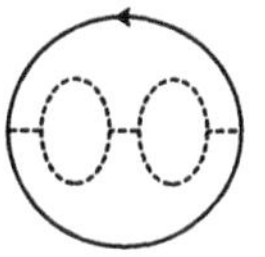 $\Big) = 8(n^3-n)$, $W_{sl_2,V_n}\Big($ 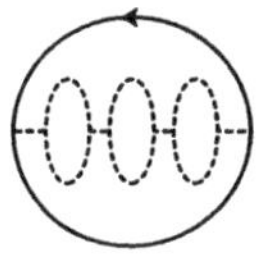 $\Big) = 32(n^3-n)$,

$W_{sl_2,V_n}\Big($ 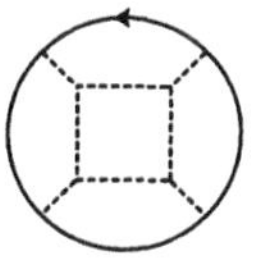 $\Big) = 2n(n^2-1)^2$.

Recovery of the quantum $(\mathfrak{g}, R)$ invariant from the framed Kontsevich invariant

Let $\mathfrak{g}$ be a simple Lie algebra, N the rank of $\mathfrak{g}$, and R a representation of $\mathfrak{g}$. For a framed oriented link L the *quantum* $(\mathfrak{g}, R)$ *invariant* $Q^{\mathfrak{g},R}(L)$ in $\mathbb{Z}[q^{\pm 1/2N}]$ is defined to be the invariant obtained from the representations of the braid groups in Theorem 5.16 by taking the quantum trace. Though we omit a complete definition of the quantum $(\mathfrak{g}, R)$ invariant for general $(\mathfrak{g}, R)$, we have already defined the quantum (sl_2, V_n) invariant, in Section 4.4, where V_n denotes the n-dimensional irreducible representation of the Lie algebra sl_2. We also define the quantum (sl_N, V) invariant, where V denotes the vector representation of the Lie algebra sl_N, later in this section. For a complete definition of the quantum $(\mathfrak{g}, R)$ invariant, see *e.g.* [Kas94, Tur89]. In this subsection, we show how the quantum invariants are recovered from the framed Kontsevich invariant.

We define a map, called the *graded weight system*,

$$\hat{W}_{\mathfrak{g},R} : \mathcal{A}(\sqcup^l S^1) \to \mathbb{C}[[\hbar]]$$

by $\hat{W}_{\mathfrak{g},R}(D) = W_{\mathfrak{g},R}(D)\hbar^d$ for a Jacobi diagram D of degree d, where $\hbar$ is an indeterminate, and $W_{\mathfrak{g},R}$ is the weight system derived from the substitution of $\mathfrak{g}$

and R into the diagram; recall that the degree of a Jacobi diagram is half the number of univalent and trivalent vertices of the Jacobi diagram.

Theorem 6.14 (see, for example, [Kas94, LeMu96a]). The quantum $(\mathfrak{g}, R)$ invariant $Q^{\mathfrak{g},R}(L)$ of a framed oriented link L is recovered from the framed Kontsevich invariant $\hat{Z}(L)$ by the graded weight system $\hat{W}_{\mathfrak{g},R}$ as

$$\hat{W}_{\mathfrak{g},R}(\hat{Z}(L)) = Q^{\mathfrak{g},R}(L)\Big|_{q=e^{\hbar}}.$$

The theorem implies universality of the framed Kontsevich invariant among the quantum $(\mathfrak{g}, R)$ invariants.

Proof. In Section 5.1 we considered the monodromy representation (5.19) of the braid group B_n along solutions of the KZ equation. As in Section 5.3 the monodromy representation is presented by the formulae obtained from the following formulae by applying the comultiplication Δ to both sides,

$\rho_{\mathrm{KZ}}\Big($  $\Big) = P \circ e^{\hbar\tau/2},$

$\rho_{\mathrm{KZ}}\Big($  $\Big) = \varphi_{\mathrm{KZ}}(\hbar\tau_{12}, \hbar\tau_{23}).$

Let $b \in B_n$ be a braid whose closure is isotopic to L. By taking the quantum trace of the monodromy representation of b we obtain the quantum $(\mathfrak{g}, R)$ invariant of L, since the monodromy representation is equivalent to the representation derived from the quantum group by Theorem 5.16.

Further, in Section 6.2 we introduce the formal KZ equation replacing $\hbar\tau$ in the KZ equation with a chord. We obtain the monodromy representation (6.5) of the pure braid group P_n along solutions of the formal KZ equation. Motivated by the monodromy representation we defined the framed Kontsevich invariant by putting

$\hat{Z}\Big($  $\Big) =$ 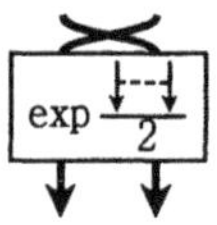 ,

$\hat{Z}\Big($ 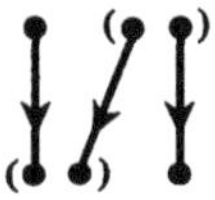 $\Big) = \Phi.$

Therefore the monodromy representation (5.19) is recovered from the framed Kontsevich invariant by substituting $\hbar\tau$ to each chord of Jacobi diagrams. The substitution is realized by the graded weight system as

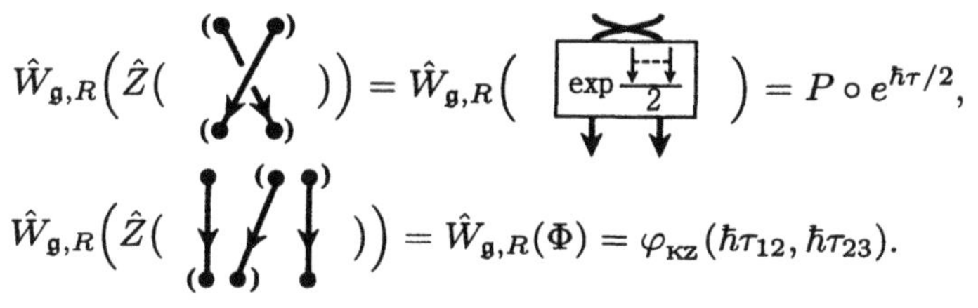

Hence,

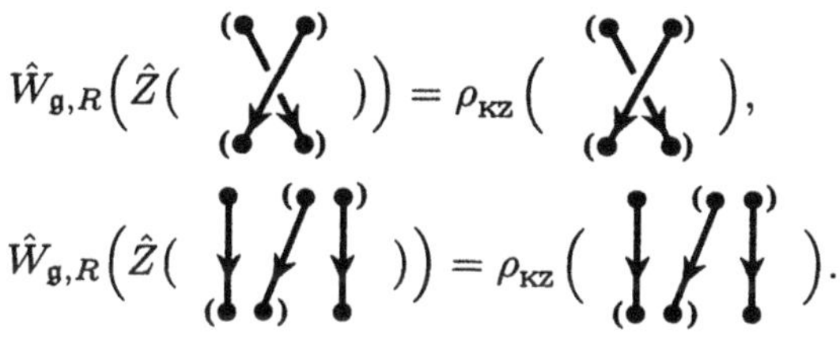

By applying the comultiplication Δ to the above formulae we obtain $\hat{W}_{\mathfrak{g},R}\big(\hat{Z}(E)\big) = \rho_{\mathrm{KZ}}(E)$ for each elementary parenthesized braid E. Hence, we obtain $\hat{W}_{\mathfrak{g},R}\big(\hat{Z}(b)\big) = \rho_{\mathrm{KZ}}(b)$ for any parenthesized braid b.

By taking the "quantum trace" of both sides of this formula we obtain the required formula. We omit the proof of this part; see [Kas94, Theorem XX.8.3] for a detailed proof. □

In [LeMu95b] J. Murakami and T.T.Q. Le gave a direct combinatorial proof, based on the skein relation, of Theorem 6.14 in the case that R is the vector representation V of sl_N. Later in this section we will give a combinatorial proof of Theorem 6.14 along the lines of the proof in [LeMu95b]. Further, based on the case of the vector representation of sl_2, at the end of this section we will give a proof of Theorem 6.14 in the case of (sl_2, R) for any representation R of sl_2.

The quantum (sl_N, V) invariant

Let V be the vector representation of the Lie algebra sl_N; as mentioned before, it is the N-dimensional irreducible representation of sl_N, which takes a matrix in sl_N to the same matrix in $\mathrm{End}(\mathbb{C}^N)$ regarding V as $\mathbb{C}^N$. It is known that the *quantum (sl_N, V) invariant* $Q^{sl_N,V}(L)$ of a framed oriented link L is determined by

the following relations,

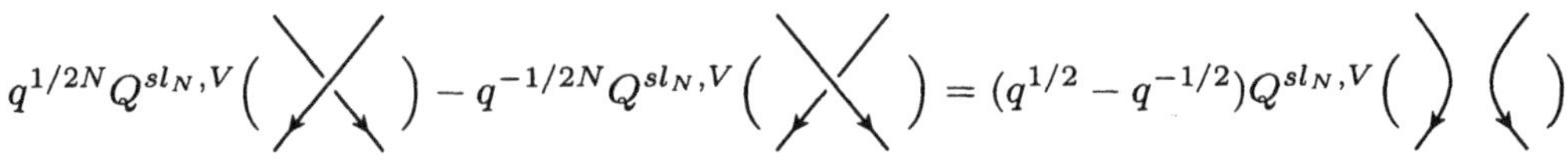

(6.36)

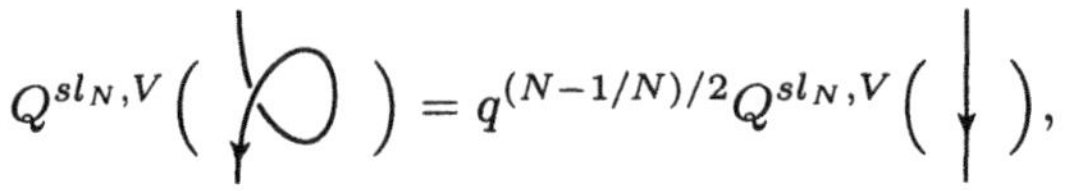

(6.37)

(6.38)

where we call the first relation the *skein relation* of the quantum (sl_N, V) invariant. Recall that we define the quantum integer $[N]$ to be $(q^{N/2} - q^{-N/2})/(q^{1/2} - q^{-1/2})$. See [MOY98] for an elementary proof of the fact that the above relations determine an invariant. When $N = 2$, we have already showed in Section 1.2 that the quantum (sl_2, V) invariant is an invariant. Further, when $N = 3$, we discuss the quantum (sl_3, V) invariant in Appendix B.1.

We show the following lemma before we prove Theorem 6.14 for (sl_N, V).

Lemma 6.15. We have that

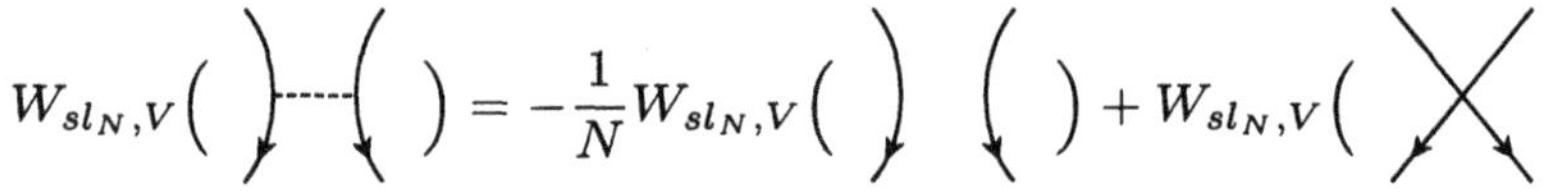

where both sides belong to $\mathrm{End}_{sl_N}(V \otimes V)$.

Proof. Since $V \otimes V$ is isomorphic to the direct sum of two irreducible representations (see, *e.g.*, [Hum72]), the dimension of $\mathrm{End}_{sl_N}(V \otimes V)$ is equal to 2. Further, since the maps $W_{sl_N,V}$ 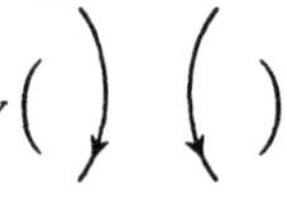 and $W_{sl_N,V}$  are linearly independent (this is shown in the following argument, as mentioned later), we can write

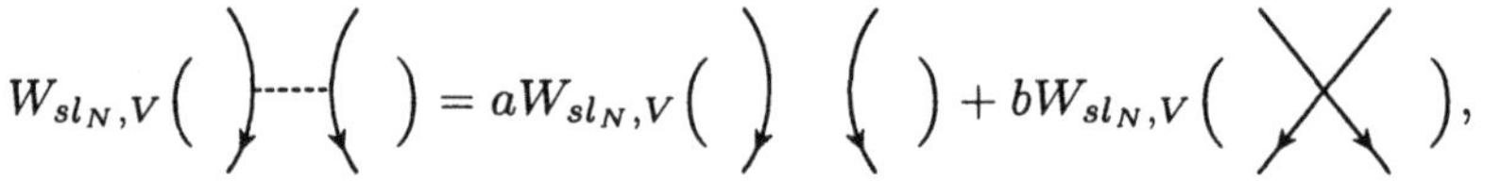

(6.39)

for some scalars a and b. We derive two relations from the above equation, as follows.

Firstly, by closing the right strand in the pictures in (6.39), we obtain

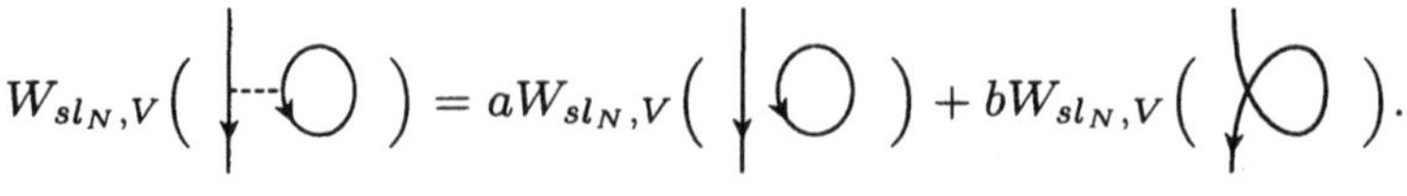

If a Jacobi diagram includes a solid circle with one dashed univalent vertex, it vanishes by the weight system $W_{sl_N,V}$, since any intertwiner

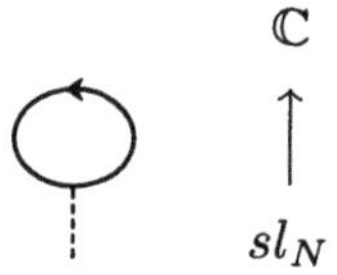

from the adjoint representation of the Lie algebra sl_N to the unit representation is equal to the zero map, because the adjoint representation of sl_N is irreducible. Hence, $W_{sl_N,V}\Big($ 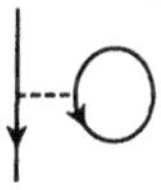 $\Big) = 0$. Further, $W_{sl_N,V}\Big($ 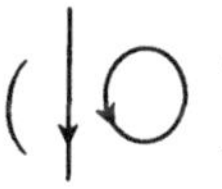 $\Big) = NW_{sl_N,V}\Big($  $\Big)$, since $W_{sl_N,V}\Big($  $\Big)$ is equal to the dimension of the vector representation V. Therefore, $aN + b = 0$.

Secondly, by closing the right strand after changing two bottom ends in the pictures in (6.39), we obtain

$W_{sl_N,V}\Big($  $\Big) = aW_{sl_N,V}\Big($  $\Big) + bW_{sl_N,V}\Big($  $\Big)$.

The left hand side is computed as $W_{sl_N,V}\Big($ 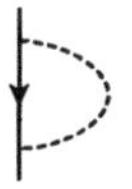 $\Big) = (N - 1/N)W_{sl_N,V}\Big($  $\Big)$, since the eigenvalue of the Casimir element of sl_N on the vector representation V is equal to $N - 1/N$. Therefore, $N - 1/N = a + bN$.

Note that the right hand sides of the above resulting formulae, $aN + b$ and $a + bN$, are linearly independent as linear sums of a and b. This implies that $W_{sl_N,V}\Big($ 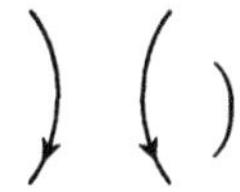 $\Big)$ and $W_{sl_N,V}\Big($  $\Big)$ are linearly independent, as mentioned before.

By the above resulting formulae, we have that $a = -1/N$ and $b = 1$. Substituting them into (6.39) we obtain the required formula. □

Proof of Theorem 6.14 in the case of (sl_N, V). Since the quantum (sl_N, V) invariant is characterized by the three relations (6.36), (6.37), and (6.38), it is sufficient to show that $\hat{W}_{sl_N,V}(\hat{Z}(L))$ also satisfies the above three relations.

Firstly, we show the skein relation (6.36) for $\hat{W}_{sl_N,V}(\hat{Z}(L))$. We put

$H =$ 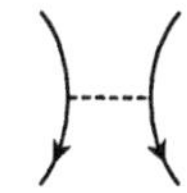 , $P =$ 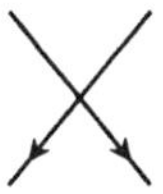 , $1 =$ 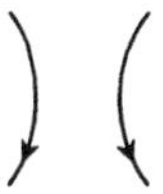 .

From the definitions of $\hat{W}_{sl_N,V}$ and $\hat{Z}$, we have that

$\hat{W}_{sl_N,V}\Big(\hat{Z}\big($  $\big)\Big) = \hat{W}_{sl_N,V}\Big(P(1+\frac{1}{2}H+\frac{1}{8}H^2+\cdots)\Big)$

$$= W_{sl_N,V}\Big(P(1+\frac{\hbar}{2}H+\frac{\hbar^2}{8}H^2+\cdots)\Big)$$

$$= W_{sl_N,V}(Pe^{\hbar H/2}).$$

By the above lemma, this is equal to $W_{sl_N,V}(Pe^{\hbar(P-1/N)/2})$. Then,

$e^{\hbar/2N}\hat{W}_{sl_N,V}\Big(\hat{Z}\big($  $\big)\Big) = W_{sl_N,V}(Pe^{\hbar P/2}).$

In the same way as above, we have that

$e^{-\hbar/2N}\hat{W}_{sl_N,V}\Big(\hat{Z}\big($  $\big)\Big) = W_{sl_N,V}(Pe^{-\hbar P/2}).$

It follows from the above two equations that the left hand side of the skein relation is equal to $W_{sl_N,V}(Pe^{\hbar P/2} - Pe^{-\hbar P/2})$. Note that every term in the expansion has an even power of P. Further, since $P^2 = 1$, the left hand side is equal to $W_{sl_N,V}(e^{\hbar/2} - e^{-\hbar/2})$. This is equal to the right hand side of the skein relation (6.36).

Secondly, we show that the invariant $\hat{W}_{sl_N,V}\big(\hat{Z}(L)\big)$ satisfies the second relation (6.37). Applying the following relation

$W_{sl_N,V}\Big($ 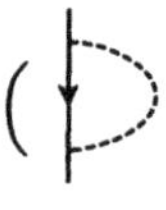 $\Big) = (N-\frac{1}{N})W_{sl_N,V}\Big($  $\Big)$

to the relation (6.14), we obtain

$W_{sl_N,V}\Big($  $\Big) = e^{\hbar(N-1/N)/2}\hat{W}_{sl_N,V}\Big(\hat{Z}\big($  $\big)\Big).$

Hence, we obtain (6.37).

Lastly, we show that the invariant $\hat{W}_{sl_N,V}(\hat{Z}(L))$ satisfies the third formula

(6.38), using the formulae obtained above. We have that

$$(q^{1/2} - q^{-1/2})\hat{W}_{sl_N,V}\Big(\hat{Z}($$  $$)\Big)\hat{W}_{sl_N,V}\Big(\hat{Z}($$  $$)\Big)$$

$$= (q^{1/2} - q^{-1/2})\hat{W}_{sl_N,V}\Big(\hat{Z}($$ 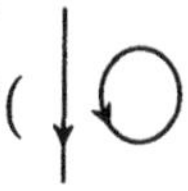 $$)\Big)$$

$$= q^{1/2N}\hat{W}_{sl_N,V}\Big(\hat{Z}($$ 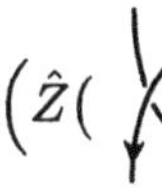 $$)\Big) - q^{-1/2N}\hat{W}_{sl_N,V}\Big(\hat{Z}($$  $$)\Big)$$

$$= q^{1/2N}q^{(N-1/N)/2}\hat{W}_{sl_N,V}\Big(\hat{Z}($$ 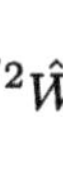 $$)\Big) - q^{-1/2N}q^{-(N-1/N)/2}\hat{W}_{sl_N,V}\Big(\hat{Z}($$  $$)\Big)$$

$$= (q^{N/2} - q^{-N/2})\hat{W}_{sl_N,V}\Big(\hat{Z}($$  $$)\Big),$$

where we obtain the second and third equalities by the relations (6.36) and (6.37) respectively. Hence, we obtain the relation (6.38) as $\hat{W}_{sl_N,V}\Big(\hat{Z}($ 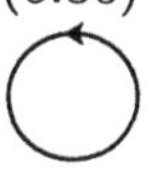 $)\Big) = [N]$.

□

The quantum* (sl_2, V_n) *invariant

In Section 4.4 we defined the quantum (sl_2, V_n) invariant, where V_n denotes the n-dimensional irreducible representation of the Lie algebra sl_2. We show Theorem 6.14 for the quantum (sl_2, V_n) invariant as follows.

***Proof of Theorem 6.14 in the case of* (sl_2, V_n).** Putting $N = 2$ in the previous proof of Theorem 6.14, in the case of (sl_N, V), we have that

$$\hat{W}_{sl_2,V}\Big(\hat{Z}(L)\Big) = Q^{sl_2,V}(L)\Big|_{q=e^{\hbar}}, \tag{6.40}$$

where V denotes the vector representation of sl_2.

We prove the theorem for (sl_2, V_n), for simplicity, in the case that L is a knot K.

When $n = 1$, since $\hat{W}_{sl_2,V_1}\Big(\hat{Z}(K)\Big) = 1$ and $Q^{sl_2,V_1}(K) = 1$ for the unit representation V_1 from the definition of the invariants, we obtain the theorem for (sl_2, V_1).

When $n = 2$, (6.40) implies the theorem for (sl_2, V_2), noting that the vector representation of sl_2 is isomorphic to the 2-dimensional irreducible representation V_2.

When $n = 3$, by Proposition 6.8 (1), we have that

$$\begin{aligned}\hat{W}_{sl_2,V}\Big(\hat{Z}(K^{(2)})\Big) &= \hat{W}_{sl_2,V}\Big(\Delta\hat{Z}(K)\Big) = \hat{W}_{sl_2,V\otimes V}\Big(\hat{Z}(K)\Big) \\ &= \hat{W}_{sl_2,V_1}\Big(\hat{Z}(K)\Big) + \hat{W}_{sl_2,V_3}\Big(\hat{Z}(K)\Big),\end{aligned}$$

noting that $V \otimes V$ is isomorphic to the direct sum of V_1 and V_3. Further, by Proposition 4.13, we have that

$$Q^{sl_2,V}\left(\hat{Z}(K^{(2)})\right) = Q^{sl_2,V\otimes V}\left(\hat{Z}(K)\right) = Q^{sl_2,V_1}\left(\hat{Z}(K)\right) + Q^{sl_2,V_3}\left(\hat{Z}(K)\right).$$

Since, by (6.40), the left hand sides of the above two formulae are equal, we have that

$$\hat{W}_{sl_2,V_1}\left(\hat{Z}(K)\right) + \hat{W}_{sl_2,V_3}\left(\hat{Z}(K)\right) = Q^{sl_2,V_1}\left(\hat{Z}(K)\right) + Q^{sl_2,V_3}\left(\hat{Z}(K)\right).$$

Since the theorem has already been shown for (sl_2, V_1), we obtain the theorem for (sl_2, V_3).

The argument for general n is similar. We show that the case of (sl_2, V_{n+1}) follows from the cases of $(sl_2, V_{\leq n})$, and the theorem follows by induction. By Proposition 6.8 (1) and Proposition 4.13, we have that

$$\begin{aligned}
\hat{W}_{sl_2;V}\left(\hat{Z}(K^{(n)})\right) &= \hat{W}_{sl_2;V}\left(\Delta^{(n-1)}\hat{Z}(K)\right) = \hat{W}_{sl_2,V^{\otimes n}}\left(\hat{Z}(K)\right) \\
&= \hat{W}_{sl_2,V_{n+1}}\left(\hat{Z}(K)\right) + n\hat{W}_{sl_2,V_{n-1}}\left(\hat{Z}(K)\right) + \cdots, \\
Q^{sl_2;V}\left(\hat{Z}(K^{(n)})\right) &= Q^{sl_2,V^{\otimes n}}\left(\hat{Z}(K)\right) \\
&= Q^{sl_2,V_{n+1}}\left(\hat{Z}(K)\right) + nQ^{sl_2,V_{n-1}}\left(\hat{Z}(K)\right) + \cdots,
\end{aligned}$$

where $V^{\otimes n}$ is isomorphic to the direct sum of V_{n+1}, n copies of V_{n-1}, and irreducible representations of dimension $\leq n-3$. The left hand sides of the above two formulae are equal, by (6.40). Further, terms after the second terms on the right hand sides are equal by the above assumption. Hence, the first terms on the right hand sides are equal. This implies the theorem for (sl_2, V_{n+1}). □

Chapter 7

Vassiliev invariants

The notion of a Vassiliev invariant (also known as a finite type invariant) was introduced by Vassiliev [Vas90]. He derived this notion from the 0-th cohomology of the space of knots, *i.e.*, the space of all smooth embeddings of S^1 into S^3. This space is a subspace of the space of all smooth immersions of S^1 into S^3, and the space of knots is the complement of the space of singular immersions. The space of singular immersions has a stratification given by the immersions having 1 double point, 2 double points, and so on. The notion of a Vassiliev invariant is derived from a computation of the 0-th cohomology of the space of knots using this stratification.

The notion of a Vassiliev invariant was given a combinatorial reformulation by Birman-Lin [BiLi93]. Our definition of a Vassiliev invariant in this chapter follows their reformulation. In this reformulation, a map

$$v : \{\text{knots}\} \longrightarrow \mathbb{C}$$

is defined to be a Vassiliev invariant of degree d if v satisfies a certain property related to singular knots with $d+1$ double points. Unlike the invariants which we have introduced so far in previous chapters, this definition of Vassiliev invariants gives, not a constructive presentation of each invariant, but instead a characterization of (the classes of) the invariants.

As mentioned in the previous chapter, we can deal with infinitely many quantum invariants of knots by unifying them into the Kontsevich invariant. Another way of studying so many quantum invariants is to study the space of Vassiliev invariants; all coefficients of quantum invariants (after a suitable change of variables) are Vassiliev invariants. Thus, a large class of quantum invariants becomes well organized when they are unified into the Kontsevich invariant and when they are characterized by Vassiliev invariants.

The sets of Vassiliev invariants of degree $d = 0, 1, 2, \cdots$ give a descending filtration of the vector space of all maps $\{\text{knots}\} \to \mathbb{C}$. A fundamental property of Vassiliev invariants is that each graded quotient space associated to the filtration is finite dimensional, and further, is isomorphic to the dual vector space of a space of Jacobi diagrams on S^1. Furthermore, this identification relates the Kontsevich invariant directly to Vassiliev invariants in the way how the Kontsevich invariant is universal among all Vassiliev invariants.

In Section 7.1 we introduce Vassiliev invariants and show their fundamental properties. In particular, we show the universality of the Kontsevich invariant among Vassiliev invariants in Section 7.2. In Section 7.3 we introduce a descending series of equivalence relations among knots to describe how finely Vassiliev invariants distinguish knots. In Section 7.4, we give a constructive presentation of each Vassiliev invariant by counting Gauss diagrams with certain properties. In Section 7.5 we present Vassiliev invariants by configuration space integrals, whose background is derived from perturbative Chern-Simons theory. See also [Bar95b] for a detailed exposition of Vassiliev invariants. Many references for articles on Vassiliev invariants at Bar-Natan's web site [Bar] would be helpful for the reader.

7.1 Definition and fundamental properties of Vassiliev invariants

In this section we introduce Vassiliev invariants (finite type invariants) of knots and framed knots. Vassiliev invariants are defined by characterizing them by some property, as opposed to the invariants we introduced earlier which have constructive definitions. We then show that the graded spaces of Vassiliev invariants are related to spaces of Jacobi diagrams. For a detailed introduction to Vassiliev invariants see also [Bar95b].

Let $\mathcal{K}$ be the vector space over $\mathbb{C}$ freely spanned by (the isotopy classes of) oriented knots in S^3. A *singular knot* is an immersion of S^1 into S^3 whose singularities are transversal double points. We regard a singular knot as an element in $\mathcal{K}$ by removing each singularity linearly by the following relation

For example,

Let $\mathcal{K}_d$ denote the vector subspace of $\mathcal{K}$ spanned by singular knots with d double points. We call a linear map $v : \mathcal{K} \to \mathbb{C}$ a *Vassiliev invariant* (or a *finite type invariant*) of *degree* d if $v|_{\mathcal{K}_{d+1}} = 0$. By definition, we can identify the set of Vassiliev invariants of degree d with the dual vector space of the quotient space $\mathcal{K}/\mathcal{K}_{d+1}$.

What is a Vassiliev invariant of degree 0? For a Vassiliev invariant v of degree 0, the value $v(K)$ is invariant under a crossing change of a knot K; for example,

since a singular knot with one double point vanishes by the definition of a Vassiliev invariant of degree 0. Since any knot K can be obtained from the trivial knot  by a finite series of crossing changes, we have that $v(K) =$ $v($  $)$ for any knot K. It follows that any Vassiliev invariant of degree 0 is a constant function on the set of knots.

What is a Vassiliev invariant v of degree 1? By the same argument as above, among singular knots K with one double point the value $v(K)$ is constant since the value is invariant under crossing changes; for example,

for a Vassiliev invariant of degree 1. Further, the constant is equal to 0, since

Hence, any Vassiliev invariant of degree 1 is a Vassiliev invariant of degree 0, *i.e.*, a constant function on the set of knots.

What is a Vassiliev invariant v of degree 2? By the same argument as above, among singular knots K with two double points the value $v(K)$ is invariant under crossing changes; note that there are two possible values, $v($  $)$ and $v($  $)$ in this case. The latter constant vanishes by

We present the former constant by $v($  $)$, where the singular knot  can be obtained as an embedding of the singular S^1 obtained from the chord diagram  by collapsing each of the chords; see also Figure 7.2. Since all singular knots obtained as embeddings of the singular S^1 corresponding to  are related to each other by crossing changes, each such singular knot is taken to the same value by a Vassiliev invariant of degree 2. It follows from an easy argument that the value $v(K)$ of a Vassiliev invariant v of degree 2 on any knot K can be presented by a linear sum,

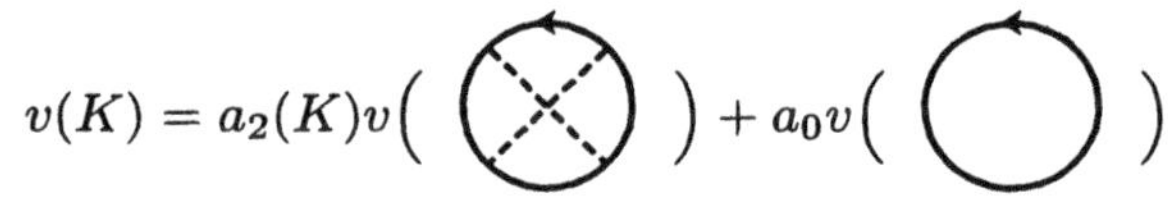

where a_0 is a constant and $a_2(K)$ is some fixed Vassiliev invariant of degree 2: in this case, the second coefficient of the Conway polynomial.

What is a Vassiliev invariant of degree d? By the same argument as above, among singular knots K with d double points, the value $v(K)$ can only take one of a finite number of values, corresponding to the different chord diagrams of degree d. That is, for a chord diagram D of degree d, we consider all the singular knots that can be obtained as embeddings of the singular S^1 that arises by collapsing each chord of D. Since such singular knots are related to each other by crossing changes, the value $v(K)$ is constant among such singular knots, by the same argument as for the above low degree cases. Hence, we present the value simply by $v(D)$.

For a Vassiliev invariant v of degree d, we have the following relation among the values of $v(D)$,

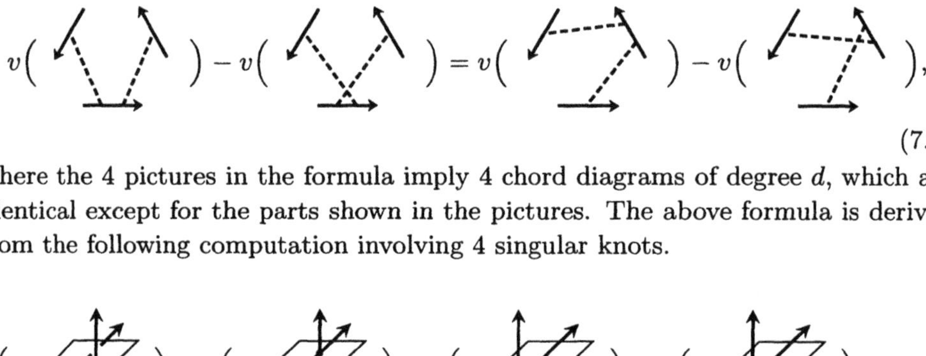

(7.1)

where the 4 pictures in the formula imply 4 chord diagrams of degree d, which are identical except for the parts shown in the pictures. The above formula is derived from the following computation involving 4 singular knots.

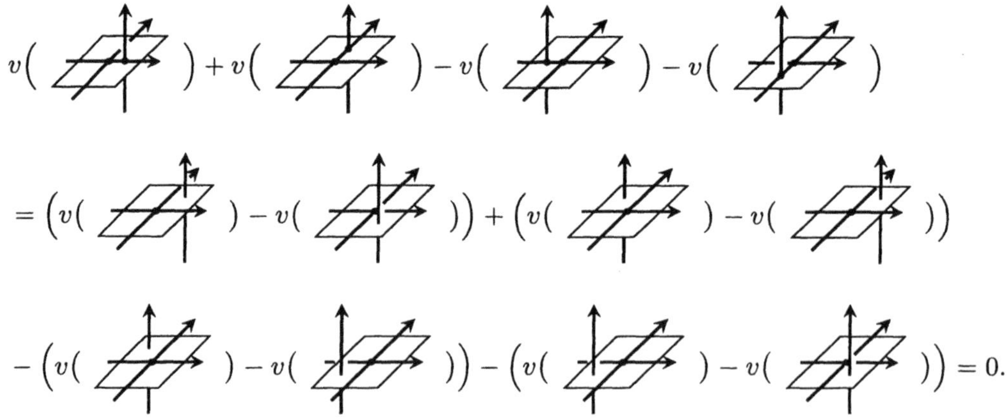

Further, we have the following relation

$v($ 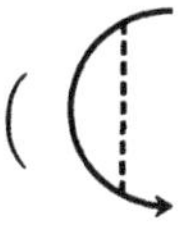 $) = 0,$ (7.2)

where the picture implies a chord diagram with an isolated chord. Here, an *isolated chord* is a chord such that the part of S^1 between the two ends of the chord has no ends of the other chords. The above formula is derived from the following computation,

$v($  $) = v($  $) - v($  $) = 0.$

Motivated by the relations (7.1) and (7.2) we consider the 4T and FI relations in the vector space spanned by chord diagrams, as shown in Figure 7.1. It is a remarkable fact that we needed the same relations in the definition of the Kontsevich invariant in Section 6.2. Recall that the Kontsevich invariant of a knot is defined to take values in the space $\mathcal{A}(S^1)$ modulo the FI relation, where $\mathcal{A}(S^1)$ is the vector space spanned by Jacobi diagrams on S^1 subject to the AS, IHX and STU relations.

The 4T relation :

The FI relation :

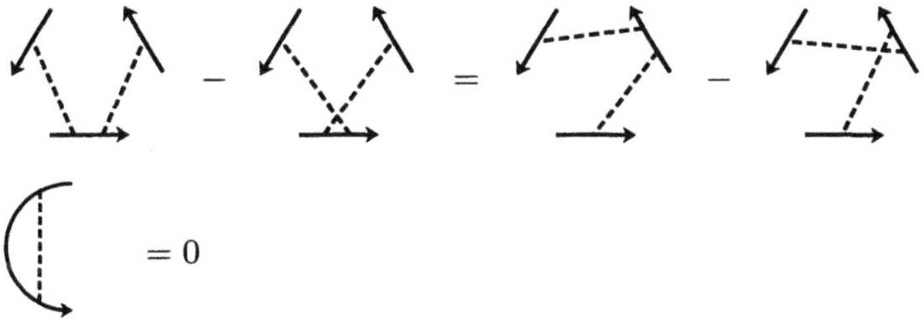

Figure 7.1 The 4T and FI relations.

Lemma 7.1 ([Bar95b]). We have the following isomorphism,

$$\mathcal{A}(S^1) \cong \mathrm{span}_{\mathbb{C}}\{\text{chord diagrams on } S^1\}/\text{the 4T relation.}$$

Proof. By the STU relation both sides of the 4T relation are equal to 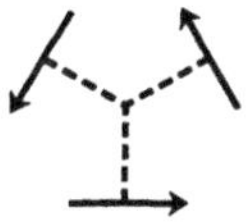 . Hence, we have the leftward map

$$\mathcal{A}(S^1) \longleftarrow \mathrm{span}_{\mathbb{C}}\{\text{chord diagrams on } S^1\}/\text{the 4T relation.}$$

This map is surjective, since any Jacobi diagram is equal to a linear sum of chord diagrams up to the STU relation. The injectivity of the map is shown by induction on the numbers of trivalent vertices of Jacobi diagrams; see [Bar95b] for a detailed proof. □

Recall that $\mathcal{A}(S^1)^{(d)}$ denotes the subspace of $\mathcal{A}(S^1)$ spanned by the chord diagrams of degree d. By the relations (7.1) and (7.2) a Vassiliev invariant v of degree d induces a linear map $\mathcal{A}(S^1)^{(d)}/\mathrm{FI} \to \mathbb{C}$; we call the map the *weight system* of v.

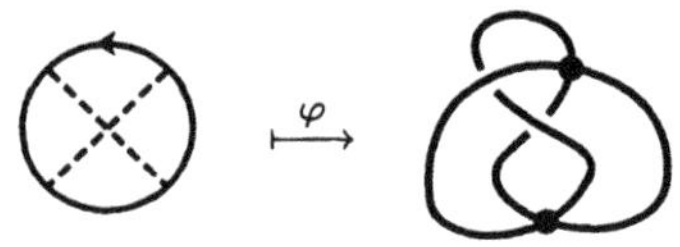

Figure 7.2 For a chord diagram D of degree d we obtain singular knots as embeddings of the singular S^1 obtained from D by collapsing each chord of D. Any Vassiliev invariant of degree d gives the same value on such singular knots, since all such are related by crossing changes.

We consider a map,

$$\varphi : \mathcal{A}(S^1)^{(d)}/\mathrm{FI} \longrightarrow \mathcal{K}_d/\mathcal{K}_{d+1}, \tag{7.3}$$

which is defined as follows. Let D be a chord diagram of degree d. By collapsing each chord of D to a point, we obtain a singular S^1. Let K_D be an embedding of the singular S^1 in S^3; see, *e.g.*, Figure 7.2. Any singular knots obtained from D in such a way are related to each other by a finite sequence of crossing changes. Note that, if two singular knots with d singular points differ by a crossing change, they are equal to each other in $\mathcal{K}_d$ modulo $\mathcal{K}_{d+1}$. Hence, the equivalence class $[K_D]$ in $\mathcal{K}_d/\mathcal{K}_{d+1}$ does not depend on the choice of embedding of the singular S^1. We define $\varphi(D)$ to be the class $[K_D] \in \mathcal{K}_d/\mathcal{K}_{d+1}$. The map φ is surjective by definition. Further, it is shown later that φ is an isomorphism between vector spaces.

The weight system of a Vassiliev invariant is presented by the map φ as follows. Let v be a Vassiliev invariant of degree d. Then, the weight system of v factors

$$\mathcal{A}(S^1)^{(d)}/\mathrm{FI} \xrightarrow{\varphi} \mathcal{K}_d/\mathcal{K}_{d+1} \subset \mathcal{K}/\mathcal{K}_{d+1} \xrightarrow{[v]} \mathbb{C}$$

where $[v]$ denotes the linear map induced by v, noting that v vanishes on $\mathcal{K}_{d+1}$ from the definition of a Vassiliev invariant of degree d.

Habiro [Habi00] introduced claspers to describe finite type invariants of knots and 3-manifolds from the viewpoint of a topological description of Jacobi diagrams; see Figure 7.3 for the definition of a clasper. Using claspers, Habiro gave a reconstruction of the map φ which replaces each dashed edge of a Jacobi diagram with

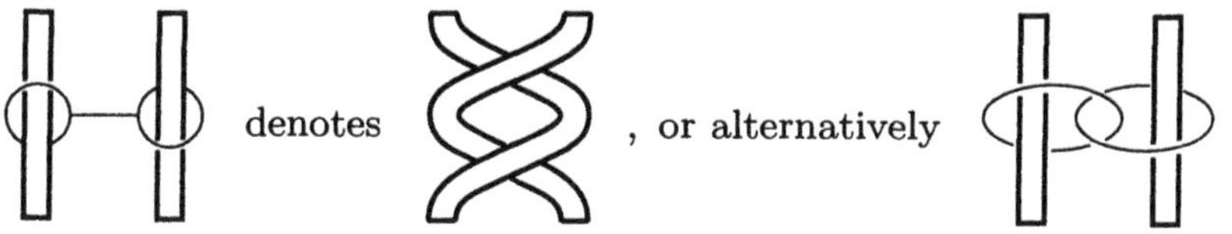

Figure 7.3 Definition of Habiro's clasper. The right picture implies the result obtained by integral surgery along the Hopf link in the picture with 0 framings. We call the embedded framed graph in the left picture a *clasper.*

a clasper and each trivalent vertex with a set of Borromean rings, such as

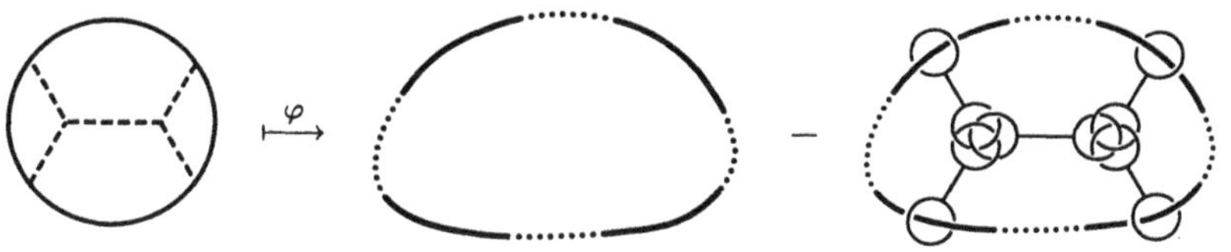
(7.4)

where the dotted lines imply strands possibly knotted and linked in some fashion. The image of a Jacobi diagram by the above map in $\mathcal{K}_d/\mathcal{K}_{d+1}$ does not depend on the choice of a knot (the middle picture) and an embedding of the edges of claspers (in the right picture). See Proposition E.26 for a proof of the fact that the above map gives a reconstruction of the map φ.

Vassiliev invariants of framed knots

The framing of a blackboard framed knot changes by ± 2 when a crossing is changed. Hence, we restrict to framed knots with even framings to define a good theory of Vassiliev invariants of framed knots. (If we only considered framed knots with odd framings, we would obtain the same results as below. For a formulation of Vassiliev invariants of framed knots, see also [KRT97].) In pictures we depict framed knots by blackboard framing as usual.

Let $\mathcal{K}'$ be the vector space over $\mathbb{C}$ freely spanned by (the isotopy classes of) the framed knots in S^3 whose framings are even. Similarly as the unframed case $\mathcal{K}'_d$ denotes the subspace of $\mathcal{K}'$ spanned by singular knots with d double points. We call a linear map $v : \mathcal{K}' \to \mathbb{C}$ a *Vassiliev invariant* of *degree* d if $v|_{\mathcal{K}'_{d+1}} = 0$.

Any Vassiliev invariant of degree 0 is a constant function on the set of framed knots in the same way as in the unframed case.

A Vassiliev invariant v of degree 1 is constant on the set of singular knots with one double point, just as in the unframed case. We write this constant

$v($ 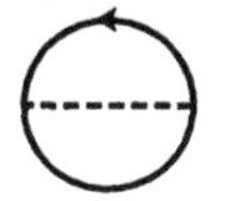 $)$, which is not necessarily equal to 0, unlike the unframed case. By an easy argument, any Vassiliev invariant v of degree 1 is presented by

$v(K) = v\Big($  $\Big) + \frac{\text{framing of } K}{2} v\Big($ 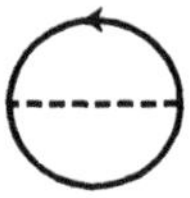 $\Big).$

For a Vassiliev invariant v of degree d we obtain the 4T relation in the same way as in the unframed case, while the FI relation does not hold in this case. Further, the surjective linear map

$$\varphi : \mathcal{A}(S^1)^{(d)} \longrightarrow \mathcal{K}'_d/\mathcal{K}'_{d+1} \tag{7.5}$$

is defined similarly; it will be shown later that φ is an isomorphism. Furthermore, the *weight system* of v is defined, just as in the unframed case, to be the following linear map

$$\mathcal{A}(S^1)^{(d)} \xrightarrow{\varphi} \mathcal{K}'_d/\mathcal{K}'_{d+1} \subset \mathcal{K}'/\mathcal{K}'_{d+1} \xrightarrow{[v]} \mathbb{C},$$

where $[v]$ denotes the linear map induced by v.

7.2 Universality of the Kontsevich invariant among Vassiliev invariants

As we showed in Theorem 6.14, the framed Kontsevich invariant is universal among quantum invariants. Remarkably, the (framed) Kontsevich invariant is also universal among Vassiliev invariants of (framed) knots. In this section we describe this latter universality of the (framed) Kontsevich invariant.

Proposition 7.2.
(1) The Kontsevich invariant Z induces the inverse of the map φ in (7.3). In particular, the map φ is an isomorphism between the space $\mathcal{A}(S^1)^{(d)}/\mathrm{FI}$ and the graded space $\mathcal{K}_d/\mathcal{K}_{d+1}$.
(2) The framed Kontsevich invariant $\hat{Z}$ induces the inverse of the map φ in (7.5). In particular the map φ is an isomorphism between the space $\mathcal{A}(S^1)^{(d)}$ and the graded space $\mathcal{K}'_d/\mathcal{K}'_{d+1}$.

Proof. We show (2) as follows. The image of $\mathcal{K}_d$ under $\hat{Z}$ lies in $\mathcal{A}(S^1)^{(\geq d)}$ by Lemma 7.3 below; note that $\mathcal{A}(S^1)$ is spanned by chord diagrams. Hence, $\hat{Z}$ induces a map $\mathcal{K}_d \to \mathcal{A}(S^1)^{(\geq d)}$. By replacing d with $d+1$, $\hat{Z}$ induces a map $\mathcal{K}_{d+1} \to \mathcal{A}(S^1)^{(\geq d+1)}$. Further, by taking the quotient of the two maps, we have $\mathcal{K}_d/\mathcal{K}_{d+1} \to \mathcal{A}(S^1)^{(d)}$. This map induces the inverse of φ by Lemma 7.3 below. Hence, we obtain (2).

We show (1) as follows. From the definitions of the φ maps, we have the following commutative diagram.

$$\begin{array}{ccc} \mathcal{A}(S^1)^{(d)} & \xrightarrow{\varphi} & \mathcal{K}'_d/\mathcal{K}'_{d+1} \\ {\scriptstyle\text{projection}}\downarrow & & \downarrow{\scriptstyle\text{forgetting framings}} \\ \mathcal{A}(S^1)^{(d)}/\mathrm{FI} & \xrightarrow{\varphi} & \mathcal{K}_d/\mathcal{K}_{d+1} \end{array}$$

Recall that the framed Kontsevich invariant $\hat{Z}$ of a framed knot modulo the FI relation is equal to the Kontsevich invariant Z of that knot with the framing forgotten. Since $\hat{Z}$ induces the inverse of the upper φ by (2), Z induces the inverse of the lower φ. Hence, we obtain (1). □

Lemma 7.3. Let D be a chord diagram with d chords, and let φ be the map in (7.5). Then,

$$\hat{Z}\Big(\varphi(D)\Big) = D + \ (\text{terms of degree} > d).$$

Proof. From the definition of $\hat{Z}$,

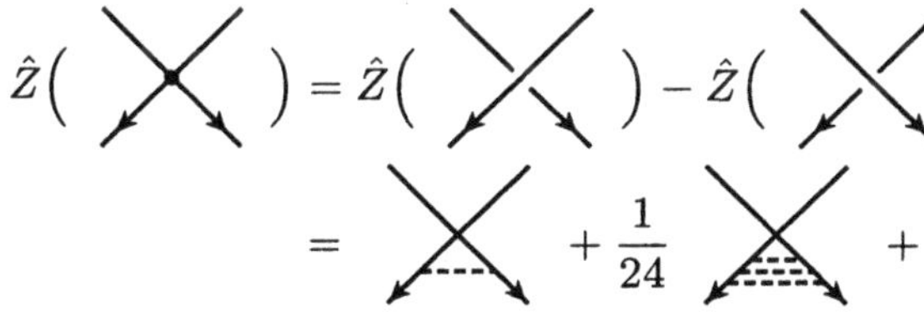

+ (terms of higher degrees).

We consider the singular S^1 obtained from the chord diagram D by collapsing each chord to a point. Recall that $\varphi(D)$ is represented by a singular knot obtained as an embedding of the singular S^1. Hence, we obtain $\hat{Z}(\varphi(D))$ as

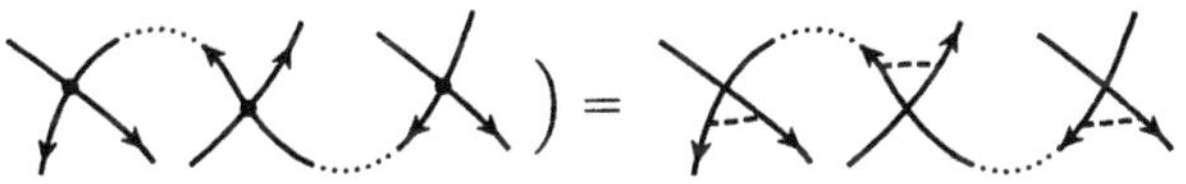

+ (terms of higher degrees),

where the chord diagram D appears again as the leading term on the right hand side, and the other terms have degrees greater than d. This implies the required formula. □

Theorem 7.4.
(1) The Kontsevich invariant is universal among Vassiliev invariants of knots in the same sense as in (2) below.
(2) The framed Kontsevich invariant is universal among Vassiliev invariants of framed knots. That is, for any positive integer d, any Vassiliev invariant v of degree d is presented by a composite map

$$v : \{\text{framed knots}\} \xrightarrow{\hat{Z}} \mathcal{A}(S^1) \xrightarrow{\text{projection}} \mathcal{A}(S^1)^{(\leq d)} \xrightarrow{W} \mathbb{C} \qquad (7.6)$$

for some linear map W. Conversely, for any linear map $W : \mathcal{A}(S^1)^{(\leq d)} \to \mathbb{C}$ the above composite map v is a Vassiliev invariant of degree d.

Proof. We show (2). (The proof of (1) is left to the reader.)

As in the proof of Proposition 7.2, $\hat{Z}(\mathcal{K}'_{d+1})$ lies in $\mathcal{A}(S^1)^{(\geq d+1)}$. Hence, the image of $\mathcal{K}'_{d+1}$ by the map (7.6) is equal to 0. Therefore, the map (7.6) is a Vassiliev invariant of degree d.

Conversely, let v be a Vassiliev invariant of degree d. We show the existence of W by induction on d. We put $W^{(d)}$ to be the weight system of v, *i.e.*, we put

$$W^{(d)} : \mathcal{A}(S^1)^{(d)} \xrightarrow{\varphi} \mathcal{K}'_d/\mathcal{K}'_{d+1} \xrightarrow{[v]} \mathbb{C}.$$

By Proposition 7.2, $\hat{Z}$ induces the inverse of φ; we denote the inverse by $\hat{Z}^{(d)}$. We put $v' = v - W^{(d)} \circ \hat{Z}^{(d)}$. Since v and $W^{(d)} \circ \hat{Z}^{(d)}$ coincide on $\mathcal{K}'_d$, the map v' vanishes on $\mathcal{K}'_d$, that is, v' is a Vassiliev invariant of degree $d-1$. By assumption, there exists a weight system W' satisfying $v' = W' \circ \hat{Z}^{(\leq d-1)}$. Hence,

$$v = W^{(d)} \circ \hat{Z}^{(d)} + W' \circ \hat{Z}^{(\leq d-1)} = (W^{(d)} + W') \circ \hat{Z}^{(\leq d)}.$$

The map W defined by $W = W^{(d)} + W'$ satisfies (7.6) for v. □

By Theorems 6.14 and 7.4 we obtain the following corollary. Historically speaking, this corollary was given a direct proof by Birman-Lin [BiLi93, Lin*91*] and Piunikhin [Piu95b] before the theorems were proved. As in Section 6.6, $Q^{\mathfrak{g},R}(K)$ denotes the quantum $(\mathfrak{g}, R)$ invariant of a framed knot K for a simple Lie algebra $\mathfrak{g}$ and a representation R of $\mathfrak{g}$. Further, as in Section 6.6, $W_{\mathfrak{g},R}$ denotes the weight system derived from the substitution of $\mathfrak{g}$ and R into Jacobi diagrams.

Corollary 7.5 (see also [Bar95b, Oht99]).
(1) For a framed knot K, the coefficient of $\hbar^d$ in $Q^{\mathfrak{g},R}(K)|_{q=e^\hbar}$ is a Vassiliev invariant of degree d (BIRMAN-LIN [BILI93, LIN*91*]).
(2) Furthermore, the weight system of this Vassiliev invariant is equal to $W_{\mathfrak{g},R}$ (PIUNIKHIN [PIU95B]).

Proof. We show (1) following the proof of [BiLi93]. (It is left to the reader to show (1) as a corollary of Theorems 6.14 and 7.4.) In the construction of quantum invariants, we associate a matrix R and its inverse R^{-1} to a positive and a negative crossing respectively. These matrices coincide when $\hbar = 0$. Hence, $R - R^{-1}$ is a matrix whose entries are divisible by $\hbar$ in $\mathbb{C}[[\hbar]]$. This difference is associated to the double point of a singular knot that occurs in the definition of a Vassiliev invariant. Therefore, if K is a singular knot with $d+1$ singular points, then $Q^{\mathfrak{g},R}(K)$ is divisible by $\hbar^{d+1}$. Hence, the coefficient of $\hbar^d$ is equal to 0 for such singular knots. This implies (1).

We show (2) by Theorems 6.14 and 7.4 as follows. By Theorem 6.14, we have that

$$Q^{\mathfrak{g},R}(\cdot)|_{q=e^\hbar} : \mathcal{K}' \xrightarrow{\hat{Z}} \mathcal{A}(S^1) \xrightarrow{\hat{W}_{\mathfrak{g},R}} \mathbb{C}[[\hbar]].$$

Further, by restricting it to $\mathcal{K}_d$ and $\mathcal{K}_{d+1}$ respectively, we obtain the following maps,

$$Q^{\mathfrak{g},R}(\cdot)|_{q=e^\hbar} : \mathcal{K}'_d \xrightarrow{\hat{Z}} \mathcal{A}(S^1)^{(\geq d)} \xrightarrow{\hat{W}_{\mathfrak{g},R}} \hbar^d \cdot \mathbb{C}[[\hbar]],$$
$$Q^{\mathfrak{g},R}(\cdot)|_{q=e^\hbar} : \mathcal{K}'_{d+1} \xrightarrow{\hat{Z}} \mathcal{A}(S^1)^{(\geq d+1)} \xrightarrow{\hat{W}_{\mathfrak{g},R}} \hbar^{d+1} \cdot \mathbb{C}[[\hbar]].$$

By taking the quotient of these maps, we obtain the following map,

$$v : \mathcal{K}'_d/\mathcal{K}'_{d+1} \xrightarrow{\hat{Z}^{(d)}} \mathcal{A}(S^1)^{(d)} \xrightarrow{W_{\mathfrak{g},R}} \mathbb{C},$$

where v is the d-th coefficient of $Q^{\mathfrak{g},R}(\cdot)|_{q=e^\hbar}$. This implies that the weight system of v is equal to $W_{\mathfrak{g},R}$. □

The relations between invariants in Chapters 4–7

In Chapters 4–7 we gave three kinds of invariants of knots; quantum invariants, Vassiliev invariants and the (framed) Kontsevich invariant. We show the relations between them in Figure 7.4.

The framed Kontsevich invariant has two universal properties. One is the universality among quantum invariants; for each simple Lie algebra $\mathfrak{g}$ and each representation R of it, the quantum $(\mathfrak{g}, R)$ invariant $Q^{\mathfrak{g},R}$ is presented by

$$Q^{\mathfrak{g},R} = \hat{W}_{\mathfrak{g},R} \circ \hat{Z}$$

with the graded weight system $\hat{W}_{\mathfrak{g},R}$ derived from the substitution of $\mathfrak{g}$ and R into Jacobi diagrams (Theorem 6.14). The other is the universality among Vassiliev invariants; each Vassiliev invariant v is presented by

$$v = W \circ \hat{Z}$$

for some weight system W (Theorem 7.4).

As a corollary of the two universalities, we obtain a relation between quantum invariants and Vassiliev invariants; the coefficients of the quantum $(\mathfrak{g}, R)$ invariant are Vassiliev invariants and their weight systems are equal to $W_{\mathfrak{g},R}$ (Corollary 7.5).

7.3 A descending series of equivalence relations among knots

How finely do Vassiliev invariants distinguish knots? In this section we consider a descending series of equivalence relations among knots, regarding two knots as equivalent if all Vassiliev invariants of a given degree take the same value on them. Further, we show that the sets of equivalence classes under the equivalence relations are equal to lattices whose ranks are equal to the dimensions of some spaces of Jacobi diagrams.

Let $\mathbb{K}$ denote the set of knots. The set $\mathbb{K}$ becomes a semigroup with respect to the connected sum of knots. In this section we will distinguish the completion

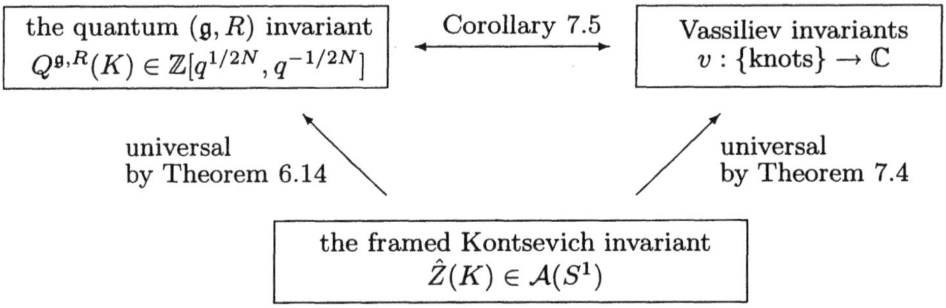

Figure 7.4 Three kinds of invariants of knots and the relations between them

(with respect to degree) of the space $\mathcal{A}(S^1)$ (spanned by finite combinations of Jacobi diagrams) by denoting it $\hat{\mathcal{A}}(S^1)$ (note that earlier sections used the notation $\mathcal{A}(S^1)$ for both spaces).

A Vassiliev invariant v is called *primitive* if $v(K_1 \# K_2) = v(K_1) + v(K_2)$ for any two knots K_1 and K_2, where $K_1 \# K_2$ denotes the connected sum of K_1 and K_2. In other words, a primitive Vassiliev invariant is a Vassiliev invariant preserving the additive structure of the semigroup $\mathbb{K}$.

Theorem 7.6. The primitive Kontsevich invariant z is universal among primitive Vassiliev invariants. That is, for any positive integer d, any primitive Vassiliev invariant v of degree d is presented by the following composite map,

$$v : \{\text{knots}\} \xrightarrow{z} \hat{\mathcal{A}}(S^1)_{\text{prim}} \xrightarrow{\text{projection}} \mathcal{A}(S^1)^{(\leq d)}_{\text{prim}} \xrightarrow{W} \mathbb{C},$$

for some linear map W. Conversely, for any linear map $W : \mathcal{A}(S^1)^{(\leq d)}_{\text{prim}} \to \mathbb{C}$ the above composite map v is a primitive Vassiliev invariant of degree d.

Proof. It follows from the definition of the Kontsevich invariant that the normalization $\hat{Z}(K) \# \nu^{-1}$ is multiplicative. Since the primitive Kontsevich invariant z is its logarithm, z is additive with respect to the connected sum of knots. Hence, any linear sum of coefficients of z is a primitive Vassiliev invariant.

Conversely, let v be a primitive Vassiliev invariant. By Theorem 7.4, v is equal to a linear combination of the coefficients of $\hat{Z}$. Hence, it is presented by a polynomial in coefficients of z. Since v is primitive, it must be a linear sum of coefficients of z. □

Corollary 7.7. Any Vassiliev invariant of degree d is equal to a polynomial in primitive Vassiliev invariants of degree $\leq d$.

Proof. The degree $\leq d$ part $\hat{Z}^{(\leq d)}$ of the Kontsevich invariant is universal among Vassiliev invariants of degree $\leq d$ by Theorem 7.4. Further, the degree $\leq d$ part $z^{(\leq d)}$ of the primitive Kontsevich invariant is universal among primitive Vassiliev invariants of degree $\le d$ by Theorem 7.6. Since each coefficient of $\hat{Z}^{(\le d)}$ is presented by some polynomial in coefficients of $z^{(\le d)}$, we obtain the corollary. $\square$

How finely do quantum invariants distinguish knots? Since the Kontsevich invariant dominates quantum invariants, this problem is reduced to the problem: how finely does the Kontsevich invariant distinguish knots? Further, noting that the Kontsevich invariant is a universal Vassiliev invariant, as in Theorem 7.4, we consider the problem: how finely do Vassiliev invariants distinguish knots? To distinguish them, we define the $C_d^{\mathbb{Q}}$-equivalence relation, denoted by $\underset{(d)}{\sim}$, among knots as follows. Two knots K and K' are called *$C_d^{\mathbb{Q}}$-equivalent*, denoted by $K \underset{(d)}{\sim} K'$, if $v(K) = v(K')$ for any Vassiliev invariant v of degree $< d$. We have an ascending series of sets of equivalence classes;

$$\mathbb{K}/\underset{(1)}{\sim} \longleftarrow \mathbb{K}/\underset{(2)}{\sim} \longleftarrow \mathbb{K}/\underset{(3)}{\sim} \longleftarrow \cdots .$$

By Theorem 7.6 and Corollary 7.7, $K \underset{(d)}{\sim} K'$ if and only if $z^{(<d)}(K) = z^{(<d)}(K')$, where $z^{(<d)}$ denotes the degree $< d$ part of the primitive Kontsevich invariant z. Hence, we have an identification between $\mathbb{K}/\underset{(d)}{\sim}$ and $z^{(<d)}(\mathbb{K})$ as

$$z^{(<d)} : \mathbb{K}/\underset{(d)}{\sim} \xrightarrow{=} z^{(<d)}(\mathbb{K}) \subset \mathcal{A}(S^1)^{(<d)}_{\text{prim}}/\text{FI}. \tag{7.7}$$

Let us consider how fine are the quotient sets $\mathbb{K}/\underset{(d)}{\sim}$. As an image of $\mathbb{K}$ the quotient set $\mathbb{K}/\underset{(d)}{\sim}$ becomes a commutative semigroup. Further, $z^{(<d)}(\mathbb{K})$ forms a commutative semigroup without torsion in the vector space $\mathcal{A}(S^1)^{(<d)}_{\text{prim}}$; the above identification preserves their additive structures. Hence, $\mathbb{K}/\underset{(d)}{\sim}$ is a commutative semigroup without torsion. For example,

$$\mathbb{K}/\underset{(1)}{\sim} \cong \mathbb{K}/\underset{(2)}{\sim} \cong \{0\}, \qquad \mathbb{K}/\underset{(3)}{\sim} \cong \mathbb{Z}, \qquad \mathbb{K}/\underset{(4)}{\sim} \cong \mathbb{Z}\oplus\mathbb{Z},$$

where the relations are verified by a concrete computation. In general, as shown below, $\mathbb{K}/\underset{(d)}{\sim}$ forms a commutative group isomorphic to a lattice in $\mathcal{A}(S^1)^{(<d)}_{\text{prim}}/\text{FI}$ whose rank is equal to the dimension of $\mathcal{A}(S^1)^{(<d)}_{\text{prim}}/\text{FI}$.

To describe it we consider the *C_d-equivalence*,* denoted by $\underset{d}{\sim}$, among knots; roughly speaking it is generated by the equivalence between two pictures in the

*The C_d-equivalence is also called the $(d-1)$-*equivalence* (due to Goussarov) in some literatures.

image of the map (7.4) for each Jacobi diagram of degree $d-1$; for example,

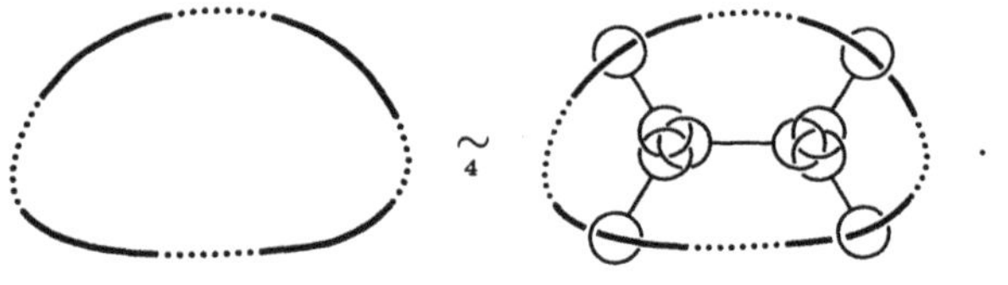

For the precise definition of the C_d-equivalence see Appendix E.2. As shown in Appendix E.2, if $K \underset{d}{\sim} K'$, then $K \underset{(d)}{\sim} K'$ by Proposition E.25. Hence, we have the following homomorphism induced by the identity map of knots,

$$\mathbb{K}_d/\underset{d+1}{\sim} \longrightarrow \mathbb{K}_{(d)}/\underset{(d+1)}{\sim},$$

where $\mathbb{K}_d$ (resp. $\mathbb{K}_{(d)}$) denotes the set of knots which are C_d-equivalent (resp. C_d^Q-equivalent) to the trivial knot. Further, by Theorem E.12, we have a surjective homomorphism,

$$\operatorname{span}_{\mathbb{Z}}\left\{\begin{array}{l}\text{primitive Jacobi diagrams}\\ \text{on } S^1 \text{ of degree } d\end{array}\right\}\Big/\text{AS, IHX, STU} \longrightarrow \mathbb{K}_d/\underset{d+1}{\sim}.$$

Recall that a primitive Jacobi diagram on S^1 is a Jacobi diagram on S^1 whose uni-trivalent graph is connected. By composing the above two maps we have that

$$\operatorname{span}_{\mathbb{Z}}\left\{\begin{array}{l}\text{primitive Jacobi}\\ \text{diagrams on } S^1\\ \text{of degree } d\end{array}\right\}\Big/\text{AS, IHX, STU} \longrightarrow \mathbb{K}_d/\underset{d+1}{\sim} \longrightarrow \mathbb{K}_{(d)}/\underset{(d+1)}{\sim} \xrightarrow{z^{(d)}} \mathcal{A}(S^1)^{(d)}_{\text{prim}}.$$

It follows from the definitions of the above maps and a computation in Section E.4 that the composition of the above sequence of maps takes each Jacobi diagram to itself. Hence, $\mathbb{K}_{(d)}/\underset{(d+1)}{\sim}$ is isomorphic to the lattice in $\mathcal{A}(S^1)^{(d)}_{\text{prim}}$ spanned by primitive Jacobi diagrams.

By Proposition E.25 we have a surjective homomorphism $\mathbb{K}/\underset{d}{\sim} \longrightarrow \mathbb{K}/\underset{(d)}{\sim}$. Since $\mathbb{K}/\underset{d}{\sim}$ is a commutative group by Corollary E.14, so is $\mathbb{K}/\underset{(d)}{\sim}$. Further, as in (7.7), $\mathbb{K}/\underset{(d)}{\sim}$ is identified with a lattice in $\mathcal{A}(S^1)^{(<d)}_{\text{prim}}$. Furthermore, by the above argument, its rank is equal to the dimension of $\mathcal{A}(S^1)^{(<d)}_{\text{prim}}$. Hence, we have

Theorem 7.8. The quotient set $\{\text{knots}\}/\underset{(d)}{\sim}$ forms a commutative group which is isomorphic to the following lattice in the space on S^1,

$$\operatorname{span}_{\mathbb{Z}}\{\text{primitive Jacobi diagrams on } S^1 \text{ with } 1 < \text{degree} < d\} \subset \mathcal{A}(S^1).$$

Further, by taking the direct limit of the sequence of isomorphisms, we have the following homomorphism,

$$\{\text{knots}\} \longrightarrow \varprojlim_{d}\{\text{knots}\}/\underset{(d)}{\sim} \cong \operatorname{span}_{\mathbb{Z}}\left\{\begin{array}{l}\text{primitive Jacobi diagrams}\\ \text{on } S^1 \text{ of degree} > 1\end{array}\right\} \subset \hat{\mathcal{A}}(S^1). \tag{7.8}$$

Furthermore, the above map gives a reconstruction of the primitive Kontsevich invariant z.

Proof. The former part of the theorem was shown above.

As in (7.7), the isomorphism at degree $< d$ is given by $z^{(<d)}$. Hence, by taking direct limit we obtain z again. □

By the above theorem we immediately obtain

Corollary 7.9. The image of the primitive Kontsevich invariant $z : \{\text{knots}\} \to \hat{\mathcal{A}}(S^1)_{\text{prim}}$ lies in a lattice in $\hat{\mathcal{A}}(S^1)_{\text{prim}}$ which is isomorphic to the lattice spanned over $\mathbb{Z}$ by primitive Jacobi diagrams on S^1 in $\hat{\mathcal{A}}(S^1)_{\text{prim}}$.

The corollary implies integrality of coefficients of the primitive Kontsevich invariant in the sense that, if $z^{(<d)}(K) = z^{(<d)}(K')$, then $z^{(d)}(K) - z^{(d)}(K')$ is equal to a linear sum of Jacobi diagrams with integer coefficients. This integrality is also verified by the fact that each Vassiliev invariant can be presented by a mapping degree on a certain configuration space as shown in Section 7.5 later.

If it could be shown that the map (7.8) (or the primitive Kontsevich invariant z) was injective, then we would identify the set of knots with a subset of the lattice, and all invariants related to quantum invariants would be understood as functions of weight systems on the lattice. Further, we would expect that there would be structures on the set of knots induced by combinatorial structures on the lattice of Jacobi diagrams.

7.4 Extending the set of knots by Gauss diagrams

Goussarov-Polyak-Viro [GPV00] introduced the idea that Gauss diagrams generalize knot diagrams. They used Gauss diagrams to enlarge the set of knots to the set of "virtual knots"; see also [Kau99a] for a definition of virtual knots. Further, Goussarov showed that any Vassiliev invariant of knots is obtained by specializing some Vassiliev invariant of virtual knots (to be precise, of virtual 1-tangles), see [GPV00]. As an application of the theory of Gauss diagrams we can give an explicit constructive presentation of each Vassiliev invariant by counting certain arrangements of crossings of knot diagrams (to be precise, of 1-tangle diagrams); for example, at the end of this section we give such a concrete presentation of a Vassiliev invariant of degree 2.

Figure 7.5 A Gauss diagram on S^1 and the corresponding surface with a knot diagram

Gauss diagrams

A *Gauss diagram* is a chord diagram, each of whose chords has an orientation and a sign $\pm$. We associate a Gauss diagram with an oriented knot diagram on an oriented surface such that the surface is a tubular neighborhood of the knot diagram; see, for example, the right picture in Figure 7.5. That is, we have the following identification,

$$\{\text{Gauss diagrams on } S^1\} = \{\text{such knot diagrams on surfaces}\}. \tag{7.9}$$

The one-to-one correspondence between Gauss diagrams on S^1 and such knot diagrams on surfaces is given as follows. Let us consider such a knot diagram on a surface. Forgetting the surface we obtain a chord diagram on S^1 whose chords correspond to the crossings of the knot diagram. Further, we associate a sign $+$ (resp. $-$) to each chord when the corresponding crossing is positive (resp. negative). Furthermore, we give each chord the orientation from the end corresponding to the over path to the end corresponding to the under path at the crossing. This gives a Gauss diagram on S^1. It is left to the reader to show that this correspondence gives a bijection between the two sides of (7.9).

By attaching discs along the boundary circles of such a surface we obtain a closed surface. In particular, when we obtain S^2 in such a way, we can construct from the knot diagram on such a surface a usual knot diagram on S^2. Hence, through the identification (7.9), we have the following inclusion,

$$\{\text{Gauss diagrams on } S^1\} \supset \{\text{knot diagrams on } S^2\}. \tag{7.10}$$

As mentioned in Section 1.1, the set of (oriented) knots can be identified with the quotient set of the set of (oriented) knot diagrams by the Reidemeister moves in Figure 1.6. Through the identification (7.9), the Reidemeister moves are taken to the Reidemeister moves among Gauss diagrams in Figure 7.6. Furthermore, by regarding the quotient of (7.10) by the Reidemeister moves, the set of knots is regarded as a subset of the quotient set of the set of Gauss diagrams by the Reidemeister moves in Figure 7.6. An element of this quotient set is called a *virtual*

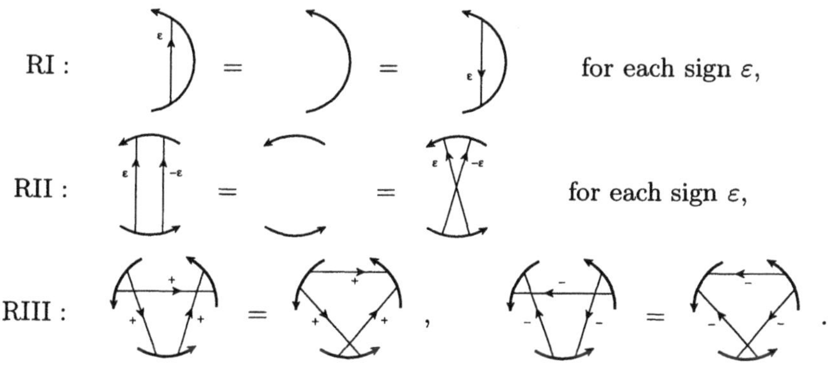

Figure 7.6 The Reidemeister moves among Gauss diagrams [GPV00]

knot. Thus, we have the following inclusion,

$$\{\text{knots}\} \subset \{\text{virtual knots}\}. \tag{7.11}$$

See also Figure 7.7 for the relations among these sets.

$$\begin{array}{rl}
\{\text{knot diagrams on } S^2\}/\text{RI, RII, RIII} & = \{\text{knots in } S^3\} \\
\cap & \\
\{\text{Gauss diagrams on } S^1\}/\text{RI, RII, RIII} & = \{\text{virtual knots}\} \\
\cap & \\
\text{span}_{\mathbb{C}}\{\text{Gauss diagrams on } S^1\}/\text{RI, RII, RIII} & \\
{\scriptstyle I}\downarrow{\scriptstyle =} & \\
\text{span}_{\mathbb{C}}\{\text{Gauss diagrams on } S^1 \text{ with dashed chords}\}/\text{RI}', \text{RII}', \text{RIII}' &
\end{array}$$

Figure 7.7 Extending the set of knots by Gauss diagrams.

Vassiliev invariants of virtual knots

We consider the vector space freely spanned by Gauss diagrams. The set of Gauss diagrams is a basis of the vector space. We give an isomorphism

$$I : \text{span}_{\mathbb{C}}\{\text{Gauss diagrams}\} \longrightarrow \text{span}_{\mathbb{C}}\{\text{Gauss diagrams}\}$$

defined by

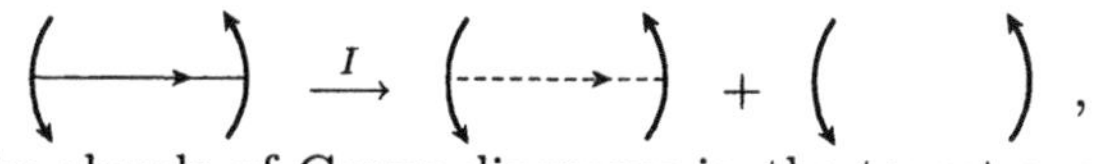

where we depict the chords of Gauss diagrams in the target space of I by dashed lines, to distinguish Gauss diagrams in the source and target spaces of I in pictures. Further, we denote by RI$'$, RII$'$, RIII$'$ the images of the Reidemeister moves under the map I. By the correspondence in Figure 7.7 we identify $\mathrm{span}_{\mathbb{C}}\{\text{virtual knots}\}$ with the vector space spanned by Gauss diagrams on S^1 with dashed chords subject to the relations RI$'$, RII$'$, RIII$'$. A linear map $v : \mathrm{span}_{\mathbb{C}}\{\text{virtual knots}\} \to \mathbb{C}$ is called a *Vassiliev invariant* of *degree* d if any Gauss diagram with more than d dashed chords is taken to 0 by v.

Proposition 7.10. Any Vassiliev invariant of virtual knots of degree d induces a Vassiliev invariant of knots of degree d by restricting the map to the set of knots. That is, we have the following map,

$$\{\text{Vassiliev invariants of virtual knots}\} \longrightarrow \{\text{Vassiliev invariants of knots}\}, \quad (7.12)$$

induced by the restriction by the inclusion (7.11).

Proof. A double point of a singular knot is expressed as a linear sum of Gauss diagrams with dashed chords by

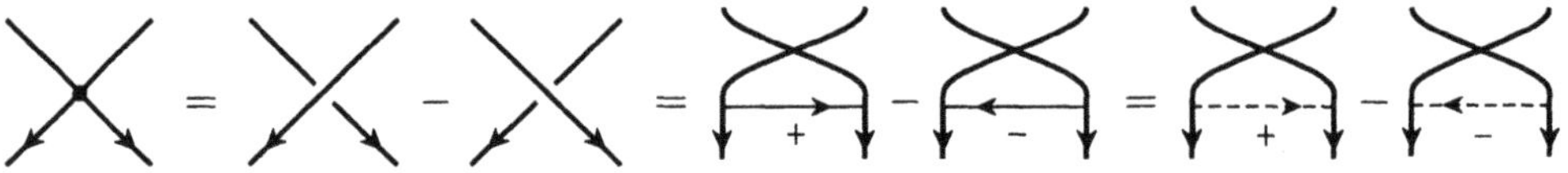

Hence, a singular knot with $d+1$ double points is presented by a linear sum of Gauss diagrams with $d+1$ dashed chords.

Let v be a Vassiliev invariant of virtual knots of degree d. Then, any Gauss diagram with $d+1$ dashed chords is taken to 0 by v. Hence, v takes any singular knot with $d+1$ double points to 0. Therefore, by restricting v to the set of knots we obtain a Vassiliev invariant of knots of degree d. □

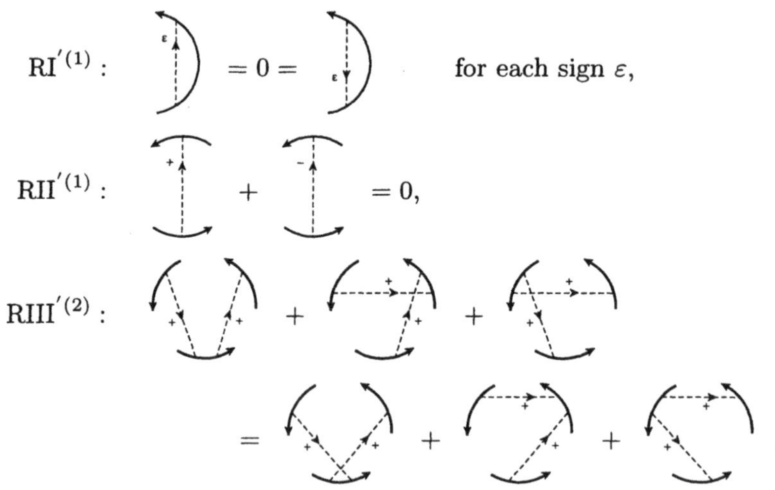

Figure 7.8 The relations RI$'^{(1)}$, RII$'^{(1)}$, and RIII$'^{(2)}$

Let v be a Vassiliev invariant of virtual knots of degree d. Through the correspondence in Figure 7.7 we have a map

$$v : \mathrm{span}_{\mathbb{C}}\{\text{Gauss diagrams on } S^1 \text{ with dashed chords}\}/\mathrm{RI}', \mathrm{RII}', \mathrm{RIII}' \longrightarrow \mathbb{C}.$$

We present such a map v by $v = v^{(d)} + v^{(d-1)} + \cdots$, where $v^{(i)}$ denotes the restriction of v to the subspace spanned by Gauss diagrams of degree i. In particular, the leading map $v^{(d)}$ induces the following linear map

$$v^{(d)} : \mathrm{span}_{\mathbb{C}}\{\text{Gauss diagrams with } d \text{ chords}\}/\mathrm{RI}'^{(1)}, \mathrm{RII}'^{(1)}, \mathrm{RIII}'^{(2)} \longrightarrow \mathbb{C}, \quad (7.13)$$

where $\mathrm{RI}'^{(1)}, \mathrm{RII}'^{(1)}, \mathrm{RIII}'^{(2)}$ denote the relations obtained by picking up the terms of maximal degrees in the images RI′, RII′, RIII′ of the Reidemeister moves under the map I; see Figure 7.8 for these relations.

We call a chord diagram with oriented chords an *arrow diagram* ([Pol00]). We identify a dashed chord with a + sign of a Gauss diagram with a dashed chord with no sign of an arrow diagram. Further, noting the relation $\mathrm{RII}'^{(1)}$, we identify a chord with a − sign with −1 times a chord with no sign. Then, the space in (7.13) is isomorphic to $\vec{\mathcal{A}}(S^1)^{(d)}/\vec{\mathrm{FI}}$. Here, the $\vec{\mathrm{FI}}$ relation is shown in Figure 7.9. Further, we define $\vec{\mathcal{A}}(S^1)$ by

$$\vec{\mathcal{A}}(S^1) = \mathrm{span}_{\mathbb{C}}\{\text{arrow diagrams on } S^1\}/\text{the 6T relation}, \quad (7.14)$$

where the 6T relation is defined in Figure 7.9. Furthermore, $\vec{\mathcal{A}}(S^1)^{(d)}$ denotes the degree d subspace of $\vec{\mathcal{A}}(S^1)$. Hence, the map (7.13) can be rewritten as

$$v^{(d)} : \vec{\mathcal{A}}(S^1)^{(d)}/\vec{\mathrm{FI}} \longrightarrow \mathbb{C}.$$

We call this map the *weight system* of a Vassiliev invariant v of virtual knots.

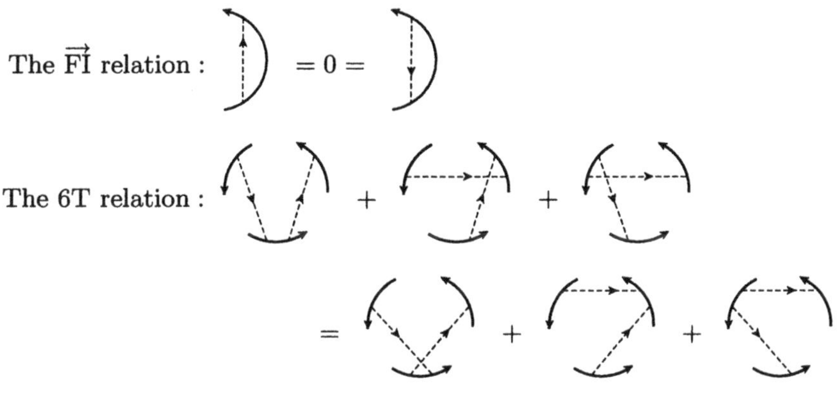

Figure 7.9 The oriented FI and the 6T relations among arrow diagrams

Further, we have the map $i : \mathcal{A}(S^1) \to \vec{\mathcal{A}}(S^1)$ defined by

where this map is well defined, because i(STU relation) is derived from the 6T relation through the map i; the proof is left to the reader, which is similar to the proof of Lemma 7.12 below.

Proposition 7.11. Let v be a Vassiliev invariant of virtual knots of degree d, and let v' be its restriction to the set of knots. Recall that v' is a Vassiliev invariant of knots of degree d, as shown in Proposition 7.10. Then, the weight system W' of v' is equal to the following composition,

$$\mathcal{A}(S^1)/\mathrm{FI} \xrightarrow{i} \overrightarrow{\mathcal{A}}(S^1)/\overrightarrow{\mathrm{FI}} \xrightarrow{W} \mathbb{C},$$

where W is the weight system of v.

Proof. As in the proof of Proposition 7.10 we have that

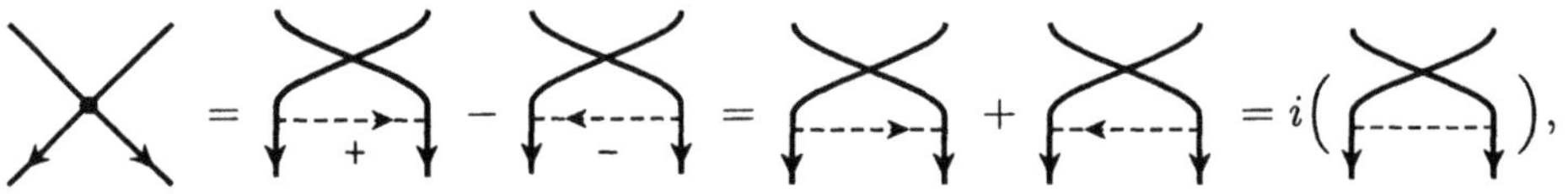

where we obtain the second equality from the identification of Gauss diagrams and arrow diagrams, and we obtain the third equality from the definition of the map i. Hence,

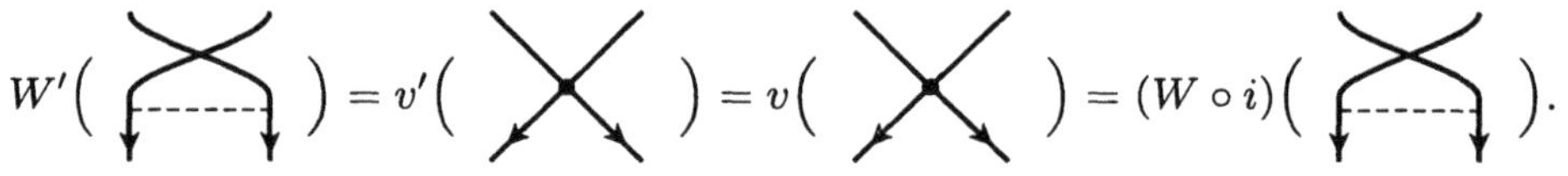

Therefore, we obtain the proposition. □

Oriented Jacobi diagrams

An *oriented Jacobi diagram* is a Jacobi diagram whose dashed edges have been equipped with orientations. Here, we do not allow a trivalent vertex with 3 inward edges or with 3 outward edges; we regard an oriented Jacobi diagram with such a trivalent vertex as 0, which is graphically shown as

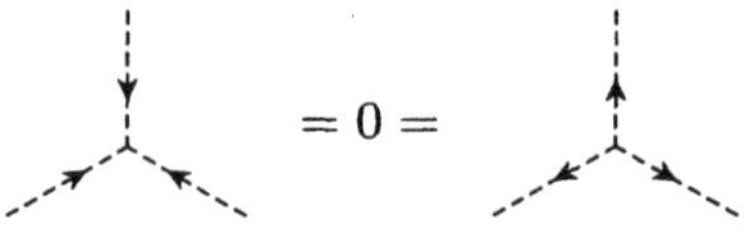

(7.15)

Further, it is convenient to introduce a thick dashed edge (with no orientation) which implies the sum of the two diagrams that arise from the possible orientations

of that edge, graphically shown as

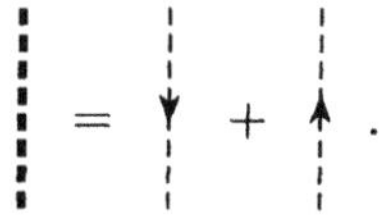

(7.16)

An *acyclic* oriented Jacobi diagram is an oriented Jacobi diagram whose uni-trivalent graph does not contain an oriented cycle (*i.e.*, a subgraph homeomorphic to an oriented circle). As an oriented version of Lemma 7.1 we have the following lemma; recall that we defined $\overrightarrow{\mathcal{A}}(S^1)$ to be the vector space spanned by arrow diagrams subject to the 6T relation in (7.14).

Lemma 7.12. We have the following isomorphism,

$$\overrightarrow{\mathcal{A}}(S^1) \cong \mathrm{span}_{\mathbb{C}}\{\text{acyclic oriented Jacobi diagrams on } S^1\}\big/\overrightarrow{\mathrm{AS}}, \overrightarrow{\mathrm{IHX}}, \overrightarrow{\mathrm{STU}},$$

where the $\overrightarrow{\mathrm{AS}}$, $\overrightarrow{\mathrm{IHX}}$, and $\overrightarrow{\mathrm{STU}}$ relations are defined in Figure 7.10.

The $\overrightarrow{\mathrm{AS}}$ relation :

The $\overrightarrow{\mathrm{IHX}}$ relation :

The $\overrightarrow{\mathrm{STU}}$ relation :

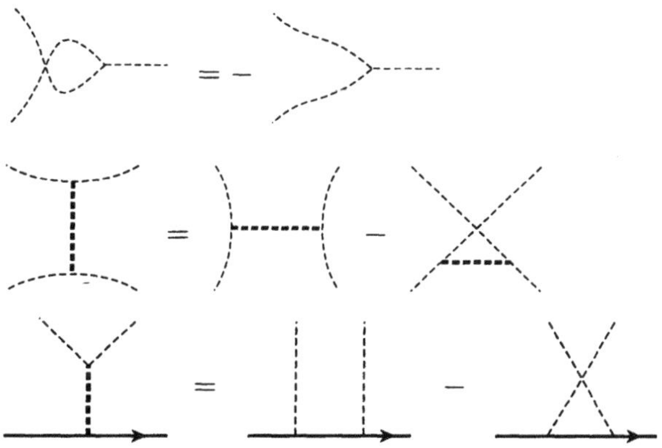

Figure 7.10 The oriented AS, IHX, and STU relations. Here, a thick dashed line implies the sum in (7.16), and thin dashed lines of different terms in the same formula which correspond (end at the same point) are given the same (arbitrarily chosen) orientation (in a vicinity of the common endpoint).

Proof. We have a natural linear map,

$$\overrightarrow{\mathcal{A}}(S^1) \longrightarrow \mathrm{span}_{\mathbb{C}}\{\text{acyclic oriented Jacobi diagrams on } S^1\}\big/\overrightarrow{\mathrm{AS}}, \overrightarrow{\mathrm{IHX}}, \overrightarrow{\mathrm{STU}}, \quad (7.17)$$

which takes an arrow diagram to the same diagram; the map is well defined because the 6T relation is obtained from the $\overrightarrow{\mathrm{STU}}$ relation as follows. We have that

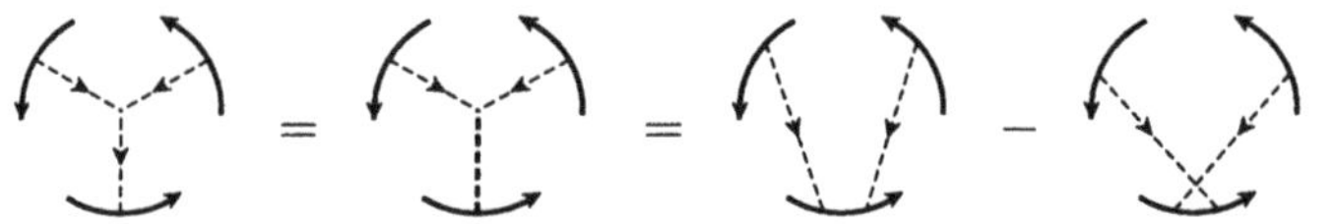 , (7.18)

where we obtain the first and second equalities by (7.15) and the $\overrightarrow{\mathrm{STU}}$ relation respectively. Similarly,

 . (7.19)

Further,

 ,

where we obtain the first and second equalities by (7.16) and the $\overrightarrow{\mathrm{STU}}$ relation respectively. By applying (7.18) and (7.19) to the two terms on the left hand side of the above formula, we obtain the 6T relation.

For the proof of the isomorphism of (7.17) see [Pol00]. □

Lemma 7.13. We have that $\vec{\mathcal{A}}(S^1)^{(2)}/\overrightarrow{\mathrm{FI}} = 0$.

Proof. By (7.16) we have that

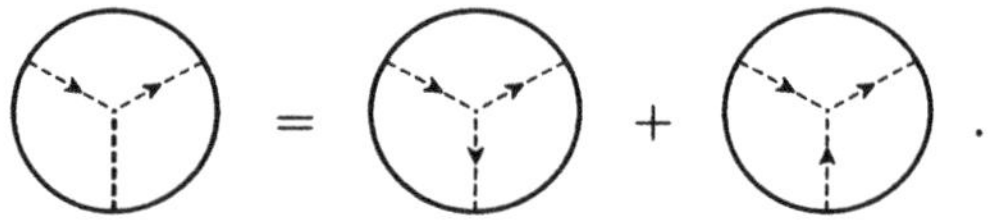 .

Each term is equal to -1 times 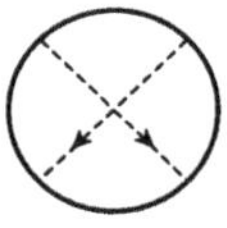 (modulo the $\overrightarrow{\mathrm{STU}}$ and $\overrightarrow{\mathrm{FI}}$ relations). Hence,  $= 0$. Since the space $\vec{\mathcal{A}}(S^1)^{(2)}$ is spanned by 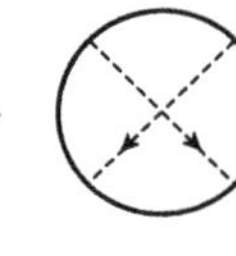 , the space is trivial. □

Recall that $\mathcal{A}(S^1)^{(2)}/\mathrm{FI}$ is one-dimensional, while $\vec{\mathcal{A}}(S^1)^{(2)}/\overrightarrow{\mathrm{FI}}$ vanishes by the lemma. This implies that the map (7.12) is *not* surjective, *i.e.*, not every Vassiliev invariant of knots is obtained from a Vassiliev invariant of virtual knots, unfortunately.

Vassiliev invariants of virtual 1-tangles

To present any Vassiliev invariant of knots by using Gauss diagrams we instead consider Vassiliev invariants of virtual 1-tangles. Unlike the above case, any Vassiliev invariant of knots is obtained from a Vassiliev invariant of virtual 1-tangles.

In this section, by a *1-tangle* we mean a tangle with 1 open strand oriented rightward at its end points, *i.e.*, an embedding of an oriented interval into $[0,1]\times\mathbb{R}^2$ whose end points are on $\{0,1\}\times\{0\}\times\{0\}$ such that it is oriented rightward at the end points. We have a map

$$\{\text{rightward oriented 1-tangles}\} \xrightarrow{\cong} \{\text{oriented knots}\},$$

which takes a 1-tangle to its closure; it is the knot obtained by gluing the two end points of the 1-tangle. This map is bijective, since a 1-tangle is recovered from a knot by cutting a point of the knot, and such a 1-tangle is uniquely determined (up to isotopy) by the knot. (Because the complement of a small ball around the cutting point is homeomorphic to a ball including the 1-tangle, and the choices of such a small ball are isotopic to each other.) Pulling back the above bijection we have the following bijection,

$$\{\text{Vassiliev invariants of knots}\} \xrightarrow{\cong} \{\text{Vassiliev invariants of 1-tangles}\}. \qquad (7.20)$$

In a similar way as we defined a virtual knot, we define a *virtual 1-tangle* to be an equivalence class of Gauss diagrams on an oriented interval modulo the relations RI, RII, and RIII shown in Figure 7.6, *i.e.*, the set of virtual 1-tangles is given by

$$\{\text{virtual 1-tangles}\} = \{\text{Gauss diagrams on an interval}\}/\text{RI, RII, RIII}.$$

Just as in the knot case, we have a natural inclusion,

$$\{\text{1-tangles}\} \subset \{\text{virtual 1-tangles}\}.$$

Figure 7.11 A Gauss diagram on an interval, and the corresponding surface with a tangle diagram

Theorem 7.14 (Goussarov, see [GPV00]**).** A Vassiliev invariant of virtual 1-tangles of degree d induces a Vassiliev invariant of 1-tangles of degree d by restricting the map to the set of 1-tangles. Further, any Vassiliev invariant of 1-tangles is induced by some Vassiliev invariant of virtual 1-tangles.

See [GPV00] for the proof of the theorem.

This theorem implies that we have a surjective map,

$$\{\text{Vassiliev invariants of virtual 1-tangles}\} \xrightarrow{\cong} \{\text{Vassiliev invariants of 1-tangles}\},$$

induced by restricting Vassiliev invariants of virtual 1-tangles to the set of 1-tangles. Further, this theorem together with the bijection (7.20) implies that any Vassiliev invariant of knots is obtained from a Vassiliev invariant of virtual 1-tangles.

Unlike Lemma 7.13 we have the following lemma, whose proof is left to the reader as an exercise.

Lemma 7.15. The space $\overrightarrow{\mathcal{A}}(\text{interval})^{(2)}/\overrightarrow{\mathrm{FI}}$ is a 2-dimensional vector space spanned by the following two diagrams,

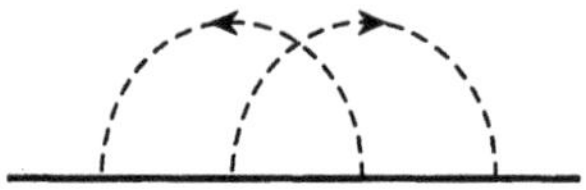 , 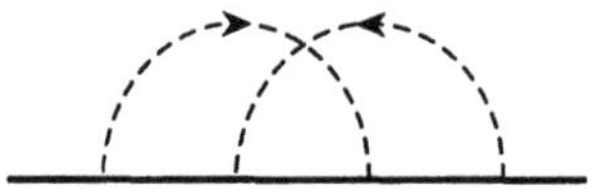 .

Counting combinations of crossings

A remarkable application of the theory of Gauss diagrams is that we obtain an explicit constructive presentation of each Vassiliev invariant, whereas the definition of a Vassiliev invariant only gives an implicit characterization.

Let us consider the construction of such an explicit presentation of a Vassiliev invariant in the case of degree 2. We consider the weight system

$$W : \overrightarrow{\mathcal{A}}(\text{interval})^{(2)}/\overrightarrow{\mathrm{FI}} \longrightarrow \mathbb{C},$$

given by $W = (D)^\star$, where D is given by

$D =$ 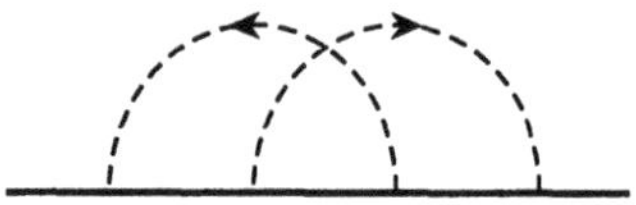 . (7.21)

That is, W is the weight system defined by $W(D) = 1$ for the above D and $W(D') = 0$ for the diagram D' other than D in Lemma 7.15. Recall that $\overrightarrow{\mathcal{A}}(\text{interval})^{(2)}/\overrightarrow{\mathrm{FI}}$ is the vector space spanned by arrow diagrams of degree 2 on an interval subject to the 6T and $\overrightarrow{\mathrm{FI}}$ relations. We regard W as lying in the following space,

$$W \in \langle 6\mathrm{T}^\star, \overrightarrow{\mathrm{FI}}^\star \rangle^\perp \subset \mathrm{span}_{\mathbb{C}}\{\text{arrow diagrams on an interval}\}^\star,$$

where $\mathrm{span}\{\cdots\}^\star$ denotes the dual vector space of $\mathrm{span}\{\cdots\}$, and $6\mathrm{T}^\star$ and $\overrightarrow{\mathrm{FI}}^\star$ denote the dual vectors of the relations 6T and $\overrightarrow{\mathrm{FI}}$ respectively. Further, $\langle 6\mathrm{T}^\star, \overrightarrow{\mathrm{FI}}^\star \rangle^\perp$ denotes the orthogonal complement of the subspace spanned by the vectors $6\mathrm{T}^\star$ and $\overrightarrow{\mathrm{FI}}^\star$.

A Vassiliev invariant v of degree 2 can be regarded as a linear map of the vector space spanned by Gauss diagrams of degree ≤ 2 which takes the relations RI$'$, RII$'$, and RIII$'$ to 0. Similarly as above, we regard v as lying in the following space,

$$v \in \langle \mathrm{RI}'^\star, \mathrm{RII}'^\star, \mathrm{RIII}'^\star \rangle^\perp \subset \mathrm{span}_{\mathbb{C}} \left\{ \begin{array}{l} \text{Gauss diagrams on an interval} \\ \text{with dashed chords} \end{array} \right\}^\star, \qquad (7.22)$$

where we mean $(\cdots)^\star$ in the same sense as above. Let us consider the case that the weight system of v is equal to the above W, *i.e.*, $W = v^{(2)}$. Noting the relation $\mathrm{RII}'^{(1)}$, we have that

$$v(D_{++}) = -v(D_{+-}) = -v(D_{-+}) = v(D_{--}) = W(D).$$

Hence, v is presented by

$$v = (D_{++})^\star - (D_{+-})^\star - (D_{-+})^\star + (D_{--})^\star + \text{ (terms of lower degrees)},$$

where $D_{\varepsilon_1\varepsilon_2}$ denotes the diagram D in (7.21) with the signs ε_1 and ε_2 on the left and right chords of D respectively. In particular, putting

$$v = (D_{++})^\star - (D_{+-})^\star - (D_{-+})^\star + (D_{--})^\star,$$

v belongs to the orthogonal complement in (7.22), that is, v takes RI$'$, RII$'$ and RIII$'$ to 0; it is left to the reader to verify that.

Let us compute the value of the above v on the 1-tangle of the trefoil knot, shown as below.

Its corresponding Gauss diagram is the following diagram.

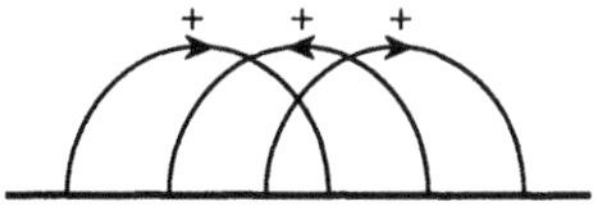

Further, this is equal to the following sum,

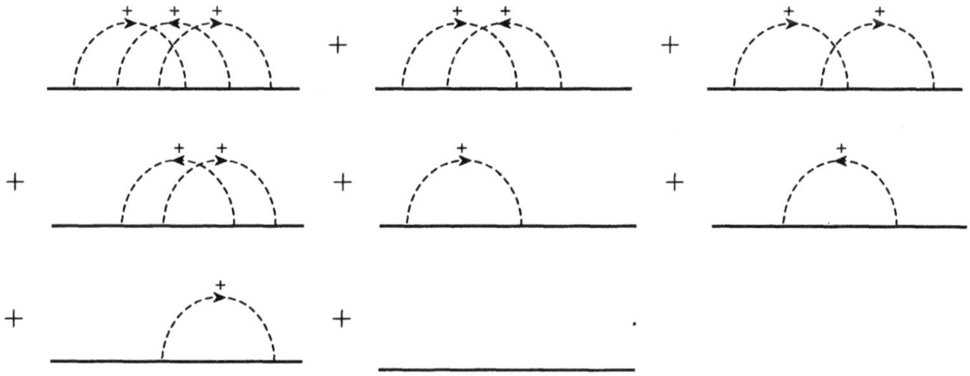

Since the above sum includes one copy of D_{++}, we have that

$v\Big($  $\Big) = 1.$ (7.23)

In other words, the above value is obtained as the number of pairs of crossings of the 1-tangle diagram which are arranged in the way described by the diagram (7.21). In the following 1-tangle diagram of the trefoil knot, such a pair of crossings is indicated. (It is unique in this case.)

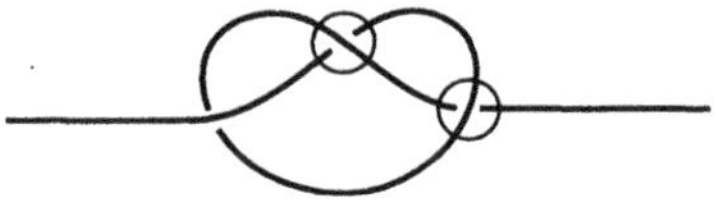

Since the two crossings in the pair are positive, we obtain (7.23).

For a diagram of the 1-tangle of the figure 8 knot we have such a pair of crossings as below.

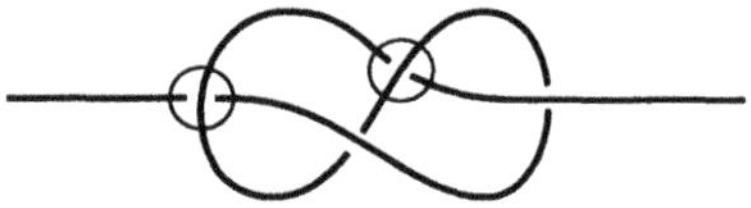

Since the two crossings are positive and negative, the corresponding Gauss diagram includes D_{-+} as a subset of it. Hence,

$v($  $) = -1,$

by the same argument as above.

In general, we obtain the value of v of any 1-tangle by counting pairs of crossings of its diagram in the same way as above. That is, we have

Theorem 7.16. Let T be a 1-tangle, and let D_T be a diagram of T. We put $v(T)$ by

$$v(T) = \sum_{(C_i, C_i')} \varepsilon_{C_i} \varepsilon_{C_i'},$$

where the sum runs over all pairs (C_i, C_i') of crossings C_i and C_i' of D_T which are arranged in the way described by the diagram D in (7.21). Here, for a crossing C we put $\varepsilon_C = 1$ if C is positive, and $\varepsilon_C = -1$ if C is negative. Then, v is a Vassiliev invariant of 1-tangles of degree 2.

7.5 Vassiliev invariants as mapping degrees on configuration spaces

Guadagnini-Martellini-Mintchev [GMM89, GMM90] and Bar-Natan [Bar91, Bar95a] showed that the coefficients of the formal perturbative expansion of the path integral in Chern-Simons field theory on S^3 in the presence of a Wilson loop around a knot could be presented by (finite dimensional) configuration space integrals, and pointed out that these integrals yield knot invariants. Further, Bott-Taubes

[BoTa94] investigated the isotopy invariance of these knot invariants derived from the configuration space integrals from a topological viewpoint which constructs suitable compactifications of the configuration spaces. (They gave a complete definition only for the invariant of degree 2.) Furthermore, Kohno [Koh94] investigated high degree cases by using the graph complex. These approaches were further investigated and completed by Thurston [Th$_{\mathrm{D}}$*99*], Altschuler-Freidel [AlFr96], Yang [Yan*97*]; see also [Poi*99a*, PoVi01].

In this section we give concrete presentations of Vassiliev invariants of degree 2 as mapping degrees of configuration spaces following these approaches. In the global viewpoint on a configuration space we present a mapping degree by a configuration space integral, while we can alternatively view a mapping degree in a local viewpoint, where we express it by the number of certain combinations of crossings as in Theorem 7.16. See Appendix F.3 for the formal perturbative expansion of the Chern-Simons path integral as the physical background of a configuration space integral.

We put two Jacobi diagrams X and Y by

$X =$ 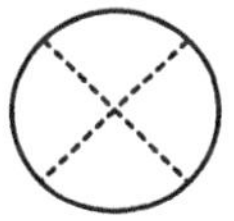 , $\quad Y =$ 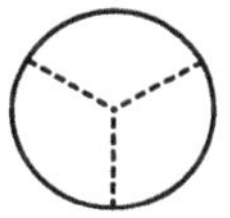 .

For an oriented knot $K \subset \mathbb{R}^3$ we define the *configuration spaces* C_X and C_Y associated to X and Y by

$$C_X = \left\{ (x_1, x_2, x_3, x_4) \in K^4 \subset (\mathbb{R}^3)^4 \;\middle|\; \begin{array}{l} x_1, x_2, x_3, x_4 \text{ are 4 distinct points} \\ \text{on } K \text{ appearing in this cyclic order} \\ \text{along the orientation of } K \end{array} \right\},$$

$$C_Y = \left\{ \begin{array}{l} (x_1, x_2, x_3, x_4) \\ \quad \in K \times K \times K \times \mathbb{R}^3 \subset (\mathbb{R}^3)^4 \end{array} \;\middle|\; \begin{array}{l} x_1,\ x_2,\ x_3 \text{ are 3 distinct points} \\ \text{on } K \text{ appearing in this cyclic order} \\ \text{along the orientation of } K, \\ \text{and } x_4 \text{ is a point in } \mathbb{R}^3 \\ \text{different from the 3 points} \end{array} \right\}.$$

Here, x_4 of C_Y can lie on K, as long as it is different from x_1, x_2, and x_3. Further, we define a map $\phi_{ij} : C_X, C_Y \to S^2$ by

$$\phi_{ij}(x_1, x_2, x_3, x_4) = \frac{x_j - x_i}{|x_j - x_i|} \in S^2, \tag{7.24}$$

where we regard S^2 as the unit 2-sphere in $\mathbb{R}^3$. Furthermore, we define the maps ϕ_X and ϕ_Y by the following direct products,

$$\phi_X = \phi_{13} \times \phi_{24} : C_X \to S^2 \times S^2,$$
$$\phi_Y = \phi_{14} \times \phi_{24} \times \phi_{34} : C_Y \to S^2 \times S^2 \times S^2.$$

They are graphically shown by the following pictures.

Further, we present them symbolically by  ,  .

We define a compactification $\overline{C_X}$ to be the closure of C_X in K^4. As a topological space, it is described as follows. Putting

$$C'_X = \left\{ (x_1, x_2, x_3, x_4) \in (S^1)^4 \;\middle|\; \begin{array}{l} x_1, x_2, x_3, x_4 \text{ are 4 distinct points} \\ \text{on } S^1 \text{ appearing in this cyclic order} \\ \text{along a fixed orientation of } S^1 \end{array} \right\},$$

we identify C'_X and C_X through the map $S^1 \to \mathbb{R}^3$ whose image is the knot K. Further, putting

$$\Delta_3 = \{(t_1, t_2, t_3, t_4) \in \mathbb{R}^4 \mid t_i > 0, \quad t_1 + t_2 + t_3 + t_4 = 1\},$$

we identify C'_X and $S^1 \times \Delta_3$ through the following homeomorphism,

$$C'_X \to S^1 \times \Delta_3, \qquad (x_1, x_2, x_3, x_4) \mapsto \big(x_1, (x_2 - x_1, x_3 - x_2, x_4 - x_3, x_1 - x_4)\big),$$

where by $x_j - x_i$ we mean the distance from x_i to x_j along the orientation of S^1. Through these identifications $\overline{C_X}$ is homeomorphic to $S^1 \times \overline{\Delta_3}$, where $\overline{\Delta_3}$ denotes the closed tetrahedra given by

$$\overline{\Delta_3} = \{(t_1, t_2, t_3, t_4) \in \mathbb{R}^4 \mid t_i \geq 0, \quad t_1 + t_2 + t_3 + t_4 = 1\}.$$

The boundary of $\overline{C_X}$ is homeomorphic to the union of direct products of S^1 and 4 faces of the tetrahedra $\overline{\Delta_3}$. From the point of view of configuration spaces, the boundary can be obtained by gluing 4 copies of the configuration space of 3 points on S^1; we present such a configuration graphically by 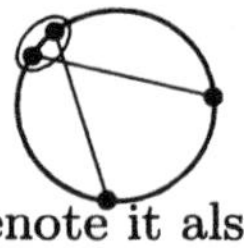 . The map ϕ_X can extend to the continuous map $\overline{C_X} \to S^2 \times S^2$; we denote it also by ϕ_X.

We define a compactification $\overline{C_Y}$ of C_Y as follows. We put a compactification $\overline{\mathbb{R}^3}$ of $\mathbb{R}^3$ to be the closed 3-ball which is the union of $\mathbb{R}^3$ and the infinite 2-sphere. Further, we put $\hat{C}_Y$ to be the closure of C_Y in $K \times K \times K \times \overline{\mathbb{R}^3}$. Though $\hat{C}_Y$ is a compactification of C_Y, the map ϕ_Y can not extend to a continuous map on $\hat{C}_Y$. To let ϕ_Y extend, we consider the graph of $\phi_Y : C_Y \to S^2 \times S^2 \times S^2$:

$$\{(\mathbf{x}, \ \phi_Y(\mathbf{x})) \mid \mathbf{x} \in C_Y\} \subset C_Y \times (S^2 \times S^2 \times S^2).$$

Further, we define a compactification $\overline{C_Y}$ of it to be the closure of the graph in $\hat{C}_Y \times (S^2 \times S^2 \times S^2)$. Then, the map ϕ_Y does extend to a continuous map on $\overline{C_Y}$. Note that $\overline{C_Y}$ and $\hat{C}_Y$ are different compactifications of C_Y and that we have a natural projection $p : \overline{C_Y} \to \hat{C}_Y$ so that we can regard $\overline{C_Y}$ as obtained by blowing-up singularities of $\hat{C}_Y$. (This compactification $\overline{C_Y}$ is a modification of the compactification in [PoVi01], which is a quotient of the compactification in [BoTa94], see also [Th$_D$*99*, Poi*99a*].)

The boundary of $\overline{C_Y}$ is the union of 5 spaces, written $\partial_1\overline{C_Y}$ through $\partial_5\overline{C_Y}$, which we will presently describe.

Firstly, $\partial_1\overline{C_Y}$ is the space consisting of the boundary points that arise when the coordinate x_4 of C_Y tends to the infinity of $\mathbb{R}^3$. As a topological space, $\partial_1\overline{C_Y}$ is homeomorphic to the direct product of S^2 and the configuration space of 3 points on S^1. The image of $\partial_1\overline{C_Y}$ under the map ϕ_Y lies in $\{(x,x,x) \mid x \in S^2\}$.

Secondly, $\partial_2\overline{C_Y}$ is the space consisting of configurations which are symbolically shown as  . The image of $\partial_2\overline{C_Y}$ under the projection p is the space in $\hat{C}_Y$ consisting of points with $x_1 = x_2$, $x_2 = x_3$ or $x_3 = x_1$. The image of $\partial_2\overline{C_Y}$ under the map ϕ_Y lies in the following space

$$\{(z_1, z_2, z_3) \in S^2 \times S^2 \times S^2 \mid z_i = z_j \text{ for some } i \neq j\}.$$

Thirdly, $\partial_3\overline{C_Y}$ is the space consisting of configurations which are symbolically shown as 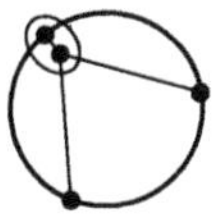 . The image of $\partial_3\overline{C_Y}$ in $\hat{C}_Y$ is the space consisting of points with $x_4 = x_1$, $x_4 = x_2$ or $x_4 = x_3$. Further, $\partial_3\overline{C_Y} \to p(\partial_3\overline{C_Y})$ is a blow up such that the 2-sphere is taken to a point. As a topological space $\partial_3\overline{C_Y}$ consists of 3 copies of the direct product of S^2 and the configuration space of 3 points on S^1.

Fourthly, $\partial_4\overline{C_Y}$ is the space consisting of configurations which are symbolically shown as  . The image of $\partial_4\overline{C_Y}$ in $\hat{C}_Y$ is the space consisting of points with $x_1 = x_2 = x_4$, $x_2 = x_3 = x_4$ or $x_1 = x_3 = x_4$. Further, $\partial_4\overline{C_Y} \to p(\partial_4\overline{C_Y})$ is a blow up such that the inverse image of a point is homeomorphic to a 2-holed 3-ball, which is a 3-manifold obtained from a 3-ball by removing two disjoint 3-balls from it. More precisely, the inverse image is described as follows. Let $\mathbf{v} \in \mathbb{R}^3$ be the tangent vector of the knot at a point with, say, $x_1 = x_2 = x_4$. Then, the inverse image is homeomorphic to the closure of the following space,

$$\left\{ \left(\frac{\mathbf{x}}{|\mathbf{x}|}, \frac{\mathbf{x}-\mathbf{v}}{|\mathbf{x}-\mathbf{v}|} \right) \in S^2 \times S^2 \;\middle|\; \mathbf{x} \in \mathbb{R}^3 - \{0, \mathbf{v}\} \right\}.$$

Hence, the closure is a compactification of $\mathbb{R}^3 - \{0, \mathbf{v}\}$, which is homeomorphic to a

2-holed 3-ball. As a topological space $\partial_4 \overline{C_Y}$ consists of 3 copies of the fiber bundle whose fiber is a 2-holed 3-ball and whose base space is the configuration space of 2 points on S^1. Further, $\partial_4 \overline{C_Y}$ has the orientation-reversing involution Θ given by

$$\Theta : (\mathbb{R}^3 - \{0, \mathbf{v}\}) \longrightarrow (\mathbb{R}^3 - \{0, \mathbf{v}\}), \qquad \mathbf{x} \longmapsto \mathbf{v} - \mathbf{x},$$

on each fiber. It satisfies that $(\phi_{14} \times \phi_{24} \times \phi_{34}) \circ \Theta = (\phi_{42} \times \phi_{41} \times \phi_{34})$.

Lastly, $\partial_5 \overline{C_Y}$ is the space consisting of configurations which are symbolically shown as 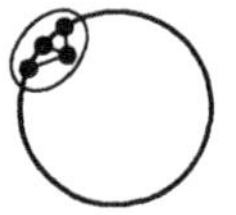 . The image of $\partial_4 \overline{C_Y}$ in $\hat{C}_Y$ is the space consisting of points with $x_1 = x_2 = x_3 = x_4$. Further, $\partial_5 \overline{C_Y} \to p(\partial_5 \overline{C_Y})$ is a blow up such that the inverse image of a point is the 4-manifold described as follows. Let $\mathbf{v} \in \mathbb{R}^3$ be the tangent vector of the knot at a point with $x_1 = x_2 = x_3 = x_4$. Then, the inverse image is homeomorphic to the closure of the following space

$$\Big\{ \Big(\frac{\mathbf{x}}{|\mathbf{x}|}, \frac{\mathbf{x} - \mathbf{v}}{|\mathbf{x} - \mathbf{v}|}, \frac{\mathbf{x} - c\mathbf{v}}{|\mathbf{x} - c\mathbf{v}|} \Big) \in S^2 \times S^2 \times S^2 \;\Big|\; c > 1, \quad \mathbf{x} \in \mathbb{R}^3 - \{0, \mathbf{v}, c\mathbf{v}\} \Big\}.$$

Since the above three vectors in S^2 are linearly dependent, the image of $\partial_5 \overline{C_Y}$ under the map ϕ_Y lies in the following space,

$$\{(z_1, z_2, z_3) \in S^2 \times S^2 \times S^2 \mid z_1, z_2, z_3 \text{ are linearly dependent as vectors in } \mathbb{R}^3\}.$$

The orientations of the configuration spaces C_X and C_Y are given as follows. The orientation of C_X is given by the order $\langle x_1, x_3, x_2, x_4 \rangle$ of the coordinates of C_X; this order is induced by the order of subscripts of $\phi_X = \phi_{13} \times \phi_{24}$. Further, the orientation of C_Y is given by the order $\langle x_1, (x_4)_1, x_2, (x_4)_2, x_3, (x_4)_3 \rangle$, where $(x_4)_1, (x_4)_2, (x_4)_3$ denote the coordinates of $\mathbb{R}^3$. This order is induced by the order of subscripts $\phi_Y = \phi_{14} \times \phi_{24} \times \phi_{34}$, by using the vertex-orientation of the trivalent vertex of the Jacobi diagram Y. Though we use temporary orientations of edges of a Jacobi diagram, say Y, to define the map ϕ_Y and the orientation of C_Y, the orientation of $\phi_Y(C_Y)$ (which is a cycle in $S^2 \times S^2 \times S^2$) does not depend on the choice of the temporary orientation, but depends only on the vertex-orientation of the trivalent vertex of Y. See [Th$_{\mathrm{D}}$*99*, Poi*99a*] for detailed descriptions of the orientations of configuration spaces associated to any Jacobi diagrams.

To define a mapping degree of a map of a configuration space, let us make a 6-dimensional space as a union of copies of $\overline{C_X} \times S^2$ and $\overline{C_Y}$ by canceling their boundaries. The boundary of $\overline{C_X} \times S^2$ consists of 4 copies of the direct product of S^2 and the configuration space of 3 points on S^1. Further, $\partial_3 \overline{C_Y}$ consists of 3 copies of the same direct product. We consider the space obtained by gluing 3 copies of $\overline{C_X} \times S^2$ and 4 copies of $\overline{C_Y}$ along the above boundaries. Further, we consider the quotient of this space by the involution Θ on the boundary $\partial_4 \overline{C_Y}$. We put the quotient space to be C. As a union of copies of the map $\phi_X \times \mathrm{id}_{S^2}$ on $\overline{C_X} \times S^2$ and the map ϕ_Y on $\overline{C_Y}$, we obtain the following map,

$$\phi : (C, \partial C) \to \Big((S^2 \times S^2 \times S^2)/(\mathfrak{S}_3 \times \mathbb{Z}_2 \times \mathbb{Z}_2), \{ \text{ linearly dependent } (z_1, z_2, z_3)\} \Big),$$

where the 3rd symmetric group $\mathfrak{S}_3$ acts on $S^2 \times S^2 \times S^2$ by permuting the 3 entries and the group $\mathbb{Z}_2 \times \mathbb{Z}_2$ acts on $S^2 \times S^2 \times S^2$ in the way that the actions of the two generators are given by $(z_1, z_2, z_3) \mapsto (-z_1, -z_2, z_3)$ and $(z_1, z_2, z_3) \mapsto (-z_1, z_2, -z_3)$. The mapping degree of ϕ is defined to be the image of the fundamental homology class of C in $H_6(C, \partial C)$ by $\phi_\star$, noting that the 6th homology group of the space of the image of ϕ is isomorphic to $\mathbb{Z}$.

Theorem 7.17.
(1) The mapping degree of ϕ is an isotopy invariant of knots.
(2) Furthermore, it is a Vassiliev invariant of degree 2.

Proof of Theorem 7.17 (1). The homeomorphism types of the configuration spaces C_X and C_Y do not change when the knot is isotoped, and neither do unions C of copies of them. Furthermore, the homotopy type of the map ϕ is invariant under isotopies of the knot. Hence, its mapping degree is an isotopy invariant. □

To show (2) (and an alternative proof of (1)) of this theorem, we present the mapping degree of ϕ by using integrals on configuration spaces. Let ω be a normalized volume form on S^2 such that the volume of S^2 is equal to 1. Then, the mapping degree of ϕ is presented by

$$I(K) = \int_C \phi^\star(\omega \times \omega \times \omega), \tag{7.25}$$

since this integral calculates the volume of the cycle which is the image of C under ϕ. (For a physical background of this integral, see Appendix F.3.) We give below a direct proof of the isotopy invariance of $I(K)$.

Alternative proof of Theorem 7.17 (1). For simplicity, we assume that C was a closed 6-manifold. We will show that $I(K_0) = (K_1)$ for any two isotopic knots K_0 and K_1. Along an isotopy between them, we obtain a 1 parameter family C_t of the configuration space C for $t \in [0, 1]$. We denote by $C_{[0,1]}$ the union of the C_t for all $t \in [0, 1]$; we assume that it was a 7-manifold. Further, we have a 1 parameter family of maps ϕ_t on the C_t, and the union $\phi_{[0,1]}$ of them. Then,

$$\begin{aligned} I(K_0) - I(K_1) = & \int_{C_0} (\phi_0)^\star(\omega \times \omega \times \omega) - \int_{C_1} (\phi_1)^\star(\omega \times \omega \times \omega) \\ = & \int_{\partial C_{[0,1]}} (\phi_{[0,1]})^\star(\omega \times \omega \times \omega) = \int_{C_{[0,1]}} d\Big((\phi_{[0,1]})^\star(\omega \times \omega \times \omega) \Big) \\ = & \int_{C_{[0,1]}} (\phi_{[0,1]})^\star \Big(d(\omega \times \omega \times \omega) \Big) = 0, \end{aligned}$$

where we obtain the third equality by Stokes' theorem.

Indeed C is not a 6-manifold, but a 6-cycle. However, we can justify the above argument for a 6 cycle, by a minor technical modification of it (which is left to the reader). □

Proof of Theorem 7.17 (2). We show that the alternating sum of $I(K)$ vanishes for knots K that arise from any singular knot with 3 double points. Fix a small ball centered at each double point, and choose a δ ball centered at each double point in the fixed ball. We denote by $K_{\varepsilon_1,\varepsilon_2,\varepsilon_3}$ (for $\varepsilon_1, \varepsilon_2, \varepsilon_3 = \pm$) the knot that arises by replacing each double point with a positive, or with a negative crossing respectively, according to $\varepsilon_1, \varepsilon_2, \varepsilon_3$, as below.

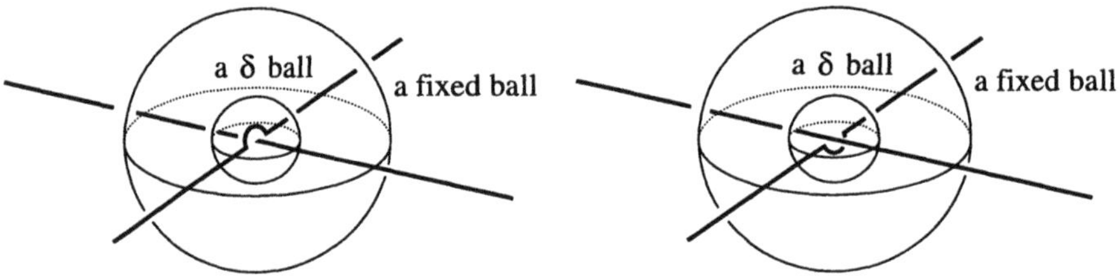

It is sufficient to show the vanishing of

$$\sum_{\varepsilon_1,\varepsilon_2,\varepsilon_3=\pm} \varepsilon_1\varepsilon_2\varepsilon_3 I(K_{\varepsilon_1,\varepsilon_2,\varepsilon_3}). \tag{7.26}$$

We denote by C_{n_1,n_2,n_3} the subset of the configuration space C (for the knot $K_{\varepsilon_1,\varepsilon_2,\varepsilon_3}$) consisting of configurations where exactly n_1, n_2, n_3 points among the coordinates x_1, x_2, x_3, x_4 lie in the 3 fixed balls respectively. Further, we put $I_{n_1,n_2,n_3}(K_{\varepsilon_1,\varepsilon_2,\varepsilon_3})$ to be the same integral (7.25) restricting its domain to C_{n_1,n_2,n_3}. Then, (7.26) is equal to

$$\sum_{n_1,n_2,n_3} \sum_{\varepsilon_1,\varepsilon_2,\varepsilon_3=\pm} \varepsilon_1\varepsilon_2\varepsilon_3 I_{n_1,n_2,n_3}(K_{\varepsilon_1,\varepsilon_2,\varepsilon_3}). \tag{7.27}$$

To show the vanishing of this formula, we estimate

$$\sum_{\varepsilon_1,\varepsilon_2,\varepsilon_3=\pm} \varepsilon_1\varepsilon_2\varepsilon_3 I_{n_1,n_2,n_3}(K_{\varepsilon_1,\varepsilon_2,\varepsilon_3}), \tag{7.28}$$

for each triple of n_1, n_2, n_3. Since the configuration space has only 4 coordinates, $n_1 + n_2 + n_3 \leq 4$. Hence, at least one of n_i is less than or equal to 1; say, $n_1 \leq 1$. When $n_1 = 0$, we have that $I_{0,n_2,n_3}(K_{+,\varepsilon_2,\varepsilon_3}) = I_{0,n_2,n_3}(K_{-,\varepsilon_2,\varepsilon_3})$ by definition. When $n_1 = 1$, the difference $I_{1,n_2,n_3}(K_{+,\varepsilon_2,\varepsilon_3}) - I_{1,n_2,n_3}(K_{-,\varepsilon_2,\varepsilon_3})$ is at most of order of δ. Therefore, so is (7.28). Since (7.27) is invariant under isotopy, it is estimated by an arbitrarily small constant, as $\delta \to 0$. This implies the vanishing of it. □

Configuration spaces for 1-tangles

In this subsection we reformulate the above argument for 1-tangles. By localizing the presentation of the mapping degree, we obtain the concrete presentation of the Vassiliev invariant which counts arrangements of crossings, as in Theorem 7.16.

We fix a vector $\mathbf{v} = (1,0,0) \in \mathbb{R}^3$. In this subsection by a *1-tangle* we mean a smooth embedding of an oriented copy of $\mathbb{R}$ into $\mathbb{R}^3$ such that its image outside a compact set is equal to a line in the direction of $\mathbf{v}$.

We put two Jacobi diagrams X' and Y' to be

$X' =$  , $Y' =$ 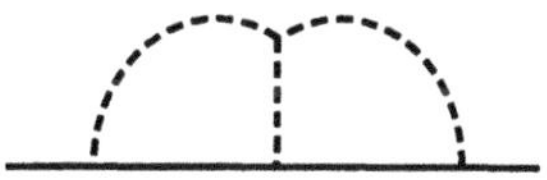 .

For a 1-tangle we define configuration spaces $C_{X'}$ and $C_{Y'}$ associated to X' and Y' by

$$C_{X'} = \left\{ (x_1, x_2, x_3, x_4) \in T^4 \subset (\mathbb{R}^3)^4 \;\middle|\; \begin{array}{l} x_1, x_2, x_3, x_4 \text{ are 4 distinct points} \\ \text{on } T \text{ appearing in this order} \\ \text{along the orientation of } T \end{array} \right\},$$

$$C_{Y'} = \left\{ \begin{array}{l} (x_1, x_2, x_3, x_4) \\ \quad \in T \times T \times T \times \mathbb{R}^3 \subset (\mathbb{R}^3)^4 \end{array} \;\middle|\; \begin{array}{l} x_1,\ x_2,\ x_3 \text{ are 3 distinct points} \\ \text{on } T \text{ appearing in this order} \\ \text{along the orientation of } T, \\ \text{and } x_4 \text{ is a point in } \mathbb{R}^3 \\ \text{different from the 3 points} \end{array} \right\}.$$

By using the maps ϕ_{ij} in (7.24), we define the maps $\phi_{X'}$ and $\phi_{Y'}$ by

$$\begin{aligned} \phi_{X'} &= \phi_{31} \times \phi_{24} : C_{X'} \to S^2 \times S^2, \\ \phi_{Y'} &= \phi_{41} \times \phi_{24} \times \phi_{43} : C_{Y'} \to S^2 \times S^2 \times S^2. \end{aligned}$$

Further, we present them symbolically by the following diagrams,

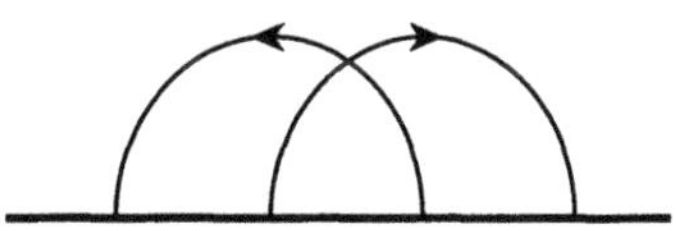 , 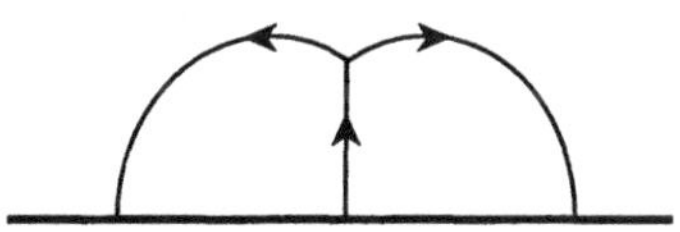 .

We define a compactification $\overline{C_{X'}}$ to be the closure of $C_{X'}$ in $(\overline{\mathbb{R}^3})^4$, where $\overline{\mathbb{R}^3}$ denotes a compactification of $\mathbb{R}^3$ defined to be the union of $\mathbb{R}^3$ and the infinite 2-sphere, as before. We denote by $\pm\infty$ the boundary points in $\overline{\mathbb{R}^3}$ in the direction $\pm\mathbf{v}$. The boundary of $\overline{C_{X'}}$ consists of 5 parts; the part with $x_1 = -\infty$, the part with $x_4 = \infty$, and the other three parts consisting respectively of configurations

symbolically presented by

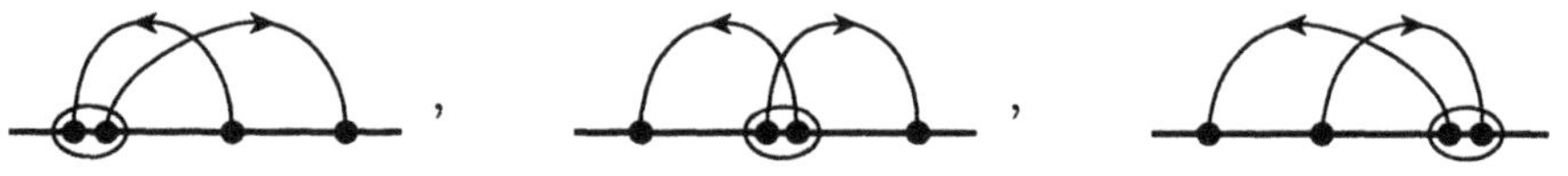

We define a compactification $\overline{C_{Y'}}$ of $C_{Y'}$ as follows. We put $\hat{C}_{Y'}$ to be the closure of $C_{Y'}$ in $(\overline{\mathbb{R}^3})^4$. Further, we define $\overline{C_{Y'}}$ to be the closure of the graph of the map $\phi_{Y'}$ in $\hat{C}_{Y'} \times (S^2 \times S^2 \times S^2)$. Though we omit a complete description of the boundary of $\overline{C_{Y'}}$, the main part of the boundary consists of configurations symbolically presented by

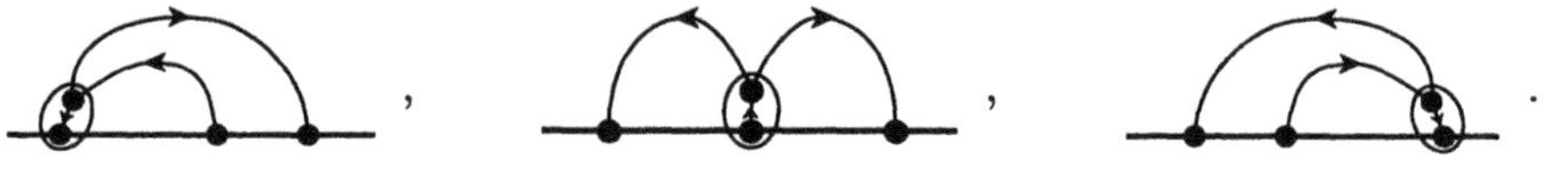

Further, another part of the boundary has an orientation-reversing involution Θ defined similarly as the involution on $\partial_4 \overline{C_Y}$ mentioned earlier.

We consider the space obtained by gluing $C_{X'} \times \mathrm{id}_{S^2}$ and $C_{Y'}$ along the boundaries symbolically presented above. We put C to be its quotient space by Θ. As the union of $\phi_{X'} \times \mathrm{id}_{S^2}$ and $\phi_{Y'}$ we obtain the following map,

$$\phi : (C, \partial C) \longrightarrow \Big((S^2 \times S^2 \times S^2)/(\mathfrak{S}_3 \times \mathbb{Z}_2 \times \mathbb{Z}_2), S\Big),$$

where S is the union of $S^2 \times S^2 \times \{\pm\mathbf{v}\}$ and the space consisting of linearly dependent (z_1, z_2, z_3). The mapping degree of ϕ is defined similarly as in the previous case.

Theorem 7.18.
(1) The mapping degree of ϕ is an isotopy invariant of 1-tangles.
(2) Furthermore, it is a Vassiliev invariant of degree 2.

Proof. We can prove (1) in the same way as we proved Theorem 7.17 (1).

We prove (2) by showing the equality of the invariant with the invariant of Theorem 7.16, following a proof in [PoVi01]. We show the equality by localizing the mapping degree as follows; it is well known that the mapping degree of a map is equal to the number of points (counted with signs) of the inverse image of a regular value of the map.

Let T be a 1-tangle, and let D be a diagram of T in a plane vertical to $\mathbf{v}$, letting $\mathbf{v}$ be the vector $(0,0,1)$ in this proof. We choose a triple $(\mathbf{v}_1, \mathbf{v}_2, \mathbf{v}_3)$ in a neighborhood of $(\mathbf{v}, \mathbf{v}, \mathbf{v})$, which does not belong to S. The inverse image of $(\mathbf{v}_1, \mathbf{v}_2, \mathbf{v}_3)$ by $\phi_{Y'}$ consists of configurations of the following sort.

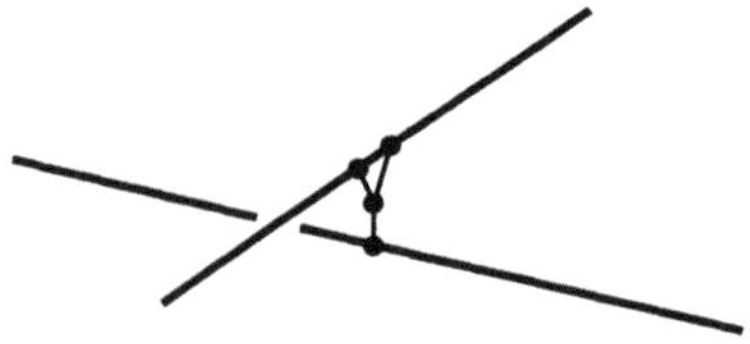

When the triple $(\mathbf{v}_1, \mathbf{v}_2, \mathbf{v}_3)$ is sufficiently close to $(\mathbf{v}, \mathbf{v}, \mathbf{v})$, such a configuration is close to a crossing of the diagram, as shown in the above picture. For simplicity, we regard the strands near the crossing as linear in the neighborhood of the crossing. Then, a tangent vector to one of the linear strands and two of $\mathbf{v}_1, \mathbf{v}_2, \mathbf{v}_3$ would be linearly dependent, if such a configuration existed. Hence, for a generic choice of $(\mathbf{v}_1, \mathbf{v}_2, \mathbf{v}_3)$, there exists no such configuration; this implies that the inverse image of such a generic triple under $\phi_{Y'}$ is empty. Furthermore, the number (counted with signs) of points of the inverse image of such a triple under $\phi_{X'}$ is equal to the number (counted with signs) of pairs of crossings arranged according to the diagram (7.21), whose configurations are of the following sort.

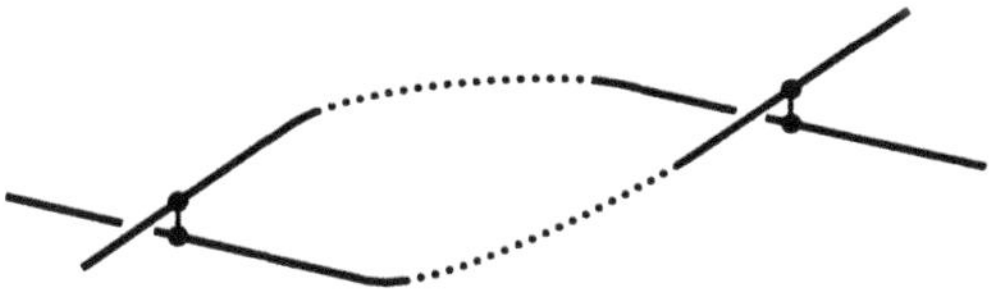

Finally, recall that the invariant in Theorem 7.16 is equal to the number of such pairs (counted with signs) from the definition of the invariant.

Since the number (counted with signs) of points of the inverse image of a regular value under ϕ is equal to the mapping degree of ϕ, the mapping degree is equal to the invariant of Theorem 7.16, which Theorem 7.16 assures us is a Vassiliev invariant of degree 2. □

Chapter 8

Quantum invariants of 3-manifolds

In the 1950-60s the great development of differential topology had led to a solution of the classification problem of high-dimensional (of dimension at least 5) simply connected manifolds, in the sense that such a problem is reduced to homotopy theory by the h-cobordism theorem and surgery theory (see, *e.g.*, [Kos93]). In low the dimensional case we know (say, [Hir94, Walla68]) that 2-manifolds are classified by their genus. Further, it is known, see [BHP68], that the classification problem of 4-manifolds is algorithmically unsolvable since the set of fundamental groups of 4-manifolds is sufficiently large to be algorithmically unsolvable as a word problem in group theory. It is still unknown whether the classification problem of 3-manifolds is solvable.

To study each 3-manifold toward the classification of 3-manifolds, Thurston [Thw97] has proposed classifying 3-manifolds in terms of 8 geometric structures; in this picture, the classification problem is reduced to the classification of discrete subgroups of the 8 Lie groups associated to the geometric structures. There are also combinatorial approaches to the classification problem of 3-manifolds; see, *e.g.*, [Hemi92, Lins95]. To study the set of 3-manifolds, toward the classification of 3-manifolds, topological invariants of 3-manifolds would be helpful. Here, for a well-known set S, a map

$$I : \{\text{3-manifolds}\} \longrightarrow S$$

is called a *topological invariant* of 3-manifolds if $I(M) = I(M')$ for any homeomorphic 3-manifolds M and M'. An invariant gives rough classification of 3-manifolds, *i.e.*, a partition of the set of 3-manifolds.

The constructive invariants of 3-manifolds which we knew in classical topology were homology groups and fundamental group. The situation changed with the interaction between low-dimensional topology and quantum field theory in the 1980s. Since the late 1980s, many topological invariants of 3-manifolds have been discovered, what are called quantum invariants of 3-manifolds. They are mostly based on the quantum field theory whose Lagrangian is the Chern-Simons functional, that was described by Witten [Wit89a] from the viewpoint of mathematical physics; see Appendix F for the theory. Reshetikhin-Turaev [ReTu91] gave the first rigorous mathematical construction of quantum invariants as linear sums of quantum invariants of framed links. After that, rigorous constructions of quantum invariants were obtained by various approaches, say, by using surgery along links [ReTu91, BHMV92, Lic97, TuWe93], Heegaard splittings [Koh92] and simplicial complexes [TuVi92], see also [KiMe91, Wen93, KoTa93, KoTa96, Yok97]. Further invariants of 3-manifolds related to quantum invariants have been discovered, which we discuss in the following chapters.

This chapter is organized as follows. In Section 8.1 we show how to present 3-manifolds by surgery along links. Here, a surgery is an operation replacing a $D^2 \times S^1$ embedded in a 3-manifold with a $S^1 \times D^2$; this operation is the simplest amongst the surgeries of manifolds in surgery theory which describe changes of manifolds by cobordisms. Different surgery presentations of a 3-manifold are related to each other by Kirby moves, or alternatively by Fenn-Rourke moves. In Section 8.2 we introduce the quantum $SU(2)$ and $SO(3)$ invariants of 3-manifolds using skein theory, following Lickorish [Lic97]; this is the most elementary approach (which is also self-contained) to introduce quantum invariants of 3-manifolds. Further, in Section 8.3 we introduce the original approach of Reshetikhin-Turaev [ReTu91] constructing the quantum $SU(2)$ and $SO(3)$ invariants again; this approach adapts to construct the other quantum invariants of 3-manifolds.

8.1 3-manifolds and their surgery presentations

In this section we express closed 3-manifolds by using framed links in S^3; such expressions are called surgery presentations of 3-manifolds. Surgery presentations of the same 3-manifold are related to each other by the Kirby moves, or equivalently by the Fenn-Rourke moves, introduced in this section.

Let K be a framed knot in S^3. We denote by $N(K)$ a tubular neighborhood of K and $S^3 - N(K)$ the 3-manifold obtained from S^3 by removing $N(K)$. We call $S^3 - N(K)$ an *exterior* of K. The boundary of the exterior is homeomorphic to $S^1 \times S^1$, which is called a *torus*. We have a simple closed curve in the boundary corresponding to the framing of K; we also call the curve the *framing* of K. It is depicted by a thick line in the picture of S^3_K appearing in Figure 8.1. Further, we consider the 3-manifold $D^2 \times S^1$, which is called a *solid torus*. We consider the simple closed curve $\partial D^2 \times \{\text{point}\}$ in the boundary of the solid torus, which is called a *meridian* of the solid torus. It is depicted by a thick line in the picture of S^3_K in appearing Figure 8.1. Furthermore, we consider the closed 3-manifold obtained by gluing the exterior $S^3 - N(K)$ and a solid torus $D^2 \times S^1$ along their boundaries such that the meridian of the solid torus is glued to the framing of the exterior. We denote the 3-manifold by S^3_K and call it the 3-manifold obtained from S^3 by *integral surgery* along the framed knot K.

For an l-component framed link L in S^3 we define S^3_L to be the closed 3-manifold

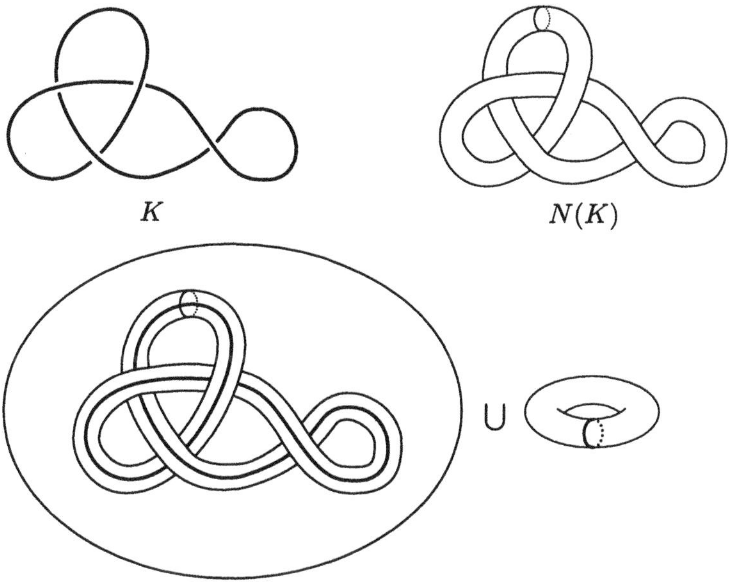

$$S^3_K = (S^3 - N(K)) \cup (D^2 \times S^1)$$

Figure 8.1 A framed knot K, its tubular neighborhood $N(K)$, and the 3-manifold S^3_K

obtained by gluing the exterior $S^3 - N(L)$ and l solid tori such that a meridian of a solid torus is glued to the framing of each component of L. We call S^3_L the closed 3-manifold obtained by *integral surgery* along the framed link L.

A *framed link* in a 3-manifold M is a link with a framing in M, *i.e.*, the image of an embedding of a disjoint union of annuli into M. For a framed link L in a 3-manifold M we denote by M_L the 3-manifold obtained by gluing the link exterior and solid tori as above, and call it the 3-manifold obtained from M by *integral surgery* along L.

For a given framed link L in a 3-manifold M the 3-manifold M_L is determined uniquely up to homeomorphism for the following reason. For simplicity we let L be a framed knot K. Let D be a neighborhood of a meridian disc in the solid torus $D^2 \times S^1$, where we call a disc $D^2 \times \{\text{point}\}$ in a solid torus a *meridian disc.* We put the remaining part to be $B = (D^2 \times S^1) - D$, which is homeomorphic to a 3-ball. Then,

$$M_K = \Big(\big(M - N(K)\big) \cup D \Big) \cup B.$$

The part $\big(M - N(K)\big) \cup D$ is uniquely determined up to homeomorphism by the framing of K. Finally we attach a 3-ball B to obtain M_K. Lemma 8.1 below assures us that this choice is unique. Hence, the 3-manifold M_K is uniquely determined up to homeomorphism.

Lemma 8.1. Let N be a 3-manifold whose boundary is homeomorphic to S^2, and let B^3 denote a 3-ball whose boundary is S^2. Further, let f and g be two homeomorphisms $\partial N \to \partial B^3$. Then, the two 3-manifolds $N \underset{f}{\cup} B^3$ and $N \underset{g}{\cup} B^3$ are homeomorphic.

Proof. We identify B^3 with $(S^2 \times [0,1])/\sim$, where "$\sim$" denotes the collapse of $S^2 \times \{1\}$. We define the homeomorphism $\varphi : B^3 \to B^3$ by $\varphi(x,t) = ((g \circ f^{-1})(x), t)$ for $x \in S^2$ and $t \in [0,1]$. The restriction of φ to the boundary of B^3 is equal to $g \circ f^{-1}$. Hence, we obtain a homeomorphism $N \underset{f}{\cup} B^3 \to N \underset{g}{\cup} B^3$ which is the identity map on N and is equal to φ on B^3. □

When a given 3-manifold M is homeomorphic to S^3_L for some framed link L in S^3, we call such a framed link L (together with S^3) a *surgery presentation* of M. Any closed connected orientable 3-manifold has a surgery presentation by the following theorem.

Theorem 8.2 ([Lic62]). Any closed connected orientable 3-manifold can be obtained from S^3 by integral surgery along some framed link.

To prove the theorem we require some lemmas.

Let F be a closed orientable surface of genus g as shown below.

$F =$ 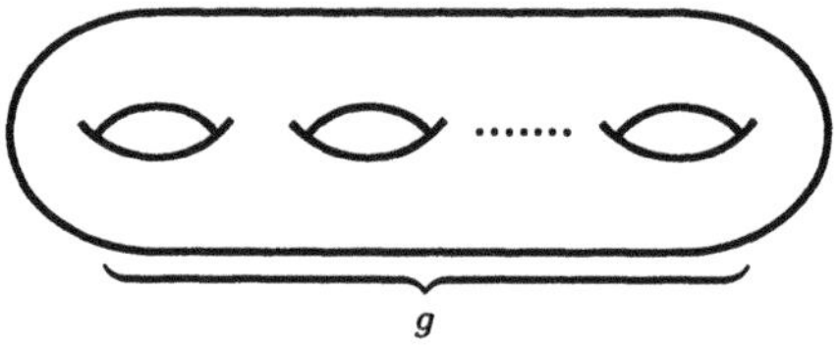

A *handlebody* H of genus g is a 3-manifold bounded by F as shown below, which is homeomorphic to a 3-manifold obtained from a 3-ball by removing tubular neighborhoods of g parallel straight lines in the 3-ball.

$H =$ 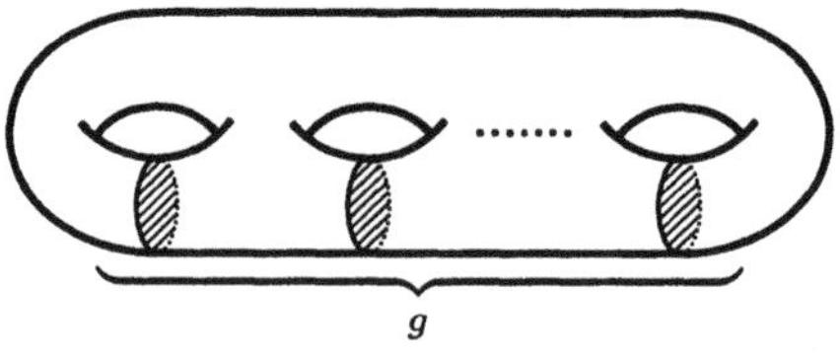

For a homeomorphism f of F to itself, we consider the 3-manifold obtained as the union of two copies of H by attaching their boundaries by f. We denote it by $H \cup_f H$. When a given 3-manifold M is homeomorphic to $H \cup_f H$ for some f, we call the union $H \cup_f H$ a *Heegaard splitting* of M.

Lemma 8.3. Any closed connected orientable 3-manifold has a Heegaard splitting.

Proof. Let M be a closed connected orientable 3-manifold. We consider a triangulation of M as a simplicial complex, *i.e.*, we express M as a union of tetrahedras by gluing their faces to each other such that two faces are glued toghther always by a linear homeomorphism between two triangles. (See [Moi52, Moi54, Mun66] for the existence of a triangulation of any 3-manifold. In fact, it is known that the piecewise linear and the differentiable categories are both equivalent to the topological category in the topology of 3-manifolds.) The *1-skeleton* of a triangulation is the union of edges and vertices of the triangulation. We consider a tubular neighborhood of the 1-skeleton. Its closure is homeomorphic to a handlebody. Further, the complement of the closure is a tubular neighborhood of the 1-skeleton of the dual triangulation of the original triangulation. Hence, the closure of the complement is also homeomorphic to a handlebody. Therefore, as the union of the two handlebodies, we obtain a Heegaard splitting of M. □

To describe homeomorphisms of a surface we introduce Dehn twists as follows. Let C be a simple closed curve in a surface F. We identify a tubular neighborhood of C in F with an annulus. A *Dehn twist* along C is a homeomorphism of F to itself which is the identity outside the tubular neighborhood, and is equal to the map shown in Figure 8.2 inside that neighborhood. In other words, we obtain a Dehn twist along C by cutting F along C, making one rotation of an end, and gluing the ends to each other.

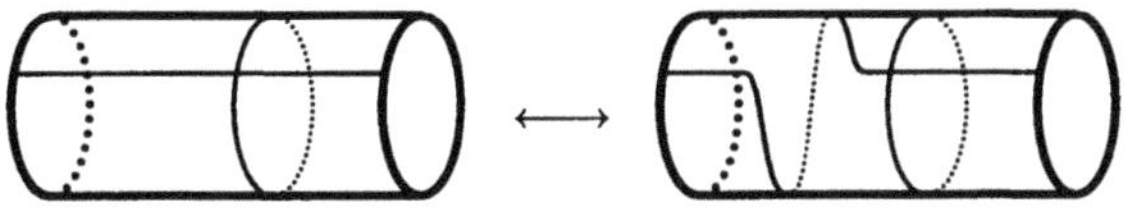

Figure 8.2 Dehn twist along a simple closed curve C, where C is the vertical thin closed curve in the picture

Lemma 8.4 ([Deh38, Lic62]). Any orientation-preserving homeomorphism of F to itself can be expressed (up to isotopy) as a composition of Dehn twists along simple closed curves in F.

The lemma is proved with a step-by-step argument. We omit the proof of the lemma; see, *e.g.*, [PrSo97] for the proof.

Proof of Theorem 8.2. Let M be a closed connected orientable 3-manifold. By Lemma 8.3, M has a Heegaard splitting, *i.e.*, M is homeomorphic to $H \cup_f H$ for some homeomorphism f of a closed surface. Further, we fix a Heegaard splitting of

S^3 of the same genus; for example, we obtain S^3 as

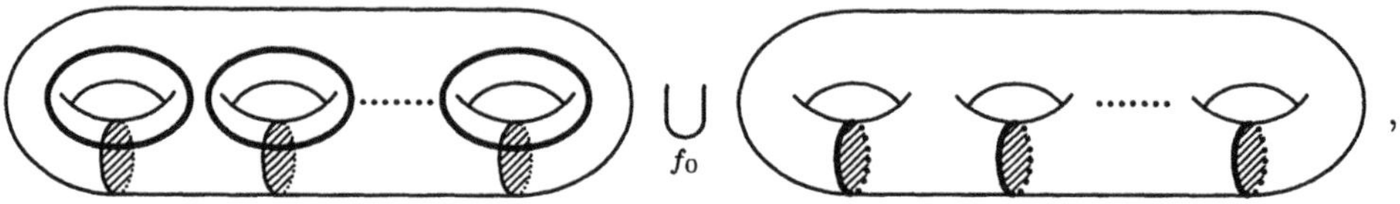

where f_0 is a homeomorphism of the surface sending thick closed curves in the boundary of the left handlebody to those of the right handlebody respectively. We express M as

$$M \cong H \underset{f_0^{-1} \circ f}{\cup} (F \times [0,1]) \underset{f_0}{\cup} H.$$

Further, by Lemma 8.4 we can express $f_0^{-1} \circ f$ by a composition of Dehn twists τ_1, τ_2, $\cdots$, τ_n. Then, M is expressed as

$$M \cong H \underset{\tau_1}{\cup} (F \times [0,1]) \underset{\tau_2}{\cup} \cdots \underset{\tau_n}{\cup} (F \times [0,1]) \underset{f_0}{\cup} H.$$

By Lemma 8.5 below, we replace each Dehn twist by an integral surgery along a framed knot. Hence, we obtain M from $H \cup_{f_0} H$ by integral surgery along a framed link with n components. Since $H \cup_{f_0} H$ is homeomorphic to S^3, we obtain the theorem. □

Lemma 8.5. Let X and Y be 3-manifolds whose boundaries are equal to the same closed surface F. Let τ be a Dehn twist along a closed curve on F. Then, $X \cup_\tau Y$ is homeomorphic to $(X \cup_{\mathrm{id}} Y)_K$, where K is a framed knot in a neighborhood of the closed curve in $X \cup_{\mathrm{id}} Y$. Here, $(X \cup_{\mathrm{id}} Y)_K$ denotes the 3-manifold obtained from $X \cup_{\mathrm{id}} Y$ (which is the union of X and Y along the common boundary F) by integral surgery along K.

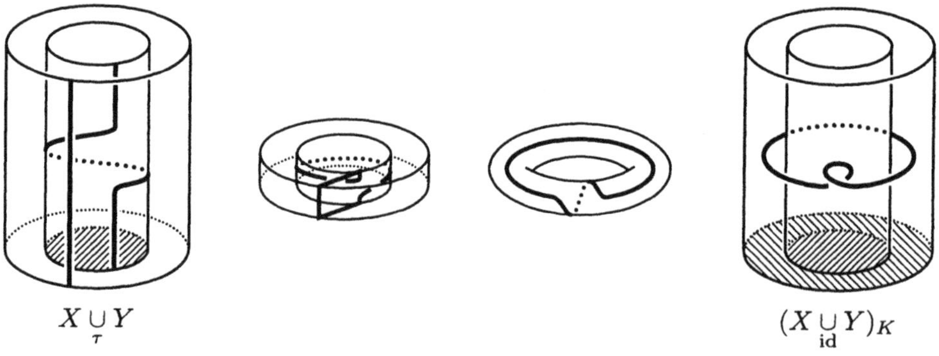

Figure 8.3 In a Heegaard splitting a Dehn twist is alternatively expressed by an integral surgery.

Proof. The proof follows the pictures in Figure 8.3. In the first picture, the outer and inner cylinders imply parts of the boundaries of X and Y respectively, and τ is a Dehn twist sending the outer thick line to the inner one. We glue the boundaries of X and Y except for the middle part of the cylinders. Then, the resulting boundary is a torus with a thick closed curve, as shown in the second picture. We then obtain $X \cup_\tau Y$ by collapsing the thick closed curve. Further, up to homeomorphism of 3-space, such a collapsing can alternatively be expressed by the attachment of the boundary of a disc along the thick closed curve in the picture. Hence, we obtain $X \cup_\tau Y$ by attaching a solid torus by gluing its boundary to the torus in the third picture (which is isotopic to the second picture), taking its meridian to the thick closed curve. This implies the integral surgery along the framed knot, denoted by K, in the last picture. Hence, $X \cup_\tau Y$ is homeomorphic to $(X \cup_{\mathrm{id}} Y)_K$ with the framed knot K. □

The KI move:

The KII move:

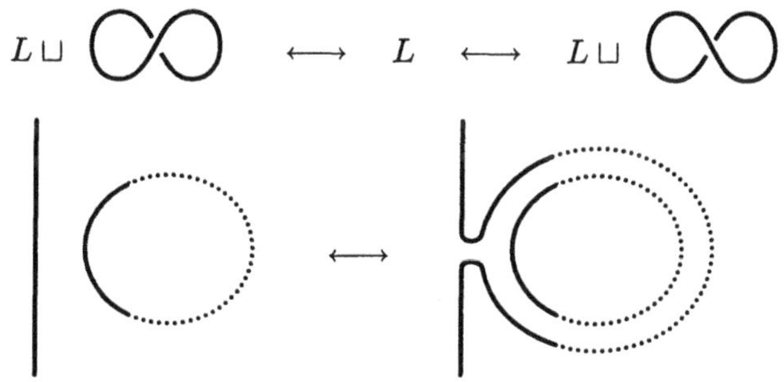

Figure 8.4 The *Kirby moves* KI and KII. A dotted line implies a strand possibly knotted and linked with other components.

To characterize the set of surgery presentations of a 3-manifold we introduce the Kirby moves KI and KII among surgery presentations of 3-manifolds, *i.e.*, among framed links in S^3. The KI move is a move adding (resp. deleting) a trivial knot with ± 1 framing to (resp. from) a framed link. The KII move is a move taking a connected sum of a component of a framed link with a parallel copy of another component of the link along its framing, as shown in Figure 8.4. For example, the two pictures below are related by the KII move.

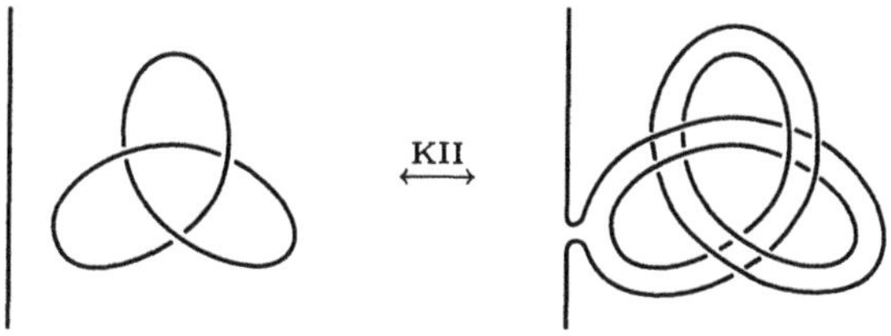

(8.1)

The KII move is also called a *handle slide*, say, of the left vertical strand over the right closed component in the above picture. The KI and KII moves do not change homeomorphism types of 3-manifolds. This is shown as follows.

As for the KI move, we consider an exterior of a trivial knot, which is the complement of a tubular neighborhood of the trivial knot. Further, we consider a disc in the exterior bounded by the 0 framing of the trivial knot. Then, the thick closed curves in the two pictures below are related by a Dehn twist along the boundary of the disc.

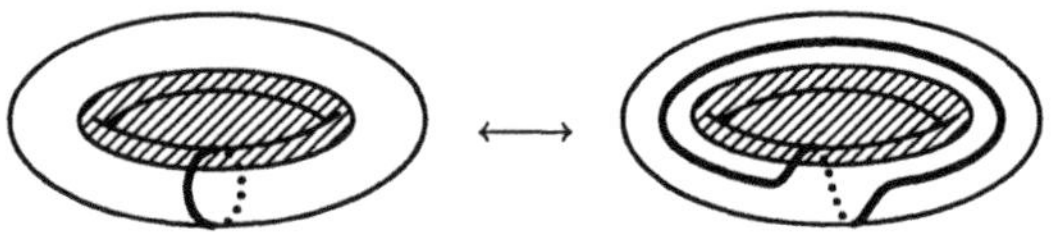

Further, the Dehn twist extends to a homeomorphism between the exteriors. By attaching solid tori to the exteriors, gluing their meridians to the thick closed curves in the above pictures, we obtain S^3 and $S^3_{U_-}$ respectively, where U_- denotes the trivial knot with -1 framing and $S^3_{U_-}$ denotes the 3-manifold obtained from S^3 by integral surgery along U_-. Furthermore, the above homeomorphism between the exteriors extends to a homeomorphism between S^3 and $S^3_{U_-}$. Hence, the KI move adding U_- does not change the homeomorphism type of a 3-manifold. We obtain the same consequence for a KI move which adds the trivial knot U_+ with $+1$ framing by the mirror image of the above argument.

As for the KII move, we consider a handle slide of a strand over a component K of a framed link. The move is expressed by an isotopy of the strand in S^3_K as follows. Recall that S^3_K is the union of the exterior of K and a solid torus. By pushing a meridian of the solid torus into the exterior of K in S^3_K, we obtain a parallel copy of K along its framing. Hence, such a parallel copy bounds a disc in S^3_K, which is an enlargement of a meridian disc in the solid torus. Therefore, the framed link after the handle slide is obtained from the original link by taking the connected sum of the strand with the boundary of a disc in S^3_K. Hence, the two framed links are isotopic in S^3_K. This implies that the KII move does not change homeomorphism types of 3-manifolds.

A surgery presentation of a 3-manifold can be described using 4-manifolds as follows. The 4-manifold $D^2 \times D^2$ is called a *2-handle*. Its boundary is expressed as the union of two solid tori,

$$\partial(D^2 \times D^2) = (\partial D^2 \times D^2) \cup (D^2 \times \partial D^2) = (S^1 \times D^2) \cup (D^2 \times S^1).$$

Let B^4 denote a 4-ball whose boundary is S^3. We consider a framed knot K in S^3. Using the framing of K we have an identification of $S^1 \times D^2$ and a tubular neighborhood $N(K)$ of K. We consider the union of B^4 and a 2-handle $D^2 \times D^2$ by gluing parts of their boundaries using the above identification. The resulting union is a 4-manifold whose boundary is homeomorphic to S^3_K. In the same way, for a framed link L with l components in S^3, we obtain a 4-manifold from B^4

by attaching l 2-handles along tubular neighborhoods of the components of L such that its boundary is homeomorphic to S^3_L. We call such a 4-manifold the *4-manifold bounded by the 3-manifold according to its surgery presentation.*

In the setting of 4-manifolds bounded by 3-manifolds according to their surgery presentations, the Kirby moves are described as follows. As for the KI move, let us consider the trivial knot $U_\pm$ with ± 1 framing, which is a surgery presentation of S^3. The 4-manifold bounded by S^3 according to the surgery presentation U_+ (resp. U_-) is homeomorphic to the 4-manifold obtained from the complex 2-dimensional projective space $\mathbb{C}P^2$ (resp. its mirror image $\overline{\mathbb{C}P^2}$) by removing a 4-ball in it; this follows from an elementary computation, which is left to the reader as an exercise. Hence, by the KI move, a 4-manifold bounded by a 3-manifold according to a surgery presentation changes by the connected sum of a $\mathbb{C}P^2$ or a $\overline{\mathbb{C}P^2}$. As for the KII move, let us consider two surgery presentations related by the KII move. The two associated 4-manifolds are homeomorphic; in fact, one of them is obtained from the other by sliding a 2-handle over another 2-handle. That is the reason why the KII move is also called a *handle slide.*

Surgery presentations of the same 3-manifold are related to each other by the following theorem.

Theorem 8.6 (Kirby [Kirb78]). For framed links L and L' in S^3, the 3-manifold S^3_L is homeomorphic to $S^3_{L'}$ if and only if L can be obtained from L' by a sequence of isotopies of framed links and the KI and KII moves.

This theorem (together with Theorem 8.2) has the following symbolic representation:

$$\begin{aligned}&\{\text{closed connected oriented 3-manifolds}\}/\text{homeomorphism}\\&= \{(\text{unoriented) framed links in } S^3\}/\text{isotopy, KI, KII}.\end{aligned}$$

This equality allows us to *define* the notion of a 3-manifold to be an equivalence class of framed links, modulo the Kirby moves. For the purpose of studying the geometry of the set of 3-manifolds, this equality is fundamental, in the sense that the left hand side of the equality is topological while we can deal with the right hand side in a combinatorial way, reducing framed links to link diagrams using Theorem 1.1.

Sketch of the proof of Theorem 8.6. It is shown above that if L and L' are related by the KI and KII moves, then S^3_L and $S^3_{L'}$ are homeomorphic.

It is rather hard to show the converse. We will give just a sketch of the proof of this part, following the original paper [Kirb78] of Kirby.

Let L and L' be two framed links in S^3 such that S^3_L and $S^3_{L'}$ are homeomorphic to the same 3-manifold M. Let W_0 and W_1 be the 4-manifolds bounded by S^3_L and $S^3_{L'}$ according to their surgery presentations. We put a closed 4-manifold V by

$$V = W_0 \cup (M \times [0,1]) \cup W_1,$$

where we glue ∂W_i and $M \times \{i\}$ $(i = 0, 1)$ by the above mentioned homeomorphisms. It is known that a 4-manifold whose signature is equal to 0 can bound a 5-manifold. As mentioned earlier, by replacing L with the split union of L and the trivial knot with ± 1 framing by the KI move, W_i can be replaced by the connected sum of W_i and either of $\mathbb{C}P^2$ and $\overline{\mathbb{C}P^2}$. Hence, we can assume, without loss of generosity, that the signature of V is equal to 0 by changing L and L' by the KI move, if necessary. Let X be a 5-manifold bounded by V. We choose a Morse function $f : X \to [0, 1]$ such that $f^{-1}(i) = W_i$ and $f|_{M \times [0,1]}$ is the projection. By some arguments, we can remove all critical points of f by making appropriate changes of W_0 and W_1 that arise from the KI and KII moves on L and L'. Thus, the new W_0 and W_1 are diffeomorphic to the same 4-manifold W bounded by M. Corresponding to the constructions of W_0 and W_1 according to the surgery presentations of M, we have two Morse functions f_i on W. Choose a homotopy f_t between f_0 and f_1 such that each f_t is a Morse function except for a finite number of t. By some argument, the change of the Morse function at each of such exceptional t can be described by a KII move on the corresponding surgery presentations. Thus, we obtain the theorem. For a complete proof see [Kirb78] (see also [Lu92]). □

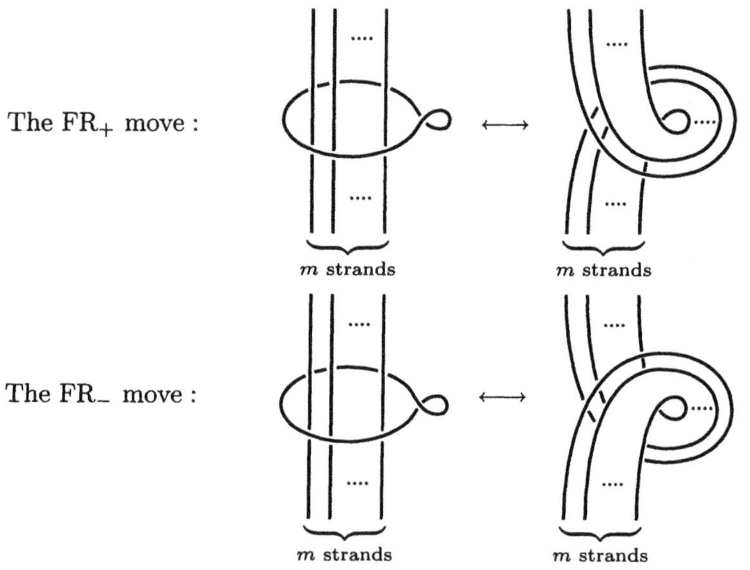

Figure 8.5 The m-strand *Fenn-Rourke moves* FR for any non-negative integer m

We introduce a certain modification of the Kirby moves, the Fenn-Rourke moves FR, shown in Figure 8.5. (We call these moves the Fenn-Rourke moves since Fenn and Rourke proved Theorem 8.7 below, though the moves were not invented by them. For a brief history of the moves see comments in [PrSo97, Chapter IV].) In Theorem 8.7 below, the FR moves play the same role as that of the Kirby moves in

Theorem 8.6, though the FR moves arise as a special case of the Kirby moves. For example, the FR_+ move is obtained

where a band implies a bundle of strands. Further, the FR_- move is obtained by the mirror image of the above procedure.

Theorem 8.7 ([FeRo79]). For framed links L and L' in S^3, the 3-manifold S^3_L is homeomorphic to $S^3_{L'}$, if and only if L can be obtained from L' by a sequence of isotopies of framed links and the FR moves.

We show the proof following [FeRo79].

Proof. By Theorem 8.6 it is sufficient to show that two framed links are related by a sequence of the FR moves if and only if they are related by a sequence of the Kirby moves.

If two framed links are related by the FR moves, then they are related by a sequence of the KI and KII moves, as shown above.

Conversely, we show that, if two framed links are related by the Kirby moves, then they are related by a sequence of the FR moves. The KI moves are the 0-strand FR moves. The KII move is obtained by a sequence of the FR moves as follows. We show the claim that any handle slide over a component C of a framed link L is expressed by a sequence of the FR moves. From the definition of the FR moves we have that

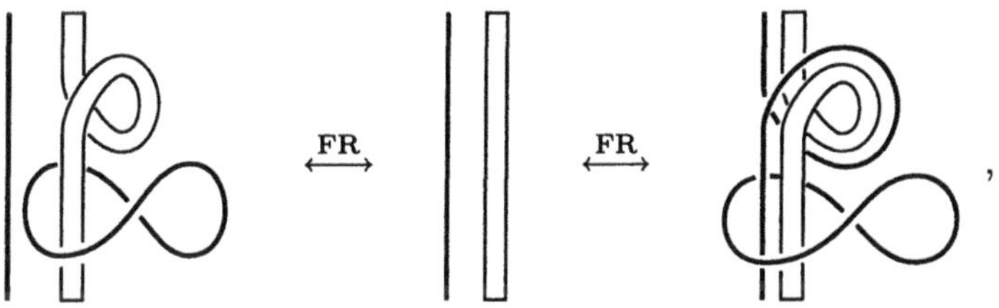

where a band implies a bundle of strands. Hence,

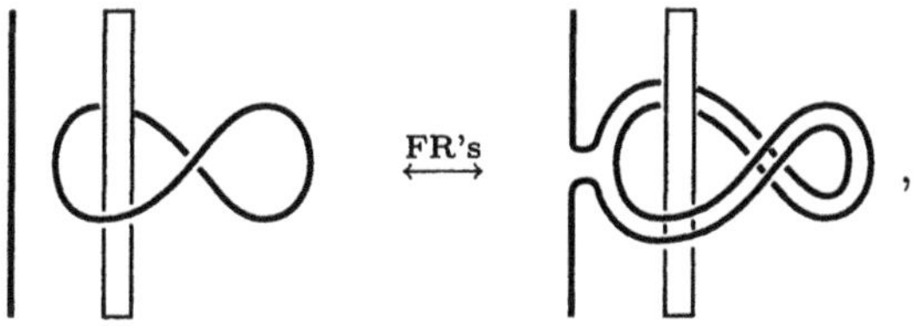

(8.2)

where we obtain both sides from sides of the previous relation (up to isotopy), by removing +1 full twist of the band. The above relation implies the required claim in the case that C is the trivial knot with +1 framing. In general, C is a framed knot. We reduce this general case to the above case as follows. By the FR move we can make a crossing change at any crossing of C as

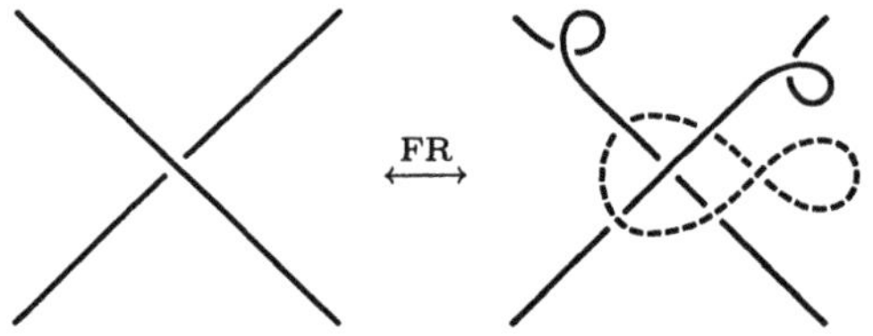

by adding an additional component (depicted in a dashed line) to L. By crossing changes we can change C to a trivial knot. Further, we can change the framing of C with

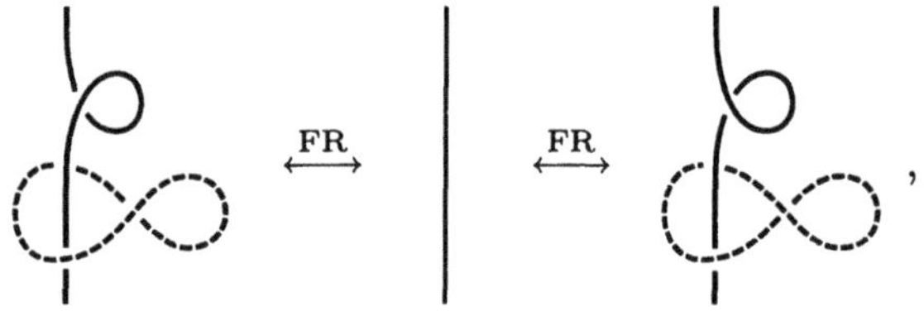

with additional components depicted in dashed lines. By such a procedure we can change C to a trivial knot with +1 framing. That is, for suitably chosen dashed components $K_1, \cdots, K_k$, the framed link L is related to $\hat{L} \cup K_1 \cup \cdots \cup K_k$ by the FR moves, such that the component $\hat{C}$ of $\hat{L}$ corresponding to C is a trivial knot with +1 framing. Let L' be a framed link obtained from L by a handle slide over the component C and let $\hat{L}' \cup K_1 \cup \cdots \cup K_k$ be the framed link obtained from $\hat{L} \cup K_1 \cup \cdots \cup K_k$ by the corresponding handle slide over the component $\hat{C}$. Since $\hat{C}$ is a trivial knot with +1 framing, $\hat{L} \cup K_1 \cup \cdots \cup K_k$ and $\hat{L}' \cup K_1 \cup \cdots \cup K_k$ are related by the FR moves, by (8.2). Further, L' and $\hat{L}' \cup K_1 \cup \cdots \cup K_k$ are related by the FR moves with the same procedure as was used to get from L to $\hat{L} \cup K_1 \cup \cdots \cup K_k$. Hence, L and L' are related by a sequence of the FR moves. □

8.2 The quantum $SU(2)$ and $SO(3)$ invariants via linear skein

In Section 1.2 we defined the quantum (sl_2, V) invariant of framed links via the Kauffman bracket of link diagrams. In this section we define a "linear skein" to be a vector space spanned by link diagrams, subject to the defining relations of the Kauffman bracket. Further, we introduce the "box over n strands" in the linear skein, which is also called the Jones-Wenzl idempotent. Using such boxes we define some invariants of a framed link, which are the quantum $SU(2)$ and $SO(3)$

invariants of the 3-manifold obtained from S^3 by integral surgery along the framed link. For further reading on the topics in this section see [Lic97, KaLi94].

For a surface F, the *linear skein* $\mathcal{S}(F)$ of F is the vector space over $\mathbb{C}$ spanned by link diagrams (admitting the empty diagram) on F subject to the relations

, (8.3)

 $D = (-A^2 - A^{-2})D \qquad$ for any diagram D. (8.4)

By the same argument as appeared following the definition of the Kauffman bracket in Section 1.2, the linear skein $\mathcal{S}(\mathbb{R}^2)$ of $\mathbb{R}^2$ is the 1-dimensional vector space spanned by the empty diagram.

Fix $2n$ points on the boundary of a disc D^2. A *tangle diagram on* $(D^2, 2n)$ is a tangle diagram on the disc D^2 such that the boundary of the tangle diagram is the union of the $2n$ fixed points on ∂D^2. Let the linear skein $\mathcal{S}(D^2, 2n)$ be the vector space spanned by tangle diagrams on $(D^2, 2n)$ subject to the relations (8.3) and (8.4). Further, by regarding D^2 as a rectangle, and regarding the $2n$ points on ∂D^2 as a set of n points on the left edge and n points on the right edge of the rectangle, we introduce an algebra structure on $\mathcal{S}(D^2, 2n)$ such that the product of two tangle diagrams on D^2 is defined to be the union of the two diagrams along a vertical edge, such as

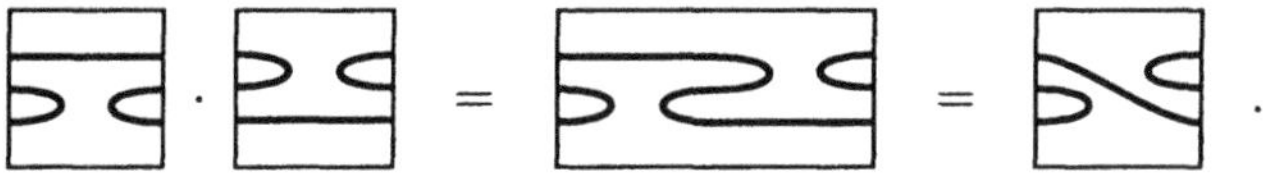

The algebra $\mathcal{S}(D^2, 2n)$ is called the *Temperley-Lieb algebra*.

Proposition 8.8. The Temperley-Lieb algebra $\mathcal{S}(D^2, 2n)$ is generated by

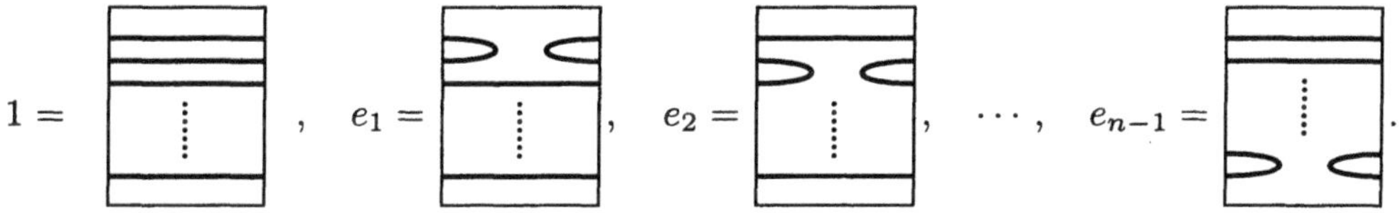

Proof. We consider a tangle diagram on $(D^2, 2n)$. By resolving crossings of the diagram using (8.3), we obtain a linear sum of diagrams without crossings. Further, it follows from an elementary argument that a diagram without a crossing can be expressed by a product of the above generators; the detailed argument is left to the reader as an exercise. □

We define the *box* over n strands in $\mathcal{S}(D^2, 2n)$ recursively by

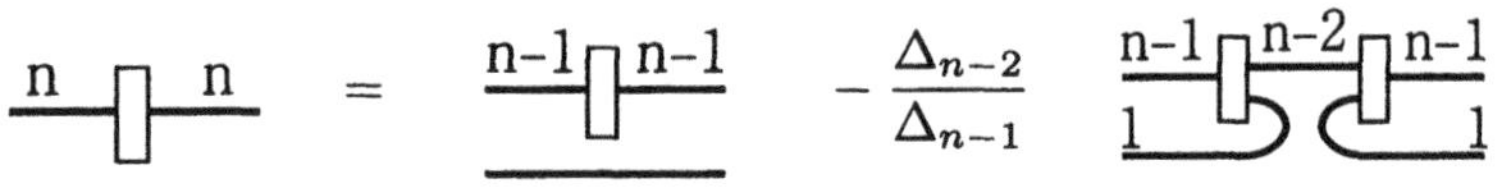

(8.5)

(8.6)

where a strand with a number n implies the union of n parallel copies of the strand, and we define Δ_k by $\Delta_k = (-1)^k(A^{2(k+1)} - A^{-2(k+1)})/(A^2 - A^{-2})$. For example,

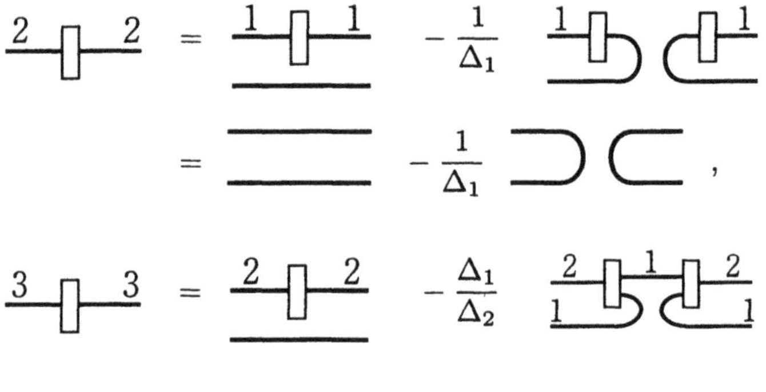

+ (linear sum of non-empty products of e_1 and e_2).

The box over n strands is called the *Jones-Wenzl idempotent* in the Temperley-Lieb algebra $\mathcal{S}(D^2, 2n)$. It is shown later, in Lemma H.2, that the operator invariant of the box over n strands corresponds to the projector to the $(n+1)$ dimensional irreducible representation of $U_q(sl_2)$. This is a reason why the box is an idempotent, *i.e.*, the composition of two copies of the box is equal to the box as an element of the linear skein. The fact that the box is an idempotent is shown, straightforwardly, as a special case of the following proposition.

Proposition 8.9.
$(1)_n$ The following formulae holds for $k = 0, 1, \cdots, n-2$,

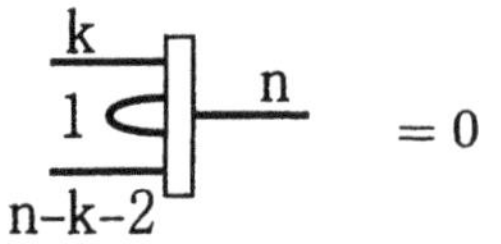

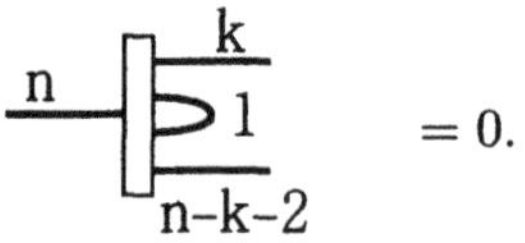

$(2)_n$ The following formula holds,

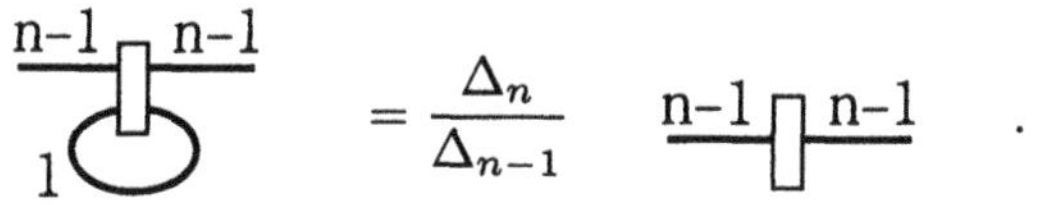

$(3)_n$ The following formula holds for $k = 0, 1, \cdots, n$,

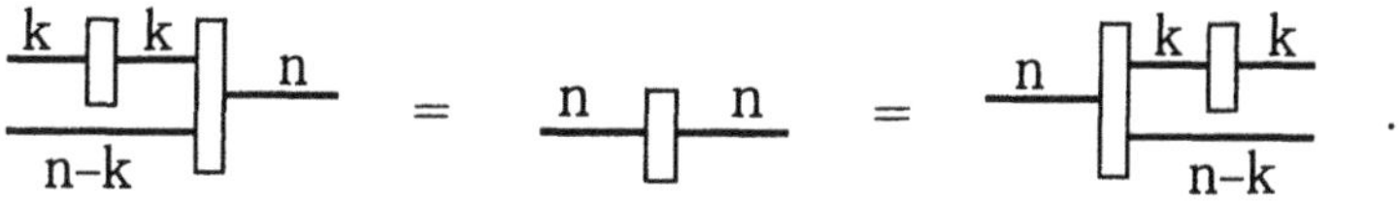

It follows from the proposition that the box is symmetric with respect to π rotation, as an element of $\mathcal{S}(D^2, 2n)$.

Proof of Proposition 8.9. We prove the proposition by induction on n.

Firstly, we show $(1)_n$, assuming $(1)_{n-1}$, $(2)_{n-1}$ and $(3)_{n-1}$, as follows. If $0 \leq k \leq n-3$, then expanding the box over n strands by the recursive formula (8.5) we reduce the proof to $(1)_{n-1}$. If $k = n-2$, then by (8.5) we have that

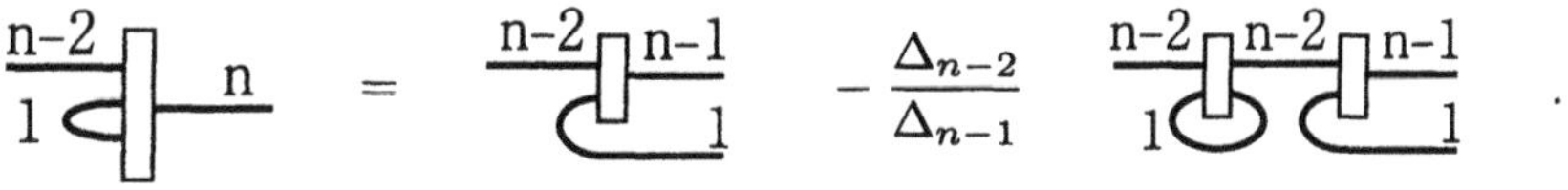

Further, we compute the last diagram as follows:

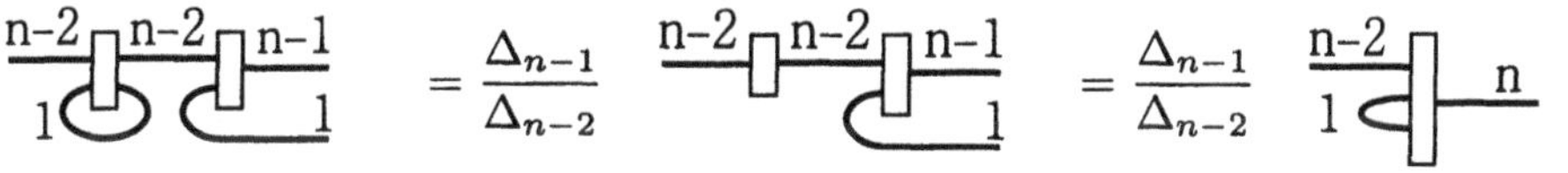

where the first and second equalities are derived from $(2)_{n-1}$ and $(3)_{n-1}$ respectively.

Hence, we obtain 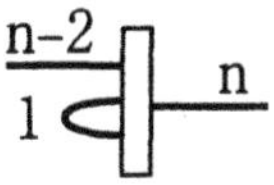 $= 0$. Further, we obtain 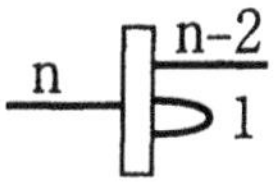 $= 0$ in the same way.

Secondly, we show $(2)_n$, assuming $(3)_{n-1}$, as follows. By (8.5) the left hand side of $(2)_n$ is equal to

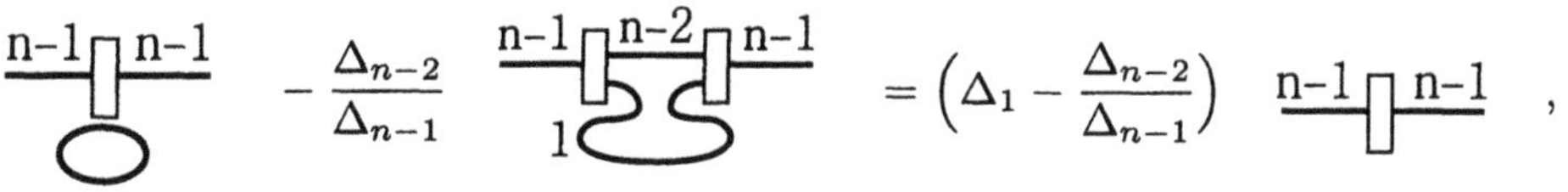

where the above equality is derived from (8.4) and $(3)_{n-1}$. Since the equality $\Delta_1 - \Delta_{n-2}/\Delta_{n-1} = \Delta_n/\Delta_{n-1}$ holds, we obtain $(2)_n$.

Thirdly, we show $(3)_n$, assuming $(1)_n$, as follows. Using (8.5) recursively we

have that

$$+\begin{pmatrix}\text{linear sum of non-empty products}\\ \text{of } e_1, e_2, \cdots, e_{k-1}\end{pmatrix}.$$

By attaching this formula to the box over n strands, the second part on the right hand side vanishes by $(1)_n$. Hence, we obtain $(3)_n$.

Summarizing the above steps, we obtain the proposition by induction on n along the following diagram.

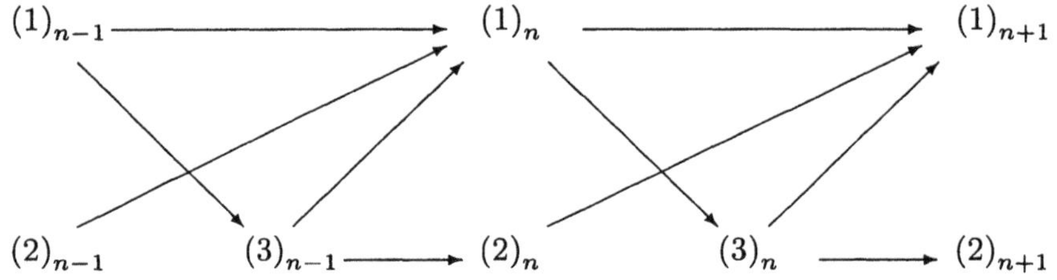

□

Fixing r to be an integer ≥ 3, we define the element $\omega \in \mathcal{S}(S^1 \times I)$ by

$$\omega = \sum_{n=0}^{r-2} \Delta_n$$

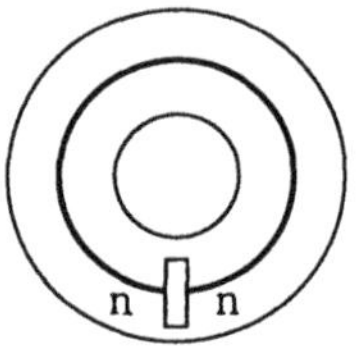

Further, for a link diagram D on a surface F we define $\langle D^\omega \rangle \in \mathcal{S}(F)$ by substituting ω into each component of D; for example, given the link diagram

we have that

$$\langle D^\omega \rangle =$$

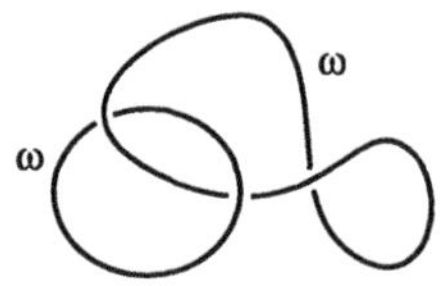

$$= \sum_{n=0}^{r-2} \sum_{m=0}^{r-2} \Delta_n \Delta_m$$

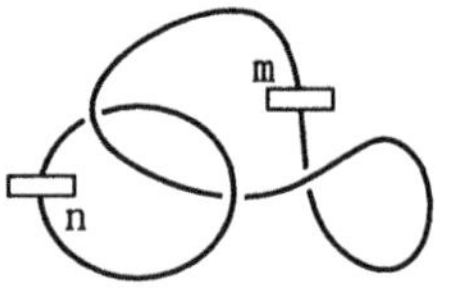

Further, by regarding the Kauffman bracket as an invariant of framed links, as in Section 1.2, we also denote $\langle D^\omega \rangle$ by $\langle L^\omega \rangle$, for a framed link L expressed by D. In the remainder of this section we assume that A^4 is a primitive r-th root of unity, *i.e.*, $A^{4r} = 1$ and $A^{4n} \neq 1$ for any n with $0 < n < r$, noting that this inequality implies that $\Delta_{n-1} \neq 0$, which is necessary for the definition of the box over n strands in (8.5). We have the following possibilities for the value of A:

- When r is odd, A is a primitive r-th, $2r$-th or $4r$-th root of unity.
- When r is even, A is a primitive $4r$-th root of unity.

Lemma 8.10. If a diagram D on $\mathbb{R}^2$ includes a box over $r-1$ strands, then $\langle D\rangle = 0$.

Proof. By using Reidemeister moves RII and RIII, the diagram D may be put in the form 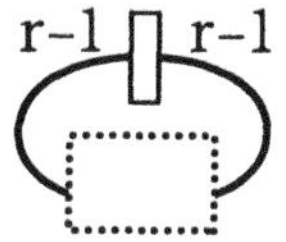

for some tangle diagram in the dotted box. By Proposition 8.8, the tangle diagram (as an element in the linear skein) is equal to a linear sum of products of $1, e_1, e_2, \cdots, e_{r-2}$. If a product includes e_i's, then it vanishes by Proposition 8.9. Otherwise, it is

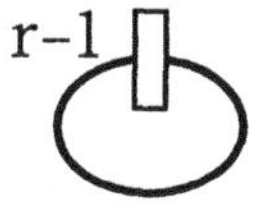

$$= \Delta_{r-1} = (-1)^{r-1}\frac{A^{2r} - A^{-2r}}{A^2 - A^{-2}} = 0.$$

Hence, we obtain the lemma. □

Let L be an oriented framed link in S^3 with l components $L_1, L_2, \cdots, L_l$. The *linking matrix* of L is the $l \times l$ matrix whose (i, j) entry is equal to the linking number $\mathrm{lk}(L_i, L_j)$ and whose (i,i) entry is equal to the framing of L_i. Since the linking matrix is symmetric, the eigenvalues of the matrix are real. Further, it is left to the reader as an exercise that the eigenvalues do not depend on the orientations of the components of L. Let W be the 4-manifold bounded by S^3_L according to the surgery presentation. It is another exercise to show that the linking matrix of L is a presentation matrix of the *intersection form*

$$\iota : H_2(W;\mathbb{R}) \times H_2(W;\mathbb{R}) \longrightarrow \mathbb{R},$$

where $\iota(F_1, F_2)$ is defined to be the algebraic intersection number of transversal closed surfaces F_1 and F_2 in W.

Theorem 8.11. Let M be a 3-manifold obtained from S^3 by integral surgery along a framed link L. Then,

$$\langle U_+^\omega\rangle^{-\sigma_+}\langle U_-^\omega\rangle^{-\sigma_-}\langle L^\omega\rangle \in \mathbb{C} \tag{8.7}$$

is invariant under the Kirby moves on L, where $U_\pm$ denotes the trivial knot with ± 1 framing and σ_+ and σ_- denote the numbers of positive and negative eigenvalues of the linking matrix of L. In particular, (8.7) is a topological invariant of M, by Theorem 8.6

When $A = -\exp(\pi\sqrt{-1}/2r)$, we denote the topological invariant presented by this theorem by $\tau_r^{SU(2)}(M) \in \mathbb{C}$, and call it the *quantum* $SU(2)$ *invariant* of a closed connected oriented 3-manifold M.

Proof of Theorem 8.11. We show the invariance of (8.7) under the KI move as follows. Let L' be a split union of L and U_+. Then, the linking matrix of L' is the block sum of the linking matrix of L and the 1×1 unit matrix (1). Hence, the numbers of eigenvalues of the linking matrix of L' are given by $\sigma'_+ = \sigma_+ + 1$ and $\sigma'_- = \sigma_-$. Further, a diagram D' of L' is obtained as the split union of D' and U_+. Hence, $\langle D'^{\omega} \rangle = \langle D^{\omega} \rangle \langle U_+{}^{\omega} \rangle$. Therefore, we obtain the invariance of (8.7) as

$$\langle U_+^{\omega} \rangle^{-\sigma'_+} \langle U_-^{\omega} \rangle^{-\sigma'_-} \langle D'^{\omega} \rangle = \langle U_+^{\omega} \rangle^{-\sigma_+} \langle U_-^{\omega} \rangle^{-\sigma_-} \langle D^{\omega} \rangle.$$

We obtain the invariance of (8.7) under the split union of a link and U_- in the same way as above. It is left to the reader to show $\langle U_{\pm}^{\omega} \rangle \neq 0$.

We show the invariance of (8.7) under the KII move as follows. Consider the KII move of the handle slide of a component L_i of L over another component L_j. For the corresponding component L'_i of the new framed link L' obtained by the handle slide we have that

$$\begin{aligned} &(\text{framing of } L'_i) = (\text{framing of } L_i) + (\text{framing of } L_j) + 2\,\mathrm{lk}(L_i, L_j), \\ &\mathrm{lk}(L'_i, L_k) = \mathrm{lk}(L_i, L_k) + \mathrm{lk}(L_j, L_k) \qquad \text{for each of the other components } L_k, \end{aligned}$$

since homologically L'_i is obtained as the sum of L_i and L_j. Hence,

$$(\text{the linking matrix of } L') = {}^tP(\text{the linking matrix of } L)P$$

where P is the matrix such that the diagonal entries and the (i, j) entry are equal to 1 and the other entries 0. Therefore, they have the same numbers of positive and negative eigenvalues of the linking matrix of L' as those of L. Hence, $\sigma_\pm$ is invariant under the KII move.

We show the invariance of $\langle D^{\omega} \rangle$ under the KII move in what follows of this proof. We fix 2 points on a component of the boundary of an annulus $S^1 \times I$. A tangle diagram in $(S^1 \times I, 2)$ is a tangle diagram on the annulus $S^1 \times I$ such that the boundary of the diagram is the pair of fixed points. Let $\mathcal{S}(S^1 \times I, 2)$ be the linear skein spanned by the link diagrams on $(S^1 \times I, 2)$. By the recursive formula (8.5) of the box we have that

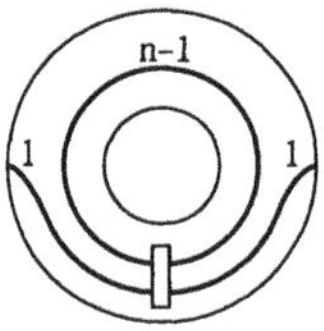

$=$
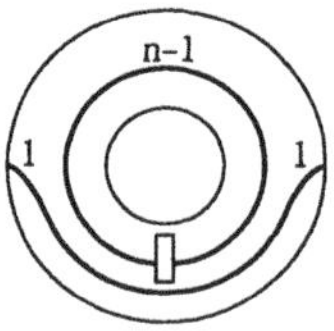

$-\dfrac{\Delta_{n-2}}{\Delta_{n-1}}$
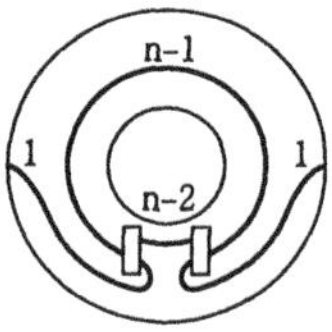

$\in \mathcal{S}(S^1 \times I, 2).$

By applying Proposition 8.9 (3) to the two boxes in the last diagram,

Δ_{n-1}
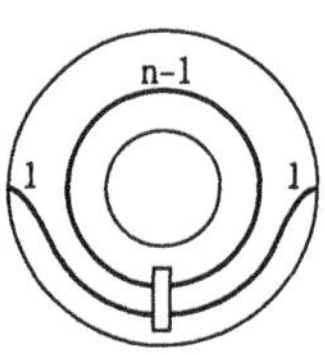

$= \Delta_{n-1}$
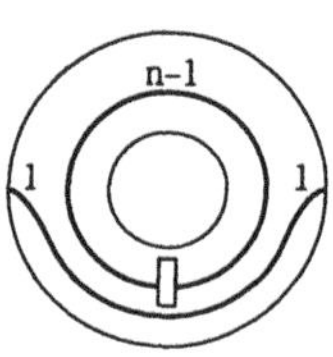

$- \Delta_{n-2}$
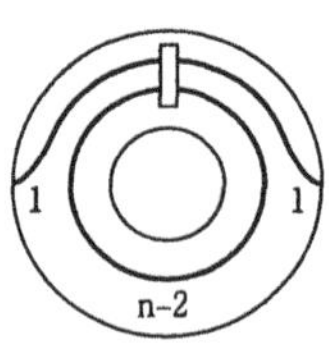

.

Letting a_n denote the difference between the above formula and the formula obtained by π rotation, that is,

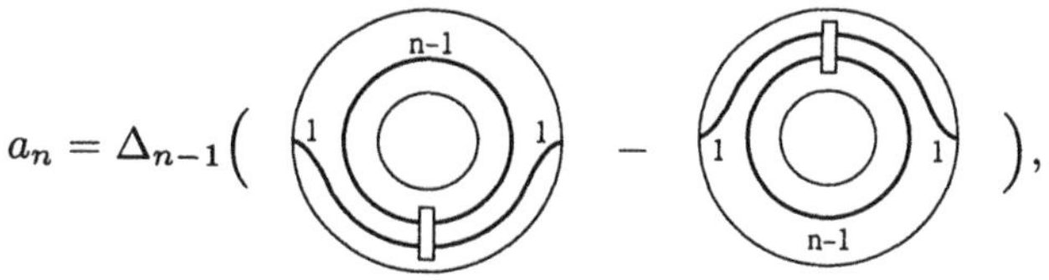

we find that

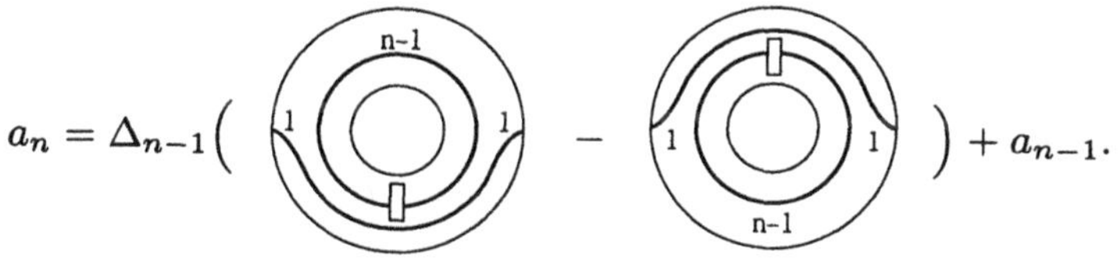

By computing the above recursive formula for a_n we obtain

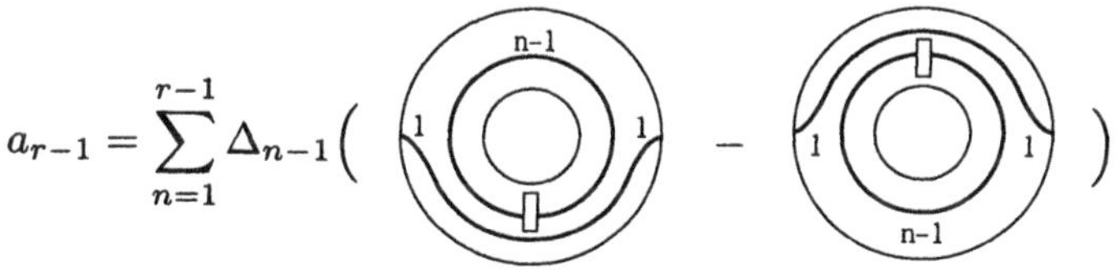

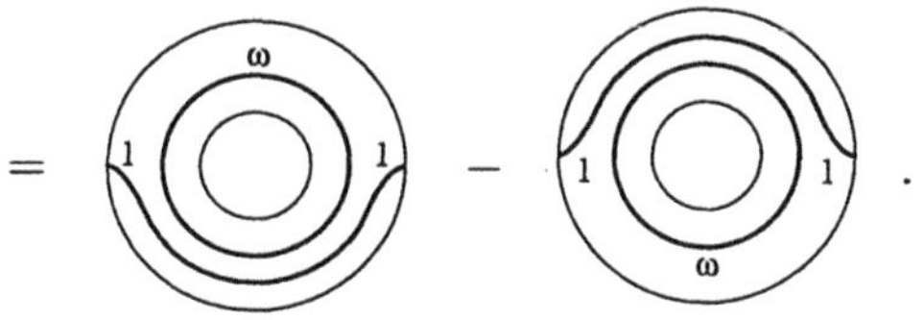

Noting that a_{r-1} includes the box over $r-1$ strands from its definition, we have that

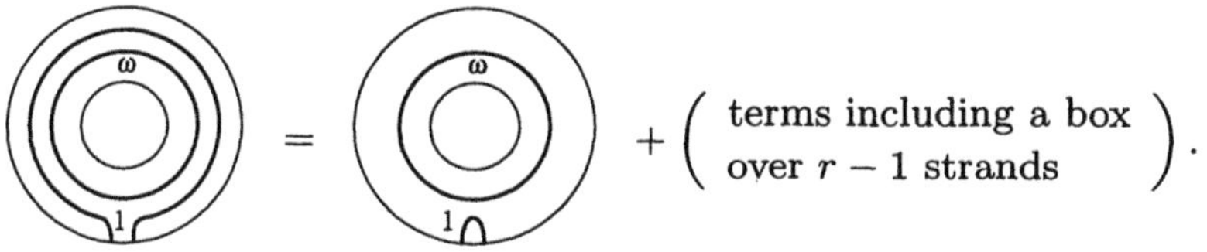

Since the element ω is locally expressed as the union of parallel strands, we obtain the following formula by applying the above formula successively to each of the parallel strands,

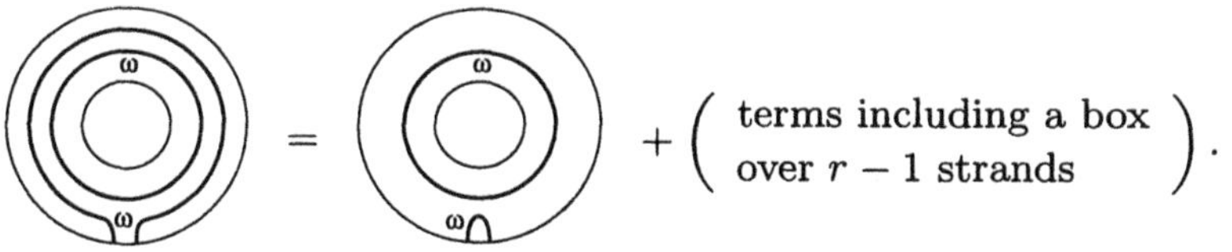

(8.8)

By embedding the above annulus into $\mathbb{R}^3$ we obtain the KII move of strands associated with the element ω; for example, substituting the annulus into the following picture, we obtain the KII move (8.1) for strands associated with ω. Note that such an embedding induces a well-defined map $\mathcal{S}(S^1 \times I) \to \mathcal{S}(\mathbb{R}^2)$.

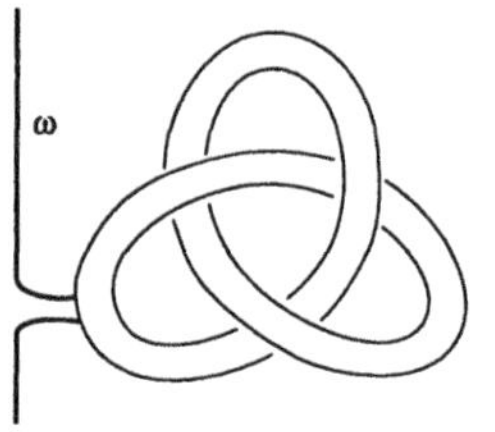

Since the last part of (8.8) vanishes by Lemma 8.10, we obtain the invariance of $\langle D^\omega \rangle$ under the KII move. □

As in the proof of Theorem 8.11 $\langle L^\omega \rangle$ is invariant under the KII move, while it changes under the KI move. Recall that a similar situation occurred when we sought to construct a link invariant from the Kauffman bracket, in Section 1.2. In that case the Kauffman bracket of link diagrams is invariant under the RII and RIII moves, while it changes under the RI move. We considered two options to make the Kauffman bracket invariant. One was to modify the Kauffman bracket of a link diagram by using the writhe of the diagram. The other was to regard the Kauffman bracket as an invariant of framed links, introducing that notion.

Similarly, in the present case we also have two options to make $\langle L^\omega \rangle$ be a topological invariant of a 3-manifold. One is to modify $\langle L^\omega \rangle$ by using the linking matrix of L, as in Theorem 8.11. The other is to regard $\langle L^\omega \rangle$ as an invariant of a "framed 3-manifold", introducing the notion of framed 3-manifolds as follows. (See [Ati90a, KiMe99] for other approaches to introduce the notion of framings of 3-manifolds).

Let M be a closed oriented 3-manifold. We regard oriented 4-manifolds W and W' bounded by M as *equivalent* if the signature of $W \cup_M (-W')$ is equal to 0, where $-W'$ denotes W' with the opposite orientation and $W \cup_M (-W')$ denotes the closed 4-manifold obtained by gluing W and $-W'$ along their common boundary M. Here, the *signature* of a closed oriented 4-manifold V is defined to be the number of positive eigenvalues minus the number of negative eigenvalues of the intersection form

$$H_2(V;\mathbb{R}) \times H_2(V;\mathbb{R}) \longrightarrow \mathbb{R}.$$

Thus, we have such an equivalence relation among oriented 4-manifolds bounded by the 3-manifold M. We call an equivalence class of the equivalence relation a *framing* of a closed oriented 3-manifold M. Further, we call a 3-manifold together with a framing a *framed 3-manifold*. When a surgery presentation of a 3-manifold is given, the 3-manifold has a natural framing which is given by the 4-manifold bounded by the 3-manifold according to the surgery presentation.

It is left to the reader to show that $\langle L^\omega \rangle$ is a topological invariant of a framed 3-manifold M with the framing given by the 4-manifold bounded by the surgery presentation of M along L.

We define the element $\omega' \in \mathcal{S}(S^1 \times I)$ by

$$\omega' = \sum_{\substack{0 \le n \le r-3 \\ n \text{ is even}}} \Delta_n$$ 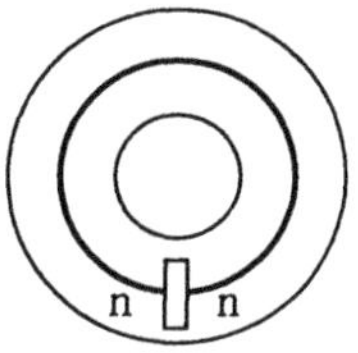 .

We obtain a topological invariant of a 3-manifold M again as

$$\langle U_+^{\omega'} \rangle^{-\sigma_+} \langle U_-^{\omega'} \rangle^{-\sigma_-} \langle D^{\omega'} \rangle \in \mathbb{C} \tag{8.9}$$

instead of (8.7). The topological invariance of (8.9) is shown by modifying the proof of Theorem 8.11, which is left to the reader as an exercise. In particular, when $A = -\exp(\pi\sqrt{-1}/2r)$, we denote the invariant by $\tau_r^{SO(3)}(M)$ and call it the *quantum $SO(3)$ invariant* of a closed oriented connected 3-manifold M.

8.3 Quantum invariants of 3-manifolds via quantum invariants of links

After Witten [Wit89a] proposed that the Chern-Simons path integral leads to topological invariants (called quantum invariants) of 3-manifolds from the viewpoint of mathematical physics, Reshetikhin and Turaev [ReTu91] gave the first rigorous construction of quantum invariants of 3-manifolds as linear sums of quantum invariants of framed links; see Appendix F.2 for a physical background of the construction. (Historically speaking, the construction of the invariants via linear skein in the previous section was obtained after [ReTu91].) Their idea, in [ReTu91], was to construct an invariant of a 3-manifold, expressed by integral surgery along a framed link, as a linear sum of quantum invariants of the framed link derived from representations of quantum groups, such that the linear sum would be invariant under the Fenn-Rourke moves. In this section, we construct quantum invariants of 3-manifolds along these lines.

For simplicity, we consider 3-manifold invariants, based on quantum sl_2 invariants of links. Let L be a framed link whose l components are associated with representations $V_{n_1}, V_{n_2}, \cdots, V_{n_l}$, where V_n denotes the irreducible n-dimensional representation of $U_q(sl_2)$. Recall that we define the quantum $(sl_2; V_{n_1}, V_{n_2}, \cdots, V_{n_l})$ invariant of L, denoted by $Q^{sl_2;V_{n_1},V_{n_2},\cdots,V_{n_l}}(L)$, in Section 4.4. Let us consider a linear sum

$$\sum_{n_1,n_2,\cdots,n_l \in J} C_{n_1} C_{n_2} \cdots C_{n_l} Q^{sl_2;V_{n_1},V_{n_2},\cdots,V_{n_l}}(L),$$

for some constants C_j $(j \in J)$, where J is a finite set of positive integers. Further, we seek constants C_j which make the linear sum invariant under the FR moves among framed links; concretely we require the following relations to be satisfied,

$\sum_{j\in J} C_j Q^{sl_2;V_j,V_{i_1},\cdots,V_{i_m}}($ 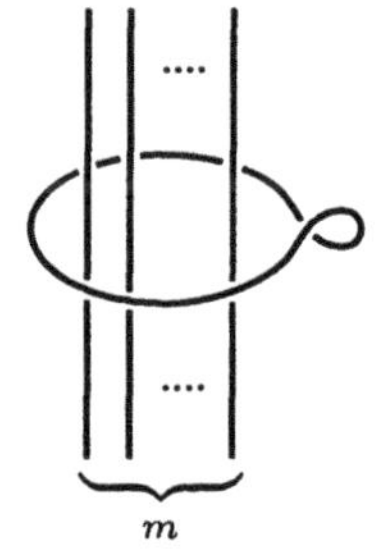 $) = c_+ Q^{sl_2;V_{i_1},\cdots,V_{i_m}}($  $),$ (8.10)

$\sum_{j\in J} C_j Q^{sl_2;V_j,V_{i_1},\cdots,V_{i_m}}($ 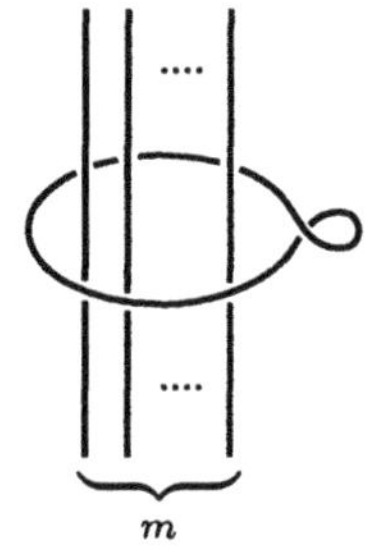 $) = c_- Q^{sl_2;V_{i_1},\cdots,V_{i_m}}($ 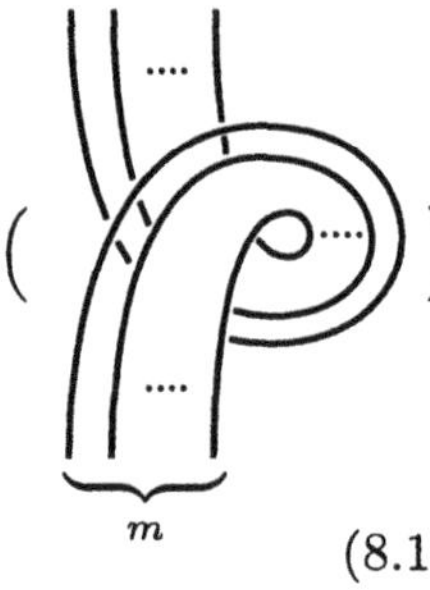 $),$ (8.11)

for some constants $c_\pm$, where V_j is associated to the closed components in the pictures on the left hand sides and $V_{i_1}, V_{i_2}, \cdots, V_{i_m}$ are associated to the m parallel strands respectively. Here, the pictures on the two sides of each formula depict two links which are identical except for a ball, where they differ as shown in the pictures.

We can solve the equations (8.10) and (8.11) by setting the C_j's and J as follows. Let r be an odd integer ≥ 3, and put $\zeta = \exp(2\pi\sqrt{-1}/r)$. Set $C_j = [j]\big|_{q=\zeta}$ and $J = \{1, 2, \cdots, r-1\}$; we discuss the reason why we choose these values later, in Remark 8.13. It is only necessary to verify the relations in the cases $m = 0$ and 1, since the other cases are derived from these cases by applying Proposition 4.13, as shown later. Further, we note that the relation (8.11) is derived from (8.10) by applying Example 4.17, which says that the quantum sl_2 invariant of the mirror image of a framed link is equal to the conjugate (the complex conjugate in this case) of the invariant of the framed link. Hence, it is sufficient to show (8.10) for $m = 0, 1$.

When $m = 0$, the relation (8.10) is expressed as

$$\sum_{1\leq n<r} [n] Q^{sl_2;V_n}(U_+)\Big|_{q=\zeta} = c_+,$$

where U_+ denotes the trivial knot with $+1$ framing. Hence, the constant c_+ is computed as

$$c_+ = \sum_{0 \le n < r} \zeta^{(n^2-1)/4} [n]^2 = \frac{\zeta^{-1/4}}{(\zeta^{1/2} - \zeta^{-1/2})^2} \sum_{0 \le n < r} \zeta^{n^2/4} (\zeta^n + \zeta^{-n} - 2)$$
$$= \frac{-2\zeta^{-3/4}}{\zeta^{1/2} - \zeta^{-1/2}} \sum_{n \in \mathbb{Z}/r\mathbb{Z}} \zeta^{n^2/4},$$

where we obtain the last equality by using the following relation,

$$\sum_{0 \le n < r} \zeta^{n^2/4 + kn/2} = \sum_{n \in \mathbb{Z}/r\mathbb{Z}} \zeta^{(n+k)^2/4 - k^2/4} = \zeta^{-k^2/4} \sum_{n' \in \mathbb{Z}/r\mathbb{Z}} \zeta^{n'^2/4}. \tag{8.12}$$

By computing c_- similarly, we obtain

$$c_\pm = \frac{\mp 2\zeta^{\mp 3/4}}{\zeta^{1/2} - \zeta^{-1/2}} \sum_{n \in \mathbb{Z}/r\mathbb{Z}} \zeta^{\pm n^2/4}. \tag{8.13}$$

When $m = 1$, we show the relation (8.10) as follows. It is known, see, *e.g.*, [Kas94, ChPr95, KiMe91], that the tensor product of two irreducible representations of $U_q(sl_2)$ splits into a direct sum of irreducible representations as

$$V_i \otimes V_j \cong \bigoplus_{\substack{|i-j|+1 \le k \le i+j-1 \\ i+j+k \text{ is odd}}} V_k. \tag{8.14}$$

Hence, by Proposition 4.13 we have that

$Q^{sl_2;V_i,V_j}\Big($  $\Big) = Q^{sl_2;V_i \otimes V_j}\Big($  $\Big) = \sum_k Q^{sl_2;V_k}\Big($  $\Big)$

$$= \sum_k q^{(k^2-1)/4}[k] = \frac{q^{-1/2}}{q^{1/2} - q^{-1/2}} \sum_k (q^{(k+1)^2/4} - q^{(k-1)^2/4}) = q^{(i^2+j^2-2)/4}[ij],$$

where each sum runs over the same k's as in (8.14). By cutting a component of the link in the first picture in the above formula we obtain

$Q^{sl_2;V_i,V_j}\Big($  $\Big) = q^{(i^2+j^2-2)/4} \dfrac{[ij]}{[i]} \mathrm{id}_{V_i}.$

Hence,

$$\sum_j [j] Q^{sl_2;V_i,V_j}\Big($$

$$\Big)\Big|_{q=\zeta} = \sum_j [j]\zeta^{(i^2+j^2-2)/4}\frac{[ij]}{[i]}\mathrm{id}_{V_i}$$

$$= c_+ \cdot \mathrm{id}_{V_i} = c_+ Q^{sl_2;V_i}\Big($$

$$\Big)\Big|_{q=\zeta},$$

where we obtain the second equality by the following computation using (8.12),

$$\sum_{0\le j<r} [j]\zeta^{(i^2+j^2-2)/4}\frac{[ij]}{[i]} = \frac{\zeta^{(i^2-2)/4}\sum_{0\le j<r}\zeta^{j^2/4}(\zeta^{j/2}-\zeta^{-j/2})(\zeta^{ij/2}-\zeta^{-ij/2})}{(\zeta^{1/2}-\zeta^{-1/2})(\zeta^{i/2}-\zeta^{-i/2})} = c_+.$$

Therefore, we obtain the relation (8.10) as

$$\sum_{1\le j<r} [j] Q^{sl_2;V_i,V_j}\Big($$

$$\Big)\Big|_{q=\zeta} = c_+ Q^{sl_2;V_i}\Big($$

$$\Big)\Big|_{q=\zeta}. \qquad (8.15)$$

Theorem 8.12 ([ReTu91]). Let L be an oriented framed link in S^3, and M the 3-manifold obtained from S^3 by integral surgery along L. We put $\zeta = \exp(2\pi\sqrt{-1}/r)$ for any fixed integer $r \ge 3$ as before. Then,

$$c_+^{\ -\sigma_+} c_-^{\ -\sigma_-} \sum_{1\le n_1,n_2,\cdots,n_l<r} [n_1][n_2]\cdots[n_l] Q^{sl_2;V_{n_1},V_{n_2},\cdots,V_{n_l}}(L)\Big|_{q=\zeta} \in \mathbb{C} \qquad (8.16)$$

is a topological invariant of M, where $c_\pm$ are the constants given by (8.13) and $\sigma_\pm$ are the numbers of positive and negative eigenvalues of the linking matrix of L. Further, the invariant is equal to the invariant given in Theorem 8.11 when we put $A = -\zeta^{1/4} = -\exp(\pi\sqrt{-1}/2r)$.

The theorem gives a reconstruction of the invariant in Theorem 8.11. As mentioned earlier, we denote the invariant by $\tau_r^{SU(2)}(M)$, and call it the *quantum* $SU(2)$ *invariant* of a closed connected oriented 3-manifold M.

We show the proof of the theorem in three parts; the former and the latter parts of showing topological invariance of (8.16) and the part of showing the equality of (8.16) to the invariant of Theorem 8.11.

Proof of Theorem 8.12; topological invariance. In order to show the topological invariance of the theorem, we must show that the value of (8.16) does not depend on the orientation of L, and that it is invariant under the m-strand FR moves for any non-negative integer m; then we obtain the topological invariance by Theorem 8.7.

The independence of the orientation of L is derived from Example 4.16 which says that the quantum invariant $Q^{sl_2;V_{n_1},V_{n_2},\cdots,V_{n_l}}(L)$ does not depend on the orientations of the components of L.

We show the invariance of (8.16) under the m-strand FR moves as follows. Under the $\mathrm{FR}_\pm$ move the value of $\sigma_\pm$ increases by 1 and the value of $\sigma_\mp$ is unchanged; it is left to the reader to show this (for a similar argument see the proof of Theorem 8.11). It is also left to the reader to show that $c_\pm \neq 0$. Hence, it is sufficient to show that the relations (8.10) and (8.11) hold for any non-negative integer m when we set C_j to $[j]\big|_{q=\zeta}$ and J to $\{0, 1, \cdots, r-1\}$. The relation (8.11) is derived from (8.10) by applying Example 4.17, as mentioned above. Further, we have already verified (8.10) for the cases $m = 0, 1$. We show (8.10) for $m \geq 2$ later, after introducing some notations below. □

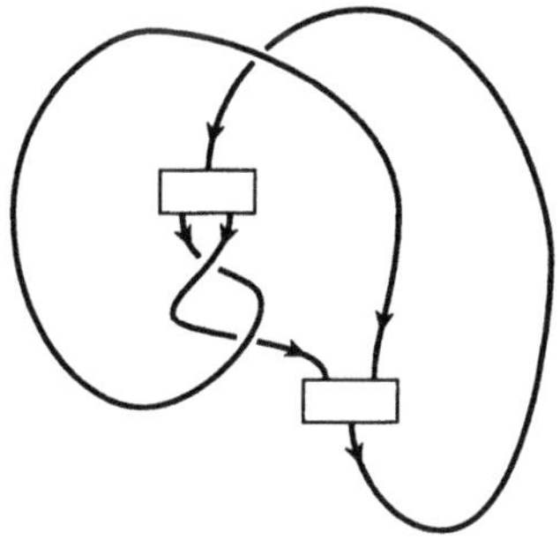

Figure 8.6 A trivalent graph embedded in $\mathbb{R}^3$ whose trivalent vertices are boxes

We consider a trivalent graph embedded in $\mathbb{R}^3$ with oriented edges such that each trivalent vertex is expressed as a box whose neighborhood is as depicted in either of the following two pictures; see Figure 8.6 for such an embedded trivalent graph.

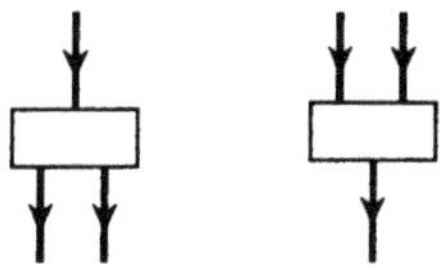

We decorate each edge of such an embedded trivalent graph with a representation V_i of $U_q(sl_2)$. Further, we decorate a box with an intertwiner as follows,

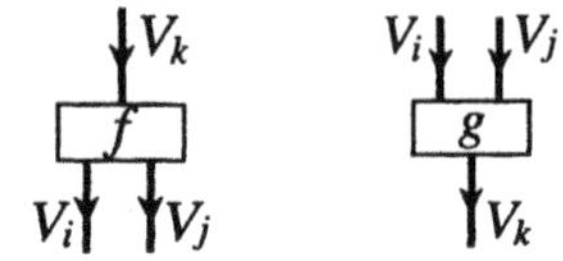

,

where f is an intertwiner in $\mathrm{Hom}(V_i \otimes V_j, V_k)$ and g is an intertwiner in $\mathrm{Hom}(V_k, V_i \otimes V_j)$.

We define the quantum sl_2 invariant of such an embedded trivalent graph decorated with representations and intertwiners by extending the definition of the quantum invariant in Section 4.4, putting the operator invariants of the boxes to

$Q^{sl_2;\cdot}\Big($ 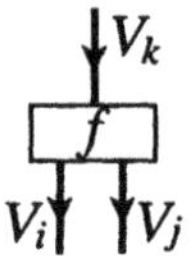 $\Big) = f \in \mathrm{Hom}(V_i \otimes V_j, V_k),$

$Q^{sl_2;\cdot}\Big($ 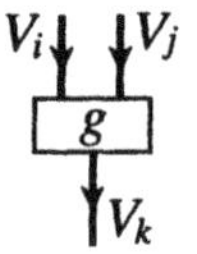 $\Big) = g \in \mathrm{Hom}(V_k, V_i \otimes V_j).$

We omit the proof of the isotopy invariance of the quantum sl_2 invariant of such embedded graphs. Instead, it is sufficient for our present purpose to show the following formula,

$c_+ Q^{sl_2;\cdot}\Big($ 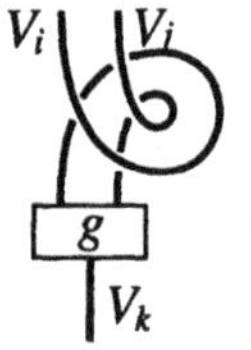 $\Big) = c_+ Q^{sl_2;\cdot}\Big($ 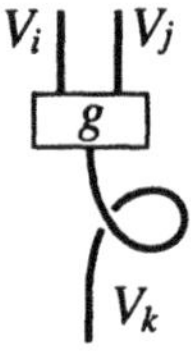 $\Big)$

$= \sum_{1\le l<r} [l] Q^{sl_2;\cdot}\Big($ 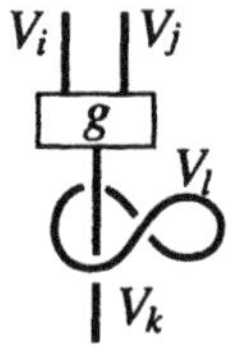 $\Big) = \sum_{1\le l<r} [l] Q^{sl_2;\cdot}\Big($ 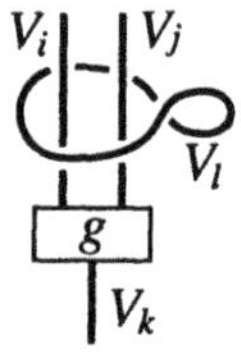 $\Big).$ (8.17)

The first equality follows from the commutativity of the following diagram, noting that the invariant of -1 full twist of one strand is given by v and that the invariant of -1 full twist of parallel two strands is given by the element obtained from v by applying Δ by Proposition 4.13.

$$\begin{array}{ccc} V_i \otimes V_j & \xrightarrow{\Delta(v)} & V_i \otimes V_j \\ {\scriptstyle g}\uparrow & & \uparrow{\scriptstyle g} \\ V_k & \xrightarrow{v} & V_k \end{array}$$

The commutativity of this diagram is derived from the fact that g is an intertwiner, noting that v acts on $V_i \otimes V_j$ by $\Delta(v)$ from the definition of a tensor representation.

The second equality follows from (8.15) (if k is not divisible by r). The third equality follows in a similar way as the first.

Proof of Theorem 8.12; topological invariance (continued). To complete the proof of the topological invariance it remains to show the invariance of (8.16) under the m-strand FR_+ moves for $m \geq 2$. We show the invariance for $m = 2$ as follows.

Fixing an isomorphism (8.14), we identify the two sides of (8.14). (There is an ambiguity of the choice of this isomorphism by a scalar multiple on each V_k. We fix it.) Let $f_{ij}^k : V_i \otimes V_j \to V_k$ be the projector to V_k and $g_k^{ij} : V_k \to V_i \otimes V_j$ the inclusion of V_k. Then, the sum $\sum_k g_k^{ij} \circ f_{ij}^k$ is equal to the identity map of $V_i \otimes V_j$. Hence, we obtain the first and the third equalities of the following formula,

$\sum_{1\leq l<r} [l] Q^{sl_2;\cdot}\Big($
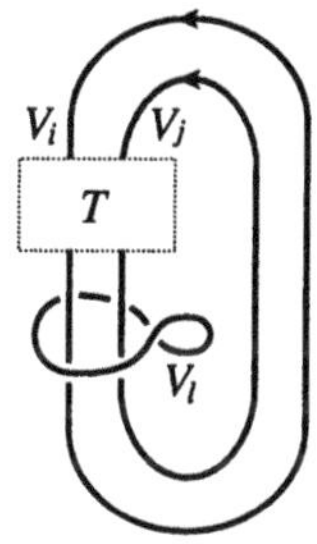

$\Big) = \sum_{1\leq l<r} [l] \sum_k Q^{sl_2;\cdot}\Big($
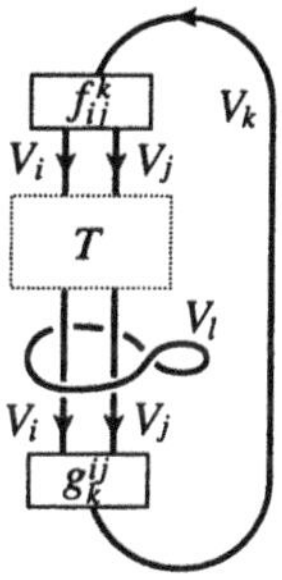

$\Big)$

$= c_+ \sum_k Q^{sl_2;\cdot}\Big($
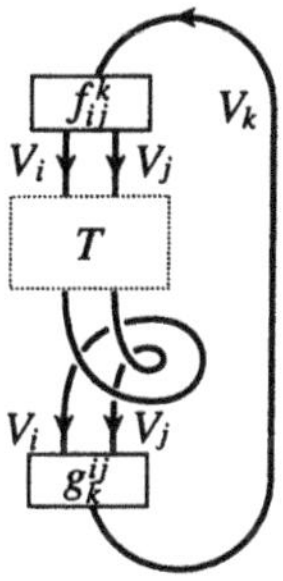

$\Big) = c_+ Q^{sl_2;\cdot}\Big($
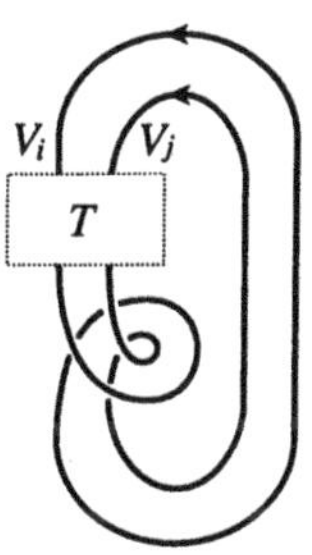

$\Big),$ (8.18)

for any 2-tangle T, where the sums over k run over the same range as in (8.14). The second equality follows from (8.17) if k is not divisible by r. Further, this equality also holds for k divisible by r; in this case the two sides of the equality vanish, since each side is equal to a multiple of the quantum dimension $[k]$ of V_k, which is equal to 0 in this case. Therefore, we obtain the invariance of (8.16) under the 2-strand FR_+ move.

The invariance under the m-strand FR_+ moves for $m \geq 3$ is shown similarly as above. For example, when $m = 3$, we consider intertwiners f and g as below.

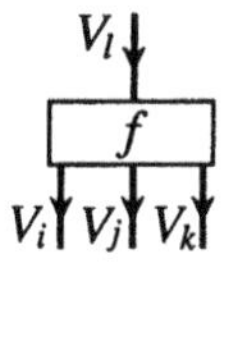

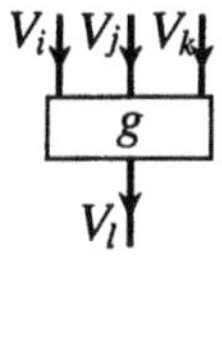

For such intertwiners we obtain a relation similar to (8.17) by Proposition 4.13 and from the fact that g is an intertwiner, just as in the proof of (8.17). Further, instead of using (8.14) we express tensor products of three representations as

$$V_i \otimes V_j \otimes V_k \cong \bigoplus_l V_{n_l}$$

for some sequence $\{n_l\}$, noting that we have repetitions in the sequence in this case, unlike the case of (8.14). Fixing the above isomorphism we obtain a relation similar to (8.18). Using such relations we can show the invariance under the 3-strand FR_+ moves in the same way as the above case of $m = 2$. We obtain the invariance under the m-strand FR_+ moves for the other m's similarly. □

Further, we prove the equality of Theorem 8.12 as follows.

Proof of Theorem 8.12; equality to the invariant in Theorem 8.11. Let L be a framed link with l components in S^3, and let M be the 3-manifold obtained from S^3 by integral surgery along L. By Theorem H.3, the quantum sl_2 invariant of L is expressed by the Kauffman bracket as

$$Q^{sl_2;V_{n_1+1},\cdots,V_{n_l+1}}(L) = (-1)^{n_1+\cdots+n_l}\big\langle L^{[n_1,\cdots,n_l]}\big\rangle\big|_{A=-q^{1/4}},$$

where $L^{[n_1,\cdots,n_l]}$ denotes the framed link obtained from L by replacing the i-th component with n_i parallel copies of the component together with a box over n_i strands (the Jones-Wenzl idempotent) for each i; to be precise it is a linear sum of framed links in the linear skein. Hence, from the definition of L^ω, we have that

$$\begin{aligned}\langle L^\omega\rangle &= \sum_{0\le n_1,\cdots,n_l\le r-2} \Delta_{n_1}\cdots\Delta_{n_l}\langle L^{[n_1,\cdots,n_l]}\rangle\big|_{A=-\exp(\pi\sqrt{-1}/2r)} \\ &= \sum_{1\le n_1,\cdots,n_l<r} [n_1]\cdots[n_l]Q^{sl_2;V_{n_1},\cdots,V_{n_l}}(L)\Big|_{q=\exp(2\pi\sqrt{-1}/r)},\end{aligned}$$

noting that $\Delta_n = (-1)^n[n+1]$ holds in the setting here. Further, by applying the above formula to the case that L is the trivial knot $U_\pm$ with ± 1 framing we have that

$$\langle U_\pm^\omega\rangle = \sum_{1\le n<r}[n]Q^{sl_2;V_n}(U_\pm)\Big|_{q=\exp(2\pi\sqrt{-1}/r)} = c_\pm.$$

By the above two formulae we obtain the equality of (8.7) and (8.16). Therefore, the invariant of Theorem 8.12 is equal to the invariant in Theorem 8.11. □

Remark 8.13. Let us discuss the idea of the construction of the invariant of Theorem 8.12 again; it remains to discuss the reason why we should have set the C_j's and J to $C_j = [j]|_{q=\zeta}$ and $J = \{1, 2, \cdots, r-1\}$. The aim of this remark is to explain, not the complete necessity of this choice, but a speculative idea of why this choice is natural.

Firstly, we explain the reason why we should set $C_j = [j]\big|_{q=\zeta}$, assuming that we have set $J = \{1, 2, \cdots, r-1\}$. The quantum sl_2 invariant (or, in general, the quantum $\mathfrak{g}$ invariant) of the following tangles decorated with irreducible representations V_i and V_j are expressed

$Q^{sl_2;V_i,V_j}\Big($  $\Big) = A_{ij} \cdot \mathrm{id}_{V_i},$ (8.19)

$Q^{sl_2;V_i}\Big($  $\Big) = \lambda_i \cdot \mathrm{id}_{V_i},$ (8.20)

for some scalars A_{ij} and λ_i, since they are intertwiners of irreducible representations. Fixing a constant c_+, we consider the following relation,

$$\sum_{j\in J} A_{ij} C_j = c_+ \lambda_i, \tag{8.21}$$

with the C_j's undetermined scalars. Note that this relation implies invariance under the 1-strand FR_+ move. Suppose that the matrix $A = (A_{ij})_{i,j\in J}$ is invertible. Then, C_j is uniquely determined by

$$C_j = c_+ \sum_{i\in J} (A^{-1})_{ji} \lambda_i,$$

where $(A^{-1})_{ji}$ denotes the (j, i) entry of the inverse matrix A^{-1}. It can be shown, by some computation, that the matrix A really is invertible in our case. Hence, the C_j's are uniquely determined by the above formula, and it follows from (8.15) that we should set $C_j = [j]\big|_{q=\zeta}$. Noting that C_j is a real number in this case, we obtain the following formula as the complex conjugate of (8.21),

$$\sum_{j\in J} \overline{A_{ij}} C_j = c_- \overline{\lambda_i}, \tag{8.22}$$

where we put c_- to be the complex conjugate of c_+. This relation implies invariance under the 1-strand FR_- move, since the quantum invariant at $q = \zeta$ of a mirror image is equal to the complex conjugate of the original invariant, by Example 4.17. Further, we obtain the invariance under the 0-strand FR moves by putting V_i in (8.19) and (8.20) to be the unit representation.

Secondly, we give a reason why we should set $J = \{1, 2, \cdots, r-1\}$. To deduce the invariance under the m-strand FR moves for $m \geq 2$ from the invariance under the 1-strand FR moves we require that the set $\{V_j \mid j \in J\}$ should be closed with respect to tensor product of representations, in some sense. In the case of the construction of Theorem 8.12 we put q to be an r-th root of unity ζ and consider the set $\{V_1, V_2, \cdots, V_{r-1}\}$, putting $J = \{1, 2, \cdots, r-1\}$. This set is actually closed with respect to tensor product of representations from the viewpoint of evaluating invariants of links. That is, for any framed link L and any positive integer k we have that

$$Q^{sl_2;V_{kr},V_{n_2},\cdots,V_{n_l}}(L)\Big|_{q=\zeta} = 0,$$

because the left hand side is equal to $[kr]$ times the invariant of the 1-tangle obtained from L by cutting the first component, and furthermore $[kr] = 0$ when $q = \zeta$. Further, by Proposition 4.13 we have that

$$Q^{sl_2;V_{kr},V_2}\big(K^{(2)}\big)\Big|_{q=\zeta} = Q^{sl_2;V_{kr-1}}(K)\Big|_{q=\zeta} + Q^{sl_2;V_{kr+1}}(K)\Big|_{q=\zeta} = 0.$$

In this sense, we can regard V_{kr+1} as $-V_{kr-1}$. Further, by similar arguments we obtain

$$Q^{sl_2;V_{kr-n}}(K)\Big|_{q=\zeta} = -Q^{sl_2;V_{kr+n}}(K)\Big|_{q=\zeta}.$$

Hence, we regard V_{kr+n} as $-V_{kr-n}$. The *representation ring* of $U_q(sl_2)$ is the ring spanned by V_i $(i = 1, 2, \cdots)$ whose addition is direct sum of representations and whose product is tensor product of representations given by (8.14) in this basis. Putting $q = \zeta$, from the viewpoint of evaluating invariants of links, we can reduce the representation ring to its quotient ring subject to the relations $V_{kr} = 0$ and $V_{kr+n} = -V_{kr-n}$ for any $k \geq 1$ and any n with $0 < n < r$. That is, when we consider the value of $Q^{sl_2;V_{n_1},V_{n_2},\cdots,V_{n_l}}(L)\big|_{q=\zeta}$, we can regard each V_{n_i} as an element in the quotient ring. Noting that $\{V_1, V_2, \cdots, V_{r-1}\}$ is a basis of the quotient ring, we should set $J = \{1, 2, \cdots, r-1\}$.

The above argument suggests another possibility in the case that r is odd: $J = \{1, 3, 5, \cdots, r-2\}$. That is, in this case, the set of odd-dimensional representations is also closed in the above sense. Hence, we obtain another invariant of 3-manifolds as below, which gives the definition of the *quantum* $SO(3)$ *invariant*,

$$\tau_r^{SO(3)}(M) = {c'_+}^{-\sigma_+} {c'_-}^{-\sigma_-} \sum_{\substack{1 \leq n_1, \cdots, n_l < r \\ n_1, \cdots, n_l \text{ are odd}}} [n_1] \cdots [n_l] Q^{sl_2;V_{n_1},\cdots,V_{n_l}}(L)\Big|_{q=\zeta} \in \mathbb{C}, \quad (8.23)$$

for any odd integer $r \geq 3$, where we set

$$c'_\pm = \sum_{\substack{1 \leq n < r \\ n \text{ is odd}}} [n] Q^{sl_2;V_n}(U_\pm)\Big|_{q=\zeta}.$$

Here, $U_\pm$ denotes the trivial knot with ± 1 framing as before. It is shown in the same way as in the proof of Theorem 8.12 that (8.23) is equal to (8.9). Further, we can give a direct proof of the topological invariance of (8.23), similar to the proof of Theorem 8.12; a detailed proof is left to the reader as an exercise.

It was shown [KiMe91] that the quantum $SU(2)$ invariant (for odd r) can be presented by using the quantum $SO(3)$ invariant,

$$\tau_r^{SU(2)}(M) = \begin{cases} \tau_3^{SU(2)}(M)\tau_r^{SO(3)}(M) & \text{if } r \equiv 3 \bmod 4, \\ \overline{\tau_3^{SU(2)}(M)}\tau_r^{SO(3)}(M) & \text{if } r \equiv 1 \bmod 4. \end{cases}$$

Here, $\tau_3^{SU(2)}(M)$ can be presented by

$$\tau_3^{SU(2)}(M) = (1+\sqrt{-1})^{\sigma_+}(1-\sqrt{-1})^{\sigma_-} \sum_{\mathbf{x}\in(\mathbb{Z}/2\mathbb{Z})^l} \sqrt{-1}^{\,{}^t\mathbf{x}A\mathbf{x}},$$

where A is the linking matrix of a framed link, which gives a surgery presentation of M, and σ_+ and σ_- denote the numbers of positive and negative eigenvalues of A. Thus, $\tau_3^{SU(2)}(M)$ is an invariant determined by a linking matrix A, *i.e.*, determined by the cohomology ring of M and the linking pairing on the torsion of the first homology group of M (see [MOO92]).

The quantum $SU(2)$ and $SO(3)$ invariants are so named for the following reason. Recall that $SU(2)$ and $SO(3)$ are Lie groups whose Lie algebra is isomorphic to sl_2. All irreducible representations of sl_2 are derived from representations of $SU(2)$, while only the odd-dimensional irreducible representations are derived from representations of $SO(3)$. Thus, the quantum $SU(2)$ and $SO(3)$ invariants are obtained by using representations whose associated classical representations (*i.e.*, representations at $q = 1$) are induced from representations of the Lie groups $SU(2)$ and $SO(3)$ respectively.

The *lens space* $L(f,1)$ of type $(f,1)$ is the 3-manifold obtained from S^3 by integral surgery along the trivial knot with f framing. As an example of the quantum $SO(3)$ invariant we compute the following example; for the general computation of invariants of lens spaces, see [Jef92, Gar92, Yam95, LiLi96].

Example 8.14. Let p be an odd prime and f a positive integer not divisible by p. Then, the quantum $SO(3)$ invariant of the lens space $L(f,1)$ of type $(f,1)$ is given by

$$\tau_p^{SO(3)}\Big(L(f,1)\Big) = \left(\frac{f}{p}\right)\zeta^{-(f-1)(f-2)/4f}\frac{\zeta^{1/2f}-\zeta^{-1/2f}}{\zeta^{1/2}-\zeta^{-1/2}},$$

where we put $\zeta = \exp(2\pi\sqrt{-1}/p)$ as before and we regard powers of ζ as lying in $\mathbb{Z}/p\mathbb{Z}$.

Proof. From the definition of the quantum $SO(3)$ invariant,

$$\tau_p^{SO(3)}\Big(L(f,1)\Big) = c'^{-1}_{+} \sum_{\substack{1\le n<p \\ n \text{ is odd}}} \zeta^{f(n^2-1)/4}[n]^2.$$

The sum on the right hand side is computed as

$$\begin{aligned}
&\frac{1}{(\zeta^{1/2}-\zeta^{-1/2})^2}\sum_{\substack{1\le n<p\\ n \text{ is odd}}}\zeta^{f(n^2-1)/4}(\zeta^n+\zeta^{-n}-2)\\
&=\frac{1}{2(\zeta^{1/2}-\zeta^{-1/2})^2}\sum_{\substack{-p<n\le p\\ n \text{ is odd}}}\zeta^{f(n^2-1)/4}(\zeta^n+\zeta^{-n}-2)\\
&=\frac{1}{(\zeta^{1/2}-\zeta^{-1/2})^2}\sum_{\substack{-p<n\le p\\ n \text{ is odd}}}\zeta^{f(n^2-1)/4}(\zeta^n-1)\\
&=\frac{1}{(\zeta^{1/2}-\zeta^{-1/2})^2}\sum_{m\in\mathbb{Z}/p\mathbb{Z}}\left(\zeta^{fm^2+(f+2)m+1}-\zeta^{fm^2+fm}\right),
\end{aligned}$$

where we obtain the third equality by putting $n = 2m + 1$. The powers of ζ in the sum are computed respectively as

$$\begin{aligned}
fm^2+(f+2)m+1 &= f\Big(m+\frac{1}{2}+\frac{1}{f}\Big)^2-\frac{f}{4}-\frac{1}{f},\\
fm^2+fm &= f\Big(m+\frac{1}{2}\Big)^2-\frac{f}{4}.
\end{aligned}$$

By replacing m with $m-\frac{1}{2}-\frac{1}{f}$ and $m-\frac{1}{2}$ respectively the above sum becomes equal to

$$\frac{\zeta^{-f/4}(\zeta^{-1/f}-1)}{(\zeta^{1/2}-\zeta^{-1/2})^2}\sum_{m\in\mathbb{Z}/p\mathbb{Z}}\zeta^{fm^2}.$$

Further, by putting $f = 1$,

$$c'_+=\frac{\zeta^{-1/4}(\zeta^{-1}-1)}{(\zeta^{1/2}-\zeta^{-1/2})^2}\sum_{m\in\mathbb{Z}/p\mathbb{Z}}\zeta^{fm^2}.$$

Hence,

$$\tau_p^{SO(3)}\big(L(f,1)\big)=\Big(\frac{f}{p}\Big)\frac{\zeta^{(1-f)/4}(\zeta^{-1/f}-1)}{\zeta^{-1}-1}.$$

This is equal to the required formula. □

Let us compute the same example by the linear skein method.

Alternative proof of Example 8.14. From the definition (8.3) of the linear skein, and by Proposition 8.9, we have that

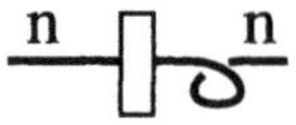

$= (-1)^n A^{n(n+2)}$
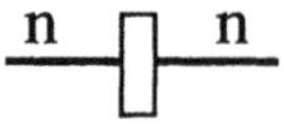

.

Hence, the invariant of Theorem 8.11 is expressed by

$$\langle U_+^{\omega'} \rangle^{-1} \sum_{\substack{0 \le n \le p-3 \\ n \text{ is even}}} A^{fn(n+2)} \Delta_n^2 .$$

By $\Delta_n = (A^{2(n+1)} - A^{-2(n+1)})/(A^2 - A^{-2})$ for even n, the sum of the above formula is computed as

$$\begin{aligned}
&\frac{1}{(A^2 - A^{-2})^2} \sum_{\substack{0 \le n \le p-3 \\ n \text{ is even}}} A^{fn(n+2)} (A^{4(n+1)} + A^{-4(n+1)} - 2) \\
&= \frac{1}{(A^2 - A^{-2})^2} \sum_{0 \le m \le \frac{p-3}{2}} A^{4fm(m+1)} (A^{4(2m+1)} + A^{-4(2m+1)} - 2) \\
&= \frac{1}{2(A^2 - A^{-2})^2} \sum_{-\frac{p-1}{2} \le m \le \frac{p-3}{2}} A^{4fm(m+1)} (A^{4(2m+1)} + A^{-4(2m+1)} - 2) \\
&= \frac{1}{(A^2 - A^{-2})^2} \sum_{m \in \mathbb{Z}/p\mathbb{Z}} A^{4fm(m+1)} (A^{4(2m+1)} - 1),
\end{aligned}$$

where we obtain the first equality by putting $n = 2m$, and obtain the second equality changing the formula in part by replacing m with $-m-1$. By putting $A^2 = \zeta^{1/2}$ the above formula is equal to

$$\frac{1}{(\zeta^{1/2} - \zeta^{-1/2})^2} \sum_{m \in \mathbb{Z}/p\mathbb{Z}} \zeta^{fm(m+1)} (\zeta^{2m+1} - 1).$$

This is equal to a formula in the previous proof of Example 8.14. In the same way as in that proof we obtain the required formula. □

Quantum G and PG invariants of 3-manifolds

An invariant of 3-manifolds associated with each simple Lie algebra $\mathfrak{g}$ was rigorously defined, by Turaev and Wenzl [TuWe93] and Andersen [And93], to be a linear sum of the quantum $\mathfrak{g}$ invariant of framed links whose components are decorated by representations of $\mathfrak{g}$ in a certain class. We call this invariant the quantum G invariant, where G denotes the simply connected Lie group whose Lie algebra is $\mathfrak{g}$. For its construction see also [KoTa93, Wen93, Tur94, KRT97, Bla00, Le*00*]. For the linear skein approach of this invariant see [OhYa97, Yok97] and Appendix B.2.

We obtain the definition of the quantum PG invariant (where PG denotes the quotient group of G by its center) by restricting the class of representations in the definition of the quantum G invariant to the subclass consisting of representations that arise from representations of PG. This invariant was defined by Kirby and Melvin [KiMe91] for $PG = PSU(2)$ $(= SO(3))$, by Kohno and Takata [KoTa96] for $PG = PSU(N)$, and by Le [Le*00*] (see also [Saw*99*]) for general PG.

In this subsection we describe a sketch of constructions of these invariants. See [Hum72, Kac90] for terminologies of the theory of Lie algebra in this subsection.

Let $\mathfrak{g}$ be a simple Lie algebra. Finite dimensional irreducible representations V_λ of $\mathfrak{g}$ (and those of $U_q(\mathfrak{g})$ for generic q) are parameterized by dominant integral weights λ. For a fixed positive integer k, let P_k denote the set of dominant integral weights λ with $0 \leq (\theta, \lambda) \leq k$, where θ is the highest root and $(\cdot,\cdot)$ denotes the scalar multiple of the Cartan-Killing form normalized by $(\theta,\theta) = 2$. We denote by $Q^{\mathfrak{g};V_{\lambda_1},\cdots,V_{\lambda_l}}(L)$ the quantum $\mathfrak{g}$ invariant of a framed link L whose components are decorated with representations $V_{\lambda_1},\cdots,V_{\lambda_l}$ of $\mathfrak{g}$; this invariant can be defined by using the quantum group $U_q(\mathfrak{g})$ as a ribbon Hopf algebra (see Chapter 4), or alternatively, by using the monodromy representation of the braid group along solutions of the KZ equation (see Chapter 5). Put $\zeta = \exp(2\pi\sqrt{-1}/r)$, where $r = k + h^\vee$ and $h^\vee$ denotes the dual Coxeter number of $\mathfrak{g}$. Let M be the 3-manifold obtained from S^3 by integral surgery along a framed link L. Then, the *quantum G invariant* of M, where G denotes the simply connected Lie group whose Lie algebra is $\mathfrak{g}$, is defined by

$$\tau_r^G(M) = c_+{}^{-\sigma_+}c_-{}^{-\sigma_-} \sum_{\lambda_1,\cdots,\lambda_l\in P_k} (\text{q-dim}\, V_{\lambda_1})\cdots(\text{q-dim}\, V_{\lambda_l})Q^{\mathfrak{g};V_{\lambda_1},\cdots,V_{\lambda_l}}(L)\Big|_{q=\zeta} \in \mathbb{C},$$

where $\text{q-dim}\, V_\lambda$ denotes the quantum dimension of V_λ, which is the trace of uv^{-1} on V_λ for the canonical elements u and v in $U_q(\mathfrak{g})$ as a ribbon Hopf algebra, and σ_+ and σ_- denote the numbers of positive and negative eigenvalues of the linking matrix of L, and the constants $c_\pm$ are given by

$$c_\pm = \sum_{\lambda\in P_k} (\text{q-dim}\, V_\lambda)Q^{\mathfrak{g};V_\lambda}(U_\pm)\Big|_{q=\zeta} \in \mathbb{C}$$

for the trivial knot $U_\pm$ with framing ± 1.

Let PG denote the quotient subgroup of G by its center. We denote by $\hat{P}_k$ the subset of P_k consisting of the highest weights of representations that arise from representations of PG. Using the above notation we define the *quantum PG invariant* of M by

$$\tau_r^{PG}(M) = c'_+{}^{-\sigma_+}c'_-{}^{-\sigma_-} \sum_{\lambda_1,\cdots,\lambda_l\in \hat{P}_k} (\text{q-dim}\, V_{\lambda_1})\cdots(\text{q-dim}\, V_{\lambda_l})Q^{\mathfrak{g};V_{\lambda_1},\cdots,V_{\lambda_l}}(L)\Big|_{q=\zeta} \in \mathbb{C},$$

where the constants $c'_\pm$ are given by

$$c'_\pm = \sum_{\lambda\in \hat{P}_k} (\text{q-dim}\, V_\lambda)Q^{\mathfrak{g};V_\lambda}(U_\pm)\Big|_{q=\zeta} \in \mathbb{C}.$$

In particular, when $\mathfrak{g} = sl_N$, the above notation is concretely presented by

$h^\vee = N$ and

$$P_k = \Big\{ \sum_{i=1}^{N-1} m_i \Lambda_i \;\Big|\; m_i \in \mathbb{Z}_{\geq 0}, \sum_{i=1}^{N-1} m_i \leq k \Big\},$$

$$\hat{P}_k = \Big\{ \sum_{i=1}^{N-1} m_i \Lambda_i \in P_k \;\Big|\; \sum_{i=1}^{N-1} i m_i \equiv 0 \bmod N \Big\},$$

for the fundamental weights Λ_i $(1 \leq i \leq N-1)$. Further, the quantum dimension of V_λ for the highest weight $\lambda = \sum_i m_i \Lambda_i$ is given by

$$\text{q-dim}\, V_\lambda = \prod_{1 \leq i \leq j \leq N-1} \frac{[m_i + \cdots + m_j + j - i + 1]}{[j - i + 1]},$$

where the quantum integer is defined to be $[n] = (q^{n/2} - q^{-n/2})/(q^{1/2} - q^{-1/2})$ as before. For example, when we put $N = 2$, the above notation becomes as follows,

$$\begin{aligned} P_k &= \{ m_1 \Lambda_1 \mid m_1 \in \mathbb{Z}_{\geq 0},\ m_1 \leq r - 2 \}, \\ \hat{P}_k &= \{ m_1 \Lambda_1 \in P_k \mid m_1 \text{ is even} \}, \\ \text{q-dim}\, V_{m_1 \Lambda_1} &= [m_1 + 1]. \end{aligned}$$

Noting that $V_{m_1 \Lambda_1}$ is the irreducible $(m_1 + 1)$ dimensional representation, denoted by V_{m_1+1} in this chapter, the definitions of $\tau_r^{SU(2)}(M)$ and $\tau_r^{SO(3)}(M)$ in (8.16) and (8.23) are derived from the above definitions of $\tau_r^G(M)$ and $\tau_r^{PG}(M)$ respectively. Further, when $N = 3$, it follows that

$$\begin{aligned} P_k &= \{ m_1 \Lambda_1 + m_2 \Lambda_2 \mid m_1, m_2 \in \mathbb{Z}_{\geq 0},\ m_1 + m_2 \leq r - 3 \}, \\ \hat{P}_k &= \{ m_1 \Lambda_1 + m_2 \Lambda_2 \in P_k \mid m_1 + 2m_2 \text{ is divisible by } 3 \}, \\ \text{q-dim}\, V_{m_1 \Lambda_1 + m_2 \Lambda_2} &= [m_1 + 1][m_2 + 1][m_1 + m_2 + 2]/[2]. \end{aligned}$$

Using these, the quantum $SU(3)$ and $PSU(3)$ invariants can be defined. These data suggest how we should construct the definitions of these invariants by the linear skein approach (see Appendix B.2).

Chapter 9

Perturbative invariants of knots and 3-manifolds

A perturbative expansion is a way to get information from a path integral in mathematical physics. In the case of the Chern-Simons field theory the path integral is an infinite dimensional Gaussian integral, and we can obtain its perturbative expansion by the formal stationary phase method (see Appendix F.3). As mentioned in Appendix F, the path integral in the Chern-Simons field theory is a part of the physical background of quantum topological invariants of 3-manifolds. Hence, a power series obtained by perturbative expansion of the path integral is also expected to be a background of a power series topological invariant of 3-manifolds.

In the previous chapter we gave a rigorous construction of quantum invariants of 3-manifolds. Motivated by the perturbative expansion of the path integral, we consider a certain number-theoretical limit of the quantum $SO(3)$ invariant $\tau_r^{SO(3)}(M)$ of a homology 3-sphere M at $r \to \infty$ in Section 9.2. We call this limit the perturbative $SO(3)$ invariant of M ([Oht96b]), which is a power series invariant of M. Rozansky [Roz96, Roz98b] derived perturbative invariants of homology 3-spheres from perturbative invariants of knots by using Gaussian integrals. We introduce the perturbative sl_2 invariant of knots in Section 9.1, and in Section 9.3 show how this invariant determines the perturbative $SO(3)$ invariant of the homology 3-sphere obtained by integral surgery along that knot, along the lines of Rozansky's approach using Gaussian integration.

9.1 Perturbative invariants of knots

In this section we consider the expansion of the quantum (sl_2, V_n) invariant $Q^{sl_2;V_n}(K)$ of a knot K with respect to n. We call this expansion the perturbative sl_2 invariant of K. We show the existence of this expansion in two ways; by using the universal sl_2 invariant and by using the Kontsevich invariant.

The perturbative sl_2 invariant as obtained from the universal sl_2 invariant

Let V_n be the irreducible n-dimensional representation of the quantum group $U_q(sl_2)$, *i.e.*, we have $\rho_{V_n} : U_q(sl_2) \to \mathrm{End}(V_n)$. See Section 4.4 for a concrete presentation of ρ_{V_n} for the generators $E, F, K^{\pm 1}$ of $U_q(sl_2)$ given in Section 4.4.

The representation ρ_{V_n} is computed, for example, as

$$\rho_{V_n}(EF) = \begin{pmatrix} [n-1][1] & & & & & 0 \\ & [n-2][2] & & & & \\ & & [n-3][3] & & & \\ & & & \ddots & & \\ & & & & [1][n-1] & \\ 0 & & & & & 0 \end{pmatrix}.$$

Hence, the (i,i) entry of $\rho_{V_n}(EF)$ is computed as

$$\begin{aligned}[n-i][i] &= \frac{(q^{(n-i)/2} - q^{-(n-i)/2})(q^{i/2} - q^{-i/2})}{(q^{1/2} - q^{-1/2})^2} \\ &= \frac{q^{n/2} + q^{-n/2} - q^{n/2-i} - q^{-n/2+i}}{(q^{1/2} - q^{-1/2})^2}.\end{aligned}$$

Further, the (i,i) entry of $\rho_{V_n}(K)$ is equal to $q^{(n+1)/2-i}$. Hence,

$$\rho_{V_n}(c) = (q^{n/2} + q^{-n/2}) \cdot \mathrm{id}_{V_n} \tag{9.1}$$

for the element c given by

$$c = (q^{1/2} - q^{-1/2})^2 EF + q^{-1/2}K + q^{1/2}K^{-1}, \tag{9.2}$$

which is central in $U_q(sl_2)$. In general, if x is a polynomial in $E, F, K^{\pm 1}$, then the (i,j) entry of $\rho_{V_n}(x)$ is a polynomial in $q^{\pm 1/2}, q^{\pm i/2}, q^{\pm j/2}, q^{\pm n/2}, n$, and $(q^{1/2} - q^{-1/2})^{-1}$.

In particular, the representation ρ_{V_n} is computed for the center of $U_q(sl_2)$ as follows. Let C be a polynomial in $E, F, K^{\pm 1}$ which belongs to the center of $U_q(sl_2)$. By Lemma G.7 in Appendix G.2 the center of $U_q(sl_2)$ is generated by the element c given by (9.2). Hence, C is equal to a polynomial in c and $q^{\pm 1/2}$. Further, by (9.1),

$$\rho_{V_n}(c^k) = (q^{n/2} + q^{-n/2})^k \cdot \mathrm{id}_{V_n}.$$

Since $\rho_{V_n}(C)$ is equal to a scalar multiple of id_{V_n} by Schur's lemma, we have that

$$\rho_{V_n}(C) = \lambda \cdot \mathrm{id}_{V_n}$$

for some $\lambda \in \mathbb{Q}[q^{\pm 1/2}, q^{\pm n/2}]$.

Let T be a 1-tangle with 0 framing. Then, the universal sl_2 invariant $Q^{sl_2;\star}(T)$ belongs to the center of $U_q(sl_2)$; this is shown by comparing the universal sl_2 invariant of the following 2-tangles.

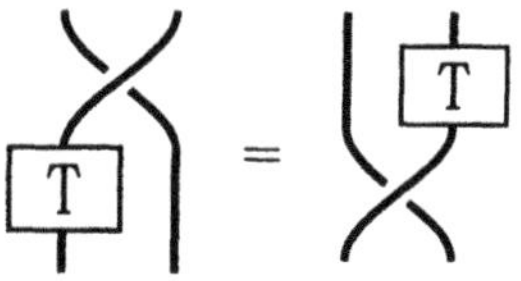

Hence, $Q^{sl_2;\star}(T)$ is equal to a power series in $q^{1/2}-q^{-1/2}$ with coefficient polynomials in c. Therefore,

$$\rho_{V_n}\left(Q^{sl_2;\star}(T)\right) = \lambda \cdot \mathrm{id}_{V_n} \tag{9.3}$$

for some $\lambda \in \mathbb{Q}[q^{\pm 1/2}, q^{\pm n/2}][[q^{1/2} - q^{-1/2}]]$.

Proposition 9.1. For any 1-tangle T with 0 framing there exists a unique series of polynomials $p_k(n) \in \mathbb{Q}[q^{\pm 1/2}, q^{\pm n/2}]$ satisfying

$$Q^{sl_2;V_{n_0}}(T) = \Big(\sum_{k=0}^{\infty} p_k(n_0)(q^{1/2} - q^{-1/2})^k\Big) \cdot \mathrm{id}_{V_{n_0}}$$

for each constant n_0 of positive integer, while n is an indeterminate.

Proof. With λ as in (9.3) we put $p_k(n)$ by

$$\lambda = \sum_{k=0}^{\infty} p_k(n)(q^{1/2} - q^{-1/2})^k,$$

where $p_k(n) \in \mathbb{Q}[q^{\pm 1/2}, q^{\pm n/2}]$. Further, by Theorem 4.11,

$$Q^{sl_2;V_{n_0}}(T) = \rho_{V_{n_0}}\left(Q^{sl_2;\star}(T)\right)$$

for each constant n_0. Hence, we obtain the proposition. □

As a corollary of Proposition 9.1 we have

Theorem 9.2. For any framed knot K with 0 framing there exists a unique power series

$$P^{sl_2;V_n}(K) \in [n] \cdot \mathbb{Q}[q^{\pm 1/2}, q^{\pm n/2}][[q^{1/2} - q^{-1/2}]],$$

satisfying

$$Q^{sl_2;V_{n_0}}(K) = P^{sl_2;V_n}(K)|_{n=n_0} \in \mathbb{Q}[q^{\pm 1/2}][[q^{1/2} - q^{-1/2}]]$$

for each positive integer constant n_0, while n is an indeterminate. Here, we regard $Q^{sl_2;V_{n_0}}(K) \in \mathbb{Z}[q^{\pm 1/2}]$ as in $\mathbb{Q}[q^{\pm 1/2}][[q^{1/2} - q^{-1/2}]]$ through the natural inclusion $\mathbb{Z}[q^{\pm 1/2}] \to \mathbb{Q}[q^{\pm 1/2}][[q^{1/2} - q^{-1/2}]]$.

We call the power series $P^{sl_2;V_n}(K)$ the *perturbative sl_2 invariant* of a framed knot K with 0 framing.

Proof of Theorem 9.2. Let T be a 1-tangle whose closure is isotopic to K. When $Q^{sl_2;V_{n_0}}(T)$ is expressed as in Proposition 9.1, the quantum invariant $Q^{sl_2;V_{n_0}}(K)$ is given by

$$Q^{sl_2;V_{n_0}}(K) = [n_0] \sum_{k=0}^{\infty} p_k(n_0)(q^{1/2} - q^{-1/2})^k.$$

Hence, putting

$$P^{sl_2;V_n}(K) = [n] \sum_{k=0}^{\infty} p_k(n)(q^{1/2} - q^{-1/2})^k,$$

we obtain the theorem by Proposition 9.1. □

Perturbative invariants obtained from the Kontsevich invariant

Let V_n be the n-dimensional irreducible representation of $U(sl_2)$, *i.e.*, we have $\rho_{V_n} : U(sl_2) \to \mathrm{End}(V_n)$. See Section 4.4 for a concrete presentation of ρ_{V_n} for the generators E, F, H of $U(sl_2)$ given in Section 4.4.

The representation V_n is computed, for example, as follows.

$$\rho_{V_n}(EF) = \begin{pmatrix} (n-1)\cdot 1 & & & & \text{\Large 0} \\ & (n-2)\cdot 2 & & & \\ & & (n-3)\cdot 3 & & \\ & & & \ddots & \\ & & & & 1\cdot(n-1) & \\ \text{\Large 0} & & & & & 0 \end{pmatrix}$$

Hence, the (i,i) entry of $\rho_{V_n}(EF)$ is equal to $(n-i)i$. By similar computation the (i,i) entry of $\rho_{V_n}(FE)$ and $\rho_{V_n}(H^2)$ are equal to $(i-1)(n-i+1)$ and $(n-2i+1)^2$ respectively. Hence,

$$\rho_{V_n}(c) = \frac{n^2-1}{2} \cdot \mathrm{id}_{V_n}, \tag{9.4}$$

for the element c given by

$$c = EF + FE + \frac{1}{2}H^2, \tag{9.5}$$

which is central in $U(sl_2)$. In general, if x is a polynomial in E, F, H, then the (i,j) entry of $\rho_{V_n}(x)$ is a polynomial in i, j and n.

In particular, the representation ρ_{V_n} is computed for the center of $U(sl_2)$ as follows. Let C be a polynomial in E, F, H which belongs to the center of $U(sl_2)$.

By Lemma G.6 in Appendix G.2 the center of $U(sl_2)$ is generated by the element c given by (9.5). Hence, C is equal to a polynomial in c. Further, by (9.4),

$$\rho_{V_n}(c^k) = (\frac{n^2-1}{2})^k \cdot \mathrm{id}_{V_n}.$$

Since $\rho_{V_n}(C)$ is equal to a scalar multiple of id_{V_n}, by Schur's lemma, we have that

$$\rho_{V_n}(C) = \lambda \cdot \mathrm{id}_{V_n}$$

for some $\lambda \in \mathbb{Q}[n]$.

Let T be a 1-tangle with 0 framing. Then, $\hat{W}_{sl_2;\star}(\hat{Z}(T))$ belongs to the center of $U(sl_2)[[h]]$, where $\hat{W}_{sl_2;\star} : \mathcal{A}(\downarrow) \to U(sl_2)[[h]]$ is the graded weight system obtained by substituting sl_2 into the uni-trivalent graphs of Jacobi diagrams. Hence,

$$\rho_{V_n}\Big(\hat{W}_{sl_2;\star}(\hat{Z}(T))\Big) = \lambda \cdot \mathrm{id}_{V_n} \tag{9.6}$$

for some $\lambda \in \mathbb{Q}[n][[h]]$

Proposition 9.3. For any 1-tangle T with 0 framing there exists a unique series of polynomials $p_k(n)$ in n of degree $\leq k$ satisfying

$$Q^{sl_2;V_{n_0}}(T)\Big|_{q=e^h} = \Big(\sum_{k=0}^{\infty} p_k(n_0)h^k\Big) \cdot \mathrm{id}_{V_{n_0}}$$

for each constant n_0 of positive integer, while n is an indeterminate.

Proof. With λ as in (9.3) we put $p_k(n)$ by

$$\lambda = \sum_{k=0}^{\infty} p_k(n)h^k,$$

where $p_k(n) \in \mathbb{Q}[n]$. Further, by Theorem 6.14,

$$Q^{sl_2;V_{n_0}}(T) = \hat{W}_{sl_2;V_{n_0}}\Big(\hat{Z}(T)\Big) = \rho_{V_{n_0}}\Big(\hat{W}_{sl_2;\star}(\hat{Z}(T))\Big)$$

for each constant n_0. Hence, we obtain the formula of the proposition with these $p_k(n)$. It is sufficient to show that $p_k(n)$ is of degree $\leq k$ for each k.

Similarly as in Section 6.5, it is shown that for a 1-tangle T, $\log \hat{Z}(T)$ is equal to a linear sum of primitive Jacobi diagrams as shown below.

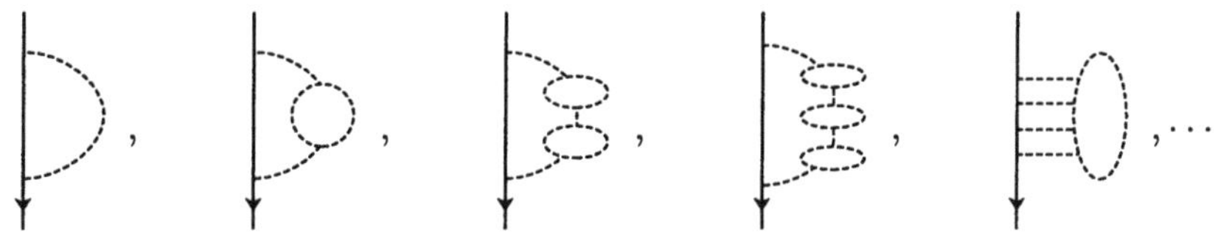

(9.7)

In the present case, since T has 0 framing, the first Jacobi diagram does not appear in the linear sum. Further, by Lemma 9.4 below,

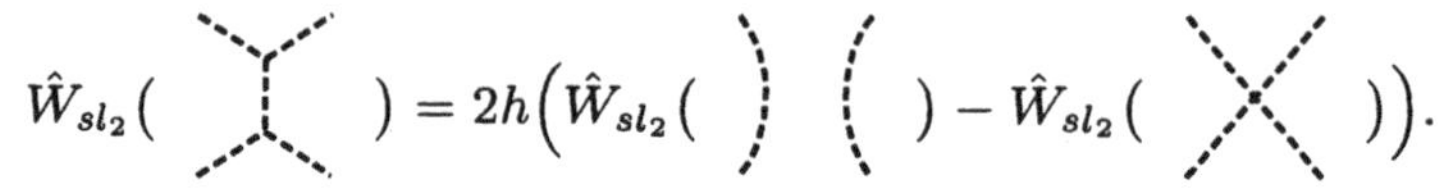

(9.8)

We consider the following congruence, which implies that the images of both sides are equal to each other when evaluated in $\hat{W}_{sl_2}$,

By decreasing trivalent vertices, using the above formula, it can be shown that any Jacobi diagram on an interval is congruent to a polynomial in  modulo the above relation. Further, any primitive Jacobi diagram on an interval (see, *e.g.*, (9.7)) except for  is congruent to a polynomial in (h^2 ). We have that

$h^2 W_{sl_2;V_n}\Big($  $\Big) = h^2 \rho_{V_n}(c) = h^2 \cdot \dfrac{n^2-1}{2} \cdot \mathrm{id}_{V_n}.$

Hence, the $\hat{W}_{sl_2;V_n}(\hat{Z}(T))$ is presented by a linear sum of polynomials in $(h^2 \cdot \frac{n^2-1}{2})$, and so is $p_k(n)$. Therefore, $p_k(n)$ is a polynomial in n of degree $\leq k$ for each k. □

One can verify the above computation for the second, third, and fourth diagrams in (9.7) by using

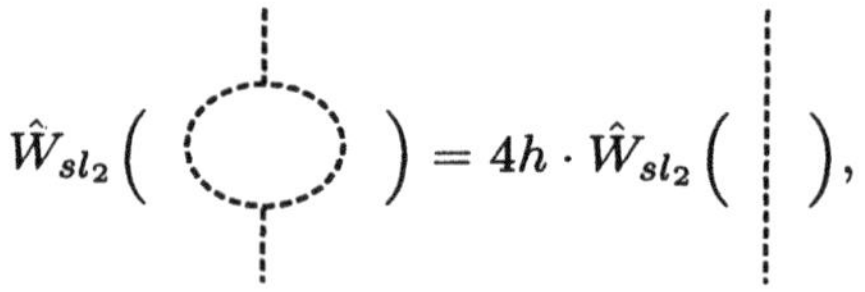

(9.9)

which is derived from (9.8).

Lemma 9.4 ([ChVa97]). We have that

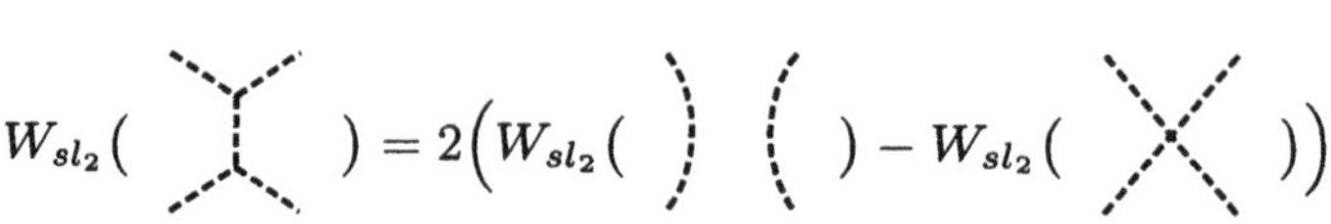

where both sides are equal as linear maps $sl_2 \otimes sl_2 \to sl_2 \otimes sl_2$.

Proof. Since the adjoint representation of sl_2 is irreducible, it is isomorphic to the 3-dimensional irreducible representation V_3. Further,

$$V_3 \otimes V_3 = V_1 \oplus V_3 \oplus V_5.$$

By Schur's lemma any intertwiner on this space is equal to $c_1 \mathrm{id}_{V_1} \oplus c_3 \mathrm{id}_{V_3} \oplus c_5 \mathrm{id}_{V_5}$ for some scalars c_1, c_3, c_5. For simplicity, we denote this map by $(c_1) \oplus (c_3) \oplus (c_5)$ in the following of this proof.

Firstly, we show that the weight system of the following diagram is equal to $(0) \oplus (-4) \oplus (0)$, as follows.

$$\begin{matrix} V_3 \otimes V_3 \\ \uparrow \\ V_3 \otimes V_3 \end{matrix}$$

Since the map factors through V_3, the map is equal to $(0) \oplus (\lambda) \oplus (0)$ for some scalar λ. Further, the composition of two copies of the diagram is equal to 4 times the diagram, in the sense of (9.9). Hence, we obtain $\lambda = 4$.

Secondly, the weight system of the following diagram is equal to $(1) \oplus (1) \oplus (1)$, from the definition of the weight system.

$$\begin{matrix} V_3 \otimes V_3 \\ \uparrow \\ V_3 \otimes V_3 \end{matrix}$$

Thirdly, we show that the weight system of the following diagram is equal to $(1) \oplus (-1) \oplus (1)$, as follows.

$$\begin{matrix} V_3 \otimes V_3 \\ \uparrow \\ V_3 \otimes V_3 \end{matrix}$$

Since the composition of two copies of the diagram is equal to the previous diagram, its weight system is equal to $(\pm 1) \oplus (\pm 1) \oplus (\pm 1)$. Further, the trace of the weight system is computed as

$\pm 1 \pm 3 \pm 5 = W_{sl_2}\Big($  $\Big) = W_{sl_2}\Big($  $\Big) = 3.$

Hence, the signs on the left hand side are $+, -, +$ respectively. Therefore, the weight system is equal to $(1) \oplus (-1) \oplus (1)$.

By computing a linear sum of the weight systems of the above three diagrams we obtain the required formula. □

As a corollary of Proposition 9.3 we have

Theorem 9.5. For any framed knot K with 0 framing there exists a unique power series

$$P^{sl_2;V_n}(K) \in [n] \cdot \mathbb{Q}[[nh, h]]$$

satisfying

$$Q^{sl_2;V_{n_0}}(K)\big|_{q=e^h} = P^{sl_2;V_n}(K)\big|_{n=n_0} \in \mathbb{Q}[[h]]$$

for each positive integer constant n_0, while n is an indeterminate.

The proof of this theorem is shown similarly as the proof of Theorem 9.2.

Theorem 9.5 give an alternative definition of the *perturbative* sl_2 *invariant* given in Theorem 9.2 by putting $q = e^h$. Here, we regard, say, $q^{n/2}$ in the original perturbative sl_2 invariant as a power series in h by

$$q^{n/2} = e^{nh/2} = 1 + \frac{1}{2}nh + \frac{1}{8}n^2h^2 + \cdots \in \mathbb{Q}[[nh]].$$

9.2 Perturbative invariants of homology 3-spheres

In this section we introduce the perturbative $SO(3)$ invariant of homology 3-spheres by an arithmetic expansion of the quantum $SO(3)$ invariant. It is known, for a certain class of 3-manifolds, that the perturbative $SO(3)$ invariant describes the asymptotic behavior of the quantum $SO(3)$ invariant, in a certain sense.

Integrality of the quantum $SO(3)$ invariant

A 3-manifold is called a *rational homology 3-sphere* if $H_\star(M;\mathbb{Q}) \cong H_\star(S^3;\mathbb{Q})$. In what follows of this section we let p be an odd prime and put $\zeta = \exp(2\pi\sqrt{-1}/p)$.

Theorem 9.6 ([Mu$_{\rm H}$95], see also [MaRo97]). For any rational homology 3-sphere M and any odd prime p the quantum $SO(3)$ invariant $\tau_p^{SO(3)}(M)$ belongs to $\mathbb{Z}[\zeta]$.

Proof. We show the theorem, for simplicity, in the case that the 3-manifold M is obtained from S^3 by integral surgery along a framed knot K with positive framing. Then, the quantum $SO(3)$ invariant of M is presented by

$$\tau_p^{SO(3)}(M) = c_+^{-1} \sum_{\substack{1 \le n < p \\ n \text{ is odd}}} [n] Q^{sl_2;V_n}(K)\Big|_{q=\zeta}, \tag{9.10}$$

where c_+ is the constant given by

$$c_+ = \sum_{\substack{1 \le n < p \\ n \text{ is odd}}} \zeta^{(n^2-1)/4}[n]^2 = \frac{-\zeta^{-3/4}}{\zeta^{1/2} - \zeta^{-1/2}} \mathbb{G}(\zeta).$$

Here, the second equality is obtained by an elementary computation. Further, $\mathbb{G}(\zeta)$ denotes the *Gaussian sum* defined by

$$\mathbb{G}(\zeta) = \sum_{k=0}^{p-1} \zeta^{k^2}.$$

By the integrality of quantum invariants of knots the sum in (9.10) belongs to $\mathbb{Z}[\zeta]$. However, it is a non-trivial question whether $\tau_p^{SO(3)}(M)$ belongs to $\mathbb{Z}[\zeta]$, that is, after the sum has been divided by the normalization factor c_+. Note that $(\zeta - 1)$ is not invertible in $\mathbb{Z}[\zeta]$. As mentioned in Appendix G.1 the Gaussian sum is divisible by $(\zeta - 1)^{(p-1)/2}$. Hence, it is necessary to show that the sum in (9.10) is divisible by $(\zeta - 1)^{(p-3)/2}$. We show this as follows.

By modifying the contribution from the framing, (9.10) is written as

$$\tau_p^{SO(3)}(M) = c_+^{-1} \sum_{\substack{1 \le n < p \\ n \text{ is odd}}} \zeta^{f(n^2-1)/4} [n] Q^{sl_2;V_n}(K_0)\Big|_{q=\zeta}, \tag{9.11}$$

where f is the framing of K and K_0 is the framed knot with 0 framing which is isotopic to K as an unframed knot.

Let T be a 1-tangle whose closure is isotopic to K_0. Since $[n]$ is invertible in $\mathbb{Z}[\zeta]$ for $0 < n < p$, the universal R matrix in $U_\zeta(sl_2)^{\otimes 2}$ can be presented by a matrix with entries in $\mathbb{Z}[\zeta]$, noting that $\zeta^{1/2} = -\zeta^{(p+1)/2} \in \mathbb{Z}[\zeta]$. Hence, the universal $U_\zeta(sl_2)$ invariant $Q^{U_\zeta(sl_2);\star}(T)$ is equal to a polynomial in c with coefficients in $\mathbb{Z}[\zeta]$, where c is a generator of the center of $U_\zeta(sl_2)$ (see Lemma G.9 in Appendix G.2) given by

$$c = (\zeta^{1/2} - \zeta^{-1/2})^2 EF + \zeta^{-1/2} K + \zeta^{1/2} K^{-1}.$$

Since

$$\rho_{V_n}(c^k) = (\zeta^{n/2} + \zeta^{-n/2})^k \cdot \mathrm{id}_{V_n},$$

the invariant $Q^{sl_2;V_n}(T)\big|_{q=\zeta}$ is equal to a linear sum of $(\zeta^{jn/2} + \zeta^{-jn/2}) \cdot \mathrm{id}_{V_n}$.

Hence, $Q^{sl_2;V_n}(K_0)\big|_{q=\zeta}$ is equal to a linear sum of $[n](\zeta^{jn/2} + \zeta^{-jn/2})$. We compute (9.10) as

$$\begin{aligned}
&\sum_{\substack{1 \le n < p \\ n \text{ is odd}}} \zeta^{f(n^2-1)/4} [n]^2 (\zeta^{jn/2} + \zeta^{-jn/2}) \\
&= \frac{1}{2} \sum_{\substack{-p < n \le p \\ n \text{ is odd}}} \zeta^{f(n^2-1)/4} [n]^2 (\zeta^{jn/2} + \zeta^{-jn/2}) \\
&= \frac{1}{2(\zeta^{1/2} - \zeta^{-1/2})^2} \sum_{\substack{-p < n \le p \\ n \text{ is odd}}} \zeta^{f(n^2-1)/4} (\zeta^n + \zeta^{-n} - 2)(\zeta^{jn/2} + \zeta^{-jn/2})
\end{aligned}$$

$$= \frac{1}{(\zeta^{1/2}-\zeta^{-1/2})^2} \sum_{\substack{-p<n\le p \\ n \text{ is odd}}} \zeta^{f(n^2-1)/4}(\zeta^n - 1)(\zeta^{jn/2}+\zeta^{-jn/2})$$
$$= \frac{1}{(\zeta^{1/2}-\zeta^{-1/2})^2} \sum_{0\le m<p} \zeta^{f(m^2+m)}(\zeta^{2m+1}-1)(\zeta^{j(m+1/2)}+\zeta^{-j(m+1/2)})$$
$$= \left(\frac{f}{p}\right)\mathbb{G}(\zeta)\zeta^{-f/4-j^2/4f}(\zeta^{(j-1)/f}+\zeta^{-(j+1)/f}-2), \tag{9.12}$$

where we obtain the last equality by such a computation as the following,

$$\sum_{0\le m<p} \zeta^{fm^2+lm} = \sum_{0\le m<p} \zeta^{f(m+\frac{l}{2f})^2-\frac{l^2}{4f}} = \zeta^{-\frac{l^2}{4f}} \sum_{0\le m'<p} \zeta^{fm'^2} = \zeta^{-\frac{l^2}{4f}}\left(\frac{f}{p}\right)\mathbb{G}(\zeta).$$

Since (9.12) is divisible by c_+ in $\mathbb{Z}[\zeta]$, so is the sum in (9.10). Hence, we obtain the theorem. □

Arithmetic expansion of the quantum SO(3) invariant

Let p be an odd prime and put $\zeta = \exp(2\pi\sqrt{-1}/p)$ as before.

As in Theorem 9.6, the quantum $SO(3)$ invariant $\tau_p^{SO(3)}(M)$ belongs to $\mathbb{Z}[\zeta]$ for any rational homology 3-sphere M. Hence, we can put

$$\tau_p^{SO(3)}(M) = a_{p,0} + a_{p,1}(\zeta-1) + a_{p,2}(\zeta-1)^2 + \cdots + a_{p,N}(\zeta-1)^N, \tag{9.13}$$

for some integers $a_{p,n}$'s. Though the expansion (9.13) is not unique, we have

Lemma 9.7. $(a_{p,n} \bmod p) \in \mathbb{Z}/p\mathbb{Z}$ is uniquely determined by $\tau_p^{SO(3)}(M)$ for $0 \le n \le p-2$.

Proof. The ring $\mathbb{Z}[\zeta]$ is isomorphic to $\mathbb{Z}[q]/T(q)$ where

$$T(q) = 1 + q + q^2 + \cdots + q^{p-1}$$
$$= \binom{p}{1} + \binom{p}{2}(q-1) + \binom{p}{3}(q-1)^2 + \cdots + \binom{p}{p}(q-1)^{p-1}.$$

Since p is an odd prime, $\binom{p}{k}$ is divisible by p for $1 \le k \le p-1$. Hence, the first $p-1$ coefficients in any expansion (9.13) are uniquely determined modulo p by $\tau_p^{SO(3)}(M)$. □

For a positive integer $m < p$, two rational numbers a and b in the ring $\mathbb{Z}[\frac{1}{2}, \frac{1}{3}, \ldots, \frac{1}{m}]$ are called *congruent modulo p* if $a - b$ is divisible by p in the ring. For an integer f not divisible by p we define the *Legendre symbol* $\left(\frac{f}{p}\right)$ by

$$\left(\frac{f}{p}\right) = \begin{cases} 1 & \text{if there exists an integer } x \text{ such that } x^2 \underset{(p)}{\equiv} f, \\ -1 & \text{otherwise.} \end{cases}$$

Theorem 9.8 ([Oht96b]). Let M be a rational homology 3-sphere, and let $a_{p,n}$ denote the coefficients in the expansion (9.13) of $\tau_p^{SO(3)}(M)$. Then, there exists a sequence of topological invariants $\lambda_n(M) \in \mathbb{Z}[\frac{1}{2}, \frac{1}{3}, \cdots, \frac{1}{2n+1}]$ such that $\lambda_n(M)$ is congruent to $\left(\frac{|H_1(M;\mathbb{Z})|}{p}\right) a_{p,n}$ modulo p for any odd prime p satisfying that $p > \max(2n+1, |H_1(M;\mathbb{Z})|)$, where $|H_1(M;\mathbb{Z})|$ denotes the order of the first homology group $H_1(M;\mathbb{Z})$.

Note that if such a $\lambda_n(M)$ exists then its topological invariance would be derived from the topological invariance of ($a_{p,n}$ mod p).

This theorem implies, in other words, that there exists a unique sequence of $\lambda_n(M) \in \mathbb{Z}[\frac{1}{2}, \frac{1}{3}, \cdots, \frac{1}{2n+1}]$ satisfying

$$\begin{aligned}\tau_p^{SO(3)}(M) &= \left(\frac{|H_1(M;\mathbb{Z})|}{p}\right) \sum_{n=0}^{N} \lambda_n(M)(\zeta-1)^n \\ &+ \left(\text{terms divisible by } (\zeta-1)^{N+1} \text{ in } \mathbb{Z}[\zeta, \tfrac{1}{2}, \tfrac{1}{3}, \cdots, \tfrac{1}{2N+1}]\right),\end{aligned}$$

for any positive integer N and any odd prime $p > \max\{2N+1, |H_1(M;\mathbb{Z})|\}$.

Using this sequence we define the following power series

$$\tau^{SO(3)}(M) = \sum_{n=0}^{\infty} \lambda_n(M)(q-1)^n \in \mathbb{Q}[[q-1]],$$

and call it the *perturbative $SO(3)$ invariant* (or the *Ohtsuki invariant*) of a rational homology 3-sphere M. It is a topological invariant of M, since each $\lambda_n(M)$ is a topological invariant.

Remark 9.9. We can define $\lambda_n(M)$ for any closed oriented 3-manifold satisfying the requirements of Theorem 9.8, but it can be shown that such $\lambda_n(M)$ vanishes unless M is a rational homology 3-sphere.

By properties of the quantum $SO(3)$ invariant we have

Proposition 9.10. For rational homology 3-spheres M_1, M_2, and M we have that

$$\begin{aligned}\tau^{SO(3)}(M_1 \# M_2) &= \tau^{SO(3)}(M_1)\tau^{SO(3)}(M_2), \\ \tau^{SO(3)}(-M)(q) &= \tau^{SO(3)}(M)(q^{-1}),\end{aligned}$$

where $M_1 \# M_2$ denotes the connected sum of M_1 and M_2, and $-M$ denotes M with the opposite orientation.

The following proposition implies that the Casson-Walker invariant (see [Walk92, Les96]) is determined by the quantum $SO(3)$ invariant. We omit the proof of the proposition; see [Mu$_{\text{H}}$95, Oht96b] for the proof.

Proposition 9.11 ([Mu$_H$95], see also [Oht96b]). As for the first two coefficients,

$$\lambda_0(M) = \frac{1}{|H_1(M;\mathbb{Z})|},$$
$$\lambda_1(M) = \frac{6\lambda(M)}{|H_1(M;\mathbb{Z})|},$$

where $\lambda(M)$ is the Casson-Walker invariant. Here we use Casson's normalization, which is half times Walker's definition.

As an example of the perturbative $SO(3)$ invariant let us compute $\tau^{SO(3)}(L(5,1))$ for the lens space $L(5,1)$. The quantum $SO(3)$ invariant of $L(5,1)$ is given (see Example 8.14) by

$$\begin{aligned}\tau_p^{SO(3)}\Big(L(5,1)\Big) &= \left(\frac{5}{p}\right)\zeta^{-3\cdot\overline{5}}\frac{\zeta^{\overline{10}}-\zeta^{-\overline{10}}}{\zeta^{\overline{2}}-\zeta^{-\overline{2}}} \\ &= \left(\frac{5}{p}\right)(\overline{5}-3\cdot\overline{5}^2(\zeta-1)+11\cdot\overline{5}^3(\zeta-1)^2+\cdots),\end{aligned}$$

where we denote the inverse of m in $\mathbb{Z}/p\mathbb{Z}$ by $\overline{m}$. By comparing the above formula with the following expansion

$$q^{-3/5}\frac{q^{1/10}-q^{-1/10}}{q^{1/2}-q^{-1/2}} = \frac{1}{5}-\frac{3}{5^2}(q-1)+\frac{11}{5^3}(q-1)^2+\cdots,$$

we obtain

$$\tau^{SO(3)}\Big(L(5,1)\Big) = q^{-3/5}\frac{q^{1/10}-q^{-1/10}}{q^{1/2}-q^{-1/2}} \in \mathbb{Q}[[q-1]].$$

The perturbative $SO(3)$ invariant $\tau^{SO(3)}(L(5,1))$ somehow describes the asymptotic behavior of $\tau_p^{SO(3)}(L(5,1))$ as $p \to \infty$ as follows. To compare $\tau^{SO(3)}(L(5,1))$ and $\tau_p^{SO(3)}(L(5,1))$ we compare $\zeta^{\overline{10}}$ and $\zeta^{1/10}$. The sequence of $\zeta^{\overline{10}}$ for $p = 7, 11, 13, \dots$ splits into subsequences, which correspond to subsequences of p consisting of primes which are congruent to 1, 3, 7 and 9 modulo 10. For a subsequence of primes p which are congruent to, say, -1 modulo 10, we can put $\overline{10} = (p+1)/10$ and obtain $\zeta^{\overline{10}} = \eta\zeta^{1/10}$ where we put $\eta = \exp(2\pi\sqrt{-1}/10)$ and regard $\exp(2\pi\sqrt{-1}/10p)$ as $\zeta^{1/10}$. Since $\left(\frac{5}{p}\right) = 1$ for these primes, the asymptotic behavior of $\tau_p^{SO(3)}(L(5,1))$ for these primes can be expressed

$$\begin{aligned}\tau_p^{SO(3)}\Big(L(5,1)\Big) &= \eta^{-6}e^{-3h/5}\frac{\eta e^{h/10}-\eta^{-1}e^{-h/10}}{\eta^5 e^{h/2}-\eta^{-5}e^{-h/2}} \\ &= \tau^{SO(3)}\Big(L(5,1)\Big)\Big|_{q^{1/10}=\eta e^{h/10}}\end{aligned}$$

for primes p which are congruent to -1 modulo 10, where $h = 2\pi\sqrt{-1}/p$. Since λ_n's appear in the expansion of $\tau^{SO(3)}(L(5,1))$ at $q^{1/10} = e^{h/10}$, the above asymptotic behavior can not be described using a finite number of $\lambda_n(L(5,1))$'s, while we *can* describe it using $\tau^{SO(3)}(L(5,1))$ at $q^{1/10} = \eta e^{h/10} \in \mathbb{Q}[[q-1]]$.

In general, we might expect the property that the perturbative $SO(3)$ invariant $\tau^{SO(3)}(M)$ might be a rational function in $t^{1/m}$ for some integer m related to the order of $H_1(M;\mathbb{Z})$ for some class of rational homology 3-spheres M. For a rational homology 3-sphere M with such a property we can describe the asymptotic behavior of the quantum invariants $\tau_p^{SO(3)}(M)$ just as above. This property has been confirmed for Seifert fibered spaces by results of [Tak97]. Lawrence [Law97] has given holomorphic expression for the perturbative $SO(3)$ invariants of rational homology 3-spheres obtained by integral surgery along $(2,n)$ torus knot (see also Appendix G.3 for the computation of the invariant).

Another property of the perturbative $SO(3)$ invariant that has been conjectured by Lawrence [Law95] is that $\tau^{SO(3)}(M)$ belongs to $\mathbb{Z}[[q-1]]$, in the case that M is an integral homology 3-sphere. For numerical examples of this conjecture see [Roz98b]. See also [Roz*98c*] for this conjecture.

For a simply connected simple Lie group G we denote by PG the quotient Lie group of G by its center; note that $PSU(2) \cong SO(3)$ in the simplest case. As mentioned in the end of Chapter 8, the quantum PG invariant $\tau_p^{PG}(M)$ was defined by Kirby and Melvin [KiMe91] for $PG = SO(3)$, by Kohno and Takata [KoTa96] for $PG = PSU(N)$, and by Le [Le*00*] for general PG. Furthermore, it is shown, by Takata and Yokota [TaYo99] (see also [MaWe98]) for $PG = PSU(N)$ and by Le [Le*00*] for general PG, that $\tau_p^{PG}(M)$ lies in $\mathbb{Z}[\zeta]$; this extends Theorem 9.6. As an extension of Theorem 9.8 Le [Le00b, Le*00*] showed that, for any rational homology 3-sphere M, there exists a unique power series $\tau^{PG}(M) \in \mathbb{Q}[[q-1]]$ obtained by arithmetic expansion of the quantum PG invariant $\tau_p^{PG}(M)$, in the sense of Theorem 9.8. We call $\tau^{PG}(M)$ the *perturbative PG invariant* of M.

What follows of this section is devoted to the proof of Theorem 9.8.

To prove this theorem we introduce the notion of a Fermat limit, defined in [LiWa99b]. Let $\mathbb{Z}_p$ be the ring of p-adic integers. We consider a complex function f on the set of odd prime numbers satisfying $f(p) \in \mathbb{Z}[\zeta]$ for each odd prime p, where we put $\zeta = \exp(2\pi\sqrt{-1}/p)$ as before. As in (9.13) we consider an expansion

$$f(p) = a_{p,0} + a_{p,1}(\zeta-1) + a_{p,2}(\zeta-1)^2 + \cdots a_{p,N}(\zeta-1)^N.$$

Then, by Lemma 9.7 ($a_{p,n}$ mod p) $\in \mathbb{Z}/p\mathbb{Z}$ is uniquely determined by $f(p)$ for $0 \le n \le p-2$. If there exists a sequence of $\lambda_n \in \mathbb{Q}$ such that each λ_n is congruent modulo p to $a_{p,n}$ in $\mathbb{Z}_p$, for all sufficiently large p, then we define the *Fermat limit* of $f(p)$ to be

$$\text{f-lim}\, f(p) = \sum_{n=0}^{\infty} \lambda_n (q-1)^n \in \mathbb{Q}[[q-1]].$$

Further, for the above $f(p)$ and λ_n's we define the Fermat limit of $(\zeta-1)^{-n_0}f(p)$, for any constant n_0, to be

$$\text{f-lim}\,(\zeta-1)^{-n_0}f(p)=\sum_{n=0}^{\infty}\lambda_n(q-1)^{n-n_0}\in\mathbb{Q}((q-1)).$$

Theorem 9.8 implies the existence of the following Fermat limit,

$$\text{f-lim}\left(\frac{|H_1(M;\mathbb{Z})|}{p}\right)\tau_p^{SO(3)}(M)=\tau^{SO(3)}(M). \tag{9.14}$$

For an odd prime p and a p-th root of unity $\zeta=\exp(2\pi\sqrt{-1}/p)$ we introduce the *Gaussian sum* $\mathbb{G}(\zeta)$ by

$$\mathbb{G}(\zeta)=\sum_{k=0}^{p-1}\zeta^{k^2}.$$

Let $P(k)$ be a polynomial in k. We define a function $\hat{P}(q)$ of q by

$$\hat{P}(q)=\text{f-lim}\,\mathbb{G}(\zeta)^{-1}\sum_{k=0}^{p-1}P(k)\zeta^{k^2}, \tag{9.15}$$

where the sum on the right hand side is called a *weighted Gaussian sum* and the existence of this Fermat limit is shown in Lemma G.2 in Appendix G.1. In particular, if $P(k)=k^l$ for a non-negative integer l, then by Lemma G.2 we have that

$$\hat{P}(q)=\begin{cases}\dfrac{(l-1)!!}{(-2\log q)^{l/2}} & \text{if } l \text{ is even,}\\ 0 & \text{if } l \text{ is odd,}\end{cases} \tag{9.16}$$

where $m!!$ denotes $m(m-2)(m-4)\cdots 1$ for an odd positive integer m, and we regard $(-1)!!$ as 1. In general, for any polynomial $P(k)$ we obtain the function $\hat{P}(q)$ as a linear sum of the functions on the right hand side of (9.16). Further, by replacing ζ in (9.15) with ζ^f we have that

$$\text{f-lim}\left(\frac{f}{p}\right)\mathbb{G}(\zeta)^{-1}\sum_{k=0}^{p-1}P(k)\zeta^{fk^2}=\hat{P}(q^f), \tag{9.17}$$

noting the relation (G.1).

Let V_n denote the n-dimensional irreducible representation of $U_q(sl_2)$ and put $[n]=(q^{n/2}-q^{-n/2})/(q^{1/2}-q^{-1/2})$ as before. Put $a=V_3-[3]V_1$ in the representation ring of $U_q(sl_2)$; recall that the representation ring is spanned by V_n's and its sum and product are the direct sum and the tensor product of representations respectively.

For simplicity, we consider the case that M is obtained from S^3 by integral surgery along a framed knot K. In this case $\tau_p^{SO(3)}(M)$ is presented by a linear sum of $Q^{sl_2;V_n}(K)$ such that the range of the sum depends on p. In order to compute the Fermat limit in (9.14) it would be convenient to reduce the sum to sums of functions of n which are independent of K. Motivated by Proposition G.11 in Appendix G.3, let us consider the expansion of V_n into a polynomial in a whose coefficients are functions of n.

From the definition of a,

$$V_n a = V_{n+2} - (q + q^{-1})V_n + V_{n-2}. \tag{9.18}$$

By using this formula repeatedly, we can present V_n for odd n by a polynomial in a, as follows. For small n,

$$\begin{aligned} V_1 &= 1, \\ V_3 &= a + [3], \\ V_5 &= a^2 + (2q + 1 + 2q^{-1})a + [5], \end{aligned}$$

where we denote the unit representation V_1 in the representation ring by 1. Further, for general n,

$$V_n = c_{(n-1)/2}(n)a^{(n-1)/2} + c_{(n-3)/2}(n)a^{(n-3)/2} + \cdots + c_1(n)a + c_0(n), \tag{9.19}$$

where the $c_k(n)$'s are polynomials in $q^{\pm 1}$ indexed by k and n, and are characterized by

$$\begin{aligned} &c_k(n+2) - (q + q^{-1})c_k(n) + c_k(n-2) = \begin{cases} c_{k-1}(n) & \text{if } k \geq 1, \\ 0 & \text{if } k = 0, \end{cases} \\ &c_k(n) = 0 \qquad \text{for } n = 1, 3, 5, \cdots, 2k-1. \end{aligned} \tag{9.20}$$

Here, the recursive formula is derived from (9.18). For example, for small k,

$$\begin{aligned} c_0(n) &= [n], \\ c_1(n) &= \frac{n(q^{n/2} + q^{-n/2}) - [2][n]}{2[2](q^{1/2} - q^{-1/2})^2}, \end{aligned}$$

since they satisfy the above characterization (9.20). Further, we can solve the equation (9.20) with functions on n by

$$c_k(n) = g_k(\frac{n}{2})q^{n/2} - g_k(-\frac{n}{2})q^{-n/2} \tag{9.21}$$

where the $g_k(n)$'s are polynomials in n of degree k; for example, for small k,

$$\begin{aligned} g_0(n) &= \frac{1}{q^{1/2} - q^{-1/2}}, \\ g_1(n) &= \frac{n}{[2](q^{1/2} - q^{-1/2})^2} - \frac{1}{2(q^{1/2} - q^{-1/2})^3}, \\ g_2(n) &= \frac{n^2}{2[2]^2(q^{1/2} - q^{-1/2})^3} - \frac{[3]n}{[2]^3(q^{1/2} - q^{-1/2})^4} + \frac{3}{8(q^{1/2} - q^{-1/2})^5}. \end{aligned}$$

In general, the sequence of polynomials $g_k(n)$'s characterized by (9.21) and (9.20) gives the required expansion (9.19) of V_n.

Let us compute the following sum which will appear later in the computation of $\tau_p^{SO(3)}(M)$. For a polynomial $g(n)$ in n,

$$\begin{aligned}
&\sum_{\substack{1\le n<p\\ n \text{ is odd}}} \zeta^{f(n^2-1)/4}[n]\Big(g(\frac{n}{2})\zeta^{n/2} - g(-\frac{n}{2})\zeta^{-n/2}\Big)\\
&= \frac{1}{2(\zeta^{1/2}-\zeta^{-1/2})} \sum_{\substack{-p<n\le p\\ n \text{ is odd}}} \zeta^{f(n^2-1)/4}\Big(g(\frac{n}{2})\zeta^{n} + g(-\frac{n}{2})\zeta^{-n} - g(\frac{n}{2}) - g(-\frac{n}{2})\Big)\\
&= \frac{1}{\zeta^{1/2}-\zeta^{-1/2}} \sum_{\substack{-p<n\le p\\ n \text{ is odd}}} \zeta^{f(n^2-1)/4}\Big(g(\frac{n}{2})\zeta^{n} - g(\frac{n}{2})\Big)\\
&= \frac{1}{\zeta^{1/2}-\zeta^{-1/2}} \sum_{0\le m<p} \Big(g(m+\frac{1}{2})\zeta^{f(m^2+m)+2m+1} - g(m+\frac{1}{2})\zeta^{f(m^2+m)}\Big). \qquad (9.22)
\end{aligned}$$

Further, for the powers in (9.22),

$$\begin{aligned}
f(m^2+m)+2m+1 &= f(m+\frac{1}{2}+\frac{1}{f})^2 - \frac{f}{4} - \frac{1}{f},\\
f(m^2+m) &= f(m+\frac{1}{2})^2 - \frac{f}{4}.
\end{aligned}$$

By replacing m with $m-\frac{1}{2}-\frac{1}{f}$ and $m-\frac{1}{2}$ in the two terms of (9.22) respectively, we have that

$$(9.22) \underset{(p)}{\equiv} \frac{\zeta^{-f/4}}{\zeta^{1/2}-\zeta^{-1/2}} \sum_{m=0}^{p-1} \Big(\zeta^{-1/f} g(m-\frac{1}{f}) - g(m)\Big)\zeta^{fm^2}.$$

Hence,

$$\begin{aligned}
&\text{f-lim}\left(\frac{f}{p}\right)\mathbb{G}(\zeta)^{-1} \sum_{\substack{1\le n<p\\ n \text{ is odd}}} \zeta^{f(n^2-1)/4}[n]\Big(g(\frac{n}{2})\zeta^{n/2} - g(-\frac{n}{2})\zeta^{-n/2}\Big)\\
&= \frac{q^{-f/4}}{q^{1/2}-q^{-1/2}}\Big(q^{-1/f}\widehat{g(m-\frac{1}{f})}(q^f) - \widehat{g(m)}(q^f)\Big), \qquad (9.23)
\end{aligned}$$

where for a polynomial $P(m)$ in m we use the function $\hat{P}(q)$ defined in (9.15).

Lemma 9.12. Let $g_k(\frac{n}{2})$ be the polynomial in n defined by (9.20) and (9.21). Then, the following function of q has a pole at $q=1$ of order at most k,

$$q^{-1/f}\widehat{g_k(m-\frac{1}{f})}(q^f) - \widehat{g_k(m)}(q^f).$$

We give the proof of this lemma in Appendix G.1.

For example, one can verify the lemma for $k = 0, 1$ with the following formulae,

$$q^{-1/f}\widehat{g_0(m-\frac{1}{f})}(q^f) - \widehat{g_0(m)}(q^f) = \frac{q^{-1/f}-1}{q^{1/2}-q^{-1/2}},$$

$$q^{-1/f}\widehat{g_1(m-\frac{1}{f})}(q^f) - \widehat{g_1(m)}(q^f) = -\frac{q^{-1/f}}{f[2](q^{1/2}-q^{-1/2})^2} - \frac{q^{-1/f}-1}{2(q^{1/2}-q^{-1/2})^3}.$$

With these computations we can now prove Theorem 9.8.

Proof of Theorem 9.8. We consider, for simplicity, some rational homology 3-sphere M that has been obtained from S^3 by integral surgery along some framed knot K with positive framing f. Then, the quantum $SO(3)$ invariant of M is given by

$$\tau_p^{SO(3)}(M) = c_+^{-1} \sum_{\substack{1\le n<p \\ n \text{ is odd}}} \zeta^{f(n^2-1)/4}[n]Q^{sl_2;V_n}(K_0), \tag{9.24}$$

where K_0 is the framed knot with 0 framing which is isotopic to K as an unframed knot. Further, c_+ is the constant given by

$$c_+ = \sum_{\substack{1\le n<p \\ n \text{ is odd}}} \zeta^{(n^2-1)/4}[n]^2 = \frac{-\zeta^{-3/4}}{\zeta^{1/2}-\zeta^{-1/2}}\mathbb{G}(\zeta),$$

where the second equality is obtained by an elementary computation. Hence,

$$\text{f-lim}\,\mathbb{G}(\zeta)^{-1}c_+ = \frac{-q^{-3/4}}{q^{1/2}-q^{-1/2}}.$$

By presenting the V_n's by polynomials in a with coefficients $c_k(n)$'s we can present the sum in (9.24) by

$$\begin{aligned}
&\sum_{\substack{1\le n<p \\ n \text{ is odd}}} \zeta^{f(n^2-1)/4}[n]Q^{sl_2;V_n}(K_0) \\
&= \sum_{n,k} \zeta^{f(n^2-1)/4}[n]c_k(n)Q^{sl_2;a^k}(K_0) \\
&= \sum_{n,k} \zeta^{f(n^2-1)/4}[n]\Big(g_k(\frac{n}{2})\zeta^{n/2} - g_k(-\frac{n}{2})\zeta^{-n/2}\Big)Q^{sl_2;a^k}(K_0).
\end{aligned}$$

Hence, by (9.23),

$$\begin{aligned}
&\text{f-lim}\left(\frac{f}{p}\right)\mathbb{G}(\zeta)^{-1} \sum_{\substack{1\le n<p \\ n \text{ is odd}}} \zeta^{f(n^2-1)/4}[n]Q^{sl_2;V_n}(K_0) \\
&= \frac{q^{-f/4}}{q^{1/2}-q^{-1/2}} \sum_{k=0}^{\infty}\Big(q^{-1/f}\widehat{g_k(m-\frac{1}{f})}(q^f) - \widehat{g_k(m)}(q^f)\Big)Q^{sl_2;a^k}(K_0),
\end{aligned}$$

noting that this infinite sum converges as a formal power series in $q-1$, since the k-th summand is divisible by $(q-1)^k$ in $\mathbb{Q}[[q-1]]$ by Lemma 9.12 and Proposition G.11. Therefore,

$$\tau^{SO(3)}(M) = \text{f-lim}\left(\frac{f}{p}\right)\tau_p^{SO(3)}(M)$$
$$= -q^{(3-f)/4}\sum_{k=0}^{\infty}\left(q^{-1/f}\widehat{g_k(m-\frac{1}{f})(q^f)} - \widehat{g_k(m)}(q^f)\right)Q^{sl_2;a^k}(K_0).$$

As mentioned in (9.14) the existence of this Fermat limit implies the existence of $\tau^{SO(3)}(M)$, *i.e.*, the fact that Theorem 9.8 holds. □

9.3 A relation between perturbative invariants of knots and homology 3-spheres

In this section we show that the perturbative $SO(3)$ invariant of a homology 3-sphere is derived from the perturbative sl_2 invariant of a knot via a Gaussian integral. This construction of the perturbative $SO(3)$ invariant is due to Rozansky [Roz96, Roz98b].

A *Gaussian integral* is an integral of the exponential of a quadratic function over Euclidean space. The simplest Gaussian integral, which we denote by $\mathcal{G}(e^{-1})$ here, is given by

$$\mathcal{G}(e^{-1}) = \int_{x\in\mathbb{R}} e^{-x^2}dx = \sqrt{\pi}.$$

Let h be a complex number whose real part is negative. Then, we obtain the following integral, denoted by $\mathcal{G}(e^h)$, from the above formula by replacing x with $\sqrt{-h}x$,

$$\mathcal{G}(e^h) = \int_{x\in\mathbb{R}} e^{hx^2}dx = \frac{\sqrt{\pi}}{\sqrt{-h}}, \tag{9.25}$$

where we choose the square root of $-h$ whose real part is positive. This integral absolutely converges. If the real part of h was 0, the integral would still converge, though it would not do so absolutely. By putting $q = e^h$ we rewrite (9.25) as

$$\mathcal{G}(q) = \int_{x\in\mathbb{R}} q^{x^2}dx = \frac{\sqrt{\pi}}{\sqrt{-\log q}}.$$

By differentiating (9.25) l times with respect to h we obtain

$$\int_{x\in\mathbb{R}} x^{2l}e^{hx^2}dx = \sqrt{\pi}\left(\frac{d}{dh}\right)^l(-h)^{-1/2} = \frac{(2l-1)!!}{(-2h)^l}\frac{\sqrt{\pi}}{\sqrt{-h}},$$

noting that the integral for each l absolutely converges if the real part of h is negative. We also note that, if the real part of h was 0, then the integral would not converge for positive l. We call the above integral a *weighted Gaussian integral.* By the above formula,

$$\mathcal{G}(q)^{-1} \int_{x \in \mathbb{R}} x^l q^{x^2} dx = \begin{cases} \dfrac{(l-1)!!}{(-2\log q)^{l/2}} & \text{if } l \text{ is even,} \\ 0 & \text{if } l \text{ is odd,} \end{cases}$$

where we obtain the equality for odd l, since $x^l q^{x^2}$ is an odd function in this case.

Recall that for a given polynomial P we introduced a function $\hat{P}$ in (9.15). By comparing the above formula with (9.16), we obtain

$$\mathcal{G}(q)^{-1} \int_{x \in \mathbb{R}} P(x) q^{x^2} dx = \hat{P}(q).$$

Further, by replacing q of the integral with q^f, we have that

$$f^{1/2} \mathcal{G}(q)^{-1} \int_{x \in \mathbb{R}} P(x) q^{f x^2} dx = \hat{P}(q^f), \tag{9.26}$$

since

$$\mathcal{G}(q^f) = \int_{x \in \mathbb{R}} q^{f x^2} dx = f^{-1/2} \int_{x' \in \mathbb{R}} q^{x'^2} dx' = f^{-1/2} \mathcal{G}(q)$$

by putting $x' = f^{1/2} x$.

In what follows of this section, we regard a weighted Gaussian integral as, not necessarily a convergent integral in the usual sense, but the correspondence of a polynomial $P(x)$ to a function $\hat{P}(q)$ in the sense given by the above formulae. Further, we re-define a weighted Gaussian integral for an indeterminate q by the above formula; thus, in what follows of this section, we need not care about the convergence of weighted Gaussian integrals.

Since affine translation of the variable x still make sense for the re-defined weighted Gaussian integral, we obtain the following lemma.

Lemma 9.13. Let $g(m, q^{m/2})$ be a polynomial in m and $q^{m/2}$. Then,

$$\text{f-lim} \left(\frac{f}{p} \right) \mathbb{G}(\zeta)^{-1} \sum_{m=0}^{p-1} \zeta^{f m^2/4} g(m, \zeta^{m/2}) = f^{1/2} \mathcal{G}(q)^{-1} \int_{x \in \mathbb{R}} q^{f x^2/4} g(x, q^{x/2}) dx.$$

Proof. The polynomial $g(x, q^{x/2})$ is equal to a linear sum of $q^{kx/2} x^l$; so, for simplicity, we let $g(x, q^{x/2})$ be $q^{kx/2} x^l$. Then, the power of q in the integral is computed as

$$\frac{f}{4} x^2 + \frac{k}{2} x = \frac{f}{4} (x + \frac{k}{f})^2 - \frac{k^2}{4f}.$$

By replacing x with $x - \frac{k}{f}$ the right hand side of the required formula is computed by (9.26) as

$$q^{-k^2/4f}\widehat{(x-\frac{k}{f})^l}.$$

In a similar way the left hand side becomes equal to the above formula by (9.17). Hence, we obtain the lemma. □

The following theorem describes the relation between the perturbative $SO(3)$ invariant of homology 3-spheres obtained from S^3 by integral surgery along a knot and the perturbative sl_2 invariant of that knot, by means of a Gaussian integral.

Theorem 9.14 (see **[Roz96, Roz98b]**). Let K_0 be a framed knot with 0 framing and let M be the 3-manifold obtained from S^3 by integral surgery along K_0 with an additional framing $f > 0$. Then,

$$\tau^{SO(3)}(M) = \frac{f^{1/2}}{c'_+} \int_{n \in \mathbb{R}} q^{f(n^2-1)/4} [n] P^{sl_2;V_n}(K_0) dn,$$

where c'_+ is the constant given by

$$c'_+ = \mathcal{G}(q)^{-1} \int_{n \in \mathbb{R}} q^{(n^2-1)/4} [n]^2 dn = \frac{-4q^{3/4}}{q^{1/2} - q^{-1/2}}.$$

Here, we regard $P^{sl_2;V_n}(K_0)$ as a power series in $q^{1/2} - q^{-1/2}$ with coefficients in $\mathbb{Q}[q^{\pm 1/2}, q^{\pm n/2}, n]$, as in Section 9.1.

We can regard $P^{sl_2;V_n}(K_0)$ in any of the senses of Theorems 9.2 and 9.5; in any such sense it is the same function of n. In fact, (in the proof of the theorem) we regard $P^{sl_2;V_n}(K_0)$ as a power series in $q^{1/2} - q^{-1/2}$ with coefficients in $\mathbb{Q}[q^{\pm 1/2}, q^{\pm n/2}, n]$.

The theorem gives an alternative definition of the *perturbative* $SO(3)$ *invariant* of a rational homology 3-sphere M. In the original definition of the perturbative $SO(3)$ invariant based on Theorem 9.8 we used the integrality of the coefficients of the quantum $SO(3)$ invariant (Theorem 9.6). On the other hand, it is not necessary to care about the integrality in the above alternative definition of the perturbative $SO(3)$ invariant, since it was a Gaussian integral instead of a Gaussian sum.

Proof of Theorem 9.14. In the proof of Theorem 9.8 we had that

$$\tau_p^{SO(3)}(M) = c_+^{-1} \sum_{n,k} \zeta^{f(n^2-1)/4} [n] \Big(g_k(\frac{n}{2}) \zeta^{n/2} - g_k(-\frac{n}{2}) \zeta^{-n/2} \Big) Q^{sl_2;a^k}(K_0), \quad (9.27)$$

where c_+ is given by

$$c_+ = \frac{-\zeta^{-3/4}}{\zeta^{1/2} - \zeta^{-1/2}} \mathbb{G}(\zeta).$$

In general, for a polynomial $g(n, q^{n/2})$ satisfying $g(-n, q^{-n/2}) = -g(n, q^{n/2})$ (for example, a factor in (9.27)), we have that

$$
\begin{aligned}
&\text{f-lim}\left(\frac{f}{p}\right)\mathbb{G}(\zeta)^{-1} \sum_{\substack{1\le n<p \\ n \text{ is odd}}} \zeta^{f(n^2-1)/4}[n]g(n,\zeta^{n/2}) \\
&= \frac{1}{2}\cdot \text{f-lim}\left(\frac{f}{p}\right)\mathbb{G}(\zeta)^{-1} \sum_{\substack{-p<n\le p \\ n \text{ is odd}}} \zeta^{f(n^2-1)/4}[n]g(n,\zeta^{n/2}) \\
&= \frac{1}{2}\cdot \text{f-lim}\left(\frac{f}{p}\right)\mathbb{G}(\zeta)^{-1} \sum_{m=0}^{p-1} \zeta^{f(m^2+m)}[2m+1]g(2m+1,\zeta^{m+1/2}) \\
&= \frac{f^{1/2}}{2}\cdot \mathcal{G}(q)^{-1} \int_{x\in\mathbb{R}} q^{f(x^2+x)}[2x+1]g(2x+1,q^{x+1/2})dx \\
&= \frac{f^{1/2}}{4}\cdot \mathcal{G}(q)^{-1} \int_{n\in\mathbb{R}} q^{f(n^2-1)/4}[n]g(n,q^{n/2})dn,
\end{aligned}
$$

where we obtain the third equality by Lemma 9.13. By applying the above formula to (9.27) we have that

$$
\begin{aligned}
&\tau^{SO(3)}(M) = \text{f-lim}\left(\frac{f}{p}\right)\tau_p^{SO(3)}(M) \\
&= \sum_{k=0}^{\infty} Q^{sl_2;a^k}(K_0)\cdot \text{f-lim}\left(\frac{f}{p}\right)c_+^{-1} \sum_{\substack{1\le n<p \\ n \text{ is odd}}} \zeta^{f(n^2-1)/4}[n]\Big(g_k(\frac{n}{2})\zeta^{n/2} - g_k(-\frac{n}{2})\zeta^{-n/2}\Big) \\
&= \sum_{k=0}^{\infty} Q^{sl_2;a^k}(K_0)\cdot \frac{f^{1/2}}{c'_+} \int_{n\in\mathbb{R}} q^{f(n^2-1)/4}[n]\Big(g_k(\frac{n}{2})q^{n/2} - g_k(-\frac{n}{2})q^{-n/2}\Big)dn \\
&= \sum_{k=0}^{\infty} Q^{sl_2;a^k}(K_0)\cdot \frac{f^{1/2}}{c'_+} \int_{n\in\mathbb{R}} q^{f(n^2-1)/4}[n]c_k(n)dn, \qquad (9.28)
\end{aligned}
$$

where we obtain the last equality from the definition of g_k. Further, from the definition of $c_k(n)$,

$$Q^{sl_2;V_n}(K_0) = \sum_{k\ge 0} c_k(n) Q^{sl_2;a^k}(K_0).$$

Since $c_k(n)$'s can be regarded as functions of n, regarding n as a variable, we can present the perturbative sl_2 invariant $P^{sl_2;V_n}(K_0)$ by

$$P^{sl_2;V_n}(K_0) = \sum_{k\ge 0} c_k(n) Q^{sl_2;a^k}(K_0). \qquad (9.29)$$

We show that this infinite sum converges as a power series in $q^{1/2} - q^{-1/2}$ as follows.

We have that

$$c_k(n)\big|_{q=1} = \binom{k+\frac{n+1}{2}}{2k+1} + \binom{k+\frac{n-1}{2}}{2k+1},$$

since the right hand side satisfy the characterization (9.20) of $c_k(n)$ putting $q = 1$. Hence, $c_k(n)\big|_{q=e^h}$ belongs to $\mathbb{Q}[n][[h]]$ for each k. Therefore, the above sum converges as a power series in $q^{1/2} - q^{-1/2}$ by Proposition G.11. Substituting (9.29) into (9.28) we obtain the required formula. □

Chapter 10
The LMO invariant

The formal perturbative expansion of the path integral in the Chern-Simons field theory (see Appendix F.3) gives an infinite sum indexed by trivalent graphs Γ such that each summand is a product of two contributions; one contribution depends on Γ and the 3-manifold M, and the other depends on Γ and a Lie algebra $\mathfrak{g}$. Furthermore, the factor which depends on Γ and $\mathfrak{g}$ is precisely the result of the weight system which substitutes $\mathfrak{g}$ into Γ. This formal perturbative expansion is the physical background of the perturbative invariants. Hence, the above infinite sum over trivalent graphs suggests the existence of a universal perturbative invariant of 3-manifolds whose values are infinite linear sums of trivalent graphs, with the property that the various perturbative invariants of 3-manifolds are recovered from it by composing with the appropriate weight systems.

In the previous chapter we rigorously constructed perturbative invariants by means of arithmetic expansions of quantum invariants of 3-manifolds. Further, as mentioned in Chapter 8, quantum invariants of 3-manifolds are expressed as linear sums of quantum invariants of links. Furthermore, the Kontsevich invariant dominates the quantum invariants of links, as shown in Chapter 6. In this chapter, motivated by these relationships, we construct a topological invariant of 3-manifolds from the Kontsevich invariant, called the LMO invariant [LMO98], and show the universality of the LMO invariant among perturbative invariants.

In order to construct a topological invariant of 3-manifolds from the Kontsevich invariant we must first consider how the Kontsevich invariant of links changes under the KII move among links; this is described in Section 10.1. Further, in Section 10.2 we construct a sequence of linear maps ι_n such that the image of (a certain modification of) the Kontsevich invariant under ι_n is invariant under the KII move. With the maps ι_n in hand, we proceed to construct the LMO invariant of 3-manifolds. In Section 10.3 we show the universality of the LMO invariant among perturbative invariants. In Section 10.4 we explain that a formal Gaussian integral, called the Aarhus integral [BGRT*97a*], gives a reconstruction of the LMO invariant.

10.1 Properties of the framed Kontsevich invariant

In this section we show properties of the framed Kontsevich invariant $\check{Z}(L)$ of a framed link L. Proposition 10.1, below, describes a property which is important for our construction of topological invariants of 3-manifolds; namely, that proposition describes the effect on the framed Kontsevich invariant of a KII move (called a handle slide move in [Kirb78]).

Let L be a framed link with l components. We define $\check{Z}(L)$ to be $\hat{Z}(L)\sharp\nu^{\otimes l}$, which is obtained by connect-summing a copy of the element ν into each component of L. We consider how $\check{Z}(L)$ changes under the KII move on a link L. Observe that the KII move on links can be described as in Figure 10.1, where we have divided the diagram into two regions with a box; the part in the box is changed by replacing the two vertical strands by the shown tangle, and the part outside the box changes by simply taking the 2-parallel of the involved component along its framing.

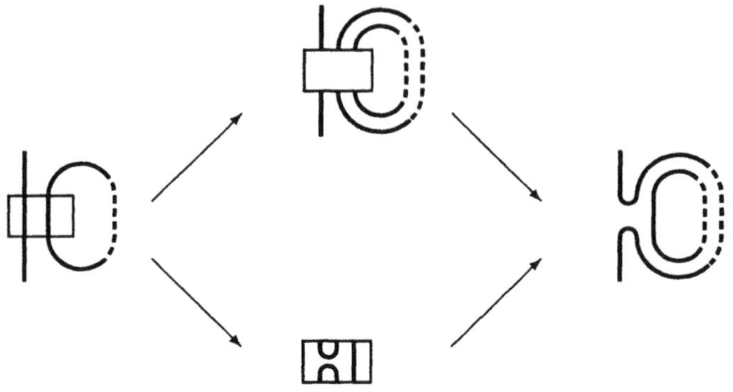

Figure 10.1 The KII move (the handle slide move) on links

Proposition 10.1 ([LeMu97, Theorem 7.3], [LMMO99]). Let L be a framed link, and let L' be a framed link obtained from L by a KII move. Then, $\check{Z}(L)$ and $\check{Z}(L')$ are related by the move shown in Figure 10.2.

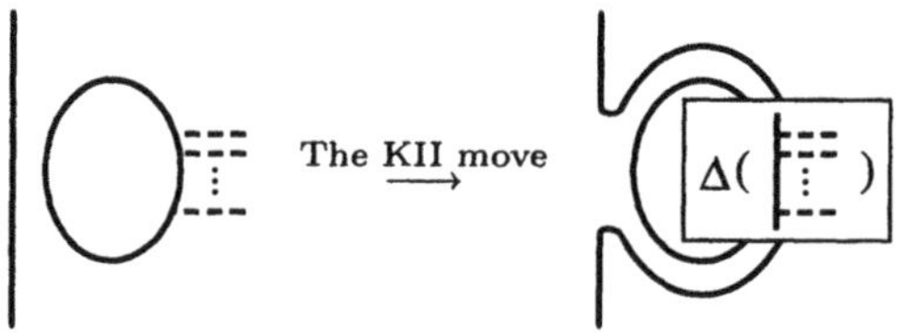

Figure 10.2 The KII move on Jacobi diagrams

Proof. The calculations $\hat{Z}(L)$ and $\hat{Z}(L')$, where L is the left and L' is the right link in Figure 10.1, are related as follows. The contribution to $\hat{Z}(L')$ from the part of L' inside the box is described by Lemma 10.2, below. The contribution to $\hat{Z}(L')$ from the part of L' outside the box is obtained from the corresponding contribution to $\hat{Z}(L)$ by an application of the comultiplication Δ (by Proposition 6.8 (1)). Hence, we have that

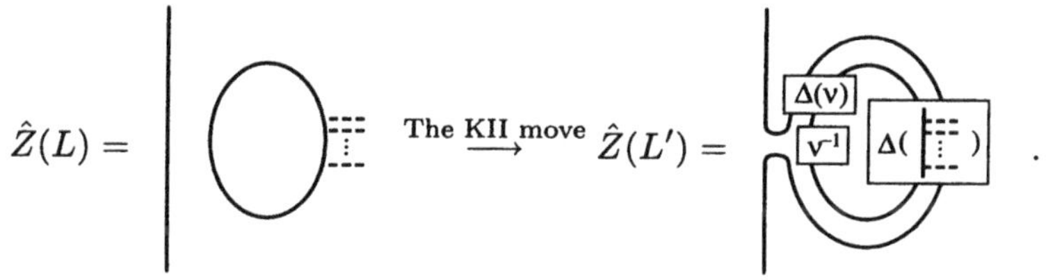

By applying the above change to the following $\hat{Z}(L)$, we have that

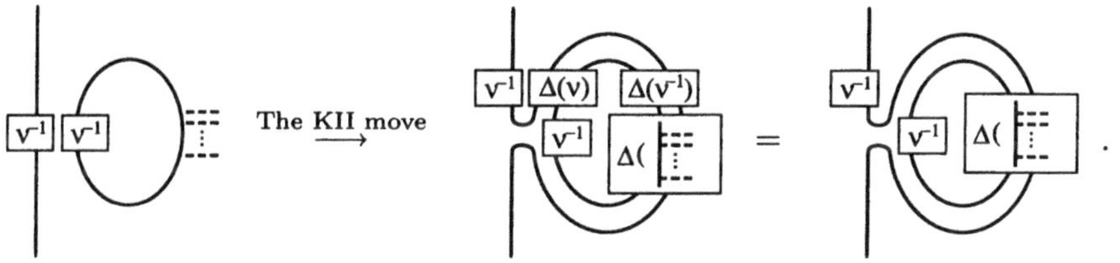

By taking a connected sum of a copy of the element ν into each component on each side of the above relation, we obtain the relation described by Figure 10.2. $\square$

Lemma 10.2. Letting $\hat{Z}$ denote the version of the framed Kontsevich invariant which uses the Drinfel'd associator, we have that

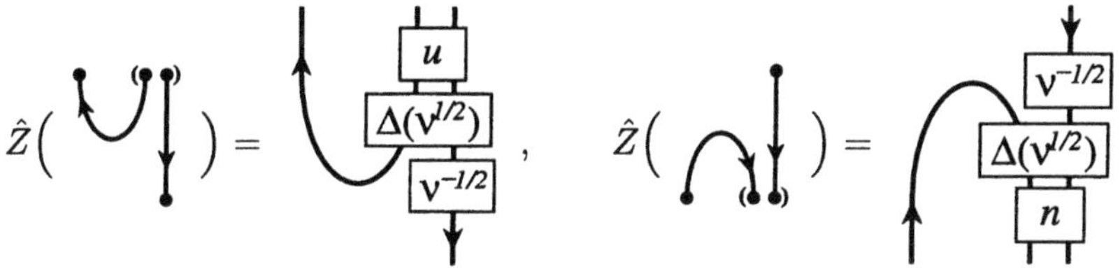

where the elements u and n were defined in Section 6.5.

Proof. We only show the first formula; the second formula can be obtained similarly. We put x by

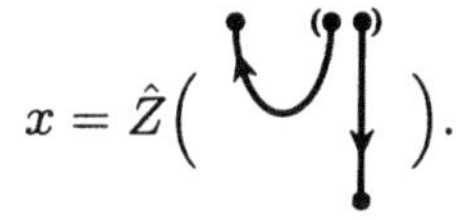

Computing the Kontsevich invariant of the following quasi-tangle in two ways, we have that

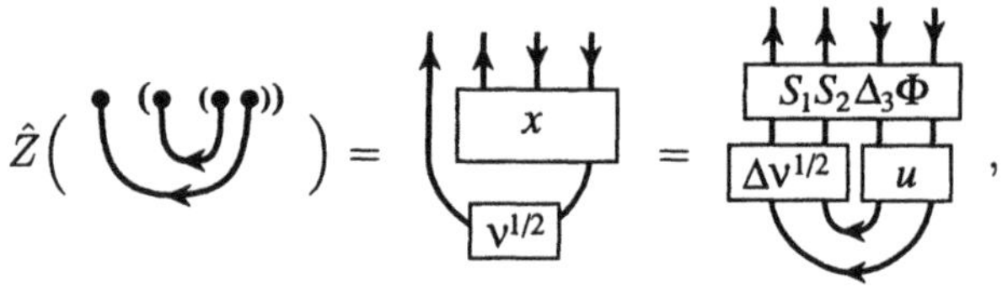

where we use (6.24) to obtain the last term. We obtain the first formula of the lemma by applying the following relation to the above formula.

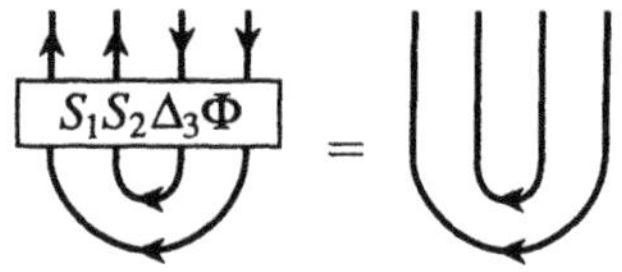

(10.1)

We show the above relation as follows. As in Section 6.4 the Drinfel'd associator Φ_{KZ} has the following expression

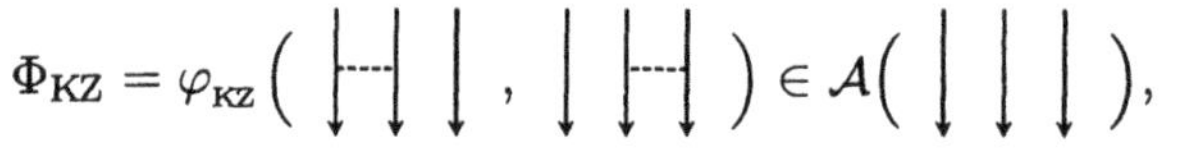

where φ_{KZ} is a formal power series in two non-commuting variables, such that $\varphi_{KZ}(A, B) = 1$ when we allow the variables A and B to commute, by Lemma D.10. Hence, we obtain (10.1), since

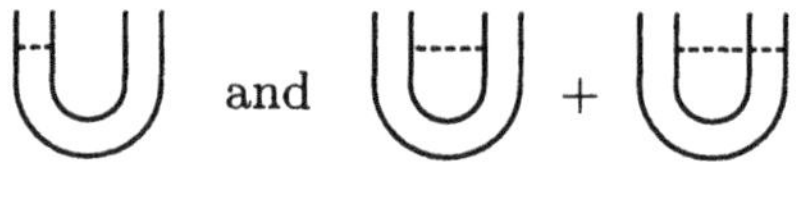

commute with each other. □

As a corollary of Proposition 6.9, we have

Proposition 10.3. The value of $\check{Z}(L)$ is group-like in $\mathcal{A}(\sqcup^l S^1)$.

Proof. By Proposition 6.9 $\hat{Z}(L)$ is group-like. In particular, when L is the trivial knot, ν ($= \hat{Z}$(the trivial knot)) is group-like. Since $\check{Z}(L)$ is defined to be the connected sum of $\hat{Z}(L)$ and ν's, $\check{Z}(L)$ is group-like, noting that $\hat{\Delta}$ is multiplicative with respect to the connected sum. □

10.2 Definition of the LMO invariant

In this section we introduce a sequence of relations P_n in the space of Jacobi diagrams. The relation P_n arises from the relationship shown in Figure 10.2 between $\check{Z}(L)$ and $\check{Z}(L')$ for L and L' two framed links related by a KII move, by assuming a certain working hypothesis (the relations $L_{<2n-2}$). As the equivalence class of $\check{Z}(L)$ under the relation P_n we obtain a topological invariant of the 3-manifold which is obtained from S^3 by integral surgery along L.

A simple case

Let us begin with the following simple case. We assume

the relations $L_{<2}$: 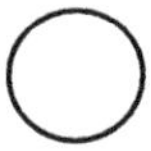 $= 0$ and  $= 0$,

as a working hypothesis. We require the equivalence of the two Jacobi diagrams shown in Figure 10.2. In particular, in the case that each of the Jacobi diagrams in Figure 10.2 has three dashed lines, we have the following relation by the relation $L_{<2}$,

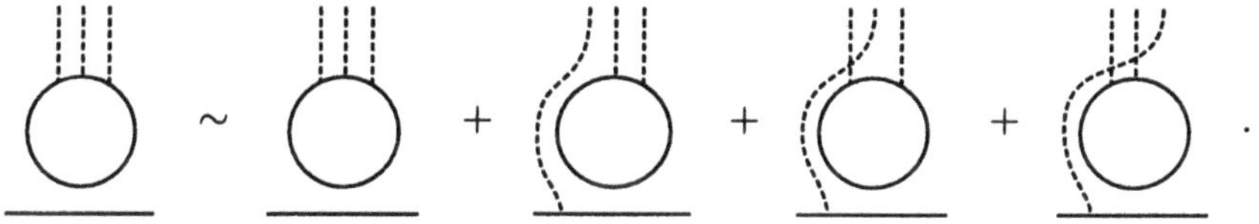

Further, temporarily assuming that 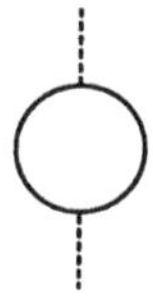 is equivalent to a constant multiple of , we obtain the following relation from the above relation:

The relation P_2 :  $\sim 0.$ (10.2)

The left hand side of the relation P_2 is the sum over disjoint unions of two dashed chords corresponding to all partitions of 4 points into 2 pairs of two points.

By closing a strand of the relation P_2, we have that

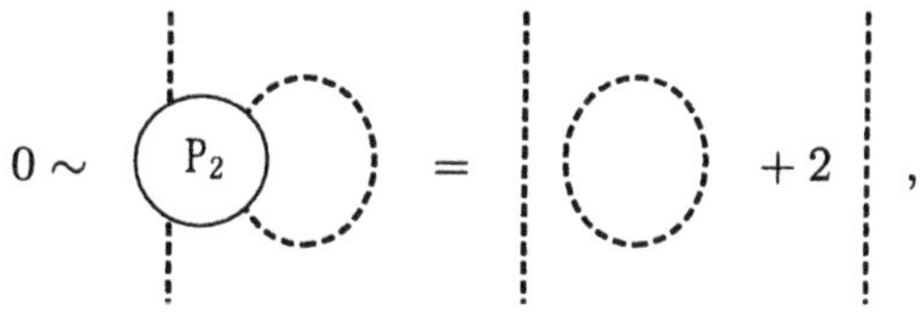

where, in the above formula, we denote the left hand side of the relation P_2 also by P_2, abusing notation. Hence, we consider the relation O_1 :  $\sim -2.$

To be precise, the relations P_2 and O_1 are relations in $\mathring{\mathcal{A}}(X)$ which is defined to be the vector space spanned by Jacobi diagrams on X allowing components of dashed loops without vertices subject to the AS, IHX and STU relations.

Later, Proposition 10.5 will show that the equivalence class $[\check{Z}(L)]$ of $\check{Z}(L)$ in the quotient space $\mathring{\mathcal{A}}(\sqcup^l S^1)/L_{<2}, P_2, O_1$ is invariant under the KII move.

Here, we continue with some further manipulations in this simple case to construct a map which removes solid circles. By the relation P_2 we have that

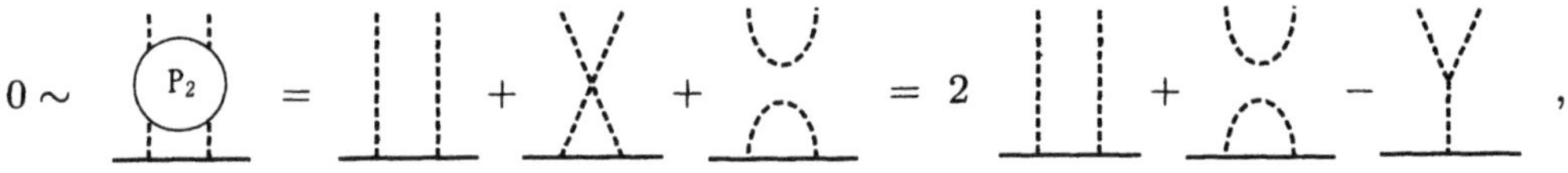

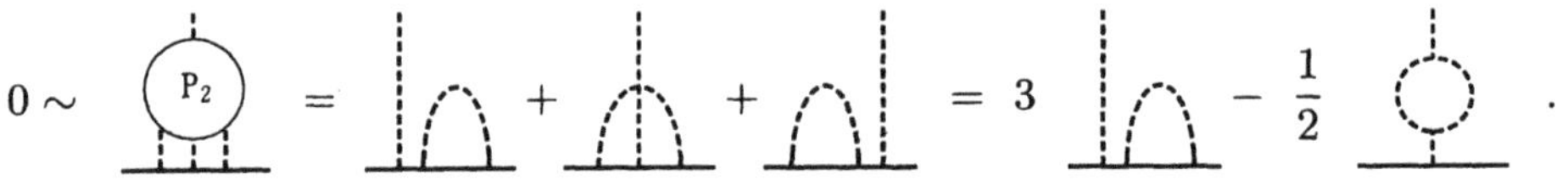

Hence,

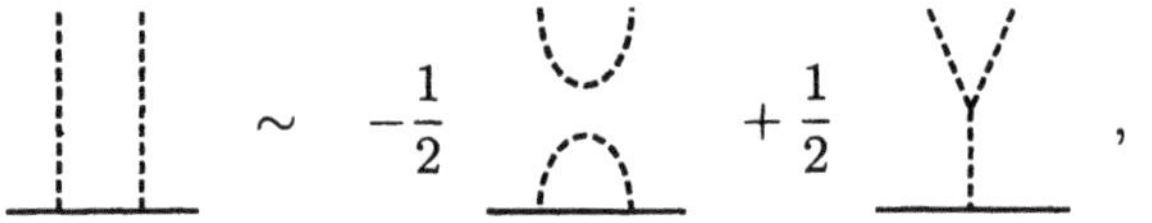 (10.3)

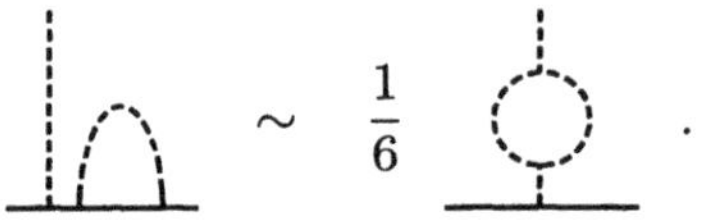 (10.4)

Using the first relation (10.3) we can replace a diagram with two adjacent legs on some solid circle by a combination of a diagram with an isolated chord in that position and a diagram with only one leg there, where by a *leg* we mean a dashed vertical line connected to a solid line. Using the second relation (10.4) we can remove an isolated chord from the solid line when there is an extra leg near the isolated chord. Hence, using the two relations together, we can always decrease the number of legs on some solid circle of a Jacobi diagram to (at most) two. By applying such procedures to a solid circle with dashed legs, we have that

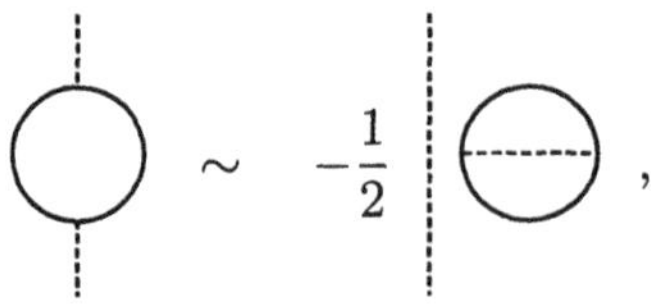

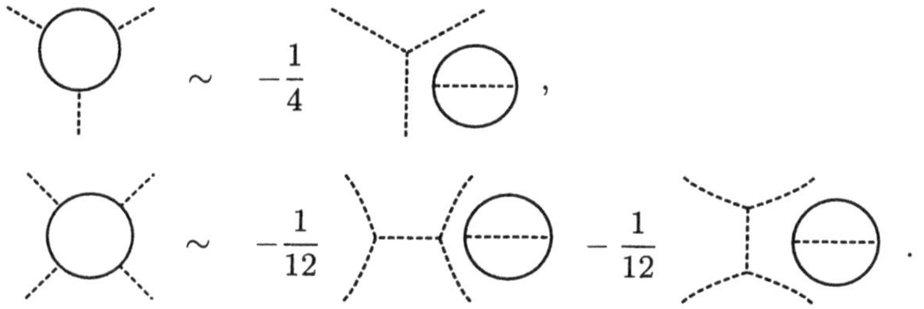

Ignoring the theta component, we consider the map given by

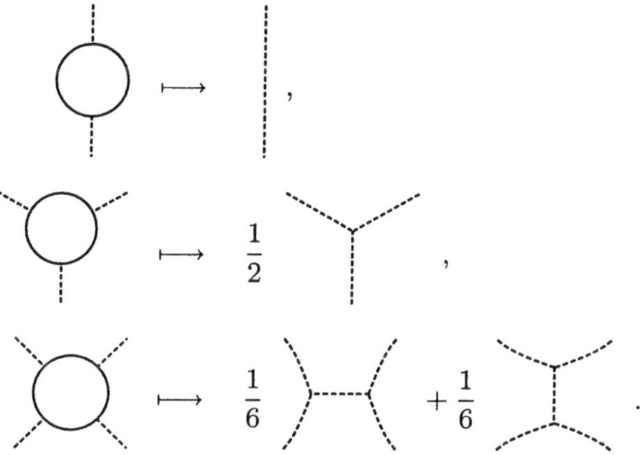

In what follows we generalize the above arguments using the relation P_2, which is the simplest case, to arguments using a relation P_n, defined below.

The relation P_n

In the formula (10.2) we defined the relation P_2 whose left hand side is the sum of all partitions of 4 points into 2 pairs. As a generalization we define the relation P_n which is defined to set to zero the sum of Jacobi diagrams which are identical except for a part, where they are disjoint unions of n dashed arcs joining $2n$ fixed points; see Figure 10.3. The relation P_n is an equivalence relation in the space $\mathring{\mathcal{A}}(X)$.

As a generalization of the procedure which uses (10.3) and (10.4), we have

Lemma 10.4 ([LMO98, Lemma 3.1]). *Let C be a component of X and let n be a positive integer. Then, a Jacobi diagram on X is equivalent modulo the relation P_{n+1} to a linear sum of Jacobi diagrams each of which has at most $2n$ univalent vertices on C.*

Proof. Let D be a Jacobi diagram on X with l legs (*i.e.*, l univalent vertices) on C. We prove the lemma by induction on l. It is sufficient to show that, if $l > 2n$, then D is equivalent modulo P_{n+1} to a linear sum of Jacobi diagrams each of which has at most $l-1$ legs on C.

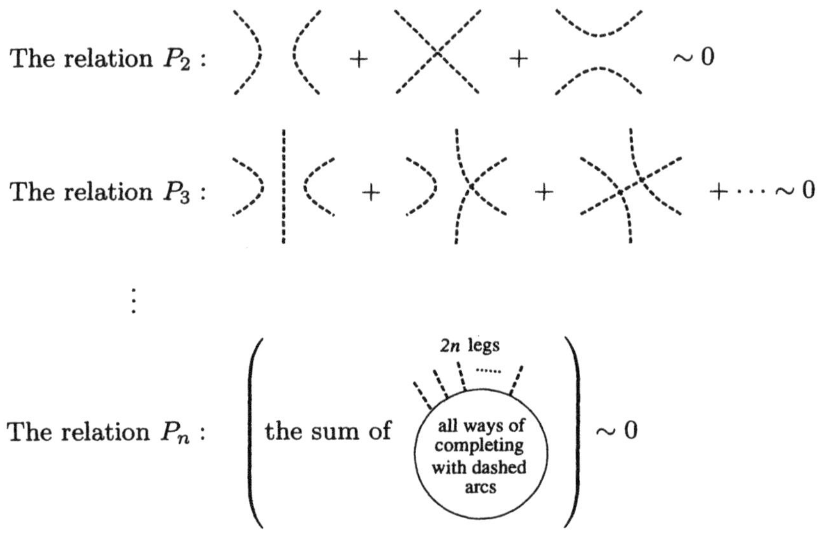

Figure 10.3 The relations P_n

When $n = 1$, the lemma follows from (10.3) and (10.4).

When $n = 2$, we can decrease the number of legs by the relation P_3 as follows. We have that

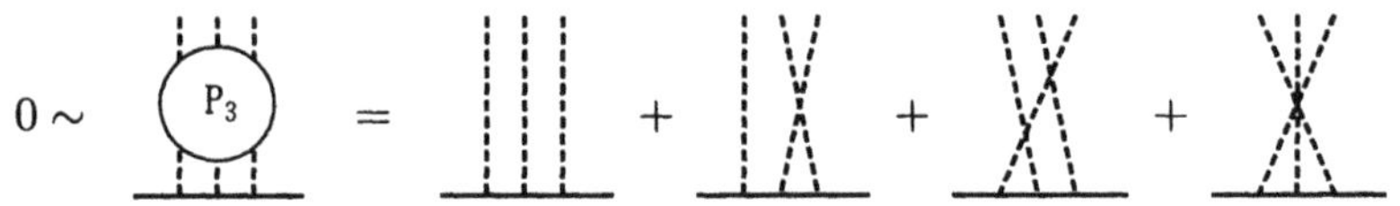

+ (the other 11 terms)

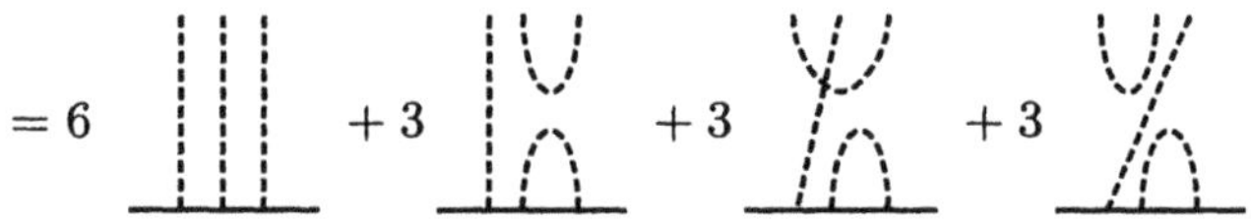

+ (terms with 2 legs in this part of the solid line)

Hence, we can express a term with (at least) 3 legs on some solid circle as a linear sum of terms with isolated chords, modulo terms with fewer legs than the original diagram. Applying the procedure again we have a term with two isolated chords modulo terms with fewer legs. Further, we can reduce a term with two isolated chords to terms with fewer legs by the following relation derived from P_3,

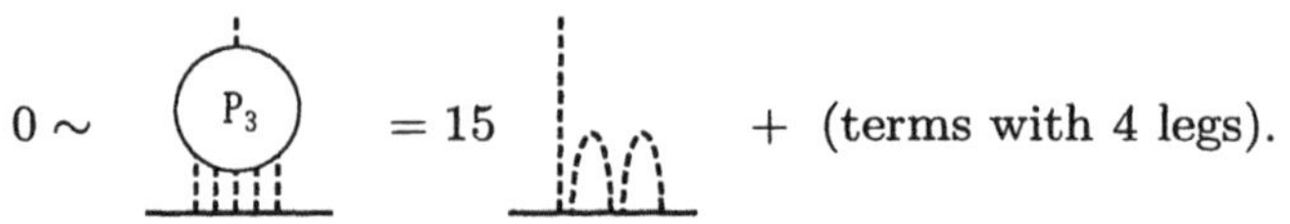

Hence, if D has at least 5 legs, we can decrease the number of legs of D by the relation P_3.

When $n \geq 3$, we can decrease the number of legs by P_{n+1} similarly as above. That is, by using the relation P_{n+1} we can replace l legs of a Jacobi diagram D with Jacobi diagrams with isolated chords modulo Jacobi diagrams with fewer legs. Repeating the procedure we obtain a Jacobi diagram with sufficiently many isolated chords. Further, we use the relation P_{n+1} again to reduce the Jacobi diagram with many isolated chords to Jacobi diagrams with fewer legs than the original Jacobi diagram D. This completes the proof; for detailed computations see [LMO98]. □

By the above lemma, we have the following proposition.

Proposition 10.5 ([LMO98, Proposition 3.1]). Let L be any oriented framed link with l components. Then, the equivalence class $[\check{Z}(L)]$ of $\check{Z}(L)$ in the quotient space $\overset{\circ}{\mathcal{A}}(\sqcup^l S^1)/L_{<2n}, P_{n+1}$ is invariant under KII moves, where we denote by $L_{<2n}$ the relation which sets to zero a Jacobi diagram containing a solid circle with less than $2n$ univalent vertices sitting on it.

Proof. By Proposition 10.1 it is sufficient to show that the two Jacobi diagrams in Figure 10.2 are equivalent modulo the relations $L_{<2n}$ and P_{n+1}. Let D and D' be the two Jacobi diagrams in Figure 10.2 respectively and let m be the number of legs (*i.e.*, dashed lines) of each of the diagrams.

If $m < 2n$, then both of D and D' vanish by $L_{<2n}$. Hence, they are equivalent modulo $L_{<2n}$.

If $m = 2n$, then D' is a sum of D and Jacobi diagrams with less than $2n$ legs on the solid circle. Since such Jacobi diagrams vanish by $L_{<2n}$, the Jacobi diagrams D and D' are equivalent modulo $L_{<2n}$.

If $m > 2n$, we can decrease the number of legs of D by Lemma 10.4 using the relation P_{n+1} as

$D =$ 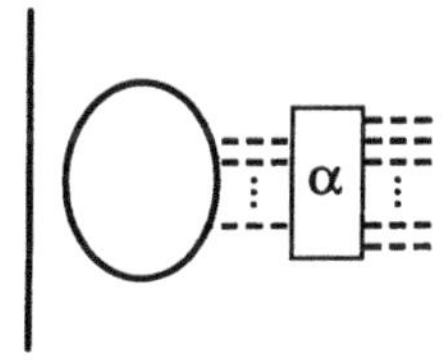 , (10.5)

where α is a linear sum of dashed uni-trivalent graphs such that the left side of α has $2n$ dashed lines. Recall that the decrease of legs that occurs in Lemma 10.4 is derived from the relation P_{n+1} and the STU relation. Since the comultiplication Δ

commutes with each of these relations, we have that

$D' =$ 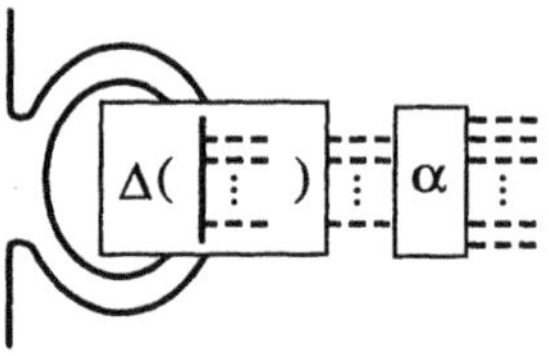 . (10.6)

Further, the left hand sides of (10.5) and (10.6) are equivalent modulo $L_{<2n}$ by the same argument as in the above case of $m = 2n$. Hence, D and D' are equivalent modulo P_{n+1} and $L_{<2n}$. □

Recall that the degree of a Jacobi diagram is half the number of univalent and trivalent vertices in the Jacobi diagram. Let $D_{>n}$ denote the relation in $\mathcal{A}(X)$ and $\mathring{\mathcal{A}}(X)$ which sets terms of degree $> n$ to 0. Further, we define the relation O_n in the space $\mathring{\mathcal{A}}(X)$ by

the relation O_n :  $\sim -2n.$

Then, we have

Lemma 10.6. The following map induced by the identity map of Jacobi diagrams is an isomorphism,

$$\mathcal{A}(\emptyset)/D_{>n} \longrightarrow \mathring{\mathcal{A}}(\emptyset)/D_{>n}, P_{n+1}, O_n.$$

Proof. Since a dashed loop can be removed by the relation O_n, we have a natural isomorphism between $\mathring{\mathcal{A}}(\emptyset)/D_{>n}, O_n$ and $\mathcal{A}(\emptyset)/D_{>n}$. Hence, it is sufficient to show that $\mathring{\mathcal{A}}(\emptyset)/D_{>n}, O_n$ is isomorphic to $\mathring{\mathcal{A}}(\emptyset)/D_{>n}, P_{n+1}, O_n$. We show that an element of $\mathring{\mathcal{A}}(\emptyset)$ including P_{n+1} vanishes by the relations $D_{>n}$ and O_n.

If there exists a chord without trivalent vertices, both of whose ends connect to P_{n+1}, then

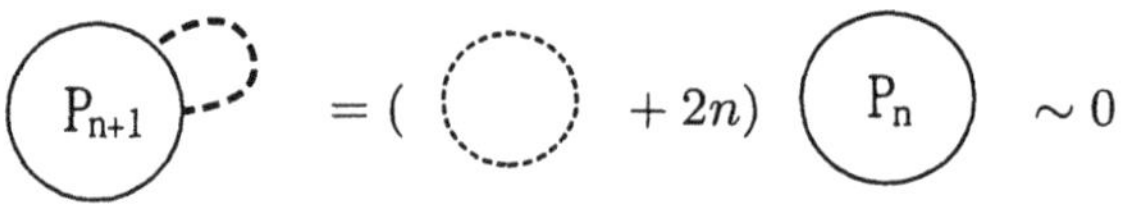 (10.7)

by the relation O_n.

If there exists a dashed arc, both of whose ends connect to P_{n+1}, and with an additional trivalent vertex on it, then

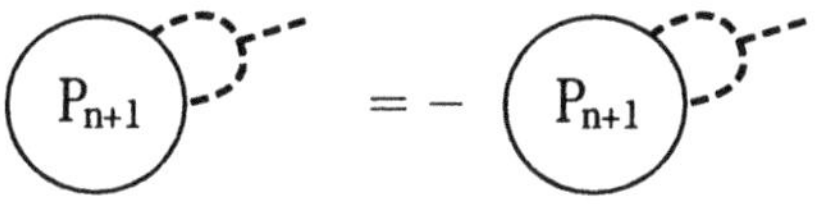

by the AS relation and the interior symmetry of P_{n+1}. Hence, such an element including P_{n+1} also vanishes.

Otherwise, we show that the degree of a Jacobi diagram including P_{n+1} is greater than n by a counting argument as follows. To each end point of P_{n+1}, we associate the trivalent vertex that is nearest to it. Corresponding to the $2n+2$ end points of P_{n+1} we find $2n+2$ associated trivalent vertices in the Jacobi diagram. Therefore, the Jacobi diagram has degree $> n$, and it vanishes by the relation $D_{>n}$. □

Replacing solid circles with dashed graphs

In this subsection we construct a sequence of maps $\iota_n : \mathcal{A}(\sqcup^l S^1) \to \mathcal{A}(\emptyset)$ for $n = 1, 2, \cdots$, which replace solid circles with dashed graphs.

We define a Jacobi diagram on m ordered points labeled by 0, 1, 2, $\cdots$, $m-1$ to be a vertex-oriented uni-trivalent dashed graph whose m univalent vertices lie on the m points, one to each. We denote by $\mathcal{A}(m)$ the vector space over $\mathbb{C}$ spanned by Jacobi diagrams on the m ordered points subject to the AS and IHX relations. In particular, we put $\mathcal{A}(0) = \mathcal{A}(\emptyset)$.

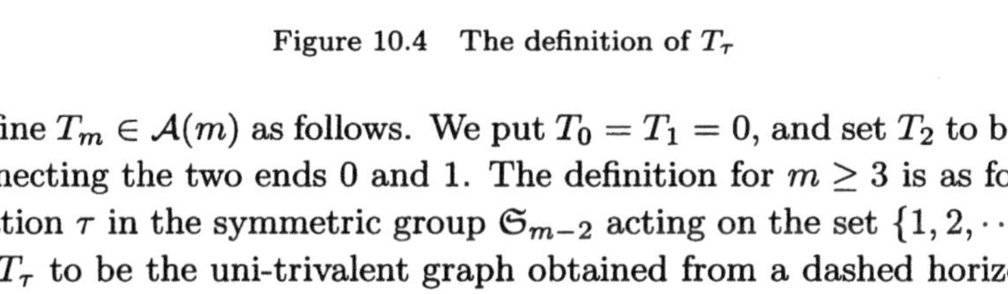

Figure 10.4 The definition of T_τ

We define $T_m \in \mathcal{A}(m)$ as follows. We put $T_0 = T_1 = 0$, and set T_2 to be a dashed chord connecting the two ends 0 and 1. The definition for $m \geq 3$ is as follows. For a permutation τ in the symmetric group $\mathfrak{S}_{m-2}$ acting on the set $\{1, 2, \cdots, m-2\}$, we define T_τ to be the uni-trivalent graph obtained from a dashed horizontal edge lying between the points 0 and $m-1$ by adding $m-2$ chords corresponding to the arrows of the permutation τ, as in Figure 10.4. Further, we define $T_m \in \mathcal{A}(m)$ by

$$T_m = \sum_{\tau \in \mathfrak{S}_{m-2}} \frac{(-1)^{r(\tau)}}{(m-1)\binom{m-2}{r(\tau)}} T_\tau, \tag{10.8}$$

where we denote by $r(\tau)$ the number of k satisfying $\tau(k) > \tau(k+1)$, and $\binom{m-2}{r(\tau)}$ the binomial coefficient. For example,

$$T_0 = 0,$$
$$T_1 = 0,$$

$T_2 =$,

$T_3 = \frac{1}{2}$  ,

$T_4 = \frac{1}{6}$  $+ \frac{1}{6}$  .

Note that we use the IHX relation to obtain the above T_4.

T_m has the following properties. For the proof of the proposition see [LMO98].

Proposition 10.7 ([LMO98]).
(1) The element $T_m \in \mathcal{A}(m)$ represents the action of the dihedral group such that T_m is invariant under rotations and each reflection takes T_m to $(-1)^m T_m$.
(2) The difference between T_m and the element of $\mathcal{A}(m)$ obtained by changing two adjacent univalent vertices is equal to T_{m-1} with one extra trivalent vertex, as shown in Figure 10.5.

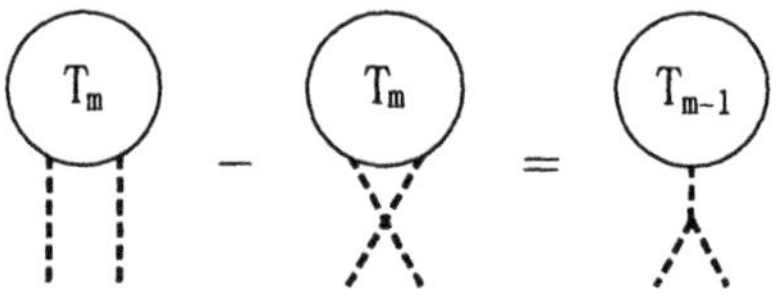

Figure 10.5 T_m satisfies a relation similar to the STU relation

The factor $(-1)^m$ in (1) of the proposition is due to the fact that reflection reverses the vertex-orientation of T_m, together with the AS relation, though T_m itself is symmetric under reflection as a linear sum of uni-trivalent graphs. The claim of (2) of the proposition is proved, for example, for $m = 4$ by a concrete

computation like

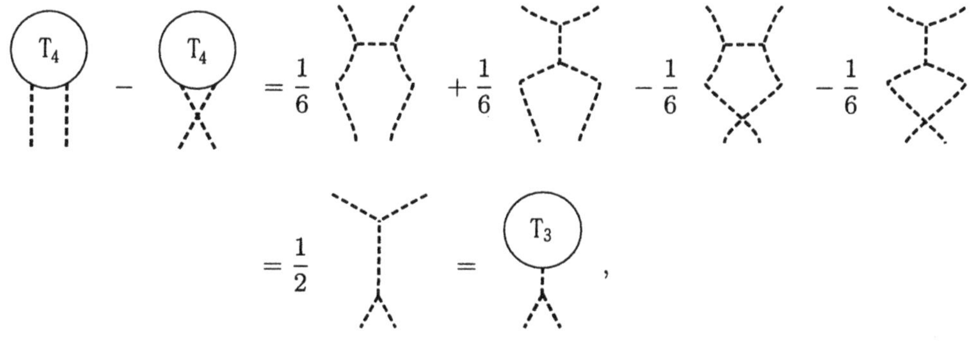

where the second equality is derived from the AS and IHX relations. For general m (2) of the proposition is proved by longer and more complicated computations; for a complete proof see [LMO98].

We define the map $j_1 : \mathcal{A}(\sqcup^l S^1) \to \mathring{\mathcal{A}}(\emptyset)$ to be the linear map which replaces a solid circle with m univalent vertices on it by gluing T_m to these vertices, as

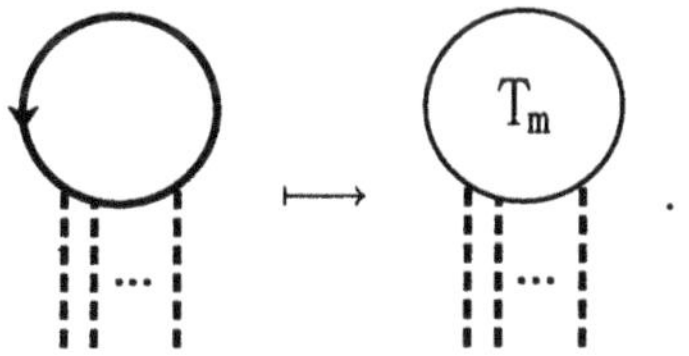

This map is well defined by Proposition 10.7. If the solid circle has the opposite orientation, we replace it by reflected T_m; it is equal to $(-1)^m T_m$ by Proposition 10.7. Further, we define the map $j_n : \mathcal{A}(\sqcup^l S^1) \to \mathring{\mathcal{A}}(\emptyset)$ as follows. We consider the map $\mathcal{A}(\sqcup^l S^1) \longrightarrow \mathcal{A}(\sqcup^{nl} S^1)$ obtained by acting with Δ on each S^1 of $\mathcal{A}(\sqcup^l S^1)$ $n-1$ times. We define j_n to be $1/n!$ times the composition of this map followed by j_1. Furthermore, by replacing each dashed loop in the image of j_n by $-2n$, we obtain a map $\mathcal{A}(\sqcup^l S^1) \to \mathcal{A}(\emptyset)$; we denote this map by ι_n.

The degree $\leq n$ part of the map ι_n is derived from the relation P_{n+1} as follows. By Lemma 10.4 a Jacobi diagram on $\sqcup^l S^1$ is equivalent modulo P_{n+1} to a linear sum of Jacobi diagrams each of which has at most $2n$ univalent vertices on each S^1. Further, by the same argument as in the proof of Lemma 10.4, we can replace each such Jacobi diagram by a combination of Jacobi diagrams with precisely n isolated chords on each solid circle modulo P_{n+1} and $L_{<2n}$ relations. Removing each solid circle with n isolated chords as in the simple case discussed above, we obtain a map $\mathcal{A}(\sqcup^l S^1) \to \mathcal{A}(\emptyset)/D_{>n}$. This map is well defined and coincides with the degree $\leq n$ part of ι_n by the following argument.

A Jacobi diagram on $\sqcup^l S^1$ which has less than $2n$ univalent vertices on a component S^1 vanishes by the map j_n because of the relation $T_0 = T_1 = 0$ and the definition of Δ. Further, if α and β are related by the relation P_{n+1}, then so are $j_n(\alpha)$ and $j_n(\beta)$. Hence, the map j_n induces the following map

$$\mathring{\mathcal{A}}(\sqcup^l S^1)/L_{<2n}, P_{n+1}, O_n \overset{j_n}{\to} \mathring{\mathcal{A}}(\emptyset)/P_{n+1}, O_n \to \mathring{\mathcal{A}}(\emptyset)/D_{>n}, P_{n+1}, O_n \overset{\cong}{\leftarrow} \mathcal{A}(\emptyset)/D_{>n},$$

where the second map is the projection, and the third map is the isomorphism given in Lemma 10.6. Further, by adding a projection $\mathcal{A}(\sqcup^l S^1) \to \mathcal{A}(\sqcup^l S^1)/L_{<2n}, P_{n+1}, O_n$ to the left of the above sequence, we obtain the map $\mathcal{A}(\sqcup^l S^1) \to \mathcal{A}(\emptyset)/D_{>n}$ which is equal to the degree $\leq n$ part of the map ι_n from the definition of ι_n. Further, it coincides with the same map obtained above by the relation P_{n+1}.

Since the equivalence class $[\check{Z}(L)]$ of $\check{Z}(L)$ under the relation P_{n+1} is invariant under the KII move on a framed link L by Proposition 10.5, we have

Proposition 10.8. The degree $\leq n$ part of $\iota_n(\check{Z}(L))$ is invariant under the KII move.

Definition and topological invariance of the LMO invariant

In this subsection we construct a sequence of topological invariants $Z_n^{\mathrm{LMO}}(M) \in \mathcal{A}(\emptyset)/D_{>n}$ of a 3-manifold M for each positive integer n. Further, we assemble the sequence of $Z_n^{\mathrm{LMO}}(M)$ into an invariant $Z^{\mathrm{LMO}}(M) \in \mathcal{A}(\emptyset)$.

We have an algebra structure in $\mathcal{A}(\emptyset)$ with the product given by the disjoint union of Jacobi diagrams. The unit of this algebra is the empty diagram; we denote it by 1.

Theorem 10.9 ([LMO98, Theorem 3.7]). Let L be a framed link, and let M be the 3-manifold obtained from S^3 by integral surgery along L. We put

$$Z_n^{\mathrm{LMO}}(M) = \big(\iota_n(\check{Z}(U_+))\big)^{-\sigma_+}\big(\iota_n(\check{Z}(U_-))\big)^{-\sigma_-}\iota_n(\check{Z}(L)) \in \mathcal{A}(\emptyset)/D_{>n}. \qquad (10.9)$$

Then, this is a topological invariant of M for any positive integer n, where $U_\pm$ denotes the trivial knot with ± 1 framing and $\sigma_\pm$ denotes the number of positive and negative eigenvalues of the linking matrix of L.

The values of the normalization constants $\iota_n(\check{Z}(U_\pm))$ are computed as follows. We have that

$\check{Z}(U_\pm) = \nu^2 \# \exp\big(\pm \frac{1}{2}$ 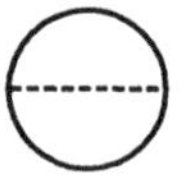 $\big)$

$=$ 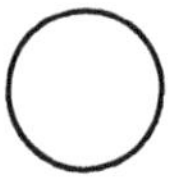 $\pm \frac{1}{2}$ 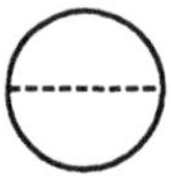 $+ \frac{1}{8}$ 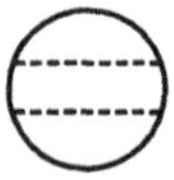 $+ \frac{1}{24}$ 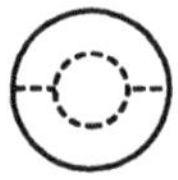 $+$ (terms of degree > 2).

Hence,

$$\iota_1\big(\check{Z}(U_\pm)\big) = \mp 1 + \frac{1}{16}$$ 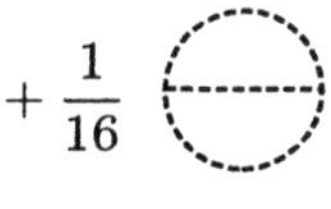 $$\in \mathcal{A}(\emptyset)/D_{>1}. \qquad (10.10)$$

Further, we obtain leading terms of the values of $\iota_n\big(\check{Z}(U_\pm)\big)$ as follows. The map $\hat{\Delta}$ induces the following map for $n = n_1 + n_2$,

$$\mathring{\mathcal{A}}(\sqcup^l S^1)/P_{n+1}, O_n, L_{<2n} \to \mathring{\mathcal{A}}(\sqcup^l S^1)/P_{n_1+1}, O_{n_1}, L_{<2n_1} \otimes \mathring{\mathcal{A}}(\sqcup^l S^1)/P_{n_2+1}, O_{n_2}, L_{<2n_2}$$

between the quotient spaces; we denote it by $\hat{\Delta}_{n_1,n_2}$. We have that

$$\hat{\Delta} \circ \iota_n = (\iota_{n_1} \otimes \iota_{n_2}) \circ \Delta_{n_1,n_2}, \qquad (10.11)$$

from the definition of ι_n. By using this formula $n-1$ times repeatedly, we have that

$$\hat{\Delta}^{(n-1)} \circ \iota_n = \iota_1^{\otimes n} \circ \hat{\Delta}^{(n-1)}, \qquad (10.12)$$

where we define $\hat{\Delta}^{(k)} : \mathcal{A}(X) \to \mathcal{A}(X)^{\otimes(k+1)}$ by the recursive formula $\hat{\Delta}^{(k)} = (\hat{\Delta}^{(k-1)} \otimes \mathrm{id}) \circ \hat{\Delta}$ and $\hat{\Delta}^{(1)} = \hat{\Delta}$. More precisely, the $\hat{\Delta}^{(n-1)}$ appearing on the right hand side is the induced map between the quotient spaces. Since $\check{Z}(U_\pm)$ is group-like, by Proposition 10.3, $\hat{\Delta}^{(n-1)}(\check{Z}(U_\pm)) = \check{Z}(U_\pm)^{\otimes n}$. Hence, by (10.12), $\hat{\Delta}^{(n-1)} \circ \iota_n\big(\check{Z}(U_\pm)\big) = \big(\iota_1 \check{Z}(U_\pm)\big)^{\otimes n}$. By using the above value (10.10) of $\big(\iota_1 \check{Z}(U_\pm)\big)$ we obtain

$$\iota_n\big(\check{Z}(U_\pm)\big) = (\mp 1)^n + \frac{(\mp 1)^{n-1}}{16}$$  $$+ \text{ (terms of degree} > 1) \in \mathcal{A}(\emptyset)/D_{>n}.$$

In particular, $\iota_n\big(\check{Z}(U_\pm)\big)$ is invertible in $\mathcal{A}(\emptyset)/D_{>n}$.

Proof of Theorem 10.9. It is sufficient to show the invariance of the right hand side of (10.9) under changes of orientation of the components of L and under the KI and KII moves.

Let L' be the link obtained from L by reversing the orientation of a component C of L. Then, by Proposition 6.8 (2), $\hat{Z}(L')$ is obtained from $\hat{Z}(L)$ by an application of the antipode $S_{(C)}$, and similarly $\check{Z}(L')$ is obtained from $\check{Z}(L)$, since $S(\nu) = \nu$. Let D be a Jacobi diagram appearing in $\check{Z}(L)$ with m univalent vertices on C. Then, the factor $(-1)^m$ arises from the definition of $S_{(C)}$. On the other hand, $(-1)^m$ also appears in the definition of the map ι_n which replaces the solid circle C with T_m and replaces C when given the opposite orientation with $(-1)^m T_m$; recall the property of T_m in Proposition 10.7. Then, the two appearances of $(-1)^m$ cancel and we obtain the invariance under the change of the orientation of L.

The invariance under the KI move is immediate, since the change of $\iota_n\big(\check{Z}(L)\big)$ under the move cancels with the change of $\sigma_\pm$.

The invariance under the KII move is derived from the invariance of the degree $\leq n$ part of $\iota_n(\check{Z}(L))$ under the KII move by Proposition 10.8. □

We now show some properties of the invariants $Z_n^{\mathrm{LMO}}(M)$, which suggest that the sequence should be assembled into a single invariant.

Proposition 10.10. For any positive integers n_1 and n_2,

$$\hat{\Delta}_{n_1,n_2}(Z_{n_1+n_2}^{\mathrm{LMO}}(M)) = Z_{n_1}^{\mathrm{LMO}}(M) \otimes Z_{n_2}^{\mathrm{LMO}}(M).$$

Proof. This formula follows from (10.11) and the property that $\check{Z}(L)$ is group-like (Proposition 10.3). □

By Proposition 10.10 we have relationships among the coefficients of the values of $Z_n^{\mathrm{LMO}}(M)$ for different n. For example, by Proposition 10.10 the values of $Z_n^{\mathrm{LMO}}(M)$ for $n \leq 3$ are presented by

$Z_1^{\mathrm{LMO}}(M) = c_0 + c_1$  ,

$Z_2^{\mathrm{LMO}}(M) = c_0^2 + c_0c_1$  $+ \frac{c_1^2}{2}$  $+ c_2$  ,

$Z_3^{\mathrm{LMO}}(M) = c_0^3 + c_0^2c_1$  $+ \frac{c_0c_1^2}{2}$  $+ c_0c_2$ 

$+ \frac{c_1^3}{6}$  $+ c_1c_2$  $+ c_3$  ,

for some scalar invariants c_i. As for the invariants c_0 and c_1 we have

Lemma 10.11 ([LMMO99]).
(1) The degree 0 part of $Z_1^{\mathrm{LMO}}(M)$, denoted by c_0 above, is equal to the order of the first homology group $H_1(M;\mathbb{Z})$ if M is a rational homology 3-sphere, 0 otherwise.
(2) The degree 1 part of $Z_1^{\mathrm{LMO}}(M)$, denoted by c_1 above, is equal to $(-1)^{b_1(M)}\lambda(M)/2$, where $\lambda(M)$ is the Casson-Walker-Lescop invariant (see [Les96] for its definition) and $b_1(M)$ denotes the first Betti number of M.

Proof. For simplicity we show the proof of (1) in the case that M is obtained from S^3 by integral surgery along a framed knot with a framing f; see [LMMO99] for the other cases and the proof of (2).

We have that

$\check{Z}(K) = \nu^2 \# \exp\big(\frac{f}{2}$ 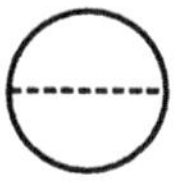 $+$ (terms of degree $> 1)\big)$

$=$ 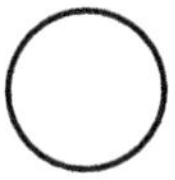 $+\frac{f}{2}$ 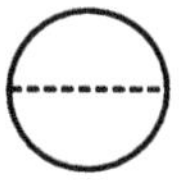 $+$ (terms of degree > 1).

Hence,

$$\iota_1\big(\check{Z}(K)\big) = -f + \text{ (a term of degree 1)} \in \mathcal{A}(\emptyset)/D_{>1}.$$

By (10.10) the leading term of the normalization constant is equal to $-\mathrm{sign}(f)$ in this case. Hence, the leading term of $Z_1^{\mathrm{LMO}}(M)$ is equal to $|f|$, which is equal to the order of $H_1(M;\mathbb{Z})$ if $f \neq 0$, and 0 if $f = 0$. □

Proposition 10.12. Let $Z_n^{\mathrm{LMO}}(M)^{(d)}$ be the degree d part of $Z_n^{\mathrm{LMO}}(M)$. Then,

$$Z_n^{\mathrm{LMO}}(M)^{(d)} = c_0^{n-d} Z_d^{\mathrm{LMO}}(M)^{(d)},$$

where c_0 is the scalar whose value is given in Lemma 10.11.

Proof. Let $\hat{\varepsilon} : \mathcal{A}(\emptyset) \to \mathbb{C}$ be the map which projects out the degree 0 part of an element of $\mathcal{A}(\emptyset)$. Since $(\hat{\varepsilon} \otimes id) \circ \hat{\Delta}_{1,n-1}$ is equal to the projection $\mathcal{A}(\emptyset)/D_{>n} \longrightarrow \mathcal{A}(\emptyset)/D_{>n-1}$, we have that

$$(\hat{\varepsilon} \otimes \mathrm{id}) \circ \hat{\Delta}_{1,n-1}(Z_n^{\mathrm{LMO}}(M)) = Z_n^{\mathrm{LMO}}(M)^{(n-1)} \in \mathcal{A}(\emptyset)/D_{>n-1}.$$

On the other hand, by Proposition 10.10,

$$\begin{aligned}(\hat{\varepsilon} \otimes \mathrm{id}) \circ \hat{\Delta}_{1,n-1}(Z_n^{\mathrm{LMO}}(M)) &= \hat{\varepsilon}(Z_1^{\mathrm{LMO}}(M)) \otimes Z_{n-1}^{\mathrm{LMO}}(M) \\ &= c_0 Z_{n-1}^{\mathrm{LMO}}(M) \in \mathcal{A}(\emptyset)/D_{>n-1}.\end{aligned}$$

Comparing the above two formulae, we obtain $c_0 Z_{n-1}^{\mathrm{LMO}}(M) = Z_n^{\mathrm{LMO}}(M)^{(\leq n-1)}$. By using this formula repeatedly, we obtain the required formula. □

By Proposition 10.12 the value of $Z_n^{\mathrm{LMO}}(M)^{(<n)}$ can be obtained from $Z_{n-1}^{\mathrm{LMO}}(M)$. Hence, the sequence of invariants $Z_n^{\mathrm{LMO}}(M)$ is determined by the sequence $Z_n^{\mathrm{LMO}}(M)^{(n)}$, together with the constant c_0.

Definition 10.13 ([LMO98]). We define an invariant $Z^{\mathrm{LMO}}(M)$ by

$$Z^{\mathrm{LMO}}(M) = 1 + \sum_{d=1}^{\infty} Z_d^{\mathrm{LMO}}(M)^{(d)} \in \mathcal{A}(\emptyset).$$

We call $Z^{\mathrm{LMO}}(M)$ *the LMO invariant* (or *the universal perturbative invariant*) of a closed oriented 3-manifold M. For a rational homology 3-sphere M, we also define $\hat{Z}^{\mathrm{LMO}}(M)$ by

$$\hat{Z}^{\mathrm{LMO}}(M) = 1 + \sum_{d=1}^{\infty} \frac{Z_d^{\mathrm{LMO}}(M)^{(d)}}{|H_1(M;\mathbb{Z})|^d} \in \mathcal{A}(\emptyset),$$

where $|H_1(M;\mathbb{Z})|$ denotes the order of $H_1(M;\mathbb{Z})$.

Using the scalar invariants c_i mentioned above, the LMO invariant $Z^{\mathrm{LMO}}(M)$ is presented by

$Z^{\mathrm{LMO}}(M) = 1 + c_1$  $+ \frac{c_1^2}{2}$  $+ c_2$ 

$+ \frac{c_1^3}{6}$  $+ c_1c_2$  $+ c_3$ 

$+$ (terms of degree > 3).

This formula let us expect the existence of the logarithm of $Z^{\mathrm{LMO}}(M)$ (see (10.13) below). In fact, the logarithm of $Z^{\mathrm{LMO}}(M)$ does exist by the following proposition.

Proposition 10.14. The value of $Z^{\mathrm{LMO}}(M)$ is group-like in $\mathcal{A}(\emptyset)$, *i.e.*,

$$\hat{\Delta}(Z^{\mathrm{LMO}}(M)) = Z^{\mathrm{LMO}}(M) \otimes Z^{\mathrm{LMO}}(M).$$

Proof. The proposition is an immediate consequence of Proposition 10.10. □

The vector space $\mathcal{A}(\emptyset)$ forms an algebra with the product given by the disjoint union of Jacobi diagrams. Further, the algebra has a Hopf algebra structure with the comultiplication $\hat{\Delta}$. As a property of a Hopf algebra (see, *e.g.*, [Abe80]), any group-like element can be presented by the exponential of some primitive element. In this case it follows from the definition of $\hat{\Delta}$ that a *primitive* element in $\mathcal{A}(\emptyset)$, *i.e.*, an element satisfying that $\hat{\Delta}(\alpha) = \alpha \otimes 1 + 1 \otimes \alpha$, is a linear sum of connected Jacobi diagrams. Hence, by Proposition 10.14 the logarithm of the LMO invariant $Z^{\mathrm{LMO}}(M)$ lies in $\mathcal{A}(\emptyset)_{\mathrm{conn}}$, where $\mathcal{A}(\emptyset)_{\mathrm{conn}}$ denotes the vector subspace of $\mathcal{A}(\emptyset)$ spanned by the connected Jacobi diagrams. We denote the logarithm by $z^{\mathrm{LMO}}(M) \in \mathcal{A}(\emptyset)_{\mathrm{conn}}$ and call it the *primitive LMO invariant.* Concretely, $z^{\mathrm{LMO}}(M)$ takes the form

$z^{\mathrm{LMO}}(M) = c_1$  $+ c_2$  $+ c_3$  $+ c_4$  $+ c_4'$ 

$+$ (terms of degree > 4), (10.13)

for some scalar 3-manifold invariants c_i. For example, for the 3-manifold $M_{n,k}$ obtained from S^3 by integral surgery along the $(2, n)$ torus knot with framing k, the primitive LMO invariant $z^{\text{LMO}}(M_{n,k})$ is given by

$$z^{\text{LMO}}(M_{n,k}) = \frac{1}{48}(3n^2 - k^2 + 3k - 5)$$

$$+\frac{1}{2^7 \cdot 3^2}(12n^4 - 12kn^3 + 3k^2n^2 - 15n^2 + 12kn - 4k^2 + 4)$$

$$+(\text{terms of degree} \geq 3).$$

See [LMO98] for a computation to obtain the above value.

Properties of the LMO invariant

Proposition 10.15.
(1) For the connected sum $M_1 \# M_2$ of two 3-manifolds M_1 and M_2, the primitive LMO invariant satisfies

$$z^{\text{LMO}}(M_1 \# M_2) = \sum_{d=1}^{\infty} (c_0(M_2)^d z^{\text{LMO}}(M_1)^{(d)} + c_0(M_1)^d z^{\text{LMO}}(M_2)^{(d)}),$$

where $c_0(M)$ denotes the scalar 3-manifold invariant c_0 of M in Lemma 10.11. In particular, for integral homology 3-spheres, the invariant z^{LMO} is additive with respect to the connected sum of 3-manifolds.
(2) Let $-M$ denote the 3-manifold M with the opposite orientation. Then,

$$z^{\text{LMO}}(-M) = \sum_{d=1}^{\infty} (-1)^{d(b_1(M)+1)} z^{\text{LMO}}(M)^{(d)}.$$

Proof. Let L_1 and L_2 be framed links which present M_1 and M_2 by integral surgery respectively. Then, the connected sum $M_1 \# M_2$ is obtained from S^3 by integral surgery along the split union $L_1 \sqcup L_2$ of L_1 and L_2. We have that $\check{Z}(L_1 \sqcup L_2) = \check{Z}(L_1) \otimes \check{Z}(L_2)$ from the definition of the Kontsevich invariant. Applying the definition of the LMO invariant to both sides of this formula, we obtain (1) of the proposition.

Let L be a framed link which presents M by integral surgery. Then, the 3-manifold $-M$ is obtained from S^3 by integral surgery along the mirror image $\overline{L}$ of L. Since $\overline{L}$ is obtained from L by exchanging positive and negative crossings, we have that $\hat{Z}(\overline{L}) = \sum_{d=0}^{\infty} (-1)^d \hat{Z}(L)^{(d)}$ from the definition of the Kontsevich invariant. By applying the definition of the LMO invariant to both sides of this formula, we obtain (2) of the proposition. □

It follows from the following proposition that the LMO invariant is not a strong invariant of 3-manifolds with positive first Betti number; we will see later, however, that the LMO invariant is indeed a strong invariant of integral homology 3-spheres. (It is known [BaLa*00*] that the LMO invariant does not separate lens spaces.)

Proposition 10.16. Let M be an oriented closed 3-manifold with positive first Betti number.
(1) When $b_1(M) = 1$, the value of $Z^{\mathrm{LMO}}(M)$ can be determined from the Alexander polynomial of M ([GaHa00, Lie00]).
(2) When $b_1(M) = 2$, the value of $Z^{\mathrm{LMO}}(M)$ can be determined from the Casson-Walker-Lescop invariant of M ([HaBe00]).
(3) When $b_1(M) = 3$, the value of $Z^{\mathrm{LMO}}(M)$ can be determined from the cohomology ring of M ([Habe*96*]).
(4) When $b_1(M) > 3$, we always have that $Z^{\mathrm{LMO}}(M) = 1$ ([Habe*96*]).

10.3 Universality of the LMO invariant among perturbative invariants

The following theorem implies the universality of the LMO invariant among perturbative invariants. It was proved for $n = 2$ in [Oht00] and for general n by Lev Rozansky and Thang Le (see also [BGRT*97a*]). The aim of this section is to show the theorem for $n = 2$.

Theorem 10.17. For a rational homology 3-sphere M, the perturbative $PSU(n)$ invariant recovers from the LMO invariant as

$$\tau^{PSU(n)}(M) = |H_1(M;\mathbb{Z})|^{-n(n-1)/2}\hat{W}_{sl_n}\big(\hat{Z}^{\mathrm{LMO}}(M)\big).$$

In particular, when $n = 2$,

$$\tau^{SO(3)}(M) = |H_1(M;\mathbb{Z})|^{-1}\hat{W}_{sl_2}\big(\hat{Z}^{\mathrm{LMO}}(M)\big)$$

noting $PSU(2) = SO(3)$. The remainder of this section is devoted to the proof of the above formula, *i.e.*, Theorem 10.17 when $n = 2$.

Expression of j_n by a map α

In this subsection, we consider a map α such that a polynomial in α is equal to the map j_n modulo the kernel of the weight system W_{sl_2}.

For a Jacobi diagram D on S^1 we define $\alpha(D)$ by

$\alpha(D) =$ (the Jacobi diagram obtained from $\Delta(D)$ by replacing one  by )

$-$ (the disjoint union of D and ).

By its linear extension we define the map $\alpha : \mathcal{A}(S^1) \to \mathcal{A}(S^1)$; note that the image of α lies, not in $\mathring{\mathcal{A}}(S^1)$, but in $\mathcal{A}(S^1)$ since $\alpha(D)$ has no dashed loops, as in the following simple cases.

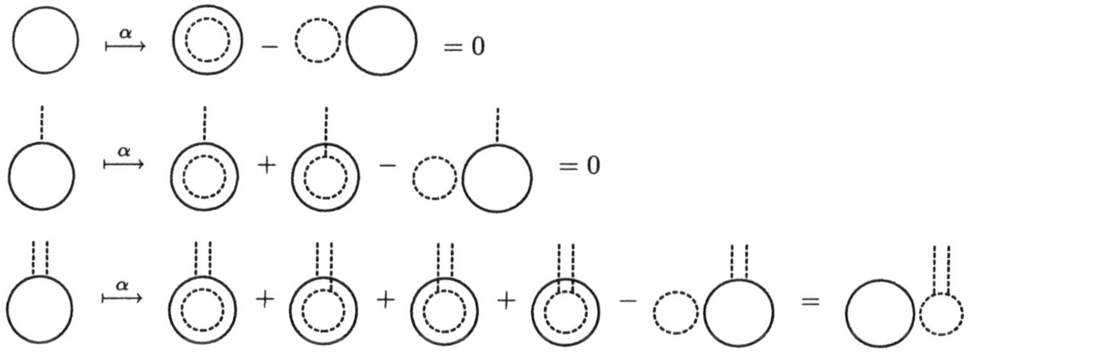

As in the above examples, the first term in an image of α cancels with the last term. Further, a dashed loop with one trivalent vertex vanishes by the AS relation. Hence, the map α decreases the number of vertices on a solid circle at least by two. Therefore, we have

Lemma 10.18 (vanishing lemma for α). If the number of univalent vertices on a solid circle of a Jacobi diagram D is less than $2m$, then $\alpha^m(D) = 0$, where α^m denotes the composition of m applications of α.

We express the map j_n by α as follows. We begin with the case of $n = 1$.

Proposition 10.19. There exists a power series $p(\alpha) = \sum_{i=1}^{\infty} c_i \alpha^i$ with coefficients $c_i \in \mathbb{Z}[\frac{1}{2}, \frac{1}{3}, \cdots, \frac{1}{2i+1}]$ satisfying the formula,

$$W_{sl_2}\big(j_1(D)\big) = W_{sl_2}\big((\varepsilon \circ p(\alpha))(D)\big),$$

for any Jacobi diagram D on S^1.

The coefficients c_i in the proposition have been concretely determined to be $c_i = (-1)^{i+1}((i-1)!)^2/2(2i)!$ by Dylan Thurston and Thang Le, though we do not use these concrete values in the proof, given in this section, that the perturbative invariants are recovered from the LMO invariant.

Proof of Proposition 10.19. Recall that a Jacobi diagram on m ordered points is a vertex-oriented uni-trivalent graph such that m univalent vertices lie on the m points, one to each, and that $\mathcal{A}(m)$ denotes the vector space spanned by such Jacobi diagrams on m ordered points subject to the AS, IHX relations. We define a Jacobi diagram on $S^1 \sqcup [m]$ to be a vertex-oriented uni-trivalent graph such that m of the univalent vertices lie on the m points and the other univalent vertices lie on the S^1. Let D_m be the Jacobi diagram on $S^1 \sqcup [m]$ consisting of m parallel dashed lines without trivalent vertices which connect the m points to S^1. Extending the maps j_1 and α to maps of Jacobi diagrams on $S^1 \sqcup [m]$, observe that $j_1(D_m)$ and $(\varepsilon \circ p(\alpha))(D_m)$ belong to $\mathcal{A}(m)$. Also, extending the definition of weight system, we have a weight system $W_{sl_2} : \mathcal{A}(m) \to (sl_2)^{\otimes m}$. Note that the image of this weight system is in the invariant subspace of the space $(sl_2)^{\otimes m}$, where sl_2 acts by the tensor product of m copies of the adjoint action of sl_2. We denote the invariant space by $\left((sl_2)^{\otimes m}\right)^{sl_2}$.

It is sufficient to show the formula

$$W_{sl_2}\big(j_1(D_m)\big) = W_{sl_2}\big((\varepsilon \circ p(\alpha))(D_m)\big) \in \left((sl_2)^{\otimes m}\right)^{sl_2} \qquad (10.14)$$

for each $m \geq 0$. We construct a power series $p(\alpha)$ solving this equation by a step-by-step procedure.

For $m = 0, 1$, the formula (10.14) always holds, since both sides of (10.14) vanish for $m = 0, 1$ from the definitions of j_1 and α.

For $m = 2$, the left hand side of (10.14) is equal to $W_{sl_2}($  $)$, from the definition of j_1. On the other hand,

$W_{sl_2}\big((\varepsilon \circ \alpha)(D_2)\big) = W_{sl_2}($  $) = 4W_{sl_2}($  $).$

Hence, we require that $c_1 = 1/4$. Since $\alpha^2(D_2) = 0$, by Lemma 10.18, the formula (10.14) holds for $m = 2$ when we put $c_1 = 1/4$, regardless of what the constants $c_2, c_3, \cdots$ are set to be.

For $m = 3$, the formula (10.14) holds when $c_1 = 1/4$ by the following argument. We put

$$x_3 = j_1(D_3) - (\varepsilon \circ c_1\alpha)(D_3) \in \mathcal{A}(3).$$

Since each of the maps j_1 and $\varepsilon \circ \alpha$ commute with the STU relation, we have that

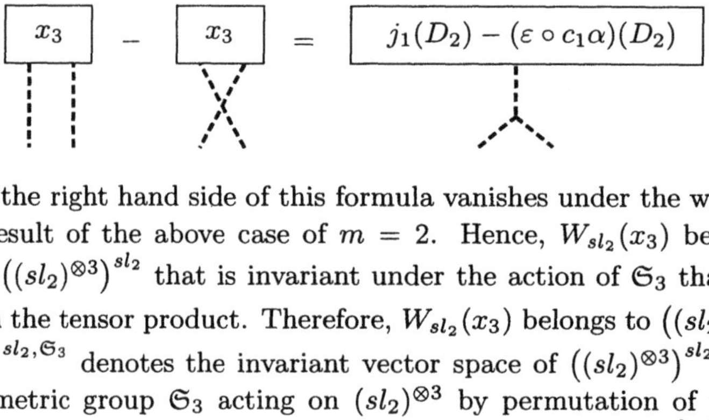

Further, the right hand side of this formula vanishes under the weight system W_{sl_2} by the result of the above case of $m = 2$. Hence, $W_{sl_2}(x_3)$ belongs to the subspace of $\left((sl_2)^{\otimes 3}\right)^{sl_2}$ that is invariant under the action of $\mathfrak{S}_3$ that permutes the 3 entries in the tensor product. Therefore, $W_{sl_2}(x_3)$ belongs to $\left((sl_2)^{\otimes 3}\right)^{sl_2, \mathfrak{S}_3}$, where $\left((sl_2)^{\otimes 3}\right)^{sl_2, \mathfrak{S}_3}$ denotes the invariant vector space of $\left((sl_2)^{\otimes 3}\right)^{sl_2}$ by the action of the symmetric group $\mathfrak{S}_3$ acting on $(sl_2)^{\otimes 3}$ by permutation of the entries. It is known (see, *e.g.*, [Hum72]) that the invariant space $\left((sl_2)^{\otimes 3}\right)^{sl_2, \mathfrak{S}_3}$ is the null vector space. Hence, $W_{sl_2}(x_3) = 0$, and the formula (10.14) holds for $m = 3$ when we put $c_1 = 1/4$.

For $m = 4$, similarly as above, we put

$$x_4 = j_1(D_4) - (\varepsilon \circ c_1\alpha)(D_4) \in \mathcal{A}(4),$$

where we put $c_1 = 1/4$, as above. By the same argument as before, $W_{sl_2}(x_4)$ belongs to the invariant space $\big((sl_2)^{\otimes 4}\big)^{sl_2, \mathfrak{S}_4}$, which is known to be one-dimensional. It follows from a concrete computation that $W_{sl_2}\big((\varepsilon \circ \alpha^2)(D_4)\big)$ is non-zero in this invariant space. We can determine the value of c_2 from the equation $W_{sl_2}(x_4) = c_2 W_{sl_2}\big((\varepsilon \circ \alpha^2)(D_4)\big)$. Then, for the values of c_1 and c_2 so determined, the formula (10.14) holds for $m = 4$, regardless of what the constants $c_3, c_4, \cdots$ are set to be, by Lemma 10.18.

For $m \geq 5$ we can continue to construct a solution $p(\alpha)$ to (10.14) by the same step-by-step procedure as above, using the fact that the invariant space $\big((sl_2)^{\otimes m}\big)^{sl_2, \mathfrak{S}_m}$ is the null vector space if m is odd, and one dimensional if m is even. We obtain the required evaluation of the denominators of c_i by computing $W_{sl_2}\big((\varepsilon \circ \alpha^k)(D_{2k})\big)$ carefully; see [Oht00] for such a computation in detail. □

Proposition 10.20. For the power series $p(\alpha)$ given in Proposition 10.19,

$$j_n(D) = \frac{1}{n!}(\varepsilon \circ p(\alpha)^n)(D),$$

for any Jacobi diagram D on S^1 and for any positive integer n.

Proof. From the definition of the comultiplication Δ,

$$(\Delta^{(k_1)} \otimes \Delta^{(k_2)}) \circ \Delta = \Delta^{(k_1+k_2+1)}, \tag{10.15}$$

recalling that $\Delta^{(k)}$ denotes the map $\mathcal{A}(S^1) \to \mathcal{A}(\sqcup^{k+1} S^1)$ that applies Δ k times, to some choice of solid circle at each step. Since $\alpha^i(D)$ is a linear sum of Jacobi diagrams obtained from $\Delta^{(k)}(D)$ by replacing some solid circles with dashed ones, we have the following formula by (10.15),

$$\big((\varepsilon \circ p_1(\alpha)) \otimes (\varepsilon \circ p_2(\alpha))\big) \circ \Delta = \varepsilon \circ \big(p_1(\alpha) p_2(\alpha)\big) \tag{10.16}$$

for any two power series $p_1(\alpha)$ and $p_2(\alpha)$, where $p_1(\alpha)p_2(\alpha)$ implies the usual product of power series. Further, since the map j_n is defined by $j_n = (1/n!) j_1 \circ \Delta^{(n-1)}$, we have the following formula by Proposition 10.19,

$$j_n(D) = \frac{1}{n!}\big((\varepsilon \circ p(\alpha))^{\otimes n} \circ \Delta^{(n-1)}\big)(D)$$

noting that there are n copies of $p(\alpha)$ because $\Delta^{(n-1)}(D)$ has n solid circles. By applying the formula (10.16) $n-1$ times, we obtain the required formula. □

Expression of α by representations

In this subsection we consider a representation a related to the map α through the weight system W_{sl_2}.

We define a to be $V_3 - 3 \cdot V_1 \in R(sl_2)$, where V_k denotes the k-dimensional irreducible representation of sl_2 and $R(sl_2)$ denotes the representation ring of sl_2 with integral coefficients. The map α is related to a by the following lemma.

Lemma 10.21. For any Jacobi diagram D on S^1,

$$W_{sl_2}\big(\varepsilon \circ \alpha^n(D)\big) = W_{sl_2;a^n}(D).$$

Proof. Let D_1 be a Jacobi diagram on $S^1 \sqcup S^1$ and let D_2 be the Jacobi diagram on S^1 obtained from D_1 by replacing the second S^1 by a dashed loop. Then,

$$W_{sl_2;R,V_3}(D_1) = W_{sl_2;R}(D_2),$$

for any $R \in R(sl_2)$, because in the definition of the weight system W_{sl_2} we substitute the adjoint representation of sl_2 (which is isomorphic to V_3) into a dashed line. Further,

$$W_{sl_2;R_1,R_2}\big(\Delta(D)\big) = W_{sl_2;R_1\otimes R_2}(D),$$

since the comultiplication Δ translates to a tensor product when a weight system is applied. Hence, by the above two formulae and the definition of α,

$$W_{sl_2;R}(\alpha(D)) = W_{sl_2;R\otimes a}(D), \tag{10.17}$$

noting W_{sl_2}(a dashed loop) $= 3$. Further,

$$W_{sl_2}\big(\varepsilon(D)\big) = W_{sl_2;V_1}(D). \tag{10.18}$$

since the counit ε translates to evaluation in the trivial representation V_1 when a weight system is applied. Applying (10.17) repeatedly to (10.18), we obtain the required formula. □

Lemma 10.22 (vanishing lemma for a). For any odd prime number r, we have the following two formulae

$$-2a^{(r-3)/2} \underset{(r)}{\equiv} \sum_{\substack{1\le m\le r-2\\ m \text{ is odd}}} mV_m,$$

$$a^{(r-1)/2} \underset{(r)}{\equiv} 0,$$

where $\underset{(r)}{\equiv}$ denotes the equivalence relation in $R(sl_2)$ generated by the following two relations; the first relation is that any element divisible by r in $R(sl_2)$ is equivalent to 0, and the second relation is that any irreducible representation whose dimension is divisible by r is equivalent to 0.

Proof. We have that

$$V_m \cdot a = V_m \otimes V_3 - 3V_m = V_{m-2} + V_m + V_{m+2} - 3V_m = V_{m-2} - 2V_m + V_{m+2},$$

where the second equality is derived from the decomposition formula for tensor products of representations for sl_2 (see, *e.g.*, [Hum72]). This formula also holds for negative integers m by regarding V_{-m} as $-V_m$. Hence, for example, in simple cases,

$$\begin{aligned} a^0 &= V_1, \\ a^1 &= V_1 \cdot a = V_{-1} - 2V_1 + V_3 \quad (= -3V_1 + V_3), \\ a^2 &= V_1 \cdot a^2 = V_{-3} - 4V_{-1} + 6V_1 - 4V_3 + V_5 \quad (= 10V_1 - 5V_3 + V_5). \end{aligned}$$

The coefficients appearing in the above formulas, say, for a^2, are equal to the coefficients appearing in the Laurent polynomial

$$t(t^{-2} - 2 + t^2)^2 = t^{-3} - 4t^{-1} + 6t - 4t^3 + t^5,$$

where the coefficient of t^m in the Laurent polynomial corresponds to the coefficient of V_m in the expansion of a^2. Similarly, in general, the coefficients appearing in the formula for a^n can be obtained as the coefficients of $t(t^{-2} - 2 + t^2)^n$. In particular, the coefficient of V_m $(= -V_{-m})$ in the expansion of a_n is equal to the sum of the coefficients of t^m and $-t^{-m}$ in $t(t^{-2} - 2 + t^2)^n$. It is, further, equal to the coefficient of t^m in

$$t(t^{-2} - 2 + t^2)^n - t^{-1}(t^2 - 2 + t^{-2})^n = (t - t^{-1})^{2n+1}.$$

Hence, the coefficient is equal to $(-1)^{n-(m-1)/2}\binom{2n+1}{n-(m-1)/2}$. Therefore,

$$a^n = (-1)^n \binom{2n+1}{n} V_1 + (-1)^{n-1} \binom{2n+1}{n-1} V_3 + \cdots - \binom{2n+1}{1} V_{2n-1} + V_{2n+1}.$$

When $n = (r-1)/2$ the coefficients of $V_1, V_3, \cdots, V_{2n-1}$ are divisible by r, by the above formula, and the remaining V_{2n+1} has a dimension divisible by r. Hence, we obtain the second formula of the lemma.

When $n = (r-3)/2$ the coefficient of V_m in the expansion of a^n is equal to

$$\begin{aligned} (-1)^{n-(m-1)/2} &\binom{2n+1}{n-(m-1)/2} = (-1)^{(r-m)/2-1} \binom{r-2}{(r-m)/2-1} \\ &= (-1)^{(r-m)/2-1} \frac{(r-2)(r-1)\cdots((r+m)/2)}{((r-m)/2-1)!} \\ &\underset{(r)}{\equiv} (-1)^{(r-m)/2-1} \frac{(-2)(-3)\cdots(-(r+m)/2)}{((r-m)/2-1)!} \\ &= -\frac{r+m}{2} \underset{(r)}{\equiv} -\frac{m}{2}. \end{aligned}$$

Hence, we obtain the first formula of the lemma. □

Proof of the universality

In this subsection, we show a sketch of the proof of Theorem 10.17 for $n = 2$, which states that the perturbative $SO(3)$ invariant recovers from the LMO invariant through the weight system W_{sl_2}.

Sketch of the proof of Theorem 10.17 for $n = 2$. For simplicity, we show the proof for the case that the 3-manifold M is obtained from S^3 by integral surgery along a framed knot K with a positive framing.

By Proposition 10.20, we have that

$$\begin{aligned} h^n \cdot \hat{W}_{sl_2}\big(j_n \check{Z}(K)\big) &= \frac{1}{n!}\hat{W}_{sl_2}\big((\varepsilon \circ p(\alpha)^n)(\check{Z}(K))\big) \\ &= \frac{1}{n!}\hat{W}_{sl_2}\big((\varepsilon \circ (c_1^n \alpha^n + n c_1^{n-1} c_2 \alpha^{n+1} + \cdots))(\check{Z}(K))\big), \end{aligned}$$

where the first h^n is derived from the fact that the map j_n decreases the degree of Jacobi diagrams by n. We consider terms of at most finite degree in the following of this proof. Then, we can reduce the power series in α on the right hand side to a finite sum by Lemma 10.18 (vanishing lemma for α). Moreover, by replacing α with a using Lemma 10.21, the right hand side of the above formula is equal to

$$\frac{c_1^n}{n!}\hat{W}_{sl_2;a^n}\big(\check{Z}(K)\big) + \frac{n c_1^{n-1} c_2}{n!}\hat{W}_{sl_2;a^{n+1}}\big(\check{Z}(K)\big) + \cdots + O(h^{n+(r-1)/2}).$$

For an odd prime r we put $n = (r-3)/2$. By Lemma 10.22 (vanishing lemma for a), the above formula is congruent modulo r to

$$\frac{c_1^n}{n!}\sum m W_{sl_2;V_m}\big(\check{Z}(K)\big) + O(h^{n+(r-1)/2}).$$

Further, by the formula $W_{sl_2;V_m}(\nu) = [m]/m$, the above formula is equal to

$$\frac{c_1^n}{n!}\sum [m] W_{sl_2;V_m}\big(\hat{Z}(L)\big) + O(h^{n+(r-1)/2}).$$

Furthermore, by the recovery of the quantum invariant from the Kontsevich invariant, the above formula is equal to

$$\frac{c_1^n}{n!}\sum [m] Q^{sl_2;V_m}(L) + O(h^{n+(r-1)/2}).$$

Recall that the maps j_n and ι_n are related by the relation  $= -2n$. Since $W_{sl_2}($  $) = 3 \equiv -2n$ modulo r, the maps j_n and ι_n are congruent modulo r in the image of W_{sl_2}. Hence, summarizing the above arguments,

$$h^n \cdot \hat{W}_{sl_2}\big(\iota_n \check{Z}(K)\big) \underset{(r)}{\equiv} \frac{c_1^n}{n!}\sum [m] Q^{sl_2;V_m}(K) + O(h^{n+(r-1)/2}). \tag{10.19}$$

By Proposition 10.12 we have that $Z_n^{\mathrm{LMO}}(M)^{(d)} = |H_1(M;\mathbb{Z})|^{n-d} Z_d^{\mathrm{LMO}}(M)^{(d)}$ for any $n \geq d$. Hence,

$$\hat{Z}^{\mathrm{LMO}}(M) = 1 + \sum_{d=1}^{\infty} \frac{Z_d^{\mathrm{LMO}}(M)^{(d)}}{|H_1(M;\mathbb{Z})|^d} \equiv \frac{Z_n^{\mathrm{LMO}}(M)}{|H_1(M;\mathbb{Z})|^n} \quad \text{modulo degree} > n \text{ part.}$$

Therefore,

$$|H_1(M;\mathbb{Z})|^n \hat{W}_{sl_2}(\hat{Z}^{\mathrm{LMO}}(M)) = \hat{W_{sl_2}}(Z_n^{\mathrm{LMO}}(M)) = \hat{W_{sl_2}}\left(\frac{\iota_n \check{Z}(K)}{\iota_n \check{Z}(U_+)}\right)$$

$$\underset{(r)}{\equiv} \frac{\sum [m] Q^{sl_2;V_m}(L)}{\sum [m] Q^{sl_2;V_m}(U_+)} + O(h^{(r-1)/2}) = \tau_r^{SO(3)}(M) + O(h^{(r-1)/2})$$

$$\underset{(r)}{\equiv} \left(\frac{|H_1(M;\mathbb{Z})|}{r}\right) \tau^{SO(3)}(M) + O(h^{(r-1)/2}),$$

where we obtain the third congruence by (10.19), obtain the fourth equality by the definition of $\tau_r^{SO(3)}(M)$ and obtain the fifth equality by Theorem 9.8. Here, $\left(\frac{\cdot}{r}\right)$ denotes the Legendre symbol. Further, it is known in number theory that $\left(\frac{f}{r}\right) \equiv f^{(r-1)/2}$ for an integer f not divisible by r. Hence,

$$|H_1(M;\mathbb{Z})|^n \hat{W}_{sl_2}(\hat{Z}^{\mathrm{LMO}}(M)) \underset{(r)}{\equiv} |H_1(M;\mathbb{Z})|^{n+1} \tau^{SO(3)}(M) + O(h^{(r-1)/2}).$$

Since this formula holds for infinitely many primes r, while both sides do not depend on r, we obtain the required formula. □

To be precise, the prime r might appear in the denominator in the expansion of $\check{Z}(K)$, though we calculated the formulae modulo r in the proof. We can avoid this technical difficulty by replacing $\check{Z}(K)$ with quantum invariants *before* we reduce modulo r. Since the quantum invariants have integral coefficients, we can then calculate the formulae modulo r. For detailed arguments see [Oht00].

10.4 Aarhus integral

In this section we describe an interpretation of the topics of this chapter by formal Gaussian integrals, along the lines of [BGRT*97a*]. We express the LMO invariant and the perturbative $SO(3)$ invariant by formal Gaussian integrals, and give another proof of the recovery of the perturbative $SO(3)$ invariant from the LMO invariant by showing the commutativity of the formal Gaussian integrals with the weight system.

Expression of the LMO invariant as a formal Gaussian integral

In this subsection we describe an expression of the LMO invariant as a formal Gaussian integral, which is defined to be a map on a subset of the space $\mathcal{B}$ of uni-trivalent graphs.

As usual we denote by sl_2 the Lie algebra spanned by a basis $\{E, F, H\}$ with Lie bracket defined by

$$[H, E] = 2E, \qquad [H, F] = -2F, \qquad [E, F] = H.$$

Further, $S(sl_2)$ denotes the symmetric algebra on sl_2, which is equal to the commutative polynomial ring in three indeterminates E, F and H. Moreover $U(sl_2)$ denotes the universal enveloping algebra of sl_2, which is the algebra generated by three non-commutative generators E, F and H subject to the relations

$$HE - EH = 2E, \qquad HF - FH = -2F, \qquad EF - FE = H. \tag{10.20}$$

The Poincare-Birkhoff-Witt theorem (see, *e.g.*, [Dix77]) implies that $S(sl_2)$ is isomorphic to $U(sl_2)$ by the map $\chi : S(sl_2) \to U(sl_2)$ defined by

$$\chi(x_1 x_2 \cdots x_n) = \frac{1}{n!} \sum_{\sigma \in \mathfrak{S}_n} x_{\sigma(1)} x_{\sigma(2)} \cdots x_{\sigma(n)} \tag{10.21}$$

where each x_i is either E, F or H and the sum runs over all elements σ in the nth symmetric group $\mathfrak{S}_n$.

Let $\mathcal{B}$ denote the vector space over $\mathbb{C}$ spanned by vertex-oriented uni-trivalent graphs subject to the AS and IHX relations. The space $\mathcal{B}$ has an algebra structure with the product of disjoint union of uni-trivalent graphs. By extending the definition of weight systems we have weight systems $W_{sl_2} : \mathcal{B} \to S(sl_2)$ and $W_{sl_2} : \mathcal{A}(S^1) \to U(sl_2)$. As a generalization of Poincare-Birkhoff-Witt isomorphism χ we have the isomorphism $\chi : \mathcal{B} \to \mathcal{A}(S^1)$ which commutes with the weight systems as

$$\begin{array}{ccc} S(sl_2) & \xrightarrow{\ \chi\ } & U(sl_2) \\ {\scriptstyle W_{sl_2}}\uparrow & & \uparrow{\scriptstyle W_{sl_2}} \\ \mathcal{B} & \xrightarrow{\ \chi\ } & \mathcal{A}(S^1) \end{array} \tag{10.22}$$

where the map $\chi : \mathcal{B} \to \mathcal{A}(S^1)$ is defined for elements of $\mathcal{B}$ as

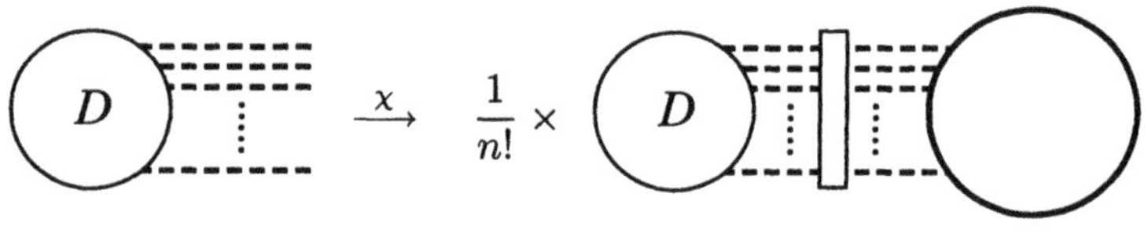

(10.23)

for a uni-trivalent graph D with n univalent vertices. Here, the rectangle over n dashed lines implies the symmetrizer of the lines such as

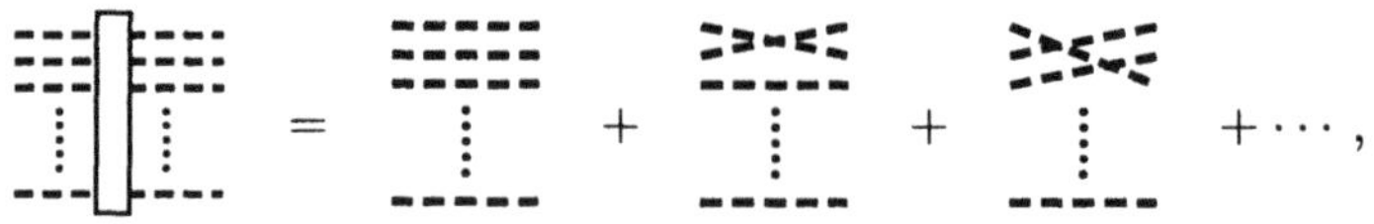

(10.24)

where the sum on the right hand side runs over the $n!$ configurations which connect the left n points to the right n points respectively with n dashed edges.

The composition

$$\mathcal{B} \xrightarrow{\ \chi\ } \mathcal{A}(S^1) \xrightarrow{\ \iota_n\ } \mathcal{A}(\emptyset)$$

is expressed by the following lemma. Here, we let P_n be the sum of the unions of n dashed arcs over all partitions of $2n$ points into n pairs. It is equal to the left hand side of the relation P_n of Figure 10.3; we denote the sum also by P_n, abusing notation as in Section 10.2.

Lemma 10.23. For any uni-trivalent graph $D \in \mathcal{B}$, we have that

$(\iota_n \circ \chi)(D) =$ 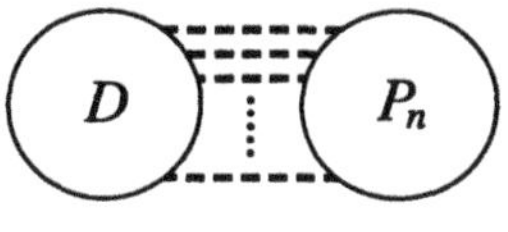 if D has just $2n$ univalent vertices,

0 otherwise.

Proof. From the definition of ι_n, $(\iota_n \circ \chi)(D)$ is obtained by gluing copies of uni-trivalent graphs T_m's of (10.8) to the univalent vertices of $\chi(D)$.

If D has less than $2n$ univalent vertices, then at least one of the involved T_m's is equal to $T_{<2}$. Hence, $(\iota_n \circ \chi)(D) = 0$, since $T_0 = T_1 = 0$.

If D has more than $2n$ univalent vertices, then at least one of such T_m's is equal to $T_{>2}$, which is a linear sum of uni-trivalent trees with at least one trivalent vertex. Hence, $(\iota_n \circ \chi)(D) = 0$, since $\chi(D)$ is symmetric with respect to permutations of univalent vertices while such trees are anti-symmetric by the AS relation.

Otherwise, D has just $2n$ univalent vertices. In the same way as in the above arguments $(\iota_n \circ \chi)(D)$ is equal to a linear sum of gluings of the univalent vertices of $\chi(D)$ to copies of T_2. This is, furthermore, equal to the gluing of $\chi(D)$ to P_n. Hence, by the symmetry of P_n, this is equal to the gluing of D to P_n. □

We define the bracket $\langle D_1, D_2\rangle$ of two uni-trivalent graphs D_1 and D_2 by

$\langle D_1, D_2\rangle =$ 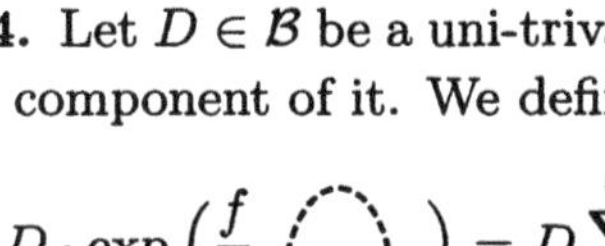 if D_1 and D_2 have the same number of univalent vertices,

0 otherwise,

where the rectangle is the symmetrizer in (10.24).

Lemma 10.24. Let $D \in \mathcal{B}$ be a uni-trivalent graph which has at least one trivalent vertex on each component of it. We define G by

$G = D \cdot \exp\Big(\frac{f}{2}$ 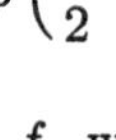 $\Big) = D\sum_{k=0}^{\infty} \frac{f^k}{2^k k!}\Big($ 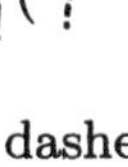 $\Big)^k \in \mathcal{B}, \qquad (10.25)$

for a non-zero scalar f, where the product of dashed graphs in $\mathcal{B}$ is the disjoint union of them, as mentioned earlier. Then,

$(\iota_n \circ \chi)(G) = (-f)^n \Big\langle D, \exp\Big(-\frac{1}{2f}$  $\Big)\Big\rangle \in \mathcal{A}(\emptyset).$

Proof. If D has $(2n-2k)$ univalent vertices, then by Lemma 10.23 we have that

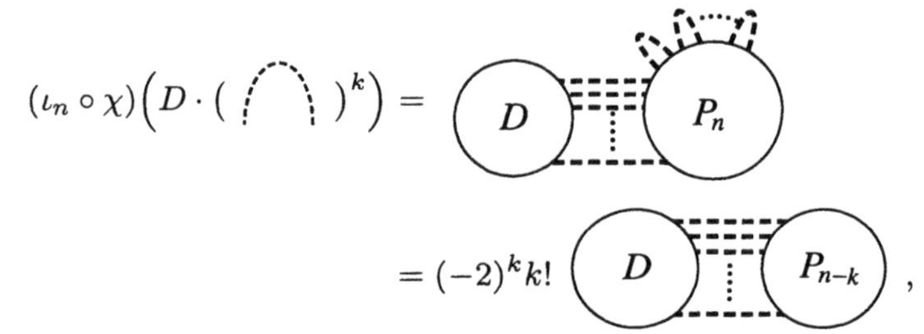

where we obtain the second equality by a computation similar to (10.7), noting that we replace a dashed circle with a factor $-2n$ in the definition of ι_n. The right hand side of the required relation is computed as

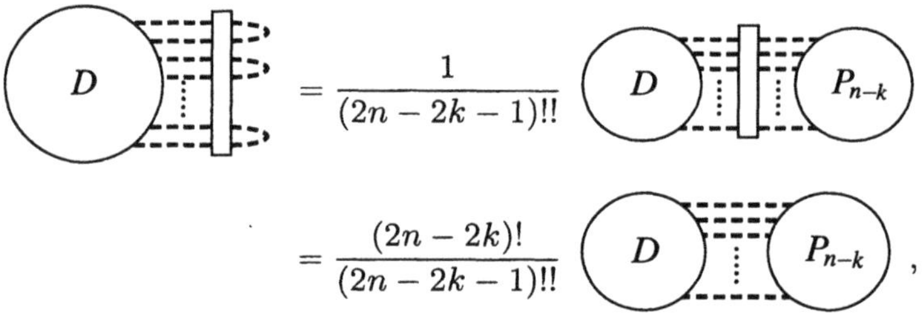

where we obtain the first equality since P_{n-k} has $(2n-2k-1)!!$ terms, also noting the symmetry of P_{n-k}. Hence, by the above two formulae,

$$\frac{f^k}{2^k k!}\cdot(\iota_n\circ\chi)\Big(D\cdot(\frown)^k\Big)=(-f)^n\cdot\frac{1}{(-2f)^{n-k}(n-k)!}\Big\langle D,(\smile)^{n-k}\Big\rangle.$$

Note that both sides vanish unless D has $(2n-2k)$ univalent vertices. By summing up both sides with respect to k we obtain the required formula. □

For G and D as in Lemma 10.24 we consider a map of a subset of $\mathcal{B}$ to $\mathcal{A}(\emptyset)$ given by

$$G\overset{\int^{\mathrm{FG}}}{\longmapsto}\Big\langle D,\exp\big(-\frac{1}{2f}\smile\big)\Big\rangle. \tag{10.26}$$

This map is regarded as a formal Gaussian integral, denoted by $\int^{\mathrm{FG}}$, in [BGRT*97a*]; see the interesting observation in [BGRT*97a*] which explains why the map can be regarded as an integral.

Lemma 10.25. Let K be a knot with a non-zero framing f. Then, $\chi^{-1}\check{Z}(K)$ can be presented by G in (10.25) for some linear sum D of uni-trivalent graphs which have at least one trivalent vertex on each connected component.

Proof. Let $\mathcal{A}(S^1)_+$ denote the subspace of $\mathcal{A}(S^1)$ spanned by Jacobi diagrams on S^1 whose uni-trivalent graphs have at least one trivalent vertex on each component. As shown in Section 6.5 $\check{Z}(K)$ is presented by $D \cdot \exp(f\Theta/2)$ for some $D \in \mathcal{A}(S^1)_+$ where Θ denotes the Jacobi diagram on S^1 with one dashed chord.

The restriction of χ to $\mathcal{B}_+$ gives the map $\mathcal{B}_+ \to \mathcal{A}(S^1)_+$. Hence, χ induces the map $\mathcal{B}/\mathcal{B}_+ \to \mathcal{A}(S^1)/\mathcal{A}(S^1)_+$. Here, $\mathcal{B}/\mathcal{B}_+$ and $\mathcal{A}(S^1)/\mathcal{A}(S^1)_+$ are isomorphic to polynomial rings generated by  and Θ respectively. Further, they are isomorphic by χ which relate  and Θ.

Therefore, χ takes $\exp\Big(-\frac{f}{2}$  $\Big)\chi^{-1}\check{Z}(K)$ into $\mathcal{A}(S^1)_+$. Hence, it belongs to $\mathcal{B}_+$. This implies the lemma. $\square$

Let K be a knot with a positive framing f. By Lemmas 10.24 and 10.25 $\iota_n\check{Z}(K)$ is presented by

$$\iota_n \check{Z}(K) = (-f)^n \Big\langle P(\chi^{-1}\check{Z}(K)), \exp\Big(-\frac{1}{2f}$$  $$\Big)\Big\rangle, \qquad (10.27)$$

where $P : \mathcal{B} \to \mathcal{B}^+ \subset \mathcal{B}$ is the natural projection onto the subspace $\mathcal{B}^+$. Therefore, from the definition of the LMO invariant, we obtain the following theorem which gives a reconstruction of the LMO invariant as a formal Gaussian integral (10.26).

Theorem 10.26 ([BGRT*97a*]). Let M be a rational homology 3-sphere obtained from S^3 by integral surgery along a knot K with a positive framing f. Then, the LMO invariant of M is presented by

$$\hat{Z}^{\mathrm{LMO}}(M) = \frac{1}{c_+}\Big\langle P(\chi^{-1}\check{Z}(K)), \exp\Big(-\frac{1}{2f}$$  $$\Big)\Big\rangle,$$

where c_+ is given by

$$c_+ = \Big\langle P(\chi^{-1}\check{Z}($$  $$)), \exp\Big(-\frac{1}{2}$$  $$\Big)\Big\rangle.$$

See [BGRT*97a*] for the general statement of the theorem for any rational homology 3-sphere M. With the formal Gaussian integral (10.26) the formula of the theorem can be rewritten

$$\hat{Z}^{\mathrm{LMO}}(M) = \frac{1}{c_+}\int^{\mathrm{FG}} \chi^{-1}\check{Z}(K).$$

Proof of Theorem 10.26. We obtain

$$\iota_n\check{Z}($$  $$) = (-1)^n\Big\langle P\Big(\chi^{-1}\check{Z}($$  $$)\Big), \exp\Big(-\frac{1}{2}$$  $$\Big)\Big\rangle$$

from (10.27) by putting K to be the trivial knot with framing $f = 1$. Taking the quotient of (10.27) by the above expression, we have that

$$\frac{1}{f^n} \frac{\iota_n \check{Z}(K)}{\iota_n \check{Z}\big(\infty \big)} = \frac{1}{c_+} \Big\langle P(\chi^{-1} \check{Z}(K)), \exp\big(-\frac{1}{2f} \smile \big) \Big\rangle,$$

for any n. Further, from the definition of the LMO invariant, the left hand side of the above formula is congruent to $\hat{Z}^{\mathrm{LMO}}(M)$ modulo terms of degree $> n$, for each n, while the right hand side is independent of n. Hence, we obtain the required formula. □

Expression of the perturbative SO(3) invariant by Aarhus integral

In this subsection we show an expression for the perturbative $SO(3)$ invariant as an Aarhus integral, which is a formal Gaussian integral over the dual of a Lie algebra.

We review the treatment of formal Gaussian integrals used in Section 9.3, as follows. For a non-zero scalar a, consider the formal expression

$$\int_{x \in \mathbb{R}} e^{ax^2} dx = \frac{1}{\sqrt{a}} \int_{x' \in \mathbb{R}} e^{x'^2} dx',$$

putting $x' = \sqrt{a} x$, even though, taken literally, the right hand side is a divergent integral. Further, we rewrite this formal expression as

$$\frac{\displaystyle\int_{x \in \mathbb{R}} e^{ax^2} dx}{\displaystyle\int_{x \in \mathbb{R}} e^{x^2} dx} = \frac{1}{\sqrt{a}}.$$

By differentiating it with respect to a, we have that

$$\frac{\displaystyle\int_{x \in \mathbb{R}} x^{2m} e^{ax^2} dx}{\displaystyle\int_{x \in \mathbb{R}} e^{x^2} dx} = \Big(\frac{d}{da}\Big)^m \frac{1}{\sqrt{a}} = \Big(-\frac{1}{2}\Big)\Big(-\frac{3}{2}\Big) \cdots \Big(-\frac{2m-1}{2}\Big) a^{-m-1/2}.$$

This formula makes sense, when we regard the integrals on the left hand side as formal Gaussian integrals, as in Section 9.3. For odd m we regard

$$\int_{x \in \mathbb{R}} x^m e^{ax^2} dx = 0.$$

Thus, for a polynomial $p(x)$ in x, we can use the above formal interpretation to define a formal Gaussian integral

$$\frac{1}{c} \int_{x \in \mathbb{R}} p(x) e^{ax^2} dx,$$

where, in this case, we formally set the normalization constant c to be $\int_{x\in\mathbb{R}} e^{x^2}dx$. This definition can be extended to the case of $Q(\mathbf{x})$ a non-degenerate quadratic polynomial in the entries of $\mathbf{x} \in \mathbb{R}^N$, and a formal Gaussian integral

$$\frac{1}{c}\int_{\mathbf{x}\in\mathbb{R}^N} p(\mathbf{x})e^{Q(\mathbf{x})}d\mathbf{x}$$

with some normalization constant c; in this case the integral is defined to be the result of diagonalizing the quadratic form Q and then employing the definition for the case $N = 1$.

The Aarhus integral [BGRT*97a*] is introduced as a formal Gaussian integral, as follows. For simplicity, we only consider the sl_2 case here (the case of general $\mathfrak{g}$ can be formulated similarly, see [BGRT*97a*]). We set a Casimir element C' to be

$$C' = EF + FE + \frac{1}{2}H^2 \in S(sl_2), \tag{10.28}$$

and regard it as a non-degenerate quadratic function on $(sl_2)^\star$. Let K be a knot with a positive framing f. Noting that $\chi^{-1}\big(W_{sl_2}\check{Z}(K)\big)$ is presented by a linear sum of $C'^d e^{fC'/2}$ for $d = 0, 1, 2, \cdots$ by Lemma 10.25, we consider a formal Gaussian integral

$$\frac{1}{c'_+}\int_{(sl_2)^\star} \chi^{-1}\big(W_{sl_2}\check{Z}(K)\big)dv,$$

for some normalization constant c'_+. Here, dv is a volume form of $(sl_2)^\star$, say, a volume form which makes $\{E^\star, H^\star, F^\star\}$ an orthonormal basis. The above integral is called the *Aarhus integral* [BGRT*97a*].

Theorem 10.27 (see [BGRT*97a*]). Let M be a rational homology 3-sphere obtained from S^3 by integral surgery along a knot K with a positive framing f. Then, the perturbative $SO(3)$ invariant of M is presented by the Aarhus integral as

$$\tau^{SO(3)}(M) = \frac{f^{1/2}}{c'_+}\int_{(sl_2)^\star} \chi^{-1}\big(W_{sl_2}\check{Z}(K)\big)dv,$$

where c'_+ is the normalization constant given by

$$c'_+ = \int_{(sl_2)^\star} \chi^{-1}\Big(W_{sl_2}\check{Z}(\ \infty\ )\Big)dv.$$

Proof. As in Section 9.3 the perturbative $SO(3)$ invariant is presented by

$$\tau^{SO(3)}(M) = \frac{f^{1/2}}{c''_+}\int_{n\in\mathbb{R}} [n]Q^{sl_2;V_n}(K)dn,$$

where c''_+ is the normalization constant given by

$$c''_+ = \int_{n\in\mathbb{R}} [n]Q^{sl_2;V_n}\big(\ \infty\ \big)dn.$$

Moreover,

$$[n]Q^{sl_2;V_n}(K) = [n]W_{sl_2;V_n}\big(\hat{Z}(K)\big) = nW_{sl_2;V_n}\big(\check{Z}(K)\big) = \mathrm{trace}_{nV_n}\big(W_{sl_2}\check{Z}(K)\big),$$

where the first and third equalities are obtained by Theorem 6.14 and from the definition of weight systems. Further, the second equality is obtained from the relation $W_{sl_2;V_n}(\nu) = ([n]/n)\cdot \mathrm{id}_{V_n}$, which is obtained by applying Theorem 6.14 to the trivial knot. Hence, the perturbative $SO(3)$ invariant is presented by

$$\tau^{SO(3)}(M) = \frac{f^{1/2}}{c''_+}\int_{n\in\mathbb{R}} \mathrm{trace}_{nV_n}\big(W_{sl_2}\check{Z}(K)\big)dn.$$

Furthermore,

$$\frac{1}{c''_+}\int_{n\in\mathbb{R}} \mathrm{trace}_{nV_n}\big(W_{sl_2}\check{Z}(K)\big)dn = \frac{1}{c'_+}\int_{(sl_2)^*} \chi^{-1}\big(W_{sl_2}\check{Z}(K)\big)dv,$$

by Proposition G.12, since $\chi^{-1}\big(W_{sl_2}\check{Z}(K)\big)$ is presented by a linear sum of $C'^d e^{fC'/2}$ for $d = 0, 1, 2, \cdots$ by Lemma 10.25. Therefore, we obtain the theorem. □

Commutativity of weight system and formal Gaussian integrals

In Theorem 10.17 we showed that the perturbative invariants are recovered from the LMO invariant. In this subsection we describe another proof of this which reduces it to the commutativity of weight systems and formal Gaussian integrals.

As before, let K be a knot with positive framing f, and let M be the rational homology 3-sphere obtained from S^3 by integral surgery along K. Then,

$$\tau^{SO(3)}(M) = \frac{f^{1/2}}{c'_+}\int_{(sl_2)^*} W_{sl_2}\big(\chi^{-1}\check{Z}(K)\big)dv,$$

by Theorem 10.27 and the commutativity of χ and W_{sl_2} that was shown in (10.22). Further, by Theorem 10.26, we have that

$$W_{sl_2}\big(\hat{Z}^{\mathrm{LMO}}(M)\big) = \frac{1}{W_{sl_2}(c_+)}W_{sl_2}\Big(\big\langle P(\chi^{-1}\check{Z}(K)), \exp\big(-\frac{1}{2f}$$  $$\big)\big\rangle\Big).$$

Hence, by Theorem 10.28 below, we obtain

$$\tau^{SO(3)}(M) = \frac{1}{f}W_{sl_2}\big(\hat{Z}^{\mathrm{LMO}}(M)\big),$$

which implies that the perturbative $SO(3)$ invariant is recovered from the LMO invariant, as shown in Theorem 10.17.

Theorem 10.28, below, implies that the weight systems commute with formal Gaussian integration. When the perturbative $SO(3)$ invariant and the LMO invariant are formulated as formal Gaussian integrals (as in Theorems 10.27 and 10.26 respectively) then the recovery of one from the other (Theorem 10.17) is explained by this commutativity.

Theorem 10.28. For a knot K with a positive framing f, we have that

$$\frac{1}{W_{sl_2}(c_+)} W_{sl_2}\Big(\big\langle P(\chi^{-1}\check{Z}(K)), \exp\big(-\frac{1}{2f}\smile\big)\big\rangle\Big) = \frac{f^{3/2}}{c'_+}\int_{(sl_2)^*} W_{sl_2}(\chi^{-1}\check{Z}(K))dv, \tag{10.29}$$

where c_+ and c'_+ are the constants in Theorems 10.26 and 10.27 respectively.

With the formal Gaussian integral $\int^{\text{FG}}$ of (10.26) the formula of the theorem can be rewritten as

$$\frac{1}{W_{sl_2}(c_+)} W_{sl_2}\int^{\text{FG}} \chi^{-1}\check{Z}(K) = \frac{f^{3/2}}{c'_+}\int_{(sl_2)^*} W_{sl_2}(\chi^{-1}\check{Z}(K))dv,$$

which implies the commutativity (up to a constant multiple) of the weight system with formal Gaussian integration, as shown in the following diagram.

$$\begin{array}{ccccc} \mathcal{B} & \supset & \mathcal{B}' & \xrightarrow{\int^{\text{FG}}} & \mathcal{A}(\emptyset) \\ {\scriptstyle W_{sl_2}}\downarrow & & {\scriptstyle W_{sl_2}}\downarrow & & \downarrow{\scriptstyle W_{sl_2}} \\ S(sl_2) & \supset & S(sl_2)' & \xrightarrow{\int_{(sl_2)^*}} & \mathbb{C} \end{array}$$

Here, $\mathcal{B}'$ denotes the subspace of $\mathcal{B}$ spanned by such G as in Lemma 10.24, and $S(sl_2)'$ denotes its image under W_{sl_2}.

Proof of Theorem 10.28. By Lemma 10.25, $\chi^{-1}\check{Z}(K)$ can be presented as a linear sum of terms of the form $D \cdot \exp\big(\frac{f}{2}\frown\big)$ for uni-trivalent graphs D. Further, when a uni-trivalent graph D has $2d$ univalent vertices, it follows from (6.32) that D is equivalent to a scalar multiple of $(\frown)^d$, where for uni-trivalent graphs D_1 and D_2 we call them equivalent, denoted by $D_1 \equiv D_2$, if $W_{sl_2}(D_1) = W_{sl_2}(D_2)$. Hence, we consider computing both sides of the required formula (10.29), temporarily assuming that $\chi^{-1}\check{Z}(K)$ is equal to

$$(\frown)^d \cdot \exp\big(-\frac{1}{2f}\frown\big).$$

The weight system W_{sl_2} takes the above formula to $C'^d e^{fC'/2}$, where C' is a Casimir element given by (10.28). Hence, the integral on the right hand side of (10.29) can be computed as

$$\int_{(sl_2)^*} C'^d e^{fC'/2}dv = 2^d\Big(\frac{d}{df}\Big)^d\int_{(sl_2)^*} e^{fC'/2}dv = 2^d\Big(\frac{d}{df}\Big)^d\Big(\frac{1}{\sqrt{f}}\Big)^3\int_{(sl_2)^*} e^{C'/2}dv.$$

By putting $I = \int_{(sl_2)^*} e^{C'/2}dv$, the above formula is, further, computed as

$$2^d\Big(-\frac{3}{2}\Big)\Big(-\frac{5}{2}\Big)\cdots\Big(-\frac{2d+1}{2}\Big)f^{-d-3/2}I = (-1)^d(2d+1)!!f^{-d-3/2}I.$$

On the other hand, the bracket on the left hand side of (10.29) is computed as

$$\left\langle \left(\frown\right)^d, \exp\left(-\frac{1}{2f}\smile\right)\right\rangle = \frac{1}{d!}\left(-\frac{1}{2f}\right)^d \left\langle \left(\frown\right)^d, \left(\smile\right)^d\right\rangle.$$

Further, the bracket on the right hand side of the above formula is computed as

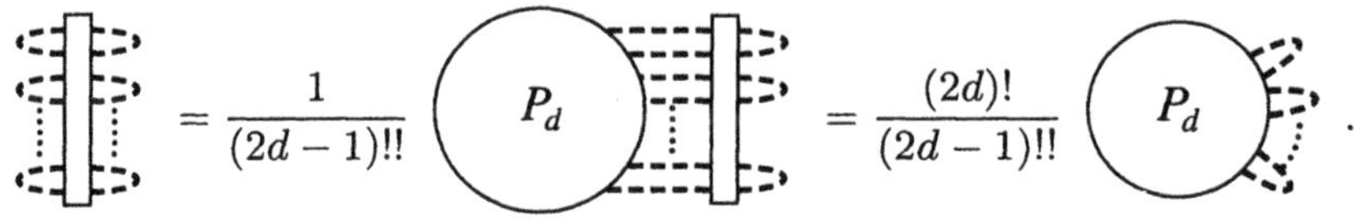

Furthermore,

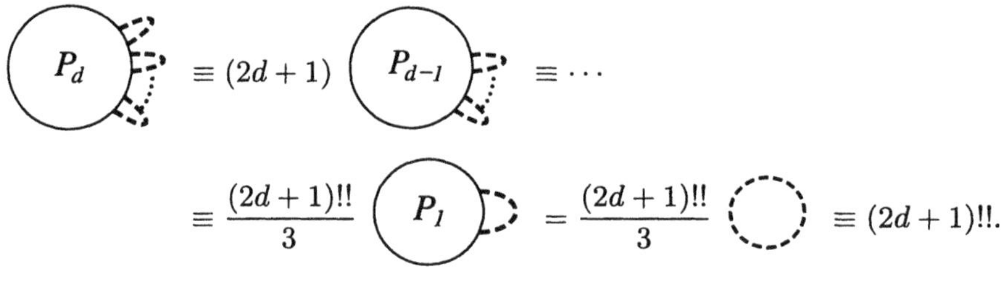

Hence,

$$W_{sl_2}\left(\left\langle \left(\frown\right)^d, \exp\left(-\frac{1}{2f}\smile\right)\right\rangle\right) = \frac{(2d+1)!}{d!}\left(-\frac{1}{2f}\right)^d$$
$$= (-1)^d(2d+1)!!f^{-d} = \frac{f^{3/2}}{I}\int_{(sl_2)^*} C'^d e^{fC'/2} dv.$$

Therefore, coming back to the case of $\chi^{-1}\check{Z}(K)$, we have that

$$W_{sl_2}\left(\left\langle P(\chi^{-1}\check{Z}(K)), \exp\left(-\frac{1}{2f}\smile\right)\right\rangle\right) = \frac{f^{3/2}}{I}\int_{(sl_2)^*} W_{sl_2}(\chi^{-1}\check{Z}(K))dv.$$

Further, by putting K to be the trivial knot U_+ with framing $f = 1$, we have that

$$W_{sl_2}\left(\left\langle P(\chi^{-1}\check{Z}(U_+)), \exp\left(-\frac{1}{2}\smile\right)\right\rangle\right) = \frac{1}{I}\int_{(sl_2)^*} W_{sl_2}(\chi^{-1}\check{Z}(U_+))dv.$$

Taking the quotient of the above two formulae, we obtain the required formula (10.29). □

Chapter 11

Finite type invariants of integral homology 3-spheres

As mentioned in Chapters 4–7, we have three kinds of isotopy invariants of knots; quantum invariants, the Kontsevich invariant, and Vassiliev invariants. There are as many quantum invariants as there are pairs of a simple Lie algebra and a representation of it. This plenitude of quantum invariants is well organized by the Kontsevich invariant and the theory of Vassiliev invariants, in the sense that the Kontsevich invariant unifies the quantum invariants and Vassiliev invariants characterize the coefficients of the quantum invariants. Further, the Kontsevich invariant is universal among Vassiliev invariants. We showed the relations between these invariants in Figure 7.4.

We expect similar relations among the topological invariants of 3-manifolds described in Chapters 8–10. As mentioned in Chapter 8, we have as many quantum invariants of 3-manifolds as there are simple Lie groups. Though they are complex-valued invariants, by their arithmetic expansion we obtain perturbative invariants (which are power series valued), as in Chapter 9. Further, as mentioned in Chapter 10, the LMO invariant unifies the perturbative invariants. Thus we are led to introduce the notion of finite type invariants of 3-manifolds, which play a similar role among the perturbative invariants as the Vassiliev invariants play among knot invariants. Specifically, in this chapter we introduce a definition where by the coefficients of perturbative invariants and the LMO invariant are finite type invariants and the LMO invariant is universal among finite type invariants. See Figure 11.3 for the relations between these invariants of 3-manifolds.

In Section 11.1 we introduce finite type invariants of integral homology 3-spheres following [Oht96a]. Recall that a fundamental property of Vassiliev invariants is that the graded spaces associated to the underlying filtration of the vector space spanned by knots are isomorphic to spaces of Jacobi diagrams on S^1. In this chapter we describe a similar property which relates finite type invariants to spaces of Jacobi diagrams on $\emptyset$. Further, we show the universality of the LMO invariant among finite type invariants in Section 11.2. Introducing, in Section 11.3, primitive finite type invariants we derive from them and investigate a descending series of equivalence relations among integral homology 3-spheres.

11.1 Definition of finite type invariants

In this section we define the notion of a finite type invariant of integral homology 3-spheres, which was introduced in [Oht96a] by replacing "crossing changes" in the definition of Vassiliev invariants of knots with "integral surgeries" on integral homology 3-spheres. (See also [CoMe00] for a definition of finite type invariants of 3-manifolds.) Further, we give a surjective map from the graded spaces of the underlying filtration to certain spaces of Jacobi diagrams; this map is shown to be an isomorphism in the next section.

A 3-manifold M is called an *integral homology 3-sphere* if $H_\star(M;\mathbb{Z}) \cong H_\star(S^3;\mathbb{Z})$. Let L be a framed link in an integral homology 3-sphere M. We call L *algebraically split* if the linking number of any two components of L is equal to 0, and call L *unit-framed* if the framing of each component of L is equal to ± 1. Note that the following two conditions are equivalent: the condition that L is algebraically split and unit-framed in M, and the condition that $M_{L'}$ is an integral homology 3-sphere for any sublink L' in L, where $M_{L'}$ denotes the integral homology 3-sphere obtained from M by integral surgery along L'.

Let $\mathcal{M}$ be the vector space over $\mathbb{C}$ freely spanned by (homeomorphism classes of) integral homology 3-spheres. For an integral homology 3-sphere M and an algebraically split and unit-framed link L in M, we put

$$[M, L] = \sum_{L' \subset L} (-1)^{\# L'} M_{L'} \in \mathcal{M},$$

where the sum runs over all sublinks L' in L including the empty link and $\# L'$ denotes the number of the components of L'. Further, we put $\mathcal{M}_{(d)}$ to be the vector subspace of $\mathcal{M}$ spanned by $[M, L]$ such that M is an integral homology 3-sphere and L is an algebraically split and unit-framed link with d components in M. Then, we have a descending filtration of M as

$$\mathcal{M} = \mathcal{M}_{(0)} \supset \mathcal{M}_{(1)} \supset \mathcal{M}_{(2)} \supset \cdots .$$

As in [Oht96a] we define a linear map $v : \mathcal{M} \to \mathbb{C}$ to be of *finite type* if $v|_{\mathcal{M}_{(d)}} = 0$ for some d. We define the degree of a finite type invariant later in Definition 11.8.

Let us compare the definition with the definition of Vassiliev invariants of knots. As mentioned in Chapter 7, a Vassiliev invariant of degree d is a linear map $\mathcal{K} \to \mathbb{C}$ which vanishes in $\mathcal{K}_{d+1}$, where $\mathcal{K}_{d+1}$ is the vector subspace spanned by linear sums of 2^{d+1} knots obtained by crossing changes at $d+1$ crossings. Here, instead of "crossing changes", we used "integral surgeries" to define the vector subspace $\mathcal{M}_{(d+1)}$ of $\mathcal{M}$.

Though the notion of finite type for integral homology 3-spheres might alternatively be defined in other ways, we are led by the expectation that the graded spaces arising from such a definition should be related to spaces of Jacobi diagrams, just as in the theory of Vassiliev invariants. In fact, as shown later in Theorems 11.7 and 11.21, the finite type invariants arising from our definition enjoy such a property. We begin the investigation of the associated graded spaces (toward the diagrammatic description mentioned above) with the following technical lemmas.

Lemma 11.1. Let $K \cup L$ be a $(d+1)$-component algebraically split and unit-framed link in an integral homology 3-sphere M such that K is a component of the link and L is the union of the other d components. Then, $[M, L] = [M_K, L]$ in $\mathcal{M}_{(d)}/\mathcal{M}_{(d+1)}$.

Proof. Since $[M, L] - [M_K, L] = [M, K \cup L]$, the difference belongs to $\mathcal{M}_{(d+1)}$. Hence, we obtain the lemma. □

Lemma 11.2. For any $[M, L]$ with $\#L = d$, there exists some framed link L' of d components in the 3-sphere S^3 such that $[M, L] = [S^3, L']$ in $\mathcal{M}_{(d)}/\mathcal{M}_{(d+1)}$.

Proof. This lemma is shown by using Lemma 11.1 for a sequence of surgeries along knots to obtain S^3 from M. □

By the above lemma $\mathcal{M}_{(d)}/\mathcal{M}_{(d+1)}$ is spanned by $[S^3, L]$ such that $\#L = d$. In the next part of this section, we will mainly be concerned with the equivalence relation among framed links in S^3 induced by congruence modulo $\mathcal{M}_{(d+1)}$. For framed links L and L' with d components, we express the equality $[S^3, L] = [S^3, L']$ in $\mathcal{M}_{(d)}/\mathcal{M}_{(d+1)}$ by $L \sim L'$, and we often represent the alternating sum $[S^3, L]$ by a picture of L.

Lemma 11.3. We have that

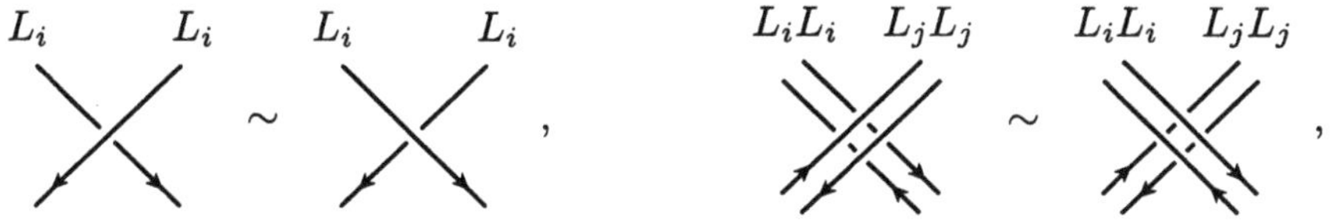

where both pictures of each relation imply two framed links which are identical except for a ball, where they differ as shown in the pictures. Further, when two strands are labeled by L_i (or L_j) in the picture, then they depict two parts of the same component of the framed link.

Proof. By Lemma 11.1, we have that

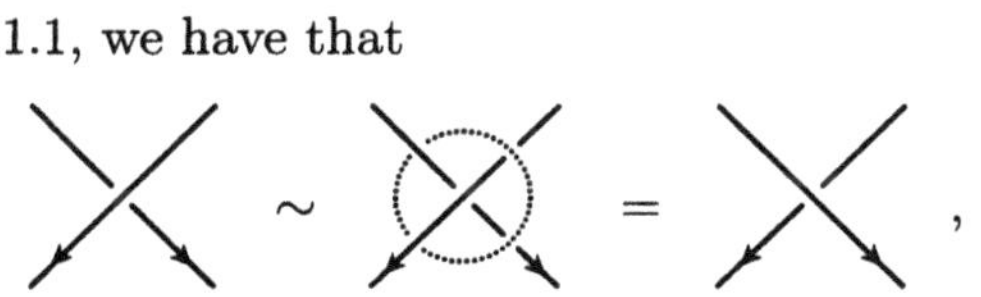

where the dotted component in the middle picture implies surgery along the component; *i.e.*, we mean the element $[S^3_K, L]$, where L is the link drawn with solid lines and K is the knot drawn with a dotted line. Hence, we obtain the first relation of the lemma.

We obtain the second relation in the same way, noting that we can change crossings of the same component by the first relation. □

Note that we can not apply the argument of this proof to a crossing of two different components of a link, because the dotted circle of the proof together with such a crossing does not form an algebraically split link.

What is the graded space $\mathcal{M}_{(0)}/\mathcal{M}_{(1)}$? The vector space $\mathcal{M}_{(0)}$ is spanned by integral homology 3-spheres. Further, its subspace $\mathcal{M}_{(1)}$ is spanned by $M - M_K$ for integral homology 3-spheres M and knots K in M. Since all integral homology 3-spheres are related to each other by a sequence of surgeries along knots, the quotient space $\mathcal{M}_{(0)}/\mathcal{M}_{(1)}$ is the one-dimensional vector space spanned by the class consisting of all integral homology 3-spheres.

What is the graded space $\mathcal{M}_{(1)}/\mathcal{M}_{(2)}$? According to Lemma 11.2, this space is spanned by $[S^3, K]$ for knots K in S^3. Further, since any knot can be obtained from the trivial knot by a sequence of crossing changes, by Lemma 11.3 any $[S^3, K]$ is congruent, modulo $\mathcal{M}_{(2)}$, to $[S^3, U_\pm]$, where $U_\pm$ denotes the trivial knot with ± 1 framing. Moreover,

$$[S^3, K] = S^3 - S^3_{U_\pm} = S^3 - S^3 = 0 \in \mathcal{M}_{(1)}/\mathcal{M}_{(2)}.$$

Hence, the graded space $\mathcal{M}_{(1)}/\mathcal{M}_{(2)}$ is the null vector space.

What is the graded space $\mathcal{M}_{(2)}/\mathcal{M}_{(3)}$? By the same argument as above, this space is spanned by $[S^3, L]$ such that L is a framed link with two components, each of which is isotopic to the trivial knot. Since L is algebraically split, such a framed link L is isotopic to the split union $U_\pm \sqcup U_\pm$ of two trivial knots. Further,

$$[S^3, U_\pm \sqcup U_\pm] = S^3 - S^3_{U_\pm} - S^3_{U_\pm} + S^3_{U_\pm \sqcup U_\pm} = S^3 - S^3 - S^3 + S^3 = 0 \in \mathcal{M}_{(2)}/\mathcal{M}_{(3)}.$$

Therefore, the graded space $\mathcal{M}_{(2)}/\mathcal{M}_{(3)}$ is the null vector space.

What is the graded space $\mathcal{M}_{(3)}/\mathcal{M}_{(4)}$? By the same argument as above, the space is spanned by $[S^3, L_1 \cup L_2 \cup L_3]$ such that $L_1 \cup L_2$ is the split union of two trivial knots. We consider an element γ in $\pi_1\big(S^3 - (L_1 \cup L_2)\big)$ such that L_3 is homotopic in $S^3 - (L_1 \cup L_2)$ to γ forgetting its base point and orientation. Since $L_1 \cup L_2 \cup L_3$ is algebraically split, γ belongs to the commutator subgroup of the fundamental group which is the normal closure of $[m_1, m_2]$, where $[m_1, m_2]$ denotes the commutator of the meridians m_1 and m_2 of L_1 and L_2 respectively. Hence, γ is presented by a product of elements of the form $g[m_1, m_2]g^{-1}$ for $g \in \pi_1\big(S^3 - (L_1 \cup L_2)\big)$. By Lemma 11.4 below we can reduce $g[m_1, m_2]g^{-1}$ to $[m_1, m_2]$ modulo $\mathcal{M}_{(4)}$. Therefore, γ is reduced to the element $[m_1, m_2]^n$ for some integer n. Further, by Lemma 11.5 below it can be reduced to a scalar multiple of a set of Borromean rings; for example, for $n = 3$, we have that

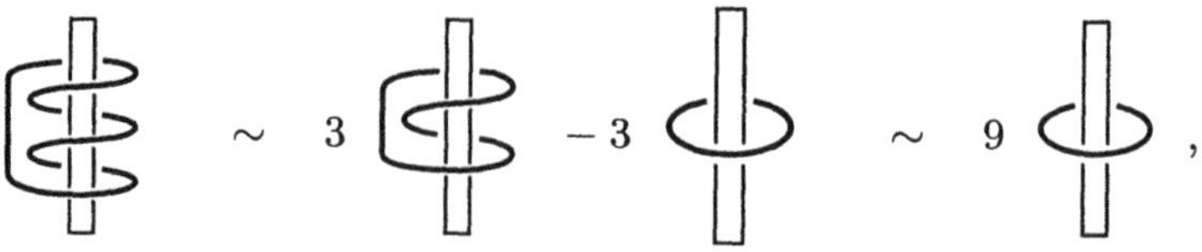

where in each picture the strand depicts L_3 and the band is a part of $L_1 \cup L_2$; recall that a picture of L implies the element $[S^3, L]$ in the formula. Hence, $[S^3, L_1 \cup L_2 \cup [m_1, m_2]^3]$ is congruent to $9 \cdot [S^3,$ Borromean rings$]$. By such arguments the space $\mathcal{M}_{(3)}/\mathcal{M}_{(4)}$ is spanned by only one element $[S^3,$ Borromean rings$]$. Further, since the Casson invariant induces a non-zero linear map $\mathcal{M}_{(3)}/\mathcal{M}_{(4)} \to \mathbb{C}$, by Proposition 11.9, the space $\mathcal{M}_{(3)}/\mathcal{M}_{(4)}$ is one dimensional.

Before investigating the graded space $\mathcal{M}_{(4)}/\mathcal{M}_{(5)}$ we show the following two lemmas, which are used both in the above argument and in the computation of $\mathcal{M}_{(4)}/\mathcal{M}_{(5)}$ later.

Lemma 11.4 (Habiro, see [Oht96a]**).** We have that

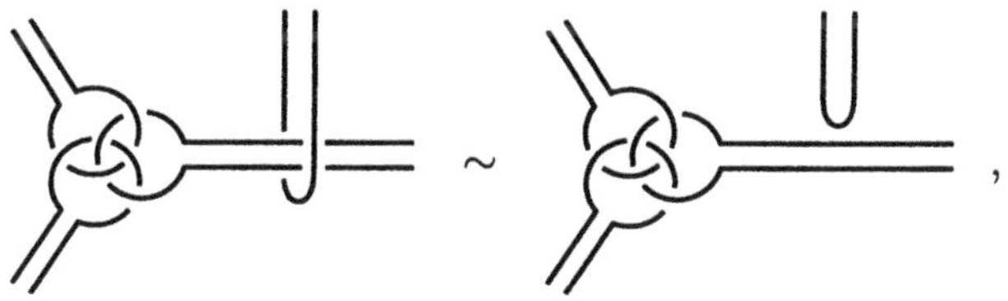

where the pictures imply two framed links which are identical except for a ball, where they differ as shown.

Proof. Applying the second relation of Lemma 11.3 to the part depicted in dotted lines in the following picture, we obtain the required relation.

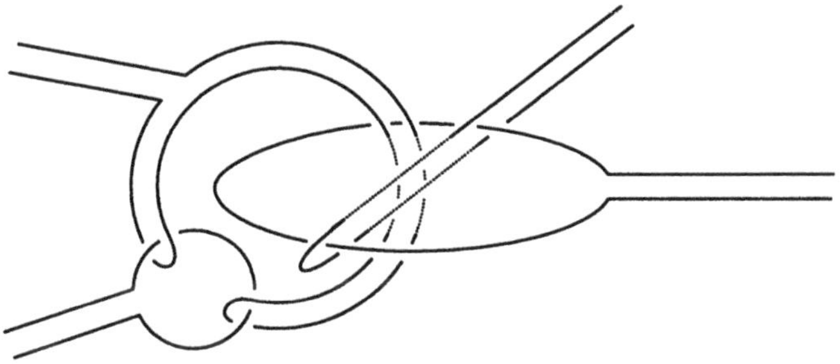

□

Lemma 11.5. We have the following two relations

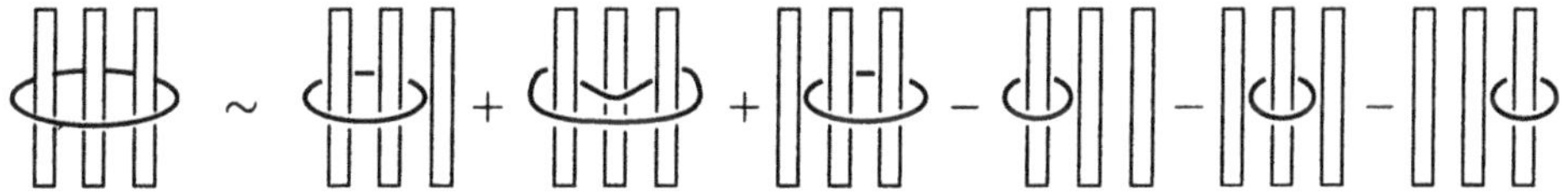

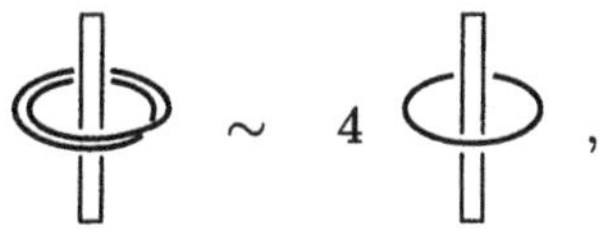

where each trivial knot in the picture has $+1$ framing and each band implies a (possibly different) bundle of pairs of parallel strands, such that each pair consists of two oppositely oriented strands of the same component, as shown in the following picture.

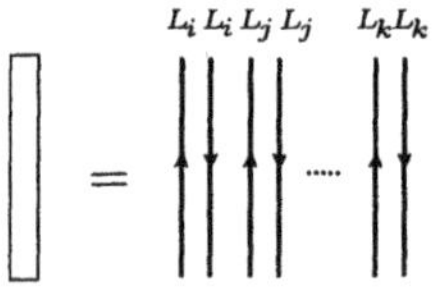

Proof. We obtain the second relation as follows. We have that

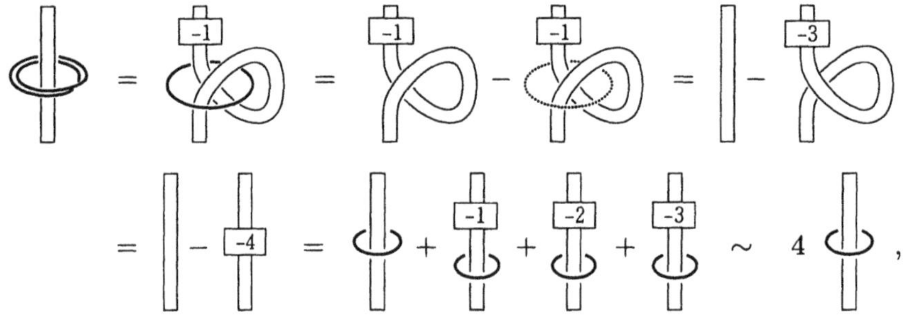

where each trivial knot in the picture has $+1$ framing and a number k in a box implies k full twists; recall that each framed link L in the picture implies the alternating sum $[S^3, L]$ and a framed link L with a dotted circle K implies $[S^3_K, L]$. Here, the third and fourth equalities are obtained by the relation $[S^3, K \cup L] = [S^3, L] - [S^3_K, L]$ for a union of a knot K and a link L, noting that a surgery along a trivial knot with $+1$ framing induces a full -1 twist of the strands intersecting a disc bounded by the trivial knot. Further, the last equivalence is obtained by Lemma 11.1. Hence, we obtain the second relation of the lemma.

We obtain the first relation in a similar way. We consider surgery along the trivial knot on the left hand side of the relation. This results in a -1 full twist of the bundle of 3 bands. Further, we resolve these full twists by a sequence of twists intersecting at most two bands. Then, we obtain the right hand side of the relation by a similar computation as the above case. □

What is the graded space $\mathcal{M}_{(4)}/\mathcal{M}_{(5)}$? By the same argument as for the case of $\mathcal{M}_{(3)}/\mathcal{M}_{(4)}$, the space is spanned by $[S^3, L_1 \cup L_2 \cup L_3 \cup L_4]$ such that $L_1 \cup L_2 \cup L_3$ is a set of Borromean rings, and L_4 is homotopic to a product of copies of commutators of the meridians of L_1, L_2 and L_3. By applying Lemmas 11.4 and 11.5 to L_4, we reduce L_4, writing the original element as a linear sum of elements, in each of which L_4 is replaced by either a commutator or the product of two commutators. Hence, the link is reduced to a framed link obtained from a uni-trivalent graph with 4 edges by the following map ϕ_d, as described in Figure 11.1.

We define the map

$$\phi_d : \mathrm{span}_{\mathbb{C}} \left\{ \begin{array}{l} \text{vertex-oriented uni-trivalent} \\ \text{graphs with } d \text{ edges} \end{array} \right\} \to \mathcal{M}_{(d)}/\mathcal{M}_{(d+1)}, \tag{11.1}$$

as follows; recall that we call a uni-trivalent graph *vertex-oriented* if a cyclic order of the three edges around each trivalent vertex of the graph is fixed. Let D be a vertex-oriented uni-trivalent graph with d edges. We associate a trivalent vertex of D with a disc and associate an edge of D with a band. We consider the orientable surface obtained by attaching the discs and the bands associated to the trivalent vertices and the edges of D such that we attach 3 bands around each disc using the vertex-orientation of the associated trivalent vertex. Further, we embed the surface in S^3 in an arbitrary way. Furthermore, by replacing each disc by a set of Borromean rings and replacing each band by its boundary, we obtain a link in S^3. By putting the framing of each component of the link to be 1, we obtain a framed link L in S^3. We define the image $\phi_d(D)$ to be $[S^3, L] \in \mathcal{M}_{(d)}/\mathcal{M}_{(d+1)}$. See Figure 11.1 for a graphical description of the definition of ϕ_d, making a framed link from a uni-trivalent graph. Note that the image $\phi_d(D)$ does not depend on the choice of embedding of the surface into S^3, by applying Lemma 11.3 to the framed link.

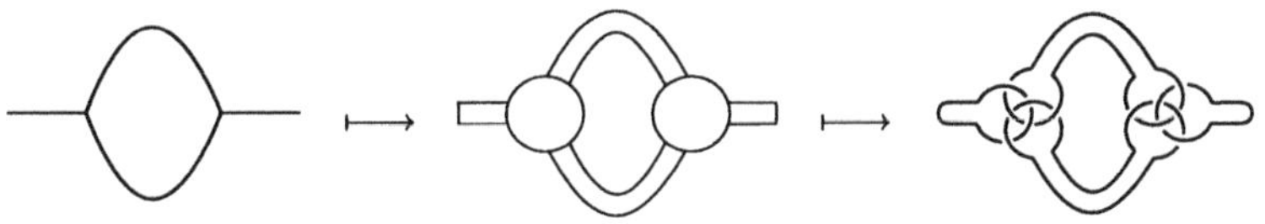

Figure 11.1 Definition of the map ϕ_d, which makes a framed link from a uni-trivalent graph

Proposition 11.6 ([Oht96a]). The map ϕ_d is surjective.

Proof. For $d \leq 4$, the proposition follows from the above argument for $\mathcal{M}_{(d)}/\mathcal{M}_{(d+1)}$.

For general d, we obtain the proposition by the following inductive argument on d. We start by assuming that the space $\mathcal{M}_{(d)}/\mathcal{M}_{(d+1)}$ is spanned by $[S^3, L]$ such that the union of the first $(d-1)$ components of L is obtained from a uni-trivalent graph by the map in Figure 11.1. By Lemmas 11.3 and 11.4 the last component of L can be replaced by a product of copies of commutators of meridians of the first $(d-1)$ components. Further, by the same argument as appears in the above case of $\mathcal{M}_{(4)}/\mathcal{M}_{(5)}$, such an element can be replaced by a linear sum of elements, in each of which the last component is replaced by either a commutator or the product of two commutators. Since such a link lies in the image of ϕ_d, the map ϕ_d is surjective. □

In the remainder of this section we will show that ϕ_{3d+1} and ϕ_{3d+2} are zero-maps, and that there exists a surjective linear map

$$\mathcal{A}(\emptyset)^{(d)} \to \mathrm{span}_{\mathbb{C}}\{\text{vertex-oriented uni-trivalent graphs with } 3d \text{ edges}\}/\mathrm{kernel}(\phi_{3d}),$$

where $\mathcal{A}(\emptyset)^{(d)}$ denotes the vector subspace of the space $\mathcal{A}(\emptyset)$ spanned by Jacobi diagrams on $\emptyset$ of degree d.

By composing this map and the map ϕ_{3d}, we obtain the map φ in the following theorem. We show the proof of the theorem at the end of this section.

Theorem 11.7 ([GaOh98]). For any non-negative integer d, we have that

$$\mathcal{M}_{(3d+1)}/\mathcal{M}_{(3d+2)} = 0,$$
$$\mathcal{M}_{(3d+2)}/\mathcal{M}_{(3d+3)} = 0.$$

Further, there exists a surjective linear map

$$\varphi : \mathcal{A}(\emptyset)^{(d)} \longrightarrow \mathcal{M}_{(3d)}/\mathcal{M}_{(3d+1)}.$$

Later, Theorem 11.21 will show that the map φ is actually an isomorphism.

Since the subscript of $\mathcal{M}_{(d)}$ is not effective, as shown in the theorem, we introduce the improved notation $\mathcal{M}_d = \mathcal{M}_{(3d)}$. Further, we define the degree of a finite type invariant by

Definition 11.8. A map $v : \mathcal{M} \to \mathbb{C}$ is called *finite type* of *degree d* if $v|_{\mathcal{M}_{d+1}} = 0$.

A finite type invariant v of degree d induces a map $\mathcal{M}_d/\mathcal{M}_{d+1} \to \mathbb{C}$. By composing this map and the map φ in Theorem 11.7 we obtain a map $\mathcal{A}(\emptyset)^{(d)} \to \mathbb{C}$. We call this map the *weight system* of v.

By the argument which followed Lemma 11.3, any finite type invariant of degree 0 is equal to a constant invariant. As for the next degree we have the following proposition, which can alternatively be obtained as a corollary of Theorem 11.19 and Lemma 10.11.

Proposition 11.9. The Casson invariant is a finite type invariant of degree 1. Further, any finite type invariant of degree 1 is equal to a linear sum of the Casson invariant and a constant.

Proof. Let M be an integral homology 3-sphere and let L be an algebraically split unit-framed link in M. By a formula of Hoste [Hos86] we have that

$$\sum_{L' \subset L} (-1)^{\# L'} \lambda(M_{L'}) = (-1)^{\# L} \psi(L) \prod f_i,$$

where $\lambda(M)$ denotes the Casson invariant of M, and f_i is the framing of the ith component of L, and $\psi(L)$ denotes the coefficient of $z^{\# L+1}$ in the Conway polynomial of L. Further, it is shown in [Hos86] that $\psi(L) = 0$ if L has more than 3 components. Hence, from the definition of finite type invariant, the Casson invariant is a finite type invariant of degree 1.

Further, by the argument which precedes Lemma 11.4, the dimension of $\mathcal{M}/\mathcal{M}_{(4)}$ is at most 2. Hence, the vector space of finite type invariants of degree 1 (which is the dual vector space of $\mathcal{M}/\mathcal{M}_{(4)}$) is spanned by the Casson invariant and a constant. □

The remainder of this section is devoted to the proof of Theorem 11.7.

To construct the map φ in Theorem 11.7 we investigate properties of the map ϕ_d given in (11.1). We consider uni-trivalent graphs with markings on their edges, and extend the definition of ϕ_d to the set of uni-trivalent graphs with markings such that ϕ_d takes a marking to a half twist as

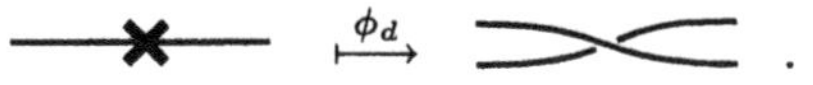

For (marked) uni-trivalent graphs D and D' with d edges, we call them equivalent, denoted by $D \sim D'$, if $\phi_d(D) = \phi_d(D')$.

Lemma 11.10. The marking can move past a trivalent vertex as

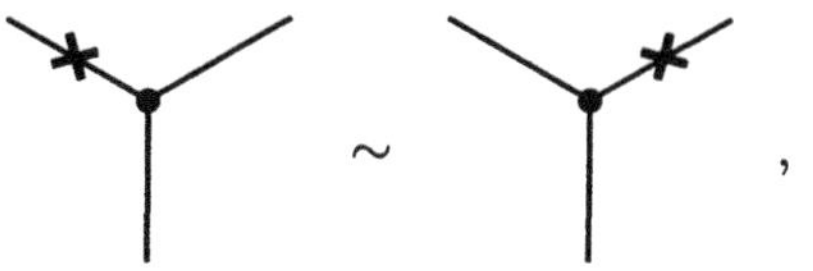

where the pictures imply two marked uni-trivalent graphs which are identical except for a neighborhood of a trivalent vertex, where they differ as shown in the pictures.

Proof. We denote by L_1, L_2, and L_3 the three components of the framed link which are the images of the three edges around the trivalent vertex by the map ϕ_d. In the same argument as in the computation of $\mathcal{M}_{(3)}/\mathcal{M}_{(4)}$ mentioned above, we can regard L_3 as the commutator $[m_1, m_2]$ of two meridians m_1 and m_2 of L_1 and L_2 respectively. If the marking appears on the first (resp. second) edge near the trivalent vertex, the commutator becomes $[m_1^{-1}, m_2]$ (resp. $[m_1, m_2^{-1}]$). Since these two commutators are conjugate in the fundamental group of $S^3 - (L_1 \cup L_2)$, they are equal in $\mathcal{M}_{(d)}/\mathcal{M}_{(d+1)}$. This implies the required relation. □

Lemma 11.11. We can remove a marking on an edge by the following relation

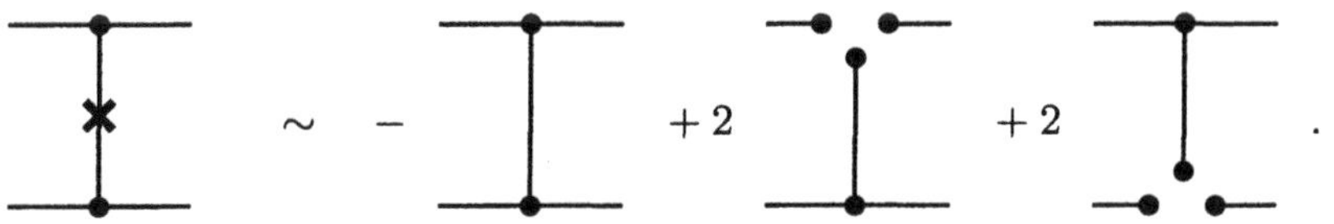

Proof. This relation is an immediate consequence of the relation

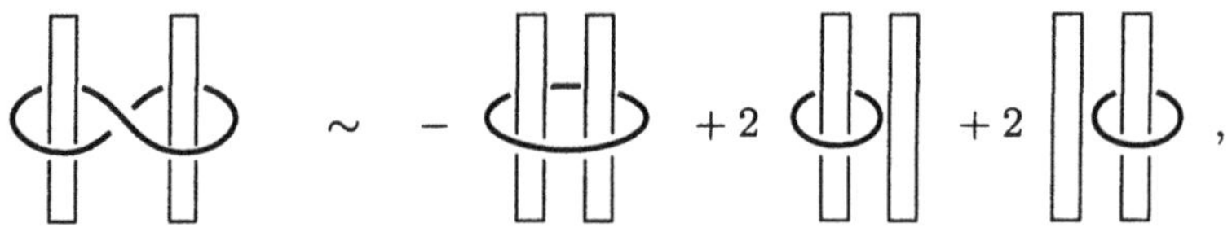

which can be obtained by a similar computation as in the proof of Lemma 11.5. □

In order to give a simpler form (Lemma 11.13 below) of Lemma 11.11, we introduce a white vertex by

By the linear extension of this formula, we regard a uni-trivalent graph with d white vertices as a linear sum of 2^d uni-trivalent graphs with the usual vertices.

Lemma 11.12. A marking can move past a white trivalent vertex.

Proof. The lemma follows immediately from Lemma 11.10 from the definition of a white vertex, noting that a marking near a univalent vertex can be removed because

□

Lemma 11.13. We have that

Proof. The lemma is obtained from Lemma 11.11 from the definition of a white vertex and the fact that a graph which includes a connected component with one edge and two univalent vertices is equivalent to zero. □

Lemma 11.14. We have that

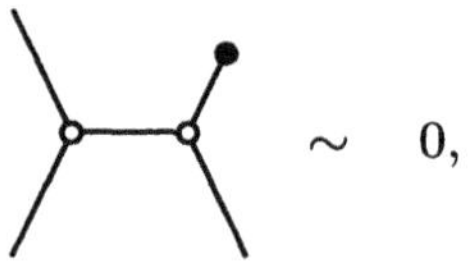

i.e., a graph containing a univalent vertex and at least two white vertices adjacent to it is equivalent to zero.

Proof. We obtain the lemma by

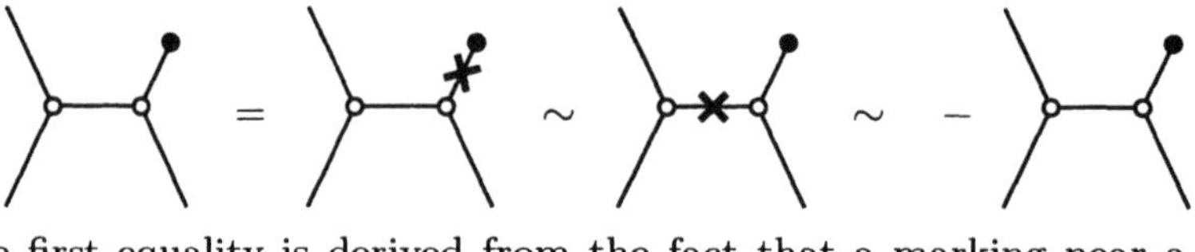

where the first equality is derived from the fact that a marking near a univalent vertex vanishes, as in the proof of Lemma 11.12, and we obtain the second and third equivalences by Lemmas 11.12 and 11.13 respectively. □

Proposition 11.15. The AS relation holds for uni-trivalent graphs with white vertices as

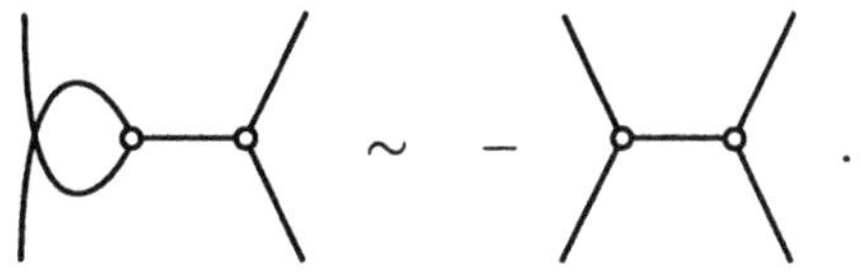

Proof. By rotating the left half of the graph on the left hand side of the relation we have that

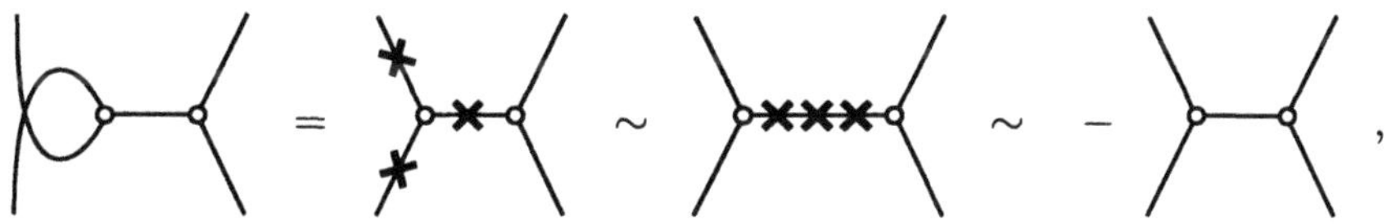

where the second and third equivalences are obtained by Lemmas 11.12 and 11.13 respectively. □

In order to prove the IHX relation, we begin with the following lemma.

Lemma 11.16. We have the following relation (similar to the STU relation),

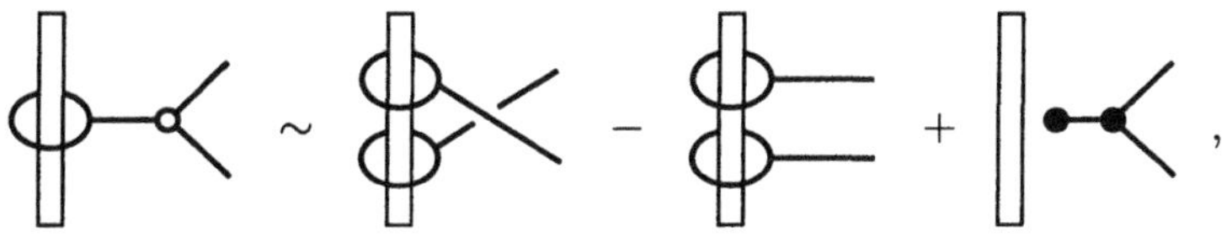

where the first and the last terms express uni-trivalent graphs with d edges and the middle two terms express uni-trivalent graphs with $d-1$ edges. Further, a loop at an end of graph winding around a band maps under the map $\phi_\star$ to a strand winding the band, as

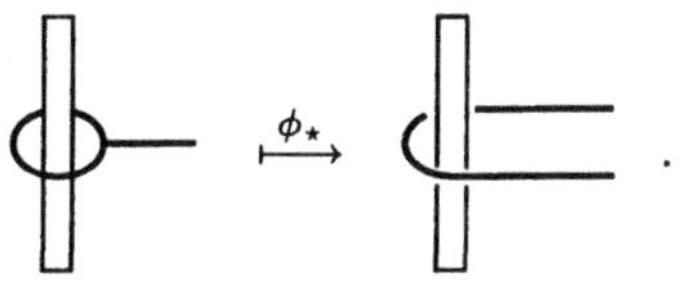

This lemma implies that the image of the difference of the middle two terms of the lemma under the map ϕ_{d-1} belongs to $\mathcal{M}_{(d)}$, even though the image of the individual terms do not, recalling that $\mathcal{M}_{(d)}$ is a vector subspace of $\mathcal{M}_{(d-1)}$.

Proof of Lemma 11.16. We consider the two framed links L and L' with d components with framings $+1$ shown in Figure 11.2.

Firstly, we calculate $[S^3, L] - [S^3, L']$ as follows. Let L_1 and L'_1 be the middle components of the framed links L and L' respectively. Note that, since L' is obtained

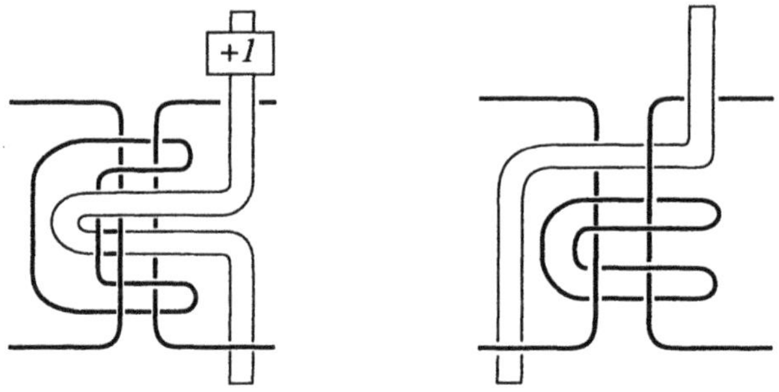

L $\qquad$ L'

Figure 11.2 Two framed links L and L'

from L (resp. L from L') by handle sliding the band over the component L_1 (resp. L'_1), the framed links $L - L_1$ in $S^3_{L_1}$ and $L' - L'_1$ in $S^3_{L'_1}$ are isotopic. Hence,

$$\begin{aligned}[S^3, L] - [S^3, L'] &= \big([S^3, L - L_1] - [S^3_{L_1}, L - L_1]\big) - \big([S^3, L' - L'_1] - [S^3_{L'_1}, L' - L'_1]\big) \\ &= [S^3, L - L_1] - [S^3, L' - L'_1].\end{aligned}$$

Further, this can be obtained as a linear sum of the images of the following uni-trivalent graphs under the map ϕ_d, as

$$[S^3, L - L_1] - [S^3, L' - L'_1] \xleftarrow{\phi_d}$$

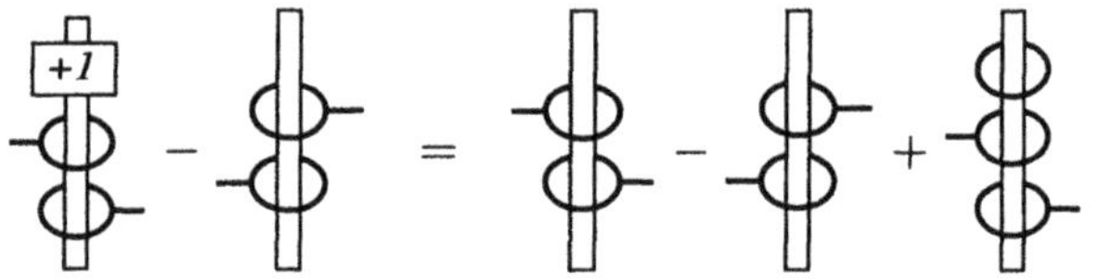

Hence,

$$[S^3, L] - [S^3, L'] \xleftarrow{\phi_d} \tag{11.2}$$

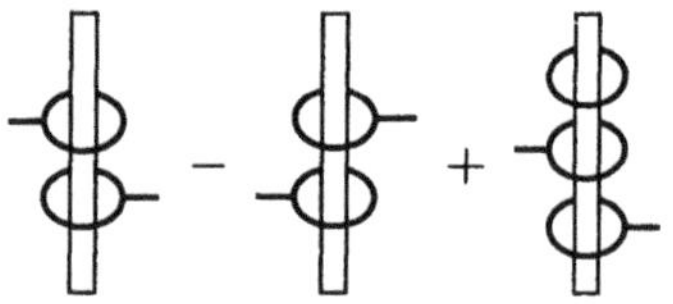

Secondly, we can calculate $[S^3, L]$ as follows. Along the lines of the proof of the surjectivity of ϕ_d in Proposition 11.6, we can find a linear sum of uni-trivalent graphs whose image is equal to $[S^3, L]$, as

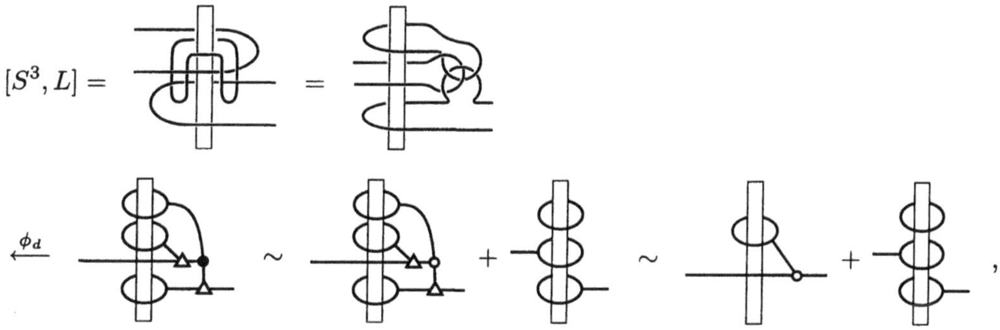

(11.3)

where we obtain the last equivalence by Lemma 11.17 below with a triangle vertex defined by

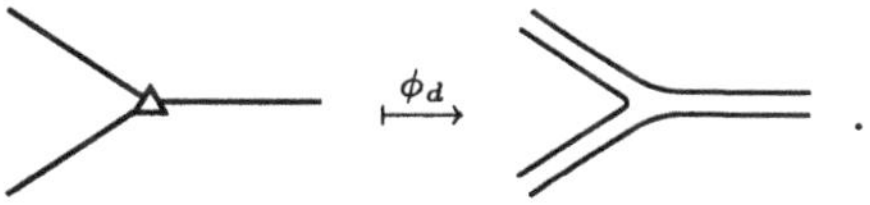

Thirdly, we calculate $[S^3, L']$ using the same argument as above, to obtain

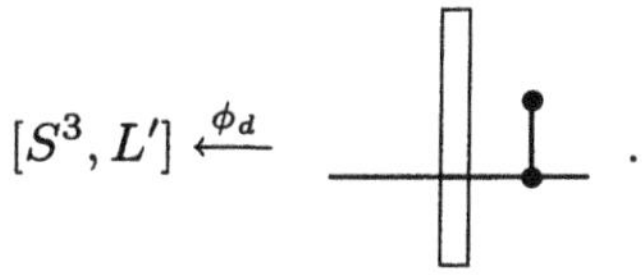

(11.4)

The detailed computation is left to the reader.

By (11.2), (11.3), and (11.4), we obtain the lemma. □

Lemma 11.17. We have the following equivalence,

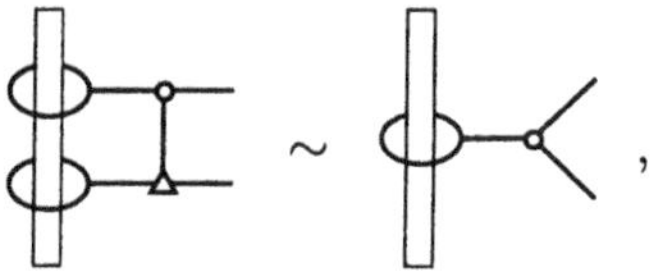

where a triangle vertex was defined in the above proof of Lemma 11.16.

Proof. Since we can change the order of two loops winding around a band by

Lemma 11.4, we have that

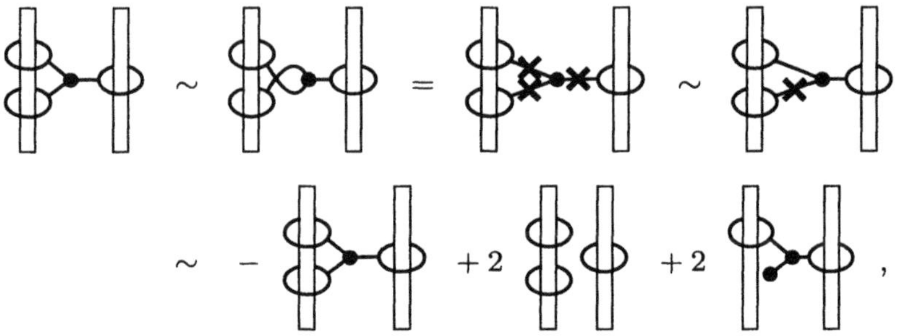

where we obtain the third equivalence by Lemma 11.10 and the fact that two markings adjacent on an edge (which imply a full twist of the two strands arising from the map ϕ_d) vanish by Lemma 11.3, and the last equivalence is derived from Lemma 11.11. Hence,

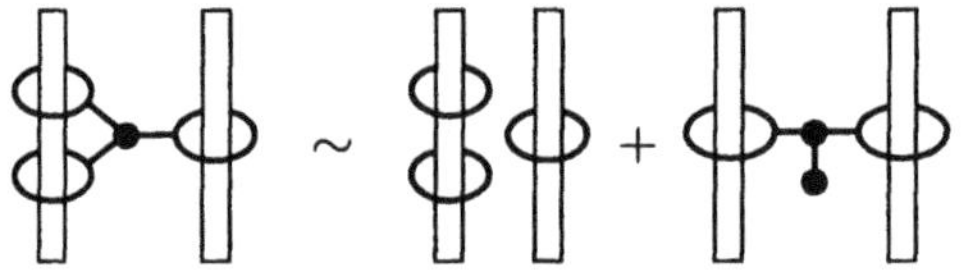

(11.5)

Further, by applying the first formula of Lemma 11.5 to the component with the triangle vertex, we have the second equivalence of the following formula,

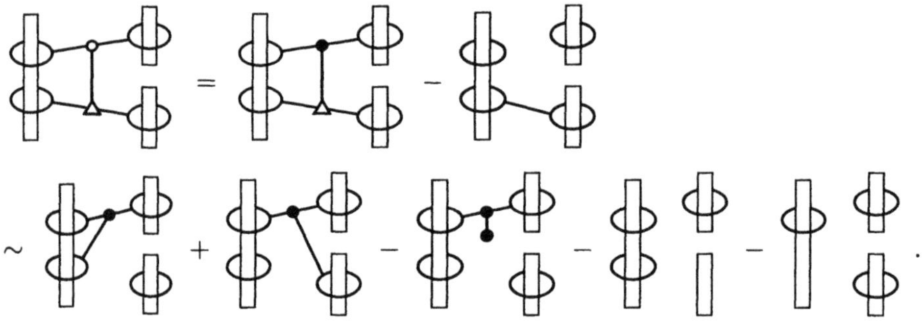

By applying (11.5) to the first term on the right hand side of the above formula, we obtain the required formula. □

Proposition 11.18. The IHX relation holds for uni-trivalent graphs with white vertices, as

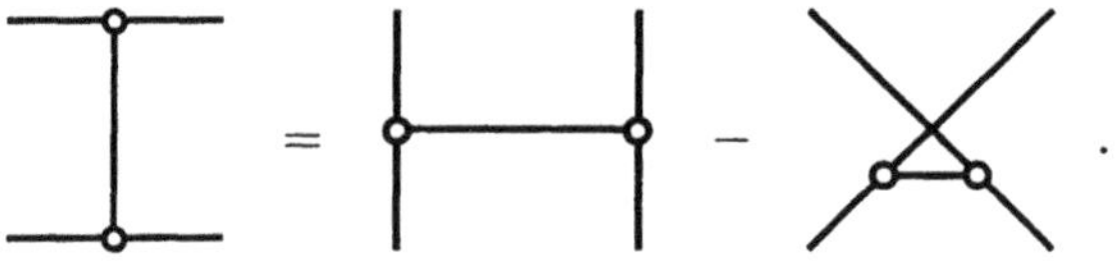

Proof. The idea of the proof is to apply Lemma 11.16 twice to two successive white vertices, as

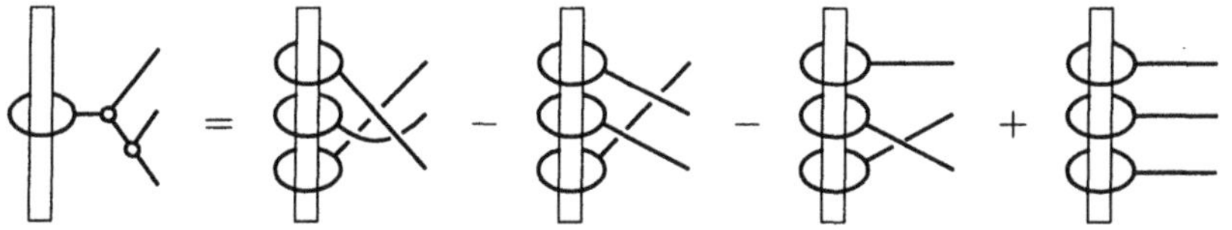

(11.6)

To be precise, we can not apply Lemma 11.16 twice in a direct way, because the two middle terms of the relation of Lemma 11.16 are uni-trivalent graphs with $d-1$ edges, though a uni-trivalent graph which we can apply Lemma 11.16 must have d edges. However, we can avoid this difficulty to obtain (11.6) along the lines of the above idea by some extra technical arguments; see [GaOh98] for detailed arguments.

By applying (11.6) to each of the following three terms, we obtain 12 terms which cancel with each other,

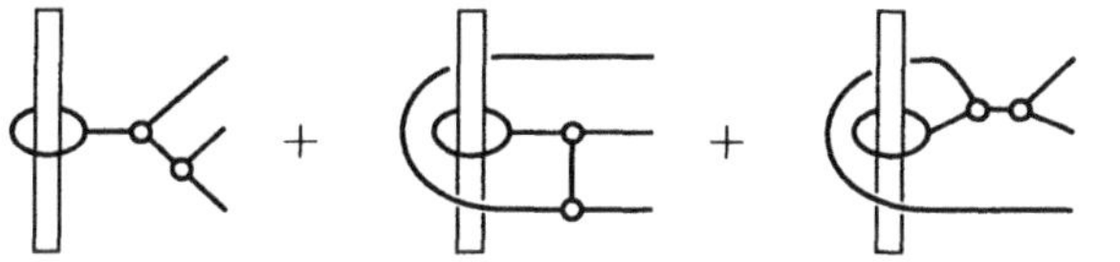

$= \text{(12 terms)} \sim 0.$

This formula implies the IHX relation. □

Proof of Theorem 11.7. The graded space $\mathcal{M}_{(d)}/\mathcal{M}_{(d+1)}$ was determined for $d \leq 2$ at the beginning of this section. Hence, we can confirm the theorem concretely for $d \leq 2$. Here, we show the proof in the cases $d \geq 3$.

Let d be not divisible by 3. Then, any uni-trivalent graph with d edges has at least one univalent vertex and at least two trivalent vertices. Hence, Lemma 11.14 implies that ϕ_d is the zero map. Since ϕ_d is surjective, $\mathcal{M}_{(d)}/\mathcal{M}_{(d+1)} = 0$.

Let d be divisible by 3. Then, by Lemma 11.14 the quotient space of uni-trivalent graphs by the kernel of ϕ_d is spanned by trivalent graphs with white vertices. By Propositions 11.15 and 11.18 the AS and IHX relations are in the kernel of ϕ_d. Hence, we can define a map from the space $\mathcal{A}(\emptyset)^{(d)}$ to the quotient space of uni-trivalent graphs by the kernel of ϕ_d by taking a trivalent vertex of a Jacobi diagram to a white vertex of a trivalent graph, as

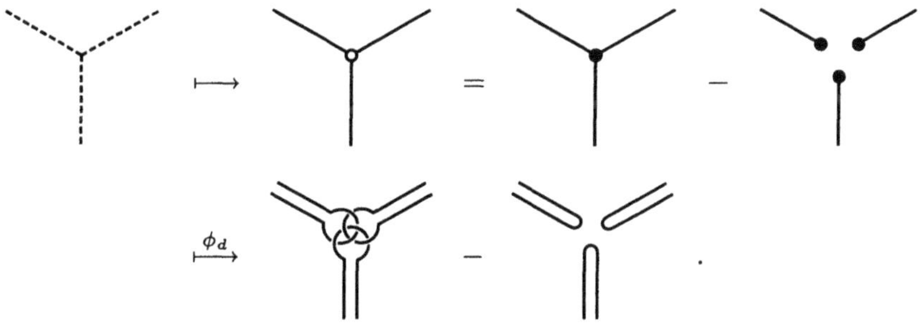

Here, for a Jacobi diagram D with $2d$ trivalent vertices we obtain a linear sum of 2^{2d} framed links L in S^3 by the above map. We define $\varphi(D)$ to be the corresponding

linear sum of $[S^3, L]$. Then, the map $\varphi : \mathcal{A}(\emptyset)^{(d)} \to \mathcal{M}_{(3d)}/\mathcal{M}_{(3d+1)}$ is a well-defined surjective linear map by the above arguments. □

Habiro [Habi00] gave a reconstruction of φ by using his claspers (see Figure 7.3 for the definition of a clasper) as

, (11.7)

where the middle picture is an integral homology 3-sphere M and the right picture is the integral homology 3-sphere obtained from M by a surgery along a *graph* of claspers, which is obtained from any embedding of the Jacobi diagram in the left picture by replacing each edge by a clasper, and each trivalent vertex by a set of Borromean rings. The image of a Jacobi diagram by φ in $\mathcal{M}_d/\mathcal{M}_{d+1}$ does not depend on the choice of an integral homology 3-sphere M, nor on the choice of an embedding of the Jacobi diagram into M. See Proposition E.28 for a proof of the fact that the above map reconstructions the map φ. This construction suggests that we can study how finely the finite type invariants distinguish integral homology 3-spheres by using a series of equivalence relations defined later in Section 11.3.

11.2 Universality of the LMO invariant among finite type invariants

The aim of this section is to show the universality of the LMO invariant Z^{LMO} among finite type invariants of integral homology 3-spheres (Theorem 11.19 below). Hence, we call the LMO invariant Z^{LMO} the "universal finite type invariant" of integral homology 3-spheres.

Theorem 11.19 ([Le97]). For any positive integer d, any finite type invariant v of degree d is presented by a composite map

$$v : \{\text{integral homology 3-spheres}\} \xrightarrow{Z^{\text{LMO}}} \hat{\mathcal{A}}(\emptyset) \xrightarrow{\text{projection}} \mathcal{A}(\emptyset)^{(\leq d)} \xrightarrow{W} \mathbb{C},$$

for some linear map W. Conversely, for any linear map $W : \mathcal{A}(\emptyset)^{(\leq d)} \to \mathbb{C}$ the above composite map v is of finite type of degree d.

Proof. The proof is reduced to Theorem 11.21 below in the same way as the proof of Theorem 7.4, which reduces the proof of the universality of the Kontsevich invariant among Vassiliev invariants to the theorem that the spaces of Jacobi diagrams are isomorphic to the associated graded spaces of the Vassiliev filtration. □

As a corollary of Theorems 10.17 and 11.19 we have the following corollary. Historically speaking, this corollary had been directly proved for $n = 2$ by Kricker-Spence [KrSp97], before the theorems were proved.

Corollary 11.20. The d-th coefficient of the perturbative $PSU(n)$ invariant is of finite type of degree d. Moreover, the weight system of the finite type invariant is equal to W_{sl_n}.

Proof. By Theorem 10.17 the d-th coefficient of the perturbative $PSU(n)$ invariant is obtained from the degree d part of the LMO invariant through the weight system W_{sl_n}. Hence, the d-th coefficient is obtained as the map in Theorem 11.19 when the linear map W is put to be the weight system W_{sl_n} for Jacobi diagrams of degree d. By Theorem 11.19 it is a finite type invariant of degree d whose weight system is equal to W_{sl_n}. □

Theorem 11.21 ([Le97]). The LMO invariant Z^{LMO} induces the inverse map of the surjective map $\varphi : \mathcal{A}(\emptyset)^{(d)} \to \mathcal{M}_d/\mathcal{M}_{d+1}$ given in Theorem 11.7. In particular, the map φ is an isomorphism, and hence the graded space $\mathcal{M}_d/\mathcal{M}_{d+1}$ is isomorphic to $\mathcal{A}(\emptyset)^{(d)}$.

We devote the remainder of this section to proving Theorem 11.21 following [Le97]. We begin with the following lemma.

Lemma 11.22. We have that

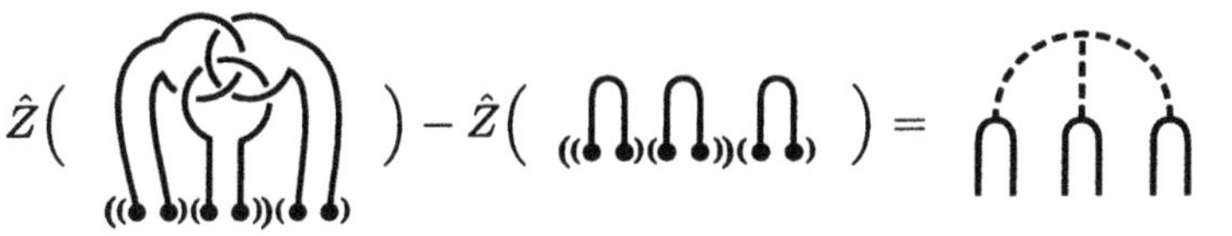

+ (remainder),

where the remainder is a linear sum of Jacobi diagrams, each of which has at least two trivalent vertices. Moreover, the uni-trivalent graph of each Jacobi diagram in the remainder is connected to every solid component.

Proof. We define two quasi-tangles b_1 and b_2 by

By the combinatorial definition of the framed Kontsevich invariant given in Section 6.4, we have that

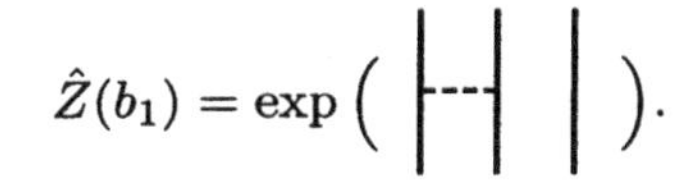

Further, $\hat{Z}(b_2)$ is obtained as the conjugate of exp (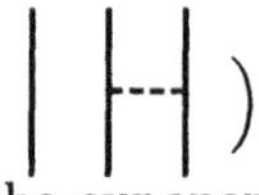) by the associator Φ. Since $\hat{Z}(b_2)$ is group-like, it can be presented as the exponential of a primitive element, as

$$\hat{Z}(b_2) = \exp\Big($$  $$+ \text{(remainder)}\Big),$$

where the remainder is a linear sum of primitive Jacobi diagrams of degree ≥ 3. Here, a primitive Jacobi diagram is a Jacobi diagram with a connected uni-trivalent graph. Hence,

$$\hat{Z}(b_2 b_1 b_2^{-1} b_1^{-1}) =$$ 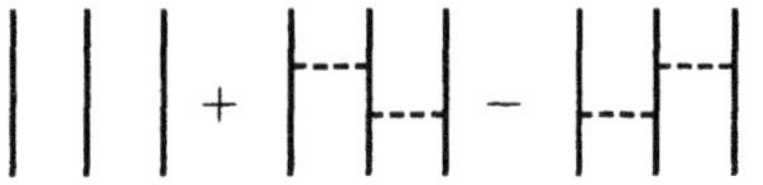 $$+ \text{ (terms of degree} \geq 3)$$

$$= \exp\Big($$  $$+ \text{ (primitive terms of degree} \geq 3)\Big)$$

$$=$$ 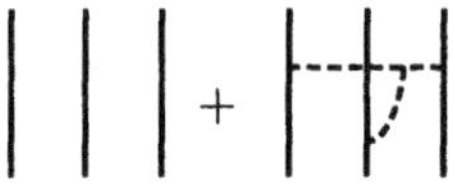 $$+ \text{(remainder)}, \qquad (11.8)$$

where the remainder is a linear sum of Jacobi diagrams, each of which has at least two trivalent vertices. The left hand side of the required formula is obtained by substituting the difference of the quasi-tangle $b_2 b_1 b_2^{-1} b_1^{-1}$ and the trivial quasi-tangle into the box in the following quasi-tangle:

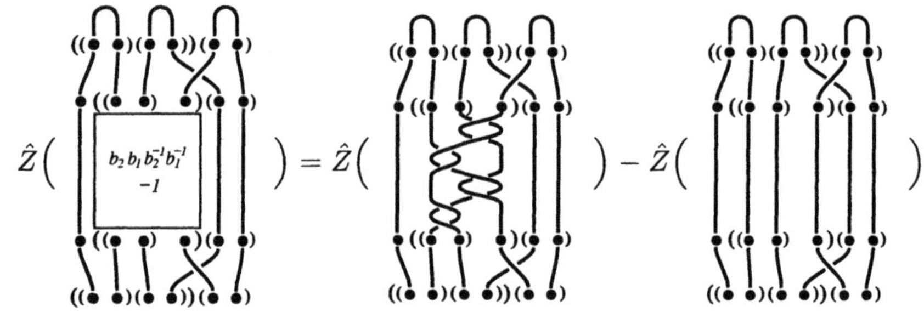

By (11.8) we obtain the above difference from

 $+ \Big($ linear sum of Jacobi diagrams, each of which has at least two trivalent vertices $\Big)$,

by conjugating by terms constructed from associators and by closing the 3 pairs of upper ends. Since the Kontsevich invariant of a critical point is equal to $\nu^{1/2}$, the difference is equal to the right hand side of the required formula.

Moreover, when we apply the counit ε to each solid component on the left hand side of the required formula, then the left hand side vanishes, by Proposition 6.8, because Borromean rings split when we remove one of its components. Hence, the right hand side of the required formula consists of Jacobi diagrams whose dashed graphs are connected to every solid component. □

By using the above lemma, we show that Z_d^{LMO} gives the left inverse of the map φ as

Lemma 11.23. Let D be a Jacobi diagram on $\emptyset$ of degree d. For the map φ in Theorem 11.7,

$$Z_d^{\text{LMO}}(\varphi(D)) = D \in \mathcal{A}(\emptyset)/D_{>d},$$

where $D_{>d}$ denotes the relation which sets Jacobi diagrams of degree $> d$ to 0.

Proof. As in the proof of Theorem 11.7, we obtain the alternating sum L_D of links from a Jacobi diagram D as

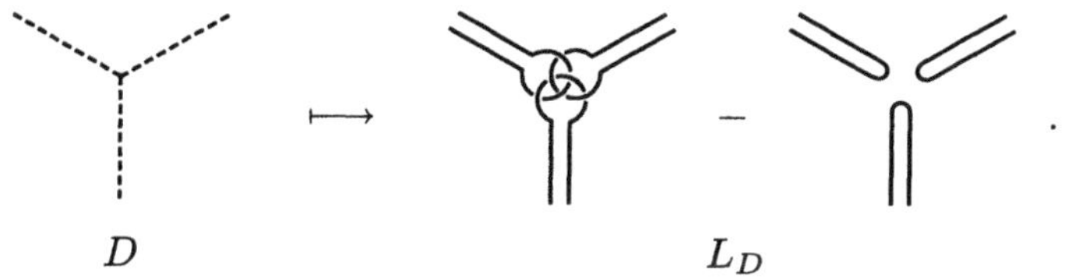

(11.9)

We put the alternating sum L_D to be $\sum_L \varepsilon_L L$ with $\varepsilon_L = \pm 1$. This alternating sum runs over 2^{2d} links L since in the definition of L_D we have 2^{2d} configurations which arise by choosing either the first or the second term in the image of (11.9) at each of $2d$ trivalent vertices of D. Then, by the construction of $\varphi(D)$, we have that

$$\varphi(D) = [S^3, L_D] = \sum_L \varepsilon_L [S^3, L] = \sum_L^{2^{2d}} \varepsilon_L \sum_{L' \subset L}^{2^{3d}} (-1)^{\# L'} S^3_{L'}, \tag{11.10}$$

where the first sum on the right hand side runs over the 2^{2d} links L in the alternating sum L_D and the second sum runs over the 2^{3d} sublinks L' of the link L with $3d$ components. In other words, the first sum is over the set of choices such that in each choice we choose either the first or the second term in the image of (11.9), and the second sum is over the set of $3d$ choices such that in each choice we determine, for each of the $3d$ edges of D, whether we let it remain or we remove it, where the edges that remain correspond to components of L'. Hence, the two sums can be regarded as independent. Thus, we can reorder them as

$$\varphi(D) = \sum_{\substack{\text{choices of} \\ 3d \text{ edges}}}^{2^{3d}} \; \sum_{\substack{\text{choices of} \\ 2d \text{ vertices}}}^{2^{2d}} \varepsilon_L (-1)^{\# L'} S^3_{L'}. \tag{11.11}$$

If we do not let all edges of D remain in the above first sum, then the second sum,

$$\sum_{\substack{\text{choices of} \\ 2d \text{ vertices}}}^{2^{2d}} \varepsilon_L (-1)^{\# L'} S^3_{L'},$$

vanishes, since the difference in the image of (11.9) is equal to 0, when one of the components is removed; note that Borromean rings split if one of Borromean rings is removed. Therefore, the double summation (11.11) reduces to the case where we choose all $3d$ edges in the first sum. This implies that the sums (11.10) reduce to the case $L' = L$. Hence, recalling that L has $3d$ components, we have that

$$\varphi(D) = \sum_L \varepsilon_L (-1)^d S^3_L.$$

From the definition of Z^{LMO}_d,

$$Z^{\mathrm{LMO}}_d(\varphi(D)) = \sum_L \varepsilon_L (-1)^d Z^{\mathrm{LMO}}_d(S^3_L) = \sum_L \varepsilon_L (-1)^d \frac{\iota_d \check{Z}(L)}{\iota_d \check{Z}(U_+)^{3d}} = (-1)^d \frac{\iota_d \check{Z}(L_D)}{\iota_d \check{Z}(U_+)^{3d}}. \tag{11.12}$$

Further, as in Section 10.2, we have that

$$\iota_d \check{Z}(U_+) = (-1)^d + (\text{terms of degree} > 0).$$

Hence, to obtain the required formula, it is sufficient to show

$$\iota_d \check{Z}(L_D) = D \in \mathcal{A}(\emptyset)^{(d)} / D_{>d}. \tag{11.13}$$

The remainder of this proof is concerned with showing (11.13). We express the alternating sum L_D as

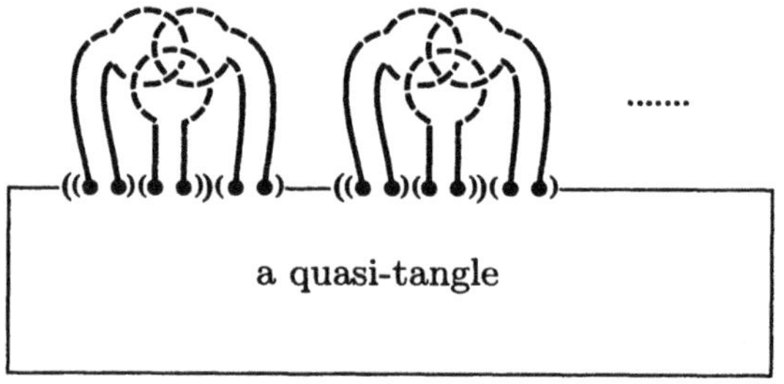

,

where a set of Borromean rings written in broken lines denotes the following difference,

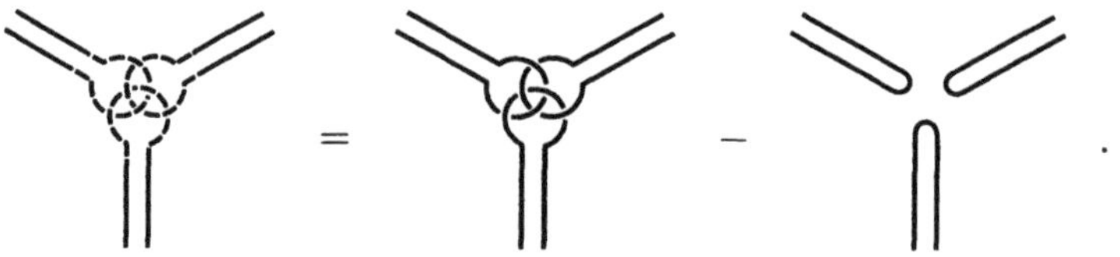

By Lemma 11.22, a Jacobi diagram D of degree d, say the diagram 

for $d = 2$, is taken to

$$\hat{Z}(L_D) =$$ 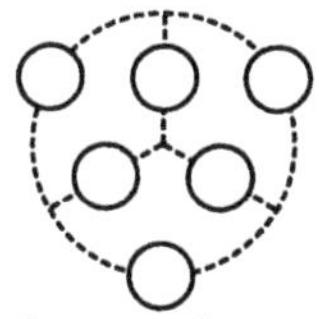 $$\# \exp\Big(\frac{1}{2}$$ 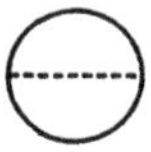 $$\Big)^{\otimes 3d} + \text{(remainder)}, \qquad (11.14)$$

where the remainder is a linear sum of Jacobi diagrams, each of which has at least $2d+1$ trivalent vertices. Here, the first Jacobi diagram is obtained from D by inserting a solid circle into each edge of D, and the $3d$ copies of the exponential term are connect-summed into the solid circles of the first Jacobi diagram; note that these exponential terms are derived from the fact that we gave a $+1$ framing to each component of the framed links in the definition of L_D. Further, $\check{Z}(L_D)$ is also presented by the same formula as the right hand side of (11.14) from the definitions of $\check{Z}$ and ν. Moreover, from the definition of ι_d, we have that

$$\iota_d\Big($$ 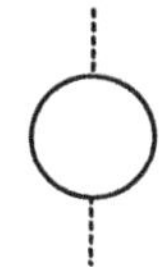 $$\# \exp\Big(\frac{1}{2}$$ 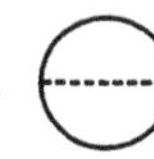 $$\Big)\Big) = (-1)^{d-1}$$ $$+ \begin{pmatrix} \text{terms with at least} \\ \text{one trivalent vertex} \\ \text{in this part} \end{pmatrix}.$$

By applying the above formula to the right hand side of (11.14) we obtain the Jacobi diagram D again. This implies the formula (11.13). □

Lemma 11.24. We have that $Z_d^{\mathrm{LMO}}(\mathcal{M}_{2d+1}) = 0$.

Proof. Recall that Theorem 11.7 asserts that $\mathcal{M}_{2d+1}$ is isomorphic to the space $\mathcal{M}_{(6d+1)}$, and further recall that that space is spanned by $[M, L]$ such that M is an integral homology 3-sphere and L is an algebraically-split unit-framed link with $6d+1$ components in M. We can obtain M from S^3 by integral surgery along a framed link K with k components in S^3. We have the link in S^3 obtained from L through inverse surgery along K. We denote the link also by L, abusing notation. We can assume, without loss of generosity, that $K \cup L$ is an algebraically-split unit-framed link with $k + (6d+1)$ components in S^3. It is sufficient to show that $Z_d^{\mathrm{LMO}}([M, L]) = 0$. From the definition of the bracket, we have that

$$Z_d^{\mathrm{LMO}}([M, L]) = \sum_{L' \subset L} (-1)^{\# L'} Z_d^{\mathrm{LMO}}(M_{L'}) = \sum_{L' \subset L} (-1)^{\# L'} Z_d^{\mathrm{LMO}}(S^3_{K \cup L'}).$$

For a framed link J with j components we denote by J° the split union of j trivial knots with the same framings as the components of J. Noting that $S^3_{K \cup L'}$ is homeomorphic to $S^3_{K \cup L' \sqcup (L-L')^\circ}$, we have that

$$Z_d^{\mathrm{LMO}}(S^3_{K \cup L'}) = Z_d^{\mathrm{LMO}}(S^3_{K \cup L' \sqcup (L-L')^\circ}) = \frac{\iota_d \check{Z}(K \cup L' \sqcup (L-L')^\circ)}{\iota_d \check{Z}(U_+)^{\sigma_+} \iota_d \check{Z}(U_-)^{\sigma_-}},$$

from the definition of Z_d^{LMO}, where $U_\pm$ denotes the trivial knot with ± 1 framing and $\sigma_\pm$ denotes the number of ± 1 framings of the components of L (and hence, of

$K \cup L \sqcup (L - L')^\circ$). Therefore, to obtain the required formula, it is sufficient to show that the alternating sum

$$\sum_{L' \subset L} (-1)^{\# L'} \iota_d \check{Z}(K \cup L' \sqcup (L - L')^\circ) \in \mathcal{A}(\emptyset)/D_{>d} \tag{11.15}$$

vanishes in $\mathcal{A}(\emptyset)/D_{>d}$. We show this vanishing in the remainder of this proof.

We show that the alternating sum

$$\sum_{L' \subset L} (-1)^{\# L'} \hat{Z}(K \cup L' \sqcup (L - L')^\circ) \in \mathcal{A}((\sqcup^k S^1) \sqcup (\sqcup^{6d+1} S^1)) \tag{11.16}$$

is equal to a linear sum of Jacobi diagrams, each of which has at least $2d+1$ trivalent vertices as follows. Let L_1 be any component of L. Then, the alternating sum (11.16) is equal to

$$\sum_{L' \subset L - L_1} (-1)^{\# L'} \Big(\hat{Z}(K \cup L' \sqcup (L - L')^\circ) - \hat{Z}(K \cup L_1 \cup L' \sqcup (L - (L_1 \cup L'))^\circ) \Big). \tag{11.17}$$

The second link $K \cup L_1 \cup L' \sqcup (L - (L_1 \cup L'))^\circ$ in the above formula is obtained from the first link $K \cup L' \sqcup (L - L')^\circ$ by removing the component L_1 and adding a trivial knot with the same framing. Since both links have the same linking matrix, they are related by a sequence of Δ-moves which involve the components of L_1. Hence, by Lemma 11.22, the difference of their values under the Kontsevich invariant is equal to a linear sum of Jacobi diagrams, each of which has a dashed component connected to L_1 with at least one trivalent vertex. Therefore, the alternating sum (11.17) is also a linear sum of such Jacobi diagrams for any component L_1 of L. Since L has $6d+1$ components, this implies that (11.16) is a linear sum of Jacobi diagrams, each of which has at least $2d+1$ trivalent vertices, by an elementary counting argument.

Since we obtain $\check{Z}$ from $\hat{Z}$ by taking a connected sum of ν into each solid component, the alternating sum obtained from (11.16) by replacing $\hat{Z}$ with $\check{Z}$ also consists of Jacobi diagrams with at least $2d+1$ trivalent vertices. Noting that the map ι_d does not decrease the number of trivalent vertices, the image of the alternating sum by ι_d consists of Jacobi diagrams with at least $2d+1$ trivalent vertices. This implies that the image (11.15) vanishes in $\mathcal{A}(\emptyset)/D_{>d}$. □

Proof of Theorem 11.21. Let D be a Jacobi diagram of degree d. Then, for the map φ given in Theorem 11.7, we have that

$$Z^{\mathrm{LMO}}(\varphi(D)) = D + (\text{terms of degree} > d), \tag{11.18}$$

by Lemma 11.23, noting that the degree $\leq d$ part of Z^{LMO} is equal to Z_d^{LMO} for integral homology 3-spheres by Lemma 10.11 and from the definition of the LMO invariant.

Further, we show that

$$Z^{\mathrm{LMO}}(\mathcal{M}_d) \subset \mathcal{A}(\emptyset)^{(\geq d)} \tag{11.19}$$

as follows. Since the map $\varphi : \mathcal{A}(\emptyset) \to \mathcal{M}_d/\mathcal{M}_{d+1}$ is surjective, $\mathcal{M}_d$ is spanned by $\mathcal{M}_{d+1}$ and $\varphi(D)$ for Jacobi diagrams D of degree d. By (11.18) for a Jacobi diagram D of degree d, the image $Z^{\mathrm{LMO}}(\varphi(D))$ belongs to $\mathcal{A}(\emptyset)^{(\geq d)}$. Hence, to obtain (11.19), it is sufficient to show the inclusion $Z^{\mathrm{LMO}}(\mathcal{M}_{d+1}) \subset \mathcal{A}(\emptyset)^{(\geq d)}$. Repeating the same argument d times we reduce (11.19) to the inclusion $Z^{\mathrm{LMO}}(\mathcal{M}_{2d-1}) \subset \mathcal{A}(\emptyset)^{(\geq d)}$, which holds by Lemma 11.24. Hence, we obtain (11.19).

By replacing d with $d+1$ in (11.19) we have that $Z^{\mathrm{LMO}}(\mathcal{M}_{d+1}) \subset \mathcal{A}(\emptyset)^{(\geq d+1)}$. Taking the quotient of (11.19) by this formula, Z^{LMO} induces the linear map

$$\mathcal{M}_d/\mathcal{M}_{d+1} \to \mathcal{A}(\emptyset)^{(\geq d+1)}/\mathcal{A}(\emptyset)^{(\geq d)} \cong \mathcal{A}(\emptyset)^{(d)}.$$

This map is the inverse of φ by (11.18). □

The relations between the invariants of Chapters 8–11

As mentioned in the introduction to this chapter, we expected certain relations between the LMO invariant, finite type invariants and the perturbative invariants, which are obtained from the quantum invariants by the arithmetic expansion. We present these relations in Figure 11.3.

The LMO invariant has two universalities. One is the universality among perturbative invariants; for a simple Lie group G the perturbative G invariant τ^G is presented by

$$\tau^G = \hat{W}_{\mathfrak{g}} \circ Z^{\mathrm{LMO}},$$

with the graded weight system $\hat{W}_{\mathfrak{g}}$ derived from the substitution of the Lie algebra $\mathfrak{g}$ of G into Jacobi diagrams (Theorem 10.17). The other is the universality among finite type invariants; each finite type invariant v is presented by

$$v = W \circ Z^{\mathrm{LMO}},$$

for some weight system W (Theorem 11.19).

As a corollary of the two universalities, we obtain a relation between perturbative invariants and finite type invariants; the coefficients of the perturbative G invariant are finite type invariants and their weight systems are equal to $W_{\mathfrak{g}}$ (Corollary 11.20).

11.3 A descending series of equivalence relations among homology 3-spheres

How finely do quantum invariants distinguish integral homology 3-spheres? Since it has been conjectured that perturbative invariants dominate quantum invariants, an answer to this problem might follow from an answer to the same problem for

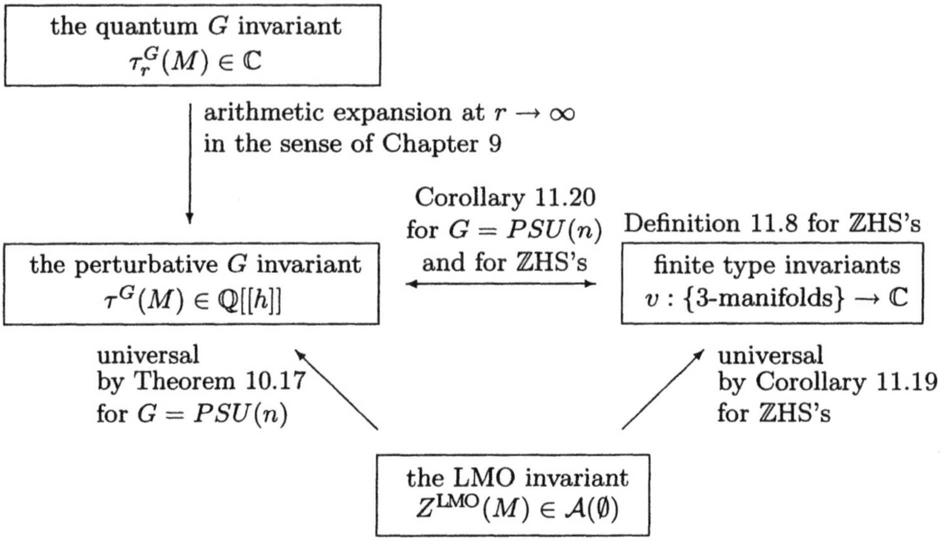

Figure 11.3 Invariants of 3-manifolds and the relations between them

perturbative invariants. Furthermore, since (as shown by Theorem 10.17) the LMO invariant is universal among perturbative invariants, this problem might be reduced to the corresponding problem for the LMO invariant $Z^{\text{LMO}}(M)$. Finally, since $Z^{\text{LMO}}(M)$ can be presented $Z^{\text{LMO}}(M) = \exp\big(z^{\text{LMO}}(M)\big)$ with the primitive LMO invariant $z^{\text{LMO}}(M)$ introduced in Section 10.2, we shall concentrate on the following problem: how finely does the primitive LMO invariant $z^{\text{LMO}}(M)$ distinguish integral homology 3-spheres? This problem is the subject of this section.

Let $\mathbb{M}$ denote the set of integral homology 3-spheres. The set $\mathbb{M}$ becomes a commutative semigroup with the sum given by the connected sum of integral homology 3-spheres. In this section we let $\hat{\mathcal{A}}(\emptyset)$ denote the completion of $\mathcal{A}(\emptyset)$ with respect to the degree of Jacobi diagrams (note that we confuse these spaces in earlier sections, using $\mathcal{A}(\emptyset)$ to denote both).

A finite type invariant v is called *primitive* if $v(M_1 \# M_2) = v(M_1) + v(M_2)$ for any two integral homology 3-spheres M_1 and M_2, where $M_1 \# M_2$ denotes the connected sum of M_1 and M_2. In other words, a primitive finite type invariant is a finite type invariant preserving the additive structure of the semigroup $\mathbb{M}$.

Theorem 11.25. The primitive LMO invariant z^{LMO} is universal among primitive finite type invariants. That is, for any positive integer d, any primitive finite type invariant v of degree d is presented by a composite map,

$$v : \mathbb{M} \xrightarrow{z^{\text{LMO}}} \hat{\mathcal{A}}(\emptyset)_{\text{prim}} \xrightarrow{\text{projection}} \mathcal{A}(\emptyset)^{(\leq d)}_{\text{prim}} \xrightarrow{W} \mathbb{C},$$

for some linear map W. Conversely, for any linear map $W : \mathcal{A}(\emptyset)^{(\leq d)}_{\text{prim}} \to \mathbb{C}$ the above composite map v is a primitive finite type invariant of degree d.

The theorem is obtained in the same way as Theorem 7.6 which asserts the universality of the primitive Kontsevich invariant among primitive Vassiliev invariants. In the same way as we obtained Corollary 7.7 we have

Corollary 11.26. Any finite type invariant of degree d is equal to a polynomial in primitive finite type invariants of degree $\leq d$.

Noting that the LMO invariant is a universal finite type invariant we consider the following problem mentioned at the beginning of this section: how finely do finite type invariants distinguish integral homology 3-spheres? To distinguish them, we define the $Y_d^{\mathbb{Q}}$-equivalence relation $\underset{(d)}{\sim}$ among integral homology 3-spheres as follows. Two integral homology 3-spheres M and M' are *$Y_d^{\mathbb{Q}}$-equivalent*, denoted by $M \underset{(d)}{\sim} M'$, if $v(M) = v(M')$ for any finite type invariant v of degree $< d$. We have the following ascending series of the sets of the equivalence classes,

$$\mathbb{M}/\underset{(1)}{\sim} \longleftarrow \mathbb{M}/\underset{(2)}{\sim} \longleftarrow \mathbb{M}/\underset{(3)}{\sim} \longleftarrow \cdots .$$

Such a series was originally studied by Goussarov and by Habiro. By Theorem 11.25 and Corollary 11.26, $M \underset{(d)}{\sim} M'$ if and only if $(z^{\text{LMO}})^{(<d)}(M) = (z^{\text{LMO}})^{(<d)}(M')$, where $(z^{\text{LMO}})^{(<d)}$ denotes the degree $< d$ part of the primitive LMO invariant z^{LMO}. Hence, we have the following identification between $\mathbb{M}/\underset{(d)}{\sim}$ and $(z^{\text{LMO}})^{(<d)}(\mathbb{M})$:

$$(z^{\text{LMO}})^{(<d)} : \mathbb{M}/\underset{(d)}{\sim} \xrightarrow{=} (z^{\text{LMO}})^{(<d)}(\mathbb{M}) \subset \mathcal{A}(\emptyset)^{(<d)}_{\text{prim}}. \tag{11.20}$$

To describe the image of the above map we consider the *Y_d-equivalence*,* denoted by $\underset{d}{\sim}$, among integral homology 3-spheres; roughly speaking it is the equivalence relation generated by the equivalence between the two pictures in the image of the map (11.7) for each Jacobi diagram with d trivalent vertices; for example,

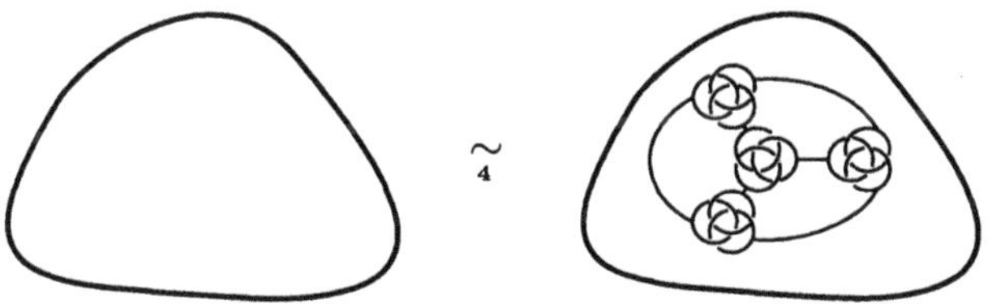 (11.21)

For the precise definition of Y_d-equivalence see Appendix E.3. As shown in Appendix E.4, if $M \underset{2d}{\sim} M'$, then $M \underset{(d)}{\sim} M'$ by Proposition E.27. Hence, we have the following

*The Y_d-equivalence is also called the $(d-1)$-*equivalence* (due to Goussarov) in some literatures.

homomorphism induced by the identity map of integral homology 3-spheres,

$$\mathbb{M}_{2d}/_{\underset{2d+2}{\sim}} \longrightarrow \mathbb{M}_{(d)}/_{\underset{(d+1)}{\sim}},$$

where $\mathbb{M}_{2d}$ (resp. $\mathbb{M}_{(d)}$) denotes the set of integral homology 3-spheres which are Y_{2d}-equivalent (resp. $Y_d^{\mathbb{Q}}$-equivalent) to S^3. Further, by Theorem E.20 we have a surjective homomorphism

$$\mathrm{span}_{\mathbb{Z}}\left\{ \begin{array}{l} \text{primitive Jacobi diagrams} \\ \text{on } \emptyset \text{ of degree } d \end{array} \right\} \Big/ \text{AS, IHX} \longrightarrow \mathbb{M}_{2d}/_{\underset{2d+2}{\sim}}.$$

Recall that a Jacobi diagram on $\emptyset$ is called *primitive* if it is connected. By composing the above three maps, we have that

$$\mathrm{span}_{\mathbb{Z}}\left\{ \begin{array}{l} \text{primitive Jacobi diagrams} \\ \text{on } \emptyset \text{ of degree } d \end{array} \right\} \Big/ \text{AS, IHX} \longrightarrow \mathbb{M}_{2d}/_{\underset{2d+2}{\sim}} \longrightarrow \mathbb{M}_{(d)}/_{\underset{(d+1)}{\sim}} \xrightarrow{z^{(d)}} \mathcal{A}(\emptyset)^{(d)}_{\mathrm{prim}}.$$

It follows from the definition of the maps and a computation in Section E.4 that the composition of the above sequence of maps takes each Jacobi diagram to itself. Hence, $\mathbb{M}_{(d)}/\underset{(d+1)}{\sim}$ is isomorphic to the lattice in $\mathcal{A}(\emptyset)^{(d)}_{\mathrm{prim}}$ spanned by primitive Jacobi diagrams.

By Proposition E.27 we have a surjective homomorphism $\mathbb{M}/\underset{2d}{\sim} \longrightarrow \mathbb{M}/\underset{(d)}{\sim}$. Since $\mathbb{M}/\underset{2d}{\sim}$ is a commutative group, by Corollary E.23, so is $\mathbb{M}/\underset{(d)}{\sim}$. Further, as in (11.20), $\mathbb{M}/\underset{(d)}{\sim}$ is isomorphic to a lattice in $\mathcal{A}(\emptyset)^{(<d)}_{\mathrm{prim}}$. Furthermore, by the above argument, its rank is equal to the dimension of $\mathcal{A}(\emptyset)^{(<d)}_{\mathrm{prim}}$. Hence, we have

Theorem 11.27. The quotient set $\mathbb{M}/\underset{(d)}{\sim}$ is a commutative group which is isomorphic to the following lattice in the space of Jacobi diagrams on $\emptyset$,

$$\mathrm{span}_{\mathbb{Z}}\{\text{primitive Jacobi diagrams on } \emptyset \text{ of degree } < d\} \subset \mathcal{A}(\emptyset).$$

Further, by taking the direct limit of the sequence of isomorphisms we have the following homomorphism

$$\mathbb{M} \longrightarrow \varprojlim_{d} \mathbb{M}/\underset{(d)}{\sim} \cong \mathrm{span}_{\mathbb{Z}}\{\text{primitive Jacobi diagrams on } \emptyset\} \subset \hat{\mathcal{A}}(\emptyset). \qquad (11.22)$$

Furthermore, the above map gives a reconstruction of the primitive LMO invariant z^{LMO}.

Since the map (11.22) gives a reconstruction of the primitive LMO invariant z^{LMO}, we have

Corollary 11.28. The image of the primitive LMO invariant z^{LMO} in $\hat{\mathcal{A}}(\emptyset)_{\mathrm{prim}}$ lies in a lattice in $\hat{\mathcal{A}}(\emptyset)_{\mathrm{prim}}$ which is isomorphic to the lattice spanned over $\mathbb{Z}$ by primitive Jacobi diagrams in $\hat{\mathcal{A}}(\emptyset)_{\mathrm{prim}}$.

The corollary implies integrality of coefficients of the primitive LMO invariant in the sense that, if $z^{\text{LMO}\,(<d)}(M) = z^{\text{LMO}\,(<d)}(M')$, then $z^{\text{LMO}\,(d)}(M) - z^{\text{LMO}\,(d)}(M')$ is equal to a linear sum of Jacobi diagrams with integer coefficients.

If it could be shown that the map (11.22) (or the primitive LMO invariant z^{LMO}) was injective, then we would be able to identify the set of integral homology 3-spheres with a subset of the above lattice, and all invariants related to quantum invariants would be understood as functions of weight systems on the lattice. Further, we would expect that there would be structures on the set of integral homology 3-spheres induced by combinatorial structures on the space of Jacobi diagrams.

Appendix A

The quantum group $U_q(sl_2)$

As mentioned in Sections 4.4 and 4.5, for a generic q, and also for q a root of unity, the quantum group $U_q(sl_2)$ forms a ribbon Hopf algebra. We prove this in Sections A.1 and A.2. Further, in Section A.3 we show some properties of certain exceptional representations of $U_\zeta(sl_2)$ at $\zeta = -1$.

A.1 $U_q(sl_2)$ at a generic q is a ribbon Hopf algebra

The aim of this section is to give a precise proof of Theorem 4.14 which says that the quantum group $U_q(sl_2)$ is a ribbon Hopf algebra. For the notations of $U_q(sl_2)$ see Section 4.4.

Computing $\mathcal{R}^{-1}$

As in Section 4.4, the universal R matrix $\mathcal{R}$ for $U_q(sl_2)$ and its inverse $\mathcal{R}^{-1}$ are given by

$$\mathcal{R} = q^{H\otimes H/4}\exp_q\Big((q^{1/2}-q^{-1/2})E\otimes F\Big),$$
$$\mathcal{R}^{-1} = \exp_{q^{-1}}\Big((q^{-1/2}-q^{1/2})E\otimes F\Big)q^{-H\otimes H/4}.$$

It is verified immediately by Lemma A.1 below that the above $\mathcal{R}^{-1}$ is the inverse of $\mathcal{R}$.

To prove the lemma, we introduce the *q-binomial coefficient* by

$$\begin{bmatrix} m \\ k \end{bmatrix} = \frac{[m]!}{[k]![m-k]!}.$$

While ordinary binomial coefficients are integers, q-binomial coefficients are polynomials in $q^{1/2}$ and $q^{-1/2}$ with integral coefficients, and the ordinary binomial coefficients are recovered from the q-binomial coefficients by putting $q = 1$. Further, as a perturbation of an ordinary recursive formula for binomial coefficients, we have the following recursive formula for the q-binomial coefficients,

$$\begin{bmatrix} m \\ k \end{bmatrix} q^{\pm k/2} = \begin{bmatrix} m-1 \\ k \end{bmatrix} + \begin{bmatrix} m-1 \\ k-1 \end{bmatrix} q^{\pm m/2}, \tag{A.1}$$

for each sign.

Lemma A.1. The following formulae hold for the q-exponential map,

$$\exp_q(x)\exp_{q^{-1}}(-x) = 1,$$
$$\exp_{q^{-1}}(x)\exp_q(-x) = 1,$$

where the q-exponential map is defined in Section 4.4.

Proof. From the definition of the q-exponential map, we have that

$$\begin{aligned}\exp_q(x)\exp_{q^{-1}}(-x) &= \sum_{n_1,n_2\ge 0} \frac{q^{n_1(n_1-1)/4-n_2(n_2-1)/4}}{[n_1]![n_2]!}(-1)^{n_2}x^{n_1+n_2} \\ &= \sum_{n\ge 0} \frac{q^{-n(n-1)/4}}{[n]!}(-x)^n \sum_{n_1=0}^{n} \begin{bmatrix} n \\ n_1 \end{bmatrix} q^{n_1(n-1)/2}(-1)^{n_1},\end{aligned}$$

where we obtain the second equality by putting $n = n_1 + n_2$. Further, we put the second sum of the last term to be ϕ_n by

$$\phi_n = \sum_{k=0}^{n} \begin{bmatrix} n \\ k \end{bmatrix} q^{k(n-1)/2}(-1)^k.$$

Then, we obtain the following recursive formula for ϕ_n by (A.1),

$$\phi_n = (1-q^{n-1})\phi_{n-1}, \qquad \phi_0 = 1.$$

It follows that $\phi_n = 0$ for any $n \ge 1$. This proves the first formula of the lemma.

The second formula of the lemma is derived from the first formula by replacing q with q^{-1}. □

$U_q(sl_2)$ *is a quasi-triangular Hopf algebra*

Proposition A.2. For the above universal R matrix $\mathcal{R}$, the pair $(U_q(sl_2), \mathcal{R})$ is a quasi-triangular Hopf algebra.

Proof. We show that the universal R matrix $\mathcal{R}$ satisfies the following defining relations of a quasi-triangular Hopf algebra,

$$(P\circ\Delta)(x) = \mathcal{R}\Delta(x)\mathcal{R}^{-1}, \qquad \text{for any } x\in A, \tag{A.2}$$
$$(\Delta\otimes \mathrm{id})(\mathcal{R}) = \mathcal{R}_{13}\mathcal{R}_{23}, \tag{A.3}$$
$$(\mathrm{id}\otimes\Delta)(\mathcal{R}) = \mathcal{R}_{13}\mathcal{R}_{12}. \tag{A.4}$$

The relation (A.3) is proved as follows. The left hand side of (A.3) is equal to

$$q^{(H\otimes 1+1\otimes H)\otimes H/4}\exp_q\Big((q^{1/2}-q^{-1/2})(E\otimes K+1\otimes E)\otimes F\Big).$$

Further, we compute the right hand side of (A.3) as

$$\begin{aligned}
&q^{(H\otimes 1\otimes H)/4}\exp_q\Big((q^{1/2}-q^{-1/2})E\otimes 1\otimes F\Big)\\
&\quad\times q^{(1\otimes H\otimes H)/4}\exp_q\Big((q^{1/2}-q^{-1/2})1\otimes E\otimes F\Big)\\
&=q^{(H\otimes 1\otimes H)/4}q^{(1\otimes H\otimes H)/4}\exp_q\Big((q^{1/2}-q^{-1/2})E\otimes K\otimes F\Big)\\
&\quad\times \exp_q\Big((q^{1/2}-q^{-1/2})1\otimes E\otimes F\Big).
\end{aligned}$$

It follows from Lemma A.3, below, that the above formulae are equal, which completes the proof of (A.3).

The relation (A.4) can be obtained similarly as above.

To prove the relation (A.2) it is sufficient to check that it is true in each case where x is set to be a generator (*i.e.*, $K^{\pm 1}$, E, or F) of $U_q(sl_2)$. That is, we prove

$$(K^{\pm 1}\otimes K^{\pm 1})\mathcal{R}=\mathcal{R}(K^{\pm 1}\otimes K^{\pm 1}), \tag{A.5}$$
$$(K\otimes E+E\otimes 1)\mathcal{R}=\mathcal{R}(E\otimes K+1\otimes E), \tag{A.6}$$
$$(1\otimes F+F\otimes K^{-1})\mathcal{R}=\mathcal{R}(F\otimes 1+K^{-1}\otimes F). \tag{A.7}$$

The relation (A.5) can be obtained immediately from the fact that $K^{\pm 1}\otimes K^{\pm 1}$ commutes with each of $q^{-H\otimes H/4}$ and $E\otimes F$. The proof of the relation (A.6) is reduced to the relations

$$(K\otimes E)\cdot q^{H\otimes H/4}=q^{H\otimes H/4}\cdot(1\otimes E), \tag{A.8}$$
$$(E\otimes 1)\cdot q^{H\otimes H/4}=q^{H\otimes H/4}\cdot(E\otimes K^{-1}), \tag{A.9}$$

and the relation

$$\begin{aligned}
&(E\otimes K^{-1}+1\otimes E)\exp_q\Big((q^{1/2}-q^{-1/2})E\otimes F\Big)\\
&\quad=\exp_q\Big((q^{1/2}-q^{-1/2})E\otimes F\Big)(E\otimes K+1\otimes E).
\end{aligned}\tag{A.10}$$

The relation (A.8) is obtained as

$$(K\otimes E)q^{H\otimes H/4}=q^{H\otimes(H-2)/4}(K\otimes E)=q^{H\otimes H/4}q^{-H/2\otimes 1}(K\otimes E)=q^{H\otimes H/4}(1\otimes E).$$

Further, the relation (A.9) is obtained in the same way. Furthermore, the relation (A.10) is proved as follows. By the second formula of Lemma A.4 below, we have that

$$EF^n-F^nE=\frac{[n]q^{-(n-1)/2}}{q^{1/2}-q^{-1/2}}(F^{n-1}K-K^{-1}F^{n-1}).$$

Hence, we obtain the following formula,

$$\frac{q^{n(n-1)/4}}{[n]!}(q^{1/2}-q^{-1/2})^n E^n \otimes (EF^n - F^n E)$$
$$= \frac{q^{(n-1)(n-2)/4}}{[n-1]!}(q^{1/2}-q^{-1/2})^{n-1} E^n \otimes (F^{n-1}K - K^{-1}F^{n-1}).$$

Summing up this formula with respect to n, we obtain (A.10), completing the proof of (A.6). We can prove (A.7) similarly. This completes the proof of (A.2) for this $\mathcal{R}$. □

Lemma A.3. For non-commutative indeterminates x and y satisfying the relation $xy = qyx$, the following equality holds for the q-exponential map,

$$\exp_q(x+y) = \exp_q(x)\exp_q(y).$$

Proof. For such x and y we have the q-binomial expansion

$$(x+y)^n = \sum_{k=0}^{n} \begin{bmatrix} n \\ k \end{bmatrix} q^{k(k-n)/2} x^k y^{n-k},$$

which can be proved by induction on n using the recursive formula (A.1) for q-binomial coefficients; for such a detailed computation see [KiMe91]. Hence,

$$\begin{aligned}\exp_q(x+y) &= \sum_{n\geq k\geq 0} \frac{q^{n(n-1)/4}}{[n]!}\begin{bmatrix} n \\ k \end{bmatrix} q^{k(k-n)/2} x^k y^{n-k} \\ &= \sum_{n\geq k\geq 0} \frac{q^{k(k-1)/4}}{[k]!}x^k \cdot \frac{q^{(n-k)(n-k-1)/4}}{[n-k]!} y^{n-k} \\ &= \exp_q(x)\exp_q(y),\end{aligned}$$

completing the proof. □

Lemma A.4 ([KiMe91]). The following formulae hold,

$$E^n F = FE^n + [n][H-n+1]E^{n-1},$$
$$F^n E = EF^n - [n][H+n-1]F^{n-1},$$

where we define $[H+m]$ to be $(q^{m/2}K - q^{-m/2}K^{-1})/(q^{1/2}-q^{-1/2})$, putting $K = q^{H/2}$.

Proof. We prove the first formula by induction on n as follows. We have that

$$\begin{aligned}E^n F &= E^{n-1}(FE + [H]) \\ &= \big(FE^{n-1} + [n-1][H-n+2]E^{n-2}\big)E + [H-2n+2]E^{n-1} \\ &= FE^n + [n][H-n+1]E^{n-1},\end{aligned}$$

where we obtain the second equality by the assumption of the induction and by the equality $[H+m]E = E[H+m+2]$, and obtain the third equality by the following relation,

$$[a][H+c+b]+[b][H+c-a]=[a+b][H+c].$$

The second formula of the lemma is obtained similarly. □

Computing u

Putting $q = e^{\hbar}$, we present $q^{H\otimes H/4}$ by

$$q^{H\otimes H/4} = 1+\frac{\hbar}{4}(H\otimes H)+\frac{\hbar^2}{32}(H^2\otimes H^2)+\cdots = \sum_i a_i(H)\otimes b_i(H),$$

for some polynomials $a_i(H)$ and $b_i(H)$ in H. Then, the universal R matrix $\mathcal{R}$ is presented by

$$\mathcal{R} = \sum_{i,n}\frac{q^{n(n-1)/4}}{[n]!}(q^{1/2}-q^{-1/2})^n a_i(H)E^n\otimes b_i(H)F^n.$$

Recall that, for $\mathcal{R} = \sum_i \alpha_i\otimes\beta_i$, we defined the element u by $u = \sum_i S(\beta_i)\alpha_i$. Hence, u is presented by

$$\begin{aligned} u &= \sum_{i,n}\frac{q^{n(n-1)/4}}{[n]!}(q^{1/2}-q^{-1/2})^n S(F)^n S(b_i(H))a_i(H)E^n \\ &= \sum_{i,n}\frac{q^{n(n-1)/4}}{[n]!}(q^{1/2}-q^{-1/2})^n(-KF)^n q^{-H^2/4}E^n, \end{aligned}$$

where we obtain the second equality by

$$\sum_i S(b_i(H))a_i(H) = \sum_i b_i(-H)a_i(H) = 1-\frac{\hbar}{4}H^2+\frac{\hbar^2}{32}H^4-\cdots = q^{-H^2/4}.$$

Further, we have that

$$KFq^{-H^2/4} = q^{H/2}Fq^{-H^2/4} = q^{-H^2/4}Fq^{-H/2} = q^{-H^2/4}FK^{-1}.$$

Hence, u is presented by

$$u = q^{-H^2/4}\sum_{n=0}^{\infty}\frac{q^{n(n-1)/4}}{[n]!}(q^{-1/2}-q^{1/2})^n(FK^{-1})^nE^n.$$

Noting that $(FK^{-1})^n = q^{n(n-1)/2}F^nK^{-n}$, we have the following expression for u, given in Section 4.4,

$$u = q^{-H^2/4}\sum_{n=0}^{\infty}\frac{q^{3n(n-1)/4}}{[n]!}(q^{-1/2}-q^{1/2})^nF^nK^{-n}E^n.$$

Noting that $U_q(sl_2)$ has the topology of a power series ring in $(q^{1/2}-q^{-1/2})$, observe that the above infinite sum converges. Further observe that when acting on the finite dimensional module V_n, given in Section 4.4, only a finite number of terms contribute.

Computing $S(u)$

Proposition A.5. We have that $S(u) = uK^{-2}$.

Proof. We define $u^{(<r)}$ to be the first r terms of the above expression of u as

$$u^{(<r)} = q^{-H^2/4} \sum_{0\leq n<r} \frac{q^{3n(n-1)/4}}{[n]!}(q^{-1/2} - q^{1/2})^n F^n K^{-n} E^n.$$

Then,

$$S(u) - uK^{-2} = \lim_{r\to\infty} \left(S(u^{(<r)}) - u^{(<r)}K^{-2} \right).$$

By Lemma A.6 below, $S(u^{(<r)}) - u^{(<r)}K^{-2}$ is divisible by $(q^{1/2} - q^{-1/2})^r$. Since $U_q(sl_2)$ has the topology of a power series ring in $(q^{1/2} - q^{-1/2})$, the above limit converges to zero, which implies the proposition. □

Lemma A.6. The following formula holds,

$$\begin{aligned} S(u^{(<r)}) - u^{(<r)}K^{-2} &= -q^{-H^2/4}(q^{-1/2} - q^{1/2})^r q^{-r(r-1)/4} \\ &\quad \times \sum_{n=0}^{r-1} \frac{q^{-n/2}}{[n]!} F^n E^n \begin{bmatrix} H \\ r-n \end{bmatrix} [r-n]! \\ &\quad \times \sum_{i=0}^{n} q^{i(r-n-1)/2} \begin{bmatrix} r-i-1 \\ r-n-1 \end{bmatrix} K^{r-2i-2}. \end{aligned}$$

Proof. We put

$$u^{(r)} = q^{-H^2/4} \frac{q^{3r(r-1)/4}}{[r]!}(q^{-1/2} - q^{1/2})^r F^r K^{-r} E^r.$$

Then,

$$u^{(r)}K^{-2} = -q^{-H^2/4}(q^{-1/2} - q^{1/2})^r \frac{q^{-r(r+3)/4}}{[r]!} F^r E^r K^{-r-2}. \tag{A.11}$$

Further,

$$S(u^{(r)}) = (q^{-1/2} - q^{1/2})^r \frac{q^{3r(r-1)/4}}{[r]!} S(E)^r S(K^{-1})^r S(F)^r q^{-H^2/4}.$$

Since we have that $S(E)^r S(K^{-1})^r S(F)^r = (-EK^{-1})^r K^r (-KF)^r = q^{-r^2} E^r F^r K^r$, the above formula can be computed

$$\begin{aligned} S(u^{(r)}) &= q^{-H^2/4}(q^{-1/2} - q^{1/2})^r \frac{q^{-r(r+3)/4}}{[r]!} E^r F^r K^r \\ &= q^{-H^2/4}(q^{-1/2} - q^{1/2})^r \frac{q^{-r(r+3)/4}}{[r]!} \sum_{n=0}^{r} \begin{bmatrix} r \\ n \end{bmatrix}^2 [r-n]!^2 \begin{bmatrix} H \\ r-n \end{bmatrix} F^n E^n K^n \\ &= q^{-H^2/4}(q^{-1/2} - q^{1/2})^r \frac{q^{-r(r+3)/4}}{[r]!} \sum_{n=0}^{r-1} \begin{bmatrix} r \\ n \end{bmatrix}^2 [r-n]!^2 \begin{bmatrix} H \\ r-n \end{bmatrix} F^n E^n K^n \end{aligned} \tag{A.12}$$

$$+ q^{-H^2/4}(q^{-1/2} - q^{1/2})^r \frac{q^{-r(r+3)/4}}{[r]!} F^r E^r K^r, \tag{A.13}$$

where we obtain the second equality by Lemma A.7 below.

We prove the lemma by induction on r. Noting that $\begin{bmatrix} r \\ n \end{bmatrix} = \dfrac{[r]!}{[n]![n-r]!}$, we have that

$$\begin{aligned} &S(u^{(<r+1)}) - u^{(<r+1)}K^{-2} \\ &= S(u^{(<r)}) - u^{(<r)}K^{-2} + (A.12) + (A.13) - (A.11) \\ &= q^{-H^2/4}(q^{-1/2} - q^{1/2})^r \frac{q^{-r(r+3)/4}}{[n]!} \sum_{n=0}^{r-1} F^n E^n \begin{bmatrix} H \\ r-n \end{bmatrix} [r-n]! K^r \\ &\quad \times \left(\begin{bmatrix} r \\ n \end{bmatrix} - \sum_{i=0}^{n} q^{(2r-n+i(r-n-1))/2} \begin{bmatrix} r-i-1 \\ r-n-1 \end{bmatrix} K^{-2i-2} \right) + (A.13) - (A.11) \\ &= -q^{-H^2/4}(q^{-1/2} - q^{1/2})^{r+1} \sum_{n=0}^{r-1} \frac{q^{-r(r+3)/4}}{[n]!} F^n E^n \begin{bmatrix} H \\ r-n \end{bmatrix} [r-n]![H-r+n] \\ &\quad \times q^{(n-r)/2} \sum_{i=0}^{n} q^{(r-n)(i+2)/2} \begin{bmatrix} r-i \\ r-n \end{bmatrix} K^{r-2i-1} + (A.13) - (A.11) \\ &= -q^{-H^2/4}(q^{-1/2} - q^{1/2})^{r+1} q^{-r(r+1)/4} \sum_{n=0}^{r} \frac{q^{-n/2}}{[n]!} F^n E^n \begin{bmatrix} H \\ r-n+1 \end{bmatrix} [r-n+1]! \\ &\quad \times \sum_{i=0}^{n} q^{i(r-n)/2} \begin{bmatrix} r-i \\ r-n \end{bmatrix} K^{r-2i-1}, \end{aligned}$$

where we obtain the third equality by Lemma A.8 below, putting $m = r - n$. Since the last formula is obtained from the right hand side of the required formula by replacing r with $r+1$, the proof is completed. □

Lemma A.7. The following formula holds,

$$E^nF^n = \sum_{i=0}^{n} \begin{bmatrix} n \\ i \end{bmatrix}^2 [i]!^2 \begin{bmatrix} H \\ i \end{bmatrix} F^{n-i}E^{n-i},$$

where $\begin{bmatrix} H \\ i \end{bmatrix} = [H][H-1]\cdots[H-i+1]/[i]!$ and $[H+m]$ is defined in Lemma A.4.

Proof. The lemma can be proved by induction on n by using Lemma A.4. □

Lemma A.8. The following formula holds,

$$\begin{bmatrix} r \\ m \end{bmatrix} - \sum_{i=0}^{r-m} q^{(r+mi+m-i)/2} \begin{bmatrix} r-i-1 \\ m-1 \end{bmatrix} K^{-2i-2}$$
$$= (q^{-m} - K^{-2}) \sum_{i=0}^{r-m} q^{m(i+2)/2} \begin{bmatrix} r-i \\ m \end{bmatrix} K^{-2i}.$$

Proof. We have that

$$\begin{bmatrix} r \\ m \end{bmatrix} = \sum_{i=0}^{r-m} q^{(r+mi+m-i)/2} \begin{bmatrix} r-i-1 \\ m-1 \end{bmatrix} q^{-m(i+1)},$$

by Lemma A.9 below, putting $l = r-m$ and $j = r-m-i$. Hence, the left hand side of the required formula is equal to

$$\sum_{i=0}^{r-m} q^{(r+mi+m-i)/2} \begin{bmatrix} r-i-1 \\ m-1 \end{bmatrix} (q^{-m(i+1)} - K^{-2i-2}).$$

Further,

$$q^{-m(i+1)} - K^{-2i-2} = (q^{-m} - K^{-2})(q^{-mi} + q^{-m(i-1)}K^{-2} + \cdots + K^{-2i})$$
$$= (q^{-m} - K^{-2}) \sum_{j=0}^{i} q^{-m(i-j)} K^{-2j}.$$

Therefore, the left hand side of the required formula is equal to

$$(q^{-m} - K^{-2}) \sum_{0 \le j \le i \le r-m} q^{(r+mi+m-i)/2 - m(i-j)} \begin{bmatrix} r-i-1 \\ m-1 \end{bmatrix} K^{-2j}$$
$$= (q^{-m} - K^{-2}) \sum_{j=0}^{r-m} q^{-m(r-m-2j-2)/2} K^{-2j} \sum_{k=0}^{r-m-j} q^{k(m+1)/2} \begin{bmatrix} m+k-1 \\ m-1 \end{bmatrix}$$
$$= (q^{-m} - K^{-2}) \sum_{j=0}^{r-m} q^{m(j+2)/2} \begin{bmatrix} r-j \\ m \end{bmatrix} K^{-2j},$$

where we obtain the first equality by putting $k = r - m - i$ and obtain the second equality by Lemma A.9 below, putting $l = r - m - j$. Hence, we obtain the required formula. □

Lemma A.9. The following formula holds,

$$\sum_{j=0}^{l} q^{j(m+1)/2} \begin{bmatrix} m+j-1 \\ m-1 \end{bmatrix} = q^{ml/2} \begin{bmatrix} m+l \\ m \end{bmatrix}.$$

Proof. We prove the lemma by induction on l. By the assumption of the induction, we have that

$$\begin{aligned}\sum_{j=0}^{l+1} q^{j(m+1)/2} \begin{bmatrix} m+j-1 \\ m-1 \end{bmatrix} &= q^{j(m+1)/2} \begin{bmatrix} m+j-1 \\ m-1 \end{bmatrix}\Bigg|_{j=l+1} + q^{ml/2} \begin{bmatrix} m+l \\ m \end{bmatrix} \\ &= q^{ml/2}\left(q^{(m+l+1)/2} \begin{bmatrix} m+l \\ m-1 \end{bmatrix} + \begin{bmatrix} m+l \\ m \end{bmatrix}\right) \\ &= q^{m(l+1)/2} \begin{bmatrix} m+l+1 \\ m \end{bmatrix},\end{aligned}$$

where we obtain the last equality by the recursive formula (A.1) for q-binomial coefficients. The above formula implies that the required formula holds for $l+1$; this completes the induction. □

$U_q(sl_2)$ is a ribbon Hopf algebra

We now show Theorem 4.14, which says that $U_q(sl_2)$ is a ribbon Hopf algebra.

Proof of Theorem 4.14. By Proposition A.2 the pair $(U_q(sl_2), \mathcal{R})$ is a quasi-triangular Hopf algebra. We show that the triple $(U_q(sl_2), \mathcal{R}, v)$ satisfies the defining relations (4.23)–(4.27) of a ribbon Hopf algebra as follows.

As for (4.23) we show that v is central as follows. We have that $S^2(x) = KxK^{-1}$ for any $x \in U_q(sl_2)$; this formula is proved by showing that it holds when x is set to be any of the generators $K^{\pm 1}$, E, F of $U_q(sl_2)$. Further, by (4.13), we have that $S^2(x) = uxu^{-1}$. Hence, $KxK^{-1} = uxu^{-1}$ holds for any $x \in U_q(sl_2)$. This implies that $K^{-1}u$ is central in $U_q(sl_2)$. By putting $v = K^{-1}u$, the element v is central. Hence, v satisfies (4.23).

The relation (4.24) is obtained as follows. By Proposition A.5, we have that $uS(u) = u^2K^{-2} = v^2$ noting the commutativity of u and K. Hence, we obtain (4.24).

The relation (4.25) is obtained as

$$\Delta(v) = \Delta(K)^{-1}\Delta(u) = (K^{-1} \otimes K^{-1}) \cdot (u \otimes u) \cdot (\mathcal{R}_{21}\mathcal{R})^{-1} = v \otimes v \cdot (\mathcal{R}_{21}\mathcal{R})^{-1},$$

where we obtain the second equality by Proposition 4.3 and from the definition of $\Delta(K)$.

The relation (4.26) is obtained as

$$S(v) = S(K^{-1}u) = S(u)S(K)^{-1} = uK^{-2} \cdot K = K^{-1}u = v,$$

where we obtain the third equality from Proposition A.5.

The relation (4.27) is obtained as

$$\varepsilon(v) = \varepsilon(K^{-1}u) = \varepsilon(K)^{-1}\varepsilon(u) = 1,$$

where we obtain the second equality by Proposition 4.3 and from the definition of $\varepsilon(K)$.

Hence, the triple $(U_q(sl_2), \mathcal{R}, v)$ is a ribbon Hopf algebra. □

A.2 $U_\zeta(sl_2)$ at a root of unity ζ is a ribbon Hopf algebra

The aim of this section is to prove Theorem 4.20 which says that $U_\zeta(sl_2)$ is a ribbon Hopf algebra; for the notations of $U_\zeta(sl_2)$ see Section 4.5. Most of the arguments in this section are similar to the arguments for $U_q(sl_2)$ in the previous section. A novelty is the computation of $S(u)$.

Throughout this section we put $\zeta = \exp(2\pi\sqrt{-1}/r)$ for a fixed integer $r \geq 2$, and we use the q-integer $[n] = (\zeta^{n/2} - \zeta^{-n/2})/(\zeta^{1/2} - \zeta^{-1/2})$.

Computing $\mathcal{R}^{-1}$

As in Section 4.5, the universal R matrix $\mathcal{R}$ for $U_\zeta(sl_2)$ and its inverse $\mathcal{R}^{-1}$ are given by

$$\mathcal{R} = \zeta^{H\otimes H/4} \exp_\zeta^{(<r)} \Big((\zeta^{1/2} - \zeta^{-1/2}) E \otimes F \Big),$$
$$\mathcal{R}^{-1} = \exp_{\zeta^{-1}}^{(<r)} \Big((\zeta^{-1/2} - \zeta^{1/2}) E \otimes F \Big) \zeta^{-H\otimes H/4}.$$

It is immediately verified, by Lemma A.10 below, that the above $\mathcal{R}^{-1}$ is the inverse of $\mathcal{R}$.

Lemma A.10. If $x^r = 0$, then the following formulae hold for the truncated q-exponential map,

$$\exp_q^{(<r)}(x) \exp_{q^{-1}}^{(<r)}(-x) = 1,$$
$$\exp_{q^{-1}}^{(<r)}(x) \exp_q^{(<r)}(-x) = 1,$$

where the truncated q-exponential map is defined in Section 4.5.

Proof. When $x^r = 0$, $\exp_q^{(<r)}(x)$ is equal to $\exp_q(x)$. Hence, we obtain the lemma by Lemma A.1. □

$U_\zeta(sl_2)$ *is a quasi-triangular Hopf algebra*

Proposition A.11. The pair $(U_\zeta(sl_2), \mathcal{R})$ is a quasi-triangular Hopf algebra.

Proof. This proposition is proved in the same way as Proposition A.2, but with Lemma A.12 (below) substituting for Lemma A.3. □

Lemma A.12. For non-commutative indeterminates x and y satisfying the relations $xy = \zeta yx$ and $x^i y^{r-i} = 0$ for any i with $0 \le i \le r$, the following equality holds for the truncated q-exponential map,

$$\exp_\zeta^{(<r)}(x+y) = \exp_\zeta^{(<r)}(x)\exp_\zeta^{(<r)}(y).$$

Proof. The proof of the lemma is obtained from the proof of Lemma A.3 by putting $x^i y^{r-i} = 0$ for each i with $0 \le i \le r$. □

Computing u

In the same way as u was computed for $U_q(sl_2)$ above, we can obtain the following expression for the u of $U_\zeta(sl_2)$,

$$u = \zeta^{-H^2/4}\sum_{n=0}^{r-1}\frac{\zeta^{3n(n-1)/4}}{[n]!}(\zeta^{-1/2}-\zeta^{1/2})^n F^n K^{-n} E^n.$$

Computing $S(u)$

Unlike the case of $U_q(sl_2)$ (Proposition A.5) we have the following proposition.

Proposition A.13. We have $S(u) = uK^{2r-2}$.

Proof. By the above expression for u, we have that

$$S(u) = \sum_{n=0}^{r-1}\frac{\zeta^{3n(n-1)/4}}{[n]!}(\zeta^{-1/2}-\zeta^{1/2})^n S(E)^n S(K^{-1})^n S(F)^n \zeta^{-H^2/4}.$$

Since we have that $S(E)^n S(K^{-1})^n S(F)^n = (-EK^{-1})^n K^n (-KF)^n = q^{-n^2} E^n F^n K^n$, the above formula can be computed

$$\begin{aligned} S(u) &= \zeta^{-H^2/4}\sum_{n=0}^{r-1}\frac{\zeta^{-n(n+3)/4}}{[n]!}(\zeta^{-1/2}-\zeta^{1/2})^n E^n F^n K^n \\ &= \zeta^{-H^2/4}\sum_{0\le i\le n<r}\frac{\zeta^{-n(n+3)/4}}{[n]!}(\zeta^{-1/2}-\zeta^{1/2})^n \begin{bmatrix} n \\ i \end{bmatrix}^2 [i]!^2 \begin{bmatrix} H \\ i \end{bmatrix} F^{n-i}E^{n-i}K^n \end{aligned}$$

$$= \zeta^{-H^2/4} \sum_{\substack{0 \le i,n \\ n+i<r}} \frac{\zeta^{-(n+i)(n+i+3)/4}}{[n+i]!} (\zeta^{-1/2} - \zeta^{1/2})^{n+i} \times \begin{bmatrix} n+i \\ i \end{bmatrix}^2 [i]!^2 \begin{bmatrix} H \\ i \end{bmatrix} F^n E^n K^{n+i}$$

$$= \zeta^{-H^2/4} \sum_{\substack{n+i<r \\ 0 \le k \le i}} \frac{\zeta^{(-(n+i)(n+i+3)+(i-2k)(i-1))/4}}{[n]!} (\zeta^{-1/2} - \zeta^{1/2})^{n} \times \begin{bmatrix} n+i \\ i \end{bmatrix} (-1)^k \begin{bmatrix} i \\ k \end{bmatrix} F^n E^n K^{n+2k},$$

where we obtain the second equality by Lemma A.7, obtain the third equality by replacing n with $n+i$, and obtain the fourth equality by Lemma A.14 below. Further, putting $i = j+k$, we have that $\begin{bmatrix} n+i \\ i \end{bmatrix}\begin{bmatrix} i \\ k \end{bmatrix} = \begin{bmatrix} n+k \\ n \end{bmatrix}\begin{bmatrix} n+k+j \\ j \end{bmatrix}$. Hence, the above formula is computed as

$$S(u) = \zeta^{-H^2/4} \sum_{\substack{0 \le n,k \\ n+k<r}} \frac{\zeta^{-((n+k)(n+k+3)+k(k-1))/4}}{[n]!} (\zeta^{-1/2} - \zeta^{1/2})^{n} (-1)^k F^n E^n K^{n+2k} \times \begin{bmatrix} n+k \\ n \end{bmatrix} \sum_{0 \le j < r-n-k} \zeta^{-j(n+k+2)/2} \begin{bmatrix} n+k+j \\ j \end{bmatrix}$$

$$= \zeta^{-H^2/4} \sum_{0 \le n < r} \frac{\zeta^{-((r-1)(r+2)+(r-n-1)(r-n-2))/4}}{[n]!} (\zeta^{-1/2} - \zeta^{1/2})^{n} (-1)^{r-n-1} \times F^n E^n K^{2r-n-2} \begin{bmatrix} r-1 \\ n \end{bmatrix},$$

where we obtain the second equality by applying Lemma A.15 below to the sum in the second line. Further, since $[r-i] = [i]$ for any i, we have that $\begin{bmatrix} r-1 \\ n \end{bmatrix} = 1$ for any n with $0 \le n < r$. Furthermore, $\zeta^{-((r-1)(r-2)+(r-n-1)(r-n-2))/4} = \zeta^{-r(r-n-1)/2}\zeta^{-n(n+3)/4} = (-1)^{r-n-1}\zeta^{-n(n+3)/4}$. Hence, the above formula is computed as

$$S(u) = \zeta^{-H^2/4} \sum_{0 \le n < r} \frac{\zeta^{-n(n+3)/4}}{[n]!} (\zeta^{-1/2} - \zeta^{1/2})^{n} F^n E^n K^{2r-n-2}.$$

This formula implies the required formula. □

Lemma A.14. The following formula holds,

$$(q^{-1/2} - q^{1/2})^i [i]! \begin{bmatrix} H \\ i \end{bmatrix} = \sum_{k=0}^{i} (-1)^k q^{(i-2k)(i-1)/4} \begin{bmatrix} i \\ k \end{bmatrix} K^{2k-i}.$$

Proof. The lemma is proved by induction on i using (A.1). □

Lemma A.15. The following formula holds

$$\sum_{j=0}^{r-m-1} \zeta^{-j(m+2)/2} \begin{bmatrix} m+j \\ j \end{bmatrix} = \begin{cases} 0 & \text{if } 0 \le m \le r-2, \\ 1 & \text{if } m = r-1 \end{cases}.$$

Proof. We denote the left hand side of the required formula by ψ_m. From (A.1) we obtain the following recursive formula.

$$\psi_{m-1} = (1 - \zeta^{-m-1})\psi_m, \qquad \psi_{r-1} = 1.$$

Hence, the required formula can be obtained by downward induction on m. □

$U_\zeta(sl_2)$ *is a ribbon Hopf algebra*

We now prove Theorem 4.20 which says that $U_\zeta(sl_2)$ is a ribbon Hopf algebra.

Proof of Theorem 4.20. By Proposition A.11 the pair $(U_\zeta(sl_2), \mathcal{R})$ is a quasi-triangular Hopf algebra. We show that the triple $(U_\zeta(sl_2), \mathcal{R}, v)$ satisfies the defining relations (4.23)–(4.27) as follows.

We show (4.23), which asserts that v is central. Just as in the proof of (4.23) that is a part of the proof of Theorem 4.14, $K^{-1}u$ is central in $U_\zeta(sl_2)$. K^r, also, is central (by definition). Writing $v = K^{r-1}u = K^r K^{-1}u$, we see that v is central.

The relation (4.24) is obtained as follows. By Proposition A.13, we have that $uS(u) = u^2K^{2r-2} = v^2$ noting the commutativity of u and K. Hence, we obtain (4.24).

The relation (4.25) is obtained as

$$\Delta(v) = \Delta(K)^{r-1}\Delta(u) = (K^{r-1} \otimes K^{r-1}) \cdot (u \otimes u) \cdot (\mathcal{R}_{21}\mathcal{R})^{-1} = v \otimes v \cdot (\mathcal{R}_{21}\mathcal{R})^{-1},$$

where we obtain the second equality by Proposition 4.3 and from the definition of $\Delta(K)$.

The relation (4.26) is obtained as

$$S(v) = S(K^{r-1}u) = S(u)S(K^{r-1}) = uK^{2r-2} \cdot K^{1-r} = K^{r-1}u = v,$$

where we obtain the third equality by Proposition A.13.

The relation (4.27) is obtained as

$$\varepsilon(v) = \varepsilon(K^{r-1}u) = \varepsilon(K)^{r-1}\varepsilon(u) = 1,$$

where we obtain the second equality by Proposition 4.3 and definition of $\varepsilon(K)$.

Hence, the triple $(U_\zeta(sl_2), \mathcal{R}, v)$ is a ribbon Hopf algebra. □

A.3 Exceptional representations of $U_\zeta(sl_2)$ at $\zeta = -1$

In this section we fix ζ to be -1 and present some properties of certain exceptional representations of $U_\zeta(sl_2)$ and certain intertwiners amongst them.

As in Section 4.5, we define a family of representations ρ_t with a complex parameter t by

$$\rho_t(E) = \begin{pmatrix} 0 & \frac{\sqrt{-1}}{2}(t^{1/2} - t^{-1/2}) \\ 0 & 0 \end{pmatrix}, \quad \rho_t(F) = \begin{pmatrix} 0 & 0 \\ 1 & 0 \end{pmatrix}, \quad \rho_t(K) = \begin{pmatrix} t^{-1/2} & 0 \\ 0 & -t^{-1/2} \end{pmatrix}.$$

Let V_t be the 2-dimensional $U_\zeta(sl_2)$ module with a $\mathbb{C}$-basis $e_{t,0}, e_{t,1}$ on which $U_\zeta(sl_2)$ acts through ρ_t.

Lemma A.16. The tensor module of two irreducible modules of $U_\zeta(sl_2)$ splits into a direct sum of irreducible modules as

$$V_{t_1} \otimes V_{t_2} \cong V_{t_1 t_2} \oplus V_{-t_1 t_2}.$$

Proof. We consider the two 2-dimensional subspaces of $V_{t_1} \otimes V_{t_2}$ spanned by $\{e_{t_1,0} \otimes e_{t_2,0}, \Delta(F)(e_{t_1,0} \otimes e_{t_2,0})\}$ and $\{e_{t_1,1} \otimes e_{t_2,1}, \Delta(E)(e_{t_1,1} \otimes e_{t_2,1})\}$ respectively. They are submodules of $V_{t_1} \otimes V_{t_2}$. Further, they are isomorphic to $V_{t_1 t_2}$ and $V_{-t_1 t_2}$ respectively. Hence, we obtain the lemma. □

As in Section 4.5, we put

$$h_t = \rho_t(K^{-1}) = \begin{pmatrix} t^{1/2} & 0 \\ 0 & -t^{1/2} \end{pmatrix}.$$

As mentioned in the proof of Theorem 3.12, the theorem can alternatively be obtained as a corollary of the following lemma. Further, the following lemma is also used in the definition of the multi-variable Alexander polynomial appearing in Section 4.5, and is used in the proof of Lemma C.5.

Lemma A.17. Let $f \in \operatorname{End}_{U_\zeta(sl_2)}(V_{t_1} \otimes V_{t_2})$ be an intertwiner with respect to the action of $U_\zeta(sl_2)$. Then,

$$\operatorname{trace}_1\Big((h_{t_1}^{-1} \otimes \operatorname{id}_{V_{t_1}}) \cdot f\Big) = c_2 \cdot \operatorname{id}_{V_{t_2}},$$
$$\operatorname{trace}_2\Big((\operatorname{id}_{V_{t_1}} \otimes h_{t_2}) \cdot f\Big) = c_1 \cdot \operatorname{id}_{V_{t_1}},$$

with the scalars c_2 and c_1 given by

$$c_2 = \frac{{t_2}^{1/2} - {t_2}^{-1/2}}{{t_1}^{1/2}{t_2}^{1/2} - {t_1}^{-1/2}{t_2}^{-1/2}}(f_+ - f_-),$$
$$c_1 = \frac{{t_1}^{1/2} - {t_1}^{-1/2}}{{t_1}^{1/2}{t_2}^{1/2} - {t_1}^{-1/2}{t_2}^{-1/2}}(f_+ - f_-),$$

where the factor $f_\pm$ is given as follows. Fixing an isomorphism as in Lemma A.16, the restriction of f on $V_{\pm t_1 t_2}$ is a scalar map for each sign. We denote this scalar by $f_\pm$.

Proof. Since the left hand sides of the lemma are intertwiners on V_{t_2} and V_{t_1} respectively, they are equal, by Schur's lemma, to scalar multiples of the identity maps on V_{t_2} and V_{t_1}. It is sufficient to show that the scalars are given as in the lemma.

We compute the vectors appearing in the proof of Lemma A.16 as

$$\Delta(F)(e_0 \otimes e_0) = e_1 \otimes e_0 + t_1^{1/2} e_0 \otimes e_1,$$
$$\Delta(E)(e_1 \otimes e_1) = \frac{\sqrt{-1}}{2}(t_1^{1/2} - t_1^{-1/2})(-t_2^{-1/2}) e_0 \otimes e_1 + \frac{\sqrt{-1}}{2}(t_2^{1/2} - t_2^{-1/2}) e_1 \otimes e_0.$$

Hence,

$$\frac{\sqrt{-1}}{2}(t_2^{1/2} - t_2^{-1/2}) \Delta(F)(e_0 \otimes e_0) - \Delta(E)(e_1 \otimes e_1) = \frac{\sqrt{-1}}{2}(t_1^{1/2} t_2^{1/2} - t_1^{-1/2} t_2^{-1/2}) e_0 \otimes e_1.$$

Therefore, the intertwiner f takes $e_0 \otimes e_0$ and $e_0 \otimes e_1$ to

$$f(e_0 \otimes e_0) = f_+ \cdot e_0 \otimes e_0,$$
$$f(e_0 \otimes e_1) = \frac{t_1^{1/2}(t_2^{1/2} - t_2^{-1/2}) f_+ + (t_1^{1/2} - t_1^{-1/2}) t_2^{-1/2} f_-}{t_1^{1/2} t_2^{1/2} - t_1^{-1/2} t_2^{-1/2}} e_0 \otimes e_1$$
$$+ \text{ (a scalar multiple of) } e_1 \otimes e_0.$$

Hence, the scalar c_1 is presented by

$$c_1 = t_2^{1/2} f_+ - t_2^{1/2} \frac{t_1^{1/2}(t_2^{1/2} - t_2^{-1/2}) f_+ + (t_1^{1/2} - t_1^{-1/2}) t_2^{-1/2} f_-}{t_1^{1/2} t_2^{1/2} - t_1^{-1/2} t_2^{-1/2}}$$
$$= \frac{t_1{}^{1/2} - t_1{}^{-1/2}}{t_1{}^{1/2} t_2{}^{1/2} - t_1{}^{-1/2} t_2{}^{-1/2}} (f_+ - f_-).$$

The scalar c_2 is presented similarly. □

Appendix B

The quantum sl_3 invariant via a linear skein

In Section 1.2 we introduced the Kauffman bracket, which gives a recursive procedure for the computation of the quantum (sl_2, V) invariant of framed links. Further, in Section 8.2 we gave a reconstruction of the quantum $SU(2)$ and $SO(3)$ invariants of 3-manifolds via a linear skein based on the Kauffman bracket. As mentioned in Section 8.3, the quantum $SO(2)$ and $SO(3)$ invariants can be defined as certain linear sums of the quantum (sl_2, V_n) invariants of framed links, which can be constructed using a "box over n strands" in the linear skein (see Appendix H). This suggests looking for constructions of other quantum invariants of 3-manifolds by similar linear sums of boxes in a linear skein.

In this appendix, in place of the Kauffman bracket, we employ a bracket first introduced by Kuperberg [Kup94], which gives a reconstruction of the quantum (sl_3, V) invariant. Motivated by the linear skein based on this bracket, we introduce a box over n rightward and m leftward strands in the linear skein, to which corresponds the quantum $(sl_3, V_{n,m})$ invariant. We proceed to define topological invariants of 3-manifolds (the quantum $SU(3)$ and $PSU(3)$ invariants) by using certain linear sums of the boxes in the linear skein, following [OhYa97].

B.1 The quantum (sl_3, V) invariant of framed links

In this section we give a construction of the quantum (sl_3, V) invariant of oriented framed links (where V denotes the vector representation of sl_3) by a bracket, due to Kuperberg [Kup94], which gives a recursive procedure for the computation of the quantum (sl_3, V) invariant.

For any oriented link diagram D, we define the bracket $\langle\langle D \rangle\rangle$ by the following

recursive formulae

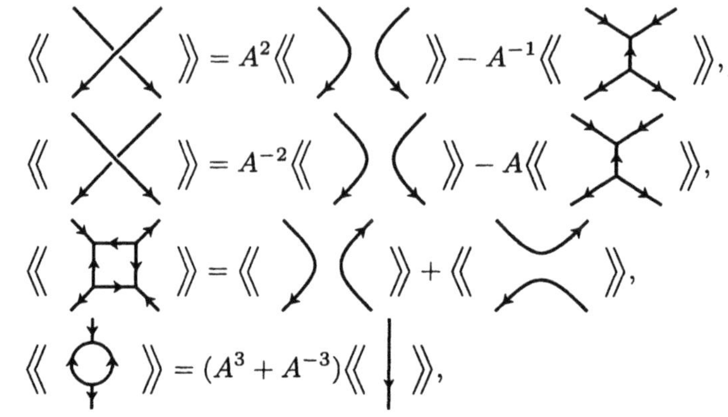

(B.1)

(B.2)

(B.3)

(B.4)

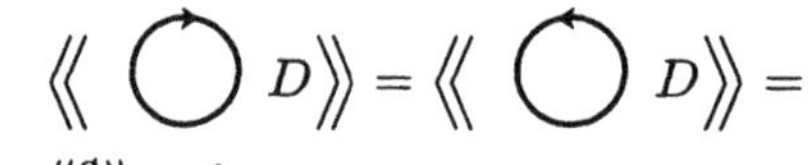

$= (A^6 + 1 + A^{-6})\langle\!\langle D \rangle\!\rangle$ for any diagram D, (B.5)

$$\langle\!\langle \emptyset \rangle\!\rangle = 1. \tag{B.6}$$

For an oriented link diagram D, we obtain its bracket $\langle\!\langle D \rangle\!\rangle \in \mathbb{Z}[A, A^{-1}]$ as follows. We resolve all crossings of D by (B.1) and (B.2) to obtain a linear sum of trivalent graphs embedded in S^2 such that each trivalent vertex of each trivalent graph has either three inward-oriented edges or three outward-oriented edges. By Lemma B.1 below, such an embedded trivalent graph has a face surrounded by 2 or 4 edges. Using the formulae (B.3) and (B.4), we can resolve such a face, decreasing the number of trivalent vertices. Continuing in this vein, we finally obtain a linear sum of disjoint unions of circles. The resulting linear sum is uniquely determined from any embedded trivalent graph by Lemma B.2 below, though the order of resolutions is not unique. Further, by (B.5) and (B.6), the bracket of the disjoint union of l circles has the value $(A^6+1+A^{-6})^l$. Hence, we obtain the value of $\langle\!\langle D \rangle\!\rangle$ as a linear sum of such values of disjoint circles.

Lemma B.1. We consider a trivalent graph embedded in S^2 such that each trivalent vertex of it has either three inward-oriented edges or three outward-oriented edges. Then, the embedded graph has a face (*i.e.*, a domain surrounded by edges) with either 2 or 4 edges.

Proof. Let v, e, and f be the numbers of vertices, edges, and faces of the embedded trivalent graph. Then, by Euler's formula, $v - e + f = 2$. Further, since the graph is trivalent, $2e = 3v$. Hence, $2e = 6f - 12 < 6f$. Then, there exists a face with less than 6 edges; this is because if each face had at least 6 edges, then we would have $2e \geq 6f$, which is a contradiction. Hence, there exists a face with either 2 or 4 edges, noting that each face has an even number of edges because any two adjacent edges have opposite orientations. □

Lemma B.2. Let T be a trivalent graph embedded in S^2 such that each trivalent vertex of it has either three inward-oriented edges or three outward-oriented edges. Then, the value of $\langle\langle T\rangle\rangle \in \mathbb{Z}[A^3, A^{-3}]$ is uniquely determined by the relations (B.3)–(B.6).

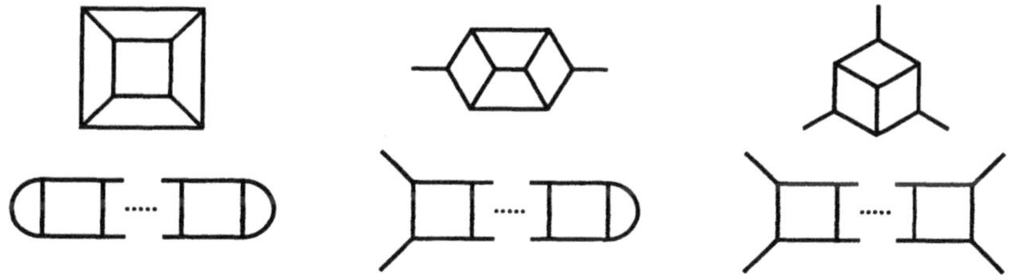

Figure B.1 Consider the domain consisting of the union of faces with 2 or 4 edges. Then, each connected component of the domain is homeomorphic to one of the pictures in this figure.

Proof. As mentioned earlier, Lemma B.1 guarantees us that we can resolve all faces of T recursively using (B.3) and (B.4). The uniqueness of the resulting value of $\langle\langle T\rangle\rangle$ is non-trivial, because there is ambiguity in the order of resolutions of faces by (B.3) and (B.4). To obtain the lemma, it is sufficient to show that we can choose a canonical process of resolutions of faces to obtain $\langle\langle T\rangle\rangle$ from T.

By Lemma B.1, T has faces with 2 or 4 edges. It can be shown, by an elementary argument, that any connected component of the union of faces with 2 or 4 edges is homeomorphic to one of the pictures shown in Figure B.1. Further, it can be shown, by concrete computations, that each graph in Figure B.1 can be uniquely resolved by (B.3) and (B.4); for example, the canonical resolution of the following graph is given by

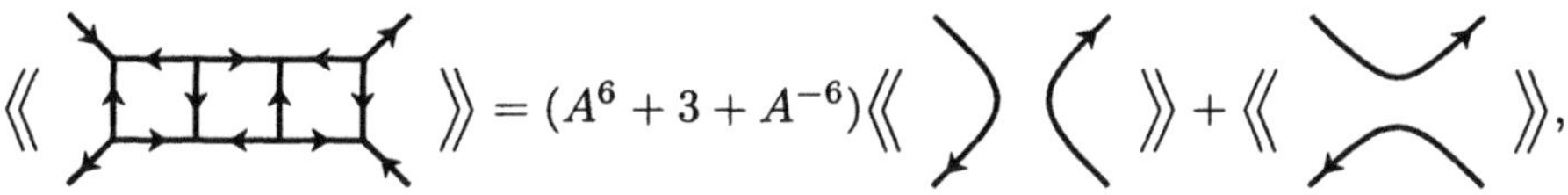

which is derived from (B.3) and (B.4). Thus, we consider a step replacing each connected component with a linear sum of trivalent graphs, say, replacing the above graph on the left hand side with the above linear sum on the right hand side. Hence, by applying such a step to each connected component of such faces of T, we uniquely obtain the resulting linear sum of trivalent graphs. Further, continuing with such steps, we uniquely obtain the resulting value of $\langle\langle T\rangle\rangle$. This gives the required canonical process to obtain $\langle\langle T\rangle\rangle$ from T. □

For example, for a trefoil knot diagram, we have that

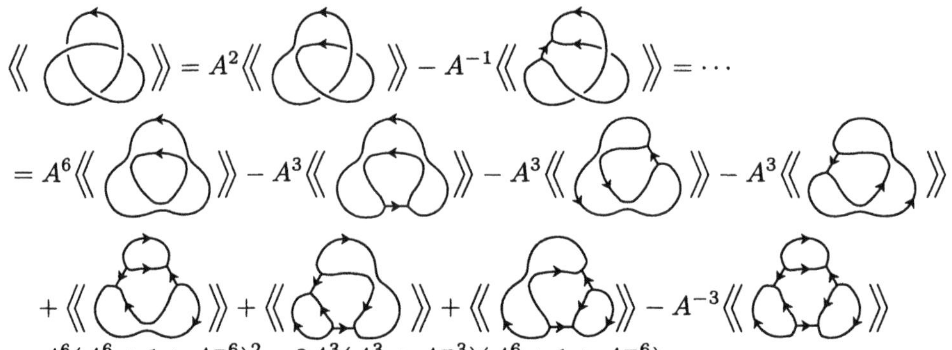

$$
\begin{aligned}
&= A^6(A^6+1+A^{-6})^2 - 3A^3(A^3+A^{-3})(A^6+1+A^{-6}) \\
&\quad + 3(A^3+A^{-3})^2(A^6+1+A^{-6}) - A^{-3}(A^3+A^{-3})^3(A^6+1+A^{-6}) \\
&= (A^6+1+A^{-6})(A^{12}+1-A^{-12}).
\end{aligned}
$$

Theorem B.3 ([Kup94]). Let L be an oriented framed link, and let D be an oriented diagram of L. Then, $\langle\!\langle D\rangle\!\rangle$ is an isotopy invariant of L.

Proof. By Corollary 1.9, the theorem can be proved by showing that $\langle\!\langle D\rangle\!\rangle$ is invariant under the RII and RIII moves on a diagram D. The details are left to the reader. □

Theorem B.4 ([Kup94]). The invariant of Theorem B.3 is equal to the quantum (sl_3, V) invariant $Q^{sl_3,V}(L)$, with $q = A^6$.

For the proof of the theorem see [Kup94] (see also [OhYa97]).

The theorem implies that the *quantum* (sl_3, V) *invariant* $Q^{sl_3,V}(L)$ of a framed link L can alternatively be defined by

$$Q^{sl_3,V}(L) = \langle\!\langle D\rangle\!\rangle\Big|_{A^6=q}.$$

For example, for the trefoil knot K with $+3$ framing, we have that

$$Q^{sl_3;V}(K) = (A^6+1+A^{-6})(A^{12}+1-A^{-12})\Big|_{q=A^6} = (q+1+q^{-1})(q^2+1-q^{-2}),$$

by the above mentioned computation.

B.2 The quantum $SU(3)$ and $PSU(3)$ invariants of 3-manifolds

In this section we sketch constructions of the quantum $SU(3)$ and $PSU(3)$ invariants of 3-manifolds via the linear skein. See [OhYa97] for detailed arguments.

For a surface F the sl_3 *linear skein* $\mathcal{S}^{sl_3}(F)$ of F for the quantum (sl_3, V) invariant is defined to be the vector space over $\mathbb{C}$ spanned by oriented link diagrams (admitting the empty diagram) on F subject to the relations (B.1)–(B.5). By the same argument as in the previous section, the linear skein $\mathcal{S}^{sl_3}(\mathbb{R}^2)$ of $\mathbb{R}^2$ is the 1-dimensional vector space spanned by the empty diagram. When F has non-empty boundary, we consider a set C of points on the boundary of F which are oriented as "inward" or "outward". The linear skein $\mathcal{S}^{sl_3}(F, C)$ is defined to be the vector space over $\mathbb{C}$ spanned by oriented tangle diagrams on F whose boundary matches C, subject to the relations (B.1)–(B.5).

We define the *box over oriented n strands* in $\mathcal{S}^{sl_3}(D^2, C)$ (where C is a set of successive inward n points and successive outward n points) recursively by

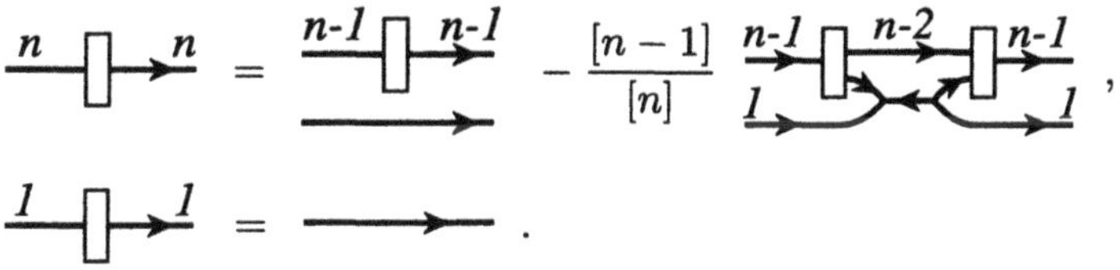

Here, a strand labeled by a number n implies the union of n parallel copies of the strand, and we define $[k]$ by $[k] = (A^{3k} - A^{-3k})/(A^3 - A^{-3})$.

As shown in [OhYa97], the boxes satisfy that

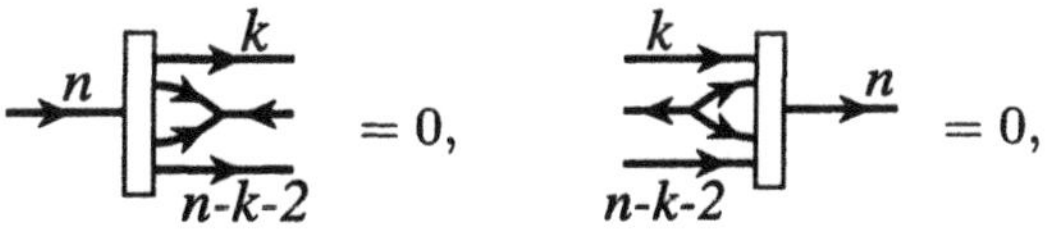

(B.7)

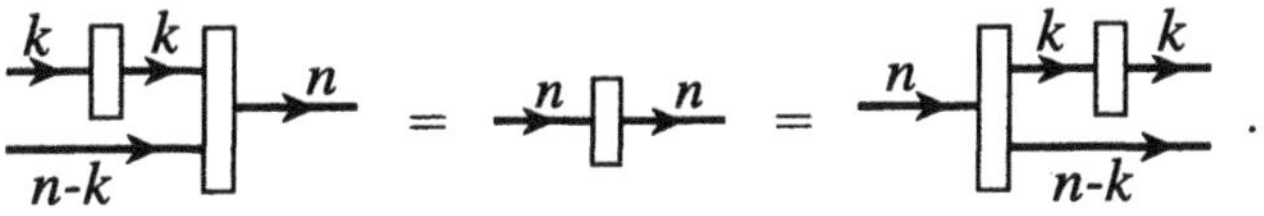

(B.8)

Further, we define the *box over oriented $n + m$ strands* by

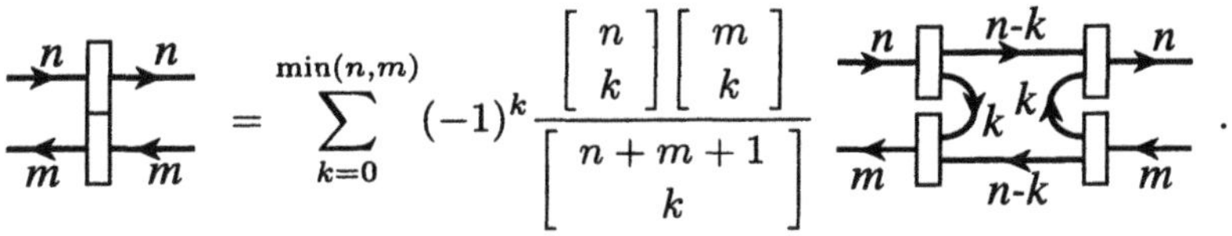

Then, as shown in [OhYa97], it satisfies that

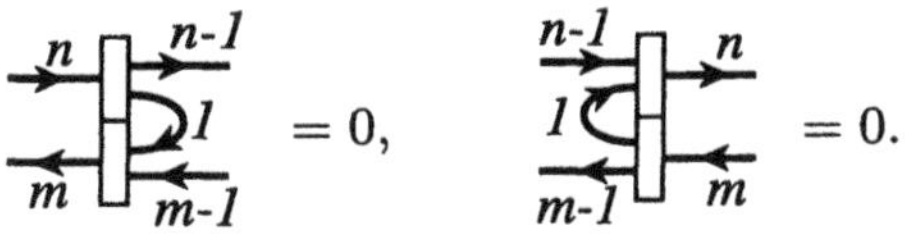

(B.9)

It is known (see, for example, [FuHa91]) that irreducible representations of sl_3 (and $U_q(sl_3)$; see, for example, [ChPr95]) are indexed by two parameters n and m, such that $V_{n,m}$ arises as the maximal irreducible representation in the decomposition of $V^{\otimes n} \otimes (V^*)^{\otimes m}$, where V denotes the vector representation of sl_3 and V^* denotes its dual representation. Consider the operator invariant derived from the representations V and V^* of $U_q(sl_3)$. Then, the relations (B.7), (B.8), (B.9) imply that the intertwiner associated to the box over $n+m$ strands is the projector to the associated irreducible representation. This is an explanation for the fact that we can obtain topological invariants of 3-manifolds (the quantum $SU(3)$ and $PSU(3)$ invariant) by considering linear sums of the boxes.

In the following of this section, we fix r to be an integer > 3, and suppose that A is a primitive $6r$-th root of unity. We define the element $\omega \in \mathcal{S}^{sl_3}(S^1 \times I)$ by

$$\omega = \sum_{\substack{n+m \leq r-3 \\ n,m \geq 0}} [n+1][m+1][n+m+2]$$ 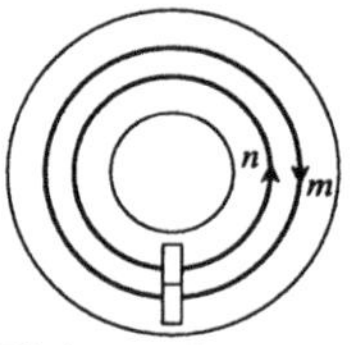

.

Further, for a framed link L, we define $\langle\!\langle L^\omega \rangle\!\rangle \in \mathcal{S}^{sl_3}(F)$ by substituting ω into each component of L in the same way as in the definition of $\langle L^\omega \rangle$ in Section 8.2.

Theorem B.5 ([OhYa97]). Let M be a 3-manifold obtained from S^3 by integral surgery along a framed link L. Then,

$$\langle\!\langle U_+^\omega \rangle\!\rangle^{-\sigma_+} \langle\!\langle U_-^\omega \rangle\!\rangle^{-\sigma_-} \langle\!\langle L^\omega \rangle\!\rangle \in \mathbb{C} \tag{B.10}$$

is invariant under the Kirby moves on L, where $U_\pm$ denotes the trivial knot with ± 1 framing, and σ_+ and σ_- denote the numbers of positive and negative eigenvalues of the linking matrix of L. In particular, Theorem 8.6 implies that (B.10) gives a topological invariant of M.

When $A = \exp(2\pi\sqrt{-1}/6r)$, we denote the topological invariant of the theorem by $\tau_r^{SU(3)}(M) \in \mathbb{C}$, and call it the *quantum $SU(3)$ invariant* of a closed connected oriented 3-manifold M.

The theorem is proved by showing the formulae

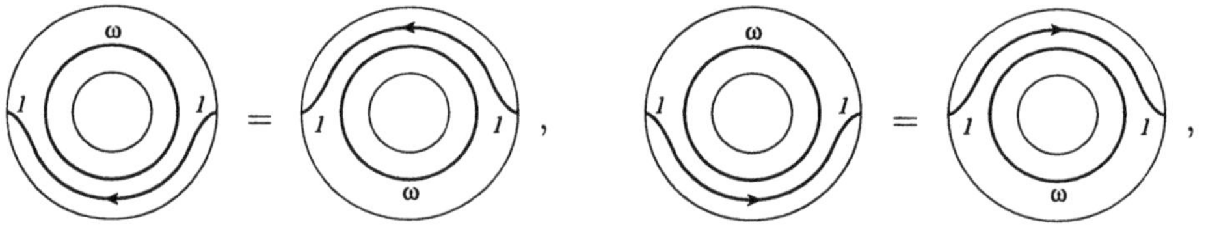

and the fact that a link diagram which includes a box over $n+m$ strands vanishes in $\mathcal{S}^{sl_3}(\mathbb{R}^2)$ if $n+m+2$ is divisible by r (see Lemma 8.10 for a similar fact in the sl_2 case). We obtain the above formulae by a computation similar to (though longer than) the computation appearing in the proof of Theorem 8.11. For a detailed proof see [OhYa97].

We define the element $\omega' \in \mathcal{S}^{sl_3}(S^1 \times I)$ by

$$\omega' = \sum_{\substack{n+m\leq r-3 \\ n,m\geq 0 \\ 3|(n-m)}} [n+1][m+1][n+m+2]$$ 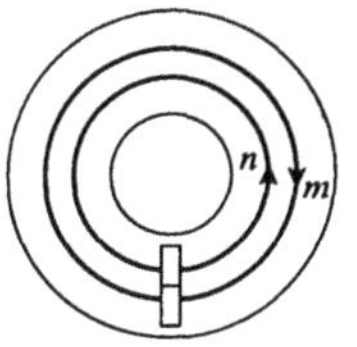 ,

where the notation $3|k$ implies that 3 divides k. We obtain a topological invariant of a 3-manifold M again as

$$\langle\!\langle U_+^{\omega'} \rangle\!\rangle^{-\sigma_+} \langle\!\langle U_-^{\omega'} \rangle\!\rangle^{-\sigma_-} \langle\!\langle D^{\omega'} \rangle\!\rangle \in \mathbb{C}, \tag{B.11}$$

instead of (B.10); the topological invariance of (B.11) is shown by modifying the proof of Theorem B.5. In particular, when $A = \exp(2\pi\sqrt{-1}/6r)$, we denote the invariant by $\tau_r^{PSU(3)}(M)$ and call it the *quantum PSU(3) invariant* of a closed connected oriented 3-manifold M. See [KoTa96] for the definition of the quantum $PSU(N)$ invariant for any integer $N \geq 2$.

Appendix C

Braid representations for the Alexander polynomial

In Sections 2.2 and 2.3 we introduced two representations of the braid group B_n in n strands which induce the Alexander polynomial of links. One is the representation on $V^{\otimes n}$ defined by using a R matrix, where V is a 2-dimensional vector space over $\mathbb{C}$. The other is the Burau representation, which is a representation on $\mathbb{C}^n$.

In this appendix we show the relation between the two representations, which induces the equality in Theorem 2.5; see also the idea of the proof mentioned after that theorem. As mentioned there, the representations are related by identifying $V^{\otimes n}$ and the exterior algebra of $\mathbb{C}^n$. In this appendix we show that the representation on $V^{\otimes n}$ is equivalent to the representation on the exterior algebra of $\mathbb{C}^n$ induced by the Burau representation. By using the equivalence we show the equality in Theorem 2.5.

C.1 Relation between two braid representations

Let W_n be an n-dimensional vector space over $\mathbb{C}$ with a basis $\{f_1, f_2, \cdots, f_n\}$. With respect to the basis of W_n the non-reduced Burau representation $\psi_{W_n} : B_n \to \mathrm{End}(W_n)$ is presented by

$$\psi_{W_n}(\sigma_i) f_j = \begin{cases} (1-t) f_i + t^{1/2} f_{i+1}, & \text{if } j = i, \\ t^{1/2} f_i, & \text{if } j = i+1, \\ f_j, & \text{if } j \neq i, i+1, \end{cases} \tag{C.1}$$

where we regard t as a complex parameter. Let Δ be the 1-dimensional vector subspace of W_n spanned by the vector $t^{-(n-1)/2} f_1 + t^{-(n-2)/2} f_2 + \cdots + t^{-1/2} f_{n-1} + f_n$. We put $\widehat{W_n} = W_n/\Delta$. Since $\psi_{W_n}(\sigma_i)$ preserves the subspace Δ, the action $\psi_{W_n}(\sigma_i)$ on W_n induces an action on the quotient space $\widehat{W_n}$. We denote the induced action by $\psi_{\widehat{W_n}}(\sigma_i)$. The action $\psi_{\widehat{W_n}}$ gives the reduced Burau representation of the braid group B_n. Theorem 2.4 implies that the Alexander polynomial $\Delta_L(t)$ of an oriented link L is determined by the reduced Burau representation as

$$\Delta_L(t) \sim \frac{1-t}{1-t^n}\det(I - \psi_{\widehat{W_n}}(b)). \tag{C.2}$$

We consider the exterior algebra $\bigwedge W_n$, which is the direct sum of the i-th exterior product $\wedge^i W_n$ for $i = 0, 1, \cdots, n$. Let V be a 2-dimensional vector space with a basis $\{e_0, e_1\}$. We define the linear map $\mathcal{I}_n : V^{\otimes n} \to \bigwedge W_n$ recursively by

$$\mathcal{I}_n : V^{\otimes n} \overset{\mathcal{I}_{n-1}\otimes \mathrm{id}_V}{\longrightarrow} (\textstyle\bigwedge W_{n-1}) \otimes V \longrightarrow \textstyle\bigwedge W_n, \tag{C.3}$$

where the second map is given by $\alpha_0 \otimes e_0 + \alpha_1 \otimes e_1 \mapsto \alpha_0 + \alpha_1 \wedge f_n$ for $\alpha_0, \alpha_1 \in \bigwedge W_{n-1}$. Here, the initial map $\mathcal{I}_1 : V \to \bigwedge W_1$ is given by $\mathcal{I}_1(e_0) = 1$ and $\mathcal{I}(e_1) = f_1$. It can be verified, by an elementary computation, that the linear map I_n is an isomorphism.

Lemma C.1. The isomorphism $\mathcal{I}_n$ is equivariant with respect to the actions of B_n on $V^{\otimes n}$ and $\bigwedge W_n$, *i.e.*, the following diagram commutes for any $b \in B_n$.

$$\begin{array}{ccc} V^{\otimes n} & \xrightarrow{\mathcal{I}_n} & \bigwedge W_n \\ \psi_{V^{\otimes n}}(b) \downarrow & & \downarrow \wedge\psi_{W_n}(b) \\ V^{\otimes n} & \xrightarrow{\mathcal{I}_n} & \bigwedge W_n \end{array}$$

Proof. It is easy to show the lemma for $n = 2$. For the general n, we show the lemma by induction on n. Suppose that $\mathcal{I}_{n-1}$ is equivariant with respect to the action of B_{n-1}. Then, $\mathcal{I}_n$ is equivariant with respect to the action of B_{n-1} by the recursive construction (C.3). Hence, it is sufficient to show that $\mathcal{I}_n$ is equivariant with respect to the action of $\sigma_{n-1} \in B_n$, where we regard $B_{n-1} \subset B_n$.

By applying (C.3) twice, the isomorphism $\mathcal{I}_n$ is presented by the following composition,

$$\mathcal{I}_n : V^{\otimes n} \overset{\mathcal{I}_{n-2}\otimes \mathrm{id}_V \otimes \mathrm{id}_V}{\longrightarrow} (\textstyle\bigwedge W_{n-2}) \otimes V \otimes V \longrightarrow \textstyle\bigwedge W_n,$$

where the second map is given by

$$\begin{aligned} &\alpha_{00} \otimes e_0 \otimes e_0 + \alpha_{01} \otimes e_0 \otimes e_1 + \alpha_{10} \otimes e_1 \otimes e_0 + \alpha_{11} \otimes e_1 \otimes e_1 \\ &\qquad \mapsto \alpha_{00} + \alpha_{01} \wedge f_n + \alpha_{10} \wedge f_{n-1} + \alpha_{11} \wedge f_{n-1} \wedge f_n. \end{aligned} \tag{C.4}$$

On one hand, from the definition (C.1), the action of σ_{n-1} on $V^{\otimes n}$ takes the left hand side of (C.4) to the image

$$\alpha_{00} \otimes e_0 \otimes e_0 + \alpha_{01} \otimes (t \cdot e_1 \otimes e_0) + \alpha_{10} \otimes (e_0 \otimes e_1 + (1-t)e_1 \otimes e_0) + \alpha_{11} \otimes (-t \cdot e_1 \otimes e_1).$$

On the other hand, from the definition of the action of σ_{n-1} on $\bigwedge W_n$, σ_{n-1} takes the right hand side of (C.4) to the image

$$\alpha_{00} + \alpha_{01} \wedge (t f_{n-1}) + \alpha_{10} \wedge \big((1-t) f_{n-1} + f_n\big) + \alpha_{11} \wedge (-t f_{n-1} \wedge f_n).$$

The above two images are related by the map (C.4). This implies that $\mathcal{I}_n$ is equivariant with respect to the actions of σ_{n-1}. □

In Section 4.5 (see also Section A.3), we considered the representation ρ_{V_t} of $U_\zeta(sl_2)$ on a 2-dimensional vector space V_t with a complex parameter t. We put $V = V_t$. For the generator K of $U_\zeta(sl_2)$, consider the action $\rho_V(K)$ of K on V. Since $\Delta(K) = K \otimes K$, the action $\rho_{V^{\otimes n}}(K)$ of K on $V^{\otimes n}$ is defined to be $\rho_V(K)^{\otimes n}$. Further, we introduce the action $\rho_{\bigwedge W_n}(K)$ of K on $\bigwedge W_n$ by $\rho_{\bigwedge W_n}(K)\alpha_\pm = \pm t^{n/2}\alpha_\pm$ for $\alpha_+ \in \bigwedge^{\text{even}} W_n$ and for $\alpha_- \in \bigwedge^{\text{odd}} W_n$.

Lemma C.2. The isomorphism $\mathcal{I}_n$ is equivariant with respect to the actions of K on $V^{\otimes n}$ and on $\bigwedge W_n$, *i.e.*, the following diagram is commutative.

$$\begin{array}{ccc} V^{\otimes n} & \xrightarrow{\mathcal{I}_n} & \bigwedge W_n \\ \rho_{V^{\otimes n}}(K) \big\downarrow & & \big\downarrow \rho_{\bigwedge W_n}(K) \\ V^{\otimes n} & \xrightarrow{\mathcal{I}_n} & \bigwedge W_n \end{array}$$

Proof. Recall that the action $\rho_V(K)$ is defined by $\rho_V(K)e_0 = t^{1/2}e_0$ and $\rho_V(K)e_1 = -t^{1/2}e_1$. That is, the action of $\rho_V(K)$ on V is given by multiplication by $\pm t^{1/2}$, where the sign is $+$ (resp. $-$) when it acts on e_0 (resp. e_1). Since $\rho_{V^{\otimes n}}(K) = \rho_V(K)^{\otimes n}$, the action $\rho_{V^{\otimes n}}(K)$ on $V^{\otimes n}$ is equal to the multiple of $\pm t^{n/2}$ where the sign is $+$ (resp. $-$) when it acts on a tensor product of some copies of e_0 with an even (resp. odd) number of copies of e_1. Such a tensor product is taken by $\mathcal{I}_n$ into $\bigwedge^{\text{even}} W_n$ (resp. $\bigwedge^{\text{odd}} W_n$). Hence, the action $\rho_{V^{\otimes n}}(K)$ is equal to the action $\rho_{\bigwedge W_n}(K)$ on $\bigwedge W_n$ through the map $\mathcal{I}_n$. □

Consider now the action $\rho_V(F)$ of F, another generator of $U_\zeta(sl_2)$, on V. The action, denoted by $\rho_{V^{\otimes n}}(F)$, of F on $V^{\otimes n}$ is defined to be $\rho_V{}^{\otimes n}(\Delta^{(n-1)}(F))$. Further, putting $\delta_n = t^{-(n-1)/2}f_1 + t^{-(n-3)/2}f_2 + \cdots + t^{-1/2}f_{n-1} + f_n$, we define an action, denoted by $\rho_{\bigwedge W_n}(F)$, of F on the space $\bigwedge W_n$ by $\rho_{\bigwedge W_n}(F)\alpha = \alpha \wedge \delta_n$.

Lemma C.3. The isomorphism $\mathcal{I}_n$ is equivariant with respect to the actions of F on $V^{\otimes n}$ and on $\bigwedge W_n$, *i.e.*, the following diagram is commutative.

$$\begin{array}{ccc} V^{\otimes n} & \xrightarrow{\mathcal{I}_n} & \bigwedge W_n \\ \rho_{V^{\otimes n}}(F) \big\downarrow & & \big\downarrow \rho_{\bigwedge W_n}(F) \\ V^{\otimes n} & \xrightarrow{\mathcal{I}_n} & \bigwedge W_n \end{array} \tag{C.5}$$

Proof. We show the lemma by induction on n. Recall that the map $\mathcal{I}_n$ is recursively defined to be the composition of $\mathcal{I}_{n-1} \otimes \mathrm{id}_V$ and the following map

$$\bigwedge W_{n-1} \otimes V \longrightarrow \bigwedge W_n \tag{C.6}$$

defined by $\alpha_0 \otimes e_0 + \alpha_1 \otimes e_1 \longmapsto \alpha_0 + \alpha_1 \wedge f_n$. The diagram (C.5) splits as the following diagram.

$$\begin{array}{ccccc} V^{\otimes n} & \xrightarrow{\mathcal{I}_{n-1}\otimes \mathrm{id}_V} & \bigwedge W_{n-1} \otimes V & \xrightarrow{(C.6)} & \bigwedge W_n \\ {\scriptstyle \rho_{V^{\otimes n}}(F)}\downarrow & & \downarrow{\scriptstyle (\rho_{\wedge W_{n-1}}\otimes\rho_V)\Delta(F)} & & \downarrow{\scriptstyle \rho_{\wedge W_n}(F)} \\ V^{\otimes n} & \xrightarrow{\mathcal{I}_{n-1}\otimes \mathrm{id}_V} & \bigwedge W_{n-1} \otimes V & \xrightarrow{(C.6)} & \bigwedge W_n \end{array} \qquad \text{(C.7)}$$

Since $\rho_{V^{\otimes n}}(F) = (\rho_{V^{\otimes(n-1)}} \otimes \rho_V)\Delta(F)$, the left square is commutative, by the assumption of the induction. We show the commutativity of the right square as follows.

The clockwise route is computed for $\alpha_0, \alpha_1 \in \bigwedge W_{n-1}$ as

$$\begin{aligned} \alpha_0 \otimes e_0 + \alpha_1 \otimes e_1 \quad &\overset{(C.6)}{\longmapsto} \quad \alpha_0 + \alpha_1 \wedge f_n \\ &\overset{\rho_{\wedge W_n}(F)}{\longmapsto} \quad \alpha_0 \wedge \delta_n + \alpha_1 \wedge f_n \wedge \delta_n. \end{aligned}$$

On the other hand, the counter-clockwise route is computed as

$$\begin{aligned} &\alpha_0 \otimes e_0 + \alpha_1 \otimes e_1 \\ \overset{(\rho_{\wedge W_{n-1}}\otimes\rho_V)\Delta(F)}{\longmapsto} \quad &(\alpha_0 \wedge \delta_{n-1}) \otimes (t^{-1/2}e_0) + \alpha_0 \otimes e_1 + (\alpha_1 \wedge \delta_{n-1}) \otimes (-t^{-1/2}e_1) \\ \overset{(C.6)}{\longmapsto} \quad &\alpha_0 \wedge (t^{-1/2}\delta_{n-1} + f_n) + \alpha_1 \wedge f_n \wedge (t^{-1/2}\delta_{n-1}). \end{aligned}$$

Since $\delta_n = t^{-1/2}\delta_{n-1} + f_n$, from the definition of δ_n, the images of the above two maps are equal. Hence, the right square of (C.7) is commutative. □

Lemma C.4. Let $X_\pm$ be the set of intertwiners in $\mathrm{Hom}(V_{\pm t^n}, V_t{}^{\otimes n})$ with respect to the action of $U_\varsigma(sl_2)$, for each choice of sign. Note that $X_\pm$ is a vector subspace of $\mathrm{Hom}(V_{\pm t^n}, V_t{}^{\otimes n})$. Then, we have the following isomorphism,

$$(X_+ \otimes V_{t^n}) \oplus (X_- \otimes V_{-t^n}) \xrightarrow{\cong} V_t{}^{\otimes n}, \qquad \text{(C.8)}$$

defined by $(\chi_+ \otimes v_+) \oplus (\chi_- \otimes v_-) \mapsto \chi_+(v_+) + \chi_-(v_-)$.

Proof. Repeatedly applying Lemma A.16, we see that $V_t{}^{\otimes n}$ is isomorphic to a direct sum of a number of copies of $V_{\pm t^n}$. Since $X_\pm$ expresses the multiplicity of the direct summand, we obtain the isomorphism of the lemma. □

For an intertwiner $f \in \mathrm{End}(V_t{}^{\otimes n})$ with respect to the action of $U_\varsigma(sl_2)$ we define $f_\pm \in \mathrm{End}(X_\pm)$ by $f_\pm(\chi_\pm) = f \circ \chi_\pm$ for $\chi_\pm \in X_\pm$, *i.e.*, we put

$$f_\pm(\chi_\pm) : V_{\pm t^n} \xrightarrow{\chi_\pm} V^{\otimes n} \xrightarrow{f} V^{\otimes n}.$$

That is, $f_\pm$ is the endomorphism of $X_\pm$ induced by the intertwiner f.

Lemma C.5. For the above f and $f_\pm$ we have the formula

$$\operatorname{trace}_{2,3,\cdots,n}\Big((\mathrm{id}_{V_t}\otimes h^{\otimes(n-1)})f\Big)=c\cdot\mathrm{id}_{V_t},$$

with the scalar c given by

$$c=\frac{t^{1/2}-t^{-1/2}}{t^{n/2}-t^{-n/2}}\Big(\operatorname{trace}(f_+)-\operatorname{trace}(f_-)\Big).$$

Proof. By Lemma C.4, $V_t^{\otimes(n-1)}$ is isomorphic to $(X'_+\otimes V_{t^{n-1}})\oplus(X'_-\otimes V_{-t^{n-1}})$ for some vector spaces X'_+ and X'_-. Hence,

$$\begin{aligned}V_t^{\otimes n}&=V_t\otimes V_t^{\otimes(n-1)}\\&\cong(X'_+\otimes V_t\otimes V_{t^{n-1}})\oplus(X'_-\otimes V_t\otimes V_{-t^{n-1}})\qquad\qquad(\mathrm{C}.9)\\&\cong\Big(X'_+\otimes(V_{t^n}\oplus V_{-t^n})\Big)\oplus\Big(X'_-\otimes(V_{-t^n}\oplus V_{t^n})\Big)\\&=\Big((X'_+\oplus X'_-)\otimes V_{t^n}\Big)\oplus\Big((X'_-\oplus X'_+)\otimes V_{-t^n}\Big),\end{aligned}$$

where we obtain the third equality by Lemma A.16. Hence, we have the isomorphisms $X_\pm\cong X'_\pm\oplus X'_\mp$. Fixing the isomorphisms we let $f_{\pm,1}$ and $f_{\pm,2}$ denote the restrictions of $f_\pm\in\operatorname{End}(X_\pm)$ to $\operatorname{End}(X'_\pm)$ and $\operatorname{End}(X'_\mp)$ respectively. The $\operatorname{trace}_{2,3,\cdots,n}$ on $V_t^{\otimes n}$ is equal to $\operatorname{trace}_{X'_+}\otimes\operatorname{trace}_2+\operatorname{trace}_{X'_-}\otimes\operatorname{trace}_2$ on (C.9). Hence, by applying Lemma A.17 to $V_t\otimes V_{\pm t^{n-1}}$, we have that

$$c=\frac{t^{1/2}-t^{-1/2}}{t^{n/2}-t^{-n/2}}\Big(\operatorname{trace}(f_{+,1})-\operatorname{trace}(f_{-,2})\Big)-\frac{t^{1/2}-t^{-1/2}}{t^{n/2}-t^{-n/2}}\Big(\operatorname{trace}(f_{-,1})-\operatorname{trace}(f_{+,2})\Big).$$

Therefore, we obtain the lemma. □

By composing the two isomorphisms (C.8) and $\mathcal{I}_n$, we have the following isomorphism,

$$(X_+\otimes V_{t^n})\oplus(X_-\otimes V_{-t^n})\overset{(C.8)}{\longrightarrow}V_t^{\otimes n}\overset{\mathcal{I}_n}{\longrightarrow}\bigwedge W_n.$$

By considering the quotient of the isomorphism by the image of the action of F, we obtain the following isomorphism,

$$X_+\otimes\Big(V_{t^n}/(\rho(F)(V_{t^n}))\Big)\oplus X_-\otimes\Big(V_{-t^n}/(\rho(F)(V_{-t^n}))\Big)\overset{\cong}{\longrightarrow}\bigwedge W_n/(\rho(F)(\bigwedge W_n)).$$

From the definition of $V_{\pm t^n}$, the quotient space $V_{\pm t^n}/(\rho(F)(V_{\pm t^n}))$ can be identified with the 1-dimensional vector space spanned by the first basis vector e_0 of $V_{\pm t^n}$. Further,

$$\bigwedge W_n/(\rho(F)(\bigwedge W_n))\;\cong\;\bigwedge W_n/(\bigwedge W_n\wedge\Delta)\;\cong\;\bigwedge(W_n/\Delta)\;\cong\;\bigwedge\widehat{W_n}.$$

Hence, the above isomorphism induces the following isomorphism,

$$X_+\oplus X_-\overset{\cong}{\longrightarrow}\bigwedge\widehat{W_n}.$$

With respect to the action of K, the above isomorphism splits into the following two isomorphisms,

$$X_+ \xrightarrow{\cong} \bigwedge^{\text{even}} \widehat{W_n}, \tag{C.10}$$

$$X_- \xrightarrow{\cong} \bigwedge^{\text{odd}} \widehat{W_n}. \tag{C.11}$$

Proof of the equality in Theorem 2.5. Let the eigenvalues of $\psi_{W_n}(b)$ be α_1, $\alpha_2, \cdots, \alpha_n$. Since the vector $\delta_n \in W_n$ is invariant with respect to the action of $\psi_{W_n}(b)$, there is a corresponding eigenvalue 1. Since $\widehat{W_n}$ is the quotient space of W_n by the subspace spanned by δ_n, the eigenvalues of $\psi_{\widehat{W_n}}(b)$ are $\alpha_2, \alpha_3, \cdots, \alpha_n$. Hence, the eigenvalues of $\psi_{\wedge \widehat{W_n}}(b)$ are their products: 1, α_i $(2 \le i \le n)$, $\alpha_i\alpha_j$ $(2 \le i < j \le n)$, $\cdots$, $\alpha_2\alpha_3\cdots\alpha_n$. Among them, the eigenvalues of $\psi_{\wedge^{\text{even}} \widehat{W_n}}(b)$ (resp. $\psi_{\wedge^{\text{odd}} \widehat{W_n}}(b)$) are the products of an even (resp. odd) number of α_i $(2 \le i \le n)$. Therefore,

$$\begin{aligned}
&\operatorname{trace}\Big(\psi_{\wedge^{\text{even}} \widehat{W_n}}(b)\Big) - \operatorname{trace}\Big(\psi_{\wedge^{\text{odd}} \widehat{W_n}}(b)\Big) \\
&= 1 - \sum_{2 \le i \le n} \alpha_i + \sum_{2 \le i < j \le n} \alpha_i\alpha_j - \cdots + (-1)^{n-1}\alpha_2\alpha_3\cdots\alpha_n \\
&= \prod_{2 \le i \le n} (1 - \alpha_i) = \det\Big(I - \psi_{\widehat{W_n}}(b)\Big).
\end{aligned} \tag{C.12}$$

As in Chapter 4, $\psi_{V^{\otimes n}}(b)$ is an intertwiner of $V_t^{\otimes n}$ to itself with respect to the action of $U_\varsigma(sl_2)$, *i.e.*, $\psi_{V^{\otimes n}}(b) \in \operatorname{End}_{U_\varsigma(sl_2)}(V_t^{\otimes n})$. Let $X_\pm$ and $\psi_{V^{\otimes n}}(b)_\pm \in \operatorname{End}(X_\pm)$ be as in Lemmas C.4 and C.5. Then, by Lemma C.5, we have the formula

$$\operatorname{trace}_{2,3,\cdots,n}\Big((1 \otimes h^{\otimes(n-1)})\psi_{V^{\otimes n}}(b)\Big) = c \cdot \mathrm{id}_V, \tag{C.13}$$

with the scalar c given by

$$c = \frac{t^{1/2} - t^{-1/2}}{t^{n/2} - t^{-n/2}}\Big(\operatorname{trace}\psi_{V^{\otimes n}}(b)_+ - \operatorname{trace}\psi_{V^{\otimes n}}(b)_-\Big). \tag{C.14}$$

By the construction, the maps (C.10) and (C.11) are equivariant with respect to the actions of the braid group B_n, *i.e.*, the following diagrams are commutative.

$$\begin{array}{ccc}
X_+ & \xrightarrow{(C.10)} & \bigwedge^{\text{even}} \widehat{W_n} \\
{\scriptstyle \psi_{V^{\otimes n}}(b)_+}\big\downarrow & & \big\downarrow{\scriptstyle \psi_{\wedge^{\text{even}} \widehat{W_n}}(b)} \\
X_+ & \xrightarrow{(C.10)} & \bigwedge^{\text{even}} \widehat{W_n}
\end{array}
\qquad
\begin{array}{ccc}
X_- & \xrightarrow{(C.11)} & \bigwedge^{\text{odd}} \widehat{W_n} \\
{\scriptstyle \psi_{V^{\otimes n}}(b)_+}\big\downarrow & & \big\downarrow{\scriptstyle \psi_{\wedge^{\text{odd}} \widehat{W_n}}(b)} \\
X_- & \xrightarrow{(C.11)} & \bigwedge^{\text{odd}} \widehat{W_n}
\end{array}$$

Hence,

$$\operatorname{trace}\Big(\psi_{V^{\otimes n}}(b)_+\Big) = \operatorname{trace}\Big(\psi_{\wedge^{\text{even}} \widehat{W_n}}(b)\Big),$$

$$\operatorname{trace}\Big(\psi_{V^{\otimes n}}(b)_-\Big) = \operatorname{trace}\Big(\psi_{\wedge^{\text{odd}} \widehat{W_n}}(b)\Big).$$

Therefore, by (C.12), the above two formulae and (C.14), we have that

$$\det\Big(I - \psi_{\widetilde{W_n}}(b)\Big) = \mathrm{trace}\Big(\psi_{V^{\otimes n}}(b)_+\Big) - \mathrm{trace}\Big(\psi_{V^{\otimes n}}(b)_-\Big) = \frac{t^{n/2} - t^{-n/2}}{t^{1/2} - t^{-1/2}} c,$$

where c is the scalar satisfying (C.13). Since the Alexander polynomial is given by (C.2), the above formula implies the equality in Theorem 2.5. □

Appendix D

Associators

The comultiplication Δ of a bialgebra A is, by definition, coassociative, which means that it satisfies the equation $(\Delta \otimes \mathrm{id})\big(\Delta(x)\big) = (\mathrm{id} \otimes \Delta)\big(\Delta(x)\big)$ for any $x \in A$. Drinfel'd introduced the notion of a quasi-bialgebra, in which the coassociativity requirement of the comultiplication is relaxed to the requirement that $(\Delta \otimes \mathrm{id})\big(\Delta(x)\big)$ is related to $(\mathrm{id} \otimes \Delta)\big(\Delta(x)\big)$ by conjugation by some fixed invertible element $\Phi \in A^{\otimes 3}$ (for all x). Drinfel'd did this in order to apply the KZ equation to the theory of quantum groups (see Section 5.5). We call the element Φ an *associator.* To be precise, in Section 5.4 we defined an associator $\Phi \in A^{\otimes 3}$ in a quasi-triangular quasi-bialgebra to be an element Φ such that a pair $(\mathcal{R}, \Phi)$ satisfying the pentagon and hexagon relations for some $\mathcal{R}$. Further, in Section 6.4 we defined an associator in the space $\mathcal{A}(\downarrow\downarrow\downarrow)$ to be an element Φ such that the pair $\Big(\exp\big(\tfrac{1}{2}\ \downarrow\text{---}\downarrow\ \big), \Phi\Big)$ satisfying the pentagon and hexagon relations.

To date, the Drinfel'd associator is the only associator we have whose coefficients have concrete expressions for all degrees. That is, the coefficients of the Drinfel'd associator can be expressed as iterated integrals (see Section D.2). On the other hand, we know a combinatorial degree-by-degree proof of the existence of solutions of the defining relations of associators by Bar-Natan (see [Bar97, Bar98, BaSt97]). However, we still do not have a direct combinatorial expression of any associator for all degrees. This implies that we still do not know a direct combinatorial expression of the Kontsevich invariant of each knot for all degrees (except for the trivial knot). Note that we already know a combinatorial direct presentation of the element ν (by Thurston [Th$_{\mathrm{D}}$00], see also [BGRT00]), which is equal to the Kontsevich invariant of the trivial knot.

In Section D.1 we introduce the notion of a Drinfel'd series and show that a Drinfel'd series induces associators both in a universal enveloping algebra and in a space of Jacobi diagrams. In Section D.2 we present the Drinfel'd associator by iterated integrals. These integrals can, moreover, be expressed in terms of multiple zeta values. Thus, the Drinfel'd associator can be regarded as a generating function of multiple zeta values. For details on associators see [Dri90] and [LeMu96a, Bar98,

Bar97, BaSt97].

D.1 Drinfel'd series

In this section we introduce the notion of a Drinfel'd series and describe properties of associators in a universal enveloping algebra, and in a space of Jacobi diagrams.

Let $\mathbb{C}\langle\langle A,B\rangle\rangle$ denote the power series ring of non-commutative indeterminates A and B. The algebra $\mathbb{C}\langle\langle A,B\rangle\rangle$ has a comultiplication Δ defined by $\Delta(A) = A\otimes 1 + 1\otimes A$ and $\Delta(B) = B\otimes 1 + 1\otimes B$. An element $\alpha \in \mathbb{C}\langle\langle A,B\rangle\rangle$ is called *group-like*, if $\Delta(\alpha) = \alpha\otimes\alpha$. A *Drinfel'd series* is defined to be a group-like power series $\varphi(A,B)$ whose leading term is 1, satisfying the following relations,

$$\varphi(\tau_{12},\tau_{23}+\tau_{24})\varphi(\tau_{12}+\tau_{23},\tau_{34}) = \varphi(\tau_{23},\tau_{34})\varphi(\tau_{12}+\tau_{13},\tau_{24}+\tau_{34})\varphi(\tau_{12},\tau_{23}), \tag{D.1}$$

$$\exp(\frac{\tau_{13}+\tau_{23}}{2}) = \varphi(\tau_{13},\tau_{12})\exp\frac{\tau_{13}}{2}\varphi(\tau_{13},\tau_{23})^{-1}\exp\frac{\tau_{23}}{2}\varphi(\tau_{12},\tau_{23}), \tag{D.2}$$

$$\exp(\frac{\tau_{12}+\tau_{13}}{2}) = \varphi(\tau_{23},\tau_{13})^{-1}\exp\frac{\tau_{13}}{2}\varphi(\tau_{12},\tau_{13})\exp\frac{\tau_{12}}{2}\varphi(\tau_{12},\tau_{23})^{-1}, \tag{D.3}$$

where τ_{ij}'s are indeterminates subject to the relations of Lemma 5.1. We call (D.1) the *pentagon relation* and call (D.2) and (D.3) the *hexagon relations* of a Drinfel'd series. In the next section we give a concrete construction of a Drinfel'd series φ_{KZ}.

Associator in a universal enveloping algebra

A Drinfel'd series gives an associator for the power series algebra $U(\mathfrak{g})[[\hbar]]$ of the universal enveloping algebra of a semi-simple Lie algebra $\mathfrak{g}$. The algebra forms a quasi-triangular quasi-bialgebra with this associator. This is shown in this subsection.

Let $\mathfrak{g}$ be a semi-simple Lie algebra, $U(\mathfrak{g})$ its universal enveloping algebra, and $U(\mathfrak{g})[[\hbar]]$ the power series algebra of an indeterminate $\hbar$. As mentioned in Chapter 5, we have the invariant 2-tensor $\tau \in U(\mathfrak{g})\otimes U(\mathfrak{g})$, dual to the Killing form. For a Drinfel'd series φ, we put

$$\Phi = \varphi\Big((\hbar\tau\otimes 1),(1\otimes\hbar\tau)\Big). \tag{D.4}$$

Then, we have

Lemma D.1. $(U(\mathfrak{g})[[\hbar]],\Delta,\varepsilon,\Phi)$ is a quasi-bialgebra for the comultiplication Δ and the counit ε of the universal enveloping algebra $U(\mathfrak{g})$.

Proof. We verify the defining relations (5.50)–(5.53) of a quasi-bialgebra. The relation (5.50) is obtained from the equality $(\text{id}\otimes\Delta)(\Delta(x)) = (\Delta\otimes\text{id})(\Delta(x))$ for any x and the commutativity between Φ and each side of this equality. The relation

(5.51) is derived from (D.1). The relation (5.52) is obtained from the definitions of Δ and ε. The relation (5.53) is obtained from the fact that a Drinfel'd series is group-like and has the leading term 1. □

Further, putting $\mathcal{R} = \exp(\hbar\tau/2)$, we have

Proposition D.2. $(U(\mathfrak{g})[[\hbar]], \Delta, \varepsilon, \Phi, \mathcal{R})$ is a quasi-triangular quasi-bialgebra.

The proposition is proved by verifying that the defining relations (5.54)–(5.56) of a quasi-triangular quasi-bialgebra follow from the relations (D.2) and (D.3); it is left to the reader to show that. Thus, by (D.4), a Drinfel'd series φ gives an associator Φ of the quasi-triangular quasi-bialgebra $U(\mathfrak{g})[[\hbar]]$.

As in Section 5.5, we call $f \in U(\mathfrak{g})^{\otimes 2}[[\hbar]]$ *symmetric* if f is invariant under the exchange of two entries of the tensor product $U(\mathfrak{g})^{\otimes 2}$. In Section 5.5 we defined twisting of Δ, $\mathcal{R}$ and Φ by f, denoted by Δ^f, $\mathcal{R}^f$ and Φ^f. We put $\mathcal{R} = \exp(\hbar\tau/2)$. Then, $\Delta = \Delta^f$ and $\mathcal{R} = \mathcal{R}^f$ for symmetric f. Hence, if $(U(\mathfrak{g})[[\hbar]], \Delta, \varepsilon, \Phi, \mathcal{R})$ is a quasi-triangular quasi-bialgebra, so is $(U(\mathfrak{g})[[\hbar]], \Delta, \varepsilon, \Phi^f, \mathcal{R})$ for symmetric f.

The following proposition is mentioned in the third step of the proof of Proposition 5.13.

Proposition D.3. Let Φ and Φ' be associators in $U(\mathfrak{g})^{\otimes 3}[[\hbar]]$ for $\mathcal{R} = \exp(\hbar\tau/2)$. Then, Φ' is obtained from Φ by twisting it by some symmetric $f \in U(\mathfrak{g})^{\otimes 2}[[\hbar]]$.

We sketch a proof of the proposition in the third step of the proof of Proposition 5.13. The proof of the proposition can also be obtained from the proof of Proposition D.6 by substituting $\mathfrak{g}$ into Jacobi diagrams.

Associator in a space of Jacobi diagrams

In Section 6.4 an associator in the space $\mathcal{A}(\downarrow\downarrow\downarrow)$ is defined to be an invertible group-like element satisfying the pentagon relation (6.11), the hexagon relations (6.12), and the relation $\varepsilon_2\Phi = 1$. For a Drinfel'd series φ, we put

$\Phi = \varphi($  , 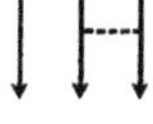 $) \in \mathcal{A}(\downarrow\downarrow\downarrow).$

Then, we have

Proposition D.4. The above Φ is an associator in the space $\mathcal{A}(\downarrow\downarrow\downarrow)$.

Proof. The proposition is proved by verifying the pentagon relation (6.11), the hexagon relations (6.12), the relation $\varepsilon_2\Phi = 1$, and the fact that Φ is group-like. These relations are derived from the pentagon and the hexagon relations (D.1)–(D.3) of a Drinfel'd series. Since a Drinfel'd series is group-like, so is Φ. Further, the equation $\varepsilon_2\Phi = 1$ holds because the leading term of a Drinfel'd series is equal to 1 and a Drinfel'd series does not contain the terms A^n and B^n (because the series is group-like). □

Lemma D.5. Any associator $\Phi \in \mathcal{A}(\downarrow\downarrow\downarrow)$ satisfies that

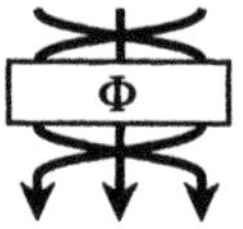

$= \Phi^{-1}.$

Proof. From the definition of the comultiplication Δ, we have that

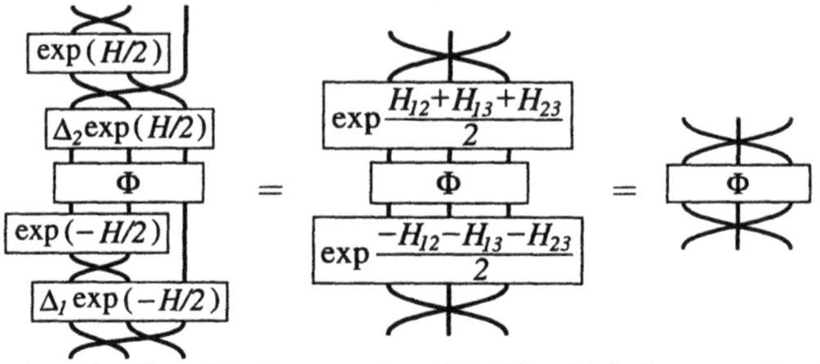

(D.5)

where H_{ij} denotes the Jacobi diagram in $\mathcal{A}(\downarrow\downarrow\downarrow)$ which has one horizontal dashed chord connecting the ith and jth vertical solid lines. Here, the second equality of the above formula is derived from the fact that $H_{12} + H_{13} + H_{23}$ commutes with any element in $\mathcal{A}(\downarrow\downarrow\downarrow)$.

Further, we show that the left hand side of the above formula is equal to Φ^{-1} as follows. This equality is described by the following parenthesized braids,

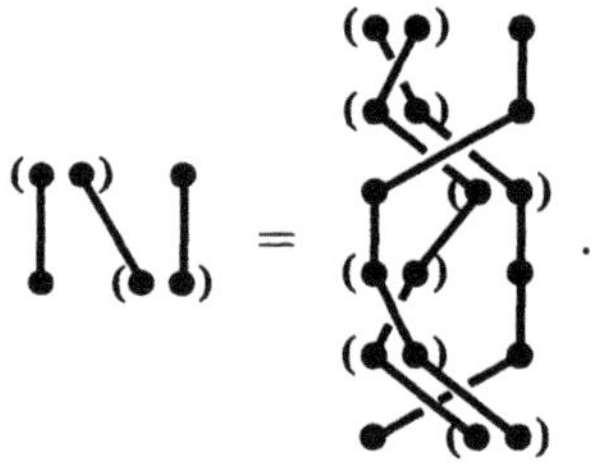

Since the above two parenthesized braids are isotopic, they are related by the hexagon relations among parenthesized braids. Hence, by the hexagon relations (6.12) and its equivalent form (6.13) among Jacobi diagrams, we obtain the equality between Φ^{-1} and the left hand side of (D.5). □

Let Φ be an associator in $\mathcal{A}(\downarrow\downarrow\downarrow)$, and f an invertible element in $\mathcal{A}(\downarrow\downarrow)$. We say that the following Φ^f is obtained from Φ by *twisting* by f.

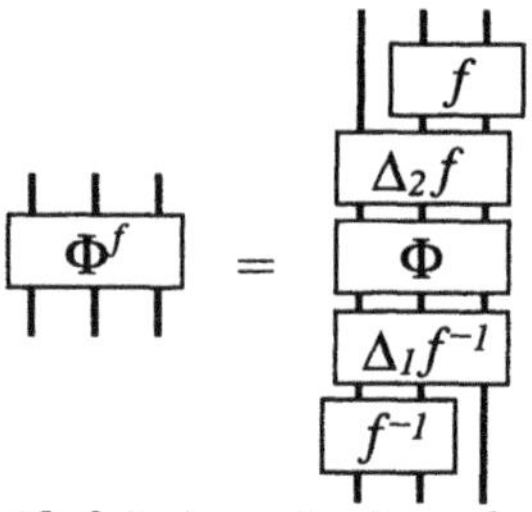

We call $f \in \mathcal{A}(\downarrow\downarrow)$ *symmetric* if f is invariant under the exchange of two vertical lines $\downarrow\downarrow$. Note that, if we defined Δ^f and $\mathcal{R}^f$ in the same way as in (5.57), $\Delta = \Delta^f$

and $\mathcal{R} = \mathcal{R}^f$ for symmetric f. Hence, in the case that f is symmetric, Φ^f is again an associator.

The following proposition implies the uniqueness of associators modulo twisting, which guarantees that the framed Kontsevich invariant $\hat{Z}(L)$ of a framed link L does not depend on the choice of an associator (see Theorem 6.7).

Proposition D.6 (see [LeMu96a]). Let Φ and Φ' be associators in the space $\mathcal{A}(\downarrow\downarrow\downarrow)$. Then, Φ' is obtained from Φ by twisting it by some symmetric $f \in \mathcal{A}(\downarrow\downarrow)$.

Sketch of the proof. We sketch a proof of the proposition which proceeds by induction on the leading degree of $\Phi - \Phi'$. Namely, assuming that $\Phi \equiv \Phi'$ modulo the degree $\geq n$ part, we show that $\Phi' \equiv \Phi^f$ modulo the degree $> n$ part for some f of the form $f = 1 + \dot{f}$ for some $\dot{f} \in \mathcal{A}(\downarrow\downarrow)^{(n)}$. Let $\dot{\Phi}$ be the degree n part of $\Phi' - \Phi$. Then, by the pentagon relations of Φ and Φ', we have that

$$1 \otimes \dot{\Phi} - \Delta_1\dot{\Phi} + \Delta_2\dot{\Phi} - \Delta_3\dot{\Phi} + \dot{\Phi} \otimes 1 = 0.$$

This equation can be written as $d_3\dot{\Phi} = 0$, with the map $d_3 : \mathcal{A}(\downarrow\downarrow\downarrow) \to \mathcal{A}(\downarrow\downarrow\downarrow\downarrow)$ defined by

$$d_3\xi = 1 \otimes \xi - \Delta_1\xi + \Delta_2\xi - \Delta_3\xi + \xi \otimes 1.$$

Further, we define the map $d_2 : \mathcal{A}(\downarrow\downarrow) \to \mathcal{A}(\downarrow\downarrow\downarrow)$ by

$$d_2\eta = 1 \otimes \eta - \Delta_1\eta + \Delta_2\eta - \eta \otimes 1.$$

It follows from the vanishing of (a modification of) the cohomology kernel(d_3)/image(d_2) that there exists $\dot{f}$ such that $d_2\dot{f} = \dot{\Phi}$. This is rewritten as

$$\dot{\Phi} = 1 \otimes \dot{f} - \Delta_1\dot{f} + \Delta_2\dot{f} - \dot{f} \otimes 1.$$

Such an $\dot{f}$ gives the required f with $\Phi' \equiv \Phi^f$. Further, by some defining relations of the associators Φ and Φ', we have that

$$\dot{\Phi}_{321} = -\dot{\Phi}, \qquad \dot{\Phi} + \dot{\Phi}_{231} + \dot{\Phi}_{312} = 0.$$

By using these relations, we can replace $\dot{f}$ with a symmetric solution. For a complete proof see [LeMu96a]. □

By modifying the proof of Proposition D.6, we obtain the following proposition, which guarantees the rationality of the coefficients of the framed Kontsevich invariant of framed links.

Proposition D.7. There exists an associator in the space $\mathcal{A}(\downarrow\downarrow\downarrow)$ whose coefficients are rational.

Sketch of the proof. We can obtain an associator by solving the defining relations of an associator step by step from low degrees. We know that there exists at least one associator Φ_{KZ}. Since the coefficients of the defining relations are rational, we obtain an associator with rational coefficients from Φ_{KZ} by twisting. For complete proofs see [Dri90] and [LeMu96a, Bar98]. □

Bar-Natan (see [Bar97, BaSt97]) gave a combinatorial degree-by-degree proof of the existence of solutions of the defining relations of associators. He [Bar97] obtained the following explicit formula for the low degrees of an associator with rational coefficients by concretely solving the defining relations degree-by-degree,

$$\begin{aligned}\log\varphi(A,B) = {}& \frac{[A,B]}{48} - \frac{8[A,[A,[A,B]]]+[A,[B,[A,B]]]}{11520} \\ & + \frac{[A,[A,[A,[A,[A,B]]]]]}{60480} + \frac{[A,[A,[A,[B,[A,B]]]]]}{1451520} \\ & + \frac{13[A,[A,[B,[B,[A,B]]]]]}{1161216} + \frac{17[A,[B,[A,[A,[A,B]]]]]}{1451520} \\ & + \frac{[A,[B,[A,[B,[A,B]]]]]}{1451520} \\ & - (\text{interchange of } A \text{ and } B) \\ & + (\text{terms of degree} \geq 8).\end{aligned}$$

D.2 The Drinfel'd associator

The Drinfel'd associator is derived from the Drinfel'd series $\varphi_{\mathrm{KZ}}(A,B)$ which is the power series in non-commutative variables A and B defined in (5.24). In this section we show some properties of φ_{KZ}, and present φ_{KZ} by iterated integrals.

The power series φ_{KZ} was defined in Section 5.2 by using solutions of the differential equation (5.23). The differential equation is rewritten as

$$G'(z) = \Big(\frac{\overline{A}}{z} + \frac{\overline{B}}{z-1}\Big)G(z), \tag{D.6}$$

putting $\overline{A} = A/2\pi\sqrt{-1}$ and $\overline{B} = B/2\pi\sqrt{-1}$, where $G(z)$ belongs to $\mathbb{C}\langle\langle A,B\rangle\rangle$ and it is analytic with respect to the variable z. In terms of the solutions $G_{(\bullet\bullet)\bullet}$ and $G_{\bullet(\bullet\bullet)}$ of the differential equation given in Lemma 5.4, the power series φ_{KZ} was defined by $G_{(\bullet\bullet)\bullet} = G_{\bullet(\bullet\bullet)}\varphi_{\mathrm{KZ}}(A,B)$ in (5.24).

Proposition D.8. $\varphi_{\mathrm{KZ}}(A,B)$ is a Drinfel'd series.

Proof. The proposition is proved by showing that $\varphi_{\mathrm{KZ}}(A,B)$ is group-like and satisfies the pentagon relation (D.1), and the hexagon relations (D.2), (D.3). The pentagon and hexagon relations are shown in the same way as in the proof of Lemma 5.10. Further, φ_{KZ} is group-like, since the solutions $G_{(\bullet\bullet)\bullet}$ and $G_{\bullet(\bullet\bullet)}$ in Lemma 5.4 are group-like by considering the differential equation obtained from (D.6) by applying the comultiplication Δ of $\mathbb{C}\langle\langle A,B\rangle\rangle$. □

From the above proposition we immediately obtain

Corollary D.9. The Drinfel'd associator Φ_{KZ} (in a quasi-triangular quasi-bialgebra and in $\mathcal{A}(\downarrow\downarrow\downarrow)$) is an associator.

Lemma D.10. For commutative variables x and y, we have that $\varphi_{\mathrm{KZ}}(x,y) = 1$.

Proof. If A and B commute, then the solutions $G_{(\bullet\bullet)\bullet}$ and $G_{\bullet(\bullet\bullet)}$ in Lemma 5.4 are given by

$$G_{(\bullet\bullet)\bullet}(z) = G_{\bullet(\bullet\bullet)}(z) = z^{A/2\pi\sqrt{-1}}(1-z)^{B/2\pi\sqrt{-1}}.$$

Hence, by (5.24), φ_{KZ} is equal to 1 when the variables commute. □

Lemma D.11. The Drinfel'd associator φ_{KZ} satisfies that $\varphi_{\mathrm{KZ}}(B,A) = \varphi_{\mathrm{KZ}}(A,B)^{-1}$.

Proof. When we exchange A and B, the solutions $G_{(\bullet\bullet)\bullet}$ and $G_{\bullet(\bullet\bullet)}$ in Lemma 5.4 are exchanged. Then, by (5.24), φ_{KZ} becomes its inverse. This implies the lemma. □

Lemma D.11 can be alternatively proved from the viewpoint of the original definition of φ_{KZ} as follows. The quasi-tangles  and  are inverses to each other and they are related (as paths in the configuration space X_3 mentioned in Section 5.3) by the change of coordinates of X_3 which takes (z_1, z_2, z_3) to (z_3, z_2, z_1). Under this change of coordinates, the variables A and B are exchanged and φ_{KZ} is changed to its inverse, since the two quasi-tangles are inverses to each other. Hence, we obtain the lemma.

Computing the Drinfel'd associator

In this subsection we compute the Drinfel'd associator φ_{KZ} by iterated integrals.

For each real number ε with $0 < \varepsilon < 1$, we denote by $G_\varepsilon(z)$ the unique solution of the differential equation (D.6) with $G_\varepsilon(\varepsilon) = 1$. Then, the power series φ_{KZ} can be expressed in these terms as

Lemma D.12. We have that

$$\varphi_{\mathrm{KZ}}(A,B) = \lim_{\varepsilon \to 0} \varepsilon^{-\overline{B}} G_\varepsilon(1-\varepsilon)\varepsilon^{\overline{A}},$$

where we regard $\varepsilon^{\overline{A}}$ and $\varepsilon^{-\overline{B}}$ as power series in $\overline{A}$ and $\overline{B}$ in such a way as

$$\varepsilon^{\overline{A}} = \exp(\overline{A}\log\varepsilon) = 1 + \overline{A}\log\varepsilon + \overline{A}^2\frac{(\log\varepsilon)^2}{2} + \cdots.$$

For the proof of the lemma see [Kas94, Lemma XIX.6.3]. By the lemma the computation of φ_{KZ} is reduced to the computation of $G_\varepsilon(1-\varepsilon)$. We compute $G_\varepsilon(1-\varepsilon)$ as follows.

We put

$$w(t) = \frac{\overline{A}}{t} + \frac{\overline{B}}{t-1}.$$

Then, the differential equation (5.23) is rewritten as $G'(t) = w(t)G(t)$. Hence, by the same argument as in Proposition 6.4, $G_\varepsilon(1-\varepsilon)$ is presented by iterated integrals as

$$G_\varepsilon(1-\varepsilon) = 1 + \sum_{m=1}^{\infty} I_m,$$

where I_m is given by

$$I_m = \int_{\varepsilon \leq t_1 \leq \cdots \leq t_m \leq 1-\varepsilon} w(t_m) \cdots w(t_1) dt_1 \cdots dt_m.$$

Therefore, we obtain a concrete expression for the power series φ_{KZ} up to any degree by computing the iterated integrals I_m's up to the degree in question. For $m = 1$ the integral is computed as

$$I_1 = \int_{\varepsilon \leq t_1 \leq 1-\varepsilon} w(t_1) dt_1 = (-\overline{A} + \overline{B}) \log \frac{\varepsilon}{1-\varepsilon}.$$

Further, for $m = 2$, we have that

$$\begin{aligned} I_2 &= \int_{\varepsilon \leq t_1 \leq t_2 \leq 1-\varepsilon} w(t_2) w(t_1) dt_1 dt_2 \\ &= \int_{\varepsilon \leq t_2 \leq 1-\varepsilon} w(t_2) \Big(\overline{A} \log \frac{t_2}{\varepsilon} + \overline{B} \log \frac{1-t_2}{1-\varepsilon} \Big) dt_2 \\ &= \overline{A}^2 \int_\varepsilon^{1-\varepsilon} \frac{1}{t} \log \frac{t}{\varepsilon} dt + \overline{AB} \int_\varepsilon^{1-\varepsilon} \frac{1}{t} \log \frac{1-t}{1-\varepsilon} dt \\ &\quad + \overline{BA} \int_\varepsilon^{1-\varepsilon} \frac{1}{t-1} \log \frac{t}{\varepsilon} dt + \overline{B}^2 \int_\varepsilon^{1-\varepsilon} \frac{1}{t-1} \log \frac{1-t}{1-\varepsilon} dt, \end{aligned}$$

where the integral of the coefficient of $\overline{A}^2$ is computed as

$$\int_\varepsilon^{1-\varepsilon} \frac{1}{t} \log \frac{t}{\varepsilon} dt = \Big[\frac{1}{2} \Big(\log \frac{t}{\varepsilon} \Big)^2 \Big]_\varepsilon^{1-\varepsilon} = \frac{1}{2} \Big(\log \frac{1-\varepsilon}{\varepsilon} \Big)^2.$$

Putting the coefficient of $\overline{AB}$ to be c_ε, we have that

$$I_2 = (\overline{A}^2 + \overline{B}^2) \frac{1}{2} \Big(\log \frac{\varepsilon}{1-\varepsilon} \Big)^2 + [\overline{A}, \overline{B}] c_\varepsilon.$$

Hence, the degree ≤ 2 part of $\varphi_{\text{KZ}}(A, B)$ is computed modulo the degree > 2 part as

$$\Big(1 - \overline{B} \log \varepsilon + \overline{B}^2 \frac{(\log \varepsilon)^2}{2} \Big) (1 + I_1 + I_2) \Big(1 + \overline{A} \log \varepsilon + \overline{A}^2 \frac{(\log \varepsilon)^2}{2} \Big)$$

$$\equiv 1 + [\overline{A}, \overline{B}]\big(\lim_{\varepsilon \to 0} c_\varepsilon\big) = 1 + \frac{1}{4\pi^2}[A, B]\big(\lim_{\varepsilon \to 0} c_\varepsilon\big).$$

Further, using the following expansion

$$\log(1 - x) = -\sum_{m=1}^{\infty} \frac{x^m}{m},$$

c_ε is computed as

$$\begin{aligned} c_\varepsilon &= \int_\varepsilon^{1-\varepsilon} \frac{1}{t} \log \frac{1-t}{1-\varepsilon} dt = \int_\varepsilon^{1-\varepsilon} \frac{\log(1-t)}{t} dt - \log(1-\varepsilon) \int_\varepsilon^{1-\varepsilon} \frac{1}{t} dt \\ &= -\int_\varepsilon^{1-\varepsilon} \sum_{m=1}^{\infty} \frac{t^{m-1}}{m} dt - \log(1-\varepsilon) \log \frac{1-\varepsilon}{\varepsilon} \\ &= -\sum_{m=1}^{\infty} \frac{(1-\varepsilon)^m - \varepsilon^m}{m^2} - \log(1-\varepsilon) \log \frac{1-\varepsilon}{\varepsilon}. \end{aligned}$$

Hence,

$$\lim_{\varepsilon \to 0} c_\varepsilon = -\sum_{m=1}^{\infty} \frac{1}{m^2} = -\zeta(2),$$

where $\zeta(2)$ is a special value of the *zeta function* $\zeta(s) = \sum_{k=1}^{\infty} k^{-s}$ which is an important function in number theory. By Euler's formula $\zeta(2) = \pi^2/6$, the coefficient of $[A, B]$ is equal to $1/24$. Continuing with similar computations, we find the degree ≤ 3 part of φ_{KZ} to be

$$\varphi_{\text{KZ}}(A, B) = 1 + \frac{1}{24}[A, B] - \frac{\zeta(3)}{(2\pi\sqrt{-1})^3}([A, [A, B]] + [B, [A, B]]) + (\text{terms of degree} \geq 4).$$

Continuing in this fashion, we find that the power series φ_{KZ} can be presented, to all degrees, as follows (see [LeMu95b]). We define the *multiple zeta function* by

$$\zeta(a_1, a_2, \cdots, a_k) = \sum_{n_1 < n_2 < \cdots < n_k \in \mathbb{N}} n_1^{-a_1} n_2^{-a_2} \cdots n_k^{-a_k}.$$

For $\mathbf{a} = (a_1, \cdots, a_l)$ and $\mathbf{b} = (b_1, \cdots, b_l)$ we put

$$\begin{aligned} &\eta(\mathbf{a}, \mathbf{b}) = \zeta(\underbrace{1, 1, \cdots, 1}_{a_1 - 1}, b_1 + 1, \underbrace{1, 1, \cdots, 1}_{a_1 - 1}, b_2 + 1, \cdots, \underbrace{1, 1, \cdots, 1}_{a_l - 1}, b_l + 1), \\ &|\mathbf{a}| = a_1 + a_2 + \cdots + a_l, \\ &\binom{\mathbf{a}}{\mathbf{b}} = \binom{a_1}{b_1}\binom{a_2}{b_2} \cdots \binom{a_l}{b_l}, \\ &(A, B)^{(\mathbf{a}, \mathbf{b})} = A^{a_1} B^{b_1} \cdots A^{a_l} B^{b_l}. \end{aligned}$$

Then, it is shown in [LeMu95b] that $\varphi_{\text{KZ}}(A,B)$ can be expressed

$$\varphi_{\text{KZ}}(A,B) = 1 + \sum_{l=1}^{\infty} \sum_{\mathbf{a},\mathbf{b},\mathbf{p},\mathbf{q}} (-1)^{|\mathbf{b}|+|\mathbf{p}|} \eta(\mathbf{a}+\mathbf{p},\mathbf{b}+\mathbf{q}) \binom{\mathbf{a}+\mathbf{p}}{\mathbf{p}} \binom{\mathbf{b}+\mathbf{q}}{\mathbf{q}} \times B^{|\mathbf{q}|}(A,B)^{(\mathbf{a},\mathbf{b})} A^{|\mathbf{p}|},$$

where the second sum runs over $\mathbf{a}, \mathbf{b}, \mathbf{p}, \mathbf{q}$ such that the sum of their lengths is equal to l and their entries are non-negative integers. For the proof of the above formula see [LeMu95b].

Thus, $\varphi_{\text{KZ}}(A,B)$ can be regarded as a generating function of multiple zeta values. It seems important to investigate the Drinfel'd associator, not only from the viewpoint of the Kontsevich invariant, but also from the viewpoint of number theory. Drinfel'd [Dri90] pointed out that the Grothendieck-Teichmuller group (which acts on the set of associators) is closely related to the Galois group $\text{Gal}(\overline{\mathbb{Q}}/\mathbb{Q})$; see also [Bar98, KRT97] for discussions of this topic.

Appendix E
Claspers

In 1994, Habiro introduced the "clasper" to describe the operation of making a "clasp", which, in other words, introduces an extra wind of one strand of a link around another. Further, he obtained many and various results, using claspers, on the finite type theories of knots and 3-manifolds, as we now describe.

As mentioned in Chapter 7, one can define Vassiliev invariants (finite type invariants) of knots using crossing changes of strands of knots. It was shown in Chapter 7 that the graded vector spaces of Vassiliev invariants are identified with the duals of the spaces spanned by chord diagrams, which are combinatorial objects consisting of solid circles and dashed chords. The association of a crossing change to a chord, which is the basis of this identification, may be reinterpreted as the forming of an extra clasp along that chord, which may be performed by introducing an extra "clasper" on the knot according to the position of that chord on the diagram. While a chord is combinatorial, a clasper is a topological object embedded in a 3-space, winding with a knot. In this sense, a clasper gives a topological realization of a chord. Using claspers we can give a reconstruction of the theory of Vassiliev invariants from the topological viewpoint.

As mentioned in Chapter 11, when studying finite type invariants of 3-manifolds, chord diagrams are generalized to Jacobi diagrams. Habiro's clasper, further, allows a direct topological realization of a Jacobi diagram. In this picture, such relations satisfied by Jacobi diagrams as the AS and IHX relations can be given topological explanations. In this sense we can reconstruct the theory of Jacobi diagrams and finite type invariants from the topological viewpoint.

This topological reconstruction of the theory of finite type invariants has another remarkable characteristic; we can give a series of descending equivalence relations in *the set of knots (or 3-manifolds)* directly, which induces the notion of a finite type invariant, while in the usual definition of finite type invariants we give a descending filtration in *the vector space formally spanned by knots (or 3-manifolds).*

Such a series of descending equivalence relations gives us another viewpoint on the theory of finite type invariants. For example, we can obtain two knots (or two 3-manifolds) which have the same value of finite type invariants up to arbitrary given degree, directly from the definition of finite type invariants using claspers. It had been difficult only by using the usual filtration of finite type invariants.

Goussarov [Gou*95*, Gou*98*, Gou99] independently developed a similar theory by using Y-graphs instead of claspers.

In Section E.1 we show basic properties of claspers which are used in the other sections of this appendix. In Sections E.2 and E.3 we introduce descending series of equivalence relations among knots and among integral homology 3-spheres respectively, which reconstruct the notions of finite type invariants of knots and integral homology 3-spheres. Results in Sections E.1, E.2, and E.3 are essentially due to Habiro [Habi00], *i.e.*, most of them are in [Habi00] and the others are obtained as minor modifications of results in [Habi00]. In Section E.4 we compute the Kontsevich and the LMO invariants of tree claspers following an idea due to Kricker. Such computations illustrate, from a different point of view, that claspers reconstruct the notion of finite type invariants.

E.1 Basic properties of claspers

In this section we introduce some notations for claspers, and show basic properties of them. All results in this section are due to Habiro [Habi00].

Habiro's clasper is defined, as in Section 7.3, as follows.

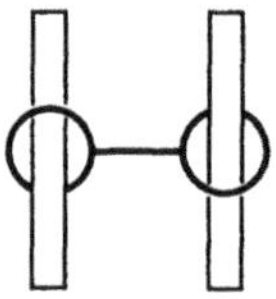 denotes  , or alternatively 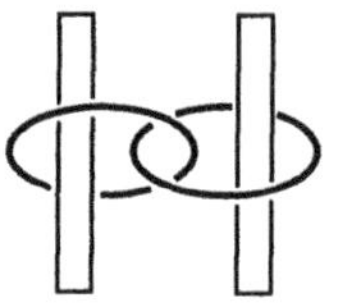 , (E.1)

where a band implies a part of a link or claspers, and the right picture implies the result obtained by surgery along the (blackboard framed) Hopf link in the picture. We call the framed embedded graph in the left picture a *clasper*. Each loop at each end of a clasper is called a *leaf* of the clasper.

Claspers satisfy the following basic relations,

 , (E.2)

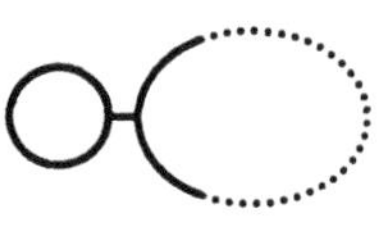 $= \emptyset$, (E.3)

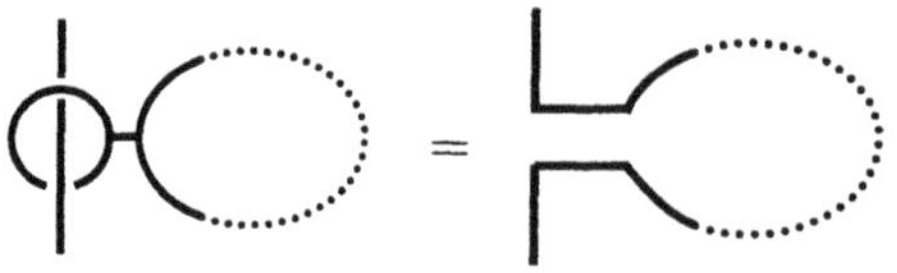 , (E.4)

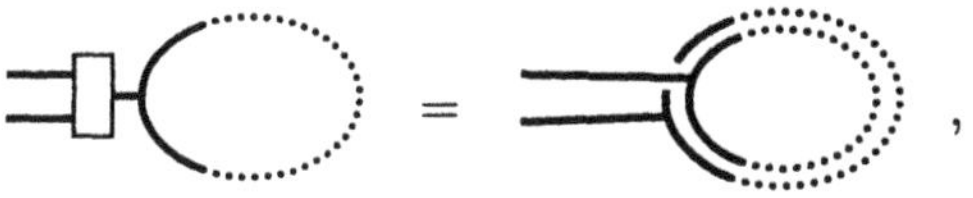 (E.5)

 (E.6)

where a box is defined in Figure E.1 and the dotted lines imply strands which are possibly knotted and linked in some fashion. We obtain (E.2) by

LHS = 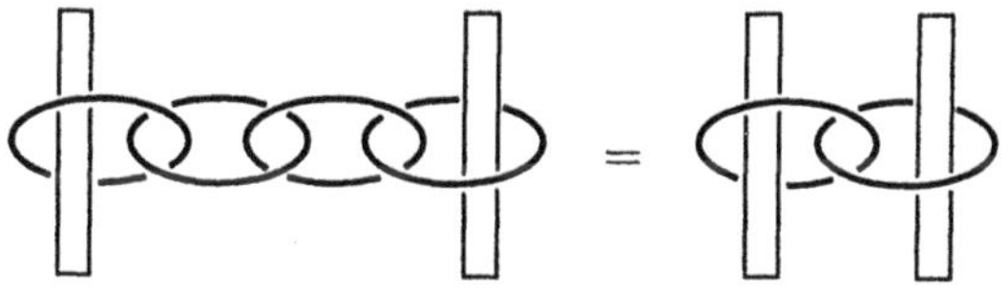 = RHS,

where the pictures imply the results obtained by surgery along the links in the pictures. We obtain (E.3), because the result obtained from a 3-space by surgery along the link  is homeomorphic to the original 3-space. We obtain (E.4) by

LHS = 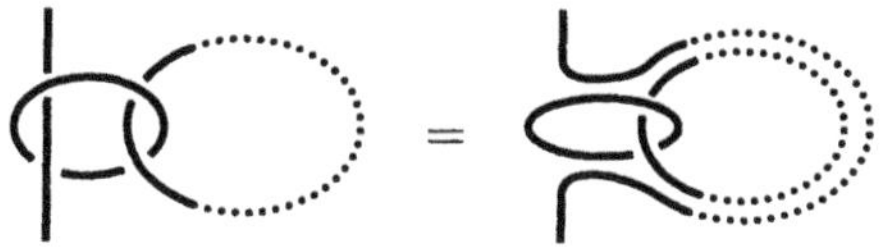 = RHS,

where we obtain the middle equality by handle slide of the vertical strand over the component of a dotted line. We obtain (E.5) by applying a similar argument to the definition of a box. We obtain (E.6) by applying (E.3) and (E.2) to the definition of a box.

Lemma E.1 (Habiro). We have that

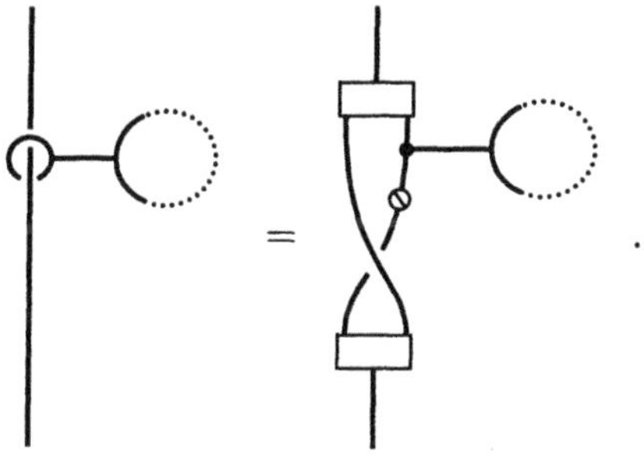

Proof. By using (E.4) the right hand side (RHS) of the required relation is computed as

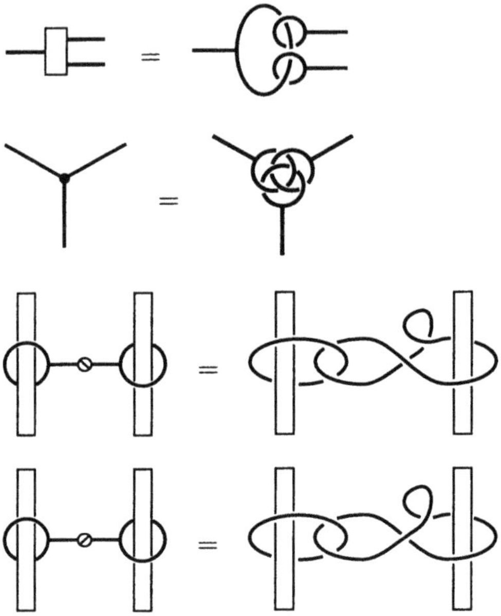

Figure E.1 Some notations for systems of claspers. The right hand side of the last two formulae imply the results obtained by surgery along the (blackboard framed) links in the pictures; note that these links are obtained from the link shown in (E.1) by adding positive and negative half twists respectively.

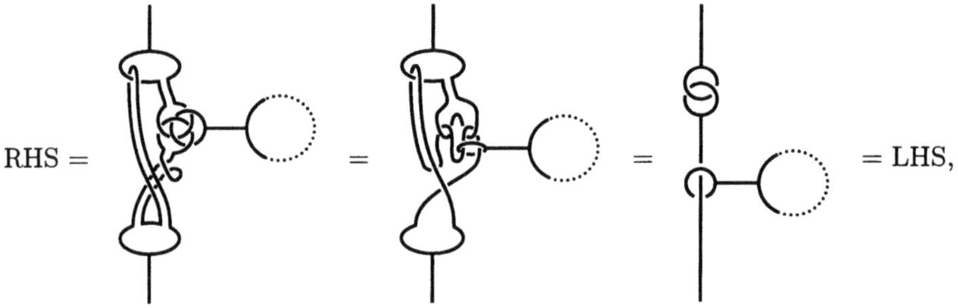

where we obtain the second and third equalities by isotopy, and obtain the last equality by (E.2). Hence, we obtain the lemma. □

Lemma E.2 (Habiro). A box moves beyond a trivalent vertex as

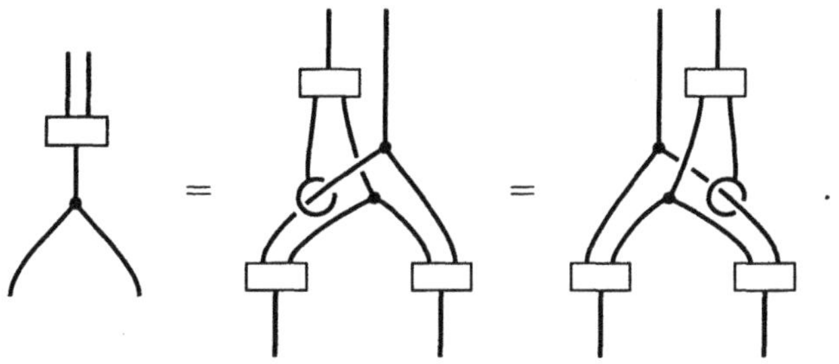

Proof. The third term is obtained from the second term by a π rotation around a vertical axis. Since the first term is symmetric under such a π rotation, the second equality is derived from the first equality.

We show the first equality as follows. The left hand side of the required formula is computed as

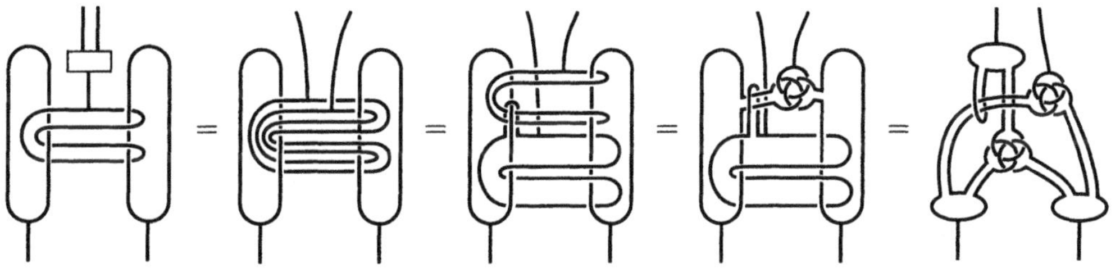

where the first equality is obtained by (E.5), and the other equalities are obtained by isotopy. The last picture is equivalent to the middle picture of the required formula. Hence, we obtain the lemma. □

E.2 A descending series of equivalence relations among knots

In this section we consider a descending series of equivalence relations among knots, similar to the series we considered in Section 7.3. The aim of this section is to give maps from lattices spanned by primitive Jacobi diagrams to the graded sets associated to the descending series (as recorded by Theorem E.12). The contents of this section are obtained as minor modifications of results due to Habiro [Habi00].

As in Section 7.3, let $\mathbb{K}$ denote the set of oriented knots in S^3. The set $\mathbb{K}$ forms a commutative semigroup with the sum given by the connected sum of knots.

A connected graph without cycles is called a *tree*. A *tree clasper* is a union of claspers obtained from a connected uni-trivalent tree embedded in a 3-dimensional space by replacing univalent vertices with leaves of claspers and trivalent vertices with sets of Borromean rings. A *tree clasper on a knot K* is a tree clasper such that at least one of leaves of the clasper is a disc-leaf and each of the other leaves bounds a disc intersecting parts of K and edges of the clasper, graphically shown as

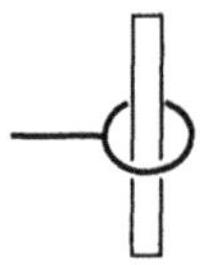 , where the band implies some bundle of strands of K and other edges of the clasper. Here, a *disc-leaf* is a leaf bounding a disc intersecting K at previsely one point.

Surgery along a tree clasper on a knot in S^3 results in another knot in S^3, for the following reason. A tree clasper on a knot has a disc-leaf by definition. We temporalily forget the knot. Then, the disc-leaf becomes a leaf bounding a disc in the complement of the tree clasper. We can remove the clasper of the leaf by (E.3). Then, the set of Borromean rings at the other end of the clasper is broken, and two new leaves bounding discs appear there. Hence, by using (E.3) again, we can remove the corresponding pair of claspers. Repeating this procedure, we can remove all the claspers of the tree clasper. Therefore, the result of surgery along the tree clasper in S^3 is S^3. Remembering the knot, we obtain a knot in S^3 after surgery along the tree clasper.

A tree clasper on a knot is called of *degree d* if it has $d-1$ trivalent vertices. We define the C_d*-equivalence*,† denoted by $\underset{d}{\sim}$, to be the equivalence relation among knots generated by the relation between two knots K and K' such that K is obtained from K' by surgery along a tree clasper on K' of degree d. Further, we put $\mathbb{K}_d$ to be the set of knots which are C_d-equivalent to the trivial knot. The set $\mathbb{K}_d$ forms a commutative semigroup as a sub-semigroup of $\mathbb{K}$. We will investigate $\mathbb{K}_d/\underset{d+1}{\sim}$ in this section.

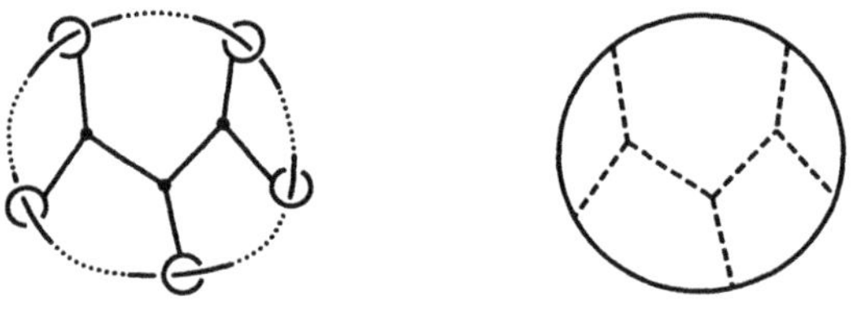

Figure E.2 A tree clasper on a knot whose leaves are disc-leaves and the Jacobi diagram on S^1 corresponding to it. Dotted lines imply strands possibly knotted and linked in some fashion.

In the following of this section a picture of tree claspers on a knot K often implies the knot obtained from K by surgery along the tree claspers. We begin with the following two lemmas.

Lemma E.3 (Habiro). Let K be a knot, and let K' be the knot obtained from K by surgery along a degree d tree clasper T on K. Then, K can be obtained from K' by surgery along some degree d tree clasper T' on K'.

†The C_d-equivalence is also called the $(d-1)$-*equivalence* (due to Goussarov) in some literatures.

Proof. For simplicity, we show the lemma in the case of a tree clasper of degree 3. (The Lemma in the general case can be proved similarly.) We express K' as

$K' = K_T =$ 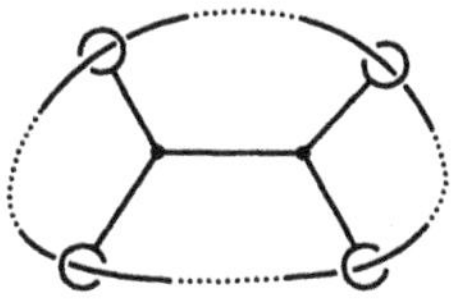 ,

where K_T denotes the knot obtained from K by surgery along T, and the picture displays K and T. We break one edge of T as shown in the next picture. Then,

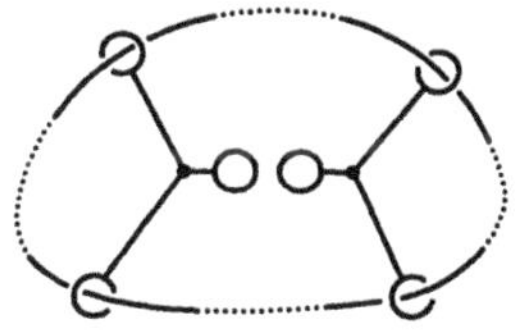 $=$ 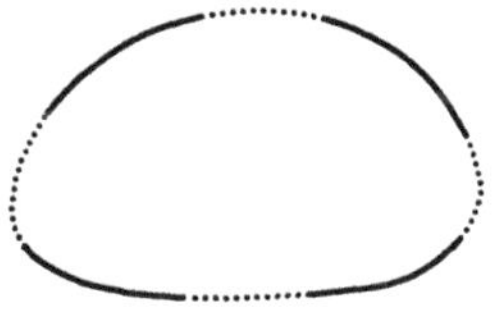 $= K,$

where we obtain the first equality by repeatedly applying (E.3) to leaves bounding discs. The above left picture is isotopic to the next picture,

$K =$ 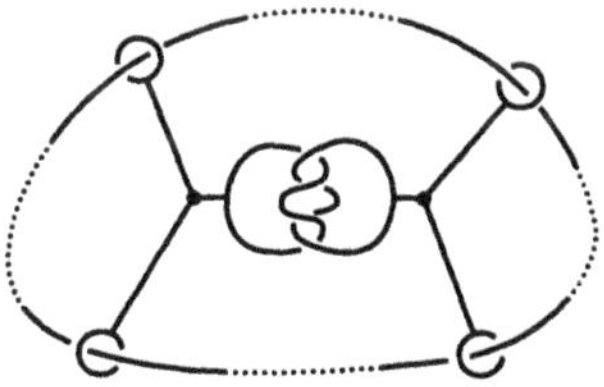 $=$ 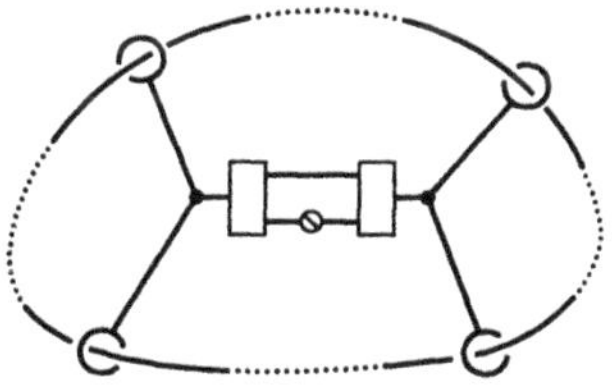 ,

where we obtain the second equality from the notations in Figure E.1. We move the introduced boxes to the knot by applying Lemma E.2 and (E.5). Then,

$K =$ 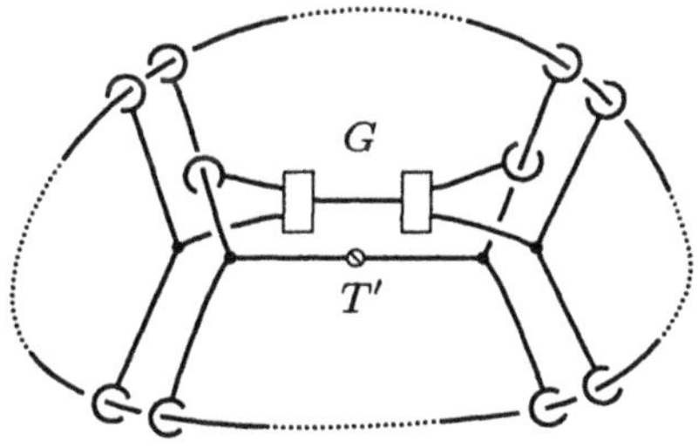

.

We denote by G the component consisting of claspers which are connected to the boxes, and denote by T' the other component, which forms a tree clasper of the same degree as T. When forgetting T', G is equal to T by (E.6). Hence, $K_G = K_T = K'$. Therefore, $K'_{T'} = K_{G \cup T'} = K$ by the above formula. This gives the required tree clasper T' on K_G $(= K')$. □

Lemma E.4 (Habiro). Let T be a degree d tree clasper in a 3-space, and let N be a tubular neighbourhood of T, as shown below. Then, the 3-manifold obtained from N by surgery along T is homeomorphic to a handlebody.

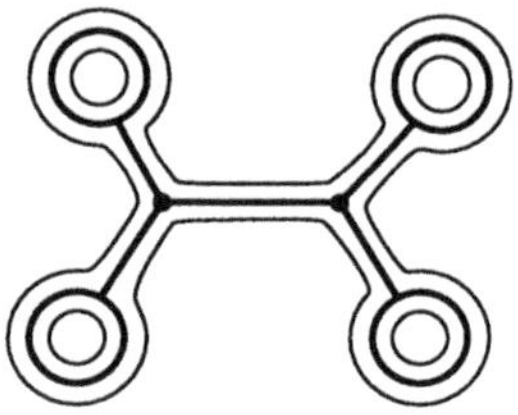

Proof. We express N by the complement of a trivial tangle in a ball, say, the following picture for $d = 3$. Then, it is sufficient (say, for $d = 3$) to show that the following two tangles are related by a homeomorphism of the ball.

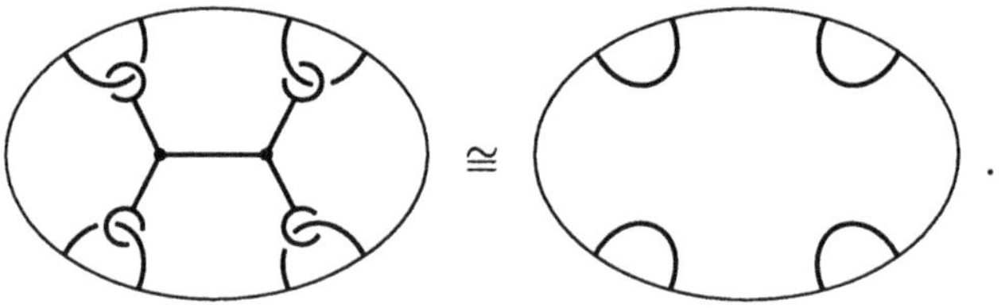

This can be shown by induction on d. We assume, say, the above homeomorphism. Then,

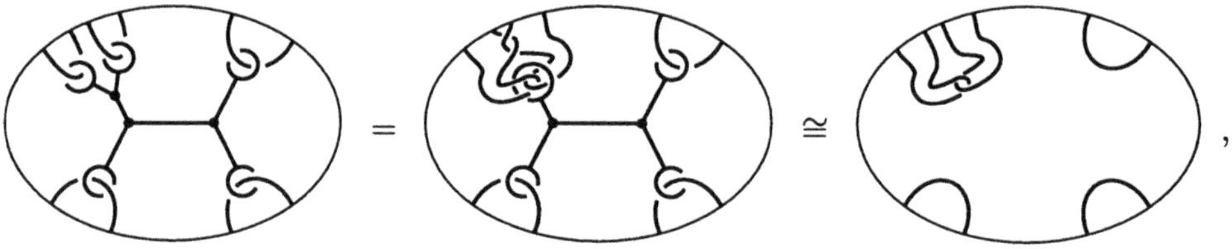

where the first equality follows from the definition of a trivalent vertex, and the second homeomorphism is obtained from a homeomorphism of a ball and the assumption of the induction. The above right tangle is homeomorphic to a trivial tangle, completing the induction. □

By Lemma E.3 we see that any knot in $\mathbb{K}_d$ can be obtained from the trivial knot by a sequence of surgeries along tree claspers of degree d. Further, by separating such tree claspers by Lemma E.4, such a knot can be obtained by surgery along a disjoint collection of degree d tree claspers decorating the trivial knot. Moreover, we can resolve the linking among such tree claspers by the following two lemmas.

Lemma E.5. We have the following relation between two knots obtained from some other knot by surgery along two degree d tree claspers which differ as shown,

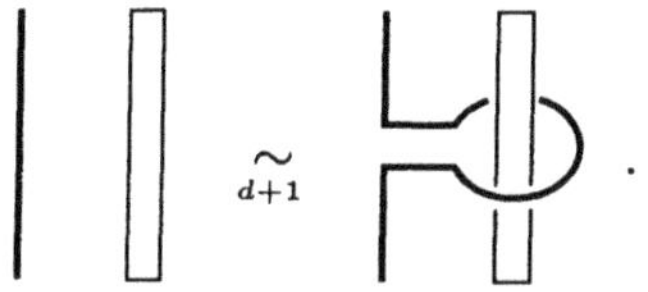

(E.7)

Here, the two pictures imply two tree claspers which are identical except for a ball, where they differ as shown in the pictures. Further, the bands imply bundles which may contain strands of the knot and edges of the claspers.

Proof. For simplicity, we focus on the proof of the following case,

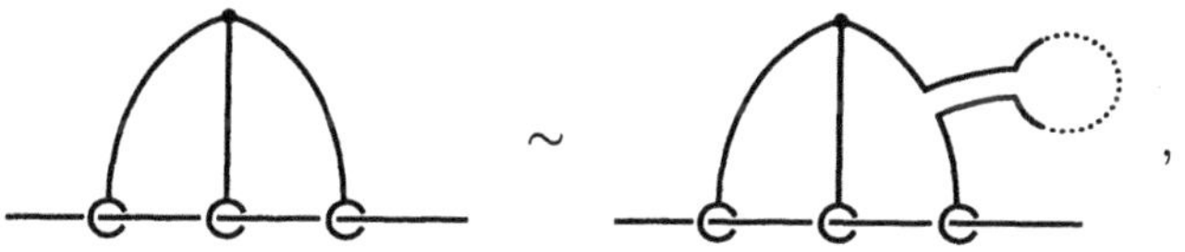

(E.8)

where the dotted line implies a strand winding the band of the lemma, which is located (possibly) outside of this picture. The right hand side of (E.8) is computed as

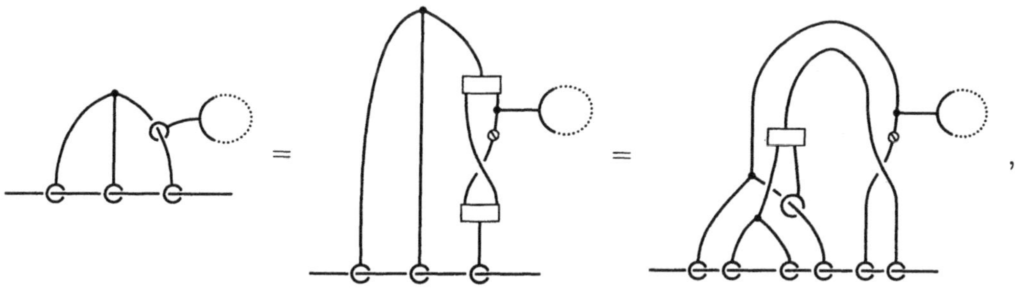

where we obtain the first equality by Lemma E.1, and obtain the second equality by Lemma E.2 and the relation (E.5). Since a connected component of the tree clasper on the right hand side has an additional trivalent vertex, the right hand side of the above formula is C_2-equivalent to the next picture,

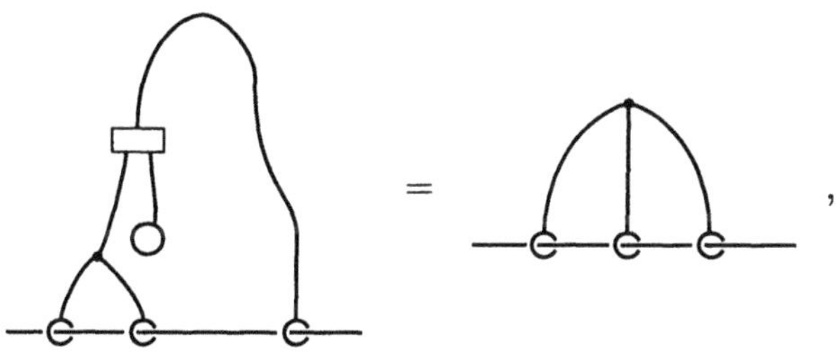

where we obtain the equality by (E.6). Hence, we obtain (E.8).

The lemma in the general case follows in a similar way. That is, we present the right hand side of (E.7) by using an additional clasper, and replace it with an additional trivalent vertex using Lemma E.1. Then, we move the introduced boxes to the leaves of the graph by repeatedly applying Lemma E.2. This results in the presence of a tree clasper with an additional trivalent vertex. Removing it by C_d-equivalence, we obtain the required formula (E.7). □

Lemma E.6. We have the following relation between two knots obtained from some other knot by surgery on collections of tree claspers.

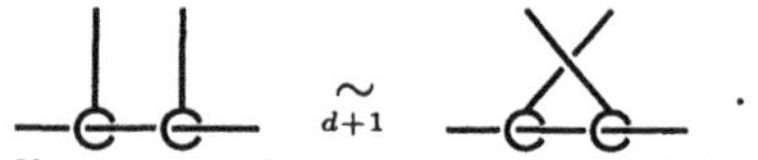

The two pictures display collections of tree claspers which are the same except inside a ball, where they differ as shown. Further, the left leaf in the left picture (and hence also the right leaf in the right picture) is a leaf of a degree d tree clasper, and the other leaves are leaves of some other clasper.

Proof. The right hand side of the relation is computed as

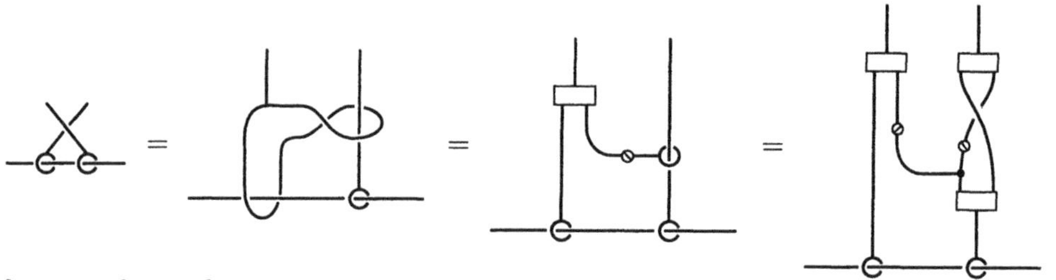

where we obtain the second equality from the notations in Figure E.1, and obtain the third equality by Lemma E.1. We show the lemma, for simplicity, in the following case. By the above formula, we have that

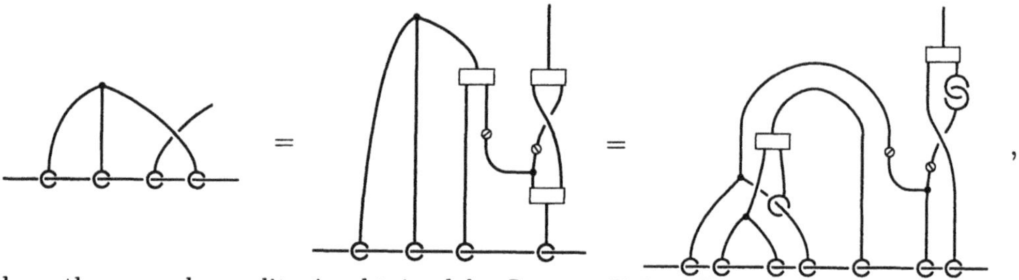

where the second equality is obtained by Lemma E.2. The right picture is C_2-equivalent to the next picture,

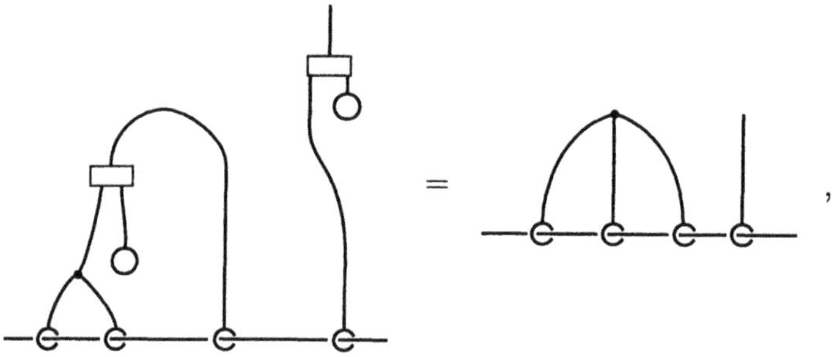

where we obtain the equality by (E.6). Hence, we obtain the lemma in this case. The lemma in the general case can be obtained in a similar way. □

By the arguments so far we see that any knot in $\mathbb{K}_d$ can be given by the connected sum of some knots, each obtained from the trivial knot by surgery along a degree d tree clasper, which implies that $\mathbb{K}_d$ is generated by such knots. Then, we focus on such knots in the following 5 lemmas.

Lemma E.7. We have the following relation between the trivial knot (denoted by 0 in the right hand side) and a knot obtained by connect-summing together a pair of knots, each of which is obtained from the trivial knot by surgery along a tree clasper. The tree claspers in the two factors are identical except for a ball, where they differ as shown:

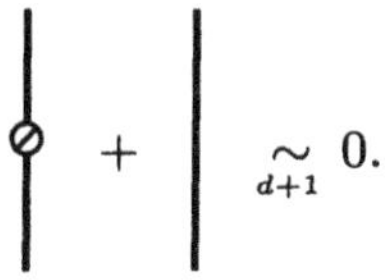

Proof. Just as in the proof of Lemma E.3, we have that

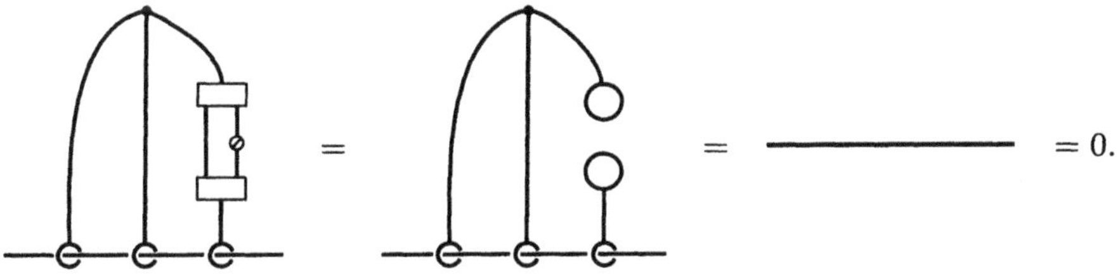

Here, the third picture implies the trivial knot; we denote it by 0 in the semigroup $\mathbb{K}$ of knots. Further, by Lemma E.2, the first picture is equal to the next picture,

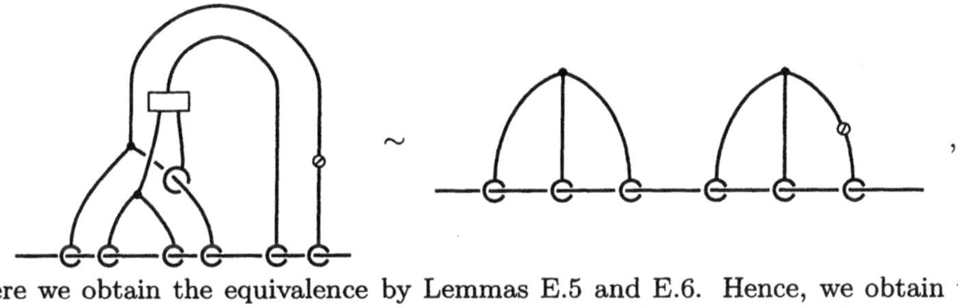

where we obtain the equivalence by Lemmas E.5 and E.6. Hence, we obtain the lemma in this simple case. The lemma in the general case can be obtained in a similar way. □

Lemma E.8. We have the following relation among three knots, each of which is obtained from the trivial knot by surgery along a tree clasper of degree d. Each tree clasper is identical except for a ball, where they differ as shown in the pictures.

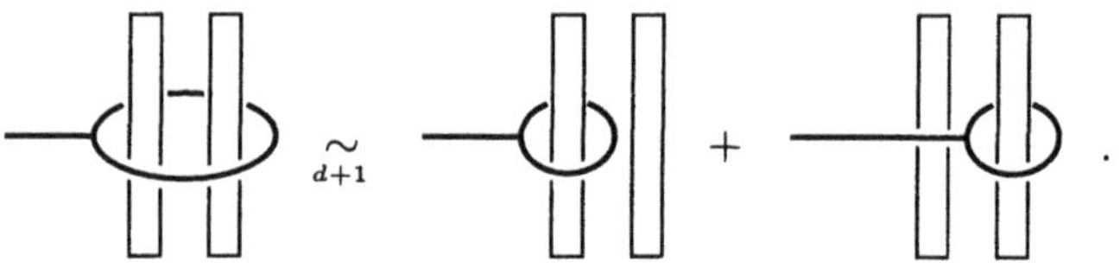

Proof. The left hand side of the required formula is equal to 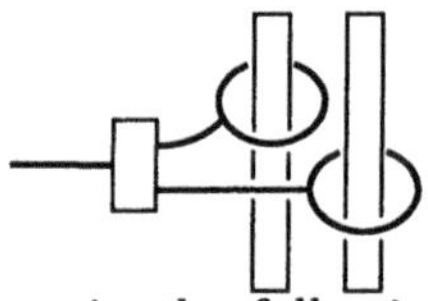 from the definition of a box. For simplicity, we show the lemma in the following simple case. We have the following left picture by expressing a leaf winding a band by a dotted line,

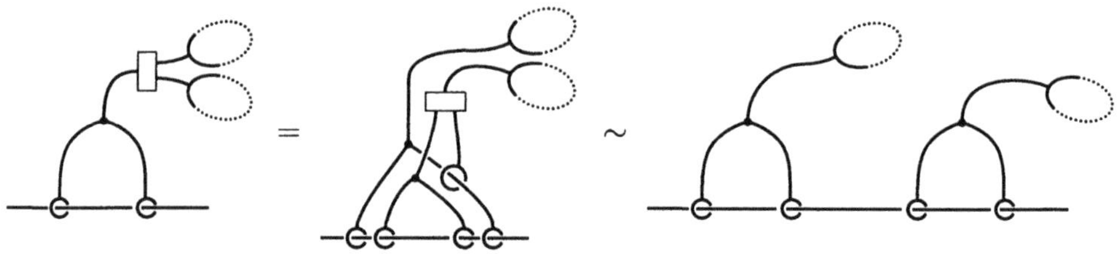

,

where we obtain the left equality by Lemma E.2 and obtain the equivalence on the right by Lemmas E.5 and E.6 and the relation (E.6). Hence, we obtain the lemma in this simple case. The lemma in the general case can be obtained in a similar way. □

Lemma E.9. We have the following AS relation, which relates the connected sum of two knots, each obtained from the trivial knot by surgery along a degree d tree clasper, to the trivial knot (denoted by 0 in the right hand side). The tree claspers are identical except for a ball, where they differ as follows.

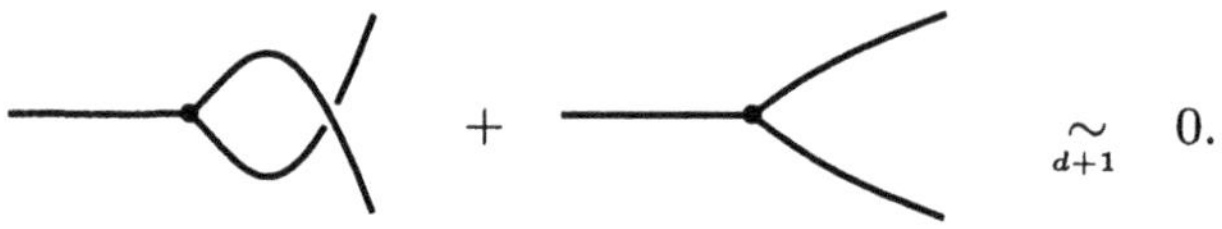

Proof. The first tree clasper of the required formula is equal to 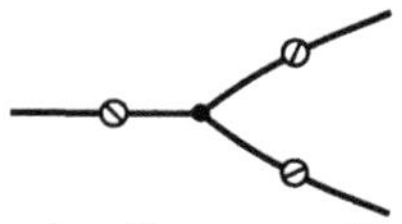 by a half twist of a neighborhood of the trivalent vertex. Hence, by three applications of Lemma E.7, we obtain the required formula. □

Lemma E.10. We have the following STU relation, which relates three knots, each obtained from the trivial knot by surgery along a degree d tree clasper. The tree claspers are identical except for a ball, where they differ as shown in the pictures.

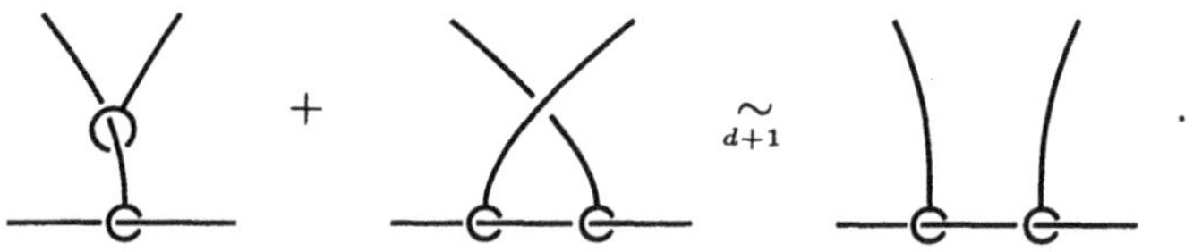

Proof. For simplicity, we show the lemma in the following simple case. We have that

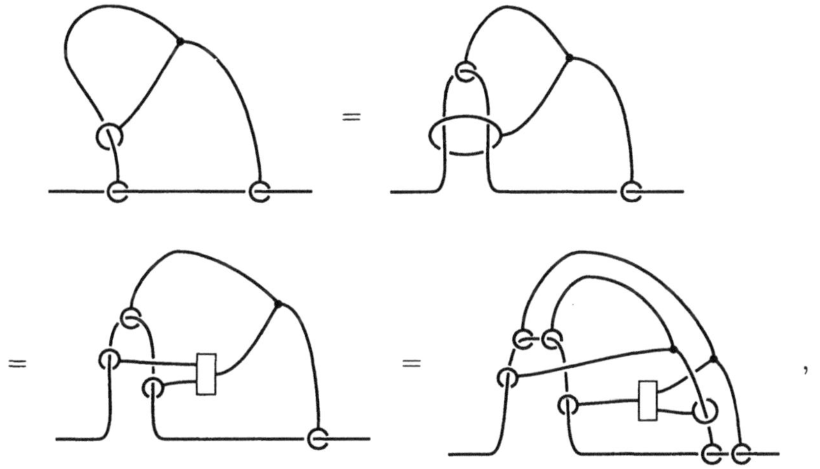

where the three equalities are obtained by isotopy, from the definition of the box, and by Lemma E.2 respectively. Further, the last picture is equivalent to the following picture by Lemmas E.5 and E.6,

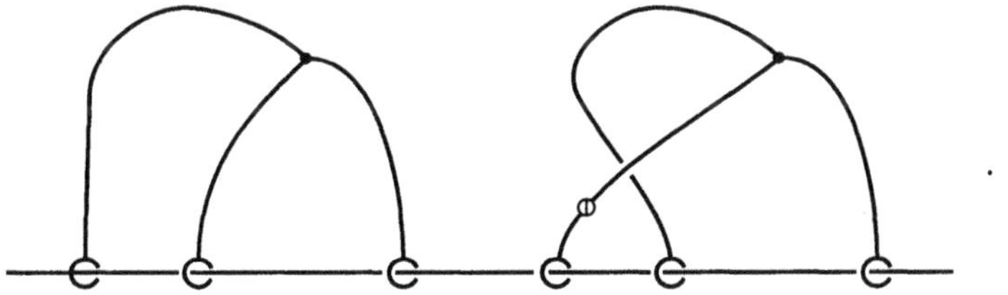

Therefore, by Lemma E.7, we obtain the required formula. □

Lemma E.11. We have the following IHX relation, which relates three knots, each obtained from the trivial knot by surgery along a degree d tree clasper. Each tree clasper is identical except for a ball, where they differ as shown.

Proof. The lemma is shown similarly as in the above proof of the STU relation. We show a sketch of the proof, for simplicity, in the following simple case. By applying Lemma E.1 to the following picture we have the following equivalence,

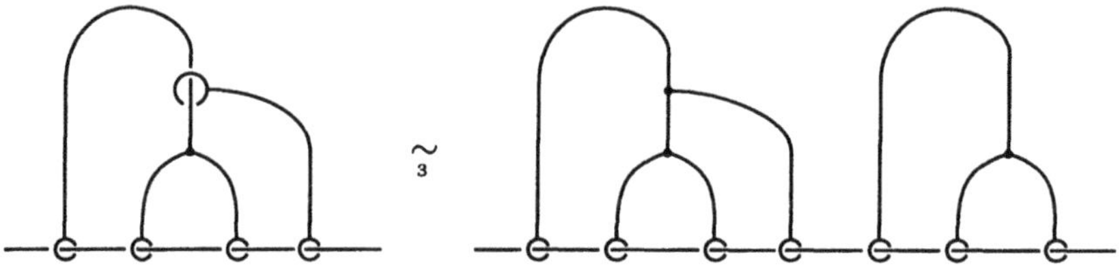

Further, similarly as in the proof of the STU relation the above left picture is equal to the next picture,

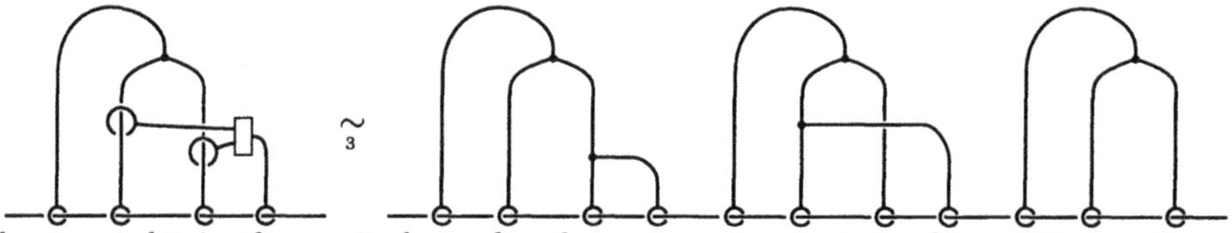

where we obtain the equivalence by the same argument as above. Comparing the right hand sides of the above relations we obtain the required relation in this simple case. The required relation in the general case can be obtained in a similar way. □

Recall that a primitive Jacobi diagram on S^1 is a Jacobi diagram whose dashed part is a connected uni-trivalent graph. We now prove the following theorem, which describes $\mathbb{K}_d/\underset{d+1}{\sim}$.

Theorem E.12. For each positive integer d, we have the following surjective homomorphism,

$$\psi : \mathrm{span}_{\mathbb{Z}} \left\{ \begin{array}{l} \text{primitive Jacobi diagrams} \\ \text{on } S^1 \text{ of degree } d \end{array} \right\} \Big/ \text{AS, IHX, STU} \longrightarrow \mathbb{K}_d/\underset{d+1}{\sim},$$

which takes a primitive Jacobi diagram to a knot obtained from the trivial knot by surgery along a tree clasper on the trivial knot; we obtain a tree clasper from a Jacobi diagram by replacing a trivalent and univalent vertices as follows,

where we choose the image of each trivalent vertex in such a way that the resulting image is a tree clasper; see, for example, Figure E.3.

In particular, $\mathbb{K}_d/\underset{d+1}{\sim}$ forms a commutative group.

Proof. The image of a Jacobi diagram under ψ is well defined, because, by Lemma E.5, the image does not depend on the choice of the embedding of the tree clasper into S^3, and, by Lemma E.13 below, does not depend on the choice of images of the trivalent vertices. Further, the AS, IHX, and STU relations are satisfied in $\mathbb{K}_d/\underset{d+1}{\sim}$ by Lemmas E.9, E.11, and E.10 respectively, noting that the STU relation in the statement of the theorem implies the STU relation among primitive Jacobi diagrams. Hence, it remains to show the surjectivity of the homomorphism.

The surjectivity is shown as follows. As mentioned earlier, $\mathbb{K}_d$ is generated by knots, each obtained from the trivial knot by surgery along a tree clasper of degree d. By Lemma E.8, a tree clasper on the trivial knot is equivalent to a sum of tree claspers, each leaf of which is a disc-leaf or a leaf bounding a disc only intersected by some edge of the tree clasper at one point. Up to equivalence, by Lemmas E.5 and E.7, such a tree clasper is equal to the image of some primitive Jacobi diagram under the map ψ. Hence, we obtain the surjectivity. □

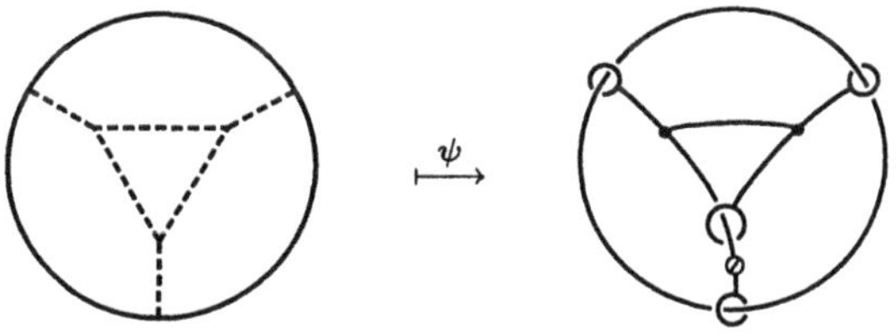

Figure E.3 Definition of the map ψ

Lemma E.13. We have the following relation between two knots, each of which is obtained from some other knot by surgery along a graph clasper.

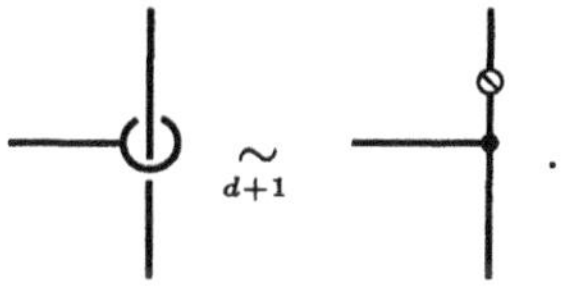

The left picture displays a leaf of a degree d tree clasper winding with an edge of the same tree clasper and the right picture is a graph clasper which is identical to the tree clasper of the left picture except for a ball, where they differ as shown in the pictures.

Proof. For simplicity, we show the lemma in the following simple case. By Lemma E.1 we have that

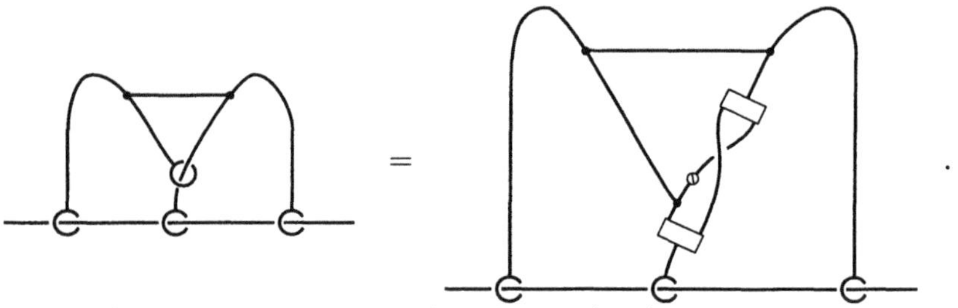

Further, by applying Lemma E.2 and the relation (E.5) to the picture on the right, we obtain the next picture,

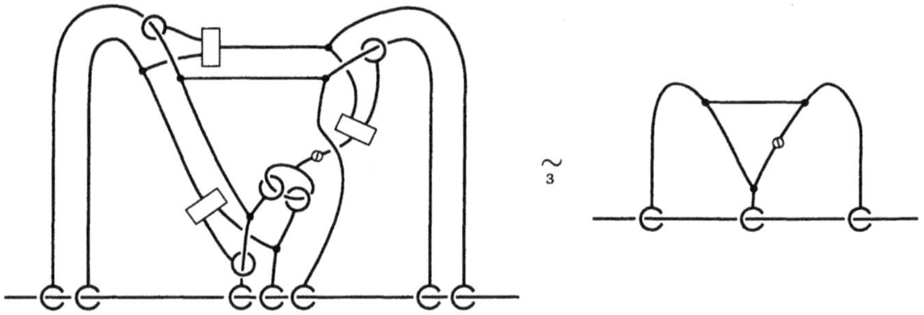

Here, we obtain the equivalence by removing the component with 3 trivalent vertices, and by using the relation (E.6). Hence, we obtain the lemma in this simple case. The lemma in the general case can be obtained in a similar way. □

We have the following corollary of Theorem E.12.

Corollary E.14. For each positive integer d, the quotient semigroup $\mathbb{K}/\underset{d}{\sim}$ forms a commutative group.

Proof. Since C_1-equivalence is the equivalence relation generated by crossing change, we have that $\mathbb{K}/\underset{1}{\sim} = \{0\}$. Hence, the corollary holds for $d = 1$.

We show the corollary by induction on d. Suppose that $\mathbb{K}/\underset{d}{\sim}$ forms a commutative group. We have the following exact sequence of commutative semigroups,

$$0 \longrightarrow \mathbb{K}_d/\underset{d+1}{\sim} \longrightarrow \mathbb{K}/\underset{d+1}{\sim} \longrightarrow \mathbb{K}/\underset{d}{\sim} \longrightarrow 0.$$

By Theorem E.12, $\mathbb{K}_d/\underset{d+1}{\sim}$ forms a commutative group. Hence, $\mathbb{K}/\underset{d+1}{\sim}$ forms a commutative group, completing the induction. □

E.3 A descending series of equivalence relations among homology 3-spheres

In this section we introduce a descending series of equivalence relations among integral homology 3-spheres and give maps of lattices spanned by primitive Jacobi diagrams to the graded sets associated to the descending series. All results in this section are due to Habiro.

As in Section 11.3, let $\mathbb{M}$ denote the set of integral homology 3-spheres. The set $\mathbb{M}$ forms a commutative semigroup with the sum given by the connected sum of integral homology 3-spheres.

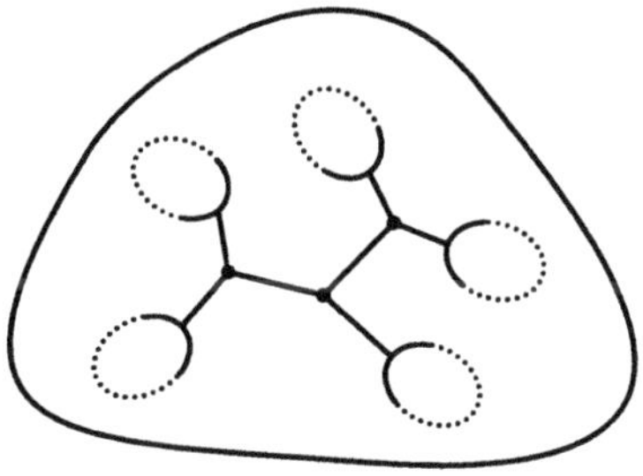

Figure E.4 A tree clasper with 3 trivalent vertices in an integral homology 3-sphere. The dotted lines imply strands possibly knotted and linked in some fashion.

A *tree clasper in an integral homology 3-sphere* M is a tree clasper obtained from a uni-trivalent tree embedded in M by replacing each trivalent vertex with a set of Borromean rings and replacing univalent vertices with framed knots possibly linking with each other and possibly linking with the edges of the embedded tree. For an example of a tree clasper see Figure E.4.

The 3-manifold obtained from an integral homology 3-sphere by surgery along a tree clasper with at least 1 trivalent vertex is an integral homology 3-sphere again, since we can, homologically, break a trivalent vertex and then remove all claspers by (E.3).

We define the *Y_d-equivalence*,‡ denoted by $\underset{d}{\sim}$, to be the equivalence relation among integral homology 3-spheres generated by the relation which relates integral homology 3-spheres M and M' where M is obtained from M' by surgery along some tree clasper in M' with d trivalent vertices. Further, we put $\mathbb{M}_d$ to be the set of integral homology 3-spheres which are Y_d-equivalent to S^3. The set $\mathbb{M}_d$ forms a commutative semigroup as a sub-semigroup of $\mathbb{M}$. We will investigate $\mathbb{M}_d/\underset{d+1}{\sim}$ in this section.

We begin with the following lemma.

Lemma E.15 (Habiro). Let M be an integral homology 3-sphere, and let M' be the integral homology 3-sphere obtained from M by surgery along a tree clasper with d trivalent vertices in M. Then, M can be obtained from M' by surgery along some tree clasper with d trivalent vertices in M'.

‡The Y_d-equivalence is also called the $(d-1)$-*equivalence* (due to Goussarov) in some literatures.

Proof. The lemma is obtained just as the proof of Lemma E.3. □

By Lemma E.15 any integral homology 3-sphere in $\mathbb{M}_d$ can be obtained from S^3 by a sequence of surgeries along tree claspers with d trivalent vertices. Further, by separating such tree claspers by Lemma E.4, such an integral homology 3-sphere can be obtained from S^3 by surgery along a disjoint collection of tree claspers with d trivalent vertices in S^3. Moreover, we can resolve the linking among such tree claspers by Lemmas E.16 and E.17 below.

In the following of this section, a picture of tree claspers in an integral homology 3-sphere M often implies the integral homology 3-sphere obtained from M by surgery along the tree claspers.

The following lemma implies that we can resolve the linking among edges of tree claspers with d trivalent vertices modulo Y_d-equivalence.

Lemma E.16 (Habiro). We have the following relation between two integral homology 3-spheres, each obtained from the same integral homology 3-sphere by surgery along a tree clasper with d trivalent vertices,

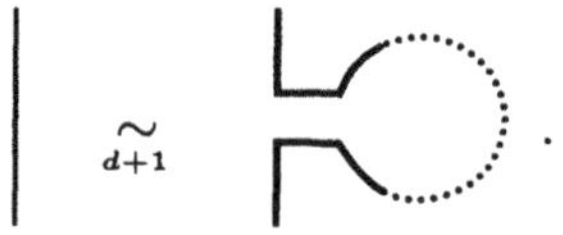

The two tree claspers are identical, except for part of one edge, which differ as shown (where it meets some ball). The dotted line denotes that part of the second tree clasper following some arbitrary path in the complement of the ball.

Proof. The lemma is shown in the same way as in the proof of Lemma E.5. □

The following lemma implies that we can resolve the linking between leaves of different tree claspers in an integral homology 3-sphere.

Lemma E.17 (Habiro). We have the following relation between two integral homology 3-spheres, each obtained from the same integral homology 3-sphere by surgery along a collection of two tree claspers with d trivalent vertices,

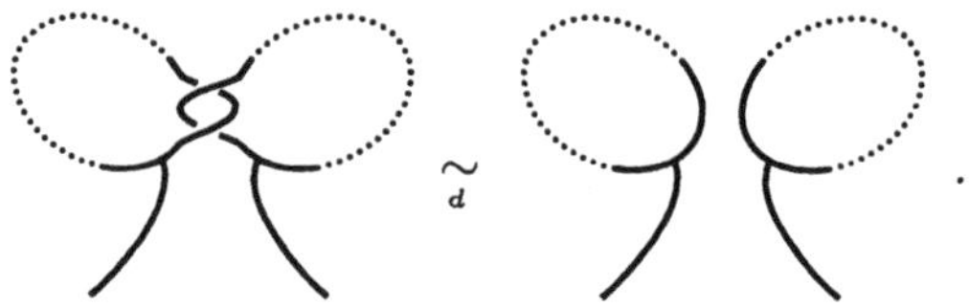

The two collection of tree claspers are identical, except for part of two leaves of different tree claspers, which differ as shown.

Proof. The left hand side of the required relation is equal to the next picture,

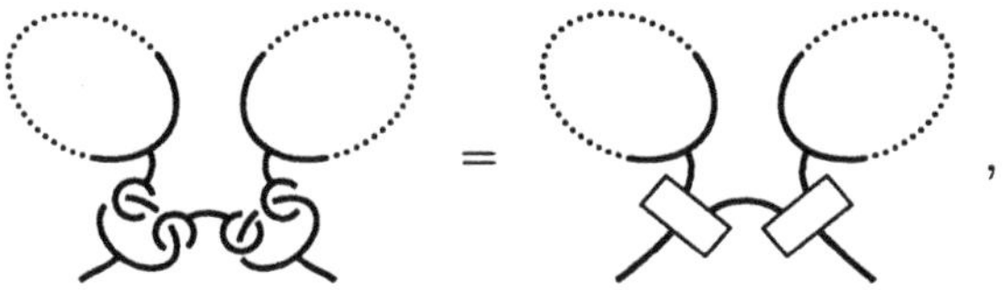

where we obtain the equality from the definition of the boxes. By moving the boxes in the right picture toward the other ends of the tree claspers, as in the proof of Lemma E.5, the arc between the two boxes becomes a tree clasper with d trivalent vertices. Removing it by Y_d-equivalence, we obtain the right hand side of the required relation. □

By the arguments so far we see that $\mathbb{M}_d/\underset{d+1}{\sim}$ is generated by integral homology 3-spheres, each of which is obtained from S^3 by surgery along a tree clasper with d trivalent vertices. To investigate $\mathbb{M}_d/\underset{d+1}{\sim}$ further, we consider relations among tree claspers in S^3. We resolve the knotting and the linking of leaves of a single tree clasper in S^3 by the following two lemmas.

Lemma E.18 (Habiro). We have the following relation between two integral homology 3-spheres, each obtained from S^3 by surgery along a tree clasper with d trivalent vertices,

The three tree claspers are identical except for part of a leaf, where they differ as shown in the pictures.

Proof. The left hand side of the required relation is equal to

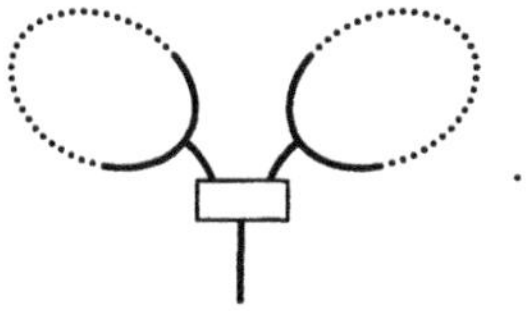

By moving the box toward the other ends, we obtain a union of two tree claspers which are possibly linked with each other. Further, by splitting the two tree claspers by Lemmas E.16 and E.17, we obtain the split union of two tree claspers shown on the right hand side of the required formula. The split union implies the connected sum of integral homology 3-spheres obtained from S^3 by surgery along the two tree claspers. Hence, we obtain the required relation. □

Lemma E.19 (Habiro). We have the following relation among three integral homology 3-spheres, each obtained from S^3 by surgery along a tree clasper with d

trivalent vertices,

The three tree claspers are identical except for part of two leaves of a tree clasper, where they differ as shown in the pictures.

Note that the pictures on the left hand sides of the relations in Lemmas E.17 and E.19 might look like same, but they mean different things; each picture in Lemma E.19 describes two leaves of a single tree clasper in S^3, while each picture in Lemma E.17 describes two leaves of two different tree claspers in an integral homology 3-sphere.

Proof of Lemma E.19. Applying Lemma E.18 to the left component on the left hand side of the required relation, it is Y_d-equivalent to

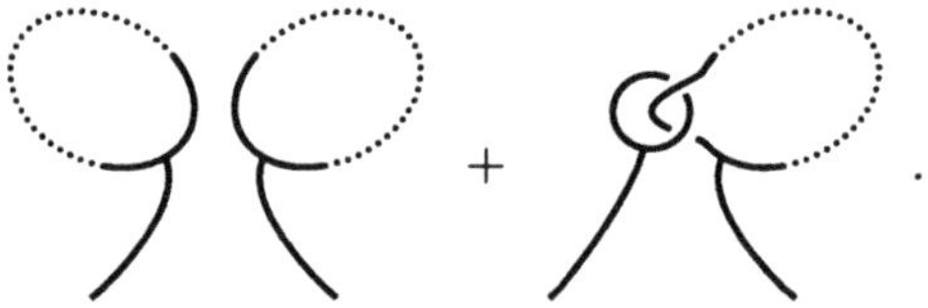

Further, applying Lemma E.18 again, now to the second term, the above sum is Y_d-equivalent to

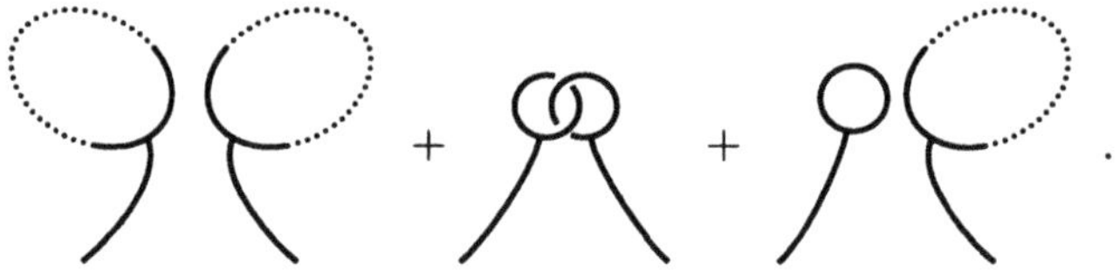

Removing the last term by (E.3), we obtain the right hand side of the required relation. □

Recall that a primitive Jacobi diagram on $\emptyset$ is a Jacobi diagram consisting of a dashed connected trivalent graph. A *graph clasper* in an integral homology 3-sphere M is a union of claspers obtained from a connected uni-trivalent graph embedded in M by replacing edges with claspers, and trivalent vertices with sets of Borromean rings in the leaves of the claspers, and univalent vertices with leaves of claspers. The map ψ in the following theorem is defined to be the map which takes a primitive Jacobi diagram to an integral homology 3-sphere obtained from S^3 by surgery along a graph clasper obtained, as just described, from some embedding of the Jacobi diagram in S^3; see, for example, Figure E.5.

Theorem E.20 (Habiro). The quotient semigroup $\mathbb{M}_d/\underset{d+1}{\sim}$ forms a commutative group. We have the following cases.

(1) We have that $\mathbb{M}_1/\underset{2}{\sim} = \mathbb{Z}/2\mathbb{Z}$. Further, for each $d \geq 2$, we have that $\mathbb{M}_{2d-1}/\underset{2d}{\sim} = 0$.

(2) For each $d \geq 1$, we have the following surjective map,

$$\psi : \mathrm{span}_{\mathbb{Z}}\{\text{primitive Jacobi diagrams on } \emptyset \text{ of degree } d\}/\mathrm{AS},\ \mathrm{IHX} \longrightarrow \mathbb{M}_{2d}/\underset{2d+1}{\sim},$$

where the map ψ is given above.

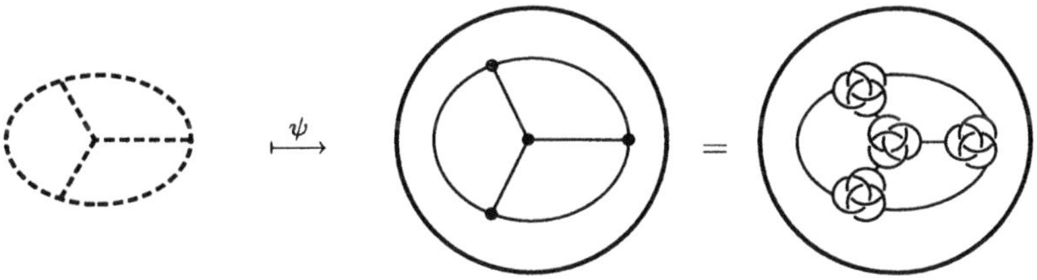

Figure E.5 Definition of the map ψ.

Proof. As mentioned earlier (see the comment which precedes Lemma E.18), the quotient semigroup $\mathbb{M}_d/\underset{d+1}{\sim}$ is generated by integral homology 3-spheres, each of which is obtained from S^3 by surgery along a tree clasper with d trivalent vertices. Further, by Lemmas E.18 and E.19, leaves of such tree claspers can be reduced to one of

 (E.9)

The leaves in the first picture can be removed by (E.2). Further, the third leaf in (E.9) is equivalent to -1 times the second leaf by

 $= 0,$

where the equivalence is obtained by Lemma E.18. Furthermore, the second leaf in (E.9) is of order 2 by

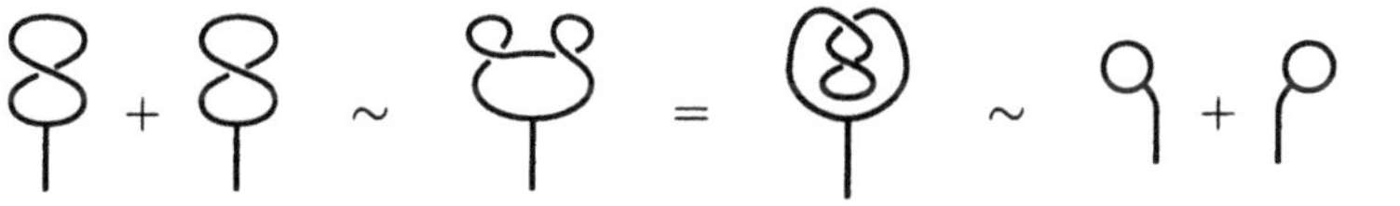 $= 0,$

where the equivalences are obtained by Lemma E.18. Moreover, if a tree clasper has the third leaf in (E.9) and at least two trivalent vertices next to the leaf, then it is equivalent to 0 by Lemma E.21 below. Hence, we obtain (1) of the theorem and obtain a surjective map,

$$\mathrm{span}_{\mathbb{Z}}\{\text{primitive Jacobi diagrams on } \emptyset \text{ of degree } d\} \longrightarrow \mathbb{M}_{2d}/\underset{2d+1}{\sim},$$

defined as in Figure E.5.

It is sufficient to show the AS and IHX relations among graph claspers with $2d$ trivalent vertices modulo Y_{2d+1}-equivalence. In a similar way as in the proof of Lemma E.7, we have that

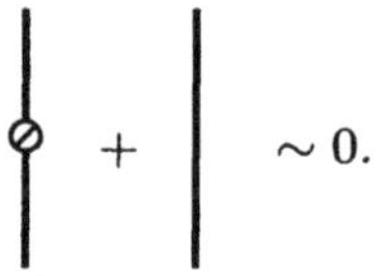

(E.10)

By using this relation we obtain the AS relation in the same way as in the proof of Lemma E.9. We show the IHX relation in Lemma E.22 below. □

We now prove the following two lemmas which were used above in the proof of Theorem E.20.

Lemma E.21 (Habiro). We have the following relation, which relates the integral homology 3-sphere obtained from S^3 by surgery along a tree clasper with d trivalent vertices that is partially shown below, to S^3 (denoted by 0 in the right hand side).

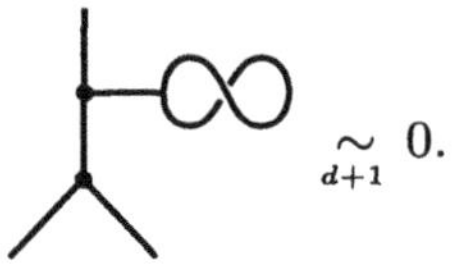

Proof. We compute a part of the picture of the required relation as follows,

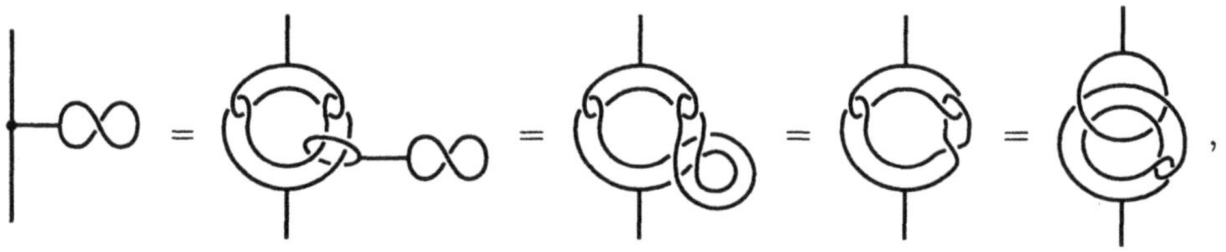

where we obtain the equalities from the definition of a trivalent vertex, by (E.4), and by isotopy. Further,

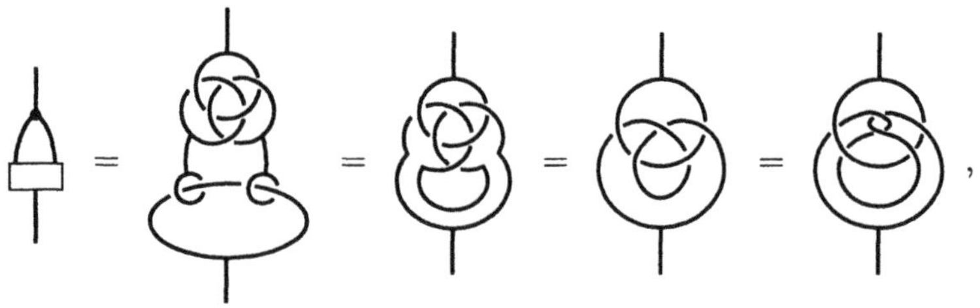

where we obtain the equalities from the definition of a box, by (E.4), and by isotopy. Since the right hand sides of the above two formulae are equal, so are the left hand sides of them. Hence,

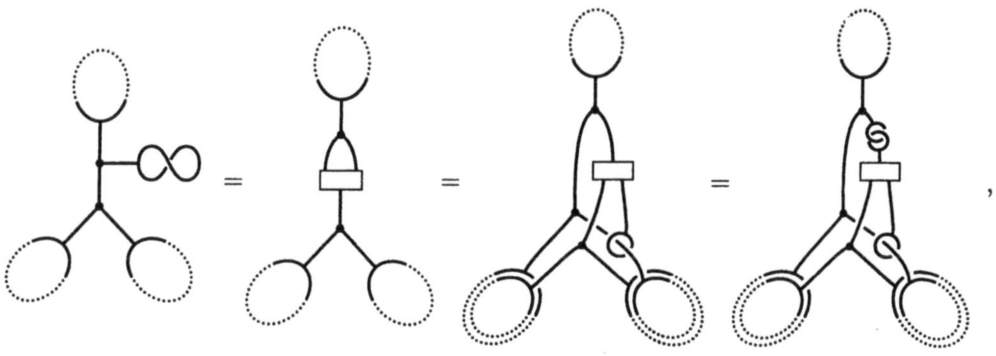

where we obtain the latter two equalities by Lemma E.2, (E.5) and (E.2). Further, this is Y_3-equivalent, by Lemma E.16, to the next picture,

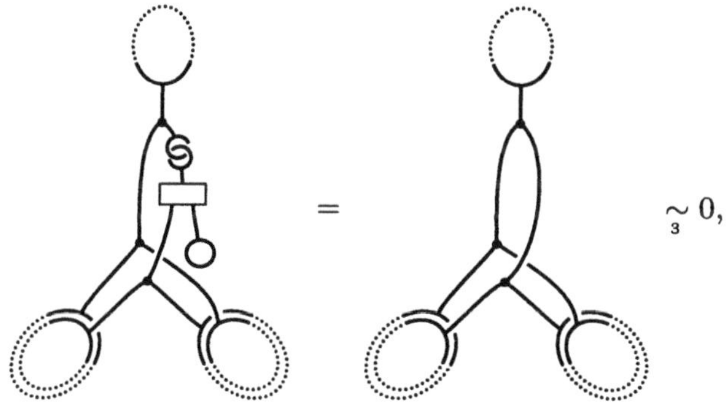

where we obtain the equality by (E.2) and (E.6) and obtain the equivalence from the definition of Y_3-equivalence. Therefore, we obtain the lemma in this simple case. In the general case we can obtain the lemma in a similar way. □

Lemma E.22 (Habiro). We have the following IHX relation, which relates three integral homology 3-spheres, each obtained from S^3 by surgery along graph clasper with d trivalent vertices and with no leaves.

These graph claspers are identical except for a ball, where they differ as shown in the pictures.

Proof. Similarly as in Lemma E.13, we have that

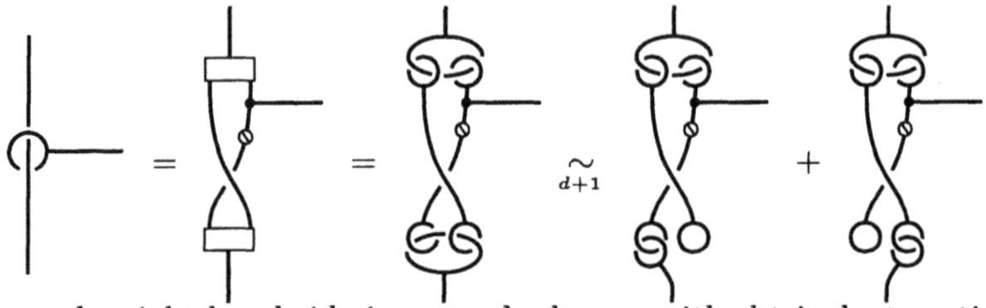

where the right hand side is a graph clasper with d trivalent vertices and with no leaves. Here, we obtain the equalities by Lemma E.1, from the definition of a box, and by (E.3), and obtain the equivalence by Lemma E.18. Hence, in a similar way as in the proof of Lemma E.11, we have that

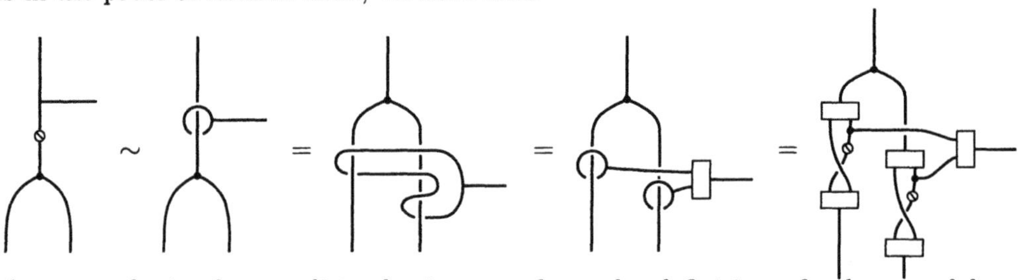

where we obtain the equalities by isotopy, from the definition of a box, and by Lemma E.1. By the same argument as in the proof of Lemma E.18, the right hand side of the above formula is equivalent to

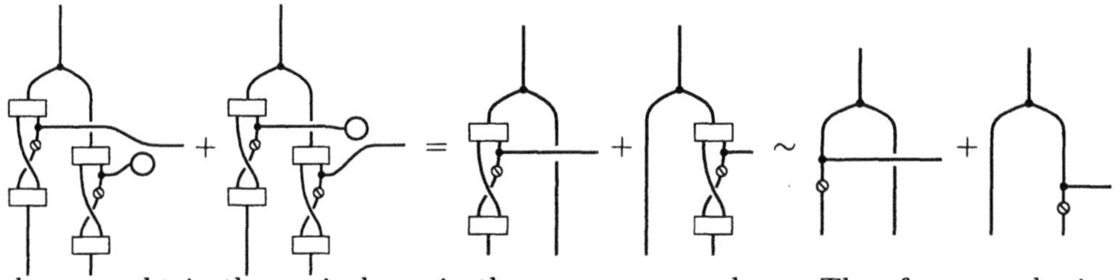

where we obtain the equivalence in the same way as above. Therefore, we obtain the required relation by (E.10). □

As a corollary of Theorem E.20 we have

Corollary E.23 (Habiro). For each positive integer d, the quotient semigroup $\mathbb{M}/\underset{d}{\sim}$ forms a commutative group.

Proof. When $d = 1$, we show that $\mathbb{M}/\underset{1}{\sim} = 0$ as follows. It is sufficient to show that any integral homology 3-sphere M is Y_1-equivalent to S^3. It is known that M can be obtained from S^3 by integral surgery along an algebraically split unit-framed link. Further, it is known [M$_{\rm H}$Na89] that such a link is related by a sequence of Δ-moves to a trivial link with the same framing. Noting that a Δ-move can be expressed by surgery along a tree clasper with one trivalent vertex, we obtain S^3 from M by a sequence of surgeries along such tree claspers, since the 3-manifold obtained from S^3 by surgery along a unit-framed trivial link is homeomorphic to S^3. Hence, we obtain the corollary for $d = 1$. See also [Hil81, Mat87] for proofs of this part.

By Theorem E.20, $\mathbb{M}_d/\underset{d+1}{\sim}$ forms a group for each d. Hence, we obtain the corollary for any d by induction on d in the same way as in the proof of Corollary E.14. □

E.4 Computing the Kontsevich and the LMO invariants of tree claspers

In this section we compute the Kontsevich and the LMO invariants of tree claspers following an idea due to Kricker [Kri*00a*]. It is shown, by the computations, that claspers give a reconstruction of the notion of finite type invariants of knots and of integral homology 3-spheres.

Let M be an integral homology 3-sphere obtained from S^3 by surgery along a framed link L. As in Chapter 10, the degree $\leq n$ part of the LMO invariant of M is given by

$$Z^{\text{LMO}}(M)^{(\leq n)} = \Big(\iota_n \check{Z}(U_+)\Big)^{-\sigma_+} \Big(\iota_n \check{Z}(U_+)\Big)^{-\sigma_-} \iota_n \check{Z}(L)$$

modulo the degree $> n$ part, where $U_\pm$ denotes the trivial knot with ± 1 framing. Further, we consider a framed knot in M obtained from a framed knot K in S^3 by surgery along L. We denote the knot in M also by K (abusing notation). Then, as in [MJOh97], we can define the LMO invariant $Z^{\text{LMO}}(M, K) \in \mathcal{A}(S^1)$ of the pair (M, K) by

$$Z^{\text{LMO}}(M, K)^{(\leq n)} = \Big(\iota_n \check{Z}(U_+)\Big)^{-\sigma_+} \Big(\iota_n \check{Z}(U_+)\Big)^{-\sigma_-} \iota_n \check{Z}(L \cup K),$$

where $\iota_n : \mathcal{A}(\sqcup^l S^1 \sqcup S^1) \to \mathcal{A}(S^1)$ is defined to be the map removing solid $\sqcup^l S^1$ in the same way as in the original definition of ι_n. In particular, when $M = S^3$, we have that

$$Z^{\text{LMO}}(S^3, K) = \hat{Z}(K). \qquad \text{(E.11)}$$

Kricker's idea is that, when K is given by surgery along claspers, the Kontsevich invariant $\hat{Z}(K)$ can be computed by (E.11) regarding the surgery along the claspers as surgery along a link obtained from the definition (E.1) of claspers. Following this idea we compute invariants of knots given by claspers in the remainder of this section.

Computation of invariants of knots given by claspers

Proposition E.24. Let D be a primitive Jacobi diagram on S^1 of degree d. Then, for the primitive Kontsevich invariant z and the map ψ of Theorem E.12, we have that

$$z\big(\psi(D)\big) = \pm D + \text{ (terms of degrees} > d)$$

for some sign.

Sketch of the proof. From the definition of the primitive Kontsevich invariant it is sufficient to show that

$$\hat{Z}\Big(\psi(D)\Big)\#\nu^{-1} = 1 \pm D + \text{ (terms of higher degrees)}.$$

We show it, say, for the following Jacobi diagram

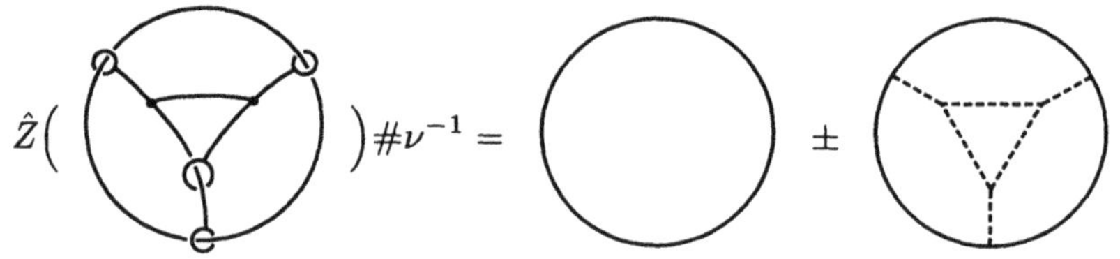

(E.12)

$+$ (terms of higher degrees).

This formula is equivalent to

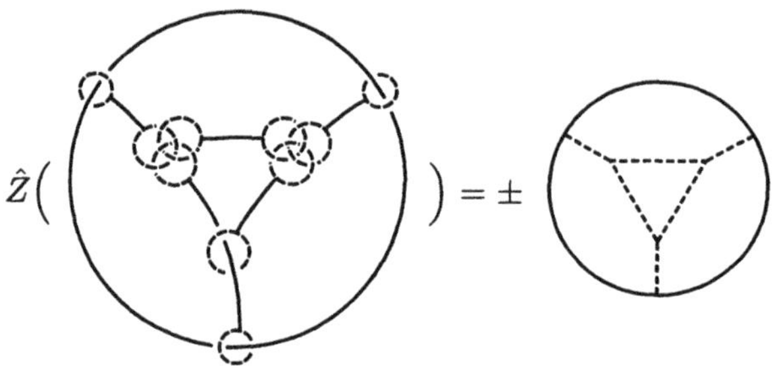

$+$ (terms of higher degrees),

(E.13)

where dashed circles imply linear sums given by

(E.14)

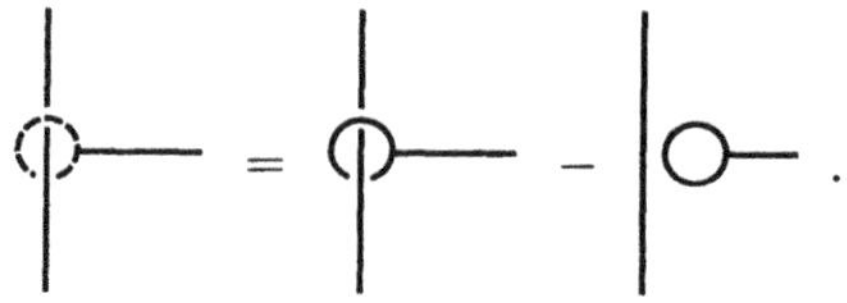

(E.15)

From the definition of claspers the knot with claspers on the left hand side of (E.12) is presented by

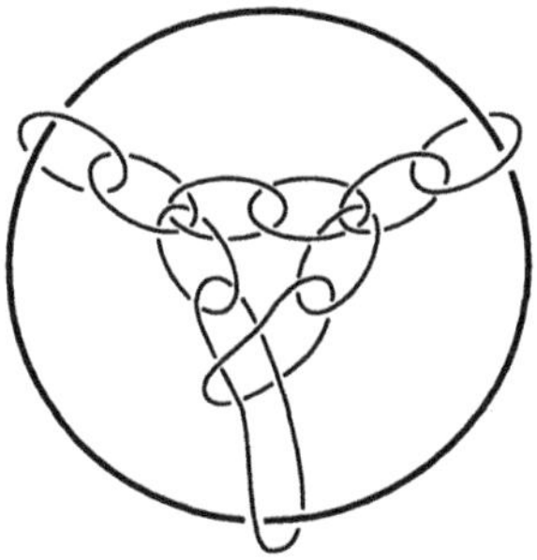

,

where the picture implies the knot obtained from the thick trivial knot by surgery along the thin link. Further, we present the linear sum of knots on the left hand side of (E.13) by

$\hat{Z}\Big($ 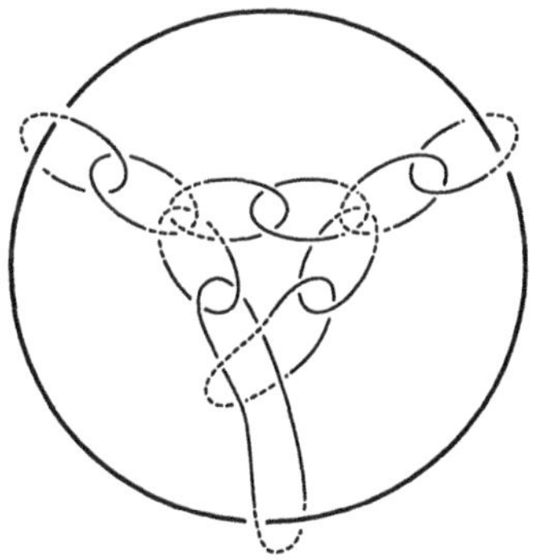 $\Big) =$ 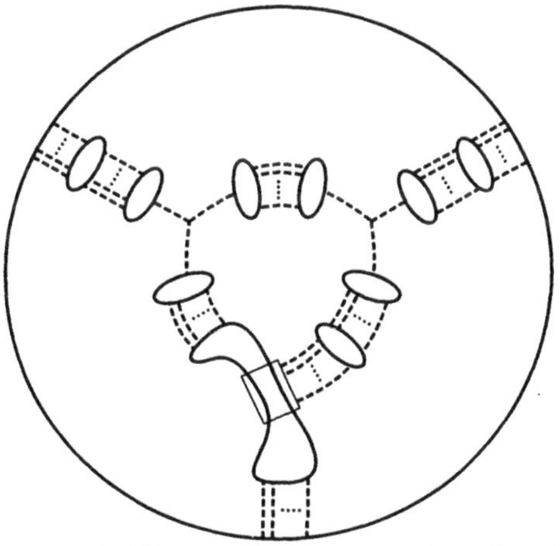 $+\cdots,$

where we obtain the equality by Lemma 11.22. Here, a box on the right hand side implies the image of the comultiplication on a solid strand in a Jacobi diagram. Applying the map ι_n in the definition of the LMO invariant to the solid circles corresponding to the thin link, we obtain the Jacobi diagram in (E.13) by (E.11). □

The C_d-equivalence in Section E.2 and the $C_d^{\mathbb{Q}}$-equivalence in Section 7.3 are related by the following proposition.

Proposition E.25. Let K and K' be two knots. If $K \underset{d}{\sim} K'$, then $K \underset{(d)}{\sim} K'$.

Proof. If $K \underset{d}{\sim} K'$, then K' is obtained from K by a sequence of surgeries along tree claspers of degree d. Suppose for simplicity that K' is obtained from K by surgery along a tree clasper on K. In the same way as in the proof of Proposition E.24 we have that

$$z(K') = z(K) \pm D + (\text{terms of higher degrees}),$$

for the primitive Kontsevich invariant z. Hence, $z(K)$ and $z(K')$ are congruent modulo the degree $\geq d$ part. This implies that $K \underset{(d)}{\sim} K'$. □

As mentioned in Section 7.1, claspers give a reconstruction of the isomorphism φ between the space $\mathcal{A}(S^1)^{(d)}$ and the graded vector space $\mathcal{K}_d/\mathcal{K}_{d+1}$ of the filtration given by Vassiliev invariants. We show that in the following proposition.

Proposition E.26. The map (7.4) gives a reconstruction of the map φ in Section 7.1.

This proposition also holds for the map φ in the theory of Vassiliev invariants over $\mathbb{Z}$,[§] though our arguments in this section are available just over $\mathbb{Q}$.

Proof of Proposition E.26. We temporarily denote the map (7.4) by φ'. In the same way as in the proof of Proposition E.24 we have that

$$\hat{Z}(\varphi'(D)) = \pm D + (\text{terms of higher degrees}).$$

Hence, φ' induces the map $\mathcal{A}(S^1) \to \mathcal{K}_d/\mathcal{K}_{d+1}$ whose inverse is given by $\hat{Z}$. Further, $\hat{Z}$ gives the inverse of φ by Proposition 7.2. Therefore, the map φ' gives a reconstruction of φ. □

Computation of invariants of homology 3-spheres given by claspers

The Y_d-equivalence in Section E.3 and the $Y_d^{\mathbb{Q}}$-equivalence in Section 11.3 are related by the following proposition.

Proposition E.27. Let M and M' be two integral homology 3-spheres. If $M \underset{2d}{\sim} M'$, then $M \underset{(d)}{\sim} M'$.

Sketch of the proof. If $M \underset{2d}{\sim} M'$, then M' is obtained from M by a sequence of surgeries along tree claspers with $2d$ trivalent vertices. Suppose, for simplicity, that $M = S^3$ and M' is obtained from S^3 by surgery along a tree clasper with $2d$ trivalent vertices. We present M', say, by

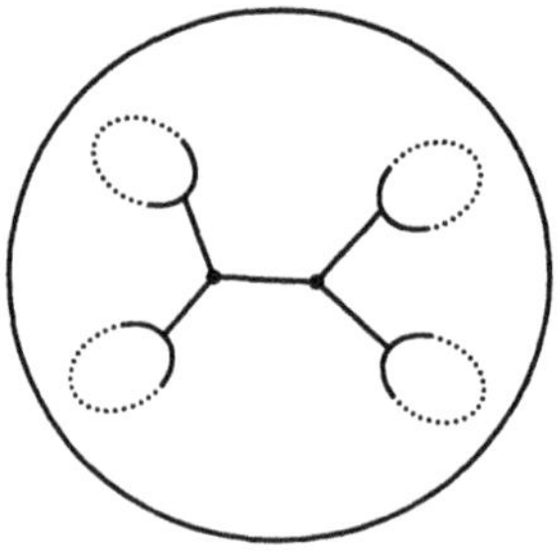,

[§]This was suggested by A. Kricker.

where dotted lines imply strands possibly knotted and linked in some fashion. Further, we consider a linear sum of integral homology 3-spheres as

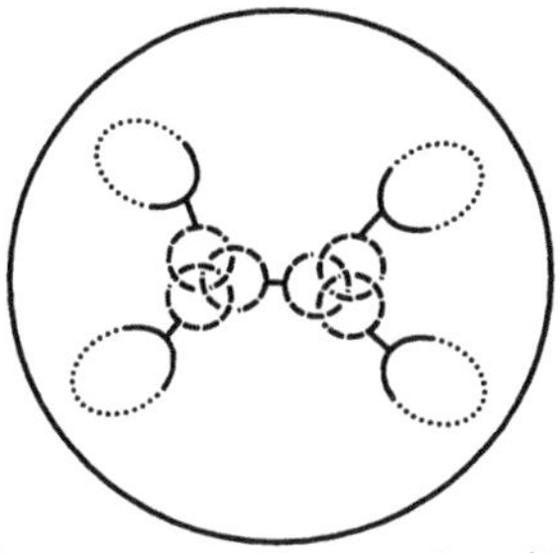

,

where dashed circles imply linear sums given by (E.14). From the definition of claspers we present the linear sum by surgery along links by

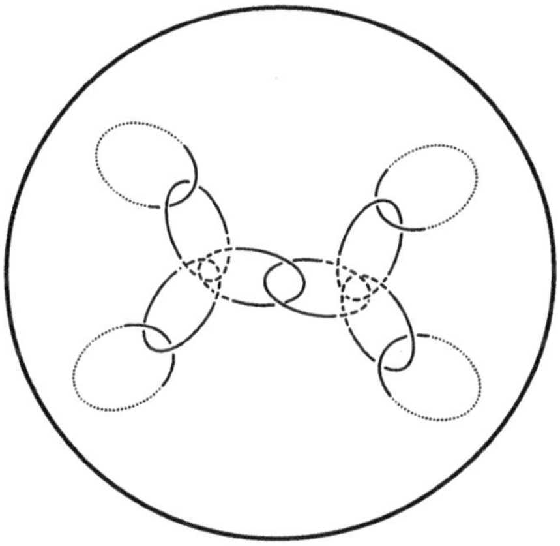

,

where the dashed parts imply linear sums in the same sense as above. By Lemma 11.22, we obtain the Kontsevich invariant of the above linear sum. Further, by applying the map ι_n to the Kontsevich invariant, we obtain the LMO invariant of M' as

$$Z^{\mathrm{LMO}}(M') = Z^{\mathrm{LMO}}(M) \pm$$ 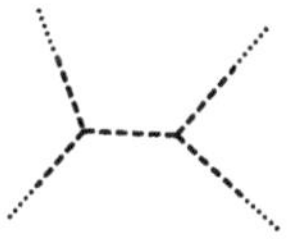 $$+ \text{ (terms with more trivalent vertices)}.$$

Since the above Jacobi diagram has at least $2d$ trivalent vertices, it is at least of degree d. Hence, $Z^{\mathrm{LMO}}(M)$ and $Z^{\mathrm{LMO}}(M')$ are congruent modulo the degree $\geq d$ part. This implies that $M \underset{(d)}{\sim} M'$. □

As mentioned in Section 11.1, claspers give a reconstruction of the isomorphism φ between the space $\mathcal{A}(\emptyset)^{(d)}$ and the graded vector space $\mathcal{M}_d/\mathcal{M}_{d+1}$ of the filtration given by finite type invariants. We show that in the following proposition.

Proposition E.28. The map (11.7) gives a reconstruction of the map φ in Section 11.1.

Proof. We temporarily denote the map (11.7) by φ'. In the same way as in the proof of Proposition E.24 we have that

$$Z^{\mathrm{LMO}}\big(\varphi'(D)\big) = \pm D + (\text{terms of higher degrees}).$$

Hence, φ' induces the map $\mathcal{A}(\emptyset) \to \mathcal{M}_d/\mathcal{M}_{d+1}$ whose inverse is given by Z^{LMO}. Further, Z^{LMO} gives the inverse of φ by Theorem 11.21. Therefore, the map φ' gives a reconstruction of φ. $\square$

Appendix F
Physical background

In classical analytical dynamics, an orbit of a moving particle is a critical point of the Lagrangian of the system, which is a functional on the space of paths. In a quantum field theory, we regard a quantum particle as traveling from one point to another point via many paths simultaneously. In the 1950s, Feynman described such moving quantum particles by using a partition function given by a formal integral (called a *path integral*) of the exponential of the Lagrangian over the infinite dimensional space of all paths. Such a path integral has not been justified mathematically, the chief difficulty being the integration over the infinite dimensional space of paths. However, from the viewpoint of mathematical physics, a path integral can be understood as follows. In the finite dimensional case it is known by the stationary phase method that the value of an oscillating integral is determined by the sum of the contributions from (neighborhoods of) critical points of the function (*i.e.*, the Lagrangian in this case), and the contribution from each critical point can be computed by expanding the integrand at the point. To construct a formal analogue of this picture in the setting of infinite dimensional integration, we start by observing that the space of (gauge equivalence classes of) critical points of the Lagrangian is finite dimensional. Hence, by an infinite dimensional formal analogue of the stationary phase method, the path integral can formally be understood as the sum of contributions presented by integrals over finite dimensional spaces.

In the late 1980s, Witten [Wit89a] (see also [Wit89b, Wit90]) considered a quantum field theory whose Lagrangian is the Chern-Simons functional, and proposed that the partition function of the theory provides a "topological invariant" of 3-manifolds. A motivation to construct this theory (expecting its topological invariance) is that the critical points of the Chern-Simons functional are flat connections, which suggests that the corresponding classical dynamical theory does not depend on the metric of the space. Further, Witten argued, using the Hamiltonian approach relating the path integral to conformal field theory, that the Chern-Simons path integral on a 3-manifold with an embedded link gives a (formal) 3-dimensional interpretation of the Jones polynomial of links; carrying Witten's argument to its logical conclusion, we are led to expect that the Chern-Simons path integral also gives a similar interpretation of quantum invariants of links (which had originally been defined by using representations of quantum groups in the 1980s before Witten's argument).

Motivated by Witten's proposal, Reshetikhin-Turaev [ReTu91] gave the first mathematically rigorous construction of such a topological invariant of 3-manifolds, as a linear sum of quantum invariants of framed links. Since then, such topological invariants (what we call *quantum invariants*) of 3-manifolds have been constructed by various mathematical approaches, as mentioned in Chapter 8.

An aim of this appendix is to give a sketch of the Chern-Simons field theory as a physical origin of quantum invariants of links and 3-manifolds. In Section F.1 we introduce the Chern-Simons path integral and sketch two approaches to it: the operator formalism (Hamiltonian formulation) and the formal perturbative expansion, which are described in Sections F.2 and F.3 respectively. Another model of the field theory, the Wess-Zumino-Witten model, is explained in Section F.4. Our aim in this appendix is to tell the stories of these theories; note that the arguments will often be sketchy, and occasionally will lack mathematical rigour.

F.1 Chern-Simons field theory

In this section we outline a sketch of the Chern-Simons field theory, which is a physical background for quantum invariants of knots and 3-manifolds. We explain this from the viewpoint of mathematical physics.

Let M be an oriented closed 3-manifold and G a semi-simple compact Lie group. Let $\mathcal{A}$ denote the set of connections on the trivial G bundle $G\times M \to M$. We identify $\mathcal{A}$ with the set $\Omega^1(M;\mathfrak{g})$ of $\mathfrak{g}$-valued 1-forms on M, *i.e.*, we regard $\mathcal{A} = \Omega^1(M;\mathfrak{g})$. The *Chern-Simons functional* $\mathrm{CS} : \mathcal{A} \to \mathbb{R}$ is defined by

$$\mathrm{CS}(A) = \frac{1}{8\pi^2}\int_M \mathrm{trace}(A \wedge dA + \frac{2}{3}A \wedge A \wedge A), \tag{F.1}$$

for a connection A. The *gauge group* $\mathcal{G}$ consisting of automorphisms of the bundle $G \times M \to M$ can be identified with the space of smooth maps $M \to G$. The gauge group $\mathcal{G}$ acts on $\mathcal{A}$ by

$$g^\star A = g^{-1}Ag + g^{-1}dg, \tag{F.2}$$

for $g \in \mathcal{G}$ and $A \in \mathcal{A}$. It follows from a concrete computation that $\mathrm{CS}(g^\star A)$ differs from $\mathrm{CS}(A)$ by an integer (with the right normalization of trace) related to the homotopy type of $g : M \to G$ (for example, when $G = SU(2)$, the integer is equal to the mapping degree of g). Hence, the Chern-Simons functional induces the map $\mathrm{CS} : \mathcal{A}/\mathcal{G} \to \mathbb{R}/\mathbb{Z}$.

The partition function of the Chern-Simons field theory is formally given by

$$Z_k(M) = \int_{\mathcal{A}/\mathcal{G}} \exp\left(2\pi\sqrt{-1}k\mathrm{CS}(A)\right)\mathcal{D}A, \tag{F.3}$$

for any positive integer k. This formal integral is called the *Chern-Simons path integral.* Note that, since $\mathcal{A}/\mathcal{G}$ is infinite dimensional, the path integral has not been defined mathematically. However, the Chern-Simons path integral gives many interesting suggestions to mathematics, as we explain in the following of this appendix.

Witten [Wit89a] proposed that the formal integral $Z_k(M)$ gives a topological invariant of the closed 3-manifold M. In the physical viewpoint the Lagrangian of the corresponding classical dynamical theory is the Chern-Simons functional. Its critical points are flat connections, which are independent of the metric of M. This suggests that the corresponding classical dynamical theory does not depend on the metric of M. This is a motivation to consider the field theory given by $Z_k(M)$, expecting that it is independent of the metric of the space. The "topological invariant" $Z_k(M)$ is called the *quantum G invariant* of M, supposing that there are indeed mathematically rigorous constructions of $Z_k(M)$.

Consider, moreover, the case that there is a link L embedded in M. To each component L_i of the link L we associate a finite-dimensional representation R_i of G. Then, the partition function for the pair (M, L) is formally given by

$$Z_k(M, L) = \int_{\mathcal{A}/\mathcal{G}} \Big(\prod_{i=1}^{l} \mathrm{trace}_{R_i} \mathrm{Hol}_{L_i}(A) \Big) \exp\big(2\pi\sqrt{-1}k\mathrm{CS}(A)\big)\mathcal{D}A,$$

for any positive integer k, where $\mathrm{Hol}_{L_i}(A)$ denotes the holonomy, described below. Witten [Wit89a] claimed that the formal integral $Z_k(M, L)$ is a topological invariant of (M, L). In particular, when $M = S^3$, we call the "invariant" the *quantum $(\mathfrak{g}; R_1, \cdots, R_l)$ invariant* of L (supposing that there is indeed a rigorous construction of it). The holonomy $\mathrm{Hol}_K(A)$ of A along K is given as follows. We express a knot K by a closed path $\gamma : [0,1] \to M$ such that $\gamma(0) = \gamma(1)$. By pulling back the $\mathfrak{g}$-valued 1-form A on M by γ we obtain a $\mathfrak{g}$-valued 1-form on $[0,1]$; we define a function $w : [0,1] \to \mathfrak{g}$ by $\gamma^\star A = w(t)dt$. Given the unique solution $f : [0,1] \to G$ of the following differential equation,

$$\frac{df(t)}{dt} = f(t)w(t), \qquad f(0) = 1 \in G,$$

we define $\mathrm{Hol}_K(A)$ to be $f(1)$. By the same argument as in Proposition 6.4 the holonomy $\mathrm{Hol}_K(A)$ can be presented by iterated integrals as

$$\begin{aligned} \mathrm{Hol}_K(A) = f(1) &= 1 + \sum_{m=1}^{\infty} \int_{0 \le t_1 \le \cdots \le t_m \le 1} w(t_1) \cdots w(t_m) dt_1 \cdots dt_m \\ &= 1 + \sum_{m=1}^{\infty} \int_{0 \le t_1 \le \cdots \le t_m \le 1} \gamma^\star A(t_1) \cdots \gamma^\star A(t_m). \end{aligned} \tag{F.4}$$

In physics, there are two approaches available to obtain observables of a partition function presented by a path integral; the operator formalism or the formal

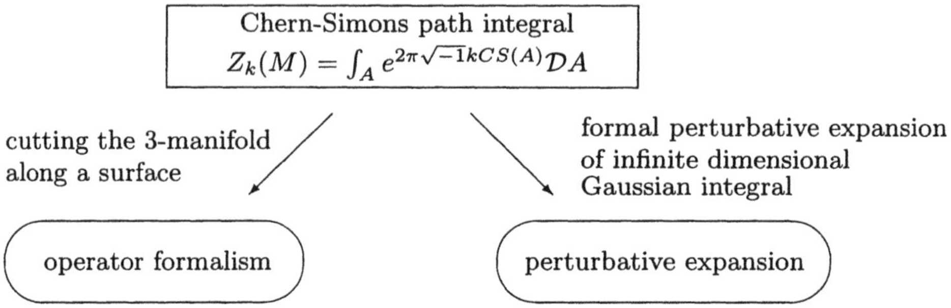

Figure F.1 How to get observables from the path integral. The observables of the path integral can be understood rigorously in mathematics, while the path integral itself can not.

perturbative expansion (see Figure F.1). They are two different ways to compute the (physically) same thing.

A sketch of the operator formalism approach (Hamiltonian formulation) for the Chern-Simons field theory is as follows. We extend the definition (F.3) of the path integral $Z_k(M)$ to the case where a 3-manifold has non-empty boundary. For a 3-manifold M with boundary $\partial M = \Sigma_1 \cup (-\Sigma_2)$, we regard $Z_k(M)$ as a linear map $\mathcal{H}_{\Sigma_1} \to \mathcal{H}_{\Sigma_2}$, where each $\mathcal{H}_{\Sigma_i}$ is a vector space of some functions on a space related to Σ_i. Physical speaking, $\mathcal{H}_{\Sigma_i}$ is a space of wave functions which represent physical states on Σ_i in the quantum theory, and $Z_k(M)$ gives a time evolution from a physical state on Σ_1 to a physical state on Σ_2. We call $\mathcal{H}_\Sigma$ the *quantum Hilbert space* of Σ. In the mathematical viewpoint, this system can be formalized by the axioms of a topological quantum field theory. Starting from a mathematically meaningful definition of the $\mathcal{H}_\Sigma$ and Z_k, we are led to a rigorous definition of quantum invariants. We explain the detailed arguments in Section F.2.

A sketch of the formal perturbative expansion approach to the Chern-Simons path integral is as follows. We consider the asymptotic behavior of the path integral (F.3) as $k \to \infty$, regarding it as a (modified) Gaussian integral. It is known as the stationary phase method that a (modified) Gaussian integral over a finite dimensional space is described by contributions from (neighborhoods of) critical points of the function in the power of the integrand. As a formal infinite dimensional analogue of the stationary phase method we consider the formal perturbative expansion of the Chern-Simons path integral. By computing the summands of the expansion we obtain an infinite sum of trivalent graphs such that the formal perturbative expansion is recovered from the infinite sum through a weight system. This is the physical background for the existence of the Kontsevich invariant of knots and the LMO invariant of 3-manifolds. We explain detailed arguments in Section F.3.

Motivated by the operator formalism of the Chern-Simons path integral, we obtain mathematically rigorous constructions of invariants of 3-manifolds as follows (see also Figure F.2). We have presentations of a 3-manifold that arise by cutting it

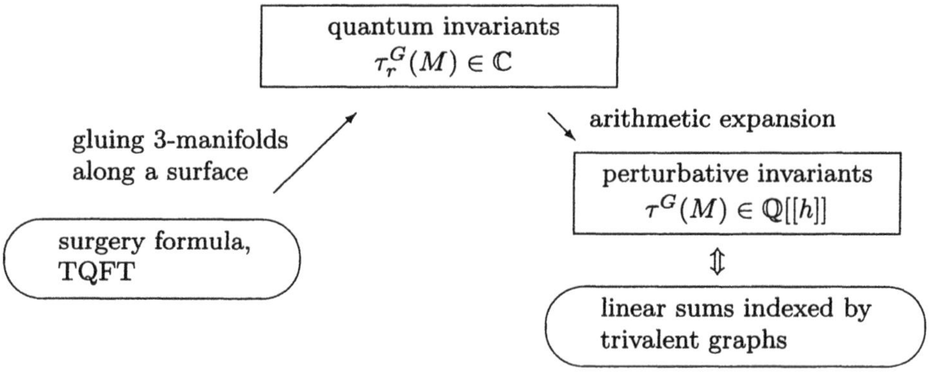

Figure F.2 Mathematically rigorous constructions of topological invariants of 3-manifolds

along surfaces; say, by a Heegaard splitting or by surgery along a link, as mentioned in Chapter 8. When a 3-manifold is presented by a Heegaard splitting, it is expressed as a union of two handlebodies along a surface. When a 3-manifold is presented by surgery along a link, it is expressed as a union of a link exterior and solid tori. Quantum invariants of the 3-manifold can be formulated as inner products of vectors in the quantum Hilbert spaces associated to the surface, or to the union of tori. See the last subsections of Section F.2 for these arguments. We denote by $\tau_r^G(M)$ the quantum G invariant constructed rigorously in Chapter 8. As mentioned in Chapter 9, we can, further, give a mathematical formulation of a perturbative invariant $\tau^G(M)$, either by an arithmetic expansion of $\tau_r^G(M)$, or by a Gaussian integral.

Based on the perturbative expansion of the Chern-Simons path integral, we obtain linear sums indexed by trivalent graphs whose coefficients are presented by finite dimensional configuration space integrals. This is a physical background for the existence of the Kontsevich invariant and the LMO invariant. It is a non-trivial mathematical problem to compare their coefficients with the configuration space integrals; note that they present physically the same thing.

F.2 Topological quantum field theory

In this section we explain how we obtain the operator formalism of the Chern-Simons field theory. Mathematically speaking, the operator formalism of a quantum theory can be understood as a functor satisfying the axioms of a topological quantum field theory. We will explain how to construct such a functor from the Chern-Simons path integral, from the viewpoint of mathematical physics. Further, again from the mathematical viewpoint, we will explain how we obtain rigorous constructions of quantum invariants of 3-manifolds based on this functor. This is the physical background for the construction of quantum invariants of 3-manifolds shown in Chapter 8.

Axioms of a topological quantum field theory

In physics a *topological quantum field theory* is a quantum field theory which does not depend on the metric of a space. Atiyah [Ati89, AHLS88] proposed the axioms below as a mathematical formulation of the notion of a topological quantum field theory.

A *topological quantum field theory* (TQFT) is a functor from the category whose objects are closed oriented surfaces and whose morphisms are oriented 3-cobordisms to the category of finite dimensional vector spaces over $\mathbb{C}$. Namely, it is a functor which assigns a finite dimensional vector space $\mathcal{H}_\Sigma$, called the *quantum Hilbert space*, to an oriented closed surface Σ, and a vector $Z(M) \in \mathcal{H}_\Sigma$ to an oriented compact manifold M with boundary Σ, satisfying the following 5 axioms.

(1) $\mathcal{H}_{-\Sigma} = \mathcal{H}_\Sigma{}^\star$, where $-\Sigma$ denotes Σ with the opposite orientation and $\mathcal{H}_\Sigma{}^\star$ denotes the dual vector space of $\mathcal{H}_\Sigma$.

(2) $\mathcal{H}_{\Sigma_1 \sqcup \Sigma_2} = \mathcal{H}_{\Sigma_1} \otimes \mathcal{H}_{\Sigma_2}$, where $\Sigma_1 \sqcup \Sigma_2$ denotes the disjoint union of two surfaces Σ_1 and Σ_2.

For a 3-cobordism M with $\partial M = (-\Sigma_1) \sqcup \Sigma_2$ the vector $Z(M)$ belongs to $\mathcal{H}_{(-\Sigma_1 \sqcup \Sigma_2)} = \mathcal{H}_{-\Sigma_1} \otimes \mathcal{H}_{\Sigma_2} = \mathcal{H}^\star_{\Sigma_1} \otimes \mathcal{H}_{\Sigma_2}$ by the axioms (1) and (2). Hence, $Z(M)$ can be regarded as a linear map $\mathcal{H}_{\Sigma_1} \to \mathcal{H}_{\Sigma_2}$.

(3) For 3-cobordisms M_1 and M_2 with $\partial M_1 = (-\Sigma_1) \sqcup \Sigma_2$ and $\partial M_2 = (-\Sigma_2) \sqcup \Sigma_3$ we have that $Z(M_1 \underset{\Sigma_2}{\cup} M_2) = Z(M_1) \circ Z(M_2)$ as linear maps $\mathcal{H}_{\Sigma_1} \to \mathcal{H}_{\Sigma_3}$.

(4) $\mathcal{H}_\emptyset = \mathbb{C}$, where $\emptyset$ denotes the empty surface.

(5) $Z(\Sigma \times I)$ is equal to the identity map of $\mathcal{H}_\Sigma$.

For a closed 3-manifold M, $Z(M)$ belongs to $\mathcal{H}_\emptyset = \mathbb{C}$ by the axiom (4).

In the next subsection we explain, from the viewpoint of mathematical physics, that the quantum Hilbert space $\mathcal{H}_\Sigma$ of the Chern-Simons field theory should be equal to the vector space of holomorphic sections of some complex line bundle over a complex manifold related to Σ. Thus, $\mathcal{H}_\Sigma$ itself can be formulated in a mathematically rigorous way independently of any path integral. In the subsection that follows the next, we explain how we obtain a mathematical construction of quantum invariants of 3-manifolds based on such $\mathcal{H}_\Sigma$.

The operator formalism for the Chern-Simons field theory

In physics a way to describe observables in a quantum field theory is to give linear maps (called operators) between quantum Hilbert spaces. Such a linear map, say, $Z(M) : \mathcal{H}_{\Sigma_1} \to \mathcal{H}_{\Sigma_2}$, can be regarded as a vector in some quantum Hilbert space, as $Z(M) \in \mathcal{H}_{(-\Sigma_1 \sqcup \Sigma_2)}$, as explained above. In this subsection, we sketch an explanation of the way how the quantum Hilbert spaces and vectors in the Hilbert spaces are obtained from the Chern-Simons field theory. For details on this topic see [Ati90b].

Let us briefly review how to compute the algebra of functions on a complex variety which are invariant under an action of a compact group on that variety. Let X be a complex projective variety, let A be its coordinate ring, and let G be a compact Lie group acting on X. Then, Mumford's geometric invariant theory [Mum65] asserts that the subring A^G of A consisting of invariant functions, with respect to the action of G, is isomorphic to the coordinate ring of (approximately) the quotient projective variety $X/G_{\mathbb{C}}$ (to be precise, the variety obtained from $X/G_{\mathbb{C}}$ by removing "bad" points; such a variety is called the *Mumford quotient*), where $G_{\mathbb{C}}$ is the complexification of G.

Further, let us briefly review the "symplectic quotient" (see [Kirw84]) to describe $X/G_{\mathbb{C}}$. A *symplectic manifold* X is a smooth manifold with a non-degenerate closed 2-form ω, called a *symplectic form*. We consider an action of a compact semi-simple Lie group G on a symplectic manifold X which preserves its symplectic 2-form ω. Then, we have a *moment map* $\mu : X \to \mathfrak{g}^\star$, where $\mathfrak{g}^\star$ is the dual of the Lie algebra $\mathfrak{g}$ of G, such that μ is G-equivariant with respect to the actions of G on X and $\mathfrak{g}^\star$, where G acts on $\mathfrak{g}^\star$ by the dual of the adjoint action, and μ satisfies the condition that $\langle d\mu_x(v), \xi\rangle = \omega_x(v, \xi_x)$ for any $x \in X$, $v \in T_xX$ and $\xi \in \mathfrak{g}$. Here, $d\mu_x : T_xX \to \mathfrak{g}^\star$ is the differential of μ at x. Further, $x \mapsto \xi_x$ is the vector field on X induced by ξ. The above condition implies that the function $x \mapsto \langle \mu(x), \xi\rangle$ is a Hamilton function for the vector field on X determined by ξ. The space $\mu^{-1}(0)/G$ is called the *symplectic quotient*. It is known that the symplectic quotient coincides with the Mumford quotient when X is a Kähler manifold, which is (roughly speaking) a symplectic complex manifold.

Let X be a Kähler manifold with a symplectic form ω. We consider an action of a compact semi-simple Lie group G on X preserving ω. Let L be the holomorphic line bundle whose curvature is equal to ω. By (a modification of) the above two arguments we can (approximately) identify the space of G-invariant holomorphic sections of $L \to X$ with the space of holomorphic sections of $\mathcal{L} \to \mu^{-1}(0)/G$, where $\mathcal{L}$ is the holomorphic line bundle obtained as the Mumford quotient of L. This is the main idea in the following interpretation of the Chern-Simons path integral, namely, to regard it as a function on an appropriate space associated with the boundary of a 3-manifold.

Let Σ be an oriented closed surface, let G be a semi-simple Lie group, and let $\mathcal{A}_\Sigma$ be the space of connections on the trivial G bundle $G \times \Sigma \to \Sigma$. By identifying $\mathcal{A}_\Sigma$ with the space $\Omega^1(\Sigma; \mathfrak{g})$ of $\mathfrak{g}$-valued 1-forms on Σ, each tangent space to $\mathcal{A}_\Sigma$ can also be identified with $\Omega^1(\Sigma; \mathfrak{g})$. A natural affine symplectic structure on $\mathcal{A}_\Sigma$ is given by the skew bilinear form,

$$(\alpha, \beta) = \frac{1}{8\pi^2} \int_\Sigma \mathrm{trace}(\alpha \wedge \beta), \tag{F.5}$$

for $\alpha, \beta \in \Omega^1(\Sigma; \mathfrak{g})$. The *gauge group* $\mathcal{G}_\Sigma$ of the trivial bundle $G \times \Sigma \to \Sigma$ is identified with the space of smooth maps $\Sigma \to G$. The action of $\mathcal{G}_\Sigma$ on $\mathcal{A}_\Sigma$ is given by (F.2); it preserves the symplectic structure of $\mathcal{A}_\Sigma$. For $a \in \mathcal{A}_\Sigma$ and $g \in \mathcal{G}_\Sigma$, we define $c(a,g) \in \mathbb{R}/\mathbb{Z}$ by

$$\begin{aligned} c(a,g) &= \mathrm{CS}(\tilde{g}^\star A) - \mathrm{CS}(A) && \text{(F.6)} \\ &= \frac{1}{8\pi^2}\int_\Sigma \mathrm{trace}(g^{-1}ag \wedge g^{-1}dg) - \int_M \tilde{g}^\star \sigma \quad \in \mathbb{R}/\mathbb{Z}, && \text{(F.7)} \end{aligned}$$

where M is a compact 3-manifold bounded by Σ, $\tilde{g}$ is a map $M \to G$ extending g, A is a connection on M extending a, and σ is a certain 3-form on G whose cohomology class is integral in $H^3(G)$. Here, $c(a,g)$ is well defined in $\mathbb{R}/\mathbb{Z}$, *i.e.*, is independent of the choices of M, $\tilde{g}$, and A in (F.6) and (F.7). We consider an action of $\mathcal{G}_\Sigma$ on $\mathbb{C} \times \mathcal{A}_\Sigma$ defined by $g(z,a) = \left(e^{2\pi\sqrt{-1}c(a,g)}z, g^\star a\right)$.

Now let us consider the operator formalism for the Chern-Simons path integral. Let M be a compact 3-manifold with boundary Σ. Fix a complex structure on Σ. This induces a natural complex structure on $\mathcal{A}_\Sigma$. For a connection a on Σ, let $\mathcal{A}_a$ be the space of connections on M whose restriction to Σ is equal to a. We consider the following path integral,

$$Z_k(M)(a) = \int_{\mathcal{A}_a/\mathrm{kernel}(\mathcal{G}\to\mathcal{G}_\Sigma)} \exp\left(2\pi\sqrt{-1}k\mathrm{CS}(A)\right)\mathcal{D}A,$$

for $A \in \mathcal{A}_a$, where $\mathcal{G}$ and $\mathcal{G}_\Sigma$ are the gauge groups of the trivial G bundles on M and Σ respectively. We formally regard $Z_k(M)(a)$ as a holomorphic function on $\mathcal{A}_\Sigma$. It satisfies $Z_k(M)(g^\star a) = \exp(2\pi\sqrt{-1}kc(a,g))Z_k(M)(a)$ by (F.6). Hence, it can be regarded as a $\mathcal{G}_\Sigma$-invariant holomorphic section of a complex line bundle $\mathbb{C}^{\otimes k} \times \mathcal{A}_\Sigma \to \mathcal{A}_\Sigma$. By a formal infinite dimensional analogue of the argument at the beginning of this subsection, the space of such $\mathcal{G}_\Sigma$-invariant sections can be formally identified with the space of holomorphic sections of a line bundle $\mathcal{L}^{\otimes k} \to \mu^{-1}(0)/\mathcal{G}_\Sigma$, where μ is a moment map on $\mathcal{A}_\Sigma$ and $\mathcal{L}$ is the holomorphic line bundle obtained as the Mumford quotient of the holomorphic line bundle $\mathbb{C} \times \mathcal{A}_\Sigma \to \mathcal{A}_\Sigma$. It is known [AtBo82] that the moment map induced by the symplectic structure on $\mathcal{A}_\Sigma$ given by (F.5) is equal to the map which takes a connection A to its curvature. Hence, $\mu^{-1}(0)$ is equal to the space of flat connections. We put $\mathcal{M}_\Sigma = \mu^{-1}(0)/\mathcal{G}_\Sigma$, which is called the *moduli space of flat connections*. The moduli space can be identified with the space of conjugacy classes of homomorphisms $\pi_1(\Sigma) \to G$. Hence, it is a finite dimensional (singular) manifold. Further, as the Mumford quotient of $\mathcal{A}_\Sigma$, $\mathcal{M}_\Sigma$ has a complex structure [AtBo82], and with respect to this structure $\mathcal{L}$ can be regarded as a holomorphic line bundle on it. We define the *quantum Hilbert space* $\mathcal{H}_\Sigma$ *for the Chern-Simons theory* to be the vector space of holomorphic sections of $\mathcal{L}^{\otimes k}$. Then, for a 3-manifold M bounded by Σ, the above formal argument shows that the function $Z_k(M)(a)$ induces (formally speaking) a vector $Z_k(M) \in \mathcal{H}_\Sigma$.

Mathematical construction of quantum invariants of 3-manifolds

Returning our viewpoint to mathematics, let us review the construction of the quantum Hilbert space $\mathcal{H}_\Sigma$. Motivated by such $\mathcal{H}_\Sigma$ we obtain rigorous constructions of quantum invariants of 3-manifolds. We sketch an explanation of this procedure in this subsection.

Let Σ be a closed oriented surface. We define $\mathcal{M}_\Sigma$ to be the set of conjugacy classes of homomorphisms $\pi_1(\Sigma) \to G$. As in [Gol84], $\mathcal{M}_\Sigma$ has a natural symplectic form ω. Moreover, if we put a complex structure J on Σ, $\mathcal{M}_\Sigma$ becomes a Kähler manifold, depending on J (see [AtBo82]). Let $\mathcal{L}$ be the holomorphic line bundle whose curvature is equal to ω. We define the quantum Hilbert space $\mathcal{H}_{\Sigma,J}$ to be the space of holomorphic sections of $\mathcal{L}^{\otimes k}$ for a fixed positive integer k.

We remove the dependence on $\mathcal{H}_\Sigma$ on the choice of a complex structure J on Σ in the following way. The set of all complex structures of Σ is called the *Teichmüller space* $\mathcal{T}$. The union of the $\mathcal{H}_{\Sigma,J}$'s for all $J \in \mathcal{T}$ can be regarded as a vector bundle on $\mathcal{T}$ whose fibers are the $\mathcal{H}_{\Sigma,J}$'s. It is known that this vector bundle has a natural (projective) flat connection. The parallel translation given by the flat connection enables us to identify each fiber $\mathcal{H}_{\Sigma,J}$ with the others. In this sense, $\mathcal{H}_{\Sigma,J}$ is independent of the choice of a complex structure J on Σ. In other words, we can rigorously define the *quantum Hilbert space* $\mathcal{H}_\Sigma$ *for the Chern-Simons theory* to be the vector space of flat sections of the vector bundle.

One way of using the above $\mathcal{H}_\Sigma$ to give a rigorous construction of quantum invariants of 3-manifolds is to apply an associated topological quantum field theory to Heegaard splittings of 3-manifolds. Kohno [Koh92] gave a mathematical construction of quantum invariants motivated by this approach. A sketch of this approach is as follows. Let $\mathfrak{M}_\Sigma$ be the mapping class group of homeomorphisms from Σ to itself. There is a natural action of $\mathfrak{M}_\Sigma$ on $\mathcal{M}_\Sigma$, which can be lifted to the action on the line bundle $\mathcal{L} \to \mathcal{M}_\Sigma$. This action naturally induces the action of $\mathfrak{M}_\Sigma$ on the above mentioned vector bundle on $\mathcal{T}$. Since this vector bundle is equipped with a (projective) flat connection, we have an action of $\mathfrak{M}_\Sigma$ on $\mathcal{H}_\Sigma$ as a monodromy representation. For a handlebody H with boundary Σ we choose a vector $Z_k(H) \in \mathcal{H}_\Sigma$ such that $Z_k(H)$ is invariant under the action of the subgroup of $\mathfrak{M}_\Sigma$ consisting of homeomorphisms which can extend to homeomorphisms from H to itself. Let M be a 3-manifold obtained by gluing two handlebodies by a homeomorphism f from Σ to itself. Then, the quantum invariant $Z_k(M)$ can be formulated to be the inner product of $Z_k(H)$'s via the automorphism of $\mathcal{H}_\Sigma$ induced by f.

Another way to give a rigorous construction of quantum invariants of 3-manifolds is to apply an associated topological quantum field theory to surgery presentations of 3-manifolds. We will explain this approach in the subsection that follows the next one, but must first introduce the quantum Hilbert space for a Riemann surface with a non-empty boundary.

The quantum Hilbert space for a Riemann surface with boundary

In this subsection we explain how to construct the quantum Hilbert space for a Riemann surface Σ with non-empty boundary, following [Ati90b]; see also [PrSe86] for details on loop groups used in this subsection. It is formulated as the quantization of a generalized symplectic quotient of the space of connections on Σ.

We review the quantization of a symplectic manifold (see [Ati90b]). Let X be a symplectic manifold with an integral symplectic form ω. Then, we can find a line bundle $\mathcal{L}$ on X whose curvature is ω. Moreover, if we can choose a complex structure on X which makes X a Kähler manifold with ω the Kähler form, then $\mathcal{L}$ is a holomorphic line bundle. We call the space of holomorphic sections of $\mathcal{L}^{\otimes k}$ the *quantization* of X at *level* k for a positive integer k (if there does indeed exist such a complex structure).

Let G denote a simple Lie group, $\mathfrak{g}$ its Lie algebra, and $\mathfrak{g}^\star$ the dual of $\mathfrak{g}$. The Lie group G acts on $\mathfrak{g}$ and $\mathfrak{g}^\star$ by the adjoint action and by its dual action respectively. We call a G-orbit in $\mathfrak{g}^\star$ a *co-adjoint orbit.* It is a symplectic manifold. Borel-Weil theory tells us that any irreducible representation of G is obtained as the quantization of the co-adjoint orbit including the highest weight corresponding to the irreducible representation.

Let us review a generalized symplectic quotient. Let X be a symplectic manifold with an integral symplectic form. As mentioned earlier, the geometric invariant theory (when applied to the symplectic quotient) asserts that the G invariant subspace of the quantization of X is isomorphic to the quantization of the symplectic quotient $\mu^{-1}(0)/G$ for the moment map

$$\mu : X \longrightarrow \mathfrak{g}^\star.$$

We consider its generalization as follows. Let R be an irreducible representation of G, and let $R^\star$ be its dual representation. We denote by W_R and $W_{R^\star}$ the co-adjoint orbits which include the highest weights corresponding to R and $R^\star$ respectively. For the above moment map μ we call $\mu^{-1}(W_R)/G$ a *generalized symplectic quotient.* We have that ${\mu'}^{-1}(0) = \mu^{-1}(W_R)$ for the following modification μ' of μ,

$$\mu' : X \times W_{R^\star} \longrightarrow \mathfrak{g}.$$

Then, the quantization of the symplectic quotient ${\mu'}^{-1}(0)/G$ is isomorphic to the G invariant subspace of the quantization $\mathcal{H} \otimes R^\star$ of $X \times W_{R^\star}$, where $\mathcal{H}$ denotes the quantization of X. Moreover, the invariant subspace can be regarded as the space, $\mathrm{Hom}_G(R, \mathcal{H})$, of intertwiners of R into $\mathcal{H}$. Hence, the quantization of the generalized symplectic quotient $\mu^{-1}(W_R)/G$ is isomorphic to $\mathrm{Hom}_G(R, \mathcal{H})$. Note that when R is the trivial representation of G this isomorphism implies the above mentioned isomorphism between the G invariant subspace of $\mathcal{H}$ and the quantization of the ordinary symplectic quotient $\mu^{-1}(0)/G$.

We consider the infinite dimensional analogue of Borel-Weil theory for the loop group LG of G; see [PrSe86] for details. Here, the *loop group* LG is the group of smooth maps $S^1 \to G$ with pointwise multiplication. Its Lie algebra is called the *loop algebra*, denoted by $L\mathfrak{g}$, which is topologically identified with the space of smooth maps $S^1 \to \mathfrak{g}$. Since we obtain projective representations of LG by the same construction as in Borel-Weil theory in this case, we consider a central extension $\widehat{LG}$ of LG, which can be defined for each positive integer k (see Section F.4). Then, we obtain any irreducible representation of $\widehat{LG}$ as the quantization of a co-adjoint $\widehat{LG}$-orbit in $(\widehat{L\mathfrak{g}})^\star$, where $(\widehat{L\mathfrak{g}})^\star$ denotes the dual of the Lie algebra of $\widehat{LG}$.

We identify LG with the gauge group $\mathcal{G}_{S^1}$ of the trivial G bundle on S^1. As in [PrSe86], the space $\mathcal{A}_{S^1}$ of connections of the trivial G bundle on S^1 can be identified with a subspace of $(\widehat{L\mathfrak{g}})^\star$ by a natural bilinear form which pairs $\mathfrak{g}$-valued 1-forms and $\mathfrak{g}$-valued smooth maps on S^1. Through this identification, an LG-orbit W in $\mathcal{A}_{S^1}$ can be regarded as a co-adjoint $\widehat{LG}$-orbit in $(\widehat{L\mathfrak{g}})^\star$, noting that an $\widehat{LG}$-orbit by the adjoint action can be regarded as an LG-orbit since $\widehat{LG}$ is a central extension of LG. We consider an LG-orbit W in $\mathcal{A}_{S^1}$ from which an irreducible representation $\hat{R}$ has been derived, in the above sense. As will be explained later in Section F.4, an irreducible representation $\hat{R}$ of $\widehat{LG}$ is related to an irreducible representation R of G. Further, the LG-orbit W in $\mathcal{A}_{S^1}$ induces a conjugacy class C of G by the bijective correspondence between LG-orbits in $\mathcal{A}_{S^1}$ and conjugacy classes of G via the monodromy along S^1.

Now we can formulate the quantum Hilbert space of a Riemann surface Σ with non-empty boundary. For simplicity, we only consider the case that Σ has one boundary component S^1. Let $\mathcal{A}_\Sigma$ be the space of connections of the trivial G bundle on Σ. The gauge group $\mathcal{G}_\Sigma$ of the bundle acts on $\mathcal{A}_\Sigma$, as mentioned earlier. As in [Ati90b] the moment map

$$\mu : \mathcal{A}_\Sigma \longrightarrow (\mathrm{Lie}\,\mathcal{G}_\Sigma)^\star$$

is defined by setting $\mu(A)$ to the curvature of A minus the image of A under the composition of the following maps,

$$\mathcal{A}_{S^1} \longrightarrow (\widehat{L\mathfrak{g}})^\star \longrightarrow (\mathrm{Lie}\,\mathcal{G}_\Sigma)^\star, \tag{F.8}$$

where the first map is the above mentioned inclusion, and the second map is the dual of the differential of the map $\mathcal{G}_\Sigma \to \widehat{LG}$, which is a lift of the restriction map $\mathcal{G}_\Sigma \to \mathcal{G}_{S^1}$. Let $R, \hat{R}, W$, and C be as above. Regarding the LG-orbit W as lying in $(\mathrm{Lie}\,\mathcal{G}_\Sigma)^\star$ through the map (F.8), we consider the generalized symplectic quotient $\mu^{-1}(W)/\mathcal{G}_\Sigma$ for the above moment map μ. From the definition of μ, the quotient $\mu^{-1}(W)/\mathcal{G}_\Sigma$ is identified with the moduli space $\mathcal{M}_{\Sigma;C}$ of conjugacy classes of representations $\pi_1(\Sigma) \to G$ which take the boundary of Σ to the conjugacy class C. Then, for a Riemann surface Σ whose boundary is associated with R, and for a fixed positive integer k, we formulate the *quantum Hilbert space* $\mathcal{H}_{\Sigma;R}$ to be the quantization of $\mathcal{M}_{\Sigma;C}$ at level k. We regard $\mathcal{H}_{\Sigma;R}$ as the space of intertwiners of $\hat{R}$ into the "quantum Hilbert space $\mathcal{H}_\Sigma$ of Σ", though we do not formulate $\mathcal{H}_\Sigma$ itself here. (There is a technical difficulty to formulate $\mathcal{H}_\Sigma$ directly such that $\mathcal{M}_\Sigma$ is not (naturally) symplectic in the present case while $\mathcal{M}_{\Sigma;C}$ is naturally symplectic.)

When a Riemann surface Σ has a boundary of n components to which has been associated irreducible representations $R_1, \cdots, R_n$ of G, we can formulate its quantum Hilbert space $\mathcal{H}_{\Sigma;R_1,\cdots,R_n}$ similarly.

Another construction of quantum invariants of 3-manifolds

In this subsection we explain how quantum invariants of 3-manifolds can be formulated as certain linear sums of quantum invariants of framed links which are surgery presentations of the 3-manifolds. This approach is based on the just introduced quantum Hilbert spaces $\mathcal{H}_{\Sigma;R_1,\cdots,R_n}$. Reshetikhin-Turaev [ReTu91] gave the first mathematical construction of quantum invariants based on this approach.

We consider the case that Σ is an n-holed S^2 with a complex structure, where the n components of the boundary of Σ are associated with irreducible representations $R_1, \cdots, R_n$ of a simple Lie group G. Then, we have the quantum Hilbert space $\mathcal{H}_{\Sigma;R_1,\cdots,R_n}$ as mentioned above. Further, we (temporarily) consider the limit as each component of the boundary of Σ tends to an infinitesimal circle. At this limit, Σ can be regarded as $\mathbb{C}P^1$ with n marked points. Hence, the Teichmüller space $\mathcal{T}$ for this Σ is naturally identified with the configuration space of n points on $\mathbb{C}P^1$. Similarly as in the case, mentioned earlier, of a Riemann surface with no marked points, the union of the $\mathcal{H}_{\Sigma;R_1,\cdots,R_n}$'s as the complex structure on Σ varies forms a vector bundle on $\mathcal{T}$ equipped with a natural flat connection (see [Ati90b]). Further, regarding a path in the configuration space as a braid, as shown in Figure 5.1, we obtain a representation of the (pure) braid group from the monodromy associated to this (projectively) flat connection in a similar way as was mentioned in Section 5.1.

It is known (see Section F.4) that the quantum Hilbert space $\mathcal{H}_{\Sigma;R_1,\cdots,R_n}$ is isomorphic to the "space of conformal blocks" $\mathcal{H}_{z;R_1,\cdots,R_n}$ described in Section F.4. Moreover, the KZ equation of Chapter 5 determines an identical flat connection of the vector bundle over the configuration space formed by the union of the $\mathcal{H}_{z;R_1,\cdots,R_n}$'s. Hence, we obtain quantum invariants of links via the above monodromy representation of the braid groups.

Let us go back to the case that Σ is an n-holed S^2. Then, instead of a braid in the above argument we have a braid exterior. Hence, by the above monodromy representation we obtain an invariant of a link exterior as a vector in the quantum Hilbert space of a disjoint union of tori such that the entries of the vector are presented by quantum invariants of links.

We consider the 3-manifold obtained from S^3 by integral surgery along the link. This 3-manifold is obtained by gluing the link exterior and an appropriate number of solid tori along their boundaries. By the gluing axiom of TQFT, the invariant of the 3-manifold is obtained as the contraction of the vector of invariants of the link exterior with the vector of invariants of the solid tori, in the (common) quantum Hilbert space of a disjoint union of tori. Hence, quantum invariants of 3-manifolds can be constructed as certain linear sums of quantum invariants of links, as mentioned in Chapter 8. This completes our explanation of the physical background of that construction.

F.3 Perturbative expansion

In this section we explain how to compute a formal perturbative expansion of the Chern-Simons path integral by an infinite dimensional analogue of the perturbative expansion of a finite dimensional Gaussian integral. We then explain how invariants of 3-manifolds and knots appear as summands in the formal perturbative expansion.

Perturbative expansion of a finite dimensional Gaussian integral

In this subsection we explain how to compute a perturbative expansion of a modified Gaussian integral in the finite dimensional case, before considering the infinite dimensional case, which will explain why trivalent graphs appear in the theory.

Let us begin with a review of a computation of a Gaussian integral. For a non-zero real constant a we have that

$$\int_{x\in\mathbb{R}} \exp(\sqrt{-1}ax^2)dx = \frac{\sqrt{\pi}}{\sqrt{|a|}} \exp\Big(\pm\frac{\sqrt{-1}\pi}{4}\Big),$$

where the sign is equal to the sign of a. In general, for a non-singular symmetric $N \times N$ matrix Q, we have that

$$\int_{\mathbf{x}\in\mathbb{R}^N} \exp(\sqrt{-1}\,{}^t\mathbf{x}Q\mathbf{x})d\mathbf{x} = \pi^{N/2}|\mathrm{det}Q|^{-1/2} \exp\Big(\frac{\sqrt{-1}\pi}{4}\mathrm{sign}Q\Big),$$

where $\mathbf{x} = \begin{pmatrix} x_1 \\ \vdots \\ x_N \end{pmatrix}$ is a variable in $\mathbb{R}^N$, and $\mathrm{det}Q$ and $\mathrm{sign}Q$ denote the determinant and the signature of Q respectively. This formula is obtained from the one that precedes it by diagonalizing the symmetric matrix Q.

We consider a modified Gaussian integral Z_k with a parameter k given by

$$Z_k = \int_{\mathbf{x}\in\mathbb{R}^N} \exp\Big(\sqrt{-1}k({}^t\mathbf{x}Q\mathbf{x} + T(\mathbf{x}\otimes\mathbf{x}\otimes\mathbf{x}))\Big)d\mathbf{x}, \qquad \text{(F.9)}$$

where Q is a non-singular symmetric $N\times N$ matrix and $T : (\mathbb{R}^N)^{\otimes 3} \to \mathbb{R}$ is a trilinear map which is symmetric with respect to the permutations of three entries in $\mathbb{R}^N$. By replacing $\mathbf{x}$ with $k^{-1/2}\mathbf{x}$ and by expanding the exponential of $\sqrt{-1}kT(\mathbf{x}\otimes\mathbf{x}\otimes\mathbf{x})$,

we have that

$$Z_k = k^{-N/2} \int_{\mathbf{x}\in\mathbb{R}^N} \exp(\sqrt{-1}\ {}^t\mathbf{x}Q\mathbf{x}) \sum_{m=0}^{\infty} \frac{\sqrt{-1}^m}{m!k^{m/2}} \big(T(\mathbf{x}\otimes\mathbf{x}\otimes\mathbf{x})\big)^m d\mathbf{x}. \qquad \text{(F.10)}$$

We compute $\big(T(\mathbf{x}\otimes\mathbf{x}\otimes\mathbf{x})\big)^m$ as follows. Using a new variable $\mathbf{u} = \begin{pmatrix} u_1 \\ \vdots \\ u_N \end{pmatrix} \in \mathbb{R}^N$ we can present a polynomial in the x_i's by

$$x_a^p x_b^q x_c^r = \frac{1}{\sqrt{-1}^{p+q+r}} \left(\frac{\partial}{\partial u_a}\right)^p \left(\frac{\partial}{\partial u_b}\right)^q \left(\frac{\partial}{\partial u_c}\right)^r \exp(\sqrt{-1}\ {}^t\mathbf{u}\mathbf{x})\Big|_{\mathbf{u}=0}.$$

Hence, by putting $T(\mathbf{x}\otimes\mathbf{x}\otimes\mathbf{x}) = \sum_{a,b,c=1}^N T_{abc}x_a x_b x_c$ we can present $\big(T(\mathbf{x}\otimes\mathbf{x}\otimes\mathbf{x})\big)^m$ by

$$\big(T(\mathbf{x}\otimes\mathbf{x}\otimes\mathbf{x})\big)^m = \frac{1}{\sqrt{-1}^{3m}} \Big(\sum_{a,b,c=1}^{N} T_{abc} \frac{\partial}{\partial u_a}\frac{\partial}{\partial u_b}\frac{\partial}{\partial u_c}\Big)^m \exp(\sqrt{-1}\ {}^t\mathbf{u}\mathbf{x})\Big|_{\mathbf{u}=0}.$$

When we substitute this formula into (F.10) a factor $\exp\big(\sqrt{-1}({}^t\mathbf{x}Q\mathbf{x} + {}^t\mathbf{u}\mathbf{x})\big)$ appears. We "complete-the-square" in the exponent, writing

$${}^t\mathbf{x}Q\mathbf{x} + {}^t\mathbf{u}\mathbf{x} = {}^t\mathbf{x}'Q\mathbf{x}' - \frac{1}{4}{}^t\mathbf{u}Q^{-1}\mathbf{u},$$

by putting $\mathbf{x}' = \mathbf{x} + \frac{1}{2}Q^{-1}\mathbf{u}$.

Hence, by the above arguments, the asymptotic behavior of (F.10) as $k \to \infty$ is given (up to a constant multiple) by

$$Z_k \underset{k\to\infty}{\sim} k^{-N/2}|\mathrm{det}Q|^{-1/2} \exp\Big(\frac{\sqrt{-1}\pi}{4}\mathrm{sign}Q\Big)$$
$$\times \sum_{m=0}^{\infty} \frac{(-1)^m}{m!k^{m/2}} \Big(\sum_{a,b,c} T_{abc} \frac{\partial}{\partial u_a}\frac{\partial}{\partial u_b}\frac{\partial}{\partial u_c}\Big)^m \exp\Big(-\frac{\sqrt{-1}}{4}{}^t\mathbf{u}Q^{-1}\mathbf{u}\Big)\Big|_{\mathbf{u}=0}. \qquad \text{(F.11)}$$

By concretely computing the derivatives with respect to the u_i's, we find that the summand for $m = 2$ in (F.11) consists of the following two sums,

$$\begin{aligned} Z_{\Gamma_1} &= \sum T_{abc}T_{a'b'c'}Q^{aa'}Q^{bb'}Q^{cc'}, \\ Z_{\Gamma_2} &= \sum T_{abc}T_{a'b'c'}Q^{ab}Q^{cc'}Q^{a'b'}. \end{aligned} \qquad \text{(F.12)}$$

We associate the trivalent graphs in Figure F.3 to the sums, where the T's (resp. the Q's) in the sums correspond to the trivalent vertices (resp. the edges) of the graphs, with the trivalent vertices and the edges connected to each other according to the subscripts in the sums.

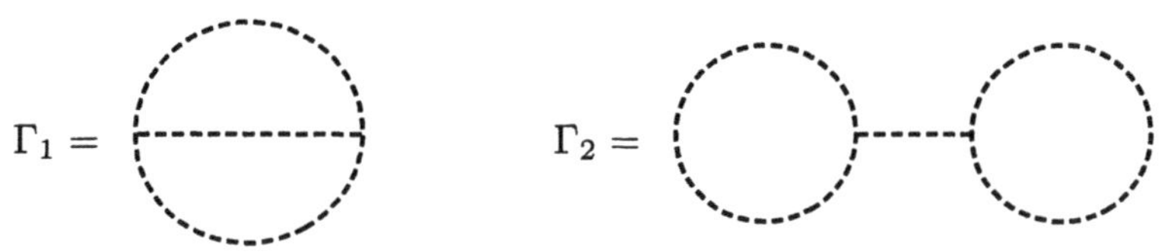

Figure F.3 Two trivalent graphs corresponding to the two sums in (F.12)

By applying similar arguments to each summand of (F.11), as m varies, we find that the sum (F.11) can be presented

$$\exp\left(\sum_{l=1}^{\infty}\frac{c^l k^{-l}}{(2l)!(3l)!}\sum_{\substack{e(\Gamma)=-l\\ \Gamma \text{ is connected}}}\frac{Z_\Gamma}{|\mathrm{Aut}(\Gamma)|}\right)$$

with a constant c, where the second sum runs over trivalent graphs whose Euler number $e(\Gamma)$ is equal to $-l$. Here, for a trivalent graph Γ the value Z_Γ is obtained by contracting $(Q^{-1})^{\otimes 3l}$ by $T^{\otimes 2l}$ according to the trivalent graph Γ, just as in the above examples of Z_{Γ_1} and Z_{Γ_2}. Further, $|\mathrm{Aut}(\Gamma)|$ denotes the order of the group $\mathrm{Aut}\Gamma$ of automorphisms of Γ.

Obtaining invariants of 3-manifolds by a perturbative expansion of the path integral

As an infinite dimensional analogue of the finite dimensional case mentioned above, we sketch, in this subsection, the formal perturbative expansion of the Chern-Simons path integral $Z_k(M)$ around the trivial connection. The expansion procedure that we now describe is known as the stationary phase approximation; see [Ati90b] and [AxSi92, AxSi94, Kon94] for such a formal perturbative expansion of the Chern-Simons path integral. By computing the summands of the formal perturbative expansion we obtain an infinite sum of trivalent graphs from which the perturbative expansion is recovered by a weight system. This is the physical background for the existence of the (primitive) LMO invariant and the recovery of the perturbative invariants from the LMO invariant through weight systems that were shown in Chapter 10.

The defining formula (F.1) of the Chern-Simons functional consists of bilinear and trilinear maps; we denote them by Q and T respectively as in the above finite dimensional case. In the present case the bilinear map $Q : \Omega^1(M;\mathfrak{g})^{\otimes 2} \to \mathbb{R}$ is given by $Q(A_1, A_2) = \int_M \mathrm{trace}(A_1 \wedge dA_2)$. Further, the trilinear map $T : \Omega^1(M;\mathfrak{g})^{\otimes 3} \to \mathbb{R}$ is given (up to a constant multiple) by $T(A_1, A_2, A_3) = \int_M \mathrm{trace}(A_1 \wedge A_2 \wedge A_3)$. As in [AxSi92, AxSi94, Kon94], after a procedure called "gauge fixing", the inverse Q^{-1} is given by $L(x,y) \otimes \tau \in \Omega^2(M \times M; \mathfrak{g} \otimes \mathfrak{g})$ for some 2-form $L(x,y)$ on $M \times M$

and the invariant 2-tensor $\tau \in \mathfrak{g} \otimes \mathfrak{g}$ which is dual to the Killing form.

The stationary phase method suggests that the asymptotic behavior of a modified Gaussian integral is determined by contributions coming from critical points of the function in the power of the integrand. Since the critical points of the Chern-Simons functional are flat connections, we expect that the asymptotic behavior of the Chern-Simons path integral $Z_k(M)$ takes the form

$$Z_k(M) \underset{k\to\infty}{\sim} \sum_{A \text{ is flat}} Z_k(M)^{(A)},$$

where the sum runs over (gauge equivalence classes of) flat connections. Further, by an infinite dimensional analogue of the finite dimensional case mentioned above, the contribution of the trivial connection is presented by

$$Z_k(M)^{(\mathrm{triv})} = \begin{pmatrix} \text{Reidemeister torsion} \\ \quad + \ \eta\text{-invariant} \end{pmatrix} \times \exp\left(\sum_{l=1}^{\infty} \frac{c^l k^{-l}}{(2l)!(3l)!} \sum_{\substack{e(\Gamma)=-l \\ \Gamma \text{ is connected}}} \frac{Z_\Gamma}{|\mathrm{Aut}(\Gamma)|} \right), \tag{F.13}$$

for some constant c, where Z_Γ is obtained by contracting $(Q^{-1})^{\otimes 3l}$ by $T^{\otimes 2l}$ according to the trivalent graph Γ as explained in the finite dimensional case.

In the first factor on the right hand side of (F.13), "Reidemeister torsion" and "η-invariant" correspond to det$Q$ and sign$Q$ in (F.11) respectively. The "η-invariant" turns out to be a factor depending on the metric of M, which is due to the fact that we need a regularization procedure to define det$Q$ and sign$Q$. Actually, this dependence was the first clue to suggest that we needed to fix a framing of a 3-manifold to rigorously define quantum invariants of it. For detailed arguments see [Ati90b].

The Z_Γ appearing in the second factor in the right hand side of (F.13) is obtained as follows. Being the term arising from the contraction of $(Q^{-1})^{\otimes 3l}$ with $T^{\otimes 2l}$ according to Γ, the summand Z_Γ is equal (up to a constant multiple) to $I_\Gamma(M)W_\mathfrak{g}(\Gamma)$, where $W_\mathfrak{g}$ is the weight system obtained by substituting $\mathfrak{g}$ into Γ (see Section 6.6 for its definition), since $W_\mathfrak{g}(\Gamma)$ is obtained by contracting $3l$ copies of the invariant 2-tensor $\tau \in \mathfrak{g} \otimes \mathfrak{g}$ with $2l$ copies of the trilinear map $\mathfrak{g}^{\otimes 3} \to \mathbb{R}$ defined by $X \otimes Y \otimes Z \mapsto \frac{1}{2}\mathrm{trace}(XYZ - ZYX)$ according to Γ. Further, $I_\Gamma(M)$ is given by

$$I_\Gamma(M) = \int_{(x_1,\cdots,x_{2l})\in M^{2l}} L(x_{i_1}, x_{j_1}) \wedge \cdots \wedge L(x_{i_{3l}}, x_{j_{3l}}). \tag{F.14}$$

This integral is obtained by contracting $3l$ copies of the 2-form $L(x,y) \in \Omega^2(M \times M)$ by $2l$ copies of the trilinear map $\Omega^1(M)^{\otimes 3} \to \mathbb{R}$ defined by $\alpha \otimes \beta \otimes \gamma \mapsto \int_M \alpha \wedge \beta \wedge \gamma$. See [BoCa98, BoCa99] for direct constructions of topological invariants of M motivated by the configuration space integral (F.14).

Axelrod-Singer [AxSi92, AxSi94] gave, in some cases, a mathematical construction for (and showed the topological invariance of) the contribution $Z_k(M)^{(A)}$ to the Chern-Simons path integral of the formal perturbative expansion around a nontrivial flat connections A. However, it is not yet known how to obtain the corresponding perturbative invariants as arithmetic limits of quantum invariants of M in the way used in Chapter 9.

Obtaining invariants of knots by a perturbative expansion of the path integral

In this subsection we explain how the formal perturbative expansion of the Chern-Simons path integral provides invariants of knots and links in $\mathbb{R}^3$. By computing summands of the formal perturbative expansion we obtain an infinite sum of graphs from which the formal perturbative expansion is recovered through a weight system. This is the physical background for the existence of the Kontsevich invariant. Further, by computing each summand of the formal perturbative expansion we obtain a configuration space integral. This is the physical background for the fact that any Vassiliev invariant can be presented by a configuration space integral, as mentioned in Section 7.5. In particular, we will explain how the linking number of a 2-component link, and also a degree 2 Vassiliev knot invariant appear as summands in the formal perturbative expansion of the Chern-Simons path integral. For references on this topic see [GMM89, GMM90, Bar91, Bar95a] for a degree 2 invariant, and [BoTa94, Koh94, ThD*99*, AlFr96, Yan*97*, Koh97, Poi*99a*, PoVi01] for further developments.

Before considering the formal perturbative expansion of the path integral for invariants of knots we consider, as a finite dimensional analogue, the following integral,

$$Z_k = \int_{\mathbf{x}\in\mathbb{R}^N} P(\mathbf{x}) \exp\Big(\sqrt{-1}k\big({}^t\mathbf{x}Q\mathbf{x} + T(\mathbf{x}\otimes\mathbf{x}\otimes\mathbf{x})\big)\Big)d\mathbf{x},$$

where we put $\mathbf{x} = \begin{pmatrix} x_1 \\ \vdots \\ x_N \end{pmatrix}$ as before, and $P(\mathbf{x})$ denotes a polynomial in the x_i's. Then, in the same fashion as the argument which obtains (F.11) from (F.9), we obtain the asymptotic behavior of the above formula as

$$\begin{aligned} Z_k \underset{k\to\infty}{\sim} k^{-N/2} \int_{\mathbf{x}'\in\mathbb{R}^N} \exp(\sqrt{-1}{}^t\mathbf{x}'Q\mathbf{x}')d\mathbf{x}' \times \sum_{m=0}^{\infty} \frac{(-\sqrt{-1})^{2m+N}}{m!k^{m/2}} \\ \times \Big(\sum_{a,b,c} T_{abc}\frac{\partial}{\partial u_a}\frac{\partial}{\partial u_b}\frac{\partial}{\partial u_c}\Big)^m P\Big(\frac{\partial}{\partial u_1},\cdots,\frac{\partial}{\partial u_N}\Big) \exp\Big(-\frac{\sqrt{-1}}{4}{}^t\mathbf{u}Q^{-1}\mathbf{u}\Big)\Big|_{\mathbf{u}=0}. \end{aligned} \tag{F.15}$$

Let us consider the summand in the asymptotic expansion corresponding to the graph Γ in Figure F.4. In the above finite dimensional case we consider the case that

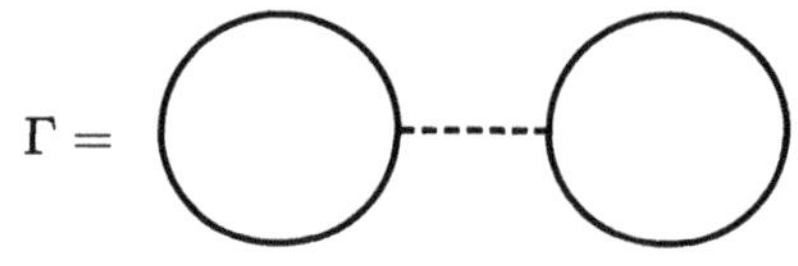

Figure F.4 The graph corresponding to the linking number

$P(\mathbf{x}) = P_1(\mathbf{x})P_2(\mathbf{x})$ for some linear maps $P_i : \mathbb{R}_N \to \mathbb{R}$, that is, the case that $P(\mathbf{x})$ is a homogeneous polynomial in x_i's of degree 2. In this case, by concretely computing the derivatives with respect to the u_i's, we find that the summand corresponding to $m = 0$ is equal, up to a constant multiple, to $(P_1 \otimes P_2)(Q^{-1})$.

Now let us consider the corresponding summand in the formal asymptotic expansion of the following path integral,

$$Z_k(\mathbb{R}^3, L) = \int_{\mathcal{A}/\mathcal{G}} \mathrm{trace}_{R_1}\mathrm{Hol}_{L_1}(A)\mathrm{trace}_{R_2}\mathrm{Hol}_{L_2}(A)\exp\left(2\pi\sqrt{-1}k\mathrm{CS}(A)\right)\mathcal{D}A,$$

for a link $L = L_1 \cup L_2$ in $\mathbb{R}^3$, by analogy with the above finite dimensional case. In this case the bilinear map $Q : \Omega^1(\mathbb{R}^3; \mathfrak{g})^{\otimes 2} \to \mathbb{R}$ is given by $Q(A_1, A_2) = \int_{\mathbb{R}^3} \mathrm{trace}(A_1 \wedge dA_2)$. Its inverse Q^{-1} is given as follows. Let ω be the normalized volume form of S^2 and let $L(\mathbf{x}, \mathbf{y})$ be the 2-form on $\mathbb{R}^3 \times \mathbb{R}^3$ obtained by pulling back ω by the map $(\mathbf{x}, \mathbf{y}) \mapsto (\mathbf{x} - \mathbf{y})/|(\mathbf{x} - \mathbf{y})|$, noting that $L(\mathbf{x}, \mathbf{y})$ is singular along the diagonal set of $\mathbb{R}^3 \times \mathbb{R}^3$. We put Q^{-1} to be $L(\mathbf{x}, \mathbf{y}) \otimes \tau \in \Omega^2(\mathbb{R}^3 \times \mathbb{R}^3, \mathfrak{g} \otimes \mathfrak{g})$ with the invariant 2-tensor $\tau \in \mathfrak{g} \otimes \mathfrak{g}$ which is dual to the Killing form. Further, we consider the linear map $P_i : \Omega^1(\mathbb{R}^3; \mathfrak{g}) \to \mathbb{R}$ given by $P_i(A) = \int_0^1 \gamma_i^\star A(t)dt$, which is the summand of degree 1 in (F.4). Here, each $\gamma_i : [0,1] \to \mathbb{R}^3$ is a closed path whose image is equal to L_i. Then, the summand $(P_1 \otimes P_2)(Q^{-1})$ in the formal asymptotic expansion of the path integral is equal to $\mathrm{trace}_{R_1 \otimes R_2}(\tau) \cdot I_\Gamma(L)$, where $I_\Gamma(L)$ is given by

$$\begin{aligned} I_\Gamma(L) &= \int_{[0,1]\times[0,1]} (\gamma_1 \times \gamma_2)^\star L(\mathbf{x}, \mathbf{y}) \\ &= \int_{S^1 \times S^1} \phi_\Gamma^\star \omega. \end{aligned} \tag{F.16}$$

Here, the map $\phi_\Gamma : S^1 \times S^1 \to S^2$ is given by

$$\phi_\Gamma(x, y) = \frac{\gamma_1(x) - \gamma_2(y)}{|\gamma_1(x) - \gamma_2(y)|}.$$

Hence, noting that (F.16) is the Gauss formula for the linking number (see, *e.g.*, [Spi79]), we find that $I_\Gamma(L)$ is equal to the linking number of L_1 and L_2, This implies that the linking number (as an isotopy invariant of 2-component links) arises as a summand in the formal perturbative expansion of the Chern-Simons path integral.

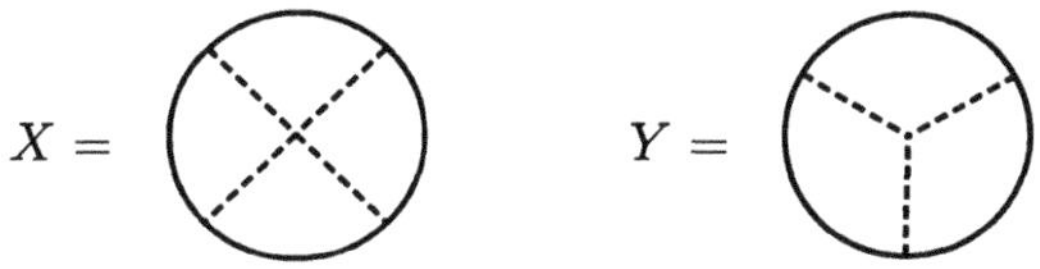

Figure F.5 Two graphs associated to the summands Z_X and Z_Y

Next we consider some other summands in the asymptotic expansion (F.15) of the finite dimensional integral again. In (F.15) we put $P(\mathbf{x}) = \sum P^{(n)}(\mathbf{x})$ where $P^{(n)}(\mathbf{x})$ denotes the homogeneous part of $P(\mathbf{x})$ of degree n, which we write $P^{(n)}(\mathbf{x}) = \sum P_{a_1 \cdots a_n} x^{a_1} \cdots x^{a_n}$. We consider the following summands in (F.15),

$$Z_X = \sum P_{abcd} Q^{ac} Q^{bd},$$
$$Z_Y = \sum T_{abc} P_{a'b'c'} Q^{aa'} Q^{bb'} Q^{cc'},$$

which are obtained by concretely computing the derivatives with respect to the u_i's in (F.15). In a similar way as the correspondence between the graphs Γ_i's in Figure F.3 and the summands Z_{Γ_i}'s mentioned previously, we associate graphs X and Y in Figure F.5 to the above summands.

Let us consider the corresponding summands in the formal asymptotic expansion of the following path integral,

$$Z_k(\mathbb{R}^3, K) = \int_{\mathcal{A}/\mathcal{G}} \text{trace}_R \text{Hol}_K(A) \exp\left(2\pi\sqrt{-1}k\text{CS}(A)\right) \mathcal{D}A, \tag{F.17}$$

for a knot K in $\mathbb{R}^3$, by analogy with the finite dimensional case. As mentioned earlier, we put Q^{-1} to be $L(\mathbf{x}, \mathbf{y}) \otimes \tau$. Further, just as in the 3-manifold case the trilinear map T is given by $T(A_1, A_2, A_3) = \text{trace} \int_{\mathbb{R}^3} A_1 \wedge A_2 \wedge A_3$ for $A_i \in \Omega^1(\mathbb{R}^3; \mathfrak{g})$. Furthermore, as the summand of degree n in (F.4) we put

$$P^{(n)}(A) = \text{trace}_R \int_{0 \le t_1 \le \cdots \le t_n \le 1} \gamma^* A(t_1) \cdots \gamma^* A(t_n),$$

where $\gamma : [0,1] \to \mathbb{R}^3$ is a closed path whose image is equal to K. Then, the summands Z_X and Z_Y in the formal asymptotic expansion of the path integral are equal to $W_{\mathfrak{g},R}(X) \cdot I_X(K)$ and $W_{\mathfrak{g},R}(Y) \cdot I_Y(K)$, where $I_X(K)$ and $I_Y(K)$ are given by

$$\begin{aligned} I_X(K) &= \int_{\Delta_4} (\gamma \times \gamma \times \gamma \times \gamma)^* (L(\mathbf{x}_1, \mathbf{x}_3) \wedge L(\mathbf{x}_2, \mathbf{x}_4)) \\ &= \frac{1}{4} \int_{C_X} \phi_X^* (\omega \times \omega), \end{aligned} \tag{F.18}$$

$$\begin{aligned} I_Y(K) &= \int_{\mathbb{R}^3\times\Delta_3} (\mathrm{id}_{\mathbb{R}^3}\times\gamma\times\gamma\times\gamma)^\star\big(L(\mathbf{x},\mathbf{x}_1)\wedge L(\mathbf{x},\mathbf{x}_2)\wedge L(\mathbf{x},\mathbf{x}_3)\big) \\ &= \frac{1}{3}\int_{C_Y} \phi_Y^\star(\omega\times\omega\times\omega), \end{aligned} \tag{F.19}$$

where Δ_n denotes the n-simplex determined by $0 \le t_1 \le \cdots \le t_n \le 1$ in $[0,1]^n$, and the notations in (F.18) and (F.19) are given in Section 7.5. As mentioned in Section 7.5, a certain linear sum of $I_X(K)$ and $I_Y(K)$ is a Vassiliev invariant of degree 2. This implies that we can obtain a Vassiliev invariant of degree 2 as a summand of a formal perturbative expansion of the Chern-Simons path integral.

The other summands in the formal perturbative expansion of (F.17) can be computed similarly as above. We obtain a summand Z_Γ for a graph Γ, and Z_Γ is equal, up to a constant multiple, to $I_\Gamma(K)W_{\mathfrak{g},R}(\Gamma)$, where $I_\Gamma(K)$ can be presented by a configuration space integral in a similar way as above. Further, we can express all Vassiliev invariants of knots as linear sums of the $I_\Gamma(K)$'s; see [Th$_{\mathrm{D}}$*99*, AlFr96, Yan*97*, Poi*99a*]. This indicates that we have an infinite sum of graphs from which the formal perturbative expansion of (F.17) is recovered by a weight system. This is the physical background for the existence of the Kontsevich invariant and the recovery of the quantum invariants of knots from the Kontsevich invariant through weight systems that was the subject of Chapter 6.

F.4 Conformal field theory by Wess-Zumino-Witten model

Conformal field theory is a 2-dimensional quantum field theory on Riemann surfaces, which is invariant under conformal (*i.e.*, holomorphic) transformations; see [Uen97, Sch97] for detailed expositions of conformal field theory. In this section we sketch an explanation of the fact (following [Gaw90a] and [Koh98]) that a conformal field theory is (formally) induced from the partition function of the Wess-Zumino-Witten model. As a differential equation satisfied by the partition function we obtain the Knizhnik-Zamolodchikov equation [KnZa84] which was introduced in Chapter 5.

The Wess-Zumino-Witten model

We call a complex 1-manifold a *Riemann surface.* We begin by reviewing differentials and differential forms on a Riemann surface. For a local coordinate z of a Riemann surface we can write $z = x + \sqrt{-1}y$ and $\overline{z} = x - \sqrt{-1}y$ for real coordinates x and y. We define differentials and differential forms on a Riemann surface by

$$\frac{\partial}{\partial z} = \frac{1}{2}\Big(\frac{\partial}{\partial x} - \sqrt{-1}\frac{\partial}{\partial y}\Big), \qquad dz = dx + \sqrt{-1}dy,$$
$$\frac{\partial}{\partial \overline{z}} = \frac{1}{2}\Big(\frac{\partial}{\partial x} + \sqrt{-1}\frac{\partial}{\partial y}\Big), \qquad d\overline{z} = dx - \sqrt{-1}dy.$$

Further, we define differential operators ∂ and $\overline{\partial}$ by

$$\partial f = \frac{\partial f}{\partial z} dz, \qquad \overline{\partial} f = \frac{\partial f}{\partial \overline{z}} d\overline{z},$$

for a smooth function f on a Riemann surface. The operators ∂ and $\overline{\partial}$ are well defined, that is, are independent of the choice of the local coordinate z.

Let Σ be a closed Riemann surface, and let G be a semi-simple complex Lie group. (The assumption of the complexity of G is not used in this subsection; it will be used later, in the subsection that follows the next one.) We define the *energy* of a smooth map $f : \Sigma \to G$ by

$$E_\Sigma(f) = \frac{\sqrt{-1}}{4\pi} \int_\Sigma \mathrm{trace}(f^{-1}\partial f \wedge f^{-1}\overline{\partial} f).$$

The energy E_Σ is a functional $\mathrm{Map}(\Sigma, G) \to \mathbb{C}$, where $\mathrm{Map}(\Sigma, G)$ denotes the space of smooth maps $\Sigma \to G$. The *Wess-Zumino-Witten functional* $S_\Sigma : \mathrm{Map}(\Sigma, G) \to \mathbb{C}/2\pi\sqrt{-1}\mathbb{Z}$ is defined by

$$S_\Sigma(f) = E_\Sigma(f) + \frac{\sqrt{-1}}{12\pi} \int_M \mathrm{trace}(\tilde{f}^{-1}d\tilde{f} \wedge \tilde{f}^{-1}d\tilde{f} \wedge \tilde{f}^{-1}d\tilde{f}) \quad \in \mathbb{C}/2\pi\sqrt{-1}\mathbb{Z},$$

for each smooth map $f : \Sigma \to G$, where M is a 3-manifold bounded by Σ, and $\tilde{f}$ is a map $M \to G$ extending f. It follows from Stokes' theorem that the second term on the right hand side is determined by the homological type of $\tilde{f}$; in fact it is equal to

$$2\pi\sqrt{-1} \int_M \tilde{f}^\star \sigma,$$

for a certain 3-form σ on G whose cohomology class is integral in $H^3(G)$. Hence, similarly as the second term in (F.7), it is independent (modulo $2\pi\sqrt{-1}\mathbb{Z}$) of the choice of M and $\tilde{f}$. We define a product of fg of $f, g \in \mathrm{Map}(\Sigma, G)$ by $(fg)(z) = f(z)g(z)$ for $z \in \Sigma$. It follows from a concrete computation [PoWi84] that $S_\Sigma(fg)$ is presented by

$$S_\Sigma(fg) = S_\Sigma(f) + S_\Sigma(g) + \frac{\sqrt{-1}}{2\pi} \int_\Sigma \mathrm{trace}(f^{-1}\overline{\partial} f \wedge \partial g g^{-1}). \tag{F.20}$$

The partition function of the Wess-Zumino-Witten model is formally given by the following path integral,

$$Z_k^{\mathrm{WZW}}(\Sigma) = \int_{f \in \mathrm{Map}(\Sigma, G)} \exp\Big(k S_\Sigma(f)\Big) \mathcal{D}f,$$

for any positive integer k. Note that, since the space $\mathrm{Map}(\Sigma, G)$ is infinite dimensional, this path integral can not yet be justified mathematically. We speculate that this path integral gives an "invariant" of a Riemann surface Σ (*i.e.*, an "invariant" of complex structures of an underlying surface).

The operator formalism for the Wess-Zumino-Witten path integral

We now consider whether there is an operator formalism (Hamiltonian formulation) for the "invariant" of the Wess-Zumino-Witten path integral, *i.e.*, we consider how to formulate the path integral $Z_k^{\rm WZW}(\Sigma)$ for Σ with non-empty boundary.

For simplicity, we will consider the case that Σ is a Riemann surface whose boundary is S^1. For a smooth map $f : \Sigma \to G$ we formulate $\exp\big(S_\Sigma(f)\big)$ as follows. Let LG denote the *loop group* $\mathrm{Map}(S^1, G)$ of G. We construct a complex line bundle $\mathcal{L}$ on LG as the quotient space of $\mathbb{C} \times \big(\bigcup_\Sigma \mathrm{Map}(\Sigma, G)\big)$, where the union runs over Riemann surfaces whose boundary is S^1, by the equivalence relation that $(\alpha, f) \sim (\alpha', f')$ for $f : \Sigma \to G$ and $f' : \Sigma' \to G$ if and only if the following two equalities hold.

- $f|_{S^1} = f'|_{S^1}$, where the sides denote the restrictions of f and f' to S^1 respectively.
- $\alpha = \alpha' \exp\big(\int_M g^\star \sigma\big)$, where M is a 3-manifold whose boundary is $(-\Sigma) \cup \Sigma'$ and g is an extension of $f \cup f'$.

Then, $\mathcal{L} \to LG$ is a complex line bundle with the projection given by $(\alpha, f) \mapsto f|_{S^1}$. We define $\exp\big(S_\Sigma(f)\big)$ to be $\Big(\exp\big(E_\Sigma(f)\big), f\Big) \in \mathcal{L}_\gamma$, where we put $\gamma = f|_{S^1}$ and let $\mathcal{L}_\gamma$ denote the fiber of $\mathcal{L}$ at $\gamma \in LG$.

The above defined $\exp\big(S_{\Sigma_1}(f_1)\big)$ for Σ_1 bounded by S^1 is well defined, in the following sense. We consider closed Riemann surfaces $\Sigma = \Sigma_1 \cup \Sigma_2$ and $\Sigma' = \Sigma_1' \cup \Sigma_2$ for Riemann surfaces Σ_1, Σ_1', Σ_2 with boundaries $\partial\Sigma_1 = \partial\Sigma_1' = S^1 = -\partial\Sigma_2$. Further, we consider maps $f : \Sigma \to G$ and $f' : \Sigma' \to G$ such that $f = f_1 \cup f_2$ and $f' = f_1' \cup f_2$ for maps f_1, f_1', f_2 on $\Sigma_1, \Sigma_1', \Sigma_2$. Suppose that $\exp\big(S_\Sigma(f_1)\big) = \exp\big(S_\Sigma(f_1')\big)$, *i.e.*, $\Big(\exp\big(E_{\Sigma_1}(f_1)\big), f_1\Big) \sim \Big(\exp\big(E_{\Sigma_1}(f_1')\big), f_1'\Big)$. Then, we show below that $\exp\big(S_\Sigma(f)\big) = \exp\big(S_\Sigma(f')\big)$; in this sense $\exp\big(S_\Sigma(f_1)\big)$ is well defined. By the above equivalence we have that

$$E_{\Sigma_1}(f_1) = E_{\Sigma_1'}(f_1') + 2\pi\sqrt{-1}\int_{M_1} \tilde{f}_1^\star \sigma, \tag{F.21}$$

where M_1 is a 3-manifold bounded by $(-\Sigma_1) \cup \Sigma_1'$ and $\tilde{f}_1 : M_1 \to G$ is a map extending $f_1 \cup f_1'$. Let M be a 3-manifold bounded by $\Sigma = \Sigma_1 \cup \Sigma_2$. We put $M' = M \cup_{\Sigma_1} M_1$, which is a 3-manifold bounded by $\Sigma' = \Sigma_1' \cup \Sigma_2$. Further, let $\tilde{f} : M \to G$ be an extension of $f = f_1 \cup f_2$. We put $\tilde{f}'^\star = \tilde{f} \cup \tilde{f}_1$, which is an extension of $f' = f_1' \cup f_2$. Then,

$$\begin{aligned} S_{\Sigma'}(f') &= E_{\Sigma'}(f') + 2\pi\sqrt{-1}\int_{M'} \tilde{f}'^\star \sigma \\ &= E_{\Sigma_1'}(f_1') + E_{\Sigma_2}(f_2) + 2\pi\sqrt{-1}\int_M \tilde{f}^\star\sigma + 2\pi\sqrt{-1}\int_{M_1} \tilde{f}_1^{\,\star}\sigma \\ &= E_{\Sigma_1}(f_1) + E_{\Sigma_2}(f_2) + 2\pi\sqrt{-1}\int_M \tilde{f}^\star\sigma \end{aligned}$$

$$= E_\Sigma(f) + 2\pi\sqrt{-1}\int_M \tilde{f}^\star\sigma = S_\Sigma(f) \quad \in \mathbb{C}/2\pi\sqrt{-1}\mathbb{Z},$$

where we obtain the third equality by (F.21). Hence, $\exp\big(S_\Sigma(f)\big) = \exp\big(S_\Sigma(f')\big)$. This completes the explanation of a reason why $\exp\big(S_{\Sigma_1}(f_1)\big)$ is well defined.

Let Σ be a Riemann surface bounded by S^1. For a positive integer k and $\gamma \in LG$ we consider the path integral

$$Z_k^{\mathrm{WZW}}(\Sigma)(\gamma) = \int_{f\in \mathrm{Map}(\Sigma,G)_\gamma} \exp\Big(kS_\Sigma(f)\Big)\mathcal{D}f,$$

where $\mathrm{Map}(\Sigma, G)_\gamma$ denotes the space of smooth maps $\Sigma \to G$ whose restrictions to S^1 are equal to γ. Here, we regard $\exp\big(kS_\Sigma(f)\big)$ as lying in $\mathcal{L}_\gamma^{\otimes k}$. Thus, we can regard $Z_k^{\mathrm{WZW}}(\Sigma)$ as a section of the line bundle $\mathcal{L}^{\otimes k} \to LG$.

Actions on the line bundle $\mathcal{L}^{\otimes k}$

In this subsection we give an action of a central extension of LG on $\mathcal{L}^{\otimes k}$. This induces actions of holomorphic maps $\Sigma \to G$ on the space of sections of $\mathcal{L}^{\otimes k} \to LG$ which leave $Z_k^{\mathrm{WZW}}(\Sigma)$ invariant.

We introduce a product $\mathcal{L}_{\gamma_1} \times \mathcal{L}_{\gamma_2} \to \mathcal{L}_{\gamma_1\gamma_2}$ for $\gamma_1, \gamma_2 \in LG$ by

$$(\alpha_1, f_1) \times (\alpha_2, f_2) = \Big(\alpha_1\alpha_2 \exp\big(\delta(f_1, f_2)\big), f_1 f_2\Big),$$

for $\alpha_1, \alpha_2 \in \mathbb{C}$ and $f_1, f_2 \in \mathrm{Map}(\Sigma, G)$ such that Σ is a Riemann surface bounded by S^1 and $f_i|_{S^1} = \gamma_i$ $(i = 1, 2)$. Here, we put

$$\delta(f_1, f_2) = -E_\Sigma(f_1) - E_\Sigma(f_2) + E_\Sigma(f_1 f_2) - \frac{\sqrt{-1}}{2\pi}\int_\Sigma \mathrm{trace}(f_1^{-1}\bar{\partial} f_1 \wedge \partial f_2 f_2^{-1}).$$

The product is well defined with respect to the equivalence relation in the definition of $\mathcal{L}$, for the following reason. Suppose that $(\alpha_i, f_i) \sim (\alpha_i', f_i')$ for $i = 1, 2$. Then,

$$\alpha_i = \alpha_i' \exp\Big(2\pi\sqrt{-1}\int_M \tilde{f}_i^\star\sigma\Big), \tag{F.22}$$

where M is a 3-manifold bounded by $(-\Sigma) \cup \Sigma$, and $\tilde{f}_i : M \to G$ is an extension of $f_i \cup f_i'$ for each i. From the definition of the Wess-Zumino-Witten functional we have that

$$2\pi\sqrt{-1}\int_M \tilde{f}_i^{\,\star}\sigma = S_{(-\Sigma)\cup\Sigma}(f_i \cup f_i') - E_{(-\Sigma)\cup\Sigma}(f_i \cup f_i'). \tag{F.23}$$

It is sufficient to show the following equivalence,

$$\Big(\alpha_1\alpha_2 \exp\big(\delta(f_1, f_2)\big), f_1 f_2\Big) \sim \Big(\alpha_1'\alpha_2' \exp\big(\delta(f_1', f_2')\big), f_1' f_2'\Big),$$

which is rewritten, using the definition of the equivalence relation, as

$$\alpha_1\alpha_2 \exp\left(\delta(f_1, f_2)\right) = \alpha_1'\alpha_2' \exp\left(\delta(f_1', f_2') + 2\pi\sqrt{-1}\int_M \tilde{f_{12}}^* \sigma\right), \tag{F.24}$$

where $\tilde{f}_{12} : M \to G$ is an extension of $f_1 f_2 \cup f_1' f_2'$. Similarly as for (F.23) we have that

$$2\pi\sqrt{-1}\int_M \tilde{f_{12}}^* \sigma = S_{(-\Sigma)\cup\Sigma}(f_1 f_2 \cup f_1' f_2') - E_{(-\Sigma)\cup\Sigma}(f_1 f_2 \cup f_1' f_2').$$

Hence, by (F.22), (F.23), and the above formula, the proof of (F.24) is reduced to showing that

$$\begin{aligned}
&S_{(-\Sigma)\cup\Sigma}(f_1 \cup f_1') - E_{(-\Sigma)\cup\Sigma}(f_1 \cup f_1') \\
&+ S_{(-\Sigma)\cup\Sigma}(f_2 \cup f_2') - E_{(-\Sigma)\cup\Sigma}(f_2 \cup f_2') + \delta(f_1, f_2) \\
&= \delta(f_1', f_2') + S_{(-\Sigma)\cup\Sigma}(f_1 f_2 \cup f_1' f_2') - E_{(-\Sigma)\cup\Sigma}(f_1 f_2 \cup f_1' f_2').
\end{aligned}$$

This is obtained by (F.20) and from the definition of δ. This completes the demonstration that the product on the fibers of $\mathcal{L}$ is well defined.

For a fixed positive integer k, let $\widehat{LG}$ denote the space obtained from $\mathcal{L}^{\otimes k}$ by removing the image of the zero section. Then, $\widehat{LG} \to LG$ is a fiber bundle whose fiber is $\mathbb{C}^\times = \mathbb{C} - \{0\}$. The product just defined induces a group structure on $\widehat{LG}$. Furthermore, the group $\widehat{LG}$ is a central extension of LG, as

$$1 \longrightarrow \mathbb{C}^\times \longrightarrow \widehat{LG} \longrightarrow LG \longrightarrow 1.$$

From the product of $\widehat{LG}$ we obtain an action of $\widehat{LG}$ on $\mathcal{L}^{\otimes k}$. It induces an action of $\widehat{LG}$ on the vector space of sections of $\mathcal{L}^{\otimes k}$.

In particular, if G is a simply connected simple Lie group, this central extension $\widehat{LG}$ can alternatively be described as follows. It is known in this case that $H_2(G;\mathbb{Z}) = 0$ and $H_3(G;\mathbb{Z}) \cong \mathbb{Z}$; hence $\pi_2(G) = 0$ and $\pi_3(G) \cong \mathbb{Z}$. As a topological space LG is homeomorphic to $G \times \Omega G$, where ΩG denotes the subspace of LG consisting of $\gamma \in LG$ with $\gamma(t_0) = 1 \in G$ for a fixed base point $t_0 \in S^1$. Then,

$$\pi_n(LG) = \pi_n(G) \times \pi_n(\Omega G) = \pi_n(G) \times \pi_{n+1}(G).$$

Hence, $\pi_1(LG) = 1$ and $\pi_2(LG) \cong \mathbb{Z}$. Therefore, $H^2(LG;\mathbb{Z}) \cong \mathbb{Z}$. The central extension $\widehat{LG}$ is equal to the central extension of LG associated with the class in $H^2(LG;\mathbb{Z})$ which is equal to k times the generator of $H^2(LG;\mathbb{Z})$; note that the generator is given by the first Chern class of $\mathcal{L}$.

Now we use the assumption that G is a complex Lie group. Let h be a holomorphic map $\Sigma \to G$. Then, $\overline{\partial} h = 0$, and hence, $E_\Sigma(h) = 0$. It follows from an elementary computation that the following formula holds,

$$(1, h) \times \left(\exp\left(E_\Sigma(f)\right), f\right) = \left(\exp\left(E_\Sigma(hf)\right), hf\right),$$

for any $f \in \mathrm{Map}(\Sigma, G)$. Hence,

$$(1, h) \times Z_k^{\mathrm{WZW}}(\Sigma)(\gamma) = Z_k^{\mathrm{WZW}}(\Sigma)\big((h|_{S^1}) \cdot \gamma\big).$$

We consider an action of such a holomorphic map h on $\widehat{LG}$ by the left multiplication of $(1, h)$, which induces an action of h on LG by the left multiplication of $h|_{S^1}$. Through such actions the holomorphic map h also acts on the space of sections of $\mathcal{L}^{\otimes k}$. The above formula leads us to speculate that $Z_k^{\mathrm{WZW}}(\Sigma)$, formally, gives a section of $\mathcal{L}^{\otimes k}$ invariant under the action of such holomorphic maps.

Consider the limit as a boundary component tends to an infinitesimal S^1. At this limit, we have an action of the Lie algebra of $\widehat{LG}$ (known as the affine algebra $\hat{\mathfrak{g}}$) on the vector space of such invariant sections of $\mathcal{L}^{\otimes k}$. Hence, the vector space splits into a direct sum of irreducible representations of $\hat{\mathfrak{g}}$. Further, we consider the case that Σ is an n-holed S^2 with a complex structure at the limit as each boundary component tends to an infinitesimal S^1. In this case Σ can be regarded as the complex projective line $\mathbb{C}P^1$ $(= \mathbb{C} \cup \{\infty\})$ with n marked points. Similarly as in the above argument we have an action of the product of n copies of $\hat{\mathfrak{g}}$ on the vector space of invariant sections of $\mathcal{L}^{\otimes k}$. Hence, the vector space splits into a direct sum of tensor products of irreducible representations of $\hat{\mathfrak{g}}$ (we put such a tensor product to be the dual of $\hat{R}_1 \otimes \cdots \otimes \hat{R}_n$ in the next subsection). The invariant subspace of such a direct summand under the action of the holomorphic maps is called a *space of conformal blocks*. We discuss its mathematical construction in the next subsection.

Mathematical construction of a space of conformal blocks

Based on the previous subsection's observation, from the viewpoint of mathematical physics, we now give a sketch of a mathematical formulation of the "space of conformal blocks".

Let G be a compact semi-simple Lie group, and let $\mathfrak{g}$ be its Lie algebra. The complexification of the loop algebra $L\mathfrak{g}$ is approximately regarded as $\mathfrak{g} \otimes \mathbb{C}((t))$, where $\mathbb{C}((t))$ denotes the algebra of Laurent power series in an indeterminate t. As a central extension of it, we define the *affine Lie algebra* $\hat{\mathfrak{g}}$ by $\hat{\mathfrak{g}} = \mathfrak{g} \otimes \mathbb{C}((t)) \oplus \mathbb{C}c$ with an indeterminate vector c, where the Lie bracket of $\hat{\mathfrak{g}}$ is given by

$$[X \otimes t^n, Y \otimes t^m] = [X, Y] \otimes t^{n+m} + B(X, Y) n \delta_{n+m,0} \cdot c \qquad \text{for any } X, Y \in \mathfrak{g}, \tag{F.25}$$

$$[c, \xi] = 0 \qquad \text{for any } \xi \in \hat{\mathfrak{g}}.$$

Here, $B(X, Y)$ denotes the Killing form of $\mathfrak{g}$, and we put $\delta_{a,b} = 1$ if $a = b$ and 0 otherwise.

We will now briefly review the theory of modules of $\mathfrak{g}$ and $\hat{\mathfrak{g}}$; see [Kac90] for the terminology used in this paragraph. For a fixed positive integer k let P_k denote the set of dominant integral weights λ with $0 \leq (\theta, \lambda) \leq k$, where θ is the highest root and $(\cdot,\cdot)$ denotes the scalar multiple of the Cartan-Killing form normalized by $(\theta,\theta)=2$. It is known (see [Kac90]) that, if R is an irreducible $\mathfrak{g}$-module with the highest weight $\lambda \in P_k$, then R induces an infinite dimensional irreducible $\hat{\mathfrak{g}}$-module $\hat{R}$ (called the *integrable highest weight module* of level k with the highest weight λ) such that the central element $c \in \hat{\mathfrak{g}}$ acts on $\hat{R}$ by multiplication by k and for any $v \in \hat{R}$ and $X \in \mathfrak{g}$ there exists a sufficiently large n such that $(X \otimes t^n)(v) = 0$.

Let $z = (z_1, \cdots, z_n)$ be an n-tuple of n distinct points $z_1, \cdots, z_n$ in $\mathbb{C}P^1$. To the n points we associate irreducible $\mathfrak{g}$-modules $R_1, \cdots, R_n$ whose highest weights are in P_k. Let $\hat{R}_1, \cdots, \hat{R}_n$ be the integrable highest weight $\hat{\mathfrak{g}}$-modules of level k induced by $R_1, \cdots, R_n$ respectively. We denote by $H^0(\mathbb{C}P^1; z)$ the vector space of meromorphic functions whose poles are (at most) at $z_1, \cdots, z_n$. For a local coordinate t_i around z_i we have the map

$$\mathfrak{g} \otimes H^0(\mathbb{C}P^1; z) \longrightarrow \mathfrak{g} \otimes \mathbb{C}((t_i)),$$

for each i, which is obtained by expanding such meromorphic functions around the z_i. Through this map we have an action of $\mathfrak{g} \otimes H^0(\mathbb{C}P^1; z)$ on each $\hat{R}_i$. This induces an action of $\mathfrak{g} \otimes H^0(\mathbb{C}P^1; z)$ on $\hat{R}_1 \otimes \cdots \otimes \hat{R}_n$. We define the *space of conformal blocks* by

$$\mathcal{H}_{z;R_1,\cdots,R_n} = \mathrm{Hom}_{\mathfrak{g}\otimes H^0(\mathbb{C}P^1;z)}(\hat{R}_1 \otimes \cdots \otimes \hat{R}_n, \mathbb{C}),$$

where the right hand side denotes the vector space of linear maps invariant under the action of $\mathfrak{g} \otimes H^0(\mathbb{C}P^1; z)$ on $\hat{R}_1 \otimes \cdots \otimes \hat{R}_n$.

Isomorphism to quantum Hilbert space

In this subsection we explain that the space of conformal blocks defined above is isomorphic to the quantum Hilbert space of $\mathbb{C}P^1$ with marked points that was mentioned in Section F.2. This was shown in [BeLa94] (as speculated in [Wit89a]).

An outline of the construction of the isomorphism is as follows. For a Riemann surface Σ with marked points, we consider *parabolic vector bundles* on Σ, which are holomorphic vector bundles on Σ whose fibers at the marked points of Σ have flag structures. Here, a *flag structure* on a vector space V is a descending series of vector subspaces of V such that the codimension of each adjacent pair of subspaces is equal to 1. We shall explain (as shown in [BeLa94]), by means of an infinite dimensional analogue of the Borel-Weil theorem, that the space of conformal blocks for Σ is isomorphic to the quantization of the moduli space of (semi-stable) parabolic vector bundles on Σ. Further, we explain (see [AtBo82]) that the moduli space of (semi-stable) parabolic vector bundles can be identified with the moduli space of flat G connections on Σ. Then, from the definition of quantum Hilbert space $\mathcal{H}_\Sigma$ we obtain the required isomorphism.

Let us review the Borel-Weil theorem. Let G be a simple complex Lie group and B a Borel subgroup of G (*i.e.*, a maximal solvable Lie subgroup of G). For example, when $G = SL(n, \mathbb{C})$, the subgroup consisting of upper half matrices is a Borel subgroup of G. The quotient G/B is called a *flag variety*. The Borel-Weil theorem asserts that each irreducible holomorphic G-module can be obtained as the space of holomorphic sections of a holomorphic line bundle on the flag variety G/B, where the line bundle is obtained from the natural principal B-bundle $G \to G/B$ by the 1-dimensional representation of B corresponding to the highest weight of the irreducible G-module.

We consider an infinite dimensional analogue of the Borel-Weil theorem, as follows. Let G be a simple complex Lie group, B a Borel subgroup of G again, and LG the loop group of G. Let B' denote the set of $f|_{S^1}$ for holomorphic maps $f : D^2 \to G$ satisfying $f(0) \in B$, where D^2 denotes the unit disc in $\mathbb{C}$ whose boundary is denoted by S^1, and $f|_{S^1}$ denotes the restriction of f to S^1. It is known (see [PrSe86]), as an infinite dimensional analogue of the Borel-Weil theorem, that any irreducible $\widehat{LG}$-module (hence, any $\hat{\mathfrak{g}}$-module) can be obtained as the space of holomorphic sections of a holomorphic line bundle on the flag variety LG/B', where $\widehat{LG}$ is a central extension of LG, and the line bundle is obtained from the character of B' corresponding to the highest weight of the $\widehat{LG}$-module.

We now explain, following [BeLa94], why the above mentioned analogue of the Borel-Weil theorem implies that the space of conformal blocks for a Riemann surface with marked points is isomorphic to the quantization of the moduli space of (semi-stable) parabolic vector bundles on the Riemann surface. For simplicity, we only consider the case that the Riemann surface Σ has one marked point P. Let D be a disc neighborhood of P whose center is P, and let Σ' be the complement of D. Then, Σ is the union of D and Σ' along S^1. Since any holomorphic vector bundle on an open Riemann surface is trivial, any holomorphic vector bundle on Σ can be obtained by gluing trivial vector bundles on D and on Σ' by an element of LG. We now consider the equivalence relation on elements of LG given by declaring elements to be equivalent of their associated parabolic vector bundles on Σ are isomorphic. Since isomorphisms among trivial vector bundles on Σ' are described by holomorphic maps $\Sigma' \to G$, two elements of LG related by an element of H give isomorphic vector bundles, where H is the subgroup of LG consisting of restrictions of holomorphic maps $\Sigma' \to G$ to S^1. A similar argument on D induces the above mentioned subgroup B' of LG, by noting that the flag at P is preserved by this Borel subgroup of G. Hence, the moduli space of (semi-stable) parabolic vector bundles on Σ can be identified with the double coset $H\backslash LG/B'$. Further, by the infinite dimensional analogue of the Borel-Weil theorem mentioned above, any $\hat{\mathfrak{g}}$-module is obtained as the space of holomorphic sections of a line bundle on LG/B'. Recall that we defined the space of conformal blocks to be the H-invariant subspace of (the dual of) the space of such holomorphic sections. Therefore, the space of conformal blocks is isomorphic to the quantization of the moduli space of (semi-stable) parabolic vector bundles on Σ.

The remaining step is to explain the identification between the moduli space of (semi-stable) parabolic vector bundles on a Riemann surface Σ and the moduli space of flat G connections on Σ. The identification is known as the Narasimhan-Seshadri theorem [NaSe65] in the case of a closed Riemann surface, and as the Mehta-Seshadri theorem [MeSe80] in the case of a closed Riemann surface with some marked points. We will explain this theorem, for simplicity, only in the case of a Riemann surface with no marked points, by sketching an explicit bijection (see [AtBo82] for details). To a flat G connection, there is canonically associated a flat (hence, holomorphic) vector bundle. Conversely, when we have a holomorphic vector bundle, we can (uniquely) choose a hermitian metric on each fiber of the vector bundle such that the curvature of the corresponding hermitian connection is equal to a constant multiple of the Kähler metric of the Riemann surface. By taking a tensor product with some line bundle, we can arrange for this curvature to become equal to 0. Thus, we obtain a flat vector bundle (up to tensor product of such a line bundle). In this way we obtain the required bijective correspondence between (semi-stable) holomorphic vector bundles and flat G connections.

Toward the Knizhnik-Zamolodchikov equation

In this subsection, we explain how the Knizhnik-Zamolodchikov equation (KZ equation) appears as a differential equation satisfied by the partition function of the Wess-Zumino-Witten model. As we shall explain, the KZ equation is obtained by using the action of the Virasoro algebra on modules of the affine Lie algebra.

Let $\mathrm{Diff}^+(S^1)$ be the space of orientation-preserving diffeomorphisms of S^1 to itself. With the product given by the composition of diffeomorphisms, it becomes an infinite dimensional Lie group, whose Lie algebra is the space of vector fields on S^1. We give (a Lie subalgebra of) the complexification of the Lie algebra of $\mathrm{Diff}^+(S^1)$ by

$$\mathcal{W} = \left\{ f(z)\frac{d}{dz} \;\middle|\; f(z) \in \mathbb{C}[z, z^{-1}] \right\},$$

which is (a subspace of) the space of vector fields on S^1, regarding S^1 as the unit circle in $\mathbb{C}$. We choose a basis $\{L_n\}_{n\in\mathbb{Z}}$ of $\mathcal{W}$ putting $L_n = -z^{n+1}\frac{d}{dz}$. With respect to this basis, the Lie bracket of $\mathcal{W}$ is presented by

$$\begin{aligned}[L_n, L_m] &= L_n L_m - L_m L_n = (-z^{n+1}\frac{d}{dz})(-z^{m+1}\frac{d}{dz}) - (-z^{m+1}\frac{d}{dz})(-z^{n+1}\frac{d}{dz}) \\ &= (m-n)z^{n+m+1}\frac{d}{dz} = (n-m)L_{n+m}.\end{aligned}$$

As a central extension of the Lie algebra $\mathcal{W}$ we introduce the *Virasoro algebra* $\mathcal{V}$ by

$$\mathcal{V} = \mathcal{W} \oplus \mathbb{C}c$$

with an indeterminate vector c, whose Lie bracket is given by

$$[L_n, L_m] = (n-m)L_{n+m} + \frac{n^3-n}{12}\delta_{n+m,0}\cdot c,$$
$$[c,\xi] = 0 \qquad \text{for any } \xi \in \mathcal{V}.$$

Let G be a semi-simple Lie group, and let LG be its loop group $\mathrm{Map}(S^1, G)$. We have a natural action of $\mathrm{Diff}^+(S^1)$ on LG such that $f \in \mathrm{Diff}^+(S^1)$ takes $\gamma \in LG$ to $\gamma \circ f$. Further, each f's action induces an automorphism of LG. Thus, we have the following homomorphism,

$$\mathrm{Diff}^+(S^1) \longrightarrow \mathrm{Aut}(LG),$$

where $\mathrm{Aut}(LG)$ denotes the group of automorphisms of LG, which is an infinite dimensional Lie group. As its derivation we consider the following homomorphism of Lie algebras,

$$\mathcal{W} \longrightarrow \mathrm{aut}(L\mathfrak{g}),$$

where $\mathrm{aut}(L\mathfrak{g})$ denotes the Lie algebra of the group of automorphisms of the loop algebra $L\mathfrak{g}$. In general, it is known that the Lie algebra of the group of automorphisms of a semi-simple Lie algebra $\mathfrak{h}$ is equal to the Lie algebra of the group of inner automorphisms of $\mathfrak{h}$ (generated by adjoint actions on $\mathfrak{h}$). Therefore, we would expect that there exists a representation

$$\mathcal{W} \longrightarrow U(L\mathfrak{g}),$$

where $U(L\mathfrak{g})$ denotes the universal enveloping algebra of $L\mathfrak{g}$, which acts on $L\mathfrak{g}$ by the adjoint action. The image L'_m of $L_m \in \mathcal{W}$ by the representation would satisfy

$$[L'_m, X\otimes t^n] = -nX\otimes t^{m+n}, \tag{F.26}$$

for any $X\otimes t^n \in L\mathfrak{g}$, since we have $-t^{m+1}\frac{d}{dt}t^n = -nt^{m+n}$. In truth, we do not have such a representation of $\mathcal{W}$, but, replacing $\mathcal{W}$ and $L\mathfrak{g}$ with their central extensions $\mathcal{V}$ and $\hat{\mathfrak{g}}$ respectively, we do have the following explicit construction (known as the *Segal-Sugawara construction*, see [Uen97]) of a linear map

$$(\text{the subspace of } \mathcal{V} \text{ spanned by } L_m\text{'s}) \longrightarrow \big(\text{infinite sums in } U(\hat{\mathfrak{g}})\big),$$

given by

$$L_m \longmapsto L'_m = \frac{1}{2(k+h^\vee)}\sum_{j\in\mathbb{Z}}\sum_{\mu} I_\mu\otimes t^{\min\{j,m-j\}}\cdot I_\mu\otimes t^{\max\{j,m-j\}}. \tag{F.27}$$

Here, k is a fixed positive integer, and $h^\vee$ denotes the dual Coxeter number of $\mathfrak{g}$ (see [Kac90] for its definition); for example, $h^\vee = N$ when $\mathfrak{g} = sl_N$. Further, $\{I_\mu\}$ denotes an orthonormal basis of $\mathfrak{g}$ with respect to the Killing form (as in Section 5.1). Note that, since $X\otimes t^j$ commutes with $X\otimes t^{m-j}$ for $m\neq 0$, by (F.25), the summand in (F.27) is equal to $I_\mu\otimes t^j\cdot I_\mu\otimes t^{m-j}$ for $m\neq 0$. The infinite sum on the right hand side of (F.27) can be justified in the sense that only finitely many summands contribute when applied to a vector of a highest weight $\hat{\mathfrak{g}}$-module. Thus, for any integrable highest weight $\hat{\mathfrak{g}}$-module $\hat{R}$ of level k we have a representation

$$\mathcal{V} \longrightarrow \mathrm{End}(\hat{R}), \tag{F.28}$$

given by (F.27) and

$$c \longmapsto \frac{k \dim \mathfrak{g}}{k+h^\vee} \cdot \mathrm{id}_{\hat{R}},$$

such that the image L'_m of L_m satisfies (F.26) in $\mathrm{End}(\hat{R})$.

Now we are ready to deduce the KZ equation. We consider the configuration space X_n consisting of n-tuples $z = (z_1, \cdots, z_n)$ of n distinct points $z_1, \cdots, z_n$ in $\mathbb{C}$. Further, we associate irreducible $\mathfrak{g}$-modules $R_1, \cdots, R_n$ to the n points respectively. In the previous subsection we defined the space $\mathcal{H}_{z;R_1,\cdots,R_n}$ of conformal blocks to be $\mathrm{Hom}_{\mathfrak{g}\otimes H^0(\mathbb{C}P^1;z)}(\hat{R}_1 \otimes \cdots \otimes \hat{R}_n, \mathbb{C})$. Then, the union of the $\mathcal{H}_{z;R_1,\cdots,R_n}$'s as $z \in X_n$ varies naturally forms a vector bundle on X_n. The vector bundle has a flat connection, as shown in [TsKa88] (see also [Hit90, ADW91]), which is obtained as follows. Earlier we speculated that the partition function $Z_k^{\mathrm{WZW}}(\Sigma)$ formally gives a vector in each $\mathcal{H}_{z;R_1,\cdots,R_n}$. Further, we speculate that these vectors collectively define a section of the above vector bundle. Noting that $L_{-1} = -\frac{d}{dz}$, we continue our speculation by observing that such a section should satisfy the following differential equations,

$$\frac{\partial \hat{W}}{\partial z_j} = -L_{-1}^{(j)} \hat{W} \qquad (j = 1, \cdots, n),$$

for functions $\hat{W} : X_n \to \mathrm{Hom}_{\mathfrak{g}\otimes H^0(\mathbb{C}P^1;z)}(\hat{R}_1 \otimes \cdots \otimes \hat{R}_n, \mathbb{C})$, where $L_{-1}^{(j)}$ denotes the action of L_{-1} on $\hat{R}_j$ given by (F.28). The above differential equations are rewritten as the following total differential equation

$$d\hat{W} = -\sum_{j=1}^{n} L_{-1}^{(j)} \hat{W} dz_j. \tag{F.29}$$

This differential equation gives a flat connection of the above vector bundle. Further, we have a natural linear injection $\mathcal{H}_{z;R_1,\cdots,R_n} \to \mathrm{Hom}_{\mathfrak{g}}(R_1 \otimes \cdots \otimes R_n, \mathbb{C})$ induced by the natural inclusions $R_i \to \hat{R}_i$. It follows from a concrete computation (see [Uen97]) that (F.29) can, further, be rewritten through the injective map as the following differential equation,

$$dW = \frac{1}{k+h^\vee} \sum_{1 \le i < j \le n} \tau_{ij} \frac{dz_i - dz_j}{z_i - z_j} W, \tag{F.30}$$

for functions $W : X_n \to \mathrm{Hom}_{\mathfrak{g}}(R_1 \otimes \cdots \otimes R_n, \mathbb{C})$. We call this equation the *Knizhnik-Zamolodchikov equation* (*KZ equation*). The formulation (5.9) of the KZ equation in Chapter 5 is derived from the above formula by putting $\hbar = 2\pi\sqrt{-1}/(k + h^{\vee})$ and $R_1 = \cdots = R_n = V^*$.

Let us see the conformal invariance of the solutions of (F.30). It is known that any conformal automorphism of $\mathbb{C}P^1$ $(= \mathbb{C} \cup \{\infty\})$ is given by the Möbius transformation associated to a matrix $\begin{pmatrix} a & b \\ c & d \end{pmatrix} \in SL(2;\mathbb{C})$, defined by

$$z \longmapsto w = \frac{az+b}{cz+d}$$

for $z, w \in \mathbb{C} \cup \{\infty\}$. We put $w_j = \frac{az_j+b}{cz_j+d}$. Then, it follows from a computation that any (local) solution W of (F.30) satisfies

$$W(z_1, \cdots, z_n) = (cz_1 + d)^{-2\Delta_{\hat{R}_1}} \cdots (cz_n + d)^{-2\Delta_{\hat{R}_n}} W(w_1, \cdots, w_n),$$

where we put $\Delta_{\hat{R}} = C_R/2(k + h^{\vee})$, which is called the *conformal weight* of $\hat{R}$. Here, C_R denotes the eigenvalue of the Casimir element of $\mathfrak{g}$ on R. Thus,

$$W(z_1, \cdots, z_n)(dz_1)^{\Delta_{\hat{R}_1}} \cdots (dz_n)^{\Delta_{\hat{R}_n}}$$

is invariant under conformal automorphisms of $\mathbb{C}P^1$. Therefore, it descends to a function on the Teichmüller space $\mathcal{T}$ of complex structures of S^2 with n marked points. This implies that the corresponding physical theory only depends on the conformal structure on Σ (which is S^2 with n marked points in the present case). Such a physical theory is called a *conformal field theory*. This completes our sketch of the induction of a conformal field theory from the Wess-Zumino-Witten model.

Appendix G

Computations for the perturbative invariant

This appendix collects some technical computations in order to prove certain properties of perturbative invariants and formal Gaussian integrals which are used in the main body of this text. In particular, we show Lemma 9.12 and Proposition G.11 in Sections G.1 and G.3, which are used in Chapter 9. Further, we show Proposition G.12 in Section G.4, which is used in Chapter 10.

G.1 Gaussian sum

In this section we show certain properties of weighted Gaussian sums and then use them to prove Lemma 9.12.

An integer a is called *congruent* modulo p to another integer b, denoted by $a \underset{(p)}{\equiv} b$, if $a - b$ is divisible by p.

Lemma G.1. Let p be an odd prime. Then,

$$\sum_{k=0}^{p-1} k^m \underset{(p)}{\equiv} \begin{cases} 0 & \text{if } m = 0 \text{ or } m \text{ is not divisible by } p-1, \\ -1 & \text{otherwise.} \end{cases}$$

Here, we regard 0^0 as 1.

Proof. If $m = 0$, then the lemma is easily shown.

If $0 < m < p - 1$, then we have that

$$\sum_{k=0}^{n-1} \binom{k}{m} = \binom{n}{m+1}$$

by induction on n. We put $n = p$. Then, the right hand side is divisible by p. Since k^m can be expressed as a linear sum of $\binom{k}{m}, \binom{k}{m-1}, \cdots$, we obtain the lemma in this case.

If $m = p - 1$, then by Fermat's little theorem we have that

$$\sum_{k=0}^{p-1} k^{p-1} \underset{(p)}{\equiv} 0 + 1 + 1 + \cdots + 1 = p - 1 \underset{(p)}{\equiv} -1.$$

Hence, we obtain the lemma in this case.

If $m > p - 1$, then we show the lemma by induction on m. By Fermat's little theorem, k^p is congruent to k. Hence,

$$\sum_{k=0}^{p-1} k^m \underset{(p)}{\equiv} \sum_{k=0}^{p-1} k^{m-(p-1)}.$$

Therefore, the case of m can be reduced to the case of $m - (p-1)$, which completes the proof. □

For an odd prime p and a p-th root of unity $\zeta = \exp(2\pi\sqrt{-1}/p)$, we introduce the *Gaussian sum* $\mathbb{G}(\zeta)$ by

$$\mathbb{G}(\zeta) = \sum_{k=0}^{p-1} \zeta^{k^2}.$$

Let f be an integer not divisible by p. It is known (see, *e.g.*, [HaWr79, Lan94]) that the Gaussian sum at ζ^f is related to the Gaussian sum at ζ as below,

$$\mathbb{G}(\zeta^f) = \sum_{k=0}^{p-1} \zeta^{fk^2} = \left(\frac{f}{p}\right)\mathbb{G}(\zeta), \tag{G.1}$$

where $\left(\frac{f}{p}\right)$ denotes the Legendre symbol. It is also known, see [HaWr79, Lan94], that

$$\mathbb{G}(\zeta) = u(\zeta - 1)^{(p-1)/2},$$
$$p = \mathbb{G}(\zeta)\mathbb{G}(\zeta^{-1}) = u'(\zeta - 1)^{p-1},$$

for some units u and u' in $\mathbb{Z}[\zeta]$. These formulae imply that $\mathbb{G}(\zeta)$ and p are divisible by $\zeta - 1$ precisely $\frac{p-1}{2}$ and $p - 1$ times in $\mathbb{Z}[\zeta]$ respectively; this is also shown in the proof of Lemma G.2 below.

Lemma G.2 ([LiWa99b]). For any non-negative integer l, we have that

$$\text{f-lim}\, \mathbb{G}(\zeta)^{-1} \sum_{k=0}^{p-1} k^l \zeta^{k^2} = \begin{cases} \dfrac{(l-1)!!}{(-2\log q)^{l/2}} & \text{if } l \text{ is even,} \\ 0 & \text{if } l \text{ is odd,} \end{cases}$$

where the Fermat limit f-lim is defined in Section 9.2.

The sum in the formula of the lemma is called a *weighted Gaussian sum*.

Proof of Lemma G.2. The lemma in the odd case is derived from the fact that $k^l\zeta^{k^2}$ is an odd function of k; this argument is left to the reader.

The lemma in the even case is shown as follows. By an elementary computation, the Taylor expansion of $\log(1+s)$ is given by

$$\log(1+s) = \sum_{i=1}^{\infty}(-1)^{i+1}\frac{s^i}{i}.$$

Putting $s = \zeta - 1$, we denote the sum of the first n terms of the above expansion by

$$h_n = s - \frac{s^2}{2} + \frac{s^3}{3} - \cdots - (-1)^n\frac{s^n}{n} \in \mathbb{Z}[\zeta, \frac{1}{2}, \frac{1}{3}, \cdots, \frac{1}{n}],$$

where n is any positive integer with $n < p$. Since h_n is congruent to $\log\zeta$ modulo $(\zeta-1)^{n+1}$, $e^{h_n k^2}$ is congruent to ζ^{k^2}. By expanding $e^{h_n k^2}$, we have that

$$\zeta^{k^2} \underset{(\zeta-1)^{n+1}}{\equiv} 1 + h_n k^2 + \frac{h_n^2}{2}k^4 + \frac{h_n^3}{3!}k^6 + \cdots + \frac{h_n^n}{n!}k^{2n}.$$

Hence, by Lemma G.1, we have that

$$\sum_{k=0}^{p-1} k^{2l}\zeta^{k^2} \equiv -\frac{h_n^{(p-1)/2-l}}{(\frac{p-1}{2}-l)!},$$

where the congruence is modulo p and $(\zeta-1)^{n+1}$. In particular, when $l = 0$,

$$\mathbb{G}(\zeta) = \sum_{k=0}^{p-1}\zeta^{k^2} \equiv -\frac{h_n^{(p-1)/2}}{(\frac{p-1}{2})!}.$$

Further,

$$\begin{aligned}\frac{(\frac{p-1}{2})!}{(\frac{p-1}{2}-l)!} &= \frac{p-1}{2}\cdot\frac{p-3}{2}\cdot\cdots\cdot\frac{p-2l+1}{2} \\ &\underset{(p)}{\equiv} \frac{-1}{2}\cdot\frac{-3}{2}\cdot\cdots\cdot\frac{-2l+1}{2} = \frac{(2l-1)!!}{(-2)^l}.\end{aligned}$$

Therefore, putting $n = p-1$,

$$\text{f-lim}\,\mathbb{G}(\zeta)^{-1}\sum_{k=0}^{p-1}k^{2l}\zeta^{k^2} = \frac{(2l-1)!!}{(-2)^l}\text{f-lim}\,{h_{p-1}}^{-l}.$$

Noting that f-lim $h_{p-1} = \log q$, we obtain the lemma. □

The remainder of this section is devoted to the proof of Lemma 9.12.

We define a family, the $C_k(m)$'s, of polynomials in m and $q^{\pm 1}$ by

$$C_k(m) = \sum_{0\le l\le (m-k-1)/2} (-1)^l \binom{m-l-1}{k}\binom{m-k-l-1}{l}(q+q^{-1})^{m-k-2l-1},$$

noting that $\binom{x}{k}$ is a polynomial in x for any non-negative integer k.

Lemma G.3. The coefficient $c_k(n)$ of a^k in the expansion (9.19) of V_n can be expressed in terms of the $C_k(m)$ as

$$c_k(n) = C_k\left(\frac{n+1}{2}\right) + C_k\left(\frac{n-1}{2}\right).$$

Proof. It is sufficient to verify that the right hand side of the above formula satisfies the characterization (9.20) of $c_k(n)$.

To verify the recursive formula of (9.20) we show the following formula,

$$C_k(m+1) - (q+q^{-1})C_k(m) + C_k(m-1) = \begin{cases} C_{k-1}(m) & \text{if } k > 0, \\ 0 & \text{if } k = 0. \end{cases}$$

If $k > 0$, then we have that

$$\begin{aligned}
&C_k(m+1) - C_{k-1}(m) - (q+q^{-1})C_k(m) \\
&= \sum_l (-1)^l \left(\left(\binom{m-l}{k} - \binom{m-l-1}{k-1} \right) \binom{m-k-l}{l} \right. \\
&\qquad\qquad \left. - \binom{m-l-1}{k}\binom{m-k-l-1}{l} \right) (q+q^{-1})^{m-k-2l} \\
&= \sum_l (-1)^l \binom{m-l-1}{k} \left(\binom{m-k-l}{l} - \binom{m-k-l-1}{l} \right) (q+q^{-1})^{m-k-2l} \\
&= \sum_l (-1)^l \binom{m-l-1}{k}\binom{m-k-l-1}{l-1} (q+q^{-1})^{m-k-2l} \\
&= -\sum_l (-1)^l \binom{m-l-2}{k}\binom{m-k-l-2}{l} (q+q^{-1})^{m-k-2l-2} = -C_k(m-1).
\end{aligned}$$

Hence, the recursive formula holds in this case. The case $k = 0$ can be verified similarly.

The relation $c_k(n) = 0$ for $n \le 2k-1$ in (9.20) is obtained from the relation $C_k(m) = 0$ for $m < k+1$, which is immediately derived from the definition of $C_k(m)$.

Hence, the left hand side of the required formula satisfies the characterization (9.20) of $c_k(n)$. Therefore, we obtain the lemma. □

We denote by $C_k(m)^{(d)}$ the coefficient of $h^d/d!$ in the expansion of $C_k(m)\big|_{q=e^h}$ as a power series in h.

Lemma G.4. $C_k(m)^{(d)}$ is an odd polynomial in m of degree $\leq 2k+d+1$.

Proof. By elementary computation, we have that

$$\left(\frac{d}{dh}\right)^d (e^h+e^{-h})^\alpha\Big|_{h=0} = \begin{cases} (\text{a polynomial in } \alpha \text{ of degree } \frac{d}{2})\cdot 2^\alpha & \text{if } d \text{ is even,} \\ 0 & \text{if } d \text{ is odd.} \end{cases}$$

Any polynomial in α of degree $d/2$ can be presented by a linear sum of $\binom{\alpha}{j}$ for $j=0,1,\cdots,d/2$. Hence, from the definition of $C_k(m)$, the coefficient $C_k(m)^{(d)}$ is equal to a linear sum of the following formula for $j=0,1,\cdots,d/2$,

$$\begin{aligned} &\sum_l (-1)^l \binom{m-l-1}{l}\binom{m-2l-1}{k}\binom{m-k-2l-1}{j} 2^{m-k-2l-1} \\ &= \sum_l (-1)^l \binom{m-l-1}{l}\binom{m-2l-1}{k+j}\binom{k+j}{j} 2^{m-k-2l-1} \\ &= 2^j \binom{m+k+j}{2k+2j+1}\binom{k+j}{j}, \end{aligned}$$

where we obtain the first and second equalities by a property of binomial coefficients and by Lemma G.5 below respectively. Fixing k and j, the above formula can be regarded as a polynomial in m of degree $2k+2j+1$. Further, since this polynomial is an odd function of m, the polynomial consists only of odd powers of m. Hence, we obtain the lemma. □

Lemma G.5. For a non-negative integer j, we have that

$$2^j\binom{m+j}{2j+1} = \sum_{0\leq l\leq (m-1)/2} (-1)^l \binom{m-l-1}{l}\binom{m-2l-1}{j} 2^{m-2l-1}.$$

Proof. By putting each side to be $\sigma_{m,j}$, it is verified that each side satisfies the following recursive formula

$$\sigma_{m,j} = 2(\sigma_{m-1,j}+\sigma_{m-1,j-1}) - \sigma_{m-2,j}.$$

Since both sides of the required formula satisfy the same recursive formula, we obtain the equality of the two sides. □

Now we prove Lemma 9.12.

Proof of Lemma 9.12. Let $g_k(\frac{n}{2})$ be the polynomial in n defined by (9.20) and (9.21). By (9.23), we have that

$$\begin{aligned} &\text{f-lim}\left(\frac{f}{p}\right)\mathbb{G}(\zeta)^{-1}\sum_{\substack{1\leq n<p \\ n \text{ is odd}}} \zeta^{f(n^2-1)/4}[n]\Big(g_k(\frac{n}{2})\zeta^{n/2} - g_k(-\frac{n}{2})\zeta^{-n/2}\Big) \\ &= \frac{q^{-f/4}}{q^{1/2}-q^{-1/2}}\Big(q^{-1/f}\widehat{g_k(m-\frac{1}{f})}(q^f) - \widehat{g_k(m)}(q^f)\Big). \qquad \text{(G.2)} \end{aligned}$$

We show that the sum in the Fermat limit is divisible by $(\zeta - 1)$ sufficiently many times. We compute the sum as follows,

$$\begin{aligned}
&\sum_{\substack{1\le n<p \\ n \text{ is odd}}} \zeta^{f(n^2-1)/4}[n]\Big(g_k(\frac{n}{2})\zeta^{n/2} - g_k(-\frac{n}{2})\zeta^{-n/2}\Big) \\
&= \sum_{\substack{1\le n<p \\ n \text{ is odd}}} \zeta^{f(n^2-1)/4}[n]c_k(n) \\
&= \sum_{\substack{1\le n<p \\ n \text{ is odd}}} \zeta^{f(n^2-1)/4}[n]\Big(C_k(\frac{n+1}{2}) + C_k(\frac{n-1}{2})\Big) \\
&= \frac{1}{2}\sum_{\substack{-p<n\le p \\ n \text{ is odd}}} \zeta^{f(n^2-1)/4}[n]\Big(C_k(\frac{n+1}{2}) + C_k(\frac{n-1}{2})\Big) \\
&\underset{(p)}{\equiv} \sum_{0\le m<p} \zeta^{f(m^2+m)}[2m+1]C_k(m), \qquad \text{(G.3)}
\end{aligned}$$

where we obtain the first and second equalities from the definition of g_k and by Lemma G.3 respectively. We now show that (G.3) is divisible by $(\zeta - 1)$ precisely $\frac{p-1}{2} - k - 1$ times, similarly as in the proof of Lemma G.2. The terms $\zeta^{f(m^2+m)}$, $[2m+1]$, $C_k(m)$ are presented by linear sums of $m^{2d_1}h^{d_1}$, $m^{d_2+1}h^{d_2}$, $m^{2k+d_3+1}h^{d_3}$ respectively, where h denotes some truncation of the expansion of $\log\zeta$, as in the proof of Lemma G.2. Hence, (G.3) can be presented by a linear sum of $m^{2d+2k+2}h^d$. Therefore, it is shown, in the same way as in the proof of Lemma G.2, that (G.3) is divisible by $(\zeta - 1)$ precisely $\frac{p-1}{2} - k - 1$ times.

Since $\mathbb{G}(\zeta)$ is divisible by $(\zeta - 1)$ precisely $\frac{p-1}{2}$ times, the formula in the Fermat limit in (G.3) belongs to $(\zeta - 1)^{-k-1} \cdot \mathbb{Z}[\zeta]$. Hence, the right hand side of (G.3) belongs to $(q-1)^{-k-1} \cdot \mathbb{Q}[[q-1]]$. This implies the lemma. □

G.2 The center of the quantum group $U_q(sl_2)$

In this section we identify the center of the universal enveloping algebra $U(sl_2)$ and the quantum group $U_q(sl_2)$.

Lemma G.6. The center of $U(sl_2)$ is the subalgebra generated by the element c given by

$$c = EF + FE + \frac{1}{2}H^2.$$

Proof. The universal enveloping algebra $U(sl_2)$ is spanned by terms of the form $H^kE^iF^j$. Further, an element $H^kE^iF^j$ commutes with H if and only if $i = j$. Hence, any central element is equal to a linear sum of such $H^kE^jF^j$'s. We put the linear sum to be $\sum_{j=0}^N a_j(H)E^jF^j$ with polynomials $a_j(H)$ in H. We now show, by induction on N, that if this linear sum is central then it is equal to a polynomial in c.

We have that

$$[E, H^k E^j F^j] = ((H-2)^k - H^k)E^{j+1}F^j + jH^k(H-j-1)E^j F^{j-1}.$$

Hence, in order for the above mentioned linear sum to be central, the leading polynomial $a_N(H)$ must be a scalar, say a_N. Therefore, the linear sum minus $a_N(c/2)^j$ is equal to another linear sum $\sum_{j=0}^{N-1} a'_j(H)E^jF^j$ for some $a'_j(H)$'s. By the assumption of the induction, we obtain the lemma. □

Lemma G.7. The center of $U_q(sl_2)$ is generated by the element c given by

$$c = (q^{1/2} - q^{-1/2})^2 EF + q^{-1/2}K + q^{1/2}K^{-1}.$$

Proof. The quantum group $U_q(sl_2)$ is spanned by terms of the form $K^kE^iF^j$. Further, an element $K^kE^iF^j$ commutes with K if and only if $i = j$. Hence, any central element is equal to a linear sum of $K^kE^jF^j$'s. Further,

$$[E, K^kE^jF^j] = (q^{-k}-1)K^kE^{j+1}F^j + \frac{[j]}{q^{1/2}-q^{-1/2}}(q^{-\frac{j+1}{2}}K^{k+1} - q^{\frac{j+1}{2}}K^{k-1})E^jF^{j-1}.$$

Hence, we obtain the lemma by induction in a similar way as in the proof of Lemma G.6. □

As a corollary of Lemma G.7, we have

Lemma G.8. Let c be as in Lemma G.7. If a polynomial in $E, F, K^{\pm1}$ with coefficients in $\mathbb{Q}[q^{\pm1/2}]$ is central in $U_q(sl_2)$, then it is equal to a polynomial in c with coefficients in $\mathbb{Q}[q^{\pm1/2}]$.

Proof. Consider a central polynomial satisfying the assumption of the lemma. Then, by Lemma G.7 it is equal to a polynomial in c, say, $\sum_{k=0}^N a_kc^k$. We show by induction on N that, if $\sum_{k=0}^N a_kc^k$ is equal to a polynomial in $E, F, K^{\pm1}$ with coefficients in $\mathbb{Q}[q^{\pm1/2}]$, then the a_k's belong to $\mathbb{Q}[q^{\pm1/2}]$.

In the expansion of the polynomial in c into a polynomial in $E, F, K^{\pm1}$ the coefficient of K^N is equal to $a_Nq^{-N/2}$. Hence, a_N belongs to $\mathbb{Q}[q^{\pm1/2}]$. Further, by the assumption of the induction, $\sum_{k=0}^{N-1} a_kc^k$ is equal to a polynomial in $E, F, K^{\pm1}$ with coefficients in $\mathbb{Q}[q^{\pm1/2}]$, and hence so is $\sum_{k=0}^N a_kc^k$. □

The following two lemmas are obtained similarly as Lemmas G.7 and G.8.

Lemma G.9. The center of $U_\zeta(sl_2)$ is generated by the element c given by

$$c = (\zeta^{1/2} - \zeta^{-1/2})^2 EF + \zeta^{-1/2}K + \zeta^{1/2}K^{-1}.$$

Lemma G.10. Let c be as in Lemma G.9. If a polynomial in $E, F, K^{\pm1}$ with coefficients in $\mathbb{Q}[\zeta]$ is central in $U_\zeta(sl_2)$, then it is equal to a polynomial in c with coefficients in $\mathbb{Q}[\zeta]$.

G.3 The quantum $(sl_2; a^k)$ invariant is divisible by $(q-1)^{2k}$

In this section we show Proposition G.11 below, which says that the quantum $(sl_2; a^k)$ invariant is divisible by $(q-1)^{2k}$. Further, we compute some examples to verify the proposition.

Let V_n denote the n-dimensional irreducible representation of $U_q(sl_2)$, and we put $[n] = (q^{n/2} - q^{-n/2})/(q^{1/2} - q^{-1/2})$ as before. We put $a = V_3 - [3]V_1$ in the representation ring of $U_q(sl_2)$. For a framed knot K we denote by $Q^{sl_2;a^k}(K)$ the linear sum of the quantum $(sl_2; V_n)$ invariant $Q^{sl_2;V_n}(K)$ corresponding to the expression of a^k as a linear sum of V_n's in the representation ring; for example we put

$$Q^{sl_2;a^2}(K) = Q^{sl_2;V_5}(K) - (2[3]-1)Q^{sl_2;V_3}(K) + ([3]^2+1)Q^{sl_2;V_1}(K),$$

since

$$a^2 = V_5 - (2[3]-1)V_3 + ([3]^2+1)V_1.$$

The aim of this section is to show the following proposition, which is used in Chapter 9.

Proposition G.11. Let K be a framed knot with 0 framing. Then, for each non-negative integer k, the quantum $(sl_2; a^k)$ invariant $Q^{sl_2;a^k}(K)$ is divisible by $(q-1)^{2k}$.

When K is the trivial knot, the proposition is trivial. In this case $Q^{sl_2;a^k}(K) = 1$ if $k = 0$, and 0 if $k > 0$.

Proof of Proposition G.11. Let T be a 1-tangle whose closure is isotopic to K. Then, $Q^{sl_2;\star}(T)$ is central in (the completion of) $U_q(sl_2)$. Hence, by Lemma G.7, it is equal to a power series in $q^{1/2} - q^{-1/2}$ with polynomials in c as coefficients, where c is given by

$$c = (q^{1/2} - q^{-1/2})^2 EF + q^{-1/2}K + q^{1/2}K^{-1}.$$

Therefore, $Q^{sl_2;\star}(T)$ is equal to a linear sum of $(q-1)^{2j}K^iE^jF^{j'}$'s.

Hence, $Q^{sl_2;\star}(K)$ is equal to a linear sum of $(q-1)^{2j}K^{i+1}E^jF^{j'}$'s from the definition of $Q^{sl_2;\star}(K)$. Therefore, $Q^{sl_2;a^k}(K)$ can be calculated by some linear sum of

$$\mathrm{trace}_{a^k}\big((q-1)^{2j}K^{i+1}E^jF^{j'}\big) = (\mathrm{trace}_a)^{\otimes k}\Big(\Delta^{(k-1)}\big((q-1)^{2j}K^{i+1}E^jF^{j'}\big)\Big),$$

where the equality is obtained from the definition of the tensor representation in terms of the comultiplication Δ. Further, $\Delta^{(k-1)}(K^{i+1}E^jF^{j'})$ is equal to a linear sum of

$$K^{i_1}E^{j_1}F^{j'_1} \otimes K^{i_2}E^{j_2}F^{j'_2} \otimes \cdots \otimes K^{i_n}E^{j_k}F^{j'_k}$$

such that $j = j_1+j_2+\cdots+j_k$ and $j' = j'_1+j'_2+\cdots+j'_k$. Noting that $\text{trace}_a(K^iE^jF^{j'}) = 0$ if $j \neq j'$, it is sufficient to show that

$$(q-1)^{2j}\text{trace}_a(K^{i_1}E^{j_1}F^{j_1})\text{trace}_a(K^{i_2}E^{j_2}F^{j_2})\cdots\text{trace}_a(K^{i_n}E^{j_k}F^{j_k}) \qquad \text{(G.4)}$$

is divisible by $(q-1)^{2k}$ when $j = j_1 + j_2 + \cdots + j_k$.

If $j \geq k$, then (G.4) is divisible by $(q-1)^{2k}$ since it is a multiple of $(q-1)^{2j}$. Hence, we consider the case that $j < k$. Since the j_l's are non-negative integers satisfying $j = j_1 + j_2 + \cdots + j_k$, at most j of the j_l's can be positive. Hence, at least $k - j$ of the j_l's are equal to 0. Further, note that $\text{trace}_a(K^i)$ is divisible by $(q-1)^2$, since it is computed as

$$\text{trace}_a(K^i) = \text{trace}_{V_3}(K^i) - [3]\text{trace}_{V_1}(K^i) = q^i + q^{-i} - q - q^{-1}.$$

Therefore, (G.4) is divisible by $(q-1)^{2k}$. □

In the remainder of this section, let us verify, by concrete computations, Proposition G.11 for the $(f, 2)$ torus knot for some small values of f and n. The results of these computations would be useful for the computation of the perturbative $SO(3)$ invariants of the homology 3-spheres obtained by integral surgeries along these knots.

To compute the quantum $(sl_2; a^k)$ invariants we introduce the following notation. Let n, m, k be positive integers whose sum is even. We define a trivalent vertex in the linear skein by

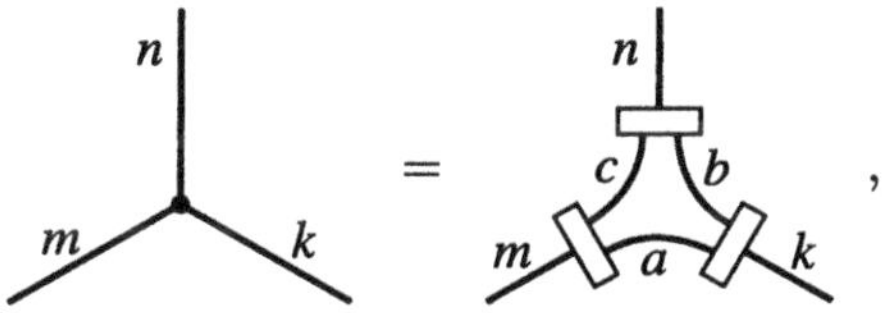

where a, b, c are integers satisfying that $n = c + b$, $m = c + a$, and $k = a + b$. As in [Lic97, KaLi94], we have that

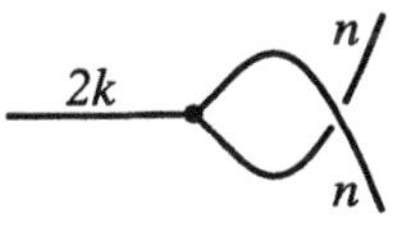

$$= (-1)^{n-k}A^{2k^2+2k-n^2-2n}$$

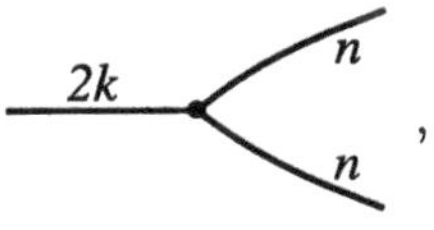

,

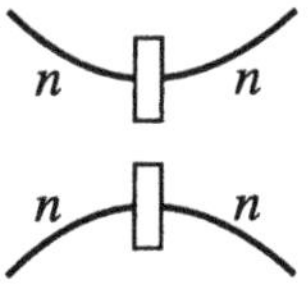

$$= \sum_{k=0}^{n} \begin{bmatrix} n \\ k \end{bmatrix}^2 \begin{bmatrix} n+k+1 \\ n-k \end{bmatrix}^{-1}$$

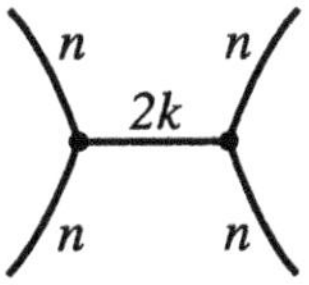

,

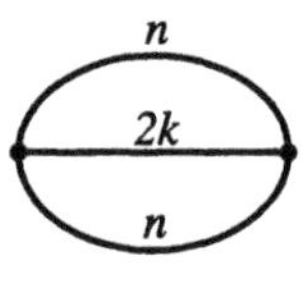

$$= \begin{bmatrix} n \\ k \end{bmatrix}^{-2} \begin{bmatrix} n+k+1 \\ n-k \end{bmatrix} [2k+1].$$

By using the above formulae the invariant of the $(f, 2)$ torus knot with f framing (where f is an odd positive integer) is computed as follows (letting n be even),

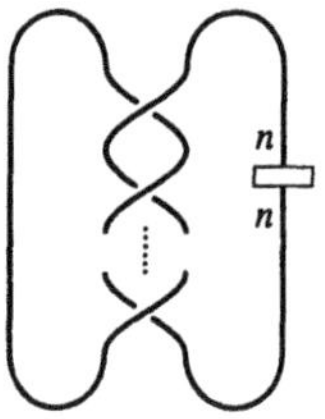

$=$

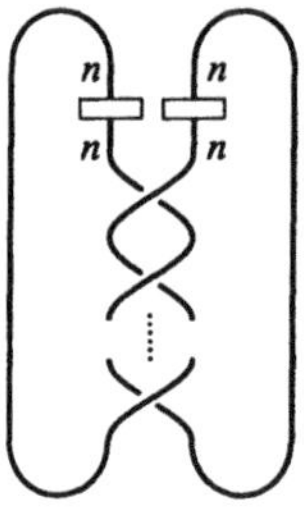

$$= \sum_{k=0}^{n} \begin{bmatrix} n \\ k \end{bmatrix}^{2} \begin{bmatrix} n+k+1 \\ n-k \end{bmatrix}^{-1}$$

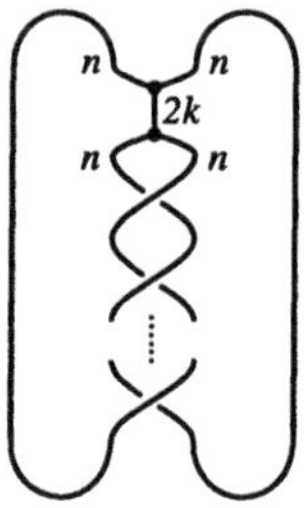

$$= q^{-fn(n+2)/4} \sum_{k=0}^{n} \begin{bmatrix} n \\ k \end{bmatrix}^{2} \begin{bmatrix} n+k+1 \\ n-k \end{bmatrix}^{-1} (-1)^k q^{fk(k+1)/2}$$

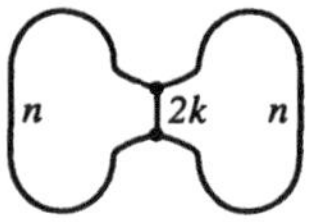

$$= q^{-fn(n+2)/4} \sum_{k=0}^{n} (-1)^k q^{fk(k+1)/2} [2k+1].$$

Let K be the $(f, 2)$ torus knot with 0 framing. By modifying the contribution from the framing, we have that

$$Q^{sl_2;V_{n+1}}(K) = q^{-fn(n+2)/2} \sum_{k=0}^{n} (-1)^k q^{fk(k+1)/2} [2k+1].$$

From the definition of a, we have that

$$\begin{aligned} a &= V_3 - [3]V_1, \\ a^2 &= V_5 - \big(2[3]-1\big)V_3 + \big([3]^2+1\big)V_1, \\ a^3 &= V_7 - \big(3[3]-2\big)V_5 + 3\big([3]^2-[3]+1\big)V_3 - \big([3]^3+3[3]-1\big)V_1. \end{aligned}$$

Hence, we can compute $Q^{sl_2;a^k}(K)$ for $k = 1, 2, 3$ as follows. When K is the $(3, 2)$ torus knot (*i.e.*, the trefoil knot),

$$\begin{aligned} Q^{sl_2;a}(K) &= (1-q)^2(1+q)^2(1+q^2)(1-q+q^2)(1+q+q^2)(1-q^2-q^3)q^{-12}, \\ Q^{sl_2;a^2}(K) &= (1-q)^4(1+q)^4(1+q^2)^2(1-q+q^2)^2(1+q+q^2)^2 \\ &\quad \times (1-q^2-q^3+q^4-q^7-q^9+2q^{12}-q^{14}-q^{15}+2q^{16}+2q^{17}+q^{18})q^{-36}, \\ Q^{sl_2;a^3}(K) &= (1-q)^6(1+q)^6(1+q^2)^3(1-q+q^2)^3(1+q+q^2)^3 \\ &\quad \times (\text{an irreducible polynomial with 42 terms}). \end{aligned}$$

When K is the $(5,2)$ torus knot,

$$\begin{aligned} Q^{sl_2;a}(K) &= (1-q)^2(1+q)^2(1+q^2)(1-q+q^2)(1+q+q^2) \\ &\quad \times (1-q^5-q^9-q^10-q^{11})q^{-20}, \\ Q^{sl_2;a^2}(K) &= (1-q)^4(1+q)^4(1+q^2)^2(1-q+q^2)(1+q+q^2) \\ &\quad \times (\text{an irreducible polynomial with 45 terms}), \\ Q^{sl_2;a^3}(K) &= (1-q)^6(1+q)^6(1+q^2)^3(1-q+q^2)(1+q+q^2) \\ &\quad \times (\text{an irreducible polynomial with 100 terms}). \end{aligned}$$

When K is the $(7,2)$ torus knot,

$$\begin{aligned} Q^{sl_2;a}(K) &= (1-q)^2(1+q)^2(1+q^2)(1-q+q^2)(1+q+q^2)(1+q^4) \\ &\quad \times (1-q^7-q^13-q^14-q^15)q^{-28}, \\ Q^{sl_2;a^2}(K) &= (1-q)^4(1+q)^4(1+q^2)^2(1-q+q^2)(1+q+q^2)(1+q^4) \\ &\quad \times (\text{an irreducible polynomial with 64 terms}), \\ Q^{sl_2;a^3}(K) &= (1-q)^6(1+q)^6(1+q^2)^3(1-q+q^2)(1+q+q^2)(1+q^4) \\ &\quad \times (\text{an irreducible polynomial with 143 terms}). \end{aligned}$$

Thus we have verified Proposition G.11 for the above knots, for $k = 1, 2, 3$.

G.4 Computation of formal Gaussian integrals

The aim of this section is to show Proposition G.12 below, which is used in Section 10.4.

As in Section 10.4 we write the Casimir element in the symmetric algebra $S(sl_2)$ of the Lie algebra sl_2 by

$$C' = EF + FE + \frac{1}{2}H^2 \in S(sl_2),$$

for the base $\{E, F, H\}$ of sl_2 used in Section 10.4.

Proposition G.12. For any non-negative integer d and any non-zero scalar f, we have that

$$\frac{\displaystyle\int_{(sl_2)^*} C'^d e^{fC'/2} dv}{\displaystyle\int_{(sl_2)^*} e^{C'/2} dv} = \frac{\displaystyle\int_{n\in\mathbb{R}} \mathrm{trace}_{nV_n}\big(\chi(C'^d e^{fC'/2})\big) dn}{\displaystyle\int_{n\in\mathbb{R}} \mathrm{trace}_{nV_n}\big(\chi(e^{C'/2})\big) dn}, \tag{G.5}$$

where the notation is as given in Section 10.4, and we regard both sides as formal Gaussian integrals in the sense mentioned in Section 10.4.

To prove the proposition we show some lemmas below.

The Casimir element in the universal enveloping algebra $U(sl_2)$ of sl_2 is

$$C = EF + FE + \frac{1}{2}H^2 \in U(sl_2).$$

As shown in Lemma G.6 the center of $U(sl_2)$ is generated by C. Since $\chi({C'}^k)$ is central, where χ is the Poincare-Birkhoff-Witt isomorphism given in (10.21), we define a polynomial Q_k in C by

$$Q_k = \chi({C'}^k) \in \mathbb{Q}[C]$$

for each non-negative integer k.

Lemma G.13. The sequence of Q_k's is characterized by $Q_0 = 1$ and the following recursive formula,

$$Q_k = \sum_{0 \le j \le k-1} \left(\binom{2k-1}{2j} C - 2 \binom{2k}{2j-1} \right) \frac{2^{k-j-1}}{2j+1} Q_j,$$

for any positive integer k.

For example, for small k's we find that

$$Q_1 = C, \qquad Q_2 = C^2 - \frac{2}{3}C, \qquad Q_3 = C^3 - 2C^2 + \frac{4}{3}C.$$

In the following of this section we often mean by a Jacobi diagram D on an interval its image $W_{sl_2}(D) \in U(sl_2)$, and write $D \equiv D'$ if $W_{sl_2}(D) = W_{sl_2}(D')$. In this sense Q_k is alternatively presented by

$Q_k = \chi\big(($ 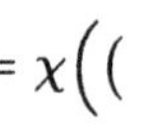 $)^k \big) \in \mathcal{A}(S^1) \cong \mathcal{A}(\downarrow),$

for the Poincare-Birkhoff-Witt isomorphism χ of Jacobi diagrams defined in (10.23).

Proof of Lemma G.13. We have that

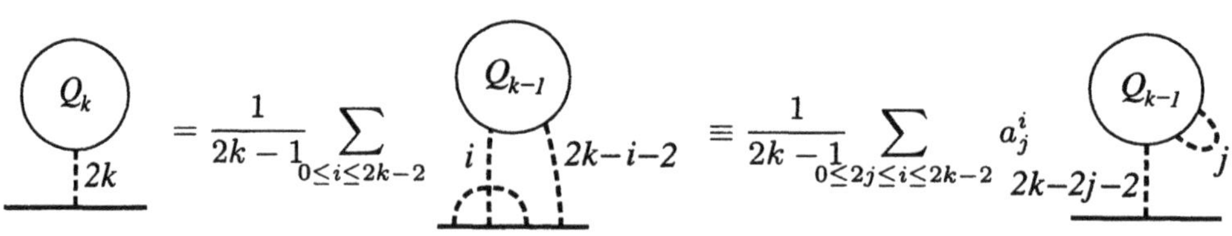

by Lemma G.14 below, where a dashed strand with a number n implies a bundle of n parallel copies of the strand. Further, similarly as in the proof of Lemma 10.6,

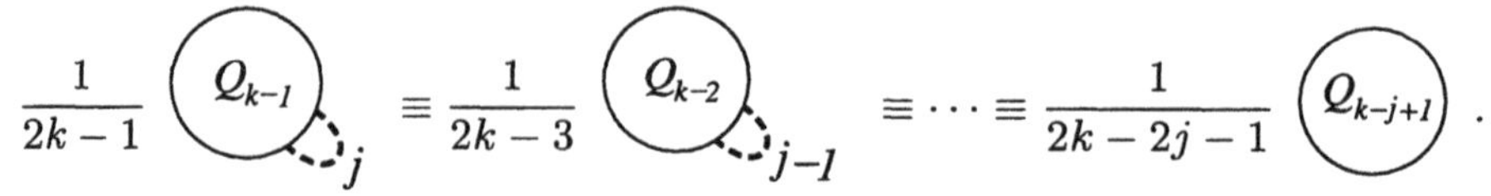

Hence, we obtain

$$\begin{aligned}Q_k &= \sum_{0\le 2j\le i\le 2k-2} \frac{a^i_j}{2k-2j-1} Q_{k-j-1} \\ &= \sum_{0\le j\le k-1} Q_{k-j-1}\frac{2^j}{2k-2j-1}\Big(\big(\sum_{2j\le i\le 2k-2}\binom{i}{2j}\big)C - 2\sum_{2j\le i\le 2k-2}\binom{i+1}{2j+2}\Big) \\ &= \sum_{0\le j\le k-1} Q_{k-j-1}\frac{2^j}{2k-2j-1}\Big(\binom{2k-1}{2j+1}C - 2\binom{2k}{2j+3}\Big),\end{aligned}$$

where the second equality is obtained by Lemma G.14 below. By replacing j with $k-j-1$ we obtain the required formula. □

We define a dotted box over n dashed strands by

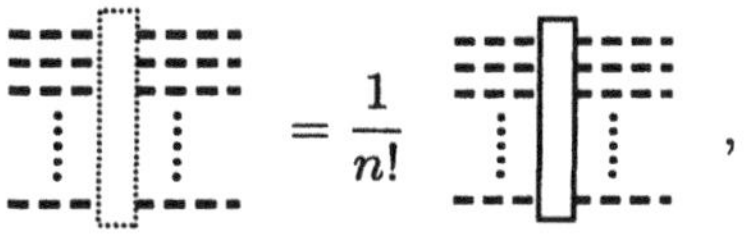

with a box of symmetrizer defined in (10.24), *i.e.*, a dotted box implies the average of the $n!$ configurations that arise by connecting the n left points to the n right points respectively.

Lemma G.14. We have that

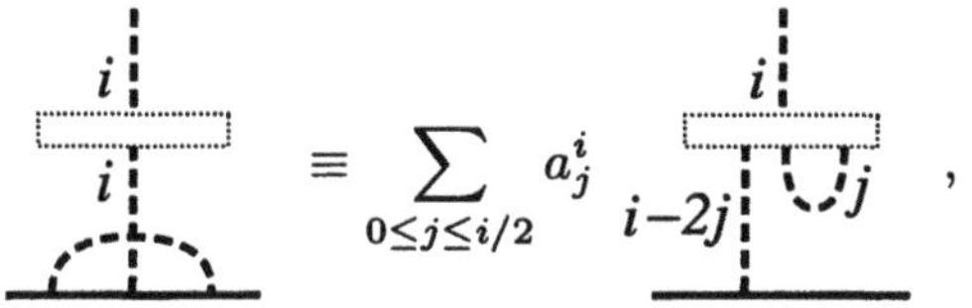

where the a^i_j's are given by

$$a^i_j = 2^j\binom{i}{2j}C - 2^{j+1}\binom{i+1}{2j+2}.$$

Proof. When $i = 1$, we have that

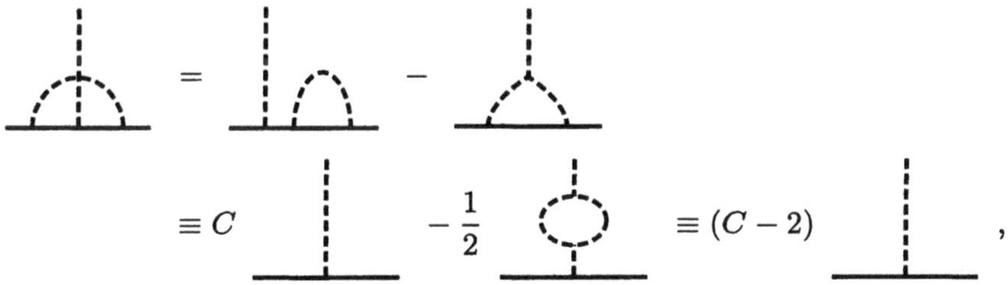

where we obtain the equalities by the STU relation and the relations (6.33) and (6.34). Hence, we obtain the required formula in this case.

When $i \geq 2$, we have that

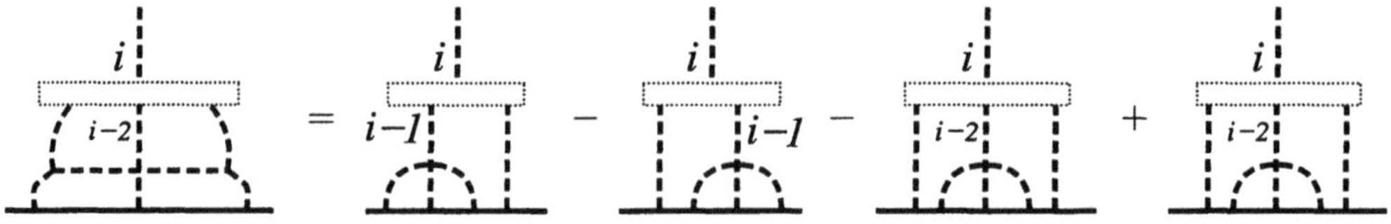

by the STU relation. Further,

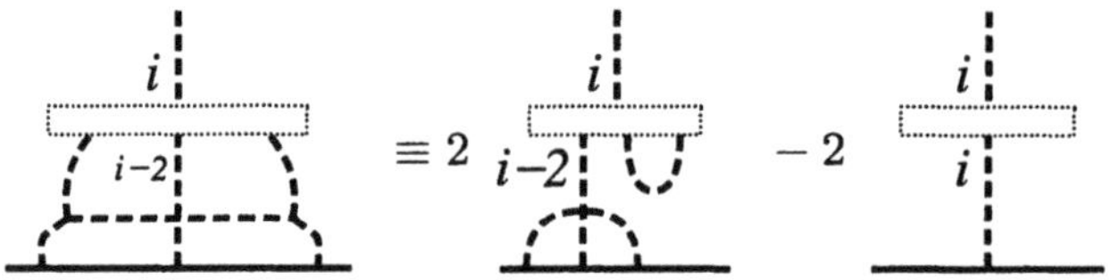

by (6.32). From the equality of the right hand sides of the above two formulae we obtain the required formula recursively, with the coefficients a^i_j satisfying the following recursive formula,

$$a^i_j = 2a^{i-1}_j - a^{i-2}_j + 2a^{i-2}_{j-1}.$$

Solving it, we obtain the required expression for the a^i_j's. □

Lemma G.15. When $C = \frac{n^2-1}{2}$, Q_k can be expressed

$$Q_k = \sum_{i=0}^{k} \frac{b_i n^{2k-2i}}{2^k(2i+1)} \binom{2k+1}{2i}$$

with the coefficients b_i's determined by $b_0 = 1$ and the following recursive formula,

$$b_k = -\sum_{j=0}^{k-1} \left(\frac{1}{4}\binom{2k-1}{2j} + \binom{2k}{2j-1} \right) \frac{2^{2k-2j}}{2j+1} b_j.$$

For example for small k's, we have that

$$b_1 = -1, \quad b_2 = \frac{7}{3}, \quad b_3 = -\frac{31}{3}, \quad b_4 = \frac{3 \cdot 127}{5}, \quad b_5 = -\frac{5 \cdot 7 \cdot 73}{3}.$$

Proof of Lemma G.15. Putting $w = 1/n^2$ we consider $Q'_k = 2^k w^k Q_k$ instead of Q_k. Then, the required formula is rewritten as

$$Q'_k = \sum_{0 \leq i \leq k} \frac{b_i w^i}{2i+1} \binom{2k+1}{2i}. \tag{G.6}$$

Further, the recursive formula for Q_k's in Lemma G.13 is rewritten as

$$Q'_k = \sum_{0 \leq j < k} \left(\binom{2k-1}{2j} \frac{w^{-1}-1}{4} - \binom{2k}{2j-1} \right) \frac{2^{2k-2j} w^{k-j}}{2j+1} Q'_j. \tag{G.7}$$

It is sufficient to show that the Q'_k's in (G.6) satisfy the above recursive formula.
By substituting Q'_k and Q'_j given by (G.6) into (G.7), we have that

$$\sum_{0\le i\le k} \frac{b_i w^i}{2i+1}\binom{2k+1}{2i} = \sum_{0\le i\le j<k} \left(\binom{2k-1}{2j}\frac{w^{-1}-1}{4} - \binom{2k}{2j-1}\right) \times \frac{2^{2k-2j}w^{k-j+i}b_i}{(2j+1)(2i+1)}\binom{2j+1}{2i}. \tag{G.8}$$

It is sufficient to show (G.8). The coefficient of w^d of (G.8) is presented by

$$\frac{b_d}{2d+1}\binom{2k+1}{2d} = \sum_{0\le i\le d}\left[\binom{2k-1}{2j}\frac{1}{4}\cdot\frac{2^{2k-2j}b_i}{(2j+1)(2i+1)}\binom{2j+1}{2i}\right]_{d=k-j+i-1}$$
$$- \sum_{0\le i<d}\left[\left(\frac{1}{4}\binom{2k-1}{2j} + \binom{2k}{2j-1}\right)\frac{2^{2k-2j}b_i}{(2j+1)(2i+1)}\binom{2j+1}{2i}\right]_{d=k-j+i}.$$

Further, by properties of binomial coefficients, (G.8) is reduced to

$$b_d\binom{2k+1}{2d} = \frac{1}{2d+2}\binom{2k-1}{2d}\sum_{0\le i\le d} 2^{2d-2i}b_i\binom{2d+2}{2i+1}$$
$$- \frac{2d+1}{2d(2d-1)}\binom{2k-1}{2d-2}\sum_{0\le i<d} 2^{2d-2i-2}b_i\binom{2d}{2i+1}$$
$$- \frac{1}{2d(2d+2)}\binom{2k}{2d-1}\sum_{0\le i<d} 2^{2d-2i}b_i(2k+2i-2d)\binom{2d+2}{2i+1}.$$

Furthermore, by using

$$(2k+2i-2d)\binom{2k}{2d-1} = (2k-2d+1)\binom{2k}{2d-1} + (2i-1)\binom{2k}{2d-1}$$
$$= 2d\binom{2k}{2d} + (2i-1)\binom{2k}{2d-1}$$

and $\binom{2k-1}{2d} - \binom{2k}{2d} = -\binom{2k-1}{2d-1}$, the required equality (G.8) is reduced to

$$b_d\left(\binom{2k+1}{2d} - \binom{2k-1}{2d}\right)$$
$$= -\frac{2d+1}{2d(2d-1)}\binom{2k-1}{2d-2}\sum_{0\le i<d} 2^{2d-2i-2}b_i\binom{2d}{2i+1}$$
$$- \frac{1}{2d(2d+2)}\binom{2k}{2d-1}\sum_{0\le i<d} 2^{2d-2i}b_i(2i-1)\binom{2d+2}{2i+1}$$
$$- \frac{1}{2d+2}\binom{2k-1}{2d-1}\sum_{0\le i<d} 2^{2d-2i}b_i\binom{2d+2}{2i+1}.$$

We obtain this formula from Lemma G.16 below, putting $m = 2k$ and $\varepsilon = 0$. Hence, we obtain (G.8). □

Lemma G.16. Let b_k's be the sequence given in Lemma G.15. Then, we have that

$$b_d\left(\binom{m+1}{2d-\varepsilon} - \binom{m-1}{2d-\varepsilon}\right)$$
$$= -\frac{2d+1}{2d(2d-1)}\binom{m-1}{2d-2-\varepsilon}\sum_{0\le i<d} 2^{2d-2i-2} b_i \binom{2d}{2i+1}$$
$$-\frac{1}{2d(2d+2)}\binom{m}{2d-1-\varepsilon}\sum_{0\le i<d} 2^{2d-2i} b_i (2i-1)\binom{2d+2}{2i+1}$$
$$-\frac{1}{2d+2}\binom{m-1}{2d-1-\varepsilon}\sum_{0\le i<d} 2^{2d-2i} b_i \binom{2d+2}{2i+1}$$

for an indeterminate m and any integers d and ε with $0 \le \varepsilon \le 2d-2$.

Proof. We put the difference of the two sides of the required formula to be $D_\varepsilon(m)$. It is a polynomial in m for each ε. We show that $D_\varepsilon(m) = 0$ by descending induction on ε.

We have a recursive formula for $D_\varepsilon(m)$ as

$$D_\varepsilon(m+1) - D_\varepsilon(m) = D_{\varepsilon+1}(m),$$

by properties of binomial coefficients, say the property $\binom{(m+1)+1}{2d-\varepsilon} - \binom{m+1}{2d-\varepsilon} = \binom{m+1}{2d-(\varepsilon+1)}$. Suppose that $D_{\varepsilon+1}(m) = 0$ by the assumption of the induction. Then, $D_\varepsilon(m+1) = D_\varepsilon(m)$ by the above recursive formula for $D_\varepsilon(m)$. Further, $D_\varepsilon(1) = 0$ for $\varepsilon < 2d-2$ from the definition of $D_\varepsilon(m)$. Hence, $D_\varepsilon(m)$ has infinitely many zeros; this implies $D_\varepsilon(m) = 0$. Therefore, by descending induction on ε we reduce the proof to the case that $\varepsilon = 2d-2$.

When $\varepsilon = 2d-2$, the required formula is written as

$$b_d(2m-1) = -\frac{2d+1}{2d(2d-1)}\sum_{0\le i<d} 2^{2d-2i-2} b_i \binom{2d}{2i+1}$$
$$-\frac{m}{2d(2d+2)}\sum_{0\le i<d} 2^{2d-2i} b_i (2i-1)\binom{2d+2}{2i+1}$$
$$-\frac{m-1}{2d+2}\sum_{0\le i<d} 2^{2d-2i} b_i \binom{2d+2}{2i+1}.$$

Since both sides are polynomials in m of degree 1, it is sufficient to show that the above formula holds for two values of m, say, $m = 0, 1$. The recursive formulae for the b_k's obtained from the above formula by putting $m = 0, 1$ give the same sequence of b_k's as given in Lemma G.15. Hence we obtain the lemma. □

Lemma G.17. When $C = \frac{n^2-1}{2}$, Q_k is presented by

$$Q_k = \sum_{i=0}^{k} b_i n^{2k-2i} \left(\frac{2^{2i-k}}{2i+1} \binom{k}{2i} + \sum_{0 \le j < i} \frac{2^{2j-k}}{i-j} \binom{i+j}{i-j-1} \binom{k}{i+j} \right)$$

with the coefficients b_i introduced in Lemma G.15.

Proof. By replacing $\binom{k}{\cdot}$ with $\binom{k+1}{\cdot}$ using a property of binomial coefficients, the required formula is rewritten as

$$Q_k = \sum_{i=0}^{k} \frac{b_i n^{2k-2i}}{(k+1)2^{k-2i}} \sum_{0 \le l \le i} \frac{1}{2^{2l}} \binom{2i-l+1}{l} \binom{k+1}{2i-l+1}.$$

We obtain the above formula by Lemma G.15 and the following equality,

$$\frac{1}{2i+1} \binom{2k+1}{2i} = \frac{2^{2i}}{k+1} \sum_{0 \le l \le i} \frac{1}{2^{2l}} \binom{2i-l+1}{l} \binom{k+1}{2i-l+1},$$

for any integers i and k with $0 \le i \le k$. In the remainder of this proof we show the above equality.

By putting $j = 2i+1$ and $m = k+1$, the above formula is rewritten as

$$\frac{1}{2^j} \binom{2m}{j} = \sum_{0 \le l \le j/2} \frac{1}{2^{2l}} \binom{j-l}{l} \binom{m}{j-l} \tag{G.9}$$

by properties of binomial coefficients. We put the difference of the two sides of this formula to be $c_j(m)$. We regard $c_j(m)$ as a polynomial in m for each positive integer j, regarding m as an indeterminate. We have the following recursive formula for $c_j(m)$,

$$c_j(m) - c_j(m-1) = c_{j-1}(m) + \frac{1}{4} c_{j-2}(m),$$

since each side of (G.9) satisfies the same recursive formula. Further, $c_1(m) = c_2(m) = 0$ from the definition of $c_j(m)$. We show that $c_j(m) = 0$ for any positive integer j by induction on j. Suppose that $c_{j-1}(m) = c_{j-2}(m) = 0$ by the assumption of the induction. Then, $c_j(m) = c_j(m-1)$ by the above recursive formula for $c_j(m)$. Since $c_j(0) = 0$ from the definition of $c_j(m)$, the polynomial $c_j(m)$ has infinitely many zeros; this implies that $c_j(m) = 0$. Hence, we obtain (G.9). □

Proof of Proposition G.12. The numerator on the left hand side of the required formula (G.5) is computed as

$$2^d \left(\frac{d}{df} \right)^d \int_{(sl_2)^*} e^{fC'/2} dv = 2^d \left(\frac{d}{df} \right)^d \left(\frac{1}{\sqrt{f}} \right)^3 \int_{(sl_2)^*} e^{C'/2} dv.$$

Hence, the left hand side of (G.5) is equal to

$$2^d d! \binom{-\frac{3}{2}}{d} f^{-d-3/2}. \tag{G.10}$$

In the remainder of this proof we compute the right hand side of (G.5).

We have that

$$\chi(C'^d e^{fC'/2}) = \sum_{k\geq 0} \frac{f^k}{2^k k!} \chi(C'^{k+d}) = \sum_{k\geq 0} \frac{f^k}{2^k k!} Q_{k+d}$$

from the definition of the polynomial Q_{k+d} in C. Further, by (9.4),

$$\mathrm{trace}_{nV_n}(C^m) = n \cdot \mathrm{trace}\Big(\big(\frac{n^2-1}{2}\big)^m \cdot \mathrm{id}_{V_n} \Big) = n^2 \Big(\frac{n^2-1}{2}\Big)^m.$$

Hence,

$$\mathrm{trace}_{nV_n}\big(\chi(C'^d e^{fC'/2})\big)$$
$$= \sum_{k\geq 0} \frac{f^k}{2^k k!} \sum_{i=0}^{k} b_i n^{2k+2d-2i+2} \Big(\frac{2^{2i-k-d}}{2i+1} \binom{k+d}{2i} + \sum_{0\leq j<i} \frac{2^{2j-k-d}}{i-j} \binom{i+j}{i-j-1} \binom{k+d}{i+j} \Big),$$

where $\binom{k+d}{2i}$ and $\binom{k+d}{i+j}$ are further replaced by

$$\binom{k+d}{2i} = \sum_{l=0}^{d} \binom{k}{2i-l} \binom{d}{l}, \qquad \binom{k+d}{i+j} = \sum_{l=0}^{d} \binom{k}{i+j-l} \binom{d}{l}.$$

Therefore, $\mathrm{trace}_{nV_n}\big(\chi(C'^d e^{fC'/2})\big)$ is presented by a linear sum of

$$\sum_{k\geq 0} \frac{f^k n^{2k-2a}}{4^k k!} \binom{k}{b} = \sum_{k\geq 0} \frac{f^k n^{2k-2a}}{4^k b!(k-b)!} = \frac{f^b n^{2b-2a}}{4^b b!} e^{fn^2/4} = \frac{f^b}{4^a b!} \Big(\frac{d}{df}\Big)^{b-a} e^{fn^2/4}$$

for some a's and b's. Integrating the above formula with respect to n, we have that

$$\frac{f^b}{4^a b!} \Big(\frac{d}{df}\Big)^{b-a} \frac{1}{\sqrt{f}} \int_{n\in\mathbb{R}} e^{n^2/4} dn.$$

By putting the above formal Gaussian integral to be I, the above formula is equal to

$$\frac{f^a (b-a)!}{4^a b!} \binom{-1/2}{b-a} \frac{1}{\sqrt{f}} I.$$

Hence,

$$\int_{n\in\mathbb{R}} \mathrm{trace}_{nV_n}\big(\chi(C'^d e^{fC'/2})\big) dn$$
$$= \sum_{0\leq j\leq i} \sum_{l=0}^{d} b_i \frac{2^{2j-d}}{i-j} \binom{i+j}{i-j-1} \binom{d}{l} \frac{f^{i-d-1}(j+d-l+1)!}{4^{i-d-1}(i+d-l)!} \binom{-1/2}{j+d-l+1} \frac{1}{\sqrt{f}} I.$$

Each summand of the above sum is equal to

$$b_i 2^{2j+d-2i+2} f^{i-d-3/2} \frac{(i+j)!}{(i-j)!(2j+1)!} \binom{d}{l} \frac{(j+d-l+1)!}{(i+d-l)!} \binom{-1/2}{j+d-l-1} I.$$

Further,

$$(j+d-l+1)! \binom{-1/2}{j+d-l+1} = \Big(-\frac{1}{2}\Big)\Big(-\frac{3}{2}\Big)\cdots\Big(-\frac{1}{2}-(j+d-l)\Big),$$

$$\frac{(2j+1)!}{2^{2j}\cdot j!} = \frac{(2j+1)!!}{2^j} = \frac{3}{2}\cdot\frac{5}{2}\cdot\cdots\cdot\frac{2j+1}{2}.$$

As the quotient of the above two formulae, we have that

$$\frac{2^{2j} j!(j+d-l+1)!}{(2j+1)!} \binom{-1/2}{j+d-l+1}$$
$$= -\frac{(-1)^{j+d-l}}{2} \Big(\frac{3}{2}+j\Big)\Big(\frac{5}{2}+j\Big)\cdots\Big(\frac{1}{2}+j+d-l\Big) = -\frac{(-1)^j}{2}(d-l)! \binom{-j-\frac{3}{2}}{d-l}.$$

Therefore,

$$\int_{n\in\mathbb{R}} \mathrm{trace}_{nV_n}\big(\chi(C'^d e^{fC'/2})\big)\,dn$$
$$= -\sum_{0\le j\le i} \sum_{l=0}^{d} (-1)^j 2^{d-2i+2} f^{i-d-3/2} \frac{d!(i+j)!}{j!(i-j)!(i+j-l)!l!} \binom{-j-\frac{3}{2}}{d-l} I$$
$$= -2^{d-2i+2} f^{i-d-3/2} \frac{d!}{i!} I \sum_{0\le j\le i} (-1)^j \binom{i}{j} \sum_{l=0}^{d} \binom{j+i}{l} \binom{-j-\frac{3}{2}}{d-l}$$
$$= -2^{d-2i+2} f^{i-d-3/2} \frac{d!}{i!} \binom{i-\frac{3}{2}}{d} I \sum_{j=0}^{i} (-1)^j \binom{i}{j}$$
$$= -2^{d+1} f^{-d-3/2} d! \binom{-\frac{3}{2}}{d} I,$$

where we obtain the third equality by Lemma G.18 below. Thus, we have showed that the above formula is equal to the numerator on the right hand side of the required formula (G.5). Further, its denominator is equal to $-2I$, since the denominator is obtained from the numerator by putting $d=0$ and $f=1$. Hence, the right hand side of (G.5) is equal to (G.10). This implies the required formula. □

Lemma G.18. For any non-negative integer d, we have that

$$\sum_{l=0}^{d} \binom{j+i}{l} \binom{-j-\frac{3}{2}}{d-l} = \binom{i-\frac{3}{2}}{d},$$

where we regard i and j as indeterminates in this lemma.

Proof. We put the difference of the two sides of the required formula to be

$$c_d(i,j) = \binom{i-\frac{3}{2}}{d} - \sum_{l=0}^{d} \binom{j+i}{l} \binom{-j-\frac{3}{2}}{d-l},$$

which is regarded as a polynomial in i and j for each d. We have that

$$\begin{aligned} c_d(i,j) &= \binom{i-\frac{1}{2}}{d} + \binom{i-\frac{1}{2}}{d-1} - \sum_{l=0}^{d} \left(\binom{j+i-1}{l} + \binom{j+i-1}{l-1} \right) \binom{-j-\frac{3}{2}}{d-l} \\ &= c_d(i-1,j) + c_{d-1}(i-1,j). \end{aligned}$$

We show that $c_d(i,j) = 0$ by induction on d. By the above formula and the assumption of the induction, we have that $c_d(i,j) = c_d(i-1,j)$. This implies that $c_d(i,j)$ is independent of i. By putting $i = -j$ in the definition of $c_d(i,j)$ we obtain $c_d(i,j) = 0$. □

Appendix H

The quantum sl_2 invariant and the Kauffman bracket

In this appendix we describe the relationship between the quantum sl_2 invariant and the Kauffman bracket of framed links. We also explain the background of the definition of the Kauffman bracket from the viewpoint of the quantum invariant.

H.1 The quantum (sl_2, V) invariant by the Kauffman bracket

In this section we relate the quantum (sl_2, V) invariant to the Kauffman bracket of framed links, where V denotes the vector representation of sl_2.

Recall that $\langle L \rangle$ denotes the Kauffman bracket of a framed link L, given in Section 1.2. We will presently exhibit a proof of Theorem 4.19 which says that

$$Q^{sl_2;V}(L) = (-1)^{\#L+f(L)} \langle L \rangle \Big|_{A=q^{1/4}}, \tag{H.1}$$

where $\#L$ denotes the number of components of L and $f(L)$ denotes the sum of the framings of the components of L. In fact, putting $A = q^{1/4}$, the Kauffman bracket $\langle L \rangle$ is equal to the invariant derived, not from the ribbon Hopf algebra $(U_q(sl_2), \mathcal{R}, v)$ given in Section 4.4, but from the closely related $(U_q(sl_2), \mathcal{R}, -v)$. That is the reason for the appearance of the sign $(-1)^{\#L+f(L)}$ in the above formula.

Proof of Theorem 4.19. According to the skein relation of Lemma 1.4, the Kauffman bracket satisfies the following relation,

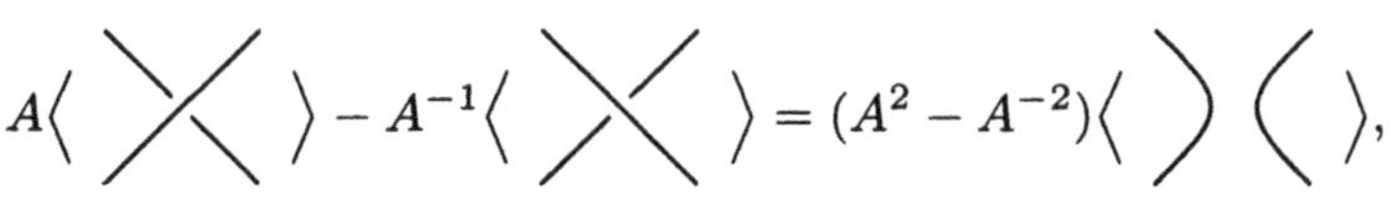

which we also call the *skein relation* of the Kauffman bracket. Note that as an invariant of framed links $(-1)^{\#L+f(L)}\langle L \rangle$ satisfies the same skein relation. Further,

by Proposition 4.18, we have that

Hence, putting $A = q^{1/4}$, both sides of (H.1) satisfy the same skein relation.

Moreover, for the trivial knot O, we have that

$$Q^{sl_2;V}(O) = q^{1/2} + q^{-1/2},$$
$$(-1)^{\#O+f(O)}\langle O\rangle = A^2 + A^{-2}.$$

Hence, putting $A = q^{1/4}$, both sides of (H.1) have the same value for the trivial knot.

Since this invariant is uniquely characterized by its skein relation and the value on the trivial knot, both sides of (H.1) must be equal. This implies the theorem. □

As a corollary of Theorem 4.19 we have the following theorem, which is used in the next section to show the second formula of Theorem H.3.

Theorem H.1. Let L be a framed link. Then,

$$Q^{sl_2;V}(L) = (-1)^{\#L}\langle L\rangle\Big|_{A=-q^{1/4}},$$

where $\#L$ denotes the number of the components of L.

This theorem gives another expression, in addition to Theorem 4.19, of the *quantum* (sl_2, V) *invariant* in terms of the Kauffman bracket.

Proof of Theorem H.1. We obtain the theorem by Theorem 4.19 and the following relation,

$$\langle L\rangle\Big|_{A=-q^{1/4}} = (-1)^{f(L)}\langle L\rangle\Big|_{A=q^{1/4}}.$$

We show the relation as follows.

Let D be a diagram expressing L by blackboard framing. Then, as shown in Section 1.2, $(-A)^{-3w(D)}\langle D\rangle$ is equal to the Jones polynomial when we set $t^{1/2} = A^{-2}$. Since the Jones polynomial of a link is a polynomial in $t^{\pm 1/2}$, $(-A)^{-3w(D)}\langle D\rangle$ is a polynomial in $A^{\pm 2}$. Hence, it is invariant under the exchange of A and $-A$. Therefore, noting the equality $\langle D\rangle = \langle L\rangle$, the Kauffman bracket $\langle L\rangle$ changes by $(-1)^{w(D)}$ under the exchange of A and $-A$. Since the writhe $w(D)$ is defined by counting crossings of D with signs, $(-1)^{w(D)} = (-1)^{f(L)}$. Hence, we obtain the above relation. □

Back to the definition of the Kauffman bracket

We now explain the background of the definition of the Kauffman bracket from the viewpoint of the quantum (sl_2, V) invariant. By similar arguments as the following, we can obtain linear relations among other quantum invariants which might give reconstructions of the invariants; for this topic see [Kup94, Kup96a] and [M$_{\text{H}}$Oh96, MOY98, OhYa97].

As mentioned in Section 4.4 the quantum (sl_2, V) invariant can be defined as an operator invariant of oriented tangles. It is known (see, *e.g.*, [Hum72]) that the tensor product $V_2 \otimes V_2$ is isomorphic as an A module to the direct sum $\mathbb{C} \oplus V_3$. From the definition of the operator invariant, we then have that

$Q^{sl_2;V_2}\Big($  $\Big) = \mathrm{id}_{\mathbb{C}} \oplus \mathrm{id}_{V_3} \in \mathrm{End}_{U_q(sl_2)}(V_2 \otimes V_2).$

Further, it follows from the definition of the operator invariant that the operator $Q^{sl_2;V_2}\Big($  $\Big)$ can be obtained by composing the linear maps $V_2 \otimes V_2 \to \mathbb{C} \to V_2 \otimes V_2$. By Proposition 4.10 this endomorphism is an intertwiner. Hence, it is equal to a scalar multiple of the projector to the direct summand $\mathbb{C}$. Moreover, we consider the operator invariant

$Q^{sl_2;V_2}\Big($  $\Big) \in \mathrm{End}_{U_q(sl_2)}(V_2 \otimes V_2).$

This also is an intertwiner, so it is equal to a linear sum of the projectors to $\mathbb{C}$ and V_3, by Schur's lemma. Hence, it can be expressed

$Q^{sl_2;V_2}\Big($  $\Big) = c_1 Q^{sl_2;V_2}\Big($  $\Big) + c_2 Q^{sl_2;V_2}\Big($  $\Big)$

for some scalars c_1 and c_2. In order to re-discover the definition of the Kauffman bracket, we employ the following ansatz,

$\Big\langle$  $\Big\rangle = c_1 \Big\langle$  $\Big\rangle + c_2 \Big\langle$  $\Big\rangle.$

Then, these scalars can be determined, for example, by using the invariance under the Reidemeister moves, and this leads to the linear relation which gives the definition of the Kauffman bracket in Section 1.2. This is a background of the definition of the Kauffman bracket.

H.2 The quantum (sl_2, V_n) invariant by the linear skein

The aim of this section is to show Theorem H.3 below, which gives an expression for the quantum (sl_2, V_n) invariant of framed links in terms of the linear skein.

Let V_m denote the m dimensional irreducible representation of $U_q(sl_2)$ as in Section 4.4. In particular, we put $V = V_2$. It is shown by using (8.14) repeatedly that $V^{\otimes n}$ includes V_{n+1} as a direct summand. We thus regard V_{n+1} as a subspace of $V^{\otimes n}$. As in Section 3.2 we consider the operator invariant of unoriented tangles whose components are associated with the representation V (see Example 4.16).

Lemma H.2. Put $A = q^{1/4}$, and consider the box over n strands (which is a linear sum of unoriented tangles defined in Section 8.2) whose components have the representation V of $U_q(sl_2)$ associated to them.

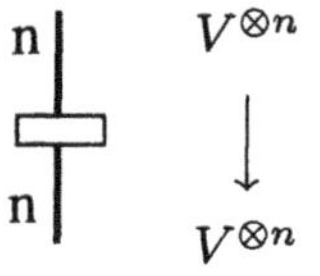

(H.2)

Then, the operator invariant of this is equal to the projector to the subspace V_{n+1} of $V^{\otimes n}$.

Proof. We show the lemma by induction on n.

We regard V_{n-1} as a subspace of $V^{\otimes(n-2)}$ and fix the following isomorphisms,

$$V_{n-1} \otimes V \otimes V \cong (V_n \oplus V_{n-2}) \otimes V \cong (V_{n+1} \oplus V_{n-1}) \oplus (V'_{n-1} \oplus V_{n-3}), \qquad \text{(H.3)}$$

where we regard each side as a subspace of $V^{\otimes n}$, and V'_{n-1} denotes a copy of V_{n-1}.

By the assumption of the induction, the operator invariant of the following tangle (to be precise, a linear sum of tangles) is equal to the projector to $V_n \otimes V$.

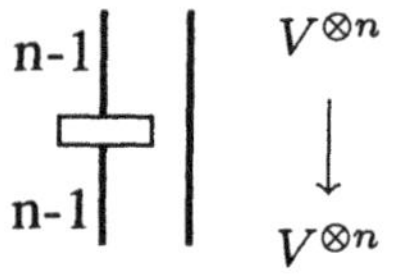

(H.4)

Fixing an isomorphism $V_n \otimes V \cong V_{n+1} \oplus V_{n-1}$ as in (H.3), the operator invariant (H.4) becomes equal to the projector to $V_{n+1} \oplus V_{n-1}$.

Further, we consider the operator invariant of the following tangle.

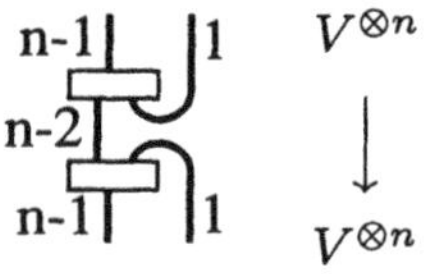

(H.5)

By Proposition 8.9 the above tangle is equal to the composition of three tangles; a copy of (H.4), the tangle below, and another copy of (H.4).

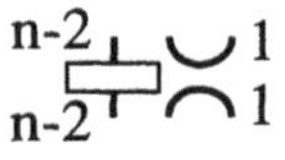

The operator invariant of the right part of the above tangle is equal to a scalar multiple of the projector to the unit representation V_1, which is a direct summand of $V \otimes V$. Hence, the operator invariant of the above tangle is equal to a scalar multiple of the projector to V_{n-1}; therefore, so is the operator invariant (H.5). Further, the scalar for the operator invariant (H.5) is equal to $[n]/[n-1]$, since it follows from Proposition 8.9 that the composition of two copies of (H.5) is equal to $[n]/[n-1]$ times (H.5). Therefore, the operator invariant (H.5) is equal to $[n]/[n-1]$ times the projector to V_{n-1}.

Hence, the following difference

$$\big(\text{the operator invariant (H.4)}\big) \quad - \quad \frac{[n-1]}{[n]}\big(\text{the operator invariant (H.5)}\big)$$

is equal to the projector to V_{n+1}. Further, by (8.5) it is also equal to the operator invariant (H.2). Therefore, we obtain the lemma. □

Let L be a framed link with l components. We denote by $L^{[n_1,\cdots,n_l]}$ the framed link obtained from L by replacing, for each i, the ith component with n_i parallel copies together with a box over n_i strands (Jones-Wenzl idempotent) connected into them. For example,

when $L =$  , we put $L^{[n,m]} =$ 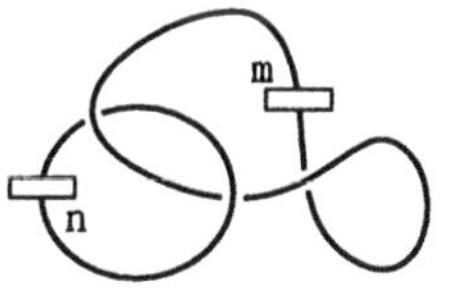

.

The following theorem is used in Section 8.3 to prove Theorem 8.12.

Theorem H.3. Let L be a framed link with l components. Then,

$$Q^{sl_2;V_{n_1+1},\cdots,V_{n_l+1}}(L) = (-1)^N \langle L^{[n_1,\cdots,n_l]}\rangle\big|_{A=q^{1/4}},$$

where we put $N = \sum_{i+1}^{l} n_i(f_i+1)$. Here, f_i is the framing of the ith component of L. Further,

$$Q^{sl_2;V_{n_1+1},\cdots,V_{n_l+1}}(L) = (-1)^{n_1+\cdots+n_l}\langle L^{[n_1,\cdots,n_l]}\rangle\big|_{A=-q^{1/4}}.$$

This theorem gives a reconstruction of the *quantum* (sl_2, V_n) *invariant* in terms of the linear skein.

Proof of Theorem H.3. Let $L^{(n_1,\cdots,n_l)}$ be the link obtained from L by replacing the ith component with n_i parallel copies of the component, for each i. By Proposition 4.13 and Theorem 4.19, we have that

$$Q^{sl_2;V^{\otimes n_1},\cdots,V^{\otimes n_l}}(L) = Q^{sl_2;V}\big(L^{(n_1,\cdots,n_l)}\big) = (-1)^{N'}\langle L^{(n_1,\cdots,n_l)}\rangle\big|_{A=q^{1/4}},$$

where $N' = \#L^{(n_1,\cdots,n_l)} + f(L^{(n_1,\cdots,n_l)})$. Further, N' is congruent modulo 2 to $N = \sum_{i+1}^{l} n_i(f_i+1)$. Since we obtain $L^{[n_1,\cdots,n_l]}$ from $L^{(n_1,\cdots,n_l)}$ by connecting in boxes (Jones-Wenzl idempotents), we obtain the first formula of the theorem by applying Lemma H.2 to the above formula.

By using Theorem H.1 instead of Theorem 4.19 in the above argument we obtain the second formula of the theorem. □

Bibliography

[Abe80] Abe, E., *Hopf algebras*, Cambridge University Press, 1980.

[Ada94] Adams, C.C., *The knot book, An elementary introduction to the mathematical theory of knots*, W. H. Freeman and Company, New York, 1994.

[AkMc90] Akbulut, S., McCarthy, J.D., *Casson's invariant for oriented homology 3-spheres, An exposition*, Mathematical Notes **36**, Princeton University Press, 1990.

[ADO92] Akutsu, Y., Deguchi, T., Ohtsuki, T., *Invariants of colored links*, J. Knot Theory Ramifications **1** (1992) 161–184.

[Ale23] Alexander, J.W., *Topological invariants of knots and links* Trans. Amer. Math. Soc. **20** (1923) 275–306.

[AlCo92] Altschuler, D., Coste, A., *Quasi-quantum group, knots, three-manifolds, and topological field theory*, Comm. Math. Phys. **150** (1992) 83–107.

[AlFr95] Altschuler, D., Freidel. L., *On universal Vassiliev invariants*, Comm. Math. Phys. **170** (1995) 41–62.

[AlFr96] ———, *Vassiliev knot invariants and Chern-Simons perturbation theory to all order*, Comm. Math. Phys. **187** (1997) 261–287.

[And93] Andersen, H.H., *Quantum groups, invariants of* 3*-manifolds and semisimple tensor categories*, Quantum deformations of algebras and their representations (Ramat-Gan, 1991/1992; Rehovot, 1991/1992), 1–12, Israel Math. Conf. Proc. **7**, Bar-Ilan Univ., Ramat Gan, 1993.

[AMR98] Andersen, J.E., Mattes, J., Reshetikhin, N., *Quantization of the algebra of chord diagrams*, Math. Proc. Camb. Phil. Soc. **124** (1998) 451–467.

[Arn69] Arnold, V.I., *The cohomology ring of the colored braid group*, Mat. Zametki **5** (1969) 227–231.

[Ati89] Atiyah, M.F., *Topological quantum field theories*, Publ. Math. IHES **68** (1989) 175–186.

[Ati90a] ———, *On framings of 3-manifolds*, Topology **29**, (1990) 1–7.

[Ati90b] ———, *The geometry and knots and physics*, Cambridge University Press, 1990.

[AtBo82] Atiyah, M.F., Bott, R., *The Yang-Mills equation over Riemann surfaces*, Phil. Trans. R. Soc. London A, **308** (1982) 523–615.

[AHLS88] Atiyah, M.F., Hitchin, N., Lawrence, R., Segal, G., Oxford seminar on Jones-Witten theory, (1988).

[ADW91] Axelrod, S., Della Pietra, S., Witten, E., *Geometric quantization of Chern-Simons gauge theory*, J. Differential Geom. **33** (1991) 787–902.

[AxSi92] Axelrod, S., Singer, I. M., *Chern-Simons perturbation theory*, Proc. XXth DGM Conference (New York, 1991), S. Catto and A. Rocha, eds., World Scientific, 3–45 (1992).

[AxSi94] ———, *Chern-Simons perturbation theory. II*, J. Diff. Geom. **39**, 173–213 (1994).

[BaKi00] Bakalov, B., Kirillov, A. Jr., *On the Lego-Teichmüller game*, Transform. Groups **5** (2000) 207–244.

[BaKi01] ———, *Lectures on tensor categories and modular functors*, University Lecture Series **21**. American Mathematical Society, Providence, RI, 2001.

[Bar91] Bar-Natan, D., *Perturbative aspects of the Chern-Simons topological quantum field theory*, PhD thesis, Princeton University, 1991.

[Bar95a] ———, *Perturbative Chern-Simons theory*, J. Knot Theory Ramifications **4** (1995) 503–547.

[Bar95b] ———, *On the Vassiliev knot invariants*, Topology, **34** (1995) 423–472.

[Bar95c] ———, *Vassiliev homotopy string link invariants*, J. Knot Theory Ramifications **4** (1995) 13-32.

[Bar97] ———, *Non-associative tangles*, Geometric topology (Athens, GA, 1993), 139–183, AMS/IP Stud. Adv. Math., 2.1, Amer. Math. Soc., Providence, RI, 1997.

[Bar98] ———, *On associators and the Grothendieck-Teichmuller group I*, Selecta Math. (N.S.) **4** (1998) 183–212.

[Bar] ———, Bibliography of Vassiliev invariants, `http://www.ma.huji.ac.il/~drorbn`.

[BaGa96] Bar-Natan, D., Garoufalidis, S., *On the Melvin-Morton-Rozansky conjecture*, Invent. Math. **125** (1996) 103–133.

[BGRT*97a*] Bar-Natan, D., Garoufalidis, S., Rozansky, L., Thurston, D.P., *The Aarhus integral of rational homology 3-spheres I: A highly non trivial flat connection on* S^3, to appear in Selecta Math., ——— *II: Invariance and universality*, to appear in Selecta Math., ——— *III: The relation with the Le-Murakami-Ohtsuki invariant*, preprint, math/9808013.

[BGRT00] ———, *Wheels, wheeling, and the Kontsevich integral of the unknot*, Israel J. Math. **119** (2000) 217–237.

[BaLa*00*] Bar-Natan, D., Lawrence, R., *A rational surgery formula for the LMO Invariant*, preprint, math.GT/0007045.

[BaSt97] Bar-Natan, D., Stoimenow, A., *The fundamental theorem of Vassiliev invariants*, Geometry and physics (Aarhus, 1995), 101–134, Lecture Notes in Pure and Appl. Math. **184**, Dekker, New York, 1997.

[Bax82] Baxter, R.J., *Exactly solved models in statistical mechanics*, Academic Press, New York, 1982.

[BeLa94] Beauville, A., Laszlo, Y., *Conformal blocks and generalized theta functions*, Commun. Math. Phys. **164** (1994) 385–419.

[BeDr82] Belavin, A.A., Drinfel'd, V.G., *Solutions of the classical Yang-Baxter equation for simple Lie algebras*, Funct. Anal. Appl. **16** (1982) 159–180.

[Bir74] Birman, J.S., *Braids, links and mapping class groups*, Ann. of Math. Studies **82**, Princeton University Press, 1974.

[Bir93] ———, *New points of view in knot theory*, Bull. Amer. Math. Soc. (N.S.) **28** (1993) 253–287.

[BiLi93] Birman, J.S., Lin, X.-S., *Knot polynomials and Vassiliev's invariants*, Invent. Math. **111** (1993) 225–270.

[BiTr98] Birman, J.S., Trapp, R., *Braided chord diagrams*, J. Knot Theory Ramifications **7** (1998) 1–22.

[Bla00] Blanchet, C., *Hecke algebras, modular categories and 3-manifolds quantum invariants*, Topology 39 (2000) 193–223.

[BHMV92] Blanchet, C., Habegger, N., Masbaum, G., Vogel, P., *Three-manifold invariants derived from the Kauffman bracket*, Topology, **31** (1992) 685–699.

[BHP68] Boone, W.W., Haken, W., Poenaru, V., *On recursively unsolvable problems in topology and their classification*, Contributions to Math. Logic (Colloquium, Hannover, 1966), North-Holland, Amsterdam, (1968) 37–74.

[BoCa98] Bott, R., Cattaneo, A.S., *Integral invariants of 3-manifolds*, J. Diff. Geom. **48** (1998) 91–133.

[BoCa99] ———, *Integral invariants of 3-manifolds II*, J. Diff. Geom. **53** (1999) 1–13.

[BoTa94] Bott, R., Taubes, C., *On the self-linking of knots*, J. Math. Phys. **35** (1994).

[Bro97] Broadhurst, D.J., *Conjectured enumeration of Vassiliev invariants*, preprint, q-alg/9709031.

[BuZi85] Burde, G., Zieschang, H., *Knots*, Studies in Math. **5** (1985) Walter de Gruyter.

[ChPr95] Chari, V., Pressley, A., *A guide to quantum groups*, Corrected reprint of the 1994 original, Cambridge University Press, Cambridge, 1995.

[Che75] Chen, K.T., *Iterated integrals, fundamental groups and covering spaces*, Trans. Amer. Math. Soc. **206** (1975) 83–98.

[ChVa97] Chmutov, S.V., Varchenko, A.N., *Remarks on the Vassiliev knots invariants coming from* sl_2, Topology **36** (1997) 153–178.

[CGO01] Cochran, T. D., Gerges, A., Orr, K., *Dehn surgery equivalence relations on 3-manifolds*, Math. Proc. Cambridge Philos. Soc. **131** (2001) 97–127.

[CoMe00] Cochran, T.D., Melvin, P., *Finite type invariants of 3-manifolds*, Invent. Math. **140** (2000) 45–100.

[CoMe01] ———, *Quantum cyclotomic orders of 3-manifolds*, Topology **40** (2001) 95–125.

[Con70] Conway, J.H., *An enumeration of knots and links, and some of their algebraic properties*, Computational Problems in Abstract Algebra (Proc. Conf., Oxford, 1967), Pergamon, Oxford, (1970) 329–358.

[CrFo77] Crowell, R.H., Fox, R.H., *Introduction to knot theory*, Reprint of the 1963 original, Graduate Texts in Math. **57**, Springer-Verlag, New York-Heidelberg, 1977.

[Deh38] Dehn, M., *Die Gruppe der Abbildungsklassen*, Acta. Math. **69** (1938) 135–206.

[DiWi90] Dijkgraaf, R., Witten, E., *Topological gauge theories and group cohomology*, Comm. Math. Phys. **129** (1990) 393–429.

[Dix77] Dixmier, J., *Enveloping algebras*, North-Holland, Amsterdam (1997).

[Dri85] Drinfel'd, V.G., *Hopf algebras and the quantum Yang-Baxter equation*, Sov. Math. Dokl. **32** (1985) 254–258.

[Dri87] ———, *Quantum groups*, In Proceedings of Int. Cong. Math. (Berkeley 86), Amer. Math. Soc., Providence (1987) 798–820.

[Dri89a] ———, *On almost cocommutative Hopf algebras*, Algebra i Analiz **1** (1989) 30–47 (in Russian), English translation in Leningrad Math. J. **1** (1990) 321–342.

[Dri89b] ———, *Quasi-Hopf algebras*, Algebra i Analiz **1** (1989) 114–148 (in Russian), English translation in Leningrad Math. J. **1** (1990) 1419–1457.

[Dri90] ———, *On quasi-triangular Hopf algebras and a group closely connected with* $Gal(\overline{\mathbb{Q}}/\mathbb{Q})$, Algebra i Analiz **2** (1990) 149–181 (in Russian), English translation in Leningrad Math. J. **2** (1991) 829–860.

[Dur77] Durfee, A.H., *Bilinear and quadratic forms on torsion modules*, Adv. in Math. **25** (1977) 133–164.

[EFK98] Etingof, P., Frenkel, I., Kirillov, A., Jr., *Lectures on representation theory and Knizhnik-Zamolodchikov equations*, Amer. Math. Soc., Providence, RI, 1998.

[Epp99] Epple, M., *Geometric aspects in the development of knot theory*, History of topology, 301–357, North-Holland, Amsterdam, 1999.

[FeRo79] Fenn, R., Rourke, C., *On Kirby's calculus of framed links*, Topology **18** (1979) 1–15.

[FrGo91] Freed, D.S., Gompf, R.E., *Computer calculation of Witten's 3-manifold invariant*, Comm. Math. Phys. **141** (1991) 79–117.

[FrYe89] Freed, P., Yetter, D., *Braided compact closed categories with applications to low dimensional topology*, Adv. in Math. **77**(2) (1989) 156–182.

[Fre*98*] Freidel, L., *From* $sl(2)$ *Kirby weight systems to the asymptotic 3-manifold invariant*, preprint, math.QA/9802069.

[F+85] Freyd, P., Yetter, D., Hoste, J., Lickorish, W.B.R., Millett, K., Ocneanu, A., *A new polynomial invariant of knots and links*, Bull. Amer. Math. Soc. (N.S.) **12** (1985) 239–246.

[Fuk96] Fukaya, K., *Morse homotopy and Chern-Simons perturbation theory*, Comm. Math. Phys. **181** (1996) 37–90.

[Fuk97] ———, *Morse homotopy and its quantization*, Geometric topology (Athens, GA, 1993), 409–440, AMS/IP Stud. Adv. Math., 2.1, Amer. Math. Soc., Providence, RI, 1997.

[FuHa91] Fulton, W., Harris, J., *Representation theory*, GTM **129** (1991) Springer-Verlag.

[Gar92] Garoufalidis, S., *Relations among 3-manifold invariants*, Thesis, Chicago University (1992).

[Gar96] ———, *On finite type 3-manifold invariants I*, J. Knot Theory Ramifications **5** (1996) 441–462.

[Gar98a] ———, *Applications of TQFT invariants to low-dimensional topology*, Topology **37** (1998) 219–224.

[Gar98b] ———, *A reappearance of wheels*, J. Knot Theory Ramifications **7** (1998) 1065–1071.

[Gar99] ———, *Signatures of links and finite type invariants of cyclic branched covers*, Tel Aviv Topology Conference: Rothenberg Festschrift (1998), 87–97, Contemp. Math. **231**, Amer. Math. Soc., Providence, RI, 1999.

[GGP01] Garoufalidis, S., Goussarov, M., Polyak, M., *Calculus of clovers and finite type invariants of 3-manifolds*, Geom. Topol. **5** (2001) 75–108 (electronic).

[GaHa00] Garoufalidis, G., Habegger, N., *The Alexander polynomial and finite type 3-manifold invariants*, Math. Annalen **316** (2000) 485-497.

[GaKr*01*] Garoufalidis, S., Kricker, A., *A rational noncommutative invariant of boundary links*, preprint, math.GT/0105028.

[GaLe96] Garoufalidis, S., Levine, J., *On finite type 3-manifold invariants II*, Math. Ann. **306** (1996) 691–718.

[GaLe97a] ———, *On finite type 3-manifold invariants IV: comparison of definitions*, Math. Proc. Camb. Phil. Soc. **122** (1997) 291–300.

[GaLe97b] ———, *Finite type 3-manifold invariants; the mapping class group and blinks*, J. Diff. Geom. **47** (1997) 257–320.

[GaLe98] ———, *Finite type* 3-*manifold invariants and the structure of the Torelli group. I*, Invent. Math. **131** (1998) 541–594.

[GaOh98] Garoufalidis, S., Ohtsuki, T., *Finite type 3-manifold invariants III: manifold weight system*, Topology **37** (1998) 227–243.

[GaRo*00*] Garoufalidis, S., Rozansky, L., *The loop expansion of the Kontsevich integral, abelian invariants of knots and S-equivalence*, math.GT/0003187.

[Gaw90a] Gawedzki, K., *Wess-Zumino-Witten conformal field theory*, Constructive quantum field theory II, G. Velo and A.S. Wightman eds., Plenum Press, 1990, 89–120.

[Gaw90b] ———, *Geometry of Wess-Zumino-Witten models of conformal field theory*, Recent advances in field theory (Annecy-le-Vieux, 1990), Nuclear Phys. B Proc. Suppl. **18B** (1990) 78–91.

[Gaw99] ———, *Lectures on conformal field theory*, Quantum fields and strings: a course for mathematicians, (Princeton, NJ, 1996/1997), 727–805, Amer. Math. Soc., Providence, RI, 1999.

[Gil99] Gilmer, P.M., *Skein theory and Witten-Reshetikhin-Turaev invariants of links in lens spaces*, Comm. Math. Phys. **202** (1999) 411–419.

[Gol84] Goldman, W.M., *The symplectic nature of fundamental groups of surfaces*, Adv. in Math. **54** (1984) 200–225.

[Gor78] Gordon, C.M., *Some aspects of classical knot theory*, Knot Theory, Proceedings 1977, J.C. Hausmann (ed.), Lect. Note in Math. **685** Springer-Verlag (1978) 1–60.

[Gou94a] Goussarov, M.N., *A new form of the Conway-Jones polynomial of oriented links*, in Topology of manifolds and varieties, edited by O Viro, Amer. Math. Soc., Providence (1994) 167-172.

[Gou94b] ———, *On n-equivalence of knots and invariants of finite degree*, in Topology of manifolds and varieties, edited by O Viro, Amer. Math. Soc., Providence (1994) 173–192.

[Gou95] ———, *Knotted graphs and a geometrical technique of n-equivalence*, POMI Sankt Petersburg preprint, circa 1995 (in Russian).

[Gou98] ———, *New theory of invariants of finite degree for 3-manifolds*, preprint, POMI, St. Petersburg, 1998.

[Gou99] ———, *Finite type invariants and n-equivalence of* 3*-manifolds*, C. R. Acad. Sci. Paris Ser. I Math. **329** (1999) 517–522.

[GPV00] Goussarov, M., Polyak, M., Viro, O., *Finite type invariants of classical and virtual knots*, Topology **39** (2000) 1045–1068.

[GrLi99] Greenwood, M., Lin, X.-S., *On Vassiliev knot invariants induced from finite type 3-manifold invariants*, Trans. Amer. Math. Soc. **351** (1999) 3659–3672.

[GMM89] Guadagnini, E., Martellini, M., Mintchev, M., *Perturbative aspects of the Chern-Simons field theory*, Phys. Lett. **B227** (1989) 111.

[GMM90] ———, *Chern-Simons field theory and link invariants*, Nucl. Phys. **B330** (1990) 575.

[Habe95] Habegger, N., *Finite type 3-manifold invariants: a proof of a conjecture of Garoufalidis*, preprint, 1995.

[Habe96] ———, *A computation of the universal quantum 3-manifold invariant for manifolds of rank greater than 2*, preprint, 1996, available at `http://www.math.sciences.univ-nantes.fr`.

[Habe01] ———, *The topological IHX relation*, J. Knot Theory Ramifications **10** (2001) 309–329.

[HaBe00] Habegger, N., Beliakova, A., *The Casson-Walker-Lescop invariant as a quantum 3-manifold invariant*, J. Knot Theory Ramifications **9** (2000) 459–470.

[HaMa00] Habegger, N., Masbaum, G., *The Kontsevich integral and Milnor's invariants*, Topology **39** (2000) 1253–1289.

[HaOr99] Habegger, N., Orr, K.E., *Finite type three manifold invariants — realization and vanishing*, J. Knot Theory Ramifications **8** (1999) 1001–1007.

[Habi97] Habiro, K., *Clasper theory and its applications*, in preparation, lecture note at Oiwake seminar, based on notes by S. Fujii, M. Hirasawa, T. Kadokami, A. Koike, B. Matsuzaka, T. Murakami, Y. Nakagawa, T. Nakatsuru, K. Nishizawa, H. Nisihara, A. Shima, T. Shintani, K. Suzuki, Y. Tsujii, Y. Tsutsumi, T. Yagi, Y. Yamada, T. Yamauchi.

[Habi00] ______, *Claspers and finite type invariants of links*, Geometry and Topology **4** (2000) 1–83.

[HaWr79] Hardy, G., Wright, E., *An introduction to the theory of numbers*, Oxford University Press, fifth edition, 1979.

[HaHa70] Hartley, R., Hawkes, T.O., *Rings, modules and linear algebra*, Chapman and Hall (1970).

[Hemi92] Hemion, C., *The classification of knots and 3-dimensional spaces*, Oxford Univ. Press, New York, 1992.

[Hemp76] Hempel, J., *3-Manifolds*, Ann. of Math. Studies **86**, Princeton University Press, Princeton, NJ; University of Tokyo Press, Tokyo, 1976.

[Hil81] Hilden, H.M., *Representations of homology 3-spheres*, Pacific J. Math. **94** (1981) 125–129.

[Hir94] Hirsch, M.W., *Differential topology*, Corrected reprint of the 1976 original, Graduate Texts in Mathematics **33**, Springer-Verlag, New York, 1994.

[Hit90] Hitchin, N.J., *Flat connections and geometric quantization*, Comm. Math. Phys. **131** (1990) 347–380.

[Hos86] Hoste, J., *A formula for Casson's invariant*, Trans. Amer. Math. Soc. **297** (1986) 547-562.

[HTW98] Hoste, J., Thistlethwaite, M., Weeks, J., *The first 1,701,936 knots*, Math. Intelligencer **20** (1998) 33–48.

[Hum72] Humphreys, J.E., *Introduction to Lie algebras and representation theory*, GTM **9** Springer-Verlag, 1972.

[Jef92] Jeffery, L.C., *Chern-Simons-Witten invariants of lens spaces and torus bundles, and the semiclassical approximation*, Comm. Math. Phys. **147** (1992) 563–604.

[Jim85] Jimbo, M., *A q-difference analogue of $U(\mathfrak{g})$ and the Yang-Baxter equation*, Lett. Math. Phys. **10** (1985) 63–69.

[Jim86] ______, *Quantum R matrix for the generalized Toda system*, Comm. Math. Phys. **102** (1986) 537–547.

[Jon83] Jones, V.F.R., *Index for subfactors*, Invent. Math. **72** (1983) 1–25.

[Jon85] ______, *A polynomial invariant for knots via von Neumann algebras*, Bull. Amer. Math. Soc. **12** (1985) 103–111.

[Jon89] ______, *On knot invariants related to some statistical mechanical models*, Pacific J. Math. **137** (1989) 311–336.

[Kac90] Kac, V.G., *Infinite-dimensional Lie algebras*, Third edition. Cambridge University Press, Cambridge, 1990.

[Kap99] Kapranov, M., *Rozansky-Witten invariants via Atiyah classes*, Compositio Math. **115** (1999) 71–113.

[Kau81] Kauffman, L.H., *The Conway polynomial*, Topology **20** (1981) 101–108.

[Kau87a] ______, *On knots*, Annals of Math. Studies **115**, 1987.

[Kau87b] ______, *State models and the Jones polynomial*, Topology **26** (1987) 395–407.

[Kau88a] ______, *Statistical mechanics and the Jones polynomial*, Contemporary Math. **78** (1988).

[Kau88b] ———, *New invariants in the theory of knots*, Asterisque **163-164** (1988) 137–219.

[Kau89] ———, *Knots, abstract tensors and Yang-Baxter equation*, in "Knots, Topology and Quantum Field Theory", Florence (1989) L. Lusanna ed., World Scientific.

[Kau91] ———, *Knots and Physics*, Series of knots and everything **1**, World Scientific, 1991.

[Kau97] ———, *Invariants of links and three-manifolds via Hopf algebras*, Geometry and physics (Aarhus, 1995), 471–479, Lecture Notes in Pure and Appl. Math. **184**, Dekker, New York, 1997.

[Kau98] ———, *Witten's integral and the Kontsevich integral*, Particles, fields, and gravitation, 368–381, AIP Conf. Proc. **453**, Amer. Inst. Phys., Woodbury, NY, 1998

[Kau99a] ———, *Virtual knot theory*, European J. Combin. **20** (1999) 663–690.

[Kau99b] ———, *Right integrals and invariants of three-manifolds*, Proceedings of the Kirbyfest (Berkeley, CA, 1998), 215–232 (electronic), Geom. Topol. Monogr. **2**, Geom. Topol., Coventry, 1999.

[KaLi94] Kauffman, L.H., Lins, S. L., *Temperley-Lieb recoupling theory and invariants of 3-manifolds*, Annals of Math. Studies **134**, Princeton Univ. Press, 1994.

[KaRa95] Kauffman, L.H., Radford. D.E., *Invariants of 3-manifolds derived from finite dimensional Hopf algebras*, J. Knot Theory Ramifications **4** (1995) 131–162.

[Kas94] Kassel, C., *Quantum groups*, Graduate Texts in Math. **155**, Springer-Verlag, 1994.

[KRT97] Kassel, C., Rosso, M., Turaev, V., *Quantum groups and knot invariants*, Panoramas et Synthéses (Panoramas and Syntheses) **5**, Société Mathématique de France, Paris, 1997.

[Kaw+90] Kawauchi, A., A survey of knot theory, Birkhäuser, 1996, English translation of "結び目理論 (Knot Theory)" (in Japanese), Springer-Verlag Tokyo, 1990, edited by A. Kawauchi, written by F. Hosokawa, T. Kanenobu, A. Kawauchi, S. Kinoshita, T. Kobayashi, T. Maeda, Y. Marumoto, K. Morimoto, H. Murakami, J. Murakami, Y. Nakanishi, M. Sakuma, T. Shibuya, J. Tao, S. Yamada and K. Yoshikawa.

[KaKo80] Kawauchi, A., Kojima, S., *Algebraic classification of linking pairings on 3-manifolds*, Math. Ann. **253** (1980) 29–42.

[Ker98a] Kerler, T., *Equivalence of a bridged link calculus and Kirby's calculus of links on non-simply connected 3-manifolds*, Topology Appl. 87 (1998) 155–162.

[Ker98b] ———, *On the connectivity of cobordisms and half-projective TQFT's*, Comm. Math. Phys. **198** (1998) 535–590.

[Ker99] ———, *Bridged links and tangle presentations of cobordism categories*, Adv. Math. 141 (1999) 207–281.

[Kirb78] Kirby, R.C., *A calculus for framed links in* S^3, Invent. Math. **45** (1978) 35–56.

[KiMe91] Kirby, R., Melvin, P., *The 3-manifold invariants of Witten and Reshetikhin-Turaev for* $sl(2,\mathbb{C})$, Invent. Math. **105** (1991) 473–545.

[KiMe99] ———, *Canonical framings for 3-manifolds*, Proceedings of 6th Gökova Geometry-Topology Conference, Turkish J. Math. **23** (1999) 89–115.

[KiRe89] Kirillov, A.N., Reshetikhin, N., *Representations of the algebra* $U_q(\mathrm{sl}(2))$, *q-orthogonal polynomials and invariants of links*, Infinite dimensional Lie algebra and groups, Proc. Conf., Marseille/Fr. 1988, Adv. Ser. Math. Phys. **7**, 285–337 (1989).

[KiRe90] ———, *q-Weyl group and a multiplicative formula for universal r-matrices*, Comm. Math. Phys. **134** (1990) 421–431.

[Kirw84] Kirwan, F.C., *Cohomology of quotients in symplectic and algebraic geometry*, Mathematical Notes **31**, Princeton University Press, 1984.

[KnPu53] Kneser, M., Puppe, P., *Quadratische Formen und Verschlingungsinvarianten von Knoten*, Math. Z. **58** (1953) 376–384.

[KnZa84] Knizhnik, V.Z., Zamolodchikov, A.B., *Current algebra and Wess-Zumino models in two dimensions*, Nucl. Phys. **B247** (1984) 83–103.

[Kod] Kodama, K., *KNOT*, computer software, available at `http://kaikei.math.kobe-u.ac.jp/~kodama/`.

[KoSu01] Kodiyalam, V., Sunder, V.S., *Topological quantum field theories from subfactors*, Chapman & Hall/CRC Res. Notes Math. **423**, Chapman & Hall/CRC, Boca Raton, FL, 2001.

[Koh87] Kohno, T., *Monodromy representations of braid groups and Yang-Baxter equations*, Ann. Inst. Fourier (Grenoble) **37** (1987) 139–160.

[Koh88] ———, *Linear representations of braid groups and classical Yang-Baxter equations*, Contemp. Math. **78** (1988) 339–363.

[Koh92] ———, *Topological invariants for 3-manifolds using representations of mapping class group I*, Topology **31** (1992) 203–230.

[Koh94] ———, *Vassiliev's invariants and de-Rham complex on the space of knots*, Contemp. Math. **179** (1994) 123–138.

[Koh97] ———, *Chern-Simons perturbative invariants*, Lectures at Knots 96, edited by Shin'ichi Suzuki, World Scientific, (1997) 235–261.

[Koh98] ———, *Conformal field theory and topology*, Tranlations of Mathmatical Monographs (to appear), American Mathematical Society, English translation of " 場の理論とトポロジー (Field theory and topology)" (in Japanese), 岩波書店 (Iwanami Shoten, Publishers), 1998.

[KoTa93] Kohno, T., Takata, T., *Symmetry of Witten's 3-manifold invariants for* $sl(n, \mathbb{C})$, J. Knot Theory Ramifications **2** (1993) 149–169.

[KoTa96] ———, *Level-rank duality of Witten's 3-manifold invariants*, Advanced Studies in Pure Mathematics **24** (1996) 243–264.

[Kon93] Kontsevich, M., *Vassiliev's knot invariants*, Adv. in Sov. Math **16(2)** (1993) 137–150.

[Kon94] ———, *Feynman diagrams and low-dimensional topology*, in First European Congress of Mathematics, II, Prof. Math. **120** (1994) 97–124.

[Kon99] ———, *Rozansky-Witten invariants via formal geometry*, Compositio Math. **115** (1999) 115–127.

[Kos93] Kosinski, A.A., *Differential manifolds*, Pure and Applied Mathematics **138**, Academic Press, Inc., Boston, MA, 1993.

[Kre98] Kreimer, D., *On the Hopf algebra structure of perturbative quantum field theories*, Adv. Theor. Math. Phys. **2** (1998) 303–334.

[Kre00] ———, *Knots and Feynman diagrams*, Cambridge Lecture Notes in Physics **13**. Cambridge University Press, Cambridge, 2000.

[Kri97] Kricker, A., *Alexander-Conway limits of many Vassiliev weight systems*, J. Knot Theory Ramifications **6** (1997) 687–714.

[Kri*99*] ———, *Covering spaces over claspered knots*, preprint, math/9901029.

[Kri*00a*] ———, *Branched cyclic covers and finite-type invariants*, preprint, math/0003035, to appear in J. Knot Theory Ramifications.

[Kri*00b*] ———, *The lines of the Kontsevich integral and Rozansky's rationality conjecture*, preprint, math/0005284.

[KrSp97] Kricker, A., Spence, B., *Ohtsuki's invariants are of finite type*, J. Knot Theory Ramifications **6** (1997) 583-597.

[KSA97] Kricker, A., Spence, B., Aitchison, I., *Cabling the Vassiliev invariants*, J. Knot Theory Ramifications **6** (1997) 327–358.

[Kup91] Kuperberg, G., *Involutory Hopf algebras and 3-manifold invariants*, Internat. J. Math. **2** (1991) 41–66.

[Kup94] ———, *The quantum G_2 link invariant*, Internat. J. Math. **5** (1994) 61–85.

[Kup96a] ———, *Spiders for rank 2 Lie algebras*, Comm. Math. Phys. **180** (1996) 109–151.

[Kup96b] ———, *Non-involutory Hopf algebras and 3-manifold invariants*, Duke Math. J. **84** (1996) 83–129.

[KuTh*99*] Kuperberg, G., Thurston, D.P., *Perturbative 3-manifold invariants by cut-and-paste topology*, preprint, math.GT/9912167.

[Lan94] Lang, S., *Algebraic number theory*, second edition, Graduate Texts in Math. **110**, Springer-Verlag, New York, 1994

[Law89] Lawrence, R.J., *A universal link invariant using quantum groups*, Differential geometric methods in theoretical physics (Chester, 1988), 55–63, World Sci. Publishing, Teaneck, NJ, 1989.

[Law90] ———, *A universal link invariant*, The interface of mathematics and particle physics (Oxford, 1988), 151–156, Inst. Math. Appl. Conf. Ser. New Ser. **24**, Oxford Univ. Press, New York, 1990.

[Law95] ———, *Asymptotic expansions of Witten-Reshetikhin-Turaev invariants for some simple 3-manifolds*, J. Math. Phys. **36** (1995) 6106–6129.

[Law97] ———, *Witten-Reshetikhin-Turaev invariants of 3-manifolds as holomorphic functions*, Geometry and physics (Aarhus, 1995), Lecture Notes in Pure and Appl. Math. **184**, Dekker, New York, (1997) 363–377.

[LaRo99] Lawrence, R.J., Rozansky, L., *Witten-Reshetikhin-Turaev invariants of Seifert manifolds*, Comm. Math. Phys. **205** (1999) 287–314.

[Le97] Le, T.T.Q., *An invariant of integral homology 3-sphere which is universal for all finite type invariants*, in "Soliton, Geometry and Topology: On the Crossroad", AMS Translations series 2 **179** (1997), Eds. V. Buchstaber and S. Novikov, 75–100.

[Le99] ——, *On denominators of the Kontsevich integral and the universal perturbative invariant of 3-manifolds*, Invent. Math., **135** (1999) 689-722.

[Le00a] ——, *Integrality and Symmetry of quantum link invariants*, Duke Math. J. **102** (2000) 273–306.

[Le00b] ——, *On perturbative $PSU(n)$ invariants of rational homology 3-spheres*, Topology **39** (2000) 813–849.

[Le*00*] ——, *Quantum invariants of 3-manifolds: integrality, splitting, and perturbative expansion*, preprint, math.QA/0004099.

[LMMO95] Le, T.T.Q., Murakami, H., Murakami, J., Ohtsuki, T., *A three-manifold invariant derived from the universal Vassiliev-Kontsevich invariant*, Proc. Japan Acad., **71**, Ser. A, 125–127 (1995).

[LMMO99] ——, *A three-manifold invariant via the Kontsevich integral*, Osaka J. Math. **36** (1999) 365–395.

[LeMu95a] Le, T.T.Q., Murakami, J., *Representations of the category of tangles by Kontsevich's iterated integral*, Comm. Math. Phys., **168**, 535–562 (1995).

[LeMu95b] ——, *Kontsevich's integral for the HOMFLY polynomial and relations between values of multiple zeta functions*, Topology Appl. **62** (1995) 193–206.

[LeMu96a] ——, *The universal Vassiliev-Kontsevich invariant for framed oriented links*, Comp. Math. **102** (1996) 41–64.

[LeMu96b] ——, *Kontsevich integral for the Kauffman polynomial*, Nagoya Math. J. **142** (1996) 39–65.

[LeMu97] ——, *Parallel version of the universal Vassiliev-Kontsevich invariant*, J. Pure and Applied Algebra **121** (1997) 271–291.

[LMO98] Le, T.T.Q., Murakami, J., Ohtsuki, T., *On a universal perturbative invariant of 3-manifolds*, Topology **37** (1998) 539–574.

[Lee90] Lee, H.C., *Tangles, links and twisted quantum groups* Plenum Press, New York, Physics, Geometry and Topology, NATO ASI Series **238** (1990) 623–659.

[Les92] Lescop, C., *Sur l'invariant de Casson-Walker: formule de chirurgie globale et generalisation aux varietes de dimension* 3 *fermees orientees* (*On the Casson-Walker invariant: global surgery formula and extension to closed oriented* 3-*manifold*), in French, C. R. Acad. Sci. Paris Ser. I, **315** (1992) 437–440.

[Les96] ——, *Global surgery formula for the Casson-Walker invariant*, Annals of Math. Studies **140**, Princeton University Press, 1996.

[Les98a] ——, *A sum formula for the Casson-Walker invariant*, Invent. Math. **133** (1998) 613–681.

[Les98b] ——, *On the kernel of the Casson invariant*, Topology **37** (1998) 25–37.

[Les*00*] ——, *About the uniqueness and the denominators of the Kontsevich integral*, prepritn, math.GT/0004094.

[Lev97] Levine, J., *The Conway polynomial of an algebraically split link*, KNOTS '96 (Tokyo), 23–29, World Sci. Publishing, River Edge, NJ, 1997.

[Lev99] ———, *Pure braids, a new subgroup of the mapping class group and finite-type invariants of 3-manifolds*, Tel Aviv Topology Conference: Rothenberg Festschrift (1998), 137–157, Contemp. Math. **231**, Amer. Math. Soc., Providence, RI, 1999.

[LiLi96] Li, B.-H., Li, T.-J., *Generalized Gaussian sums: Chern-Simons-Witten-Jones invariants of lens-spaces*, J. Knot Theory Ramifications **5** (1996) 183–224.

[Lic62] Lickorish, W.B.R., *A representation of orientable combinatorial 3-manifolds*, Ann. Math. **76** (1962) 531–540.

[Lic91a] ———, *Invariants for 3-manifolds from the combinatorics of the Jones polynomial*, Pacific J. Math. **149** (1991) 337–347.

[Lic91b] ———, *Three-manifolds and the Temperley-Lieb algebra*, Math. Ann. **290** (1991) 657–670.

[Lic92] ———, *Calculations with the Temperley-Lieb algebra*, Comment. Math. Helv. **67** (1992) 571–591.

[Lic93] ———, *The skein method for three-manifold invariants*, J. Knot theory Ramifications **2** (1993) 171–194.

[Lic97] ———, *An Introduction to Knot Theory*, Graduate Texts in Math. **175**, Springer-Verlag, 1997.

[LiMi88] Lickorish, W.B.R, Millet, K.C., *The new polynomial invariants of knots and links*, Math. Mag. **61** (1988) 3–23.

[Lie00] Lieberum, J., *The LMO-invariant of 3-manifolds of rank one and the Alexander polynomial* Math. Ann. **318** (2000) 761–776.

[Lin*91*] Lin, X.-S., *Vertex models, quantum groups and Vassiliev's knot invariants*, preprint, 1991, available at `http://math.ucr.edu/~xl/cv-html/pub.html`

[Lin94] ———, *Finite type link invariants of 3-manifolds*, Topology **33** (1994).

[Lin98] ———, *Finite type invariants of integral homology 3-spheres: a survey*, Knot theory (Warsaw, 1995), 205–220, Banach Center Publ., **42**, Polish Acad. Sci., Warsaw, 1998.

[Lin96] ———, *Finite type invariants and the non-invertibility of links*, Math. Res. Lett. **3** (1996) 405–417.

[LiWa99a] Lin, X.-S., Wang, Z., *On Ohtsuki's invariants of integral homology 3-spheres*, Acta Math. Sin. (Engl. Ser.) **15** (1999) 293–316.

[LiWa99b] ———, *Fermat limit and congruence of Ohtsuki invariants*, Proceedings of the Kirbyfest (Berkeley, CA, 1998), 321–333 (electronic), Geom. Topol. Monogr., **2**, Geom. Topol., Coventry, 1999.

[Lins95] Lins, S., *Gems, computers and attractors for* 3*-manifolds*, Series on Knots and Everything **5**, World Scientific Publishing Co., Inc., River Edge, NJ, 1995.

[Lu92] Lu, N., *A simple proof of the fundamental theorem of Kirby calculus on links*, Trans. Amer. Math. Soc. **331** (1992) 143–156.

[Lus93] Lusztig, G., *Introduction to quantum groups*, Progress in Mathematics **110**, Birkhauser Boston, Inc., Boston, MA, 1993.

[Lyu86] Lyubashenko, V.V., *Superanalysis and solutions of the triangle equation*, Candidate's Dissertation Phys.-Mat. Sciences, 1986.

[Mac71] Mac Lane, S., *Categories for the working mathematician*, Graduate Texts in Math., **5**, Springer-Verlag, New-York, 1971.

[Maj95] Majid, S., *Foundations of quantum group theory*, Cambridge University Press, Cambridge, 1995.

[MaRo95] Masbaum, G., Roberts, J., *On central extensions of mapping class groups*, Math. Ann. 302 (1995) 131–150.

[MaRo97] ———, *A simple proof of integrality of quantum invariants at prime roots of unity*, Math. Proc. Camb. Phil. Soc. **121** (1997) 443–454.

[MaWe98] Masbaum, G., Wenzl, H., *Integral modular categories and integrality of quantum invariants at roots of unity of prime order*, J. Reine Angew. Math. **505** (1998) 209–235.

[Mat87] Matveev, S.V., *Generalized surgeries of three-dimensional manifolds and representations of homology spheres*, Mat. Zametki **42** (1987) 268–278 (in Russian), English translation in Math. Notes **42** (1987) 651–656.

[MeSe80] Mehta, V. B., Seshadri, C. S., *Moduli of vector bundles on curves with parabolic structures*, Math. Ann. **248** (1980) 205–239.

[MeMo95] Melvin, P., Morton, H., *The coloured Jones function*, Comm. Math. Phys. **169** (1995) 501–520.

[Moi52] Moise, E.E., *Affine structures in 3-manifolds. V: The triangulation theorem and Hauptvermutung*, Ann. of Math. (2) **56** (1952) 96–114.

[Moi54] ———, *Affine structures in 3-manifolds. VII: Invariance of the knot-types; local tame imbedding*, Ann. of Math. (2) **59** (1954) 159–170.

[MoSe88] Moore, G., Seiberg, N., *Polynomial equations for rational conformal field theories*, Phys. Lett. **B212** (1988) 451–460.

[MoSe89] ———, *Classical and quantum field theory*, Comm. Math. Phys. **123** (1989) 177–254.

[Mor93a] Morton, H.R., *Invariants of links and 3-manifolds from skein theory and from quantum groups*, Topics in knot theory (Erzurum, 1992), 107–155, NATO Adv. Sci. Inst. Ser. C Math. Phys. Sci. **399**, Kluwer Acad. Publ., Dordrecht, 1993.

[Mor93b] ———, *Quantum invariants given by evaluation of knot polynomials*, J. Knot Theory Ramifications **2** (1993) 195–209.

[Mor99] ———, *The Burau matrix and Fiedler's invariant for a closed braid*, Topology Appl. **95** (1999) 251–256.

[Mum65] Mumford, D., *Geometric invariant theory*, Springer-Verlag, Berlin, 1965.

[Mun66] Munkres, J.R., *Elementary differential topology*, Lectures given at Massachusetts Institute of Technology, Fall, 1961. Revised edition. Annals of Mathematics Studies **54**, Princeton University Press, Princeton, NJ, 1966.

[Mu_H94] Murakami, H., *Quantum $SU(2)$-invariants dominate Casson's $SU(2)$-invariant*, Math. Proc. Camb. Phil. Soc. **115** (1994) 253–281.

[Mu_H95] ———, *Quantum $SO(3)$-invariants dominate the $SU(2)$-invariant of Casson and Walker*, Math. Proc Camb. Phil. Soc. **117** (1995) 237–249.

[M_HNa89] Murakami, H., Nakanishi, Y., *On a certain move generating link-homology*, Math. Ann. **284** (1989) 75–89.

[MOO92] Murakami, H., Okada, M., Ohtsuki, T., *Invariants of three-manifolds derived from linking matrices of framed links*, Osaka J. Math. **29** (1992) 545–572.

[M_HOh96] Murakami, H., Ohtsuki, T., *Quantum Sp(n) invariant of links via an invariant of colored planar graphs*, Kobe J. Math. **13** (1996) 191–202.

[M_HOh01] ———, *Finite type invariants of knots via their Seifert matrices*, Asian J. Math. **5** (2001) 379–386.

[MOY98] Murakami, H., Ohtsuki, T., Yamada, S., *HOMFLY polynomial via an invariant of colored planar graphs*, l'Enseignement mathematique **44** (1998) 325–360.

[Mu_J93] Murakami, J., *A state model for the multi-variable Alexander polynomial*, Pacific J. Math. **157** (1993) 109–135.

[Mu_J96] ———, *Representation of mapping class groups via the universal perturbative invariant*, Proc. of Knots 96, edited by S. Suzuki, World Scientific, 573–586.

[Mu_J00] ———, 結び目と量子群 (*Knots and quantum groups*), in Japanese, 朝倉書店, 2000.

[M_JOh97] Murakami, J., Ohtsuki, T., *Topological quantum field theory for the universal quantum invariant*, Comm. Math. Phys. **188** (1997) 501–520.

[NaSe65] Narasimhan, M. S., Seshadri, C. S., *Stable and unitary vector bundles on a compact Riemann surface*, Ann. of Math. (2) **82** (1965) 540–567.

[Oht93] Ohtsuki, T., *Colored ribbon Hopf algebras and universal invariants of framed links*, J. Knot Theory Ramifications **2** (1993) 211–232.

[Oht95] ———, *A polynomial invariant of integral homology 3-spheres*, Math. Proc. Camb. Phil. Soc. **117** (1995) 83–112.

[Oht96a] ———, *Finite type invariants of integral homology 3-spheres*, J. Knot Theory and Its Rami. **5** (1996) 101-115.

[Oht96b] ———, *A polynomial invariant of rational homology 3-spheres*, Invent. Math. **123** (1996) 241–257.

[Oht98] ———, *A filtration of the set of integral homology 3-spheres*, Proceedings of ICM, Berlin, 1998, Documenta Math. Extra Volume ICM 1998 II, invited lectures, 473–482.

[Oht99] ———, *Combinatorial quantum method in 3-dimensional topology*, lecture note of Oiwake Seminar 1996, MSJ Memoirs **3**, Math. Soc. Japan, 1999.

[Oht00] ———, *The perturbative SO(3) invariant of rational homology 3-spheres recovers from the universal perturbative invariant*, Topology **39** (2000) 1103–1135.

[Oht+99] Ohtsuki, T., *et al.*, 量子不変量 — 3 次元トポロジーと数理物理の遭遇 (*Quantum invariants, — 3-dimensional topology encounters mathematical physics*), in Japanese, editted by T. Ohtsuki, written by T. Deguchi, T. Ohtsuki, T. Ohyama, J. Murakami, H. Murakami, T. Takata and M. Wakui, 日本評論社 (Nippon Hyoronsha Co., Ltd. Publishers), 1999.

[OhYa97] Ohtsuki, T., Yamada, S., *Quantum SU(3) invariants via linear skein theory*, J. Knot Theory Ramifications **6** (1997) 373–404.

[Piu92] Piunikhin, S., *Turaev-Viro and Kauffman-Lins invariants for 3-manifolds coincide*, J. Knot Theory Ramifications **1** (1992) 105–135.

[Piu93] ———, *Reshetikhin-Turaev and Crane-Kohno-Kontsevich 3-manifold invariants coincide*, J. Knot Theory Ramifications **2** (1993) 65–95.

[Piu95a] ———, *Combinatorial expression for universal Vassiliev link invariant*, Comm. Math. Phys. **168** (1995) 1–22.

[Piu95b] ———, *Weights of Feynman diagrams, link polynomials and Vassiliev knot invariants*, J. of Knot Theory Ramifications **4** (1995) 163–188.

[Poi*99a*] Poirier, S., *Rationality results for the configuration space integral of knots*, preprint, math.GT/9901028.

[Poi*99b*] ———, *The limit configuration space integral for tangles and the Kontsevich integral*, preprint, math.GT/9902058.

[Pol*99*] Polyak, M., *Skein relations and Gauss diagram formulas for Milnor's mu-invariants*, preprint 1999.

[Pol00] ———, *On the algebra of arrow diagrams*, Lett. Math. Phys. **51** (2000) 275–291.

[PoVi94] Polyak, M., Viro, O., *Gauss diagram formulas for Vassiliev invariants*, Internat. Math. Res. Notices 1994, no. 11, 445ff., approx. 8 pp. (electronic).

[PoVi01] ———, *On the Casson knot invariant*, J. Knot Theory Ramifications **10** (2001) 711–738.

[PoWi84] Polyakov, A.M., Wiegmann, P.B., *Goldstone fields in two dimensions with multivalued actions*, Phys. Lett. **B141** (1984) 223–228.

[PrSo97] Prasolov, V.V., Sossinsky, A.B., *Knots, links, braids and 3-manifolds, — an introduction to the new invariants in low-dimensional topology*, Trans. of Math. Monographs **154**, Amer. Math. Soc., 1997.

[PrSe86] Pressley, A., Segal, G., *Loop groups*, Oxford Mathematical Monographs. Oxford Science Publications. The Clarendon Press, Oxford University Press, New York, 1986.

[Prz92] Przytycki, J.H., *History of the knot theory from Vandermonde to Jones*, XXIVth National Congress of the Mexican Mathematical Society (Spanish) (Oaxtepec, 1991), 173–185, Aportaciones Mat. Comun., **11**, Soc. Mat. Mexicana, Mexico, 1992.

[PrTr88] Przytycki, J.H., Traczyk, P., *Invariant of links of Conway type*, Kobe J. Math. **4** (1988) 115–139.

[Res*88*] Reshetikhin, N.Yu., *Quantized universal enveloping algebras, the Yang-Baxter equation and invariants of links I, II*, LOMI preprints E-4-87, E-17-87, 1988.

[Res89] ———, *Quasitriangular Hopf algebras and invariants of links*, Algebra and Analysis **1** (1989), Leningrad Math. J. **1** (1990) 491–513.

[Res91] ———, *Invariants of links and 3-manifolds related to quantum groups*, Proceedings of the International Congress of Mathematicians, Vol. I, II (Kyoto, 1990), 1373–1375, Math. Soc. Japan, Tokyo, 1991.

[ReTu90] Reshetikhin, N., Turaev, V.G., *Ribbon graphs and their invariants derived from quantum groups*, Comm. Math. Phys. **127** (1990) 1–26.

[ReTu91] ———, *Invariants of 3-manifolds via link polynomials and quantum groups*, Invent. Math. **103** (1991) 547–597.

[Rob95] Roberts, J., *Skein theory and Turaev-Viro invariants*, Topology **34** (1995) 771–787.

[Rol90] Rolfsen, D., *Knots and links*, Math. Lecture Series, **7**, Publish or Perish, Inc., Houston, Texas, 1990, Second printing, with corrections.

[Roz96] Rozansky, L., *Witten's invariants of rational homology spheres at prime values of K and trivial connection contribution*, Comm. Math. Phys. **180** (1996) 297–324.

[Roz97a] ———, *The trivial connection contribution to Witten's invariant and finite type invariants of rational homology spheres*, Comm. Math. Phys. **183** (1997) 23–54.

[Roz97b] ———, *Higher order terms in the Melvin-Morton expansion of the colored Jones polynomial*, Comm. Math. Phys. **183** (1997) 291–306.

[Roz97c] ———, *On finite type invariants of links and rational homology spheres derived from the Jones polynomial and Witten-Reshetikhin-Turaev invariant*, in "Geometry and physics", (J. E. Andersen, J. Dupont, H. Pedersen, and A. Swann, eds.), lecture notes in pure and applied mathematics **184** (1997) 379–397, Marcel Dekker, New-York.

[Roz98a] ———, *The universal R-matrix, Burau representation, and the Melvin-Morton expansion of the colored Jones polynomial*, Adv. Math. **134** (1998) 1–31.

[Roz98b] ———, *On p-adic convergence of perturbative invariants of some rational homology spheres*, Duke Math. J. **91** (1998) 353–380.

[Roz*98a*] ———, *On the structure of the colored Jones polynomial of a link I: R-matrix and Burau representation*, preprint, math/9806004.

[Roz*98b*] ———, *A contribution of a U(1)-reducible connection to quantum invariants of links II: Links in rational homology spheres*, preprint, math/9806066.

[Roz*98c*] ———, *On p-adic properties of the Witten-Reshetikhin-Turaev invariant*, preprint, math/9806075.

[RoWi97] Rozansky, L., Witten, E., *Hyper-Kahler geometry and invariants of three-manifolds*, Selecta Math. (N.S.) **3** (1997) 401–458.

[Sav99] Saveliev, N., *Lectures on the topology of 3-manifolds, an introduction to the Casson invariant*, Walter de Gruyter, Berlin, 1999.

[Saw*99*] Sawin, S.F., *Jones-Witten invariants for nonsimply-connected Lie groups and the geometry of the Weyl alcove*, preprint, math.QA/9905010.

[Sch97] Schottenloher, M., *A mathematical introduction to conformal field theory*, Based on a series of lectures given at the Mathematisches Institut der Universität Hamburg. Translated from the German. Lecture Notes in Physics. New Series m: Monographs **43**, Springer-Verlag, Berlin, 1997.

[Shi00] Shirokova, N., *The space of 3-manifolds*, C. R. Acad. Sci. Paris Ser. I Math. **331** (2000) 131–136.

[Spi79] Spivak, M., *A comprehensive introduction to differential geometry*, Second edition. Publish or Perish, Inc., Wilmington, Del., 1979.

[Sta96] Stanford, T., *Finite type invariants of knots, links, and graphs*, Topology **35** (1996) 1027–1050.

[Sta97] ———, *Computing Vassiliev's invariants*, Topology Appl. **77** (1997) 261–276.

[Swe69] Sweedler, M.E., *Hopf Algebras*, W.A. Benjamin Inc, New York, 1969.

[Tai1898] Tait, P.G., *On knots I, II, III*, Scientific papers **1** (1898) 273–347, London, Cambridge University Press.

[Tak92] Takata, T., *Invariants of 3-manifolds associated with quantum groups and Verlinde's formula*, Publ. Res. Inst. Math. Sci. **28** (1992) 139–167.

[Tak95] ———, *Universal R-matrices for the quantum group* $U_q(\mathrm{sl}(N+1,\mathbb{C}))$*: the root of unity case*, Osaka J. Math. **32** (1995) 959–986.

[Tak96] ———, *On quantum* $PSU(n)$*-invariants for lens spaces*, J. Knot Theory Ramifications **5** (1996) 885–901.

[Tak97] ———, *On quantum* PSU(n) *invariants for Seifert manifolds*, J. Knot Theory Ramifications **6** (1997) 417–426.

[TaYo99] Takata, T., Yokota, Y., *The* $PSU(N)$ *invariants of 3-manifolds are polynomials*, J. Knot Theory Ramifications **8** (1999) 521–532.

[$\mathrm{Th_D}$99] Thurston, D., *Integral expressions for the Vassiliev knot invariants*, preprint, math.QA/9901110.

[$\mathrm{Th_D}$00] ———, *Wheeling: A diagrammatic analogue of the Duflo isomorphism*, Ph.D. Thesis, U.C. Berkeley, Spring 2000, math.QA/0006083.

[$\mathrm{Th_W}$97] Thurston, W.P. *Three-dimensional geometry and topology, Vol. 1*, Edited by Silvio Levy. Princeton Mathematical Series **35**, Princeton University Press, Princeton, NJ, 1997.

[TsKa88] Tsuchiya, A., Kanie, Y., *Vertex operators in conformal field theory on* P^1 *and monodromy representations of braid groups*, in Conformal Field Theory and Solvable Lattice Models (Kyoto, 1986) ed. by A. Tsuchiya, T. Eguchi, M Jimbo, Sdv. Studies in Pure Math. **16** (1988) 297–372.

[TUY90] Tsuchiya, A., Ueno, K., Yamada, Y., *Conformal field theory on universal family of stable curves with gauge symmetries*, Advanced Studies in Pure Math. **19** (1990) 459–566.

[Tur88] Turaev, V.G., *The Yang-Baxter equation and invariants of links*, Invent. Math. **92** (1988) 527–553.

[Tur89] ———, *Operator invariants of tangles, and R-matrices*, (in Russian) Izv. Akad. Nauk SSSR Ser. Mat. **53** (1989); English translation in Math. USSR-Izv. **35** (1990) 411–444.

[Tur92] ———, *Modular categories and 3-manifold invariants*, Internat. J. Modern Phys. **B6** (1992) 1807–1824.

[Tur94] ———, *Quantum invariants of knots and 3-manifolds*, Studies in Math. **18**, Walter de Gruyter, 1994.

[TuVi92] Turaev, V., Viro, O., *State sum invariants of 3-manifolds and quantum 6j-symbols*, Topology **31** (1992) 865–902.

[TuWe93] Turaev, V., Wenzl, H., *Quantum invariants of 3-manifolds associated with classical simple Lie algebras*, Internat. J. Math. **4** (1993) 323–358.

[TuWe97] ———, *Semisimple and modular categories from link invariants*, Math. Ann. **309** (1997) 411–461.

[TuGr96] *History and science of knots*, Edited by J. C. Turner and P. van de Griend. Series on Knots and Everything **11**. World Scientific Publishing Co., Inc., River Edge, NJ, 1996.

[Uen97] Ueno, K. *Introduction to conformal field theory with gauge symmetries*, Geometry and physics (Aarhus, 1995), 603–745, Lecture Notes in Pure and Appl. Math. **184**, Dekker, New York, 1997.

[Vas90] Vassiliev, V.A., *Cohomology of knot spaces*, "Theory of singularities and its applications" (V.I. Arnold ed.) Adv. Soviet Math. **1**, Amer. Math. Soc., Providence, RI (1990) 23–69.

[Vas92] ———, *Complements of discriminants of smooth maps*, Trans. of Math. Mono. **98** Amer. Math. Society., Providence, 1992.

[Vog90] Vogel, P., *Representation of links by braids: a new algorithm*, Comment. Math. Helv. **65** (1990) 104–113.

[Vog97] ———, *Algebraic structures on modules of diagrams*, preprint, 1997, available at `http://www.math.jussieu.fr/~vogel/`

[Walk92] Walker, K., *An extension of Casson's invariant*, Annals of Math. Studies **126**, Princeton University Press, 1992.

[Wall64] Wall, C.T.C., *Quadratic forms on finite groups, and related topics*, Topology **2** (1964) 281–298.

[Walla68] Wallace, A.H., *Differential topology: first steps*, Second printing. Mathematics Monograph Series. W. A. Benjamin, Inc., Reading, Mass.-London-Amsterdam, 1968.

[Wen93] Wenzl, H., *Braids and invariants of 3-manifolds*, Invent. Math. **114** (1993) 235–275.

[Wit89a] Witten, E., *Quantum field theory and the Jones polynomial*, Comm. Math. Phys. **121** (1989) 351–399.

[Wit89b] ———, *Gauge theories and integrable lattice models*, Nuclear Phys. **B322** (1989) 629–697.

[Wit90] ———, *Gauge theories, vertex models, and quantum groups*, Nuclear Phys. **B330** (1990) 285–346.

[Wit95] ———, *Chern-Simons gauge theory as a string theory*, The Floer memorial volume, 637–678, Progr. Math., **133**, Birkhäuser, Basel, 1995.

[Yam87] Yamada, S., *The minimal number of Seifert circles equals the braid index of a link*, Invent. Math. **89** (1987) 347–356.

[Yam89] ———, *An invariant of spatial graphs*, J. Graph Theory **13**, 531–551 (1989).

[Yam95] ———, *The absolute value of the Chern-Simons-Witten invariants of lens spaces*, J. Knot Theory Ramifications **4** (1995) 319–327.

[Yan97] Yang, S.-W., *Feynman integral, knot invariant and degree theory of maps*, preprint, q-alg/9709029.

[Yok92] Yokota, Y., *Polynomial invariants of positive links*, Topology **31** (1992) 805–811.

[Yok93] ———, *The Kauffman polynomial of periodic knots*, Topology **32** (1993) 309–324.

[Yok95] ———, *The Kauffman polynomial of alternating links*, Topology Appl. **65** (1995) 229–236.

[Yok96] ———, *Topological invariants of graphs in 3-space* Topology **35** (1996) 77–87.

[Yok97] ———, *Skeins and quantum $SU(N)$ invariants of 3-manifolds*, Math. Ann. **307** (1997) 109-138.

Notation

$Z^{\mathrm{LMO}}(M)$	LMO invariant of a closed 3-manifold M, 286
$z^{\mathrm{LMO}}(M)$	primitive LMO invariant of a closed 3-manifold M, 286
$\mathcal{K}$	the vector space spanned by knots, 176
$\mathbb{K}$	the set of knots, 185
$\mathcal{M}$	the vector space spanned by integral homology 3-spheres, 306
$\mathbb{M}$	the set of integral homology 3-spheres, 328
$\underset{d}{\sim}$	C_d-equivalence, Y_d-equivalence, $(d-1)$-equivalence, 380, 391

Index

SERIES ON KNOTS AND EVERYTHING

Editor-in-charge: Louis H. Kauffman *(Univ. of Illinois, Chicago)*

The Series on Knots and Everything: is a book series polarized around the theory of knots. Volume 1 in the series is Louis H Kauffman's Knots and Physics.

One purpose of this series is to continue the exploration of many of the themes indicated in Volume 1. These themes reach out beyond knot theory into physics, mathematics, logic, linguistics, philosophy, biology and practical experience. All of these outreaches have relations with knot theory when knot theory is regarded as a pivot or meeting place for apparently separate ideas. Knots act as such a pivotal place. We do not fully understand why this is so. The series represents stages in the exploration of this nexus.

Details of the titles in this series to date give a picture of the enterprise.

Published:

Vol. 1: Knots and Physics (3rd Edition)
L. H. Kauffman

Vol. 2: How Surfaces Intersect in Space — An Introduction to Topology (2nd Edition)
J. S. Carter

Vol. 3: Quantum Topology
edited by L. H. Kauffman & R. A. Baadhio

Vol. 4: Gauge Fields, Knots and Gravity
J. Baez & J. P. Muniain

Vol. 5: Gems, Computers and Attractors for 3-Manifolds
S. Lins

Vol. 6: Knots and Applications
edited by L. H. Kauffman

Vol. 7: Random Knotting and Linking
edited by K. C. Millett & D. W. Sumners

Vol. 8: Symmetric Bends: How to Join Two Lengths of Cord
R. E. Miles

Vol. 9: Combinatorial Physics
T. Bastin & C. W. Kilmister

Vol. 10: Nonstandard Logics and Nonstandard Metrics in Physics
W. M. Honig

Vol. 11: History and Science of Knots
edited by J. C. Turner & P. van de Griend

Vol. 12: Relativistic Reality: A Modern View
edited by J. D. Edmonds, Jr.

Vol. 13: Entropic Spacetime Theory
J. Armel

Vol. 14: Diamond — A Paradox Logic
N. S. Hellerstein

Vol. 15: Lectures at KNOTS '96
S. Suzuki

Vol. 16: Delta — A Paradox Logic
N. S. Hellerstein

Vol. 19: Ideal Knots
A. Stasiak, V. Katritch & L. H. Kauffman

Vol. 20: The Mystery of Knots — Computer Programming for Knot Tabulation
C. N. Aneziris

Vol. 24: Knots in HELLAS '98 — Proceedings of the International Conference on Knot Theory and Its Ramifications
edited by C. McA Gordon, V. F. R. Jones, L. Kauffman, S. Lambropoulou & J. H. Przytycki

Vol. 26: Functorial Knot Theory — Categories of Tangles, Coherence, Categorical Deformations, and Topological Invariants
by David N. Yetter

Vol. 27: Bit-String Physics: A Finite and Discrete Approach to Natural Philosophy
by H. Pierre Noyes; edited by J. C. van den Berg

Vol. 29: Quantum Invariants — A Study of Knots, 3-Manifolds, and Their Sets
by Tomotoda Ohtsuki